Ultra-Wideband, Short-Pulse Electromagnetics

Ultra-Wideband, Short-Pulse Electromagnetics

Edited by

Henry L. Bertoni
Lawrence Carin
Leopold B. Felsen

Weber Research Institute
Polytechnic University
Brooklyn, New York

Springer Science+Business Media, LLC

Library of Congress Cataloging-in-Publication Data

Ultra-wideband, short-pulse electromagnetics / edited by Henry L.
Bertoni, Lawrence Carin, Leopold B. Felsen.
p. cm.
"Proceedings of an International Conference on Ultra-Wideband,
Short-Pulse Electromagnetics, held October 8-10, 1992, at WRI,
Polytechnic University, Brooklyn, New York"--CIP t.p. verso.
Includes bibliographical references and index.
ISBN 978-1-4613-6244-9 ISBN 978-1-4615-2870-8 (eBook)
DOI 10.1007/978-1-4615-2870-8
1. Radar--Congresses. 2. Electromagnetic devices--Congresses.
3. Signal processing--Congresses. 4. Antennas (Electronics)-
-Congresses. I. Bertoni, Henry L. II. Carin, Lawrence.
III. Felsen, Leopold B. IV. International Conference on Ultra
-Wideband, Short-Pulsed Electromagnetics (1992 : Brooklyn, N.Y.)
TK6573.U59 1993
621.3848--dc20 93-6075
CIP

Proceedings of an International Conference on Ultra-Wideband, Short-Pulse Electromagnetics, held October 8–10, 1992, at WRI, Polytechnic University, Brooklyn, New York

Originally published by Plenum Press, New York in 1993
Softcover reprint of the hardcover 1st edition 1993
A Division of Plenum Publishing Corporation
233 Spring Street, New York, N.Y. 10013

PREFACE

In 1945, Dr. Ernst Weber founded, and was the first Director of, the **Microwave Research Institute (MRI)** at POLYTECHNIC UNIVERSITY (at that time named the Polytechnic Institute of Brooklyn). MRI gained world-wide recognition in the 50's and 60's for its research in electromagnetic theory, antennas and radiation, network theory and microwave networks, microwave components and devices. It was also known through its series of topical symposia and the widely distributed hard bound MRI Symposium Proceedings. Rededicated as the **Weber Research Institute (WRI)** in 1986, the research focus today is on such areas as electromagnetic propagation and antennas, ultra-broadband electromagnetics, pulse power, acoustics, gaseous electronics, plasma physics, solid state materials, quantum electronics, electromagnetic launchers, and networks. Following the MRI tradition, WRI has launched its own series of in-depth topical conferences with published proceedings. The first conference was held in October, 1990 and was entitled **Directions in Electromagnetic Wave Modeling.** The proceedings of the conference were published under that title by Plenum Press.

This volume constitutes the Proceedings of the second WRI International Conference dealing with **Ultra-Wideband Short-Pulse Electromagnetics.** The conference was held October 8-10, 1992 at the Polytechnic University in Brooklyn, NY in cooperation with the Antennas and Propagation Society , and the Microwave Theory and Techniques Society of the IEEE. The conference was co-sponsored by the Air Force Office of Scientific Research and the Rome Laboratory of the Air Force Systems Command, who convened a workshop on **Short Pulse Research and Technology -- Open Questions and Future Trends** at the end of the conference.

Henry L. Bertoni
Lawrence Carin
Leopold B. Felsen

January, 1993

CONTENTS

SCANNING THE CONFERENCE

L. B. Felsen

Weber Research Institute, Polytechnic University
Brooklyn, New York 11201

MOTIVATION AND PERSPECTIVES

This second WRI international conference was convened two years after the first conference which introduced this series by broad topical coverage of **Directions in Electromagnetic Wave Modeling**. This broad theme was chosen intentionally to provide an overview of cutting edge issues -- both analytic and numerical -- that confront electromagnetic wave modelers in diverse natural and manmade environments, responding to needs driven by new technologies, system requirements and other missions, with an awareness of recent developments in processing of data. The relevance of these issues was underscored by presentations from program managers in various government agencies. The hard-copy Proceedings of the first conference, published by Plenum Press in 1991, attests to the variety of strategies, techniques and models that are being explored to cope with these problems.

In contrast, the second WRI conference was intentionally focused on a specific topic of current concern within the electromagnetics community -- Ultra-Wideband, Short-Pulse Electromagnetics. This topic was chosen not only because of the steadily increasing importance of time domain techniques and applications per se, but because of the general trend toward wider signal bandwidths, culminating in ultrawideband (UWB) -- short pulse (SP) inputs.

The UWB-SP regime has stimulated debate and controversy centered essentially around the question of whether one should approach that regime by "conventional" synthesis from the frequency domain (FD), or whether new advantages and insights accrue from "nonconventional" treatment directly in the time domain (TD). The FD vs. TD issue permeates all aspects of UWB-SP methodology: signal generation and detection (multiband FD or impulse sources and receivers); processing the data (wavenumber spectral filtering in

Ultra-Wideband, Short-Pulse Electromagnetics
Edited by H. Bertoni *et al.*, Plenum Press, 1993

sources and receivers); processing the data (wavenumber spectral filtering in FD, after Fourier transform from TD, or directly in TD); numerical modeling (FD or TD finite difference schemes); analytic modeling (parametrizations in terms of FD or TD basis elements); etc. Full exploitation of the direct-TD approach requires "direct-TD thinking" wherein the FD plays an auxiliary role rather than the dominant role that is traditional in electromagnetic science and engineering. Thus, TD thinking must address dispersion which is well understood in the FD, anomalous behavior triggered in materials by the initial impact of SP excitation, and other TD phenomenologies that manifest themselves in an entirely different manner when transformed to the FD. A major TD discriminant is the temporal-spatial resolution of progressive events which is translated into FD phase differences, without time gating, that may be more difficult to resolve.

While there are clear instances where one approach may be argued to have distinct advantages over the other, practical problems usually do not belong to this category and can arguably be addressed in either way, due to the FD-TD (configuration-spectrum) complementarity. This is, in fact, the dilemma confronting FD and TD protagonists, and its imprint can be discerned throughout the contributions to this volume. As with any controversial issue, the FD-TD debate has not always been confined to sober balancing of technical aspects but has led to partitioning into establishment (FD) and non-establishment (TD) constituencies. This has culminated in the formation of a DARPA (Defense Advanced Research Projects Agency) Review Panel to examine impulse radar and related issues. The panel report was met with charges and countercharges based not only on technical but also on political considerations. While the dust has settled now, memories still linger on among those who were directly involved.

The present conference was organized in cogniance of this setting. The hallmark of the WRI conferences -- hand-picked invited speakers, selected contributed and poster papers, no parallel sessions -- was adhered to in the assemblage of a program centered around leading scientists from universities, government and industry, who are active in various areas of UWB-SP electromagnetics; this group includes several who have been participants in the DARPA proceedings. One of the DARPA panel members, Dr. E. Brookner of Raytheon Co., Sudbury, MA, shared with the conference some reflections on that episode, laced with pointed humor concerning substance vs. politics. One conclusion extracted from a report by Charles A. Fowler relates to this conference and is worthy of note:

> "An honest-to-God technical meeting on this subject has been scheduled: International Conference on Ultra-Wideband Short-Pulse Electromagnetics, October 8-10, Polytechnic University, NY. For a mere $200, one can listen for 2 1/2 days to the results of substantive technical work in the field."

The WRI organizers were fortunate to gain co-sponsorship of this conference from the Air Force Office of Scientific Research and from Rome to convene a workshop on <u>Short Pulse Research and Technology -- Open Questions and Future Trends</u>. Because the proposed timing and topic of this workshop competed with the WRI conference, it was decided to combine both events, and to include the workshop at the end of the conference so that the participants could discuss open questions and future trends with perspectives gained from the conference presentations. The workshop, which was co-chaired by Dr. A. Nachman of the Air Force Office of Scientific Research and by Dr. F. J. Zucker of Rome Laboratory, was structured around a panel, with strong participation by the lively audience.

THE TECHNICAL PROGRAM

The invited and contributed papers, a total of 66, have been grouped into twelve sessions. Ten of these are topical and involved oral presentations. One of the remaining two sessions was devoted to the poster papers which cover the entire subject area of the conference, and the other -- the last session -- was devoted to the workshop. Topical content rather than numerical balance guided the grouping so that there were large sessions and small sessions. Each session had a chairman who was well versed in his session topic and would therefore stimulate discussion. Scanning the Table of Contents gives a quick overview of the session topics and the contributions to that topic. The coverage includes:

- Pulse Generation and Detection
  - Electro-Optical Methods
  - Electronic Methods
- Broadband Electronic Systems and Components
- Antennas and Arrays
- Pulse Propagation and Guidance
- Scattering Theory and Computation
  - Analytic Methods
  - Numerical Methods
- Signal processing Techniques.

The last item deserves special mention because it addressed the need for combining UWB wave techniques with modern pre and post signal processing so as to systematize the interpretation of data and extraction of desired information.

Concerning format and venue, many of the conference participants underscored the timeliness and relevance of the conference theme; praised the high caliber of the speakers presentations and discussions; expressed approval of the absence of parallel sessions; and enjoyed the congenial atmosphere as well as the new facilities at Polytechnic's Brooklyn campus where the conference took place. However, the majority of attendees would have preferred a conference hotel closer to the meeting site. In retrospect, the conference may be said to have met its objectives and can therefore be regarded as a success.

PULSE GENERATION AND DETECTION

AN ULTRA-WIDEBAND OPTOELECTRONIC THz BEAM SYSTEM

N. Katzenellenbogen and D. Grischkowsky

IBM Watson Research Center
Yorktown Heights, NY 10598

ABSTRACT

A unique optoelectronic THz beam system is described in this article. The system can generate well collimated beams of freely propagating fsec THz pulses, which can be detected with a time-resolution better than 150 fsec and with a signal-to-noise ratio of more than 1000. The useful bandwidth of the system extends from 0.2 to more than 6 THz. An experimental and theoretical analysis of the time-domain response function describing the photoconductor shows that performance of the system is now fundamentally limited only by the ballistic acceleration of the optically generated photocarriers.

INTRODUCTION

The generation of THz radiation via the photoconductive excitation of semiconducting and semi-insulating materials by ultrashort laser pulses has recently been at the focus of a great deal of work. Modern integrated circuit techniques have made possible the precise fabrication of micron-sized dipoles, which when photoconductively driven by fsec laser pulses can radiate well into the THz regime[1,2]. In an alternative and complimentary approach optoelectronic antennas can be used to extend radio and microwave techniques into the THz range[3-12]. Most recently, radiation has been generated by use of the surface field of semiconductors photoconductively driven with ultrafast laser pulses[13]. A new and quite efficient source of broadband THz radiation involves the generation of photocarriers in trap-enhanced electric fields with ultrafast laser pulses[14-16].

Some of these sources are based on an optical type approach whereby a transient point source of THz radiation is located at the focus of a dielectric collimating lens, followed by an additional paraboloidal focusing and collimating mirror[7,9-11]. A well collimated beam of THz radiation can thus be produced. An identical receiver is matched to the transmitter, to produce a THz system that has an extremely high collection efficiency. The optoelectronic THz system that is described in this article is such a system and has a demonstrated signal-to-noise ratio of 1000, a time resolution of less than 150 fsec and a frequency range from 0.2 THz to more than 6 THz. One of the most useful versions of the system is based on repetitive, subpicosecond optical excitation of a Hertzian dipole antenna imbedded in a charged coplanar transmission line structure[2,9-11]. The burst of radiation emitted by the resulting transient dipole is collimated by a THz optical system and subsequently focused onto a similar receiver structure, where it induces a transient voltage that is detected by photoconductively shorting the receiver. The THz optical system gives exceptionally tight coupling between transmitter and receiver, and the excellent focusing properties preserves the sub-picosecond time dependence of the source.

Exceptional sensitivity for repetitively pulsed beams of THz radiation is realized by the combination of THz optics with the synchronously-gated, optoelectronic detection process. Two stages of collimation are used to obtain a THz beam with a frequency independent divergence from the THz transmitter. The THz receiver with identical optical properties essentially collects all of this beam. The resulting tightly coupled system of the THz transmitter and receiver can give strong reception of the transmitted pulses of THz radiation after many meters of propagation. Gating of the THz receiver is another reason for the exceptional sensitivity. The gating window of approximately 0.6 psec is determined by the laser pulsewidth and the carrier lifetime in ion-implanted silicon-on-sapphire (SOS). The noise in the comparatively long time interval (10 nsec) between the repetitive THz pulses is not seen by the receiver. Therefore when the repetitive signal is synchronously detected, the total charge (current) from the signal increases linearly with the number of sampling pulses, while the charge (current) from noise increases only as the square root of the number of pulses. An important feature of the detection method is that it is a coherent process; the electric field of a repetitive pulse of THz radiation is directly measured.

THE OPTOELECTRONIC THz BEAM SYSTEM

The Experimental Set-Up

We begin by describing a THz system for which the transmitting and receiving antennas, used to generate and detect beams of short pulses of THz radiation are identical. Each antenna is imbedded in a coplanar transmission line[2,9-11], as shown in Fig. 1a. The antennas are fabricated on ion-implanted silicon-on-sapphire (SOS) wafers. The 20-μm-wide antenna structure is located in the middle of a 20-mm-long coplanar transmission line consisting of two parallel 10-μm-wide, 1-μm-thick, 5 Ω/mm, aluminum lines separated from each other by 30 μm. A colliding-pulse mode-locked (CPM) dye laser, produces 623 nm, 70 fsec pulses at a 100 MHz repetition rate in a beam with 5 mW average power. This beam is focused onto the 5-μm-wide photoconductive silicon gap between the two antenna arms. The 70 fsec laser creation of photocarriers causes subpsec increases in the conductivity of the antenna gap by more than 10,000 times. When a DC bias voltage of typically 10 V is applied to the transmitting antenna, these pulses of increased conductivity result in corresponding pulses of electrical current through the antenna. Consequently, bursts of electromagnetic radiation are produced. A large fraction of this radiation is emitted into the sapphire substrate in a cone normal to the interface; the radiation pattern is presented in Ref. (10). The radiation is then collected and collimated by a dielectric lens attached to the backside (sapphire side) of the SOS wafer[2,9-11]. For the work reported here, the dielectric lenses were made of high-resistivity (10 kΩcm) crystalline silicon with a measured absorption of less than 0.05 cm^{-1} in our frequency range. The silicon lens is a truncated sphere of 10 mm diameter with a focal point located at the antenna gap when attached to the back side of the chip. After collimation by the silicon lens, the beam diffracts and propagates to a paraboloidal mirror, where the THz radiation is recollimated into a highly directional beam, as shown in Fig. 1b. For the measurements to be described in the first part of this article, even though the 70 mm aperture paraboloidal mirrors have a 12 cm focal length, a 16 cm distance was used between the silicon lenses and the mirrors to optimize the response of the system at the peak of the measured spectrum. After recollimation by the paraboloidal mirror, beam diameters (10-70mm) proportional to the wavelength were obtained; thereafter, all the frequencies propagated with the same 25 mrad divergence. The combination of a paraboloidal mirror, a silicon lens (THz optics) and the antenna chip comprise the transmitter, a source of highly-directional freely-propagating beams of subpicosecond THz pulses. After a 50 cm propagation distance this THz beam is detected by an identical combination, the THz receiver, where the paraboloidal mirror focuses the beam onto a silicon lens, which focuses it onto a SOS antenna chip, similar to the one used for the emission process. The electric field of focused incoming THz radiation induces a transient bias voltage across the 5 μm gap between the two arms of this receiving antenna, directly connected to a low-noise current amplifier. The amplitude and time dependence of this transient voltage is obtained by measuring the collected charge (average current) versus the time delay between the THz pulses and the delayed CPM laser pulses in the 5 mW detection beam. The detection process with gated integration can be looked upon as a sub-picosecond boxcar integrator, whereby the CPM laser pulses synchronously gate the receiver, by driving the photoconductive switch defined by the 5 μm antenna gap.

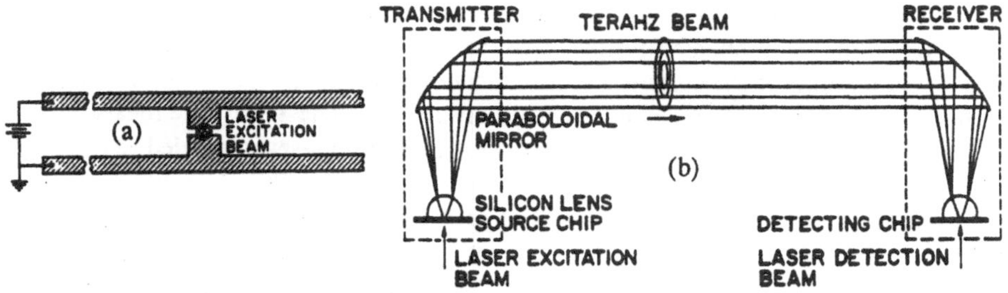

Figure 1. (a) Ultrafast dipolar antenna. (b) THz transmitter and receiver.

Signal-to-Noise Measurements

The setup described in the previous section was used to generate and detect subpicosecond radiation pulses. Figure 2a shows a typical time-resolved measurement[11]. The signal-to-noise ratio in this 4 minute scan is more than 10,000:1. The clean pulseshape is a result of the fast action of the photoconductive switch at the antenna gap, the broadband response of the ultrafast antennas, the broadband THz optical transfer function of the lenses and paraboloidal mirrors, and the very low absorption and dispersion of the silicon lenses. The measured pulsewidth of 0.54 psec (FWHM) is only an upper limit to the true pulsewidth, because no deconvolution has been applied to the measurement to take out the response time of the antenna gap. This time-response will be determined in the next section of this article.

In Fig. 2b, the Fourier transform of the measured signal (Fig. 2a) is shown to span the range from about 0.1 to 2.0 THz. This represents only a lower limit to the true extent of the emitted radiation as it contains the frequency response of the receiver. At the low frequency end, the efficiency of both emitter and receiver has been shown to be propor-

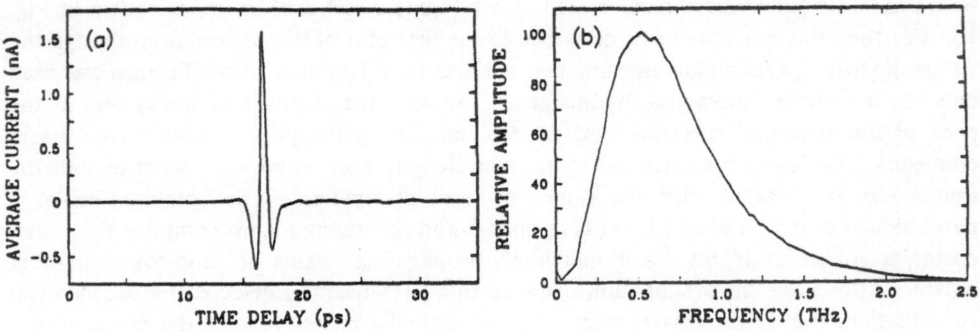

Figure 2. (a) THz pulse measured to 35 psec by scanning the time delay between the optical gating pulses and the incident THz pulses, while monitoring the current induced in the THz receiver. (b) Amplitude spectrum to 2.5 THz of the measured pulse shape.

tional to the length of the antenna, i.e., proportional to the separation between the two lines of the coplanar transmission line. At extremely low frequencies the size of the paraboloidal mirrors will also limit the efficiency. At the high frequency limit the efficiency of the antenna is strongly reduced when the half-wavelength (in the dielectric) of the emitted radiation is no longer small compared to the antenna length. The high frequency part of the spectrum is also limited by the finite risetime of the current transient and the 'less than ideal' imaging properties of the THz optics.

The detection limit of the system was investigated by reducing the intensity of the pump laser beam from the 6 mW normally used to only 15 μW. The THz pulse measured under this condition is shown in Fig. 3. This 400-fold reduction in laser power led to a reduction in the transient photocurrent of 320, instead of the expected 400. The discrepancy indicates a slight nonlinearity due to the onset of saturation, related to the fact that the electrical pulses generated on the transmission line are quite strong (almost 1 V in either direction). This 320-fold reduction in photocurrent led to a reduction in the power of the THz beam by the factor 1/100,000. However, despite this enormous reduction in power, the peak amplitude is still more than 30 times larger than the rms noise. Based on previous calculations[11], the average power in the THz beam during this measurement was about 10^{-13} W. If the power of the THz beam were even further reduced, the detection limit of

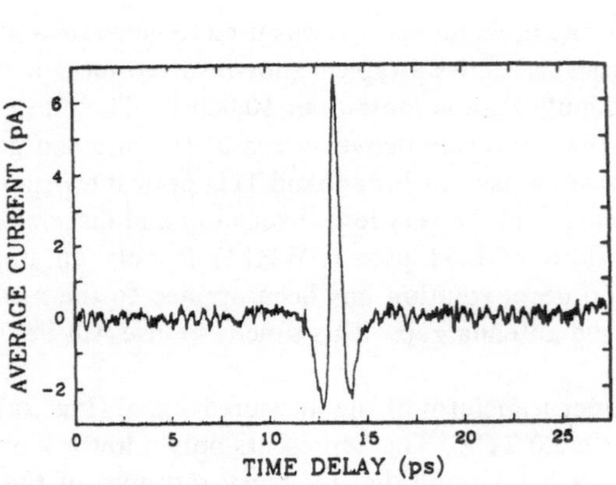

Figure 3. Measured THz pulse to 27.5 psec with a 100,000 times reduction (compared to Fig. 2a) of the THz beam power.

the THz receiver would be reached at $1x10^{-16}$ W, for a signal-to-noise-ratio of unity and a 125 ms integration time. Because the generation and detection of the THz (far-infrared) radiation is coherent, the THz receiver is inherently much more sensitive than the incoherent bolometer. The above receiver is approximately 1000 times more sensitive than a helium cooled bolometer[17].

Experimental Time-Dependent Response Function

In this section the experimental study of the THz beam system is extended to smaller 10 μm-long antennas. From the calculated THz optical transfer function together with the known THz absorption, the limiting bandwidth of the system is extracted following the procedure of Ref. (18). Given that the transmitter and receiver are identical, identical transmitter and receiver bandwidths are obtained. The time-domain response function for the antenna current is obtained by comparing the measured result to the calculated radiation spectrum from a Hertzian dipole driven by the current pulse determined by the laser pulsewidth, the current risetime, and the carrier lifetime.

The optoelectronic THz beam system is the same as previously described and as shown in Fig. 1, except that here smaller antennas are used. The antenna structure, again fabricated on ion-implanted SOS, is located in the middle of a 20-mm-long coplanar transmission line consisting of two parallel 5-μm-wide aluminum lines separated from each other by 10 μm. The performance of the colliding-pulse, mode-locked (CPM) dye laser was improved to provide 60 fsec excitation pulses in a beam with an average power of 7 mW on the excitation spot.

For these 10-μm-long antennas the measured transmitted THz pulse is shown in Fig. 4a. This pulse is shown on an expanded time scale in Fig. 4b, where the measured FWHM pulsewidth of 420 fsec (with no deconvolution) is indicated. This pulsewidth is significantly shorter than the 540 fsec pulse (Fig. 2a) obtained from the same experimental arrangement, but with the 30-μm-long antennas described in the previous section. The use of even still smaller antennas did not significantly shorten the THz pulses. The numerical Fourier transform of Fig. 4a is shown in Fig. 4c, where the amplitude spectrum is seen to extend beyond 3 THz. The sharp spectral features are water lines, from the residual water vapor present in the apparatus.

Two effects reduce the spectral extent of the measured pulse (Fig. 4c). These are the frequency-dependent transfer function[19] of the THz optical system (Fig. 1b) and the THz absorption in the sapphire (SOS) chips. The absorption of sapphire has been previously measured[20]. The transmission function describing these two effects is presented in Fig. 4d, for our focusing geometry and the SOS chip thickness of 0.46 mm. Dividing the measured spectrum in Fig. 4c by this transmission function we obtain Fig. 4e, which is the product of the receiver and transmitter spectral response. Because the transmitter and receiver are identical, by the reciprocity theorem[21] the transmitted spectrum is identical to the receiver response, and is given by the square root of Fig. 4e, shown in Fig. 4f.

Theoretical Time-Dependent Response Function

In the small antenna limit corresponding to the Hertzian dipole, the generated radiation field is proportional to the time-derivative of the current pulse. Based on our study we conclude that the current in the antenna is mainly determined by the intrinsic response of the semiconductor itself. As a first-order approximation, we consider the complex conductance of ion-implanted silicon to be described by Drude theory with the scattering time and carrier lifetime reduced due to the ion implantation. Because we excite "hot" carriers, there will be some variation in the effective mass during carrier cooling, which in

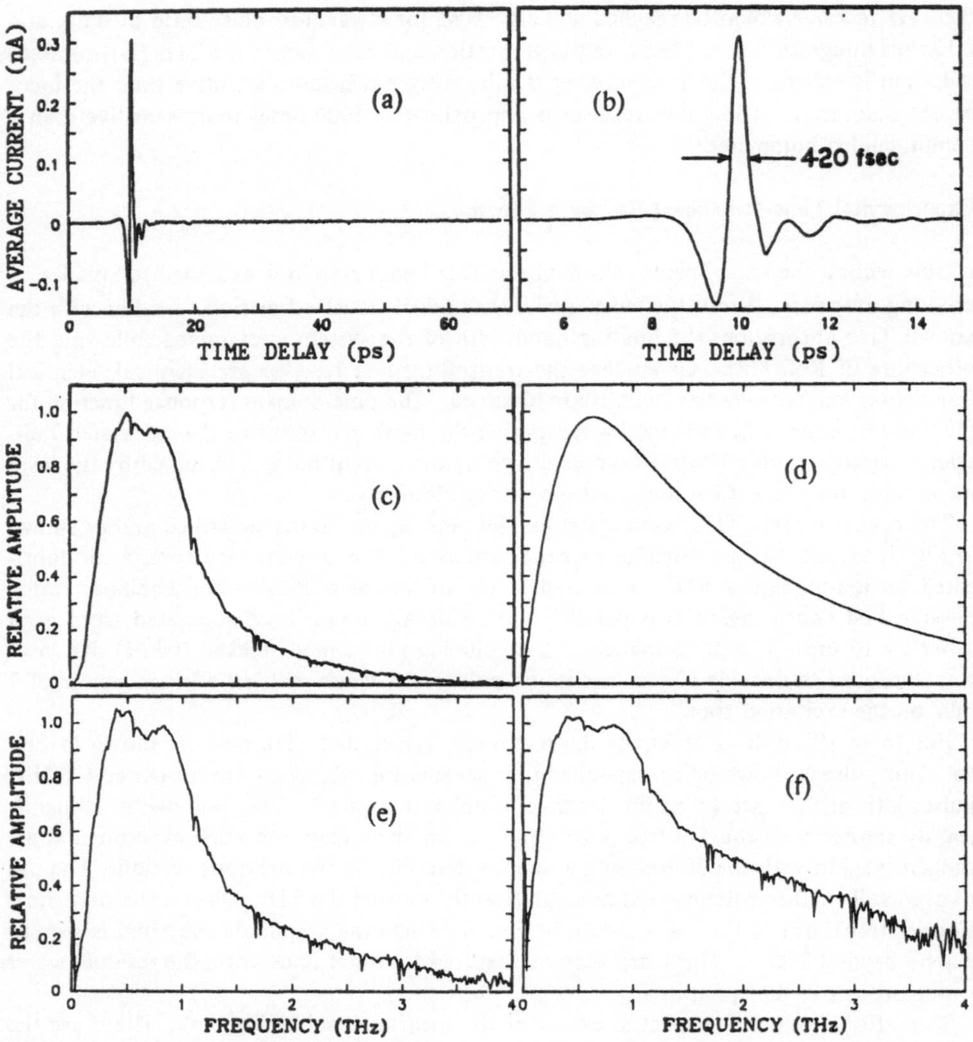

Figure 4. (a) Measured THz pulse to 70 psec. (b) Measured THz pulse on an expanded time scale from 5 to 15 psec. (c) Amplitude spectrum to 4 THz of Fig. 4b. (d) Transmission function to 4 THz. (e) Amplitude spectrum of Fig. 4c divided by transmission function. (f) Amplitude spectral response of transmitter and receiver.

our approximation can be neglected. The intrinsic time-domain response function will now be derived for a semiconductor described by the simple Drude formalism. For this case the free carriers are considered as classical point charges subject to random collisions. Here the simplest version of this model is assumed, for which the collision damping is independent of the carrier energy and for which the frequency dependent complex conductivity $\sigma(\omega)$ is given by

$$\sigma(\omega) = \sigma_{dc} \frac{i\Gamma}{\omega + i\Gamma}, \tag{1}$$

where $\Gamma = 1/\tau$ is the damping rate and τ is the average collision time. The dc conductivity is given by $\sigma_{dc} = e\mu_{dc}N$, where e is the electron charge, μ_{dc} is the dc mobility and N is the carrier density. Recent time-domain spectroscopy measurements[22] on lightly doped silicon from low frequencies to beyond 2 THz, are in good agreement with this relationship. The following procedure is similar to that of Ref. (18). It is helpful to recast the formalism into a frequency dependent mobility as

$$\mu(\omega) = \mu_{dc}\frac{i\Gamma}{\omega + i\Gamma}\,. \tag{2}$$

The dc current density is given by $J_{dc} = \sigma_{dc}E$, or equivalently $J_{dc} = eE\mu_{dc}N$, where for the simple case considered here E is a constant electric field. Since the current is linear in N, for a time dependent carrier density N(t), the time dependent current density can be written as

$$J(t) = eE\int_{-\infty}^{t} \mu(t-t')N(t')dt', \tag{3}$$

where μ(t-t′) is the time-domain response function for the mobility. This function is determined by the inverse transform of the frequency dependent mobility to be the causal function

$$\mu(t-t') = \mu_{dc}\Gamma e^{-\Gamma(t-t')} \tag{4}$$

which vanishes for negative (t-t′).

To understand the operation of the photoconductive switch it is useful to rewrite the basic Eq. (3) in the equivalent form,

$$J(t) = eEA\int_{-\infty}^{t} \mu(t-t')\int_{-\infty}^{t'} R_c(t'-t'')I(t'')dt''dt', \tag{5}$$

where I(t″) is the normalized intensity envelope function of the laser pulse, A is a constant giving the conversion to absorbed photons/volume and R_c is the response function describing the decay of the photogenerated carriers. By defining a new photocurrent response function j_{pc}(t-t′), we can rewrite Eq.(5) in the following way

$$J(t) = \int_{-\infty}^{t} j_{pc}(t-t')I(t')dt', \tag{6}$$

where j_{pc}(t-t′) is obtained by evaluating Eq.(5) with a delta function δ(t″) laser pulse. Given the causal function R_c(t′-t″)= exp-(t′-t″)/τ_c, describing a simple exponential decay of the carriers with the carrier lifetime τ_c (significantly longer than the average collision time τ) for positive (t′-t″) and vanishing for negative (t′-t″), and that μ(t-t′) is given by the Drude response of Eq. (4), the causal response function j_{pc}(t*) is then evaluated to be

$$j_{pc}(t^*) = \frac{\mu_{dc}eEA}{1-\tau/\tau_c}\left(e^{-t^*/\tau_c} - e^{-t^*/\tau}\right) \tag{7}$$

for positive t* = (t-t′) and shown to vanish for negative t*. In the short pulse limit of the ultrafast excitation pulses, the time dependence of the photocurrent J(t) is approximately equal to that of the photocurrent response function $j_{pc}(t^*)$. For a long carrier lifetime, the time dependence of $j_{pc}(t^*)$ is described by a simple exponential rise with a risetime of the order of $\tau = 1/\Gamma$, which in lightly doped silicon is equal to 270 fsec and 150 fsec for the electrons and holes, respectively[22]. As these results show, the material response can be slow compared to the duration of the ultrafast laser excitation pulses which are typically of the order of 60 fsec.

The time-dependent response function described by Eq. (7) is calculated in Fig. 5 for the two cases; τ = 270 fsec and $\tau_c = \infty$, and τ = 270 fsec and τ_c = 600 fsec. The result for infinite carrier lifetime has the following intuitive interpretation. After the instantaneous creation of carriers, the initial current and mobility is zero. The carriers then accelerate ballistically, as determined by the applied electric field, their charge and effective mass. This acceleration continues for approximately a time equal to the scattering time τ, after which the velocity and current equilibrate to their steady-state value. This discussion will now be shown to accurately describe the mathematical dependence of Eq. (7). With $\tau_c = \infty$, Eq. (7) is equal to

$$j_{pc}(t^*) = \mu_{dc} eEA(1 - e^{-t^*/\tau}), \tag{8}$$

which for times short compared to τ reduces to

$$j_{pc}(t^*) = \mu_{dc} eEAt^*/\tau. \tag{9}$$

Remembering that for Drude theory $\mu_{dc} = e/(m^*\Gamma)$, Eq. (9) is equivalent to

$$j_{pc}(t^*) = Aet^*(eE/m^*), \tag{10}$$

which describes the ballistic acceleration eE/m*.

The receiver response is identical to that of the transmitter, as demanded by the reciprocity principle[21]. Consequently the photoconductive response function is the same for both the transmitter and receiver, and their performance is both limited by the ballistic acceleration of the carriers.

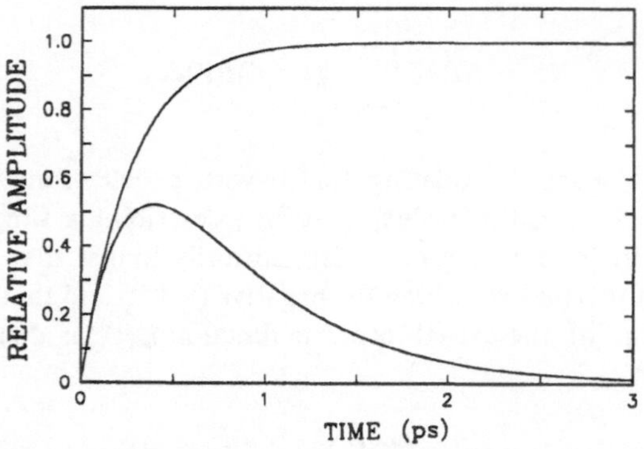

Figure 5. Calculated photoconductive response function to 3 psec with the scattering time τ = 270 fsec and infinite carrier lifetime τ_c (upper curve) and with τ_c = 600 fsec (lower curve).

Comparison of Theory with Experiment

For the photoconductive switches considered here, we assume the time-domain response function $j_{pc}(t^*)$ to be given by Eq. (7). This response function is then convolved with a Gaussian shaped laser pulse with a FWHM of 60 fsec, as prescribed by Eq. (6). The carrier lifetime τ_c has been measured[23] to be 600 fsec for ion-implanted SOS. As will be shown, good agreement with experiment is obtained with an average collision time τ = 190 fsec. With these parameters the calculated shape of the current pulse in the photoconductive switch and the Hertzian dipole antenna is presented in Fig. 6a. The time derivative of this pulse is given in Fig. 6b, where an extremely fast transient, corresponding to the rising edge of the current pulse, is seen. The numerical Fourier transform of Fig. 6b, presented in Fig. 6c, is the predicted amplitude spectrum of the transmitter. In Fig. 6d, this spectrum is compared with the amplitude spectrum of the transmitter/receiver from Fig. 4f; the agreement is excellent. Thus, we have determined an experimentally self-consistent time-domain response function describing the current in the Hertzian dipole antenna for both the transmitter and receiver. For longer antennas for which the radiated pulse is no longer the time-derivative of the current pulse, the calculated current pulse is Fourier analysed and the resulting spectral amplitudes are put into the antenna response to determine the emitted pulse.

In summary, we have shown that the 10-μm-long antenna imbedded in the coplanar transmission line has electrical properties much faster than the semiconductor. Consequently, the performance of both the transmitter and receiver is completely determined (and limited) by the intrinsic response time of the semiconductor. With the 10-μm-long antennas, it is now possible to directly study the dynamical response of free carriers in a variety of semiconductors.

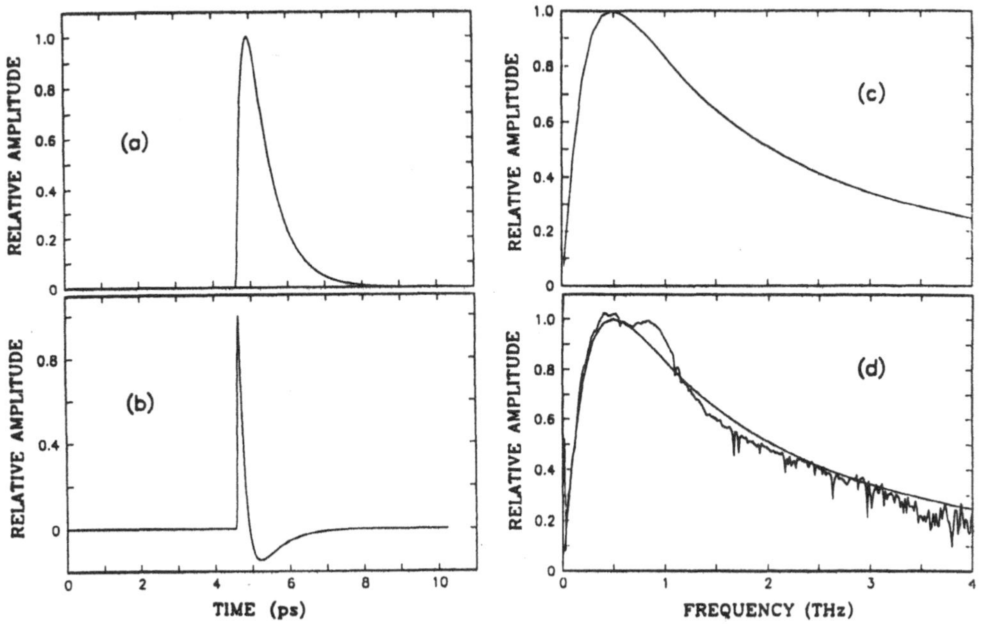

Figure 6. (a) Calculated current pulse (11 psec timescale) in semiconductor and antenna. (b) Time-derivative of current pulse. (c) Amplitude spectrum to 4 THz of Fig. 6b. (d) Comparison of Figs. 4f and 6c.

A High-Performance GaAs THz Source

A different-type, high-performance optoelectronic source chip[14], fabricated on semi-insulating GaAs is shown in Fig. 7. The simple coplanar transmission line structure consists of two 10-μm-wide metal lines separated by 80 μm. Irradiating near the metal-semiconductor interface (edge) of the positively biased line with focused ultrafast laser pulses produces synchronous bursts of THz radiation. This occurs because each laser pulse creates a spot of photocarriers in a region of extremely high electric field, the trap enhanced field[15] (TEF). The consequent acceleration of the carriers generates the burst of radiation. The CPM dye laser provides 60 fsec excitation pulses with an average power of 5 mW at the 5μm diameter excitation spot. A major fraction of the laser generated burst of THz radiation is emitted into the GaAs substrate in a cone normal to the interface and as before is collected and collimated by the crystalline silicon lens attached to the back side of the chip. This GaAs TEF source chip is completely compatible with the previously described optoelectronic THz beam system.

Because of its long carrier lifetime, the use of this chip is limited only to the transmitter. The receiver operates in conjunction with a current amplifier, whose output signal-to-noise ratio is determined by the ratio of the on-time of the photoconductive switch to its off-time. Stated more precisely the noise on the current amplifier is determined by the average resistance of the photoconductive switch. Upon illumination by the laser pulse this resistance typically changes from MegaOhms to a few hundred Ohms and recovers with the carrier lifetime. Therefore, the switch must reopen relatively rapidly, otherwise the average resistance is lowered and the system noise increases. For semi-insulating

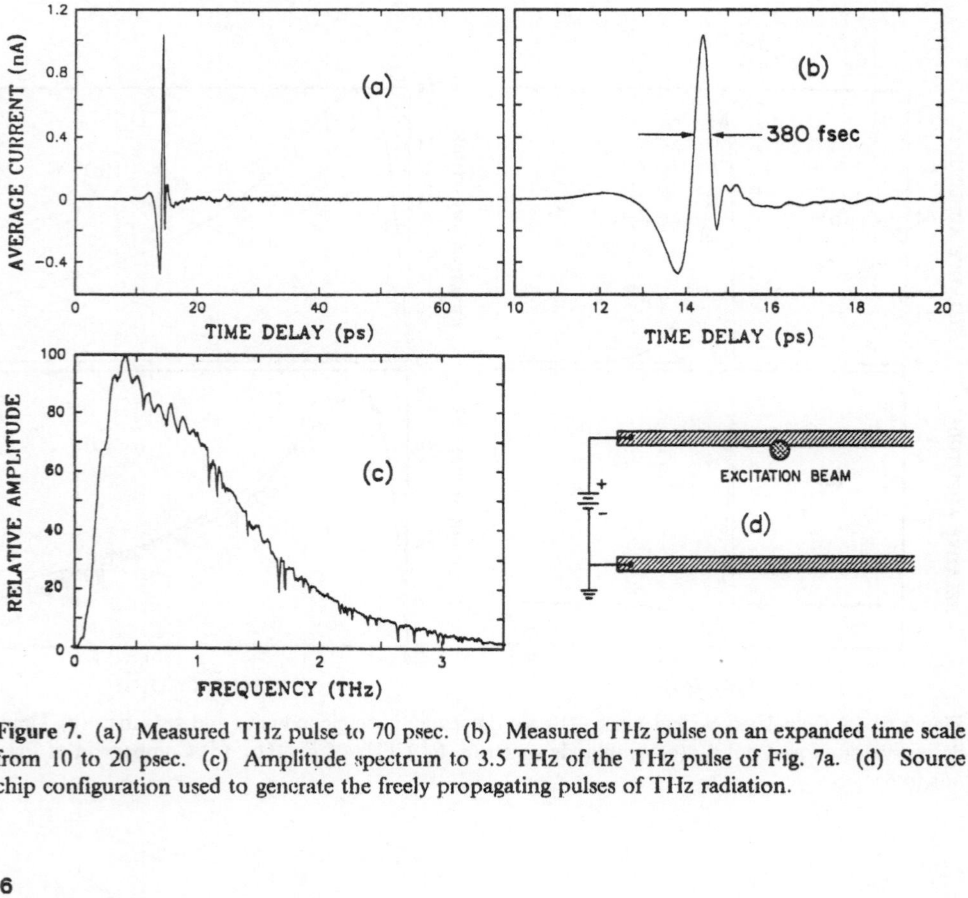

Figure 7. (a) Measured THz pulse to 70 psec. (b) Measured THz pulse on an expanded time scale from 10 to 20 psec. (c) Amplitude spectrum to 3.5 THz of the THz pulse of Fig. 7a. (d) Source chip configuration used to generate the freely propagating pulses of THz radiation.

GaAs the carrier lifetime is too long to allow for any measurement capability, resulting in an effective signal-to-noise ratio of zero. Thus, again the THz radiation detector of choice is the ion-implanted SOS detection chip with the antenna geometry shown in Fig. 1a, but with the smaller and thus faster 10-μm-long antenna.

The measured THz pulse emitted from the laser excited GaAs TEF source chip with +60V bias across the transmission line is shown in Fig. 7a, and on an expanded time scale in Fig. 7b. The measured pulsewidth with no deconvolution is seen to be 380 fsec. At the time these results were obtained[14], they were the shortest directly measured THz pulses; the dip on the falling edge was the sharpest feature ever observed with an ion-implanted detector and indicated a response time faster than 190 fsec. The numerical Fourier transform of the pulse of Fig. 7a, as presented in Fig. 7c, extends to beyond 3 THz; the sharp line structure is due to residual water vapor present in the system.

Measurements to 6 THz with an Ultrafast Flip-Chip Receiver

Using a newly developed receiver together with the same GaAs TEF source described above, we now measure a TEF source spectrum with a FWHM bandwidth almost twice as broad as the initial characterization[14] (shown in Fig. 7) and exceeding that obtained by a recent interferometric characterization of this same source[16]. The system changes enabling the doubling of the measured bandwidth will now be described. Firstly, the positions of the paraboloidal mirrors were set to have a unity transfer function for the THz radiation. For a plane-wave, 5-mm dia., Gaussian beam exiting the silicon collimating lens, the THz transfer function of the paraboloidal mirror system is unity with the faces of the lenses at the foci of the mirrors separated from each other by two focal lengths[19]. Secondly, exceptional care was taken to match the focus of the silicon lens to the antenna position. For the on-axis focusing a series of observations were made using lenses of the same curvature, but with thicknesses varying in steps of 50 μm. In the plane of the chip the position of the focus was adjusted to +/- 20 microns. Thirdly, as shown in Fig. 8, a

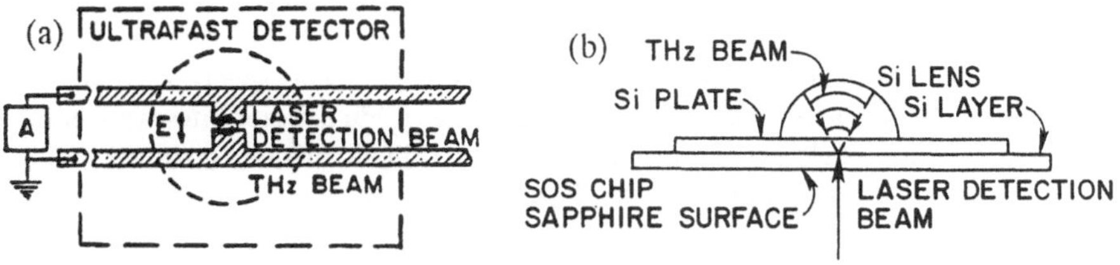

Figure 8. (a) Receiving antenna on ion-implanted SOS. The overlaying Si plate is indicated by the dashed line. (b) Cross-section of the flip-chip.

new flip-chip arrangement was used to eliminate the absorption of the incoming THz radiation by the sapphire substrate of the SOS detection chip. For this case the antenna on the SOS chip is covered by contacting a 0.5 mm thick polished plate of high-resistivity (10kΩcm) silicon. The collimating silicon lens is then contacted to this silicon plate. In contrast to the previous approach, the antenna is now driven by the ultrashort laser sampling pulses entering the sapphire side (back-side) of the chip and focusing onto the antenna on the Si layer.

The THz pulse measured with this ultrafast flip-chip receiver is shown in Fig. 9a. The feature at 17 psec is a reflection of the THz pulse from the sapphire/air interface. On the trailing edge of the pulse, rise and fall times faster than 135 fsec are observed, as shown

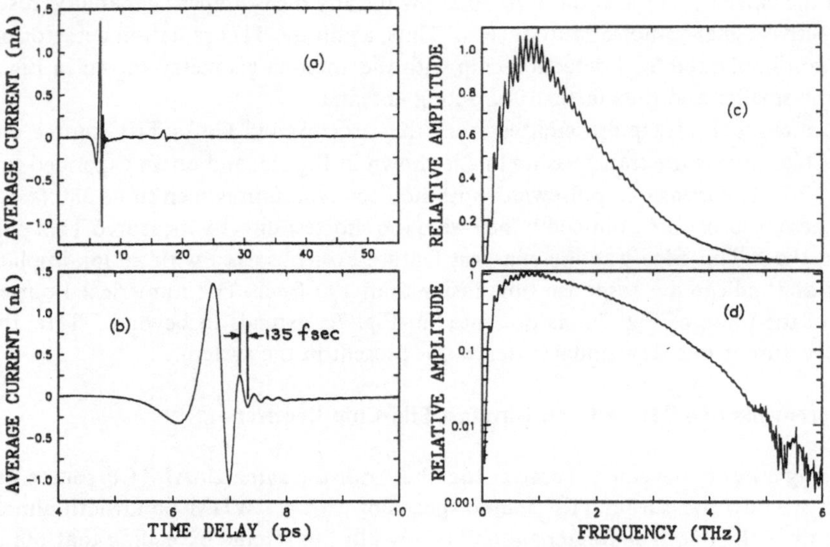

Figure 9. (a) THz pulse measured by the ultrafast flip-chip receiver. (b) Measured THz pulse on an expanded time scale from 4 to 10 psec. (c) Numerical Fourier transform to 6 THz of Fig. 9a. (d) Logarithmic presentation of Fig. 9c.

in the expanded view of Fig. 9b. This exceptional time resolution demonstrates that the ion-implanted SOS photoconductive receiver can be much faster than is generally realized, but is consistent with the earlier prediction of a 150 fsec time resolution[18]. This fast time resolution is illustrated by the 150 fsec calculated radiation pulse shown in Fig. 6b. The numerical Fourier transform of the measured pulse of Fig. 9a peaks at approximately 1 THz and extends to 6 THz, as shown in Figs. 9c and 9d. The spectral oscillations are caused by the reflection at 17 psec. For these data the residual water lines were eliminated by enclosing the entire THz system in an air-tight box filled with dry nitrogen vapor.

As described earlier and illustrated in Fig. 4, the initial characterization of the THz receiver[18] compared theory with measurements for which the THz optical transfer function peaked at 0.6 THz and significantly attenuated the higher frequency components. In addition, THz absorption by the sapphire substrate further attenuated the higher frequency components. When these effects were factored out of the data, reasonable agreement between the measured and the calculated spectrum was obtained as shown in Fig. 6d. This factorization is now unnecessary for the results presented in Fig. 9, because the transfer function is unity and independent of frequency, and the absorption in the sapphire substrate of the receiver has been eliminated by the flip-chip geometry. Therefore, in the following discussion we consider the calculated ballistic receiver response to describe the actual receiver.

The calculated amplitude spectrum of the ballistic receiver response of Fig. 6c is displayed as the upper smooth curve in Fig. 10a; the corresponding ballistic power spectrum, equal to the square of this curve, is the lower curve. Because the measured spectrum is the product of the transmitter and receiver amplitude spectra, it is meaningful to compare the measured spectrum with the ballistic receiver power spectrum shown in Figs. 10a and 10b. It is quite significant that the (FWHM) measured spectrum is broader than the ballistic power spectrum, and that it is higher than the power spectrum until 4.5 THz, after which it drops significantly with increasing frequency. Two contributions to this high frequency drop-off are the increasing THz absorption of the 0.5mm-thick GaAs source chip, and that at the highest frequencies the antenna may be too large to be in the Hertzian dipole limit. The broad measured bandwidth demonstrates the high performance

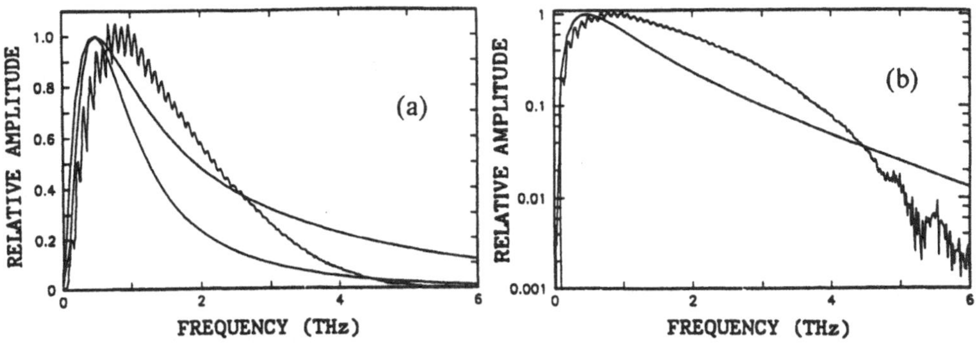

Figure 10. (a). Calculated ballistic receiver amplitude spectrum (upper curve) and power spectrum (lower curve), together with the measured spectrum. (b) Logarithmic presentation of Fig. 10a.

of the entire THz system and experimentally confirms the predicted broadband response of the ballistic photoconductive receiver.

The fact that the measured spectrum exceeds the ballistic power spectrum indicates that the frequency response of the GaAs TEF source is broader and the time dependence is faster than that of the ion-implanted SOS ballistic receiver. A possible explanation for the extremely rapid time dependence of the photocurrent for the GaAs TEF source could be the rapid and large changes in the mobility of the photocarriers associated with inter-valley scattering and carrier cooling. Clearly, the calculation of the proper photoconductive response function for the GaAs TEF source remains an important problem. Also, the relative strength and time dependence of THz radiation from the inverse Franz-Keldysh effect needs to be understood for this source[24,25]. In summary, we have experimentally confirmed the broad-band response of the SOS ballistic photoconductive receiver, and we have used the receiver to infer even faster carrier dynamics in GaAs.

REFERENCES

1. D.H. Auston, K.P. Cheung and P.R. Smith, Picosecond photoconducting Hertzian dipoles, Appl. Phys. Lett. 45:284-286 (1984).
2. Ch. Fattinger and D. Grischkowsky, Point source THz optics, Appl. Phys. Lett. 53:1480-1482 (1988); THz beams, Appl. Phys. Lett. 54:490-492 (1989).
3. G. Mourou, C.V. Stancampiano, A. Antonetti and A. Orszag, Picosecond microwave pulses generated with a subpicosecond laser-driven semiconductor switch, Appl. Phys. Lett. 39:295-297 (1981).
4. R. Heidemann, Th. Pfeiffer and D. Jager, Optoelectronically pulsed slot-line antennas, Electronics Lett. 19:317 (1983).
5. A.P. DeFonzo, M. Jarwala and C.R. Lutz, Transient response of planar integrated optoelectronic antennas, Appl. Phys. Lett. 50:1155-1157 (1987); Optoelectronic transmission and reception of ultrashort electrical pulses, Appl. Phys. Lett. 51:212-214 (1987).

6. Y. Pastol, G. Arjavalingam, J.-M. Halbout and G.V. Kopcsay, Characterization of an optoelectronically pulsed broadband microwave antenna, Electron. Lett. 24:1318 (1988).
7. Y. Pastol, G. Arjavalingam, J.-M. Halbout, Characterization of an optoelectronically pulsed equiangular spiral antenna, Electron. Lett. 26:133 (1990).
8. P.R. Smith, D.H. Auston and M.C. Nuss, Subpicosecond photoconducting dipole antennas, IEEE J. Quantum Elect. 24:255-260 (1988).
9. M. van Exter, Ch. Fattinger and D. Grischkowsky, High brightness THz beams characterized with an ultrafast detector, Appl. Phys. Lett. 55:337-339 (1989).
10. Ch. Fattinger and D. Grischkowsky, Beams of Terahertz electromagnetic pulses, "OSA Proc. on Psec. Elect. and Optoelect.," T.C.L. Gerhard Sollner and D.M. Bloom, Eds. (Optical Society of America, Washington, DC), Vol.4 (1989).
11. M. van Exter and D. Grischkowsky, Characterization of an optoelectronic teraHz beam system, IEEE Trans. Microwave Theory and Techniques, 38:1684-1691 (1990).
12. D.R. Dykaar, B.I. Greene, J.F. Federici, A.F.J. Levi, L.N. Pfeiffer and R.F. Kopf, Log-periodic antennas for pulsed terahertz radiation, Appl. Phys. Lett. 59:262-264 (1991).
13. X.-C. Zhang, B.B. Hu, J.T. Darrow and D.H. Auston, Generation of femtosecond electromagnetic pulses from semiconductor surfaces, Appl. Phys. Lett. 56:1011-1013 (1990).
14. N. Katzenellenbogen and D. Grischkowsky, Efficient generation of 380 fs pulses of THz radiation by ultrafast laser pulse excitation of a biased metal-semiconductor interface, Appl. Phys. Lett. 58:222-224 (1991).
15. S.E. Ralph and D. Grischkowsky, Trap-enhanced electric fields in semi-insulators; the role of electrical and optical carrier injection, Appl. Phys. Lett. 59:1972-1974 (1991).
16. S.E. Ralph and D. Grischkowsky, THz spectroscopy and source characterization by optoelectronic interferometry, Appl. Phys. Lett. 60:1070-1072 (1992).
17. C. Johnson, F.J. Low and A.W. Davidson, Germanium and germanium-diamond bolometers operated at 4.2K, 2.0K, 1.2K, 0.3K, and 0.1K, Optical Engr. 19:255 (1980).
18. D. Grischkowsky and N. Katzenellenbogen, Femtosecond pulses of terahertz radiation: physics and applications, "Proc. of the Psec. Elect. and Optoelect. Conf.," Salt Lake City, Utah, March 13-15, 1991, T.C.L. Gerhard Sollner and Jagdeep Shah, Eds. (Opt. Soc. of Am., Washington D.C. 1991), Vol.9, pp. 9-14.
19. J.C.G. Lesurf, "Millimetre-Wave Optics, Devices and Systems," Adam Hilger, Bristol, England (1990).
20. D. Grischkowsky, S. Keiding, M. van Exter and Ch. Fattinger, Far-infrared time-domain spectroscopy with terahertz beams of dielectrics and semiconductors, J.Opt.Soc.Am.B. 7:2006-2015 (1990).
21. G.D. Monteath, "Applications of the Electromagnetic Reciprocity Principle," Oxford:Pergamon Press (1973).
22. M. van Exter and D. Grischkowsky, Optical and electronic properties of doped silicon from 0.1 to 2 THz, Appl. Phys. Lett. 56:1694-1696 (1990); Carrier dynamics of electrons and holes in moderately doped silicon, Phys. Rev. B15. 41:12,140-12,149 (1990).
23. F.E. Doany, D. Grischkowsky and C.-C. Chi, Carrier lifetime versus ion- implantation dose in silicon on sapphire, Appl. Phys. Lett. 50:460-462 (1987).
24. B.B. Hu, X.-C. Zhang, and D.H. Auston, Terahertz radiation induced by subband-gap femtosecond optical excitation of GaAs, Phys. Rev. Lett. 67:2709 (1991).
25. S.L. Chuang, S. Schmitt-Rink, B.I. Greene, P.N. Saeta, and A.F.J. Levi, Optical rectification at semiconductor surfaces, Phys. Rev. Lett. 68:102-105 (1992).

TERAHERTZ RADIATION FROM ELECTRO-OPTIC CRYSTALS

X.-C. Zhang, Y. Jin, T.D. Hewitt, T. Sangsiri

Physics Department
Rensselaer Polytechnic Institute, Troy, NY

L. Kingsley, and M. Weiner

US Army LABCOM, Pulse Power Center
Electronic Technology and Devices Laboratory
Fort Monmouth, NJ 07703

INTRODUCTION

With illumination by femtosecond laser pulses at an oblique incident angle, a semiconductor sample emits subpicosecond submillimeter-wave radiation[1]. This phenomenon was interpreted as radiation from accelerated photocarriers in the surface depletion field[1] and the dependence of the radiation on crystal rotation symmetry was explained as strong evidence of $\chi^{(3)}$ related optical rectification[2]. More recently, quantum well structures have been used to limit the carrier transport. This allows examination of the mechanism of optical rectification as well as the study of electron coherent oscillations in a double-well potential[3]. To distinguish between the contribution from carrier transport and from optical rectification is crucial for a physical understanding of the origin of the radiation, as well as for realizing an intense THz source application.

In this paper we report the coherent measurement (phase and amplitude) of optically induced femtosecond electromagnetic radiation from a variety of inorganic and organic electro-optic crystals. In semiconductor bulk crystals with the bandgap less than the incident photon energy, the optically induced submillimeter-wave radiation has contributions from both the ultrafast photocarrier acceleration in static field region and the second order nonlinear optical rectification. Our results clearly demonstrate the contribution from both the subpicosecond optical rectification (SOR) and the photocarrier transport effect. When the subpicosecond optical rectification was separated from the background radiation from carrier transport and the semiconductor was excited at resonance, we observed an enhancement of the submillimeter-wave nonlinear susceptibility. We also observed that when the laser energy was tuned above the bandgap the change of the polarity of the radiated electromagnetic field was due to the resonant behavior of the subpicosecond nonlinear susceptibility. In crystals with a bandgap greater than the incident photon energy, the photon induced submillimeter-wave radiation is mainly due to the nonlinear optical rectification. In organic crystals, the strength of the rectified signal from an organic salt showed two orders stronger than that from inorganic crystals $LiTaO_3$ and $LiNbO_3$. We also report the use of a magnetic field to control a THz signal.

EXPERIMENTAL SETUP

The experimental setup is a time-resolved optoelectronic coherent sampling arrangement, and it has been described elsewhere[3]. Figure 1 schematically illustrates this arrangement. A cw Ar laser pumped mode-locked Ti:sapphire laser (Coherent MIRA) was used as the source of optical pulses. The MIRA laser produced an output pulse energy greater than 10 nJ at a repetition rate of 76 MHz and a pulse duration less than a 200 femtosecond.

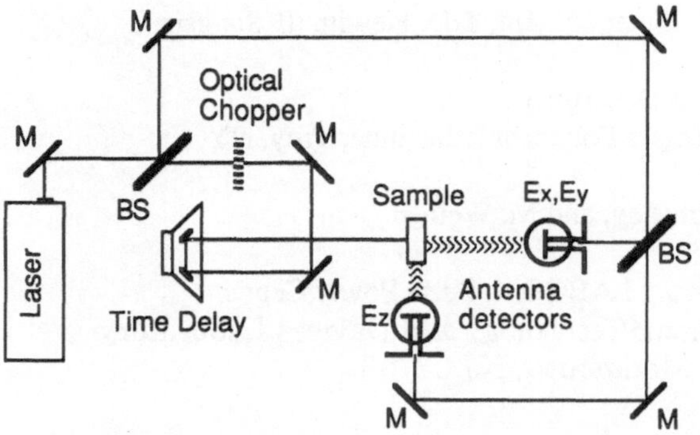

Figure 1. Experimental setup of optically induced terahertz electromagnetic radiation.

The energy spectrum of the laser beam (full width at half maximum) is about 14 meV at a wavelength about 820 nm. The laser beam was split into two parts by a beamsplitter with a 0.05/0.95 reflection/transmission ratio. The stronger optical beam passed through a variable time delay stage, and illuminated the sample crystals with a spot diameter of approximately 6 millimeters. The weaker optical beam, typically less than 20 mW, was used for optical gating at a photoconducting detector. The radiated submillimeter-wave beam in the forward direction was focused on a photoconductor attached on a 50 μm dipole antenna. This dipole antenna is a submillimeter-wave polarization sensitive detector. A photocurrent was developed at the antenna when the submillimeter-wave radiation was spatially and temporally overlapped with the optical gating pulse. The temporal measurement was achieved by varying the time-delay between the excited laser pulse illuminated on the crystals (strong optical beam) and the trigger laser pulse focused on the detector (weak optical beam). The measurements were taken at room temperature.

OPTICAL RECTIFICATION

Normal optical incidence on the GaAs surface was used to distinguish the subpicosecond optical rectification signal from the carrier transport effect. Under normal incidence the THz signal from the accelerated carriers driven by the surface depletion field can not be measured in the forward direction because no radiation is emitted along the charge acceleration direction. However, subpicosecond optical rectification can radiate in the forward direction under normal incident illumination[4]. Figure 2 shows the THz radiation (in the forward direction) from a <111> GaAs bulk sample for three incident angles (ϕ=0 and ϕ= $\pm$50 degrees). The photon energy is 1.52 eV.

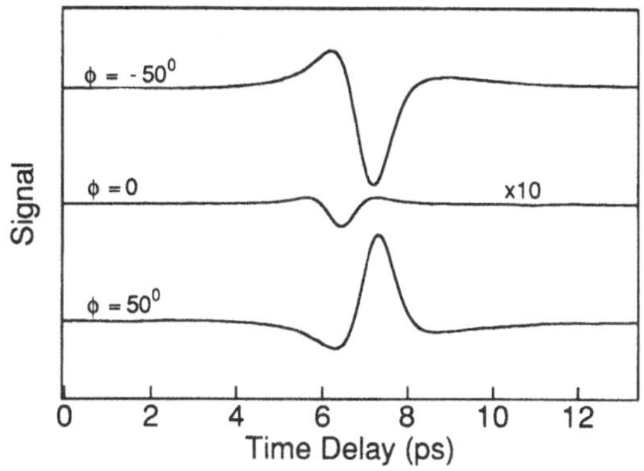

Figure 2. THz radiation from a <111> oriented GaAs with normal (center trace) and ±50 degrees (lower and upper traces) optical incident angles. Temporal broadening of the THz radiation waveform is pronounced with the oblique optical beams.

The center trace (ϕ=0) is due to subpicosecond optical rectification only, while the upper and lower traces (ϕ=±50) combine radiation from both the SOR and carrier transport effects. The timing shifts of the waveforms taken at oblique angles are due to an increase in the THz beam path from tilting the sample.

Since the carrier transport process may last much longer than the polarization dephasing time, this relatively slow feature should be reflected in time-resolved measurements. The temporal broadening of the THz signal due to the transient photocurrent, can be well resolved when oblique optical incident angles are used. We also investigated the dependence of the temporal broadening on photocarrier density (with the photocarrier densities ranging from $2.2x10^{13}/cm^3$ to $4.4x10^{16}/cm^3$), but did not observe any change in the temporal broadening within our temporal resolution. However, at higher optical fluence, a waveform change is expected, due to transition saturation, signal absorption and the screening of the static electric field. In principle, if the optical pulse duration is longer (or shorter) than the polarization dephasing time the duration of the SOR should be close to the optical pulse duration (or polarization dephasing time) since the SOR is proportional to the negative second time derivative of the nonlinear polarization. It is possible to observe the polarization dephasing time if a much shorter optical pulse is used. Due to the finite bandwidth of the detector we currently can not resolve the actual waveform of the SOR.

The SOR signal changes dramatically when the laser energy is tuned near the GaAs bandgap. The rectified signal is nearly unchanged for small photon energy tuning around 1.35 eV (70 meV lower than the bandgap). When the photon energy is tuned to the bandgap the amplitude of SOR signal increases, reaching a peak when the laser energy is just above the bandgap. Tuning the laser energy further above the bandgap results in a polarity change of the radiation signal. Figure 3a is a plot of the peak value of the SOR signal from a <111> GaAs sample vs optical excitation energy (normal incident angle). The energy resolution is limited by the laser spectrum which is about 14 meV. The resonance enhancement near the bandgap is pronounced. We measured about a 106 fold increase in the rectified field when the laser energy was tuned from 1.40 eV to 1.46 eV. Figure 3b contains two temporal waveforms at photon energies of 1.43 eV and 1.49 eV clearly demonstrating the polarity change of the rectified field. During laser energy tuning the temporal duration of the rectified field remained essentially unchanged. We measured the SOR signal from a <111> InGaAs/GaAs superlattice sample vs optical excitation energy at normal incidence. Besides the resonance due to GaAs (barriers), an additional resonance, which is due to the bound states in the InGaAs quantum wells, is observed at 1.40 eV. We also observed similar resonant behavior from <111> oriented CdTe samples. Both the resonance enhancement and sign reversal near the bandgap of CdTe (E_g is about 1.50 eV) are measured.

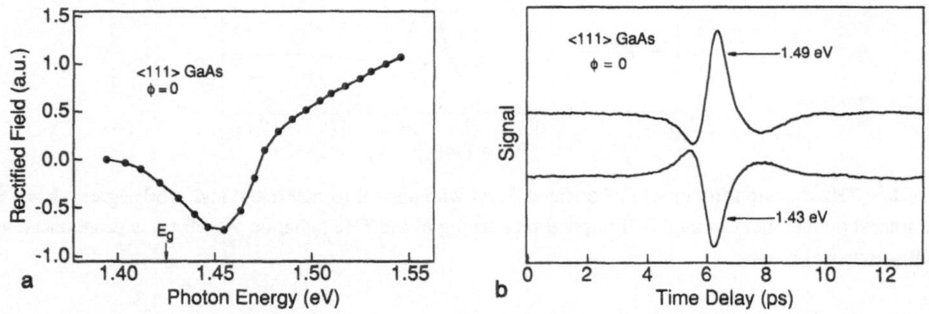

Figure 3. (a) The peak value of the rectified signal vs laser energy at normal incidence, from <111> GaAs; (b) Temporal waveforms of the radiated field at a photon energy of 1.49 eV (upper trace) and 1.43 eV (lower trace), respectively.

Resonance enhancement of the SOR signal from a <100> GaAs sample with off-normal optical illumination was observed previously, and the temporal waveform flip-over was explained as evidence of virtual photoconductivity[5]. However, the virtual current model can not generate forward radiation (longitudinal) under normal incidence. Our new result suggests that this polarity flip-over is due to a change in sign of the resonant susceptibility at the bandgap.

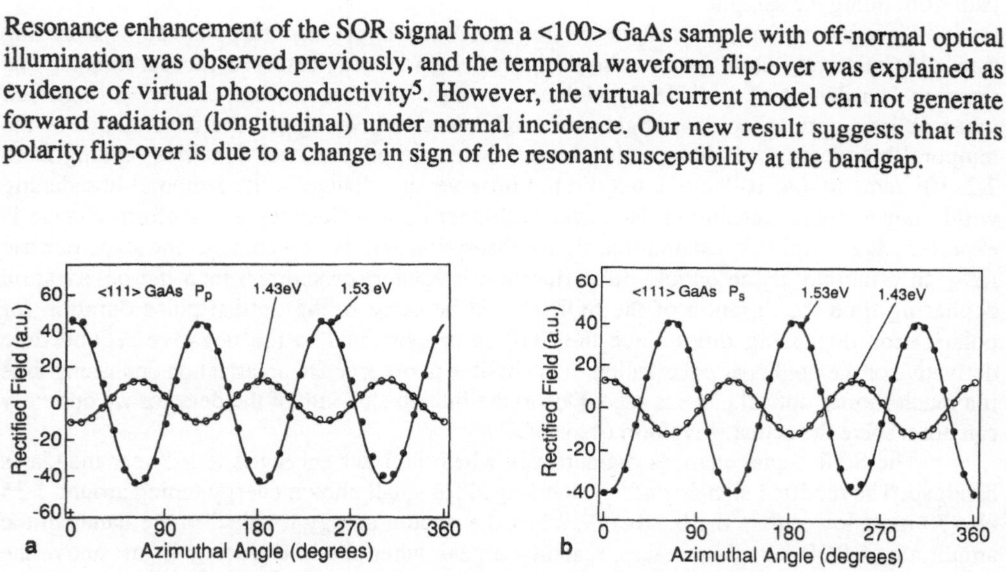

Figure 4. Rectified field vs azimuthal angle at two optical polarizations (a) parallel and (b) perpendicular to the detector axis. The open and solid dots are the experimental data at two photon energies (1.43 eV and 1.53 eV respectively). The curves are plots of the rectified field calculated from the nonlinear polarizations (a) P_p and (b) P_s.(the sign is reversed).

We measured the dependence of the SOR signal on crystallographic orientation. In this measurement a <111> oriented GaAs sample was rotated around its surface normal. Figure 4a and 4b show the SOR signal vs crystal azimuth angle at two optical energies (1.43 eV and 1.53 eV) and two optical polarizations (incident light polarized parallel and perpendicular to the detector axis), respectively. The solid and open dots are the experimental data. The SOR signal has opposite polarity at these two energies. A three-fold rotation symmetry was observed at both photon energies. Since the submillimeter-wave radiation is proportional to the negative second time derivative of the low frequency polarization, we can calculate the angular dependence of the nonlinear polarization (low frequency) to obtain information about the rectification. The optical rectification from a <111>B oriented GaAs under normal incidence, can be calculated from the nonlinear polarizations P_p and P_s corresponding to different incident optical polarization geometries (light polarization parallel and perpendicular to the dipole detector axis respectively). The nonlinear polarizations P_p and P_s (SI) are given by

$$P_p = -P_s = \frac{4d_{14}}{\sqrt{6}}\cos(3\theta)EE^* \quad (1)$$

Where the azimuthal angle θ is measured with respect to the [11-2] direction and d_{14} is a nonlinear optical susceptibility tensor element having a frequency response from dc to several terahertz (mainly characterized by the optical pulse duration[6]). Figure 4a and 4b also show the reversed rectification fields calculated from the nonlinear polarizations P_p and P_s respectively (curves are fit to the experimental data). The plots of the calculation from equation (1) fit the experimental data at 1.43 eV well. However, to fit the experimental data at 1.53 eV the sign of the susceptibility needs to be changed. In Figure 3 the rectification signal shows a sign reversal at the bandgap. After we changed the sign of d_{14} the data at 1.53 eV fit well with the calculations. We also calculated the nonlinear polarization from a <100> oriented GaAs. The calculation showed no SOR signal in the forward direction under normal incidence, regardless of the azimuthal rotation of the crystal. This calculated result has been verified in our experiments.

We also found intense THz radiation from an organic crystal (Dimethyl Amino 4-N-Methylstilbazolium Tosylate, or DAST). DAST is a member of the Stilbazolium salt family With optical excitation at a wavelength of 820 nm and a 150 fsec pulse duration, the magnitude of the rectified field from the this organic salt is 185 times larger than that from a $LiTaO_3$ crystal, and 42 times greater than the rectified signal from our best unbiased GaAs and InP crystals. Figure 5 shows the temporal waveform of the radiated field from a 1.5 mm thick DAST sample.

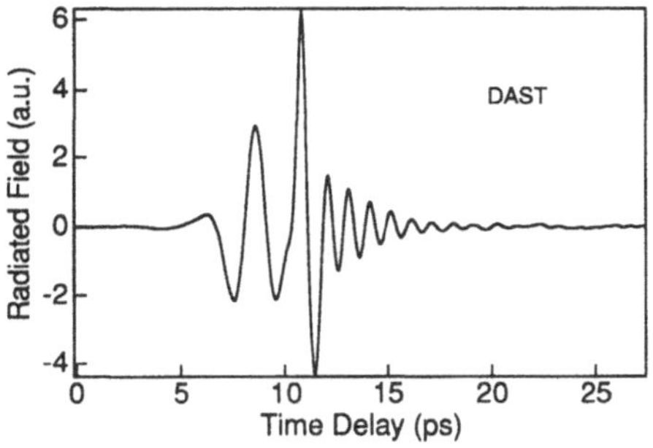

Figure 5. Temporal waveform of the rectified field from an organic crystal (DAST).

A persistent oscillation with a period of 1.04 ps following the main peak of the waveform was repeatable in all of the DAST samples, but disappeared when the DAST sample was replaced by a semiconducting (GaAs) or insulating ($LiTaO_3$) sample. The origin of this oscillation is not clear at present, but is possibly due to intrinsic molecular vibrational contributions to the subpicosecond submillimeter-waves. This organic crystal presently provides the most intense Terahertz radiated field among all of the natural non-externally biased materials we know. Due to the large second-order nonlinear susceptibilities and ease of use in a practical experimental setup, these organic crystals are excellent candidates for generating THz radiation.

MAGNETIC CONTROL OF TERAHERTZ BEAM

We have used an external magnetic field to control the direction and polarization of a THz beam radiated from a photoconductive semiconductor emitters. When an unbiased <100> GaAs wafer is illuminated with a femtosecond optical pulse at normal incidence and a magnetic field is applied perpendicular to the static electric field (perpendicular to the semiconductor surface normal) of the GaAs emitter, a THz beam radiates in the forward direction due to the Lorentz force on the moving charges (optically induced carriers). A potential application of this technique is the switching of intense THz radiation using a noncontact approach.

Figure 6 schematically illustrates the experimental configuration for the magnetic switching of a THz beam. The magnetic field is applied on a semiconductor emitter (unbiased <100> GaAs wafer).

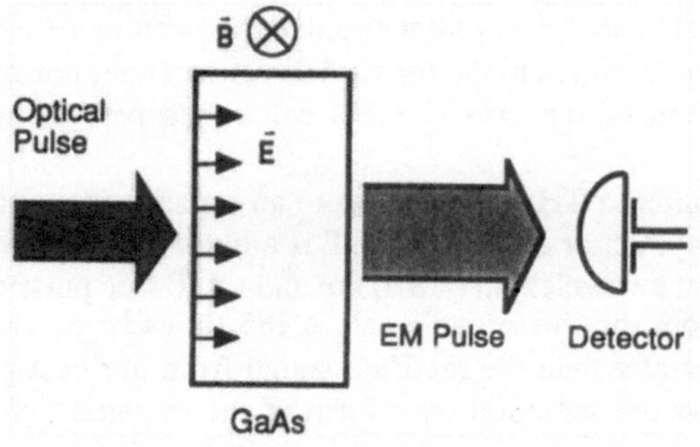

Figure 6. Experimental configuration for the magnetic control of a THz beam.

Figure 7 shows the THz beam radiation from this <100> GaAs wafer under an external magnetic field and normal incident optical angle.

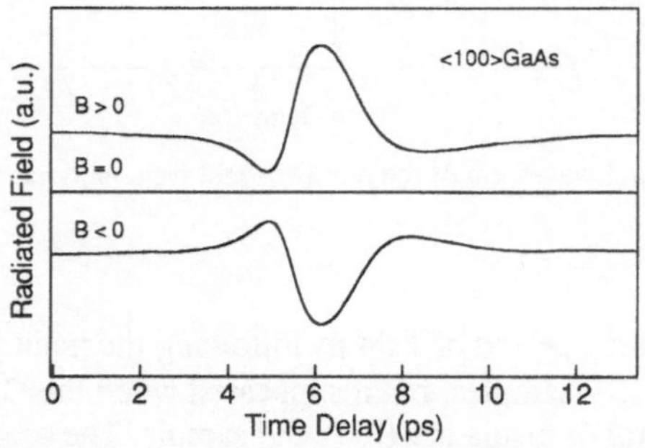

Figure 7. The terahertz radiation from an unbiased <100> GaAs wafer under an external magnetic field and normal incident optical excitation angle.

In this configuration the magnetic field is perpendicular to the surface electric field (static depletion field) and the THz beam detector axis (photoconducting antenna). The three curves are the temporal radiation waveforms at magnetic field B=0 (center), B>0 (top curve) and B<0 (reverse direction of the magnetic field, bottom curve), respectively. When the direction of the magnetic field is reversed the polarity of the radiation is also reversed (top and bottom curves). Figure 8 is a plot of the amplitude of the radiation field (peak value) versus the strength of the magnetic field. The solid dots are the experimental data and the line is the fit from a simple calculation. The magnitude of the radiated electric field component in the forward direction is found to be linearly proportional to the strength of the moderate magnetic field.

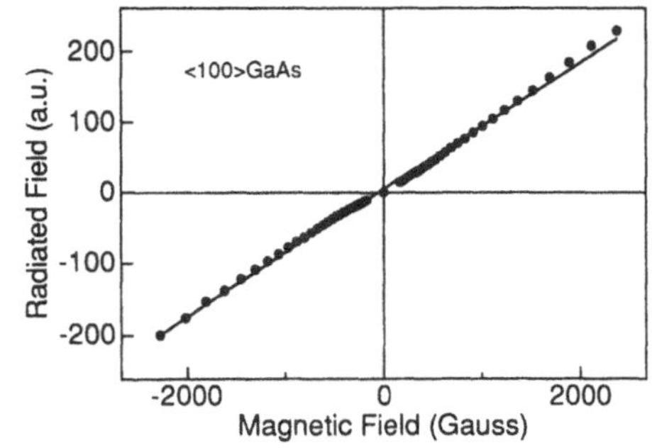

Figure 8. The peak value of terahertz radiation versus the strength of the external magnetic field.

We also applied an external magnetic field to the biased GaAs emitter. The control of the amplitude of the THz radiation and direction by the magnetic field has been confirmed also. It is possible to switch large photocurrents in the THz photoconducting antenna, which is in impractical with conventional contact methods.

ACKNOWLEDGMENTS

This work was supported by the National Science Foundation under grant number ECS-9211566 and by the U.S. Army Research Office Scientific Services Program under contract number DAAL03-91-C-0034. We would like to acknowledge D. Grischkowsky (IBM) for his support of this work.

REFERENCES

1 X.-C. Zhang, B.B. Hu, J.T. Darrow, and D.H. Auston, Appl. Phys. Lett. **56**, 1011 (1990); X.-C. Zhang and D.H. Auston, J. Appl. Phys. **71**, 326 (1992)

2 S.L. Chuang, S. Schmitt-Rank, B.I. Greene, P.N. Saeta, and A.F.J. Levi, Phys. Rev. Lett. **68**, 102 (1992)

3 H.G. Roskos, M.C. Nuss, J. Shah, K. Leo, D.A.B. Miller, A.M. Fox, S. Schmitt-Rink, and K. Kohler, Phys. Rev. Lett., **68**, 2216 (1992)

4 X.-C. Zhang, Y. Jin, K. Yang, and L.J. Schowalter, Phys. Rev. Lett., **69**, 2303 (1992)

5 B.B. Hu, X.-C. Zhang, and D.H. Auston, Phys. Rev. Lett., **67**, 2709 (1991)

6 When the optical pulse duration is less than or comparable with the crystal dielectric relaxation time or carrier momentum dephasing time T_2, the nonlinear polarization can hold longer than the optical pulse duration, and d_{14} can be very dispersed, especially under a resonant excitation.

PHOTOCONDUCTIVE SEMICONDUCTOR SWITCHES FOR HIGH POWER RADIATION

G. M. Loubriel, F. J Zutavern, G. J. Denison,
W. D. Helgeson, D. L. McLaughlin, and M. W. O'Malley

Sandia National Laboratories
Albuquerque, NM 87185

J. A. Demarest

Kaman Sciences Corporation
Albuquerque, NM 87110

ABSTRACT

In this paper we present the results of experiments on Si and GaAs Photoconductive Semiconductor Switches (PCSS). Our goal is to improve their performance for high power electromagnetic pulse generation. For Si, we show ways to alter carrier lifetime to achieve higher repetition rates, improvements in switch lifetimes to over 10^7 pulses at high field, and methods that reduce or eliminate thermal runaway and heating. For GaAs, the effect of focused trigger radiation was studied and a further reduction (by a factor of 100) in the required light energy was observed. The gain in these switches is now about 100,000 electrons generated per absorbed or trigger photon. It was further demonstrated that light can be piped through fiber optics to trigger multiple current filaments in GaAs. These results show the ability to control the location of the current filaments.

INTRODUCTION

Photoconductive semiconductor switches (PCSS) offer improvements over existing pulse power technology: higher voltage, higher current, more efficient, faster turn-on, faster turn-off, faster recovery for higher repetition rates, easier control, more precise timing control, longer device lifetime, more reproducible operation, simpler, smaller, lighter, and/or less expensive. The most significant possibilities are: 100 ps rise times and jitter, kilohertz (continuous) and megahertz (burst) repetition rates, scalable or stackable to hundreds of kilovolts and tens of kiloamps, optical control, and solid state reliability. The switching properties that have been achieved with our lateral Si (in the linear mode) and GaAs (in the non-linear, high gain switching mode[1]) are summarized in Table I. In this paper, we present experimental results which demonstrate some of these improvements and discuss the potential for future improvements.

Along with the technological improvements offered by PCSS also come trade-offs. Probably the most significant of these is the optical energy required to activate a

---

*This work supported was supported by the U.S. Dept. of Energy under Contract No. DE-AC04-76DP00789 and by the U.S. Air Force, Air Force Systems Command, Phillips Laboratory, Kirtland AFB, NM 87117-6008 under prime contract F29601-91-C0046.

large PCSS in the "normal" linear photoconductive switching mode. There are two ways to reduce the optical energy required to trigger a PCSS: reduce its size, or use a gain mechanism. To achieve the first we have increased the switched fields from 10-20 kV/cm to up to 100 kV/cm, although the most reliable operation (10^7 tests for silicon and 10^5 for GaAs) is achieved at fields of 20 to 50 kV/cm. A gain mechanism has been observed[1] in a high field, non-linear switching mode of GaAs and InP which is frequently referred to as "lock-on."

TABLE I. Best switching properties of Si and high gain GaAs PCSS (results are not simultaneous)

Parameter	Si Flash Lamp Pumped Laser	GaAs Flash Lamp Pumped Laser	GaAs Laser Diode Array
ELEC. FIELD (kV/cm)	82	57-100	37-100
VOLTAGE (kV)	123	155	74.5 kV
CURRENT (kA)	2.8	4.0	4.2
POWER (MW)	65	120	40
RISETIME (ps)	200	300	600
DEVICE LIFE (#)	10^7	not tested	100,000
REP. RATE (Hz)	540	not tested	1,000
GAIN	1	10^5	10^3

LINEAR SWITCHING

The concept of linear photoconductive semiconductor switching is described in several references[2]. By linear switching with visible light, we mean a switching mode where one electron-hole pair is created for nearly every photon absorbed in the semiconductor. The first demonstration of fast risetimes with high voltages was carried out[3] with a 100 ps Nd:YAG laser pulse which triggered a Si switch. This size switch has been used to switch over 100 kV. For Si, our highest switched field and current are[4]: 82 kV/cm (123 kV) and 2.8 kA (not simultaneous). Figure 1 shows the voltage waveform for a Si PCSS that switched 123 kV.

Thermal runaway is detrimental to PCSS systems because: 1) energy is lost in the switches that results in extra cooling needs, 2) the energy lost must be accounted for in a higher charge voltage, and 3) the increased switch temperature reduces switch lifetime. Figure 2a shows voltage and current waveforms for a 0.25 cm long by 0.25 cm wide Si switch. The voltage waveform that peaks at 8.1 kV results in thermal runaway. The current at peak voltage is 1.2 A corresponding to 6.8 kΩ, and the voltage at current peak (2.2 A) is less than 500 V corresponding to about 200 Ω. For the voltage waveforms that peak at 4 kV and 6 kV, the current looks ohmic with a resistance (at peak voltage) of about 8 kΩ. There are different ways to reduce or eliminate thermal runaway. The simplest is to reduce the charging time. A second way is to use different Si. Figure 2b shows voltage and current waveforms that are ohmic. This high switch resistance (62 kΩ) negates the currents that result in heating

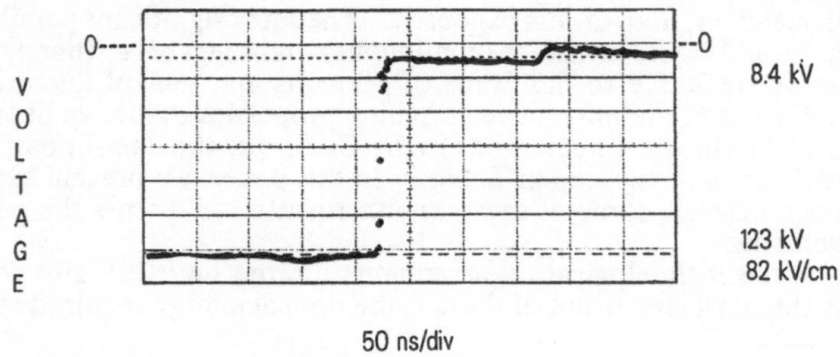

Figure 1. The voltage across through a Si switch. The switch (1.5 cm long) was triggered with a 10 ns long Nd:YAG laser. The voltage switched was 123 kV. The current (not shown) peaks at 2.0 kA and lasts about 160 ns.

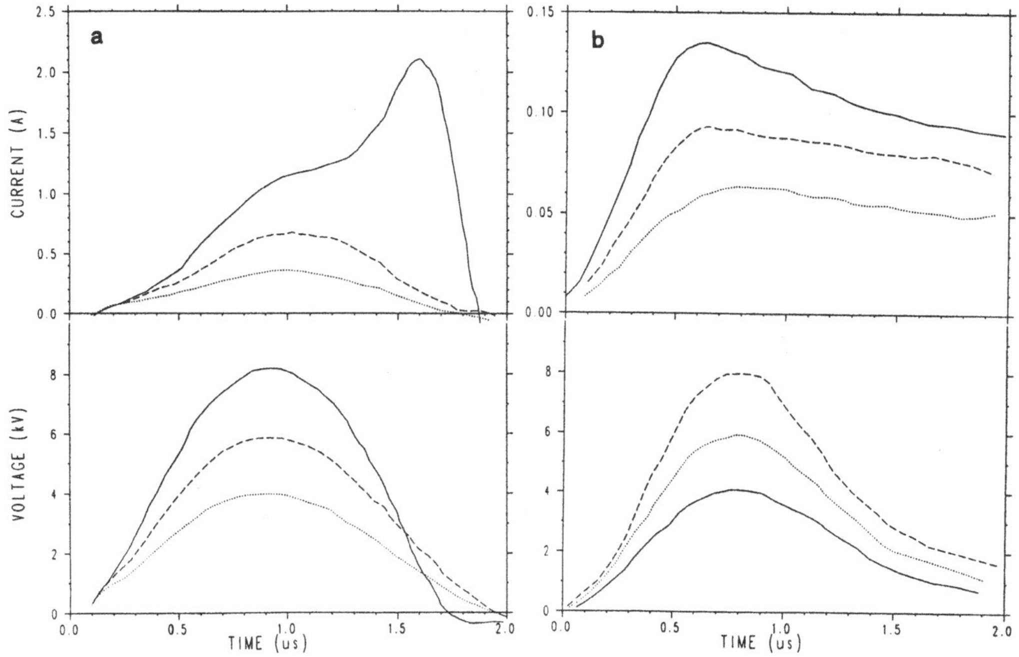

Figure 2. Pairs of current and voltage waveforms for 0.25 cm by 0.25 cm by 0.06 cm Si switch. Figure 2A: Thermal runaway is most evident in the peak at 2 A. Figure 2B: Pairs of current and voltage waveforms for a different 0.25 cm by 0.25 cm by 0.028 cm Si switch. Note the change in scale of the current waveforms. There is no thermal runaway.

and thermal runaway. A third way to reduce thermal runaway is to use longer switches. DC current is reduced because switch resistance increases with length. Transient current is reduced because the carriers require longer to cross the switch and the initiation of charge injection is controlled by the space charge limit until the carriers of opposite polarities begin to mix. The current-voltage curve for a 0.25 cm long by 0.25 cm wide switch is non-linear (the waveforms show thermal runaway) showing a resistance that starts at 500 kΩ and ends at 67 kΩ (4 kΩ-cm, at 4 kV and 16 kV/cm). Using the same type of material and voltage pulse, we then tested a 1.5 cm long by 2.0 cm wide switch. Its current voltage curve is almost linear. The switch resistance is 207 kΩ (16.6 kΩ-cm) at fields of up to 20 kV/cm.

For repetitive switching we need a short carrier lifetime so that the switch resistance recovers rapidly. We have studied the carrier lifetime in Si by applying laser pulses and measuring the switch resistance after a known delay. The conductivity is given by the intrinsic switch dark conductivity ($1/R_0$) plus that due to the laser pulse, whose decay is exponential. Figure 3 shows $1/R - 1/R_0$ vs. delay time for a switch that was not neutron bombarded. Note that there are two slopes: a fast decay of 4.3 μs and a slow one of 335 μs. By studying how these slopes were affected by neutron bombardment we were able to attribute the fast slope to recombination via defects and the slow decay to deep level (impurity) decay. In previous tests we have doped the Si with Au, and achieved opening times of 14 ns. The disadvantage of neutron irradiation and gold doping is that the carrier mobility is reduced and this increases the laser energy requirement.

The ability to obtain short carrier lifetime Si allows repetitive operation and lifetime testing at a laser-limited repetition rate of 540 Hz. Figure 4 shows typical voltage and current waveforms used in testing the 1 μm Al switch described below. The switched voltage was 8.5 kV (34 kV/cm), and the switch current was 77 A (308 A/cm) which lasted for about 20 ns. The switches with Cr-Cr-Mo-Au metallization have survived 10^7 pulses at up to 32 kV/cm. Although the switch is still running after these pulses, there is severe degradation of the metallization. To reduce the damage, we have concentrated our efforts in reducing current bunching at the edge of the contacts. We created two types of Al diffused contacts: deposition of a 1 μm thick layer of Al, followed by sintering for 20 minutes at 600 C, and deposition of 0.1 μm of Al, heating at 400 C for 20 minutes. In both cases, the Al diffused layer was covered with Cr-Mo-Au. The 1 μm switches survived 10^7 pulses at 34 kV/cm. The 0.1 μm switches survived 32 kV/cm for 2.4 million pulses, 36 kV/cm for 1.94 million, 40 kV/cm for 1.46 million, and only about 100,000 pulses at 44 kV/cm at which point the switch failed. Both the Al diffused switches fared better than the non-diffused switches. Even better results were obtained with an ion implanted contact[5] (Phosphorous) that was capped with Ti-Pt metallization. This type of switch has now been run at 48 kV/cm for 0.9 million pulses, 44 kV/cm for 2.2 million pulses, and at 32 kV/cm for 10^7 pulses. In the last test the switch was removed before it failed and showed only minor evidence of contact degradation.

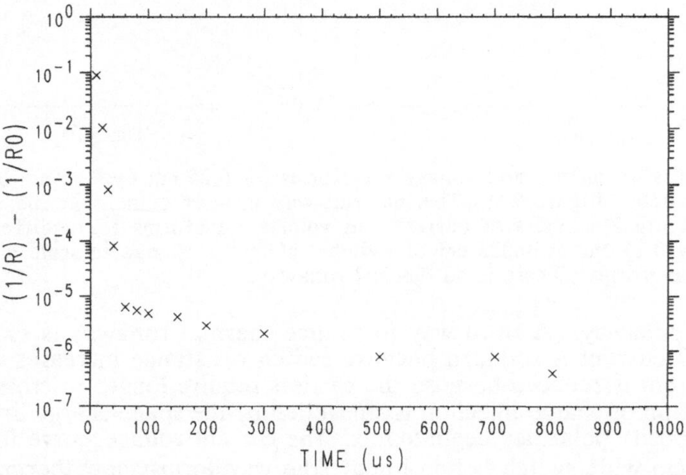

Figure 3. **Plot of switch resistance versus time. To obtain the carrier lifetime we have plotted $(1/R)-(1/R_0)$.**

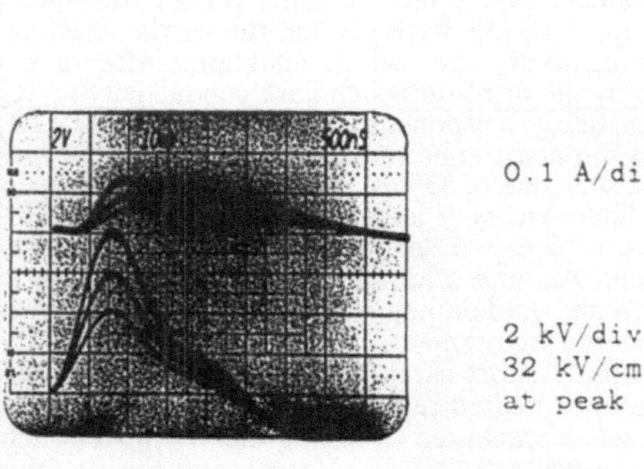

Figure 4. **Voltage and current waveforms for switch lifetime tests carried out with the 1 μ Al switch. The voltage, current, and time were recorded at 2 kV/div, 20 A/div, and 20 ns/div. Over 10^7 pulses were obtained under these conditions.**

Cr and P are donors to Si and Al is an acceptor. The damage to the Cr diffused and the ion implanted P switches is mostly on the positive contact, while the Al diffused switches show most damage on the negative contact. This correlation of damage with type of contact together with Auger analysis of the damaged regions seem to imply that the damage occurs from local heating at the semiconductor-metal interface.

NON-LINEAR SWITCHING

At low fields, linear switching was observed using a 7 ns wide optical pulse from a Q-switched Nd:YAG laser and a Cr:GaAs PCSS, which had a carrier lifetime of about 2 ns. Since carrier recombination is faster than the optical pulse, the current pulse follows the optical pulse. Above 8 kV/cm, a new effect was observed[(1)]. Our observations about this non-linear switching mode are grouped into three different phases and listed in Table II. These observations have been compiled after many experiments with different types of GaAs and one type of InP. Similar observations have also been reported by other authors (6). At present, this collective list of switching characteristics has only been observed with GaAs and InP.

The most significant characteristic of this type of switching is gain. Optical trigger gain (OTG) is, effectively, the number a carriers generated for current conduction per photon absorbed by the switch. With uniform switch illumination, OTGs as high as 1000 have been observed. This allows the use of laser diode arrays for triggering. Figure 5 shows the current through a 1.5 cm long GaAs switch triggered with a small laser diode array (50 μJ of energy, T05 package) that illuminated the switch uniformly. This corresponds to switching 40 MW.

Carrier recombination in GaAs normally results in the emission of a 1.4 eV photon. To monitor the location of high carrier densities, we recorded images of the infrared photoluminescence (PL) which was emitted from the PCSS. Previous experiments have shown that during high gain switching, the current is concentrated in filaments which extend across the normally insulating region (gap) of the switch[7-9]. The tests show that the location of current filaments can be controlled and that significantly lower minimum optical trigger energies can be obtained by focusing the light to small regions near a switch contact. Controlling filament locations will probably allow the fabrication of more reliable switches with higher average current densities. All tests were performed with a 1.5 cm long by 2 cm wide lateral PCSS fabricated from a 1" diameter by 0.025" thick wafer of Cr:GaAs. The first study[9]

Table II. Characteristics of high speed, non-linear switching in large (1-30 mm long) GaAs PCSS.

I. <u>Initiation Phase</u>
   A. Triggering depends on electrical and optical thresholds which are inversely related.
      E-field: 5-100 kV/cm; Trigger energy: 1-800 μJ.
   B. Triggering is most efficient near the contacts.
   C. Low energy triggering implies gain ($\rightarrow$100,000).
   D. One or more current filaments are always observed.
   E. Initiation is too fast for carrier injection and drift.

II. <u>Sustaining Phase</u>
   A. In a few nanoseconds, the average field drops to an "on"-state value which we call the lock-on field: 3.5 - 9.5 kV/cm.
   B. The PCSS is on after the optical pulse and drops to R = V(lock-on)/I(circuit), like a Zener diode.
   C. Carrier regeneration would imply gain.

III. <u>Recovery Phase</u>
   A. The switch resistance returns when the field drops below the lock-on field.
   B. Consistent recovery is observed after 1 ms, the fastest single pulse recovery observed is 30 ns.

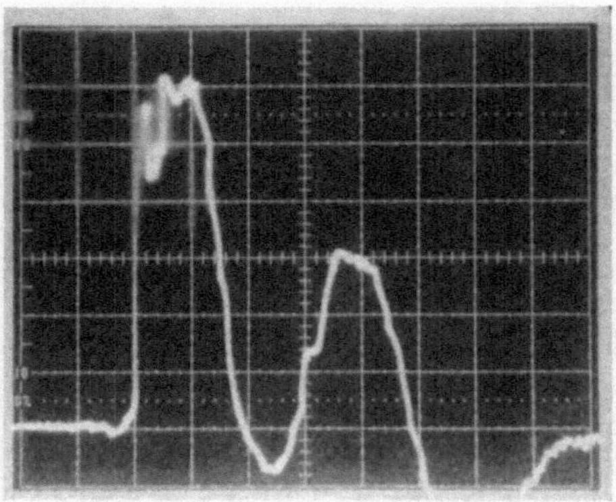

Figure 5. The current through a high gain switch charged to 75 kV, triggered by a laser diode array whose optical power was 166 W. The scales are 200 A/div. and 5 ns/div. The peak current is 1200 A. The power delivered to an ideal 30 Ω load would be 40 MW.

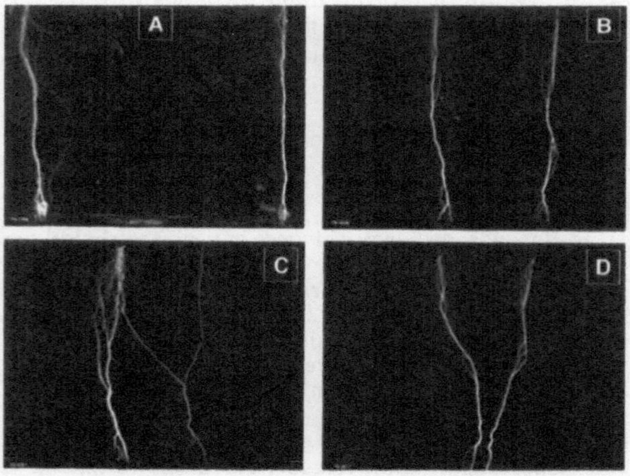

Figure 6. Twin groups of current filaments are shown in these photographs. They were produced with two 1 mm dia. fibers that were positioned with different separations along the bottom contact of the PCSS and illuminated simultaneously.

concentrated the light to a stripe or a small spot (~1 mm dia.). At low optical intensities (< 10 μJ), the filament always intersected the spot and even less light, as low as 0.5 μJ near the cathode, was required to activate the high gain mode. Since triggering at these fields with uniform illumination required several hundred microjoules, this represents a reduction in the laser triggering energy by another two orders of magnitude. Previously reported triggering gains of 1,000 may eventually be extended to 100,000. Recent studies[10] show that high gain switching can be initiated by light that is coupled through 1 mm diameter fibers. We were also able to trigger multiple current filaments and control their spacing by using two fibers near the negative contact. This last result is shown in Figure 6 where the separation between the fibers is varied and the resulting current filaments follow. We have also triggered filaments with the laser focused to a 200 μm spot. The sizes of the filaments appeared to range from 50-300 μm. A remarkable aspect of these measurements was that triggering near a contact did not change the delay and rise time (about 2 ns for a 1.5 cm long switch) to high gain switching. These data, coupled with the saturation velocity of carriers, and the observation of filaments implying the presence of electrons and holes across the entire length of the switch means that models that only use carrier injection and drift cannot explain high gain switching.

CONCLUSIONS

In this paper we presented the results of experiments on Si and GaAs PCSS. For Si, we presented ways of reducing or eliminating thermal runaway, ways of altering carrier lifetime, and improvements in switch lifetimes to obtain 10^7 pulse life. For GaAs, the effects of focusing the triggering radiation was studied and another reduction (factor of 100) in the required light energy was observed. It was further determined that light can be piped through fiber optics to trigger the GaAs and that this results in the ability to control the location of the current filaments. These data cannot be explained by models that use only carrier injection and drift.

REFERENCES

1 G. M. Loubriel, M. W. O'Malley, and F. J. Zutavern, "Toward Pulsed Power Uses for PCSS: Closing Switches," 6th IEEE Pulsed Power Conf., Arlington, VA, 1987, pp. 145-148.

2 Picosecond Optoelectronic Devices, edited by C. H. Lee (Academic Press, NY, 1984), pp 73-117.

3 F. Zutavern, M. O'Malley, M. Buttram, and W. Stygar, "Photoconductive Semiconductor Switching," in Proc. 5th IEEE Pulsed Power Conf., Arlington, VA, pp. 246-249.

4 G. M. Loubriel, M. W. O'Malley, F. J Zutavern, B. B. McKenzie, W. R. Conley, H. P. Hjalmarson, "High Current PCSS," 18th IEEE Power Modulator Symposium, Hilton Head, SC, 1988, p. 312.

5 For details on the ion implanted switch (from by David Sarnoff Research Center and Grumman Aerospace) see: G. M. Loubriel, F. J. Zutavern, G. J. Denison, W. D. Helgeson, D. L. McLaughlin, M. W. O'Malley, C. H. Sifford, L. C. Beavis, C. H. Seager, A. Rosen, and R. G. Madonna, "Long Lifetime Si PCSS," to be published in Proc. OELASE93, Los Angeles, CA.

6 Optically Activated Switching II, G. M. Loubriel (ed.), SPIE Proc. Vol. 1632, Los Angeles, CA, 1992.

7 F. J. Zutavern, G. M. Loubriel, M. W. O'Malley, W. D. Helgeson, and D. L. McLaughlin, "High Gain PCSS," 8th IEEE Pulsed Power Conf., San Diego, CA, 1991, pp. 23-28.

8 R. Aaron Falk, Jeff C. Adams, and Gail L. Bohnhoff-Hlavlacek, "Optical Probe Techniques for Avalanching Photoconductors" in Proc. 8th IEEE Pulsed Power Conference, San Diego, CA, 1991, pp. 29-36.

9 F. J. Zutavern, G. M. Loubriel, D. L. McLaughlin, W. D. Helgeson, M. W. O'Malley, "Electrical and Optical Properties of High Gain GaAs Switches", pp. 152-159 of ref. 6.

10 F. J Zutavern, G. M. Loubriel, M. W. O'Malley, W. D. Helgeson, D. L. McLaughlin, and G. J. Denison, "Charac-teristics of Current Filamentation in High Gain PCSS," IEEE Power Modulator Symposium, Myrtle Beach, SC, 1992.

TRANSIENT SCATTERING MEASUREMENTS USING PHOTOCONDUCTIVELY SWITCHED PLANAR ANTENNAS

Lawrence Carin

Department of Electrical Engineering/ Weber Research Institute
Polytechnic University
Brooklyn, NY 11201

INTRODUCTION

In most electromagnetic scattering and radar applications, the scattering process is linear, and therefore time and frequency domain data are linked via the Fourier transform. One can therefore perform a scattering measurement in the time or frequency domain and, as long as the time and frequency domain measurements cover the same bandwidth, frequency and time-domain measured data are interchangeable via the Fourier transform (assuming good measurements were performed in both domains). The choice of whether to perform the initial measurement in the time or frequency domain is therefore dictated by such issues as cost and measurement quality (immunity to noise).

There are several advantages to performing the measurements in the frequency domain. An intrinsic benefit of frequency-domain measurements is that one can usually control the amplitude of the incident fields over the available frequency band and therefore high quality measurements are possible at all frequencies. This should be compared with time-domain measurements, for which the shape of the transmitted waveform is difficult to vary, being dictated by the (usually invariant) properties of the transmitter. Thus, the frequency spectrum of a time-domain system will generally be fixed by the transmitter characteristics, with large energies transmitted around the pulse's center frequency but with power levels usually degrading quickly away from this frequency. Combining this issue with the mature nature of the technology involved in frequency-domain measurements, it would appear that at this time the quality of frequency-domain measurements is superior to that of time-domain measurements.

Over the last several years, however, many researchers have performed impressive scattering measurements directly in the time domain[1-5]. These time-domain measurements suggest that with further development, the quality of time-domain scattering measurements may approach that of frequency-domain measurements. This, combined with the fact that the equipment needed for direct time-domain measurements is expected to be far less costly than equipment associated with conventional frequency-domain measurements[6], indicates that the development of time-domain scattering systems warrants further research.

The purpose of this paper is to describe and present results for a new type of short-pulse, time-domain scattering range which is capable of generating freely propagating waveforms with bandwidths extending from a few to nearly 80 GHz. The system uses ultrashort optical pulses generated from a picosecond laser to switch planar antennas photoconductively. Although the measurements cover microwave and millimeter-wave frequencies, no conventional microwave or millimeter-wave equipment is employed in our measurements.

Over the last several years, picosecond and femtosecond lasers have been used to switch various planar antennas photoconductively, generating short bursts of microwave, millimeter-wave, and terahertz radiation[7-16]. Because the measurements are performed in the time domain, ultra-wideband (UWB) data can be obtained in a single measurement. These antennas have been used to characterize the UWB properties of dielectrics[12], semiconductors[13], and superconductors[14]. As was shown first by Robertson *et al.*[5] and later by us[15,16], these antennas can also be useful for performing short-pulse, time-domain electromagnetic scattering measurements on scaled three-dimensional targets.

EXPERIMENTAL SYSTEM

We use a mode-locked, pulse-compressed, frequency-doubled Nd-YLF laser to produce 4-5 ps green (527 nm) pulses at a 76 MHz repetition rate and 180-200 mW average power. The optical pulses are used to switch aluminum coplanar-strip (CS) horn antennas fabricated on silicon-on-sapphire substrates[8]. The transmitting CS horn antenna is connected to a 45 V battery and the receiving CS horn antenna is connected to a current pre-amplifier and then to a lock-in amplifier. Two electric current pulses are produced when the transmitting antenna is switched. One current pulse travels to the horn and is partially radiated, the second pulse travels away from the horn (along the feed), is reflected at the bonding pads, and then is subsequently partially radiated when it travels through the horn. The time window in our measurements is therefore determined by the time delay between these two radiated waveforms, and is dictated by the length of the CS feed line. The radiation from the transmitter is collimated with a lens, but no lens is used on the receiving antenna. This is because the theory calculates the scattered signal at a point, and the CS horn antenna, without a lens, acts approximately like a point detector.

In the scattering experiments, a measurement is made first for signal propagation from the transmitter to the receiver directly, without a target present. Call this signal $s_1(t)$. The waveform $s_1(t)$ can be expressed as a cross correlation[12] between the current induced on the receiver and the effects of the photoconductive sampling process. A second measurement is then taken with a scatterer placed between the transmitter and receiver. The second measured signal $s_2(t)$ can be expressed as a convolution of $s_1(t)$ with the impulse response $h(t)$ of the target at the point of observation. Once $s_1(t)$ and $s_2(t)$ have been measured, several different techniques are available to deconvolve $h(t)$ from $s_2(t)$. The simplest approach is to divide the Fourier transform of $s_2(t)$ by that of $s_1(t)$. This gives the impulse response in the frequency domain, which can be converted back to the time domain if desired.

In the above considerations, it has been assumed that the waveform incident on the scatterer is the same signal that was incident on the receive antenna for the measurement of $s_1(t)$. To achieve this, the distance between the transmitter and receiver for the measurement of $s_1(t)$ must be the same as the distance from the transmitter to the target when measuring $s_2(t)$. However, if all significant frequency components in the pulse decay approximately at the same rate with distance, this distance criterion may not be critical. In this case, if the distances discussed above are not adhered to, one would measure a scaled (in amplitude) version of $h(t)$. A measurement was performed to determine the decay rate of the EM pulse with distance. It was found that as the distance from the transmitter to receiver was varied, the pulse shape remained essentially unchanged, but the amplitude scaled. The peak amplitude of each measured pulse is plotted in Fig. 1 as a function of distance. It is seen that the pulse exhibits approximately a $1/r$ dependence, with r the distance from the transmitter. Although this is what is expected of an EM beam, it is important that it be verified experimentally in light of recent theoretical studies of EM missiles[17].

It should be pointed out that the use of a lens on the transmitter significantly increased the signal strength measured on the receiver. The use of the lens produces a collimated pulsed beam, and in the far field of the lens a $1/r$ signal decay is expected. The use of the lens effectively increases the aperture from which the energy radiates, thus increasing the distance from the transmitter to the Fraunhofer far field. If one were to use a lens on the receiver, and the rate of diffraction was such that the beam energy existed predominantly within the diameter of the lens, then the measured pulse would be independent of distance (assuming the lens efficiently focuses all the EM energy onto the antenna). Examples such as this, for which the use of a focusing lens on the receiver camouflages physical effects associated with EM radiation (diffraction in this case), are why a lens was only used on the transmitter. It should be noted that in material measurements[12-14] the problems caused by a lens on the receiver are not an issue, and the significant enhancement in signal-to-noise afforded by such a lens is obviously desirable.

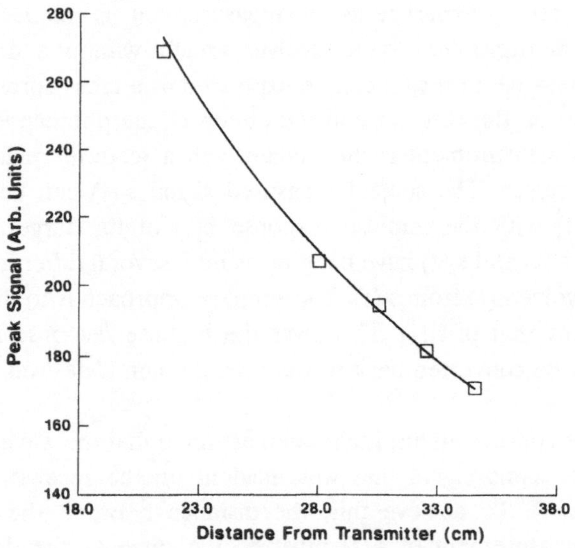

Figure 1. Peak signal as a function of distance from the transmitter. The points represent measured data and the line represents a theoretical 1/r variation, where r is the distance from the transmitter. The line was computed by using one measured point (at 34.5 cm), and assuming a 1/r variation for all other distances.

With a knowledge of the EM pulse decay rate, the signals $s_1(t)$ and $s_2(t)$ are measured as follows. The transmitter and receiver are placed at the positions desired for a given scattering measurement. A good reflector (an aluminum plate in our experiments) is positioned just in front of the location at which the scatterer will be placed. The pulse specularly reflected off the plate is measured at the receiver. Using the distances from the plate to the receiver and from the transmitter to the receiver, this measured pulse is easily scaled in amplitude to its value at the plate's surface. Therefore the scaled signal (now used as $s_1(t)$) will approximate very closely the signal actually incident on the scatterer cross correlated with the response of the receiver. The plate is then replaced with the scatterer, and $s_2(t)$ is measured.

Results

Measured data is presented for scattering from one and two conducting strips and from conducting and dielectric spheres. The experimental results are compared with theoretical data. The theoretical data was computed in the frequency domain, and a unit amplitude linearly polarized plane wave is assumed incident. We therefore compute the impulse response in the frequency domain, which is then multiplied by the Fourier transform of $s_1(t)$ (which is measured). This frequency-domain product is converted into the time via a Fourier transform, arriving at a theoretical calculation of $s_2(t)$. For scattering from conducting strips, a spectral domain moment method procedure was used[18]; for scattering from the conducting and dielectric spheres, a standard spherical harmonic expansion[19] was applied. The experimental results represent data that can be measured routinely, and the agreement shown between theory and experiment is typical but not the best (or worst) that can be achieved.

a. Strip Scattering

The strips are 5 mm wide and are separated by 5 mm (for the two strip case). In the experiment, the strips are made of thin aluminum and are sufficiently long such that the strip length is much larger than the beam cross section. The antennas are arranged such that the angle of incidence and observation are both 42°, with the angles measured from a (fictitious) line normal to the strip width. The electric field is approximately polarized parallel to the length of the strips (TE). The transmitter and receiver are placed 12.5 cm and 9 cm, respectively, from the front surface of the strips. Figure 2 shows a typical reference pulse measured at the receiver after reflecting off the aluminum plate. Also shown in Fig. 2 is the Fourier transform of the reference pulse. It is evident that the pulse contains significant EM energy at frequencies spanning from 5 to over 70 GHz. This bandwidth is limited by the laser, and can be extended by using shorter duration optical pulses[10,11]. Figures 3 and 4 show the pulse after scattering from one and two aluminum strips, respectively.

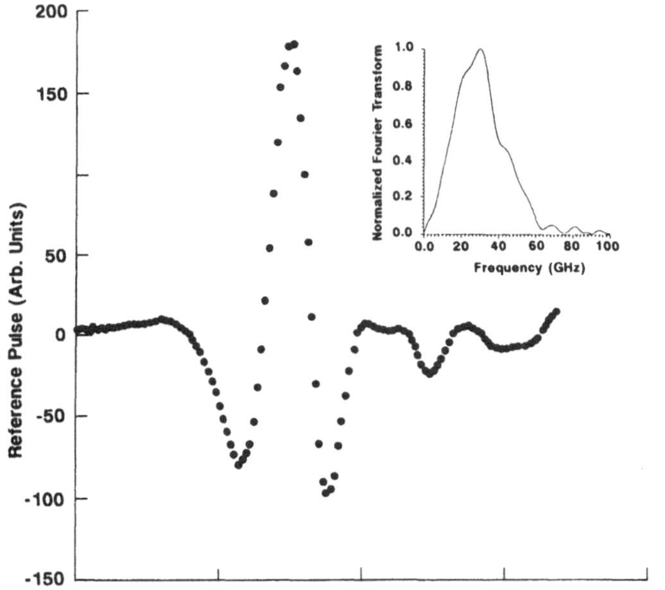

Figure 2. The measured reference pulse and its associated numerical Fourier transform.

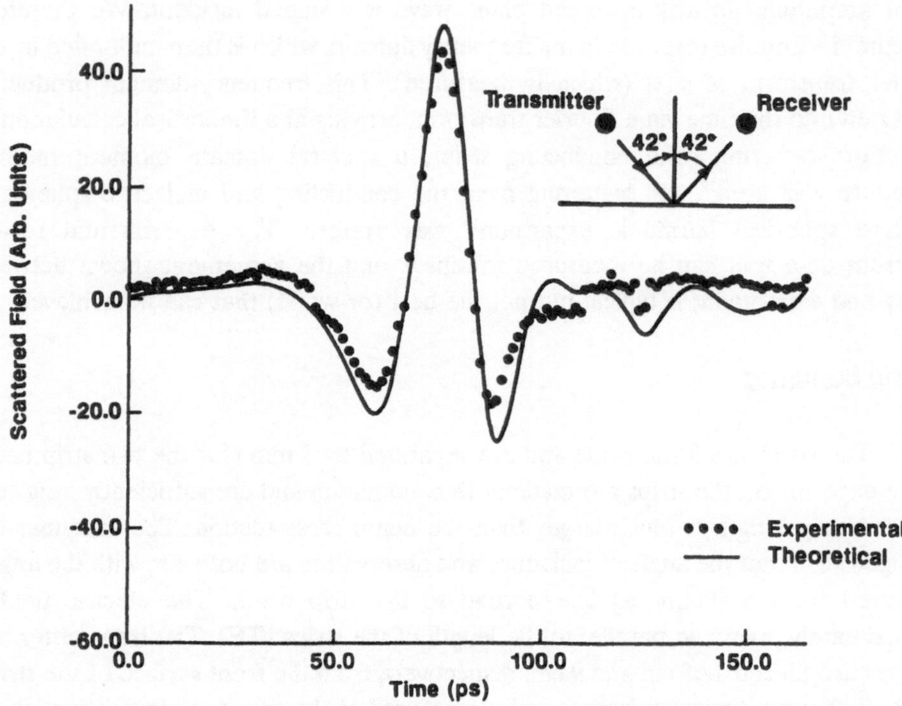

Figure 3. Measured and computed scattered signal, $s_2(t)$, from a single aluminum strip of 5 mm width. The points represent measured data and the line represents computed results. The transmitter and receiver are 12.5 cm and 9 cm, respectively, from the strip. The transmitter and receiver are both positioned at 42° angles with respect to the strip (see inset).

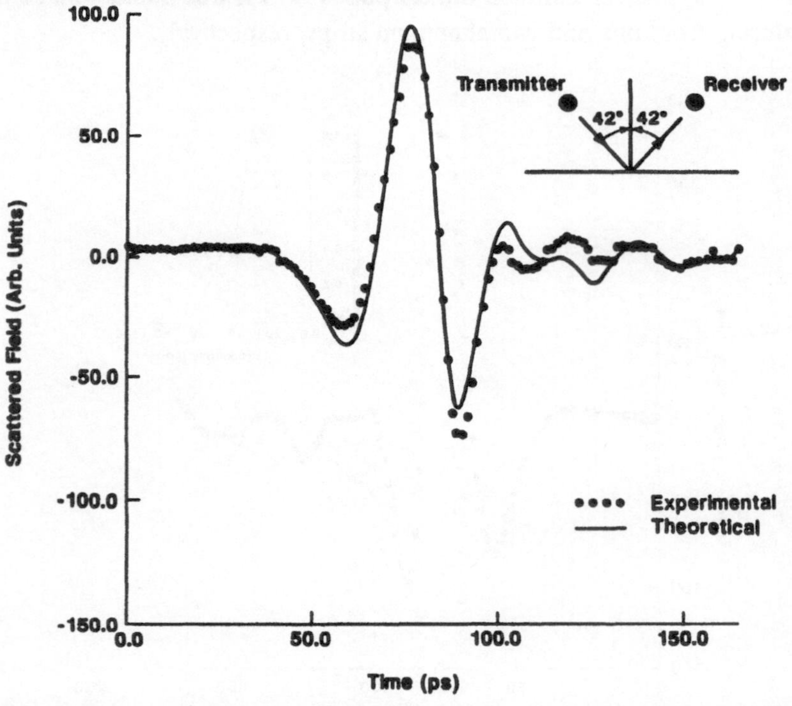

Figure 4. Measured and computed scattered signal from two coplanar aluminum strips of 5 mm width separated by 5 mm. The points represent measured data and the line represents computed results. The transmitter and receiver are positioned as in Fig. 3.

In Figs. 3 and 4, the points represent measured data and the curves represent computed results. Notice that the scattered signal off a single strip is approximately a scaled version of the reference pulse at the angle of observation chosen. This can be understood by realizing that since the angles of incidence and observation are equal, we are observing predominantly specular reflection (which is weakly dispersive). For the two strip case, destructive and constructive interference is expected from the pulses reflected from each individual strip. Notice that this results (both experimentally and theoretically) in an enhanced second dip in the scattered waveform.

b. Conducting Sphere

The steel sphere used in the experiments had a 2.5 cm diameter. The transmitting and receiving antennas were placed 7.0 cm and 5.5 cm, respectively, from the center of the sphere, separated by an angle of 75°. The antennas were situated such that the coplanar strips of the two antennas were facing one another (not in the same plane). In this transmitter-receiver arrangement, the creeping wave excited is very small in magnitude, much smaller than the initial wavefront. Figure 5 shows the theoretical and experimental data for this structure. The agreement between theory and

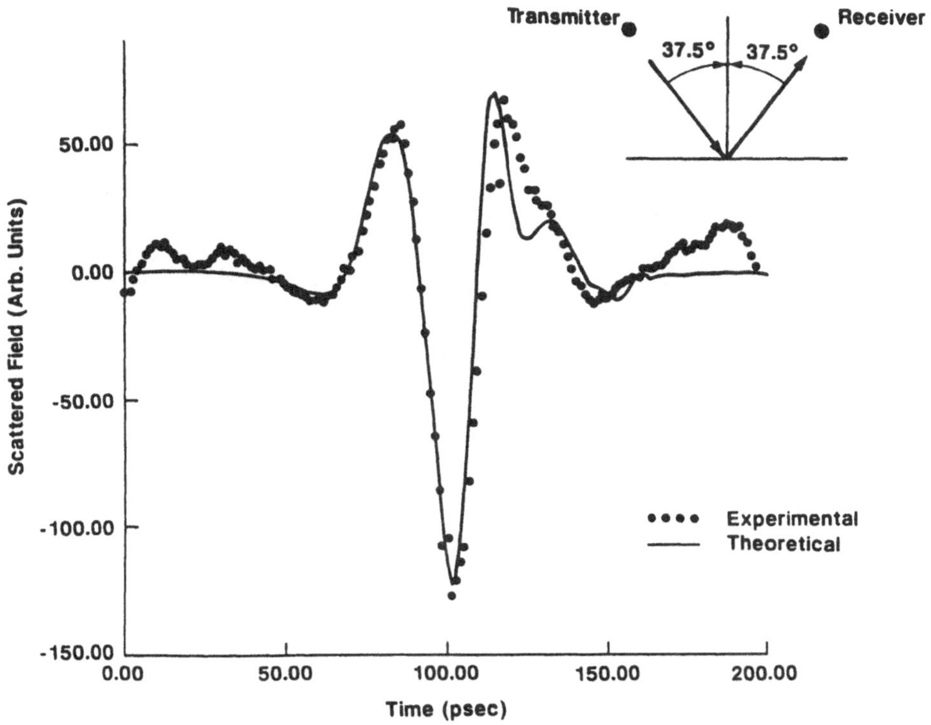

Figure 5. Measured and computed scattered signal from a 2.5 cm diameter steel sphere. The transmitter and receiver are 7.0 cm and 5.5 cm, respectively, from the center of the sphere; and are separated by an angle of 75°. The points represent measured data and the line represents computed results.

experiment is quite respectable, although noise causes a slight corruption in the early- and late-time response of the experimental data. The reference pulse used for the sphere (not shown) is different than that for the strip measurements since a different reference pulse is measured for each experiment.

c. Dielectric Sphere

The time-domain scattered signal from a given target consists of two components: (i) the early-time response and (ii) the late-time response[4]. When the incident waveform first hits the target, there is scattered radiation from localized points on the target, as these points become illuminated. This part of the time-domain scattered signal, produced before the entire target is illuminated by the incident waveform, is termed the early-time response. The early-time response is usually referred to as being aspect dependent, because it is strongly dependent on the position of the target with respect to the incident radiation. After the entire target has been illuminated, the induced fields in or on the target bounce back and forth between the target's various scattering centers. As the fields bounce back and forth, energy is radiated from the target. This part of the scattered signal, termed the late-time response, is dependent on the global character of the target and is efficiently represented in terms of the target's natural modes. The late-time response is termed aspect independent because, independent of what part of the target is hit first by the incident radiation, once the entire target is illuminated, the scattered signal can always be described in terms of the targets natural modes. The late-time response therefore provides important information about the global properties of a given target.

The measurements in (a) and (b) concentrated on the early-time response. We now consider scattering from dielectric spheres in an effort to measure both the early and late-time response. The antennas used for the measurements in (a) and (b) and those used in our initial dielectric-sphere measurements were CS antennas with a usable time window of approximately 185 ps (1 cm long CS feed). In order that the late-time signal occurred in this time window, for the dielectric sphere measurements, we were required to use a low-dielectric-constant sphere with a small radius (such that the multiple reflections inside the sphere did not take too long to evolve). Although this type of sphere has, for our purposes, the advantage of having the multiple reflections in the antenna's time window, it has a very small radar cross section (RCS) and therefore will be susceptible to noise.

In Fig. 6 are experimental and theoretical results for a 1.7 cm diameter dielectric sphere. Our short-pulse system was used to characterize[12] the dielectric slabs from which the spheres were machined, and it was determined that the index of refraction was nearly lossless and dispersionless over the pulse bandwidth, with $n=1.5$. One can see that, in spite of the very small scattered signal, we are able to accurately measure the early-time response and the beginning of the late-time response. One notices that the agreement between theory and experiment degrades after approximately 185 ps, due to the interference caused by the second radiated waveform. It is interesting to note that both the theory and experiment predict that the late-time response, for this particular configuration, is stronger than the early-time response.

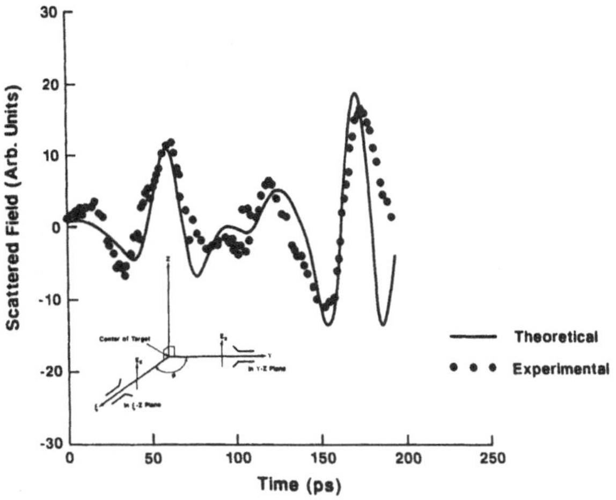

Figure 6. Measured (points) and calculated (curve) scattered signal from a 1.7 cm dielectric sphere with index of refraction n=1.5. The transmitting and receiving antenna were positioned 4 cm and 5 cm, respectively, from the center of the sphere, and they were separated by an angle ϕ=75° angle (see inset). The antenna time window was 185 ps.

We next considered scattering from a sphere of 2.2 cm diameter, with a measured index of refraction of n=3.56 (also measured to be virtually lossless and dispersionless over the pulse bandwidth). The agreement between theory and experiment shown in Fig. 7 is quite good in this case, although the limited time window of our antenna precluded our ability to measure the late-time response.

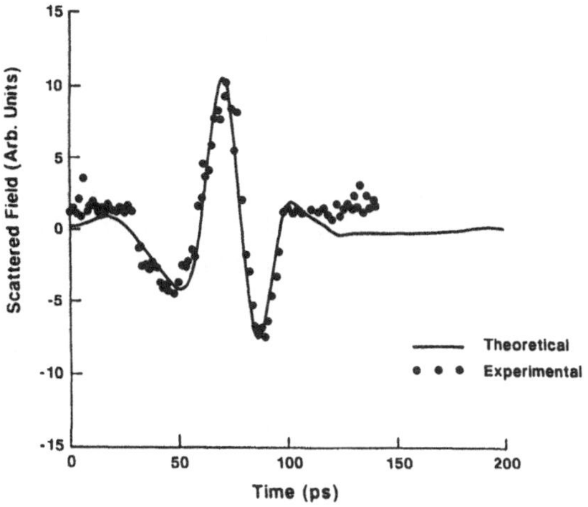

Figure 7. Measured (points) and calculated (curve) scattered signal from a 2.2 cm dielectric sphere with index of refraction n=3.56. The transmitting and receiving antenna were positioned 4.5 cm and 5 cm, respectively, from the center of the sphere, and they were separated by an angle ϕ=75° (see inset of Fig. 6). The antenna time window was 185 ps.

To achieve a longer time window, we increased the length of the CS feed to 2 cm. Inset in Fig. 8 is a typical waveform generated by our new antennas, showing a usable time window of over 300 ps. With these antennas we can investigate the late-time response from high-dielectric-constant spheres, for which the late-time response comes later in time. Figure 8 shows a typical result for scattering from the n=3.56 dielectric sphere of 2.2 cm diameter. Again, the late-time response is stronger than the early-time response, and now, with our enhanced time window, we accurately measure much of the late-time response.

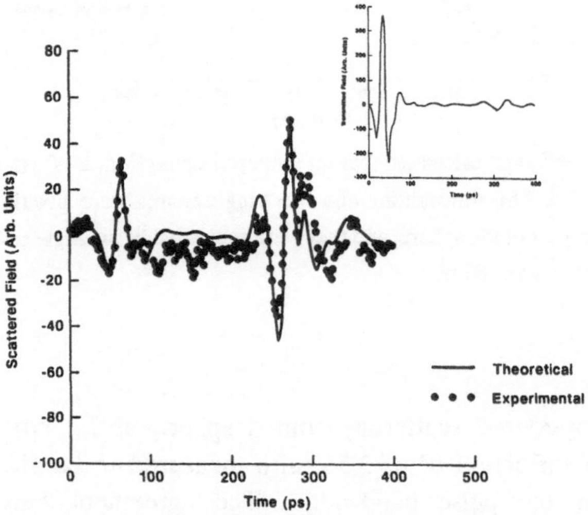

Figure 8. Measured (points) and calculated (curve) scattered signal from a 2.2 cm dielectric sphere with index of refraction n=3.56. The transmitting and receiving antenna were positioned 5.5 cm and 3.5 cm, respectively, from the center of the sphere, and they were separated by an angle ϕ=80° (see inset of Fig. 6). The antenna time window is nearly 300 ps, as indicated by a transmission measurement between the transmitter and receiver, with no target (shown inset).

DISCUSSION

The experimental results are in good agreement with the theoretical computations. There are some discrepancies, however, and it is important that their sources are understood. In the computations, it has been assumed that the incident field was polarized linearly. Although the radiation produced by the coplanar strip antennas is known to be highly linearly polarized[12], the incident radiation will have some cross-polarization components. Additionally, the theory assumes the strips and conducting sphere are perfectly conducting; and in the case of the strips, it is assumed that the conductors are infinitesimally thin. Obviously, these conditions can only be approximated experimentally. Perhaps the largest source of discrepancy, however, involves the fact that in the theory a pulsed plane wave is assumed incident while in the experiments a pulsed beam is produced. In fact, the theoretical calculations for the case of the strips assume the problem is two-dimensional, while in the experiment (due to the diffracting beam, which is only incident on a subsection of the strips) we clearly have three-dimensional scattering.

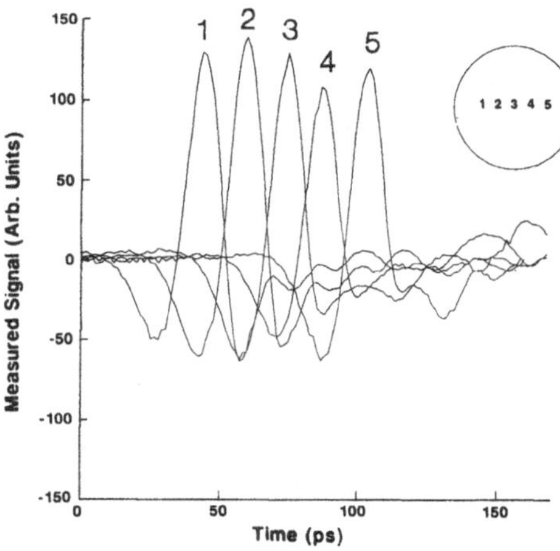

Figure 9. Experimental characterization of the spatial dependence of the pulsed beam. The measured waveform is shown at five different positions along an axis in the beam's cross section. The distance between each sample location is 0.5 cm.

An experiment was performed to evaluate the spatial dependence of the pulsed beam. In Fig. 9 is shown the measured waveform at different places along the beam cross section. Although the pulse is similar at the different points there is a spatial dependence to the measured waveform. It took over 1 hour to measure all the waveforms in Fig. 9; therefore, due to drifts in the laser power, it is expected that the spatial variation indicated in Fig. 9 is more severe than what would be expected of a single pulse.

CONCLUSIONS

An experimental system has been developed for performing UWB scattering measurements directly on an optical table. The UWB EM radiation is generated and detected by optically switching coplanar strip horn antennas with optical pulses generated by a picosecond laser system. The table-top scattering range was used to study transient scattering from conducting strips and from conducting and dielectric spheres. Upon comparison with theoretical results, good agreement is evident. To further characterize the scattering system, the transverse and longitudinal variation of the pulsed beam have been investigated experimentally. It was found that the pulsed beam decayed in the usual 1/r manner, and that the beam cross section is approximately uniform.

The facility discussed above provides a unique technique for performing ultra-wideband transient scattering measurements. The results presented here are in good agreement with theory, but this technology is still relatively new and will require further development such that this technique can compete with conventional scattering facilities.

ACKNOWLEDGEMENTS

Most of the measurements were performed by the author's students, Kamil Agi and David Kralj. This work is supported in part by the National Science Foundation under grant ECS-9211353 and by the Raytheon Company.

REFERENCES

1. C. L. Bennett and G. F. Ross, "Time-domain electromagnetics and its applications," *Proc. IEEE*, Vol. 66, pp. 299-318, March 1978.
2. F.-I Tseng and T. P. Sarkar, "Experimental determination of resonant frequencies by transient scattering from conducting spheres and cylinders," *IEEE Trans. Antennas Propagat.*, vol. AP-32, pp. 914-918, Sep. 1984.
3. M. A. Morgan and N. J. Walsh,"Ultra-wideband transient electromagnetic scattering laboratory," *IEEE Trans. Antennas Prop.*, vol. AP-39, pp. 1230-1233, Aug. 1991.
4. E. J. Rothwell and W. Sun, "Time domain deconvolution of transient radar data," *IEEE Trans. Antennas Propagat.*, vol. AP-38, pp. 470-475, Apr. 1990.
5. W. M. Robertson, G. V. Kopcsay, and G. Arjavalingam, "Picosecond time-domain electromagnetic scattering from conducting cylinders," *IEEE Microwave and Guided Wave Lett.*, vol. 1, pp. 379-381, 1991.
6. "Assessment of Ultra-Wideband (UWB) Technology," prepared by OSD/DARPA UWB Radar Review Panel, July 1990.
7. D. H. Auston, K. P. Cheung, and P. R. Smith, "Picosecond photoconducting Hertzian dipoles," *Appl. Phys. Lett.*, vol. 45, pp. 284-286, Aug. 1984.
8. A. P. DeFonzo and C. R. Lutz, "Optoelectronic transmission and reception of ultrashort electrical pulses," *Appl. Phys. Lett.*, vol. 51, pp. 212-214, July 1987.
9. D. H. Auston, "Ultrafast optoelectronics," in Ultrafast Optical Pulses, W. Kaiser (Ed.), New York: Springer-Verlang, Ch. 5, 1987.
10. Y. Pastol, G. Arjavalingam, and J.-M. Halbout, "Characterization of an optoelectronically pulsed equiangular spiral antenna," *Elect. Lett.*, vol. 26, pp. 133-135, Jan. 1990.
11. M. van Exter and D. R. Grischkowsky, "Characterization of an optoelectronic terahertz beam system," *IEEE Trans. Microwave Theory Tech.*, vol. MTT-38, pp. 1684-1691, Nov. 1990.
12. G. Arjavalingam, Y. Pastol, J.-M. Halbout, and G. V. Kopcsay, "Broad-band microwave measurements with transient radiation from optoelectronically pulsed antennas," *IEEE Trans. Microwave Theory Tech.*, vol. MTT-38, pp. 615-621, 1990.
13. M. van Exter and D. Grischkowsky, "Carrier dynamics of electrons and holes in moderately doped silicon," *Phys. Rev. B*, vol. 41, pp. 140-149, June 1990.
14. J. M. Chwalk, J. F. Whitaker, and G. A. Mourou, "Submillimeter wave response of superconducting $YBa_2Cu_3O_{7-x}$ using coherent time-domain spectroscopy," *Electon. Lett.*, vol. 27, p. 447, 1991.

15. L. Carin and K. Agi, "Ultra-wideband transient scattering measurements using optoelectronically switched antennas," *IEEE Trans. Microwave Theory Tech.*, Feb. 1993.
16. D. Kralj and L. Carin, "Short-pulse scattering measurements from dielectric spheres using optoelectronically switched antennas," *Appl. Phys. Letts.*, Mar. 15, 1993
17. T. T. Wu, "Electromagnetic missiles," *J. Appl. Phys.*, vol. 57, pp. 2370-2372, April 1985.
18. L. Carin and L. B. Felsen, "Efficient analytical-numerical modeling of ultra-wideband pulsed plane wave scattering from a large strip grating," to appear in *Int. J. of Numerical Modeling.*
19. R. F. Harrington, "Time Harmonic Electromagnetic Fields", McGraw-Hill Book Co., New York, 1961, pp. 292-298.

PHOTOCONDUCTIVE SWITCHING-REVISITED

T. Sarkar, R. Adve
Electrical and Computer Engineering Department
Syracuse University; Syracuse, NY 13244

M. Wicks
Rome Laboratory
Rome, NY 13441

INTRODUCTION

In many electromagnetic applications, it is necessary to generate waveforms of high amplitude and of picoseconds duration. The present state of the art commercial equipment is capable of delivering 10 volt amplitude pulses of 70 picoseconds duration. In this paper, we present a method to produce pulses which are about a kilovolt in amplitude and hundreds of picoseconds in duration. This technology, capable of producing high amplitude narrow pulses is derived from laser research.

Historically, high power electrical pulses of picosecond duration were first produced by laser-induced photoconductivity in high resistivity semiconductors. First demonstrated by Auston [1], many other researchers have produced similar results [2-4]. In this paper we present a method, developed by earlier researchers, for the design of a novel device capable of generating kilovolt amplitude pulses of picoseconds duration.

A mode-locking Q-switching Nd:YAG laser is used to generate a train of light pulses of approximately 100 picoseconds duration. This pulse train is passed through a switchout device such that a single pulse is selected, striking a Chromium doped GaAs semiconductor switch which then conducts. With a high voltage source applied to one end of the switch and a load to the other, an electrical pulse of half amplitude is observed. Using this device it is possible to routinely generate 1 kilovolt amplitude pulses of 100 to 300 picoseconds duration.

Ultra-Wideband, Short-Pulse Electromagnetics
Edited by H. Bertoni *et al.*, Plenum Press, 1993

MEASUREMENTS

A single Q-switched optical pulse is incident on a chromium doped GaAs switch as shown in Figure 1. One end of the switch is connected to a high voltage dc source and the other to a 50 ohm load.

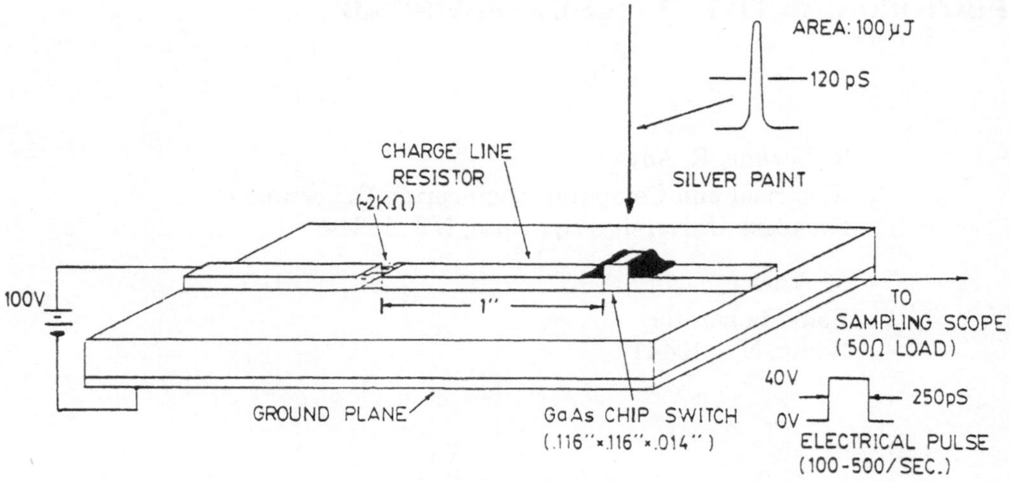

Figure 1: Experimental Setup for the Charge Line Switch.

A typical one kilovolt electrical pulse produced by the system is shown in Figure 2(a) for the single shot case, and 2(b), for the 100 shot average. The waveforms are recorded directly using a Tektronix 7854 sampling scope. It is seen that the pulses are 300 picoseconds wide. This pulse width is determined by the length of the transmission line between the semiconductor switch and the 2 kohm charging resistor, due to the fact that the switch remains on, even after the optical energy is turned off, since the laser energy saturates the switch. Also considering that there is a bulk conduction current through the device, because the pulses have flat tops.

By reducing the laser energy it is possible to sharpen the pulses as shown in Figures 3(a) for the single shot case and in Figure 3(b) for 100 shot average.

CONCLUSIONS

An experimental set up is described for generating high voltage subnanosecond pulses. The waveform generator is a laser activated switch. A laser beam turns a semiconducting GaAs switch on and off. The GaAs switch is located between the source and the load. The waveform at the load closely follows that of the fast optical pulse. In this way it is possible to routinely generate 1 kilovolt amplitude pulses of 100-300 picoseconds duration at a repetition rate of 250 Hz, (depending on how much the switch is saturated.) Also a Blumlein configuration of the switch may increase the switching efficiency from 40% to about 90%.

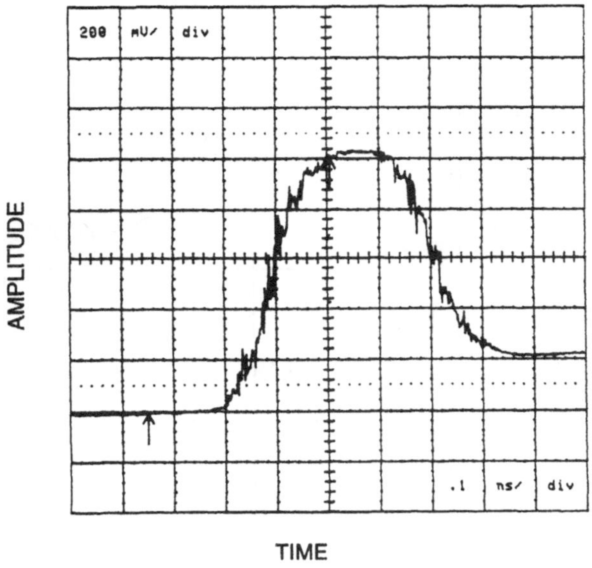

(a) No Averaging

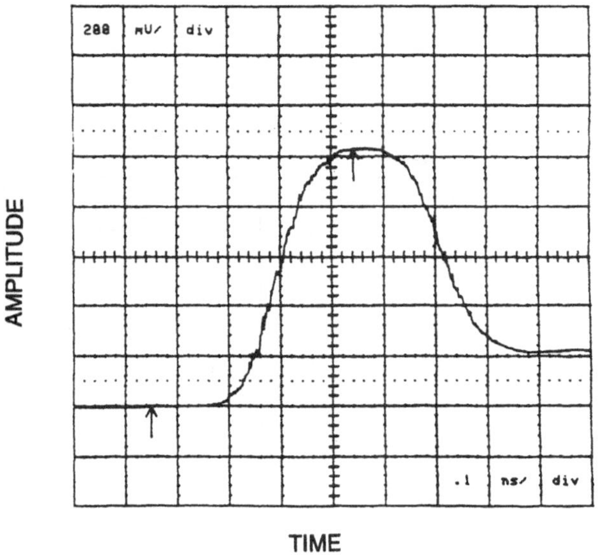

(b) Averaging 100 Points

Figure 2: Electrical Signal Output to a 50 Ohm Load.

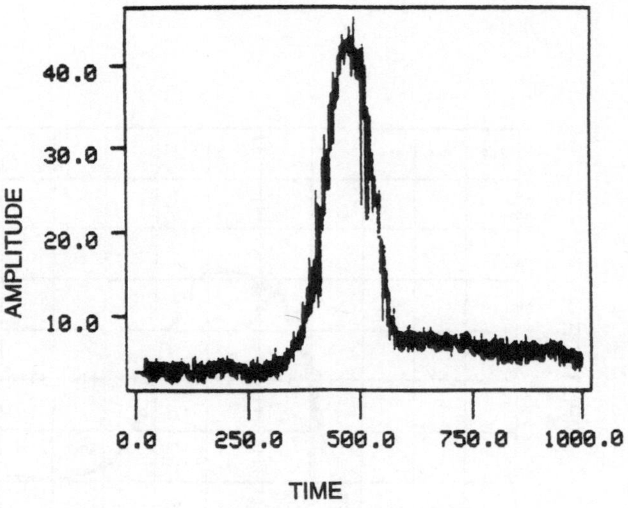

(a) No Averaging

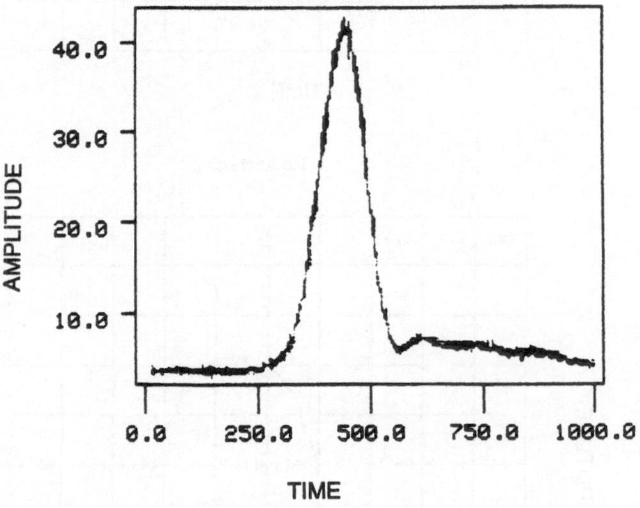

(b) Averaging 100 points

Figure 3: Electrical Signal Output to a 50 Ohm Load Under Reduced Optical Intensity.

REFERENCES

1. Auston, D. H. (1975), Appl. Phys. Lett., 26, p. 101.

2. Antonetti, et al (1977), Opt. Comm., 23, p. 434.

3. Lee, C. H. (1977), Appl. Phys. Lett., 30, p. 84.

4. Margulis, W. and Libbet, W. (1981) Ops. Commun. 37, p. 224.

5. Schoenbach, K. H. et al (1988) Optically and Electron Beam Contolled Semiconductor Switches, p. C3.

BROADBAND ELECTRONIC SYSTEMS AND COMPONENTS

ALL-ELECTRONIC SUBPICOSECOND PULSES FOR A 3-TERAHERTZ FREE-SPACE SIGNAL GENERATION AND DETECTION SYSTEM

D. W. Van Der Weide, J. S. Bostak, B. A. Auld, and D. M. Bloom

Edward L. Ginzton Laboratory
Stanford University
Stanford, CA 94305

INTRODUCTION

We report the first subpicosecond shock-waves ever generated and measured by electronic circuits. We have used these circuits with integrated antennas to generate and detect freely-propagating THz radiation. The circuits are monolithic nonlinear transmission lines fabricated on GaAs and, for peak efficiency, they operate at liquid nitrogen temperatures. Nonlinear transmission lines (NLTL's) have been used by several researchers for generating electronic shock-waves with picosecond transition times by compressing the wavefront generated by a ~0.5 W, ~5-10 GHz microwave power source.[1,2,3] While the literature reflects continual decreases in these transition times, as measured both by diode sampling bridges and by electro-optic sampling, the fastest reported 10%-90% fall time thus far has been 1.4 ps, with a ~5 V amplitude.[4] However, generating and measuring a sub-picosecond transition with an all-electronic device has been an elusive goal because a fundamental limitation has been the NLTL diode series resistance. By immersing a packaged NLTL into liquid nitrogen, we were able to lower this resistance significantly, thereby producing voltage shock-waves with 880 fs fall times and 3.5 V amplitudes, as measured by an on-chip diode sampling bridge (Figure 1). We have used these circuits with integrated slot antennas to generate freely propagating THz radiation, and we have observed measurable radiation beyond 3 THz.

NONLINEAR TRANSMISSION LINES

Nonlinear transmission lines are synthetic structures of series inductors (approximated by sections of high-impedance transmission line) with varactor diodes periodically placed as shunt elements. On this structure a voltage shock-wave develops from a sinusoidal input because the propagation velocity varies with the nonlinear capacitance-voltage relationship of the diodes,

$$V_p = 1/\sqrt{LC_{tot}(V)} \quad (1)$$

where L is the inductance and $C_{tot}(V)$ the sum of the varactor and parasitic capacitance of the line, all per unit length. Limitations of the NLTL arise from its periodic cutoff frequency,

$$\omega_{per} = 2/\sqrt{LC_{tot}(V)} \quad (2)$$

waveguide dispersion, interconnect metallization losses, and diode resistive losses.

MEASUREMENT OF SUBPICOSECOND FALL TIME

To make the fall time measurement shown in Figure 1, we mounted a monolithic NLTL/sampler chip[5] into a suitable microwave package with coaxial-to-coplanar waveguide transitions. After the shock-wave at the output of the test NLTL has propagated through a 20 dB attenuator, it is presented to the two-diode sampling bridge, itself strobed by a nonlinear transmission line. It then propagates down a ~100 µm length of coplanar waveguide to the opposite port of the sampler. We shorted this port with a gold bond ribbon to get a known broadband termination since a 50 Ω load with extremely broadband characteristics is difficult to fabricate on-wafer. While the short distorts the back edge of the pulse, it does not affect the initial falling edge.

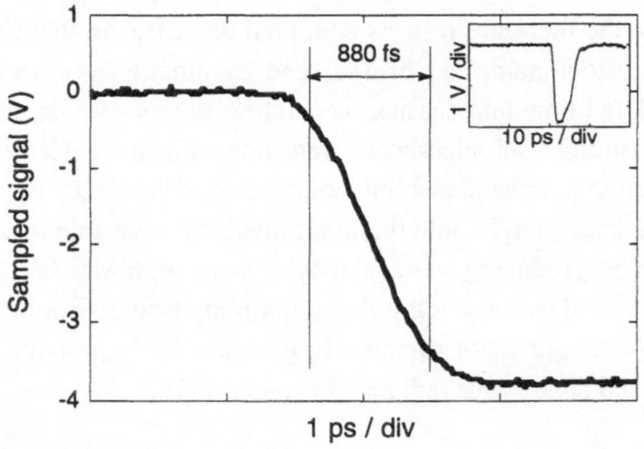

Figure 1. Time-domain reflectometry measurement of electronically-generated waveform having 880 fs, 3.5 V (10 - 90 %) fall time.

At room temperature, we measured a ~1 V, 1.8 ps fall time with a +25 dBm, 6.56 GHz input. Test port and strobe inputs to the sampler were offset by 0.5 Hz by splitting the output of a microwave signal source and running the strobe arm through a rotating phase shifter to allow the sampled waveform to be traced out in equivalent time and viewed on an oscilloscope with < 20 fs of rms jitter. We then lowered the packaged assembly into a dewar flask of liquid nitrogen. Steady improvement in both amplitude and fall time was noted until the packaged circuit was completely immersed, resulting in a minimum fall time of 880 fs with a 3.5 V amplitude. The liquid nitrogen does not significantly load the circuits because its relative permittivity is low, $\varepsilon_r = 1.4$. A root-sum-squares deconvolution of the measured fall time with the sampler aperture time (assuming that the two times are equal) gives 620 fs. This means that the -3 dB bandwidth of the sampling bridge is ~560 GHz. It also indicates that the periodic cutoff frequency

$$\omega_{per} = 2/\sqrt{LC_{tot}(V)} \tag{3}$$

($\geq 2\pi \times 500$ GHz for circuits described here) is now a significant limitation to circuit performance.

ANALYSIS OF DIODE IMPROVEMENTS

To support our analysis of this improved performance, we measured diode test structures from the same wafer at T = 300 K and 77 K. The NLTL diodes are formed on N^- material grown by molecular-beam epitaxy, doped to $1.2 \times 10^{17} \text{cm}^{-3}$, with a buried N+ layer 0.8 μm thick to provide the cathode contact. From parametric measurements, we propose that diode resistive losses are the fundamental limitation on NLTL fall time. They give rise to the dynamic RC cutoff frequency of the diodes,

$$\omega_{RC} = (S_{\max} - S_{\min})/R_d \tag{4}$$

where $S = 1/C$ and R_d is the diode series resistance.[6]

The dynamic resistance of a diode is found from the dynamic conductance,

$$g_{dynamic} = \partial I_d / \partial V_d \cong (q/kT) I_d \tag{5}$$

where I_d and V_d are the diode current and voltage. Then,

$$R_{dynamic} = 1/g_{dynamic} \cong \Delta V_d / \Delta I_d \tag{6}$$

which is measured with a Hewlett-Packard 4145A Semiconductor Parameter Analyzer. The measurement is made at $V_d = kT/q$ (where k is Boltzmann's constant, T is the absolute temperature, and q is the electronic charge), and at a current such that 1 Ω is due to the Schottky barrier alone; the remaining resistance corresponds to R_d. This quantity is area-independent, so the measurement can be made on an entire NLTL structure. We measured $R_{dynamic,300\,K} / R_{dynamic,77\,K} \approx 2.5$, for an improvement in the dynamic RC cutoff frequency of the smallest NLTL diodes from 2.3 THz to ~5.7 THz, in fairly good agreement with the ratio of fall time improvement when cutoff effects of the periodic structure are ignored. We note that the conductivity of the gold interconnections used in the

NLTL also improves from 2.4 μΩ·cm to 0.5·μΩ cm going from T = 300 to T = 77 K,[7] but its direct effect on minimizing fall time is negligible compared to that of the diode series resistance.

APPLICATIONS TO A SYSTEM

We have applied the ultrafast pulse generating capability of the NLTL to drive antennas and a sampler (Figure 2) in the coherent, all-electronic free-space signal generation system of Figure 3, the first generation of which is described in more detail elsewhere.[8,9] In addition to the increased spectral content of the driving NLTL's due to cryogenic operation, we made other improvements to this system.

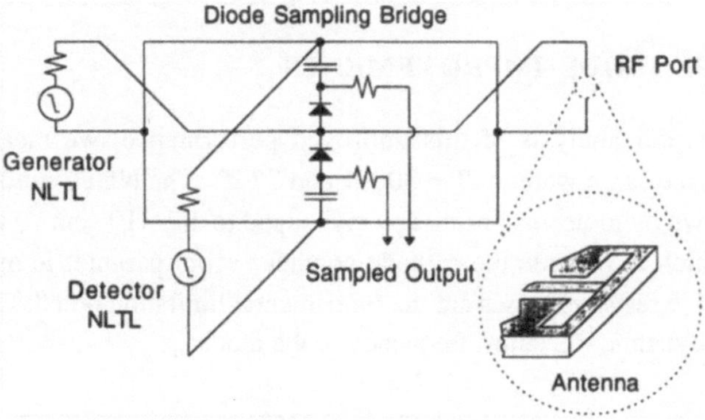

Figure 2. Generalized circuit diagram of all-electronic THz generator or detector.

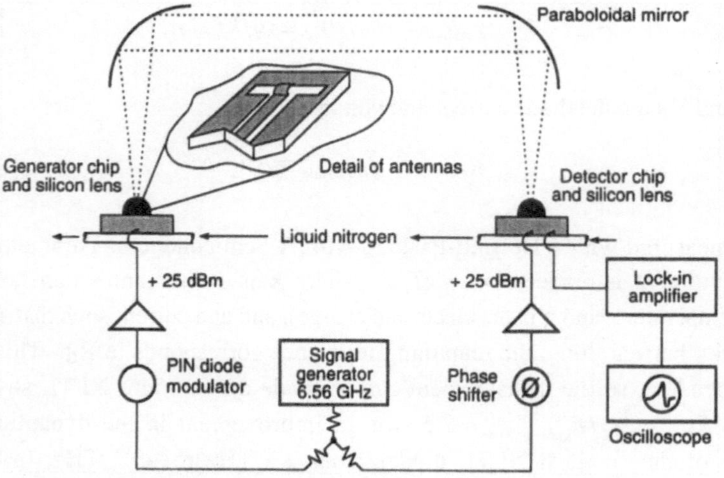

Figure 3. Diagram of 3-terahertz free-space signal generation and detection system.

We used integrated slot antennas[10] measuring 5 μm wide and 190 μm long, resonant at ~330 GHz. We also used high resistivity (10 kΩ·cm) silicon hyper-hemispheres[11] to collect and focus the quasi-optical beam. Finally, we chopped the signal with a PIN diode modulator to avoid spurious electrical reflections from a mechanical chopper used previously. With this improved system we measured the freely propagating THz pulse shown in Figure 4. This waveform has ringing associated with the resonance of the antennas; its shape was well-predicted by scale modeling using 113x antenna models (Figure 5.) The response of these scale models was measured with the low-pass time-domain option of the HP 8510B network analyzer.

By blocking the beam, we could determine the useful spectral range of the system. The Fourier transform of the time-domain signal detected both unblocked and blocked is shown in Figure 6 and shows measurable radiation beyond 3 THz. The distance between

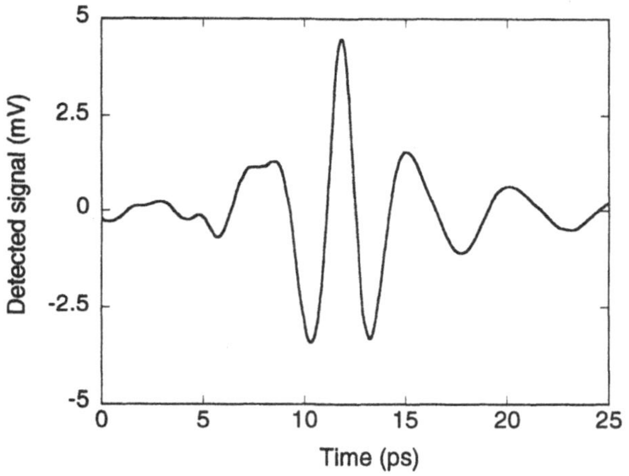

Figure 4. Pulse detected with THz free-space generation and detection system.

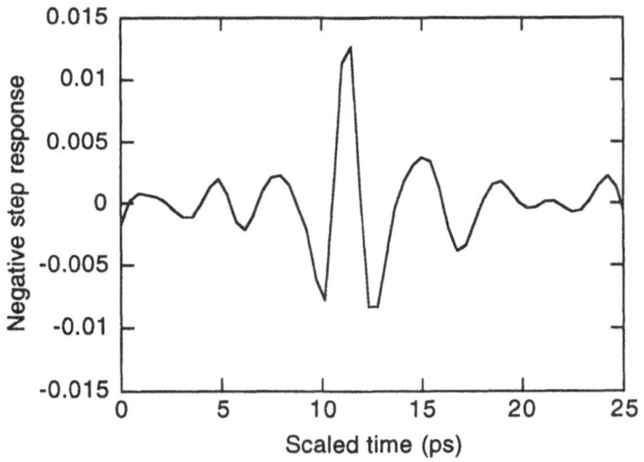

Figure 5. Pulse measured on scale-model antennas with negative-going unit step function.

the system generator and detector is 48 cm. Figure 7 demonstrates the improvement in performance gained with cryogenic treatment versus room-temperature operation of the system.

SYSTEM NOISE CONSIDERATIONS

The 3 dB bandwidth of the lock-in amplifier was 160 Hz, and we used 64 averages for an equivalent measurement bandwidth of 2.5 Hz. At T = 77 K, the measured data with the beam blocked gave the noise power shown in Fig. 4 for a 50 Ω system. Neglecting shot noise, the output resistance of the sampler (R = 160 Ω) divided by its duty factor, 0.62 ps/152 ps = 0.0041, gives an equivalent resistance of ~40 kΩ. This resistance develops $4kTRB = 425\times10^{-18}\ V^2$ mean-squared noise voltage at 77 K. In the same bandwidth, the lock-in amplifier generates $122\times10^{-18}\ V^2$, referred to the input. The sum of these mean-squared voltages corresponds to -140 dBm in a 50 Ω system, as shown in Figure 6.

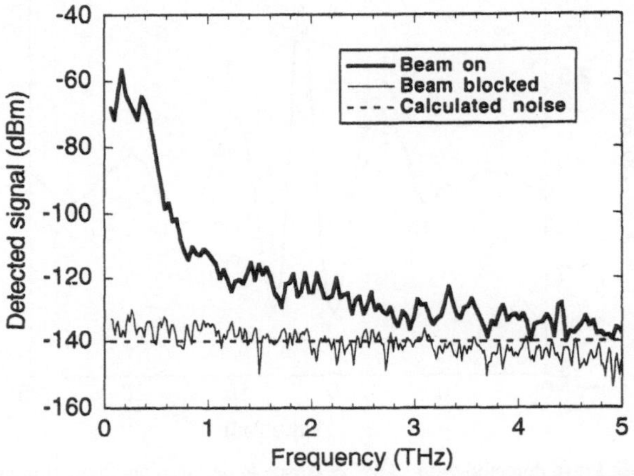

Figure 6. Frequency-domain response of THz signal generation system operating at T = 77 K, both with and without the beam blocked. The spectrum is not continuous: individual frequency points are connected by a line as a guide to the eye.

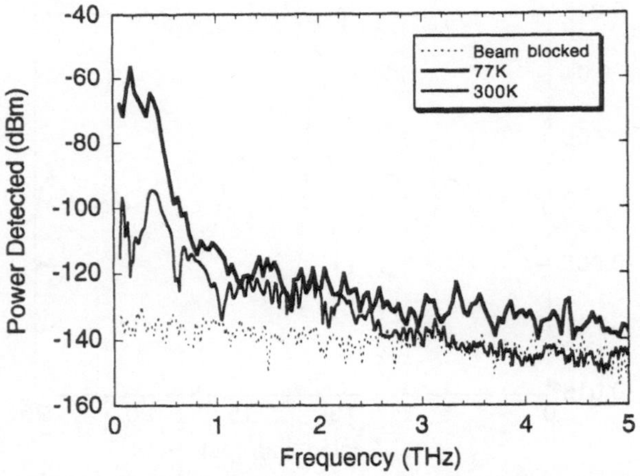

Figure 7. Frequency-domain response of THz signal generation system operating at T = 300 K, at T = 77 K, and with the beam blocked at T = 77 K. The spectrum is not continuous: individual frequency points are connected by a line as a guide to the eye.

CONCLUSIONS

In conclusion, we have generated and measured the first sub-picosecond voltage shock-waves by all-electronic means, using nonlinear transmission lines fabricated on GaAs immersed in liquid nitrogen. The improvement over room-temperature performance is due mainly to lower diode series resistance. Finally, we have described the first application of this improvement, a coherent, broadband, electronic signal generation system which has measurable radiation beyond 3 THz.

ACKNOWLEDGMENTS

The authors wish to thank J. Martin for circuit assembly. They also acknowledge discussions with P.E. Brunemeier, W.A. Strifler, D. Grischkowsky, E. Özbay, A. Black, M.S. Shakouri, I Aoki, and P.M. Solomon. Thanks are also due to Watkins-Johnson Co. for loans of equipment. This work was supported by the Joint Services Electronics Program under Contract N00014-91-J-1050.

REFERENCES

1. M.J.W. Rodwell, C.J. Madden, B.T. Khuri-Yakub, D.M. Bloom, Y.C. Pao, N.S. Gabriel, and S.P. Swierkowski, Generation of 7.8 ps electrical transients on a monolithic nonlinear transmission line, *Electron. Lett.* 24:100 (1988).
2. C.J. Madden, M.J.W. Rodwell, R.A. Marsland, Y.C. Pao, and D.M. Bloom, Generation of 3.5-ps fall-time shock waves on a monolithic GaAs nonlinear transmission line, *IEEE Electron. Device Lett.* 9:303 (1988).
3. C.J. Madden, R.A. Marsland, M.J.W. Rodwell, and D.M. Bloom, Hyperabrupt-doped GaAs nonlinear transmission line for picosecond shock-wave generation, *Appl.Phys. Lett.* 54:1019 (1989).
4. M.J.W. Rodwell, M. Kamegawa, R. Yu, M. Case, E. Carman, K.S. Giboney, GaAs nonlinear transmission lines for picosecond pulse generation and millimeter-wave sampling, *IEEE Trans. Microwave Theory Tech.* 39:1194 (1991).
5. R.A. Marsland, C.J. Madden, D.W. Van Der Weide, M.S. Shakouri, D.M. Bloom, Monolithic integrated circuits for mm-wave instrumentation, *IEEE GaAs IC Symp. Tech. Dig.* 19 (1990).
6. P. Penfield and R.P. Rafuse. "Varactor Applications," MIT Press, Boston (1962).
7. A. Goldsmith, T.T. Waterman, and H.J. Hirschorn. "Handbook of Thermophysical Properties of Solid Materials," Pergamon Press, New York (1961).
8. D.W. Van Der Weide, J.S. Bostak, B.A. Auld, D.M. Bloom, All-electronic free-space picosecond pulse generation and detection, *Electron. Lett.* 27:1412 (1991).
9. D.W. Van Der Weide, J.S. Bostak, B.A. Auld, D.M. Bloom, All-electronic free-space picosecond pulse generation and detection, *in* : "Optical Millimeter-Wave Interactions: Measurements, Generation, Transmission and Control," IEEE/LEOS, Newport Beach, CA (post-deadline appendix) (1991).
10. P.R. Smith, D.H. Auston, and M.C. Nuss, Subpicosecond photoconducting dipole antennas, *IEEE J. Quantum Electron.* 24:255 (1988).
11. Katzenellenbogen, N. and Grischkowsky, D., Efficient generation of 380 fs pulses of THz radiation by ultrafast laser pulse excitation of a biased metal-semiconductor interface, *Appl.Phys. Lett.* 58:222 (1991).

PULSE GENERATION AND COMPRESSION ON A TRAVELLING-WAVE MMIC SCHOTTKY DIODE ARRAY

Mircea Dragoman, Ralf Kremer, and Dieter Jäger

FG Optoelektronik, Universität Duisburg, D-4100 Duisburg, FRG

INTRODUCTION

An increasing interest has been paid in recent years to the study of the properties of wave propagation along nonlinear and dispersive transmission lines in electronics. In particular, due to a balance between nonlinearity and dispersion soliton waves can be generated [1,2] which preserve their shape and velocity during propagation representing a compressed pulse of excitation. These unique properties of solitons make them of considerable interest in the area of generation of picosecond pulses and millimeter waves with frequencies up to and above 100GHz. [1–5] The purpose of this paper is to study nonlinear wave propagation on special periodic Schottky contact coplanar transmission lines in order to obtain picosecond pulses for high-speed measuring techniques. Special emphasis is further laid upon the parametric generation and amplification of solitons in a resonator configuration and the behavior of gap solitons as a final example for nonlinear pulse compression techniques.

PERIODIC SCHOTTKY CONTACT TRANSMISSION LINES

The structure of a Schottky contact coplanar transmission line is sketched in Fig. 1, where the center conductor forms a rectifying metal-semiconductor junction. In this manner, a depletion layer is generated where the width depends on the applied voltage - the sum of a reverse dc bias V_0 and ac voltage $V(t)$. In the large signal regime and using a hyperabrupt doping profile, huge nonlinearities are obtained. [6–8] In order to implement the required dispersion properties of this nonlinear transmission line two types of periodic Schottky coplanar lines are realized as shown in Fig. 2. In these periodic nonlinear transmission lines solitons are generated and propagate along the structures. [7–9,1]

Ultra-Wideband, Short-Pulse Electromagnetics
Edited by H. Bertoni *et al.*, Plenum Press, 1993

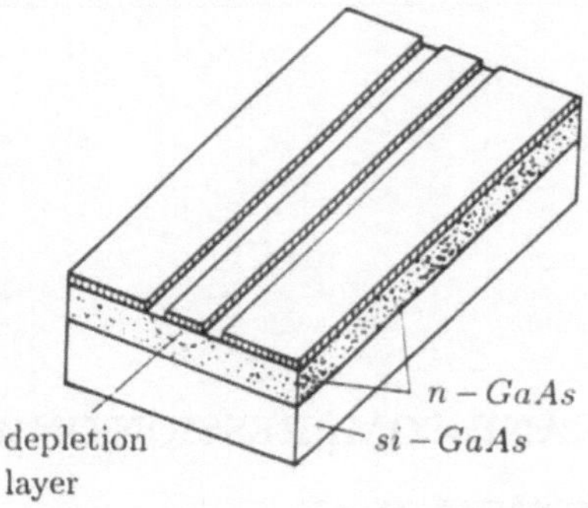

Fig.1. A coplanar Schottky contact transmission line.

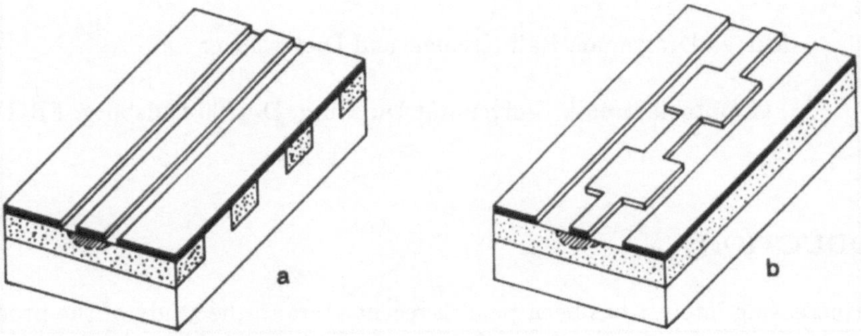

Fig.2. A periodic coplanar Schottky contact transmission line realized by (a) periodic doping or periodic ion implantation and (b) a periodic variation of the width of the center conductor.

EQUIVALENT CIRCUITS AND WAVE EQUATIONS

The equivalent circuits of the periodic structures presented in Fig. 2 are shown in Figs. 3 and 4, see Ref. 1. The wave equations for the lumped circuit displayed in Fig. 3 are

$$V_{k+1} - 2V_k + V_{k-1} = L\frac{d^2Q_k}{dt^2} + R\frac{dQ_k}{dt}, \tag{1}$$

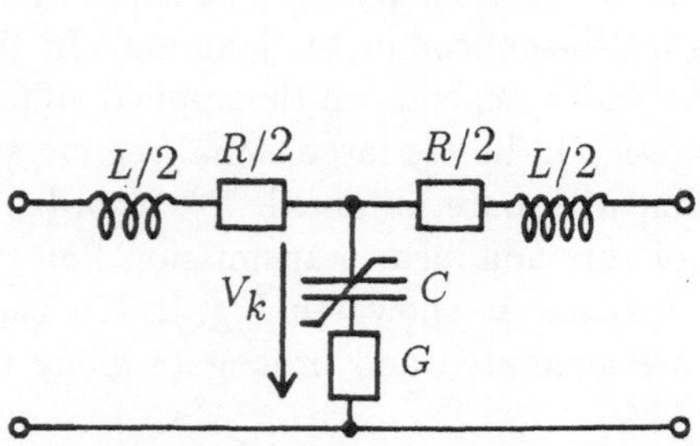

Fig.3. The equivalent circuit with lumped elements.

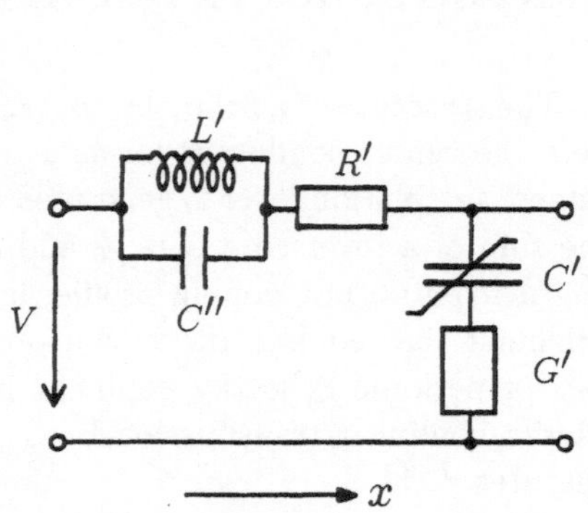

Fig.4. The equivalent circuit with distributed elements.

$$\frac{dQ_k}{dt} = G\left[V_k - V(Q_k)\right], \tag{2}$$

where C = C(V) = dQ/dV holds. Equations (1) and (2) can be combined to obtain a generalized Toda equation which exhibits soliton solutions. [7–9,1] The wave equation for the equivalent circuit with distributed elements as displayed in Fig. 4 is

$$V_\xi = g(V)V_\tau + \chi V_{\tau\tau\tau} - aV + bV_{\tau\tau}, \tag{3}$$

where $\xi = x/2u_0$, $\tau = t - x/u_0$, $u_0 = (L'C_0')^{-1/2}$, $\chi = 1/3\omega_c^2$, $\omega_c = (L'C'')^{-1/2}$, $a = R'C_0'$, $b = C_0'/G'$ and $C'(V) = C_0'(1 - g(V))$. Equation (3) represents a generalized Korteweg de Vries equation where a and b are lossy parameters, i.e., a describes the frequency independent dissipation, while b characterizes the semiconductor losses with a quadratic frequency dependence. χ is the measure of dispersion and ω_c the cut-off frequency. Neglecting the dissipative effects the one soliton solution of eq. (3) is [1,10], using $g(V) = \delta V$.

$$V(\xi,\tau) = \hat{V} sech^2 \left\{ \left[\frac{\delta\hat{V}}{12\chi}\right]^{1/2} \left[\tau + \frac{\delta}{3}\hat{V}\xi\right] \right\}. \tag{4}$$

In the damped case the soliton amplitude ($\hat{V}$) and its velocity ($v_s = -3/\delta\hat{V}$) will vary as a function of ξ. [10] Note that eq. (3) can be derived from eqs. (1) and (2) in the case of small nonlinearity and weak dispersion.

GENERATION OF SOLITON WAVES

In the time domain the evolution of an input wave along the considered nonlinear and dispersive transmission line reveals remarkable properties when the losses are of second order. The propagation of a voltage wave along the devices presented in Fig. 2 has been studied by using eqs. (1) and (2) with V_k as a function of k - the number of the elementary cell from Fig. 3. The results for $C = C_0(1 - \delta V)$ are displayed in Fig. 5 for a rectangular input wave and in Fig. 6 for a sinusoidal excitation. As can be seen, a set of solitons with different amplitudes is generated in both cases. In Fig. 7 the experimental data obtained by modelling the structure by a periodic chain of lumped elements [1] are compared with the analytical results given by eq. (4). The solitons which are formed (see Figs. 5 to 7) exhibit a temporal width as given by, see eq. (4),

$$t_w = 3.5(\delta\hat{V})^{-1/2}/\omega_c. \tag{5}$$

For example, from Fig. 5 a compression factor of more than 15 can be estimated. As a result, a soliton with a duration of $t_w < 1ps$ can only be obtained when the cut-off frequency exceeds $1THz$ and the nonlinearity $g(V) = \delta V$ is larger than 0.25. However, these conditions can now be accomplished using the technological fabrication procedures dedicated to $MMIC$ Schottky contacts with Terahertz cut-off frequencies. [3–5]

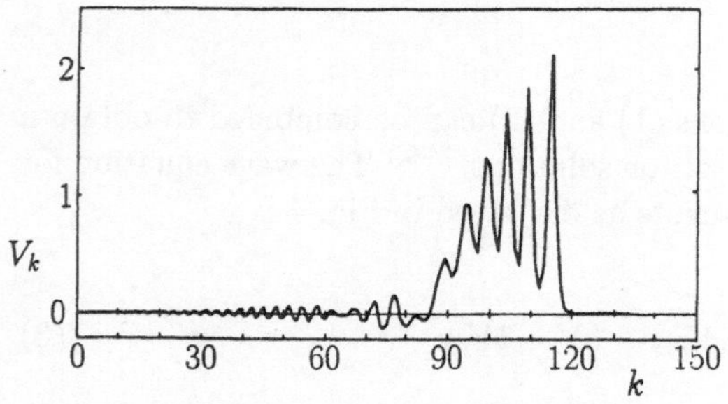

Fig.5. Numerical results for a rectangular input signal with a width of $\Delta k = 30$.

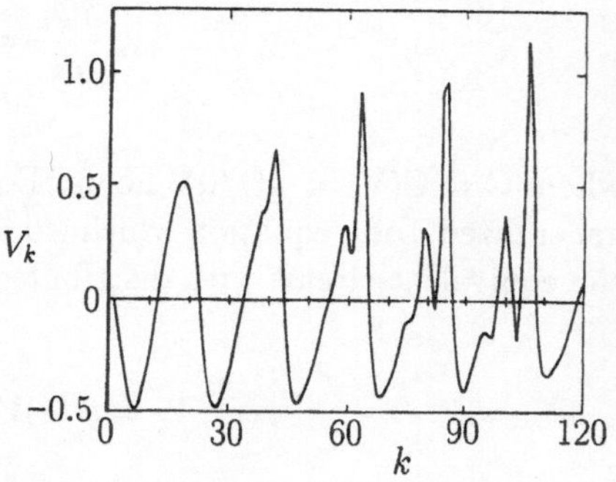

Fig.6. Numerical results for a sinusoidal input signal with a period of $\Delta k = 20$.

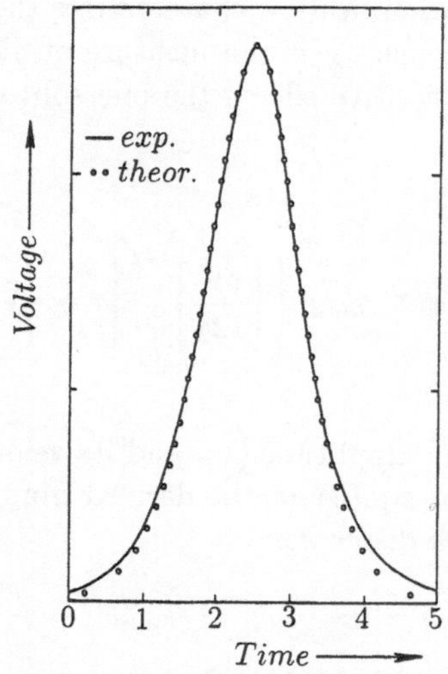

Fig.7. Comparison of a soliton between theory and experiment. [1]

In the case where the losses a and b are considered, the soliton amplitude decreases with x according to [10]

$$\hat{V}(x) = \hat{V}(0)exp(-\alpha_R x)/(1+\lambda x), \tag{6}$$

where

$$\alpha_R = \beta R'/2\omega L' \tag{7}$$

and

$$\lambda = 2\delta\hat{V}(0)\omega_c^2/9\pi\omega_r u_0 \tag{8}$$

with

$$\omega_r = R'/L' + G'/C_0'. \tag{9}$$

As a result, the minimum width of the soliton is determined by the losses leading to

$$t_w \geq 8(\delta\hat{V}\omega_r)^{-1}. \tag{10}$$

The results obtained so far reveal that the generation of solitons is accompanied by the following undesired effects: (a) The input wave is generally compressed into a set of several pulses and (b) due to losses the amplitude gradually decreases leading to an increase of the width. To overcome these drawbacks, a ring resonator configuration (see Fig. 8) can be used where the inherent feedback leads to a parametric generation and amplification of solitons.[11] Accordingly, the parametric generation of solitons is theoretically described by

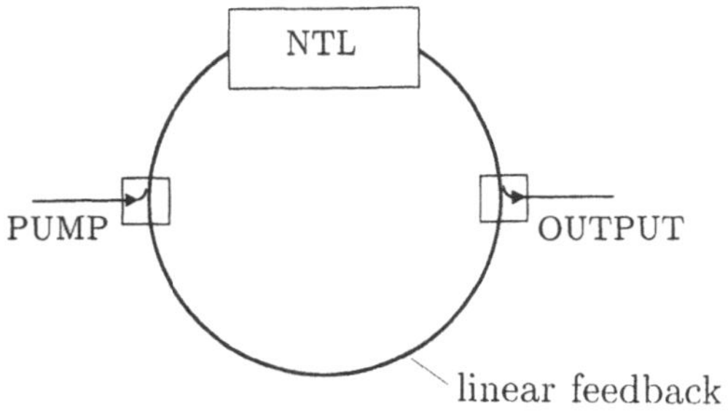

Fig.8. The ring resonator configuration. NTL denotes a nonlinear transmission line.

$$V_\xi = g(V)V_\tau + \chi V_{\tau\tau\tau} - aV + bV_{\tau\tau} + (fV)_\tau, \tag{11}$$

where the source term $(fV)_\tau$ represents the pump wave delivering energy into the system. In the case of a sinusoidal wave, f is given by

$$f(\xi,\tau) = \delta V_p cos\left[\omega_p(\tau - \Delta\xi)\right]. \tag{12}$$

Here V_p and ω_p are amplitude and frequency of the pump and Δ is a parameter of nonlinear dispersion [11]. Equation (11) has been studied numerically and the results are presented in Figs. 9 and 10. In Fig. 9 the influence of the pump is shown, i.e., in the case of $V_p \neq 0$ an unattenuated soliton wave is generated. In Fig. 10, the parametric generation of a soliton is sketched where the pump amplitude is used as a parameter. As can be seen, beyond a certain threshold (here $\delta V_p > 0.05$) unattenuated soliton propagation occurs revealing also the phenomenon of microwave bistability (see Ref. 1 for further details). In this respect, the behavior of the ring resonator is analogous to an optically bistable element. Here the different states correspond to different numbers of solitons in the system. Finally, it should be noted that graded structures [12] can also be used to limit the number of solitons generated from a general input wave.

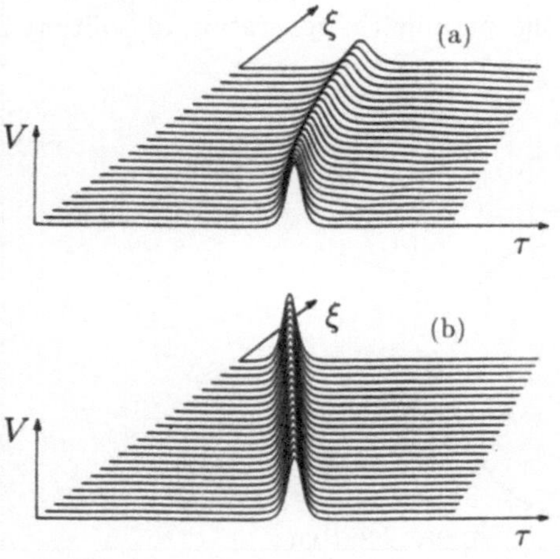

Fig.9. Compensation of losses. (a) $\delta V_p = 0$; (b) $\delta V_p \neq 0$

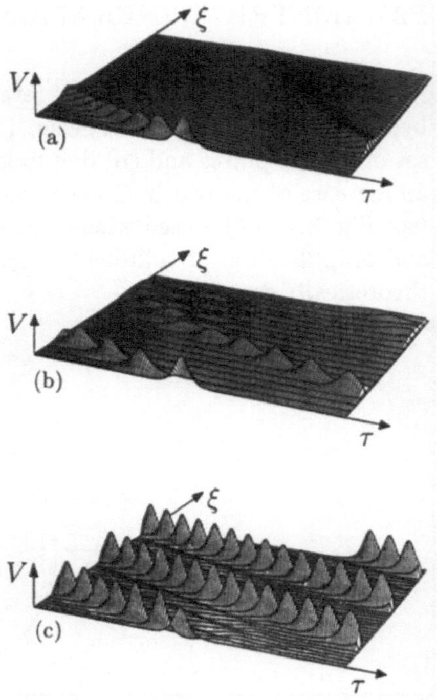

Fig.10. Parametric soliton generation. $\delta = 0.1V^{-1}$, $\xi = 3.67 \cdot 10^{-20}s^2$, $a = 4.8 \cdot 10^{-4}s^{-1}$, $b = 9.6 \cdot 10^{-12}s$, $\omega_p = 5.8 \cdot 10^8 s^{-1}$, $\Delta = 0.06$, $\delta V_p = 0, 0.01, 0.05$ for (a), (b), (c)

GAP SOLITONS

It is well known that the excitation field with a frequency within the stop band of a periodic structure is strongly reflected and the amplitude of the excitation decays exponentially with propagation distance. On the other hand, an excitation with a frequency outside the stop band, here $\omega \ll \omega_c$, can pass through the structure, a condition which has been assumed up to now in this paper. Since the periodic structure under investigation exhibits a clear nonlinearity, the cut-off frequency ω_c depends also on the wave amplitude. Thus the following situation can occur. Although the structure is excited in the stop band, a perfect transmission is induced provided that the input signal exceeds a certain threshold. This unexpected behavior is due to the excitation of a nonlinear resonance - a gap soliton [13] - which is a compressed envelope wave packet generated by nonlinear interaction of forward and backward waves within the stop band and which is caused by a shift of the dispersion curve with the applied voltage amplitude. The dynamical behaviour of the gap solitons can be described by using the equivalent circuit of Fig. 3 leading to the following nonlinear Schrödinger equation for the wave envelope $\bar{V}exp(j\omega t)$ [14]

$$j\left[\bar{V}_t + u_g \bar{V}_x\right] + (u_g'/2)\bar{V}_{xx} - (\delta'\omega/8) \mid \bar{V} \mid^2 \bar{V} = 0. \tag{13}$$

Here $u_g = (\partial\omega/\partial\beta)$ and $u_g' = (\partial^2\omega/\partial\beta^2)$ are the group velocity and group velocity dispersion, respectively, and $\delta' = \delta^2/4$. The soliton solution of eq. (13) is $V =$

$Re\{\bar{V}(x,t)exp(j\omega t)\}$ having an envelope given by [11]

$$\hat{V}(x,t) = \hat{V}_0 sech\left[\hat{V}_0\sqrt{\frac{\delta'\omega}{8u'_g}}(x - u_g t)\right], \tag{14}$$

where $\hat{V}_0$ is a constant amplitude. The temporal width of the soliton is roughly obtained from

$$t_w \sim \left[\frac{u'_g}{\delta'\hat{V}_0^2\omega u_g}\right]^{1/2}. \tag{15}$$

Relation (15) shows that picosecond wavepackets can be generated using an excitation in the stop band combined with a weak group velocity dispersion and a large nonlinearity.

CONCLUSIONS

In this paper, wave propagation along periodic nonlinear Schottky contact transmission lines is investigated, and from equivalent circuits wave equations are derived which lead to the formation of ultrashort solitons. As a result, the generation and compression of picosecond pulses is demonstrated numerically and analytically as well as experimentally. Additionally picosecond wave packets as gap solitons are predicted for the first time. In this manner, the travelling-wave *MMIC* Schottky diode array is in general found to be an interesting and novel device for *MMIC* applications in the millimeter wave region.

ACKNOWLEDGMENT

One of us (M.D.) would like to thank the Alexander von Humboldt Foundation for financial support.

REFERENCES

1. D. Jäger, Characteristics of travelling waves along the non-linear transmission lines for monolithic integrated circuits: a review, Int. J. Electron. 58: 649(1985).
2. M. Dragoman, R. Kremer, and D. Jäger, Travelling wave MMIC Schottky diodes for pulse generation and compression, 2nd Int. Workshop of the German IEEE MTT/AP Joint Chapter on Integrated Nonlinear Microwave and Millimeterwave Circuits - INMMC'92, 283(1992).
3. C.J. Madden, M.J.W. Rodwell, R.A. Marsland, Y.C. Pao, and D.M. Bloom, Generation of 3.5ps falltime shockwaves on a monolithic GaAs nonlinear transmission line, IEEE Electron Dev. Lett. EDL-9:303(1988).

4. M. Case, M. Kamegawa, R.Y. Yu, M.J.W. Rodwell, and J. Franklin, Impulse compression using soliton effects in a monolithic GaAs circuit, Appl. Phys. Lett. 58:173(1991).

5. E. Carman, M. Case, M. Kamegawa, R.Y. Yu, K. Giboney, and M.J.W. Rodwell, V-band and W-band broadband monolithic distributed multipliers, IEEE Microwave and Guided Wave Lett. 2:253(1992).

6. D. Jäger and J.P. Becker, Distributed variable-capacitance microstrip lines for microwave applications, Appl. Phys. 12:203(1977).

7. D. Jäger, Nonlinear slow-wave propagation on periodic Schottky coplanar lines, Proc. IEEE Microwave and Millimeter Wave Monolithic Circuits Symp.,15, St. Louis(1985).

8. D. Jäger, M. Block, D. Kaiser, M. Welters, and W.von Wendorff, Wave propagation phenomena and microwave-optical interactions in coplanar lines on semiconductor substrates, J. Electrom. Waves Appl. 5: 337(1991).

9. D. Jäger, Soliton propagation along periodic loaded transmission line, Appl. Phys. 16:35(1978).

10. D. Jäger, Experiments on KdV solitons, J. Phys Soc. Jpn. 51:1686(1982).

11. A. Gasch, T. Berning and D. Jäger, Generation and parametric amplification of solitons in a nonlinear resonator with a Korteweg de Vries medium, Phys. Rev. A34 :4528(1986).

12. K. Muroya and S. Watanabe, Experiment on soliton in inhomogeneous electric circuit. I. Dissipative case, J. Phys. Soc. Jpn. 50:1359(1981).

13. C.M. de Sterke, Gap soliton simulation, Phys. Rev. A45:2012(1992).

14. D. Jäger, A. Gasch, and K. Moser, Intrinsic optical bistability and collective nonlinear phenomena in periodic coupled microstructures: model experiments in: "Optical switching in low-dimensional systems", H. Haug and L. Banyai(eds.), Plenum Publ. Corp., New York(1989).

ULTRA-WIDEBAND IMPULSE ELECTROMAGNETIC SCATTERING LABORATORY

Michael A. Morgan

Electrical and Computer Engineering Department
Naval Postgraduate School
Monterey, CA 93943

INTRODUCTION

The Transient Electromagnetic Scattering Laboratory (TESL) employs a unique dual-channel impulsive illuminating source whose instantaneous frequency spans 1–12 GHz. Impulse scattering from scale model targets are performed in a shielded anechoic chamber. The TESL has facilitated research into the control of scattering mechanisms and radar target identification using natural resonances[1,2]. Both of these applications are demanding on accuracy and dynamic range; improvement in the fidelity of measured scattering signatures has been an ongoing theme in the TESL's evolution since 1980.

Impulse scattering ranges have been operational since the 1960's[3] and were made possible by the development of fast pulse sources and sampling oscilloscopes. Impulse measurements can offer an attractive alternative to the more conventional coherent stepped-frequency method, both in terms of performance and cost. One advantage is the direct observation of the temporal signature, which is useful for locating and controling local response mechanisms on scatterers. Time-domain measurements are even essential when non-linear or time-varying effects are significant.

The original TESL[4] was developed in 1980 at the Naval Postgraduate School using an outdoor $12m \times 12m$ horizontal ground-plane. This arrangement required use of bisected mirror-symmetric scattering objects illuminated by a vertically polarized field. A high power mechanically switched pulser was employed and the system had a bandwidth of 50 MHz to 2 GHz. The outdoor location also resulted in severe interference from natural noise sources, as well as local and distant RF emitters. A large nonstationary noise floor to measurements was the result.

In 1983, the free-field TESL[5] was constructed using a fully shielded anechoic chamber within which scattering targets were supported at any orientation on a low density styrofoam column. Bisectable targets were no longer required and the restriction to a single incident field polarization was removed. Of at least equal importance was a significant reduction in noise and interference brought about by the shielded enclosure.

Ultra-Wideband, Short-Pulse Electromagnetics
Edited by H. Bertoni *et al.*, Plenum Press, 1993

Another upgrade of the TESL[6] came in 1985, with the development and testing of a highly stable GaAs FET amplified impulse-generator. The effective bandwidth was 1–6 GHz, and although the pulse amplitude was smaller than provided by the earlier mechanical pulser, long-term averaging could be employed to yield much improved signal fidelity. There were limitations, however, to the duration of signal averaging that could be employed because of slow drifts in the sampling system.

A further 10 dB improvement in signal to noise ratio (SNR) was made in 1988, with the installation of a new 20 GHz (HP54120T) digital processing oscilloscope (DPO)[7]. This DPO provided improved stability for long-term averaging along with a lower noise figure in its sampling head.

A significant enhancement of the impulse-generator was made by Walsh[8,9] in 1989, whereby the $30pS$ rise-time step waveform provided by the DPO's four-channel test set was fed into two parallel GaAs FET power amplifier sections whose complementary passbands slightly overlap. The latest such configuation, as developed by Bresani[10] in 1991, uses a 5–12 GHz passband section formed from a 20 dB gain preamplifier (Avantek AWT-13533) and a 28 dB gain, 1 watt power amplifier (Avantek APT-12066). The other section has a 1–7 GHz passband and uses a 25.5cm long coaxial delay-line followed by a 35 dB gain, 1 watt power amplifier (Avantek APT-6065).

Each pulse amplifier section feeds a separate 1–12 GHz bandwidth double-ridged tapered TEM horn antenna (AEL H-1479). Signal recombination is accomplished by superposition of the incident fields at the target location. The delay line in the 1–7 GHz section compensates for the extra preamplifier time delay in the 5–12 GHz section and thus aligns the leading edges of the two bandpass impulses which illuminate the target. This alignment is important to ensure that phase interference is minimized for the spectral components in the 5–7 GHz overlap region of the two sections.

Experimental procedures and signal processing will now be discussed, followed by illustrations of attainable measurement quality. Ongoing improvements which offer the promise of a 1–50 GHz bandwidth in the near future will then be considered.

LABORATORY DESCRIPTION

The free-field TESL incorporates a shielded anechoic enclosure having inside dimensions of $6.2m$ long by $3.1m$ square, as depicted in Figure 1. Scattering targets are placed at a center distance of $2.18m$ from the antennas on a low density styrofoam column which has no discernible scattering.

The chamber shielding significantly reduces the effects of outside interference, making it insignificant compared to either thermal noise provided by the sampling head or antenna noise generated by ambient radiation of the chamber's absorber. The back-wall is composed of 46 cm long pyramidal absorber, which has a reflection coefficient of −30 dB at 500 MHz (decreasing further with frequency throughout the system's passband). The source-wall uses 21 cm long pyramids while side-walls, floor and ceiling are covered by 21 cm longitudinal wedged material, which acts to channel energy towards the back-wall.

TESL operation can be understood by referring to the system diagram shown in Figure 2. Transfer functions in the frequency-domain represent transmitting, scattering and receiving processes, which are all assumed to be linear and time-invariant. Some simplifications are being used by combining signal paths from the dual transmitting horns and by employing scalar signal notation. Although the various transfer functions will have dyadic forms to reflect polarization sensitivity, only one polariza-

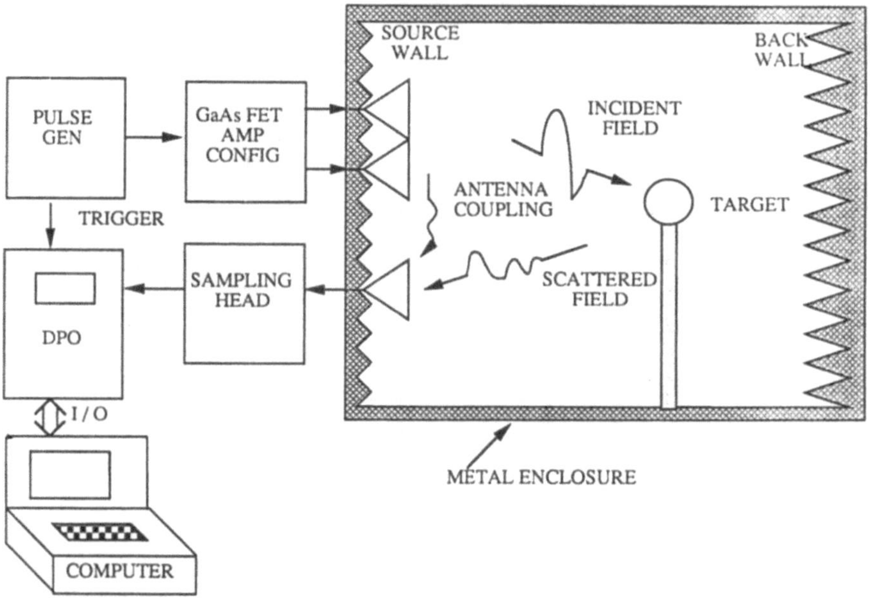

Figure 1. Transient Electromagnetic Scattering Laboratory (TESL) configuration.

tion is being transmitted and received during any measurement.

The input step pulse, with Fourier transform $X(f)$, drives the parallel amplifier and transmitting antenna sections, whose composite transfer function is denoted by $H_T(f)$. The transmitted field is incident upon the scatterer and the absorber in the chamber, which have respective transfer functions of $H_S(f)$ and $H_C(f)$. Multiple scattering between the target and the chamber structure, including absorber and antennas, is symbolized by the two-way interaction arrow between $H_C(f)$ and $H_S(f)$. Signal energy that is directly coupled from the dual transmitting horns to the adjacent receiving antenna is accounted for by the antenna coupling transfer function, $H_A(f)$. The receiving antenna transfer function, $H_R(f)$, converts the received field into the signal being sampled by the DPO. Finally, the combined effects of antenna noise and receiver noise, are lumped together at the DPO input as an equivalent noise source $N(f)$.

SIGNAL PROCESSING

TESL measurements are used to estimate, via deconvolution, the transient field scattering response at the receiving antenna location, $e_s(t)$, due to a user-specified incident plane wave signal, $e_i(t)$, which is not actually transmitted. The specified incident waveform must have a power spectrum which is essentially contained within the 1–12 GHz measurement passband.

In performing deconvolution signal processing, target scattering, as embodied in $H_S(f)$, is extracted from the combined effects of the other frequency-dependent elements of the system (e.g. amplifiers, antennas, chamber clutter, etc.). This is a multi-

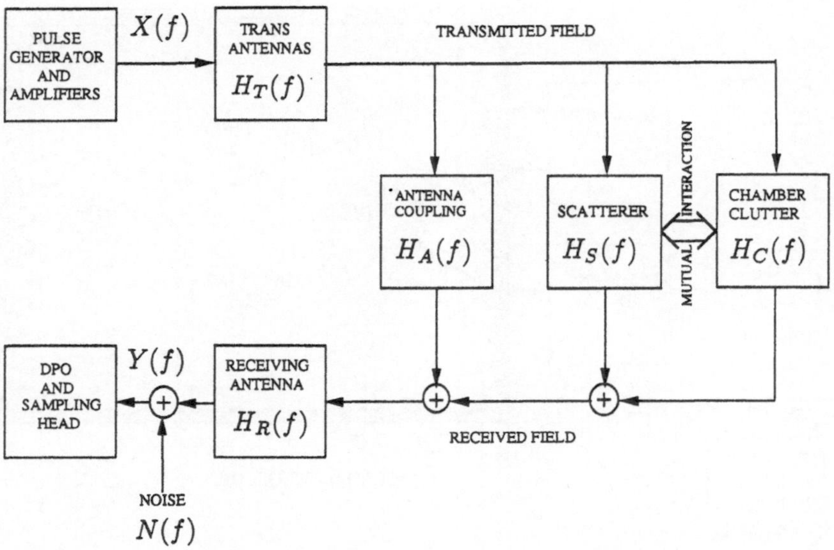

Figure 2. TESL transfer function system representation.

step procedure that requires the acquisition of three time-sampled and ensemble-averaged waveforms: (1) ***target***, with selected scatterer in place; (2) ***background***, with an empty chamber, and; (3) ***calibration***, using a metallic sphere of radius 7.62cm. Usually 1024 time-sampled data points are acquired in a 20 nsec time-window, each digitized to 12 bit accuracy. A measured waveform is typically assembled from the ensemble-average of 2048 individually sampled waveforms. Assuming uncorrelated noise, a 33 dB increase in signal-to-noise ratio (SNR) results from this process.

Measurements for background and broadside backscattering from a 10 cm long brass wire, with radius 1.18 mm, are overlaid in Figure 3. The scattering waveform with the background subtracted is displayed in Figure 4. Background subtractions eliminate the signal components due to direct clutter and antenna coupling and result in the spectra

$$Y_S(f) = X(f)\, H_T(f)\, \{H_S^T(f) + H_{SC}^T\}\, H_R(f) + N_S(f) \tag{2a}$$
$$Y_C(f) = X(f)\, H_T(f)\, \{H_S^C(f) + H_{SC}^C\}\, H_R(f) + N_C(f) \tag{2b}$$

where $N_S(f)$ and $N_C(f)$ are the respective noise differences of target minus background and calibration minus background.

The next step is deconvolution, whereby the antenna transfer functions are compensated for and the target response is estimated for a specified incident field waveform. This process employs an extension of the "optimal compensation deconvolution" procedure developed by Riad[11]. The target scattering spectral estimator is given by

$$E_i(f)\, \widehat{H}_S^T(f) = \frac{Y_S(f)\, Y_C^*(f)}{|Y_C(f)|^2 + L}\, \{E_i(f)\, H_S^C(f)\} \tag{3}$$

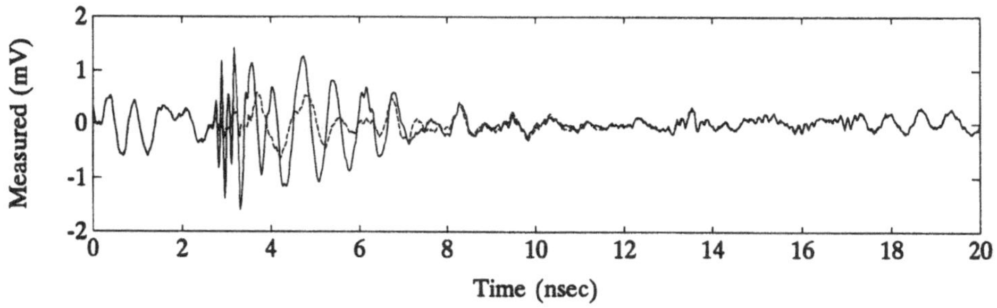

Figure 3. Measured antenna voltages for thin-wire target (broadside incidence) and background with empty chamber. Using 2048 ensemble averages. Brass wire length=10cm, radius=1.18mm.

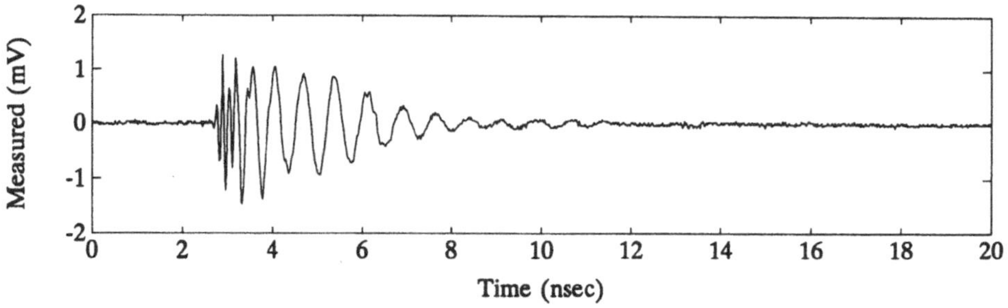

Figure 4. Target signal with background subtracted for the 10cm wire of Figure 3.

with the estimated time-domain target signature, $\widehat{e}_s(t)$, obtained by an inverse FFT of this function. The smoothing parameter L, is selected so that the effective passband of the deconvolved response in (3) fits within that of the measurement system.

Computation of the calibration sphere scattered field, as produced by the specified incident waveform, is denoted by $E_i(f) \cdot H_S^C(f)$ in (3). This computation is performed using a stepped-frequency Mie series, where the complex phasor scattered field at the location of the receiving antenna is evaluated. To increase accuracy, far-field approximations are not employed and the small bistatic angle between the transmitting and receiving antennas is incorporated.

The incident field waveform to be used is somewhat arbitrary, although its spectrum must fit within the system passband. A standardized signal shape which has been used in the TESL for several years is composed of a fast, positive-polarity Gaussian function and a slower, negative-polarity Gaussian, giving a composite double-Gaussian (DG)waveform

$$e_i(t) = A_1 e^{-\alpha_1 (t-t_0)^2} - A_2 e^{-\alpha_2 (t-t_0)^2} \tag{4}$$

This waveform allows control of both lower and upper frequency rolloffs so that the effective bandwidth of $E_i(f)$ can be confined to 1–12 GHz. Peak amplitude of the DG waveform is set at $10V/m$.

TARGET SIGNATURES

The deconvolved broadside backscattering signature for the 10 cm thin-wire previously considered is displayed in Figure 5. Computation responses are overlaid from two numerical approaches: a thin-wire time-domain integral equation[12] (TDIE) and the inverse FFT of a stepped frequency computation based on Hallén's frequency-domain integral equation[13] (FDIE). A short 3 nsec time-window is used to show detailed comparisons.

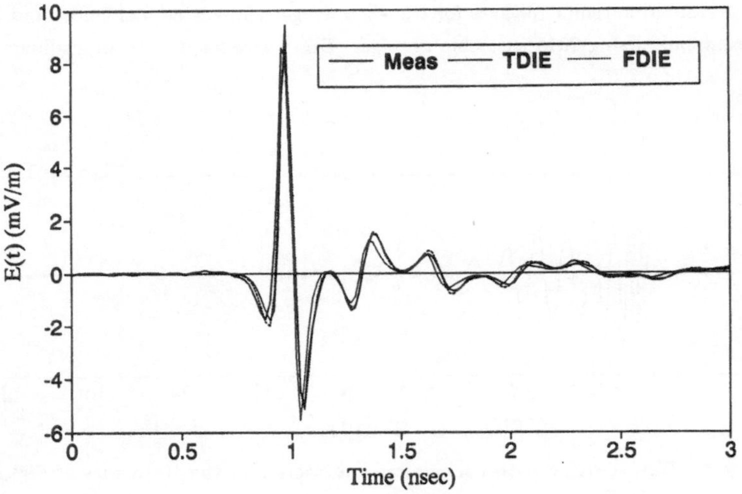

Figure 5. Comparison of deconvolved DG transient 90° backscattering with TDIE and FDIE computations for a 10 cm wire with radius 1.18 mm.

A second validation is shown in Figure 6, where the deconvolved DG transient backscattering from a metallic sphere of radius 4.05 cm is compared to a Mie series based computation which employs an inverse FFT. Both the main specular reflection and the first creeping wave return are apparent.

Providing impulse signatures of scale model radar targets has been a significant activity of the TESL. Example nose-on backscattering signatures for 1/72 scale models of two 1950's vintage tactical fighter aircraft are shown in Figure 7 for the case of $\overline{E}^{inc}$ parallel to the wings. The F-102A "Delta Dagger" is a delta-wing design of length $L = 20.8m$ while the MIG-17 "Fresco" has a much shorter length $L = 11.15m$ and uses conventional rear stabilizers. These scale models have respective "early-time" durations $T_e = 2L/c + T_p$ of $2.23nS$ and $1.33nS$ for the F-102A and MIG-17, where c is the velocity of light and $T_p = 0.3nS$ is the incident impulse duration for the DG waveform used.

Effects of structure are clearly discernible in these aircraft signatures, beginning in both cases with a small return from the nose. For the Delta Dagger, a second return begins around $0.5nS$ due to the air inlets. As the illuminating field travels over the swept back delta wing there is little forward scattering until the induced traveling wave current reaches the abrupt trailing edge, resulting in strong backscatter near $1.5nS$. A return is then seen near $2nS$ due to the rear of the fuselage. Considering the smaller Fresco, we see backscattering due to the leading and trailing edges of the wings between 0.5 and $1.0nS$, followed by strong rear stabilizer return starting just before $1nS$.

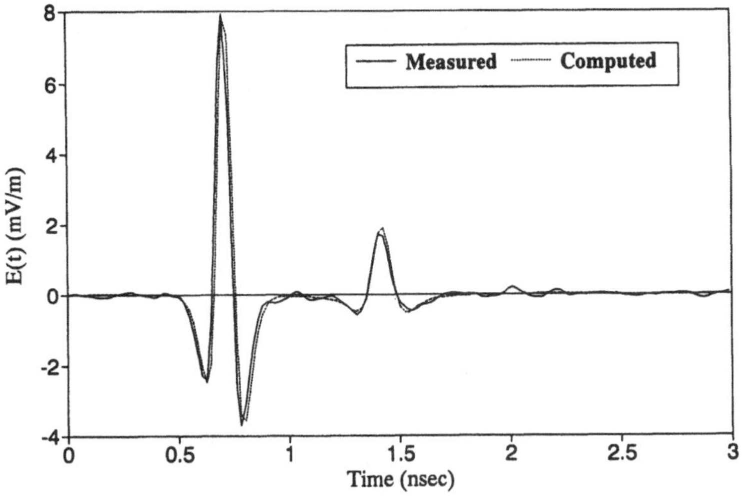

Figure 6. Comparison of deconvolved DG transient backscattering and Mie series based computation for a 4.05 cm radius metal sphere.

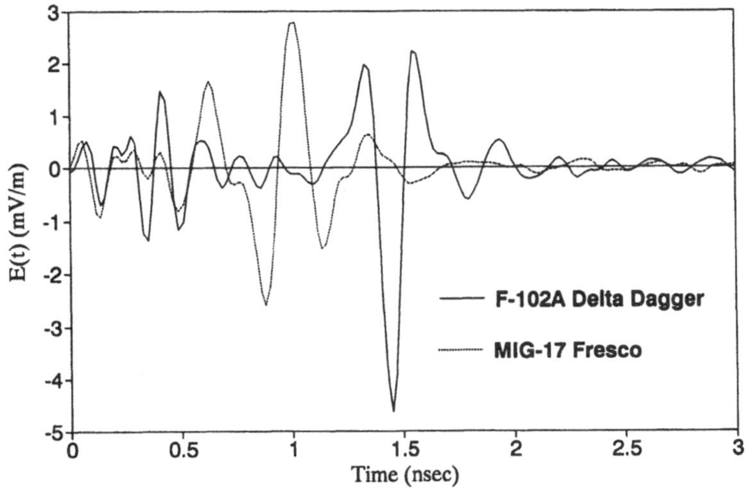

Figure 7. Nose-on backscattering signatures of two 1/72 scale model fighter aircraft.

FUTURE PLANS

There is a continuing quest to improve both the SNR and the frequency bandwidth of TESL measurements. In the realm of increased bandwidth, the dual amplifier system is employing essentially the entire available spectrum of the input pulse source provided by the 0.25 volt DPO $30pS$ step-generator. Further, the HP54121 DPO test set has a 20 GHz passband.

To remove these spectral bottlenecks, a recently developed 9 volt, $12pS$ step-generator (PSPL 4015B) made by PicoSecond Pulse Labs has been acquired along with a new HP54124 DPO four-channel 50 GHz test set. The 4015B uses the DPO step pulse as a trigger and shows excellent timing stability in our initial tests. With proper antennas and deconvolution, the potential exists for impulse scattering measurements in the TESL with bandwidths approaching 1–50 GHz. Further improvements in processed signal fidelity will also be sought using alternate deconvolution procedures[14].

REFERENCES

1. M.A. Morgan, Scatterer discrimination based upon natural resonance annihilation, *J. Electromag. Waves Appl.*, 2:481–502 (1988).
2. M.A. Morgan and P.D. Larison, Natural resonance extraction from ultra-wideband scattering signatures, *in:* "Ultra-Wideband Radar: Proceedings of the First Los Alamos Symposium," B. Noel, ed., CRC Press, Boca Raton (1991)
3. J.D. DeLorenzo, A range for measuring the impulse response of scattering objects, *1967 NEREM Record* 9:80–81 (1967).
4. C.W. Hammond, The development of a bistatic electromagnetic scattering laboratory, M.S. Thesis, Electrical Engineering Department, Naval Postgraduate School (1980).
5. M.A. Morgan, Time-domain scattering measurements, *IEEE Antennas and Propagation Newsletter* 6:5–9 (1984).
6. M.A. Morgan and B.W. McDaniel, Transient electromagnetic scattering: data acquisition and signal processing, *IEEE Trans. Instrum. Meas.* IME-37:263–267 (1988).
7. S. Sompaee, Computer algorithms for measurement control and signal processing of transient scattering algorithms, M.S. Thesis, Electrical & Computer Engineering Department, Naval Postgraduate School (1988).
8. N.J. Walsh, "Bandwidth and signal-to-noise ratio enhancement of the NPS transient electromagnetic scattering laboratory," M.S. Thesis, Electrical & Computer Engineering Department, Naval Postgraduate School (1989).
9. M.A. Morgan and N.J. Walsh, Ultra-Wideband transient electromagnetic scattering laboratory, *IEEE Trans. Antennas Propagat.* AP-39:1230–1234 (1991).
10. A.E. Bresani, Performance enhancement of the NPS Transient Electromagnetic Scattering Laboratory, M.S. Thesis, Electrical & Computer Engineering Department, Naval Postgraduate School (1991).
11. S.M. Riad, Impulse response evaluation using frequency-domain optimal compensation deconvolution, *23rd Midwest Symposium on Circuits and Systems* University of Toledo (1981).
12. E.P. Sayre and R.F. Harrington, Time-domain radiation and scattering by thin-wires, *Applied Sci. Res.*, 26:413–444 (1972).
13. R.S. Elliott, "Antenna Theory and Design," Prentice-Hall, Englewood Cliffs (1981).
14. F.I Tseng and T.K. Sarkar, Deconvolution of the impulse response of a conducting sphere by the conjugate gradient method, *IEEE Trans. Antennas Propagation*, AP-35:105–109 (1987).

SCATTERING, RESONANCE, CREEPING WAVE, TRAVELING WAVE AND ALL THAT: UWB MEASUREMENTS OF VARIOUS TARGETS

R. G. Madonna[1], P.J. Scheno[1], G.H. Vilardi[1], C. Hom[2], J. Scannepieco[2]

1 Grumman Corporate Research Center
2 Grumman Aircraft Systems Group
Grumman Aerospace Corp.
Bethpage, NY 11714

ABSTRACT

Ultra Wide Band (UWB) impulse radar has the potential to perform target identification. In order to realize this potential, target signatures must be measured and understood from the stand point of electromagnetic scattering. This paper discusses measurements of target "echoes" using an impulse UWB system.

We first examine the interaction of the transmitted pulse with conductive plates. At small incidence angles, we observe both specular scattering and a late time resonance. At large incidence angles, interactions with the leading edge and traveling waves are observed. In order to enhance the traveling wave interaction, we fabricated and test "bullets" which are 5 foot long, 6 inch wide aluminum strips whose front ends are tapered to a point. The bullets are oriented approximately 20° to the incident wave. The returns from the bullets are due almost completely to traveling wave.

We then look at data gather from a sphere. One clearly sees the specular return followed by the creeping wave. Resonance is observed at late times. We also have examined a dipole, which has a very large cross section at its resonance frequency. The dipole clearly exhibits ringing at its fundamental frequency.

We have fabricated and are in the process of measuring a resonant cavity. The cavity consists of a 27 inch diameter sphere with a 20° cap removed and a conducting cone (13.5 inches high) inserted into the sphere. We are observing resonance, and scattering from the cone. We will have more details to report at a later time as the measurements are in progress.

1. INTRODUCTION

Ultra Wide Band (UWB) radar is attracting a great deal of interest due to many of its potential applications (Noel, 1991; LaHaie, 1992). Target identification is one such application for which UWB appears well suited. However, in order to realize the potential of UWB for this application, the scattering of electromagnetic pulses from simple and complicated shapes must be well understood. A "return" from a complicated target, such as an aircraft, will be made up of a specular component as well as other phenomena such as resonance and traveling wave. Early research on this topic was performed by Kennaugh and Cosgreff (1958).

In this paper, we look at time domain UWB returns from a variety of targets. Each target exhibits a different interaction with the electromagnetic pulse. Studying these interactions in simple targets is the first step in developing target identification techniques. Understanding the simple targets' interactions also aids in the interpretation of returns from a complicated, extended target, such as a airplane. We begin by describing our measurement set-up. We then examine returns from conductive plates. The plates are large enough to be specular reflectors but we note the presence of late time ringing. In order to see resonance effects more clearly, Section 4 examines the response of a thin dipole to the UWB pulse. The dipole exhibits resonant behavior and we see the presence of harmonies of the fundamental frequency. In this section, we also look at the response of a sphere. The sphere, however, resonates via a creeping wave which is induced on the sphere by the incident UWB pulse. Our pulse is short enough so that the creeping wave can be resolved from the specular return. Section 5 looks at traveling waves on a simple structure. We show that one can measure the length of the structure based on the time difference of arrival of the specular return and the traveling wave return from the structure.

2. EXPERIMENTAL SETUP

Our UWB radar consists of a high voltage pulser, a pair of TEM (Transverse Electric Magnetic) horn antennas, a broad band digital sampling oscilloscope (HP54120T, $\Delta f = 20$ GHz), a fast digitizer (Tektronix SCD5000, $\Delta f = 4.5$ GHz), an optional 30 dB, 4 GHz wide amplifier, and a 386 laptop computer controller. The selection of pulsers is shown in Table 1. These are all commercially available. The TEM horns were designed and built by Grumman for use in our UWB testing program. These horns are described in Cermignani et. al. 1992.

Our radar is typically setup in a quasi-monostatic configuration. In the Grumman indoor RCS range, we have a 10.8° angle at the target between the direct transmit and receive paths (See Figure 1). All of the measurements were made in this range.

Table 1. Selection of available pulsers used in Grumman UWB Measurements.

Pulser	Output Voltage (kV)	Risetime (ns)	Pulse width (ns)	PRF (Hz)
Tektronix 109	0-0.4	~0.2	~0.8	500
Grant Applied Physics HYPS	2.0	~0.18	~0.5	0-2000
Grant Applied Physics HPFM	2.4	~0.15	~1.0	0-2000

We employ background subtraction in all our measurements. This eliminates any returns due to down range scatterers that are inherent to the facility, such as the pylon. It also eliminates any cross talk between the antennas. We also use spheres to calibrate the system so that RCS measurements are absolute (Madonna, 1992). We select time gates for our HP54120T and SCD5000 so that the specular return and late time effects are captured. During the course of our measurement series, we routinely shut down the equipment every hour in order to minimize the amount of time base drift due to heating. Every 3 hours, we re-measure the chamber background and the calibration object in order to track any long time scale drifts. We also perform a post calibration measurement in order to compare the system performance over the entire run.

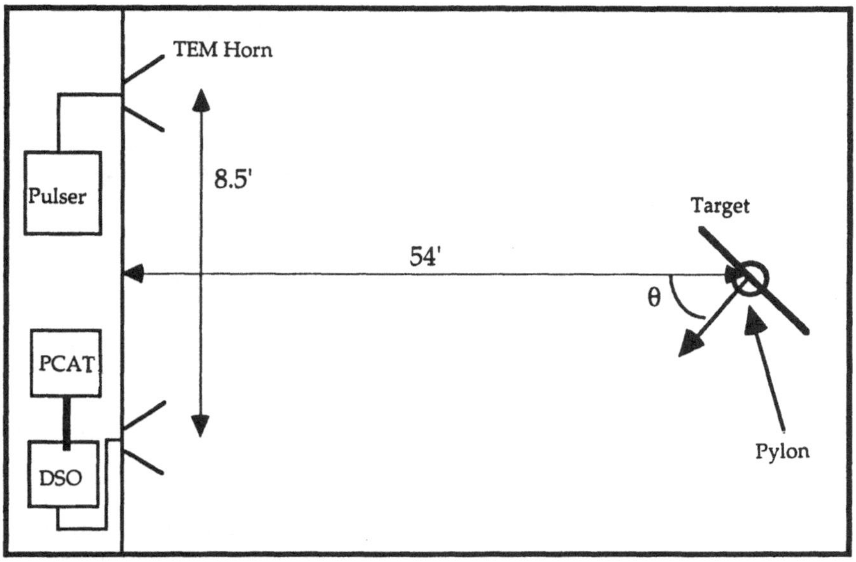

Figure 1. Schematic representation of Grumman's indoor RCS range configured for bistatic UWB tests. Note the septums between the two TEM horns are not shown in this figure.

3. UWB RETURN FROM FLAT PLATES

Flat plates are among the easiest targets to measure. They have large radar cross sections (RCS) at most frequencies, and for large (bigger than 2'x2') plates, act as specular reflectors across most of our transmitted band (Madonna et. al., 1992). However, when illuminated with pulses that are 1-2 ns in duration, the plates display a "ringing" that is not observed in cw measurements. Figure 2 shows the time domain return from a 4'x4' plate. Note the large specular return followed by some quickly damped ringing. This late time feature is dependent on the size of the plate. Figure 3 is a plot of the return from a 2' x 2' aluminum plate. Notice that the late time features are different than those in Figure 2 and that the ringing does not persist as long as in the case of the 4' x 4' plate.

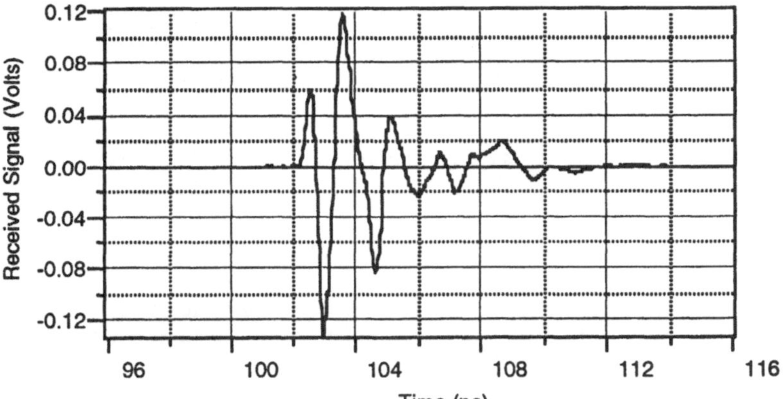

Figure 2. Measured return from a 4'x4' aluminum plate. Note the large amplitude of the specular return and the late time ringing.

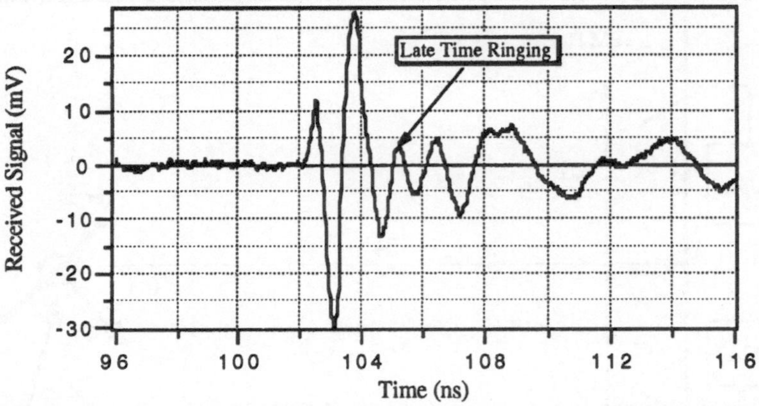

Figure 3. Return from a 2' x 2' aluminum plate. Again one sees late time ringing but at a different frequency than the 4' x 4' plate.

There is another interesting feature that we observe when measuring plates. If the plate is rotated so that the incident angle is large, say 40°, then we see scattering from the leading edge of the plate as well as a traveling wave launched from the tracking edge (Madonna et. al., 1992). Figures 4a and b show this for a 2'x2' plate and 4'x4' plate respectively. We also observe what appears to be back scattering from the plate as the incident wave sweeps along the plate's surface. The plates are not optimal structures for observing traveling waves or ringing, so we measured other simple targets that better display these mechanisms.

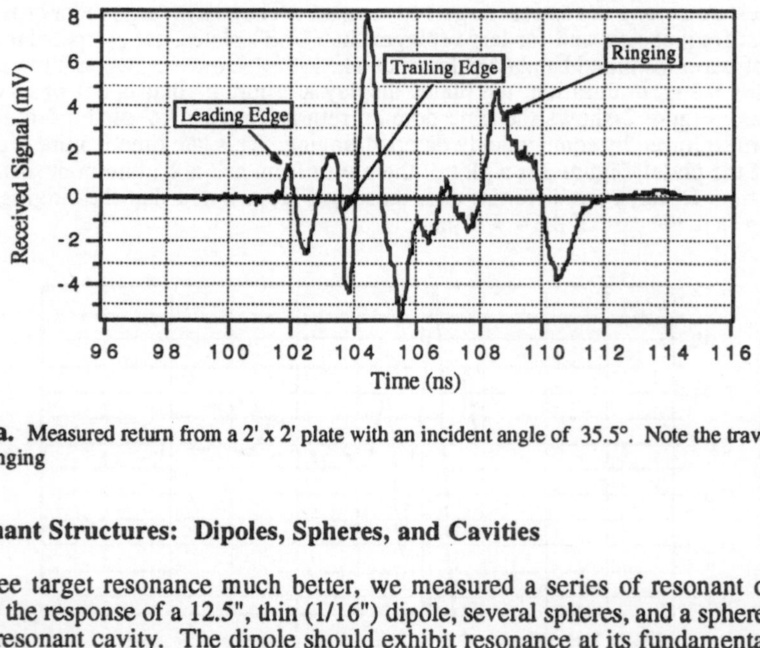

Figure 4a. Measured return from a 2' x 2' plate with an incident angle of 35.5°. Note the traveling wave and late time ringing

4. Resonant Structures: Dipoles, Spheres, and Cavities

To see target resonance much better, we measured a series of resonant objects. We measured the response of a 12.5", thin (1/16") dipole, several spheres, and a sphere modified to mimic a resonant cavity. The dipole should exhibit resonance at its fundamental frequency. The spheres resonant via a creeping surface wave. The cavity should display a complicated interaction between creeping waves, edge diffraction and internal reflections.

We mounted the thin dipole on a loseless foam block in horizontal polarization. The incident wave was horizontally polarized as was the receiving antenna. There was a 10° angle between the incident and return paths. Figure 5 shows the measured time domain return from the dipole. The average frequency is 440 MHz while the computed fundamental frequency is

472.4 MHz. Figure 6 shows the RCS of the dipole base on the received signal. Note the contributions at the fundamental and first few harmonics. This explains why the average frequency of the time domain signal is not at the fundamental frequency. The time domain

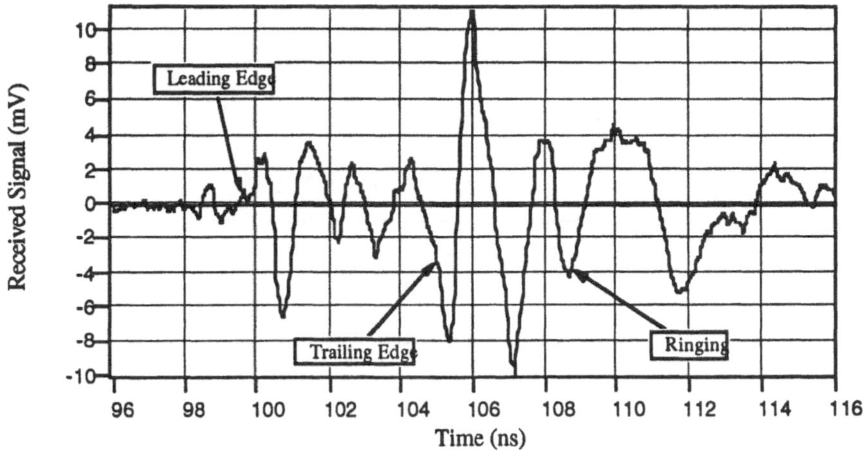

Figure 4b. Return from a 4' x 4' plate at 35.5° incident angle. The traveling wave appears later in time than for the 2' x 2' plate.

signal is a superposition of the fundamental frequency with the first few harmonics. Each harmonic is damping at a different rate which causes the time domain signal to have a lower frequency at late times. This is very consistent with the expected behavior of an impulse excited thin wire dipole (Ruck et. al., 1970).

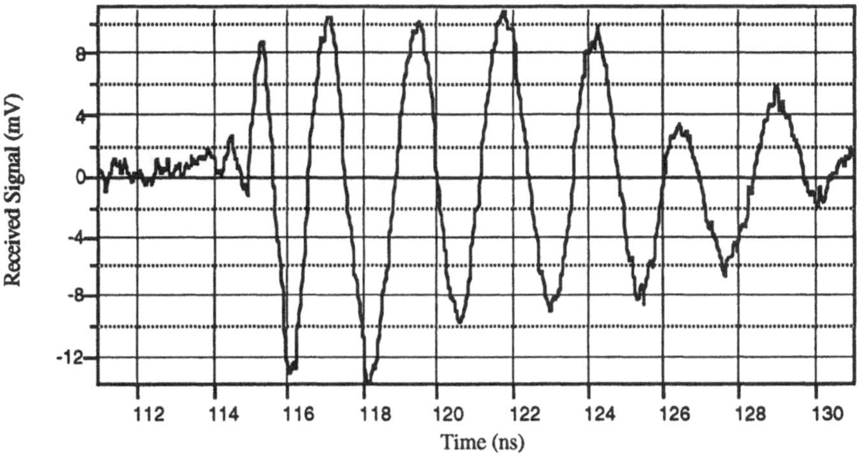

Figure 5. The time domain return from a thin, 12.5" dipole. Note the change in frequency with time as the dipole rings. and damps.

We now turn to spheres. Aside from being excellent calibration targets, spheres display a creeping surface wave which can be observed using an impulse UWB radar. This wave, which is induced by the incident impulse, will make several orbits of the sphere and radiate a signal at a frequency equal to $2\pi a/c$ where a is the sphere's radius and c is the speed of light (Kennaugh, 1961).

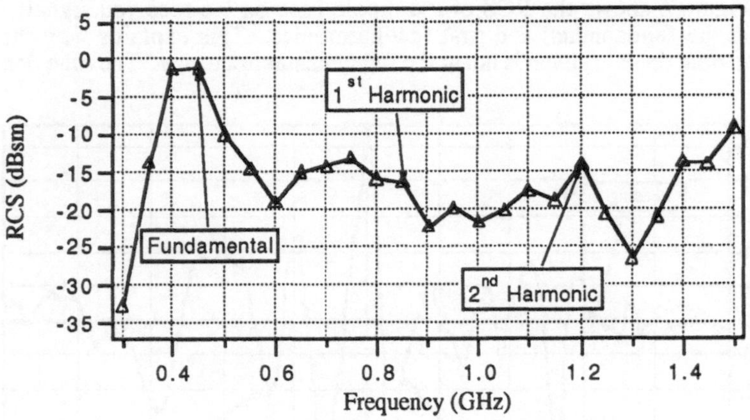

Figure 6. RCS of a thin, 12.5" dipole. Notice that the peaks in the spectrum are at the fundamental frequency and the first few harmonics.

Figures 7 and 8 are time domain returns from 20" and 27" spheres, respectively. As can be seen, both spheres "ring" with the first creeping wave appearing at $(2+\pi)a/c$ seconds after the specular return where a is the sphere's radius and c is the speed of light. The first creeping wave is identified in both figures. Note the average frequencies of both returns correspond well to the $2\pi a/c$ ideal model. There is some deviation from this due to have a pulse of finite duration and resistive losses in the spheres. We do not have calibrated spectra for these waveforms as these spheres are our calibration objects. Still one sees that the temporal behavior is in agreement with theory (Kennaugh and Moffat, 1965).

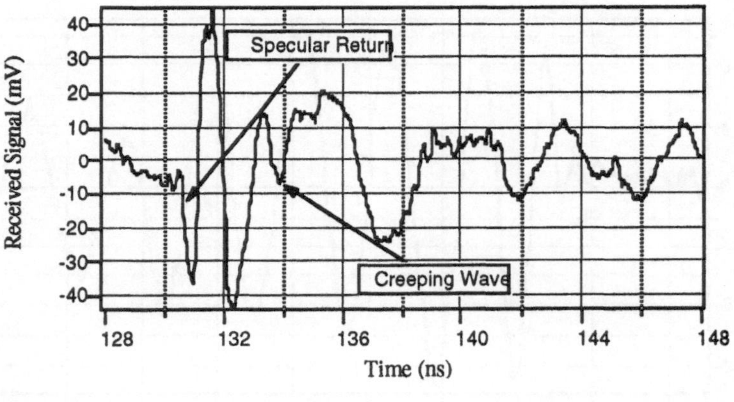

Figure 7. Return from a 20" sphere. The background subtraction was not perfect for this measurement. Still, one can see the first creeping wave radiating at approximately 4.4 ns after the specular return.

We conclude this section with a discussion of a resonant cavity created from a sphere. The cavity consists of a 27" diameter sphere with 5" cap cut out and a cylindrical cone, 13.5" in length inserted into the hole (Rolf, 1991). This target was mounted with the opening at 0° relative to the range centerline (Refer to Figure 1). Measurements were made and the target was rotated. Figure 9 shows the recorded waveforms when the aperture was at 0°, 45° and 90° respectively. One can see a difference in all three waveforms from the 27" diameter sphere in Figure 8. However, there is very little difference between any of the three waveforms shown in Figure 9. We expected to see a difference in the waveforms as the aspect angle changed due to

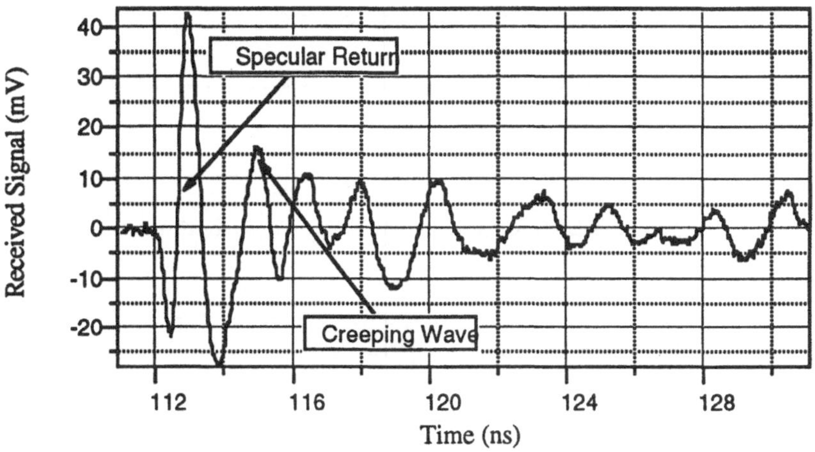

Figure 8. Return from a 27" sphere. Again, we see the creeping wave radiate well into late times.

where some of the interactions take place (i.e. where the cone is joined to the sphere). We believe that the aperture is too small to interact with the lower frequency components of our transmitted waveform. We are enlarging the aperture to 10" diameter and will be re-measuring the target.

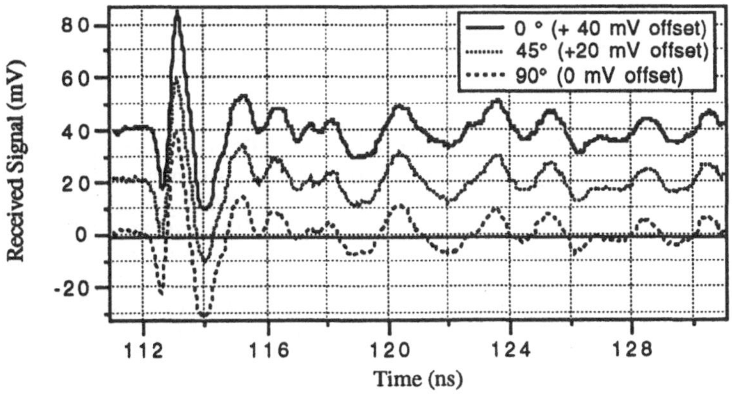

Figure 9. Returns from a resonant cavity at three different aspect angles. There appears to be very little difference in the returns.

5. Traveling Waves

In Section 3, we described traveling waves on square plates. Plates are not ideal targets for inducing and observing traveling waves. In order to observe this phenomena better, we measured the UWB radar return from a 5' long aluminum "bullet". The "bullet" has a rounded front end which flares out to a 6" wide strip (See Figure 10). The bullet is mounted at a 20° angle relative to the range centerline and is illuminated with vertically polarized radiation. The measured return is shown in Figure 11. We see diffraction from the top of the bullet and a large return from the traveling wave when it reached the end of the bullet. We also observe some reflection of the traveling wave that quickly die out. The return due to the traveling wave is quite a bit larger than the return due to tip diffraction (approximately 17 dB greater). This is not unexpected as the bullet is shaped in such a way as to reduce interactions at the tip. We also observe some late time ringing which may be due to reflected current on the bullet.

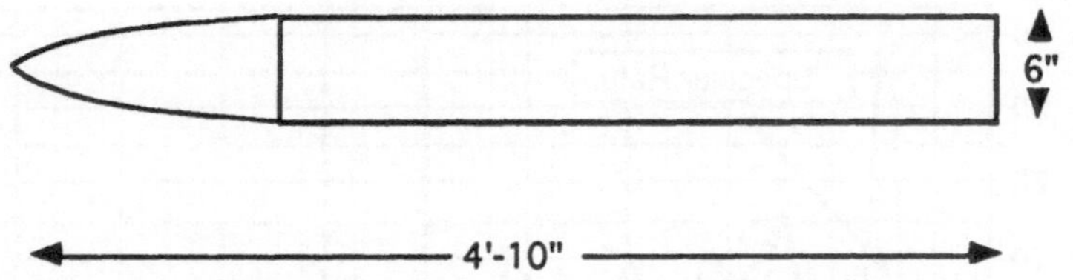

Figure 10. Schematic drawing of the bullet used to study traveling waves.

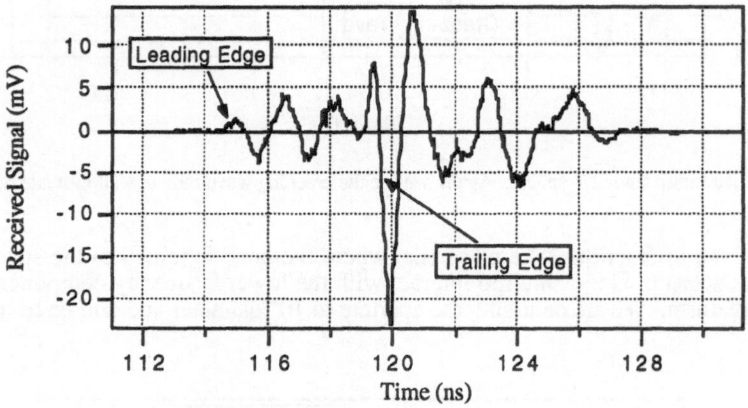

Figure 11. Measured return from the bullet. Note the portion of the return due to tip diffraction and the large return from the traveling wave.

Figure 12 shows the calibrated radar cross section (RCS) of the bullet. The RCS is obtained by methods detailed in reference Madonna, 1992. However, an FFT is done on the data in Figure 11, which results in a spectral resolution of 50 MHz per point for the RCS. We see multiple peaks in the RCS corresponding to harmonics of the fundamental resonance of the bullet. The round trip time for a wave on the bullet is approximately 10.16 ns, which corresponds to a frequency of 98.4 MHz. This frequency is below our antennas' low frequency response. However, there are harmonics of the fundamental frequency contained in the illuminating wave. Thus, we see in Figure 12 responses to harmonics at several different frequencies. The harmonics do not fall on multiples of 50 MHz (the spectral resolution of the data) and so we see a rather "choppy" RCS. However, if one uses peak finding algorithms on the data in Figure 12, one finds that the 3rd, 4th, 7th, and 11th harmonics are present in the data.

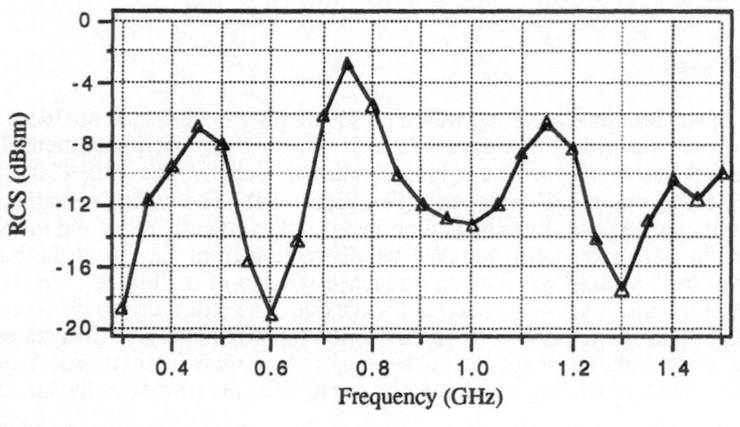

Figure 12. RCS of the bullet. Note the excitations of harmonics of the bullet's resonant frequency.

CONCLUSION

We have measured the UWB radar returns from a variety of simple targets. Each target was chosen to exhibit a specific behavior (e.g. traveling wave). We are able to clearly observe phenomena such as resonance, traveling waves, and creeping waves.

The measurements have an implication for UWB radar system development. UWB radar has the potential to provide the user with a target identification capability. This capability is strongly dependent on one's ability to process the return waveforms and compare them to a set of waveforms stored in the radar's computer. The measurements presented here imply that a large amount of information for each target of interest must be stored in the radar's processor if the system is to successfully perform target identification. Target waveforms will change with aspect angle and elevation. The target's resonances and the locations for the launching of traveling waves will change as the target configuration changes. For example, an aircraft with its flaps down will have different time domain signature than with its flaps up. Such target information must be carefully examined. The salient features that will permit rapid and accurate target identification must be extracted. This extraction will, hopefully, reduce the amount of information that the radar must store.

The above remarks are not to be taken as a discouragement for developing UWB systems, but rather as a challenge. System level considerations, such as how to implement target identification must be addressed. This is an area for future research.

ACKNOWLEDGEMENTS

This work was supported by the Grumman Aerospace Corporation IR&D funds.

REFERENCES

Cermignani, J.D., Madonna, R.G., Scheno, P.J., Anderson, J., 1992, Measurement of the performance of a cavity backed, exponentially flaired TEM horn, *Ultrawideband Radar,* SPIE Proceedings, Vol. 1631, 1992, pp146 - 155

Kennaugh, E.M., Cosgreff, R.L., 1958, The use of impulse response in electromagnetic scattering problems, *IRE Nat'l Conv. Rec., part 1,* pp. 72 -77

Kennaugh, EM,, 1961, The scattering of short electromagnetic pulses by a conducting sphere, *Proc. IRE (Correspondence),* Vol. 49 p. 380, Jan. 1961

Kennaugh, E.M., Moffat, D.L., 1965, Transient and impulse response approximations, *Proc. IEEE,* Vol. 53, No. 8, pp. 893 - 901, August 1965

LaHaie, I.J., 1992, "Ultrawideband Radar", SPIE Proceedings, Vol. 1631, Bellingham, Washington

Madonna, R.G., Scheno, P.J., Scannepicco, J., 1992, Diffraction of ultrawide band radar pulses, *Ultrawideband Radar,* SPIE Proceedings, Vol. 1631, 1992, pp. 165 - 174

Madonna, R.G., 1992, Measurement and calibration issues for RCS measurements using UWB techniques, *1992 IEEE MTT-S International Microwave Symposium*

Noel, B, 1991, " Ultra-Wideband Radar: Proceedings of the First Los Alamos Symposium", CRC Press, Boca Raton, Fl

Rolf, B, 1991, Private communication

Ruck, G.T., Barrick, D.E., Stuart, W.D., Kirchbaum, C.K., 1970, " Radar Cross Section Handbook", Plenum Press, NY

AN OVERVIEW OF SANDIA NATIONAL LABORATORIES' PLASMA SWITCHED, GIGAWATT, ULTRA-WIDEBAND IMPULSE TRANSMITTER PROGRAM

R. S. Clark, L. F. Rinehart,
M. T. Buttram, and J. F. Aurand

Sandia National Laboratories
Albuquerque, NM 87185

ABSTRACT

Sandia National Laboratories has developed several repetitive, ultra-wideband (UWB), impulse transmitters to address impulse source technology and to support experimental applications. The sources fall into two different classes, pulse peaking and pulse shorting depending on how the UWB frequency components are generated. The frequency spectrum of the radiated pulse from these sources include the spectrum of 100-MHz to 3-GHz. Depending upon the source, repetitive operation from single shot to 5-kHz (1-kHz nominal) has been obtained with excellent reliability and repeatability.

SNIPER (Sub-Nanosecond ImPulsE Radiator) is a source which uses an oil peaking switch to obtain a fast risetime (250-pS) pulse of 2-nS duration. The output voltage ranges between few tens of kilovolts to 250-kV. EMBL (EnantioMorphic Blumlein) is a similar device (presently under development) which uses a gas switch to sharpen the trailing edge of a 2-nS pulse to approximately 100-pS. To date, an output voltage of approximately 600-kV has been obtained (700- kV is the design goal). Since the frequency spectra are identical between sources with sharpened leading or trailing edges, alternatively, one can use parallel switches to short the pulse at its peak voltage.

The pulse is generated externally and then injected into the antenna. Due to the high powers involved and the need to radiate a broad spectrum of frequencies, Sandia has concentrated on TEM horn antennas with special high voltage feed adapters. Several TEM horns have been built and used during this program. In those cases where higher gains are desired for the higher frequencies, TEM horn-fed, dish antennas have been employed.

An overview of the UWB transmitters, including design and operation of the modulators, the PFN's, the pulse sharpening switches and the antennas will be presented.

---

*This work was supported by the U. S. Department of Energy under contract DE-AC04-76DP00789.

INTRODUCTION

All the UWB sources which are addressed here use the same basic component structure for pulse compression. Pulse compression is the method whereby the pulse length is shortened while the peak power is increased. Referring to Figure 1, the high-voltage, d-c power supply is energized by the a-c line and charges the energy storage capacitors to 30-kV in about 700-us, the capacitors are discharged through a 10:1 step-up transformer into the Blumlein pulse forming line (PFL) with the thyratron switch. The time to charge the Blumlein is on the order of 300 ns. The Blumlein produces a 2-ns (FWHM) pulse to the oil peaking gap which sharpens the leading edge (risetime) or falling edge. This pulse is then used to feed an antenna and the impulse wave is radiated to the target.

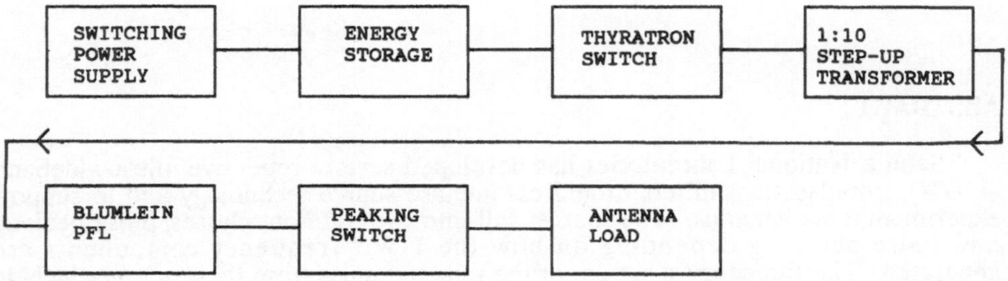

Figure 1. UWB Transmitter Flow Diagram

ASP-C

The ASP class of pulsers, shown in Figure 2, was originally developed at Sandia for single shot experimental observations. Later, it was used by the authors as a test bed for UWB source development. ASP-C, a later version of ASP, is a coaxial Blumlein 33-cm in diameter by 274-cm long mounted to a rectangular frame. The pulse width is variable (by changing the length of the inner cylinder) from 2-nS to 10-nS. A modulator was later added to ASP-C which gave the machine the capability to operate up to approximately 200-Hz. The output voltage is 200-kV maximum. ASP-C has been primarily used to conduct research into oil-insulated peaking switches, with the results being incorporated into the SNIPER project. It has also been employed in several UWB tests.

Figure 2. ASP-C Impulse Generator

SNIPER

SNIPER is a spark-gap based, high-voltage pulser which generates a subnanosecond-risetime pulse for impulse irradiation experiments.

The SNIPER pulse forming line is an 11.4-cm diameter, 40-cm long oil-filled Blumlein, a type of vector inversion generator. A cut-away view of the SNIPER Blumlein is shown in Figure 3. The Blumlein consists of three concentric cylinders isolated by dielectric supports. The outer cylinder is grounded, the intermediate cylinder is charged to approximately 250-kV by the charging transformer, and the

Table 1. SNIPER I PARAMETERS

Pulse Width	2 NS
Risetime	150 PS
Power	> 1 GW
Repetition Rate	1.2 KHZ *

*Has been operated to 5 kHz at reduced power (Power Supply Limited)

center cylinder is held to ground through an inductor (d-c ground). The load (antenna) is connected between the inner and the outer cylinders. The oil peaking gap is shown between the center line and the load.

The Blumlein is switched (main switch) by a low volume, unpurged hydrogen spark gap. When the gas gap is overvolted, it breaks down and launches a wave, between the intermediate and outer cylinder, toward the output end of the Blumlein PFL. This wave plus the static voltage between the intermediate cylinder and the center cylinder constitute the output voltage. The voltage appears across the load for twice the electrical length of the Blumlein. The oil gap sharpens the pulse before it appears across the load. The pre-pulse stripper, the annulus device of conical cross-section surrounding the peaking gap shown in Fig. 3, is a means of preventing a precursor voltage from appearing across the load during the pulse sharpening process. Displacement current through the capacitance between the oil gap electrodes is the cause for the pre-cursor voltage. The stripper is a means of minimizing the effects of this capacitance.

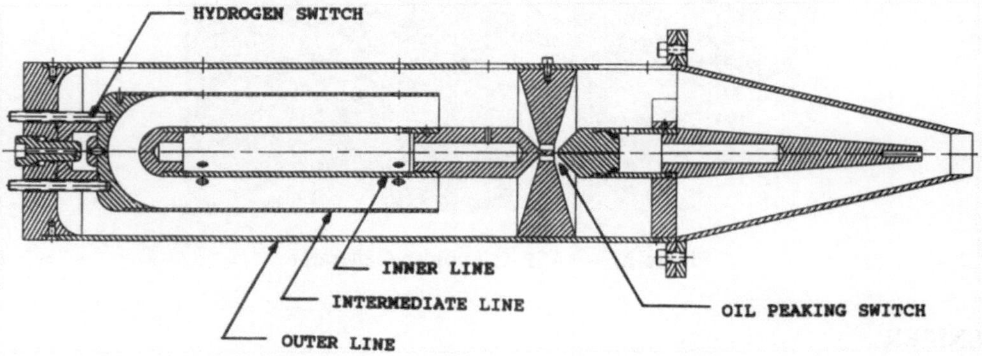

Figure 3. Cut-A-Way View of the SNIPER Blumlein PFL

The main gas switch must recover it's dielectric strength in approximately 100-us to allow high rep-rate operation. Although other gases have been used for low rep-rate operation, hydrogen appears to be the gas of choice for the higher rep-rates. The switch must be able to endure the high static pressure of the gas at upwards to a 1,000-PSI and the extreme heat from the repetitive spark channel. At the same time it must be able to resist hydrogen embrittlement. The first material to be used was polymethylmethacrylate. It showed signs of hydrogen embrittlement (stress crazing) after only a short time and a few shots. A polycarbonate material was substituted and the switch housing was able to survive weeks of firing and millions of shots at rates to 1 kHz. Brass electrodes have proved to be suitable for this application and show minimal erosion. The switch has proven to be very stable and long-lived.

The oil spark gap peaking switch with a pre-pulse stripper operates at several MV/cm (10 to 15-MV/cm) stress to keep the resistive phase time and statistical lag time very short. The peaking switch gap is on the order of 0.2 to 0.4-mm, so a modest oil flow purges it about 10 volumes/shot at 1-kHz. This gap is subject to a moderate rate of erosion and must be re-gapped occasionally to ensure stable and repetitive operation.

Figure 4 shows three SNIPER voltage waveforms. First is the un-peaked Blumlein output at 2-ns per division. The second graph shows the same pulse after passing through the peaking switch. This waveform is presented at 200-ps per division in the third photograph. The frequency spectrum of the SNIPER source covers the range to 3-GHz (the electrical pulse has a DC component that is lost on radiation). Operating the source without the peaking switch will result in a predominate frequency spectrum of approximately 200-MHz.

All of the high voltage components, from the capacitors to the SNIPER PFL, are housed in a container of insulating oil. The container is 48-in. long by 20-in. wide by 20-in. high containing approximately 75 gallons of insulating oil and the high voltage components.

The rate of repetitive operation is dependent upon the ability of the power supply to charge the primary capacitors after each discharge. For 1-kHz operation of SNIPER, the power supply must charge the capacitors in 1-mS. Switching power supplies are used which can be tailored to the source operation. This type of supply eliminates the use of charging resistors or inductors.

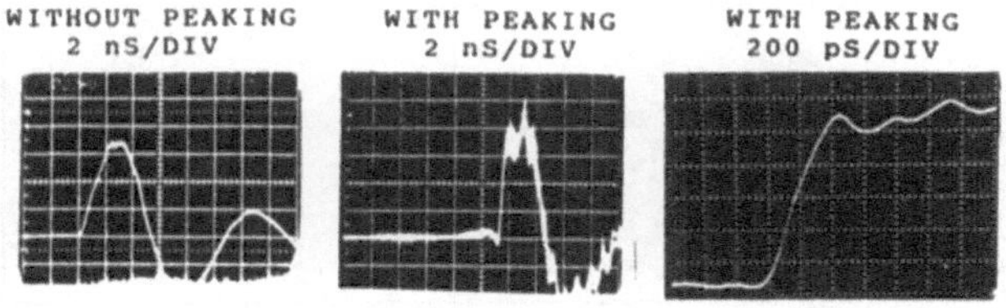

Figure 4. SNIPER Waveforms

EMBL

EMBL is the latest plasma switched UWB transmitter to be developed at Sandia National Labs. It employs a shorting switch to generate high frequency components in the transmitted wave. Following a long tradition of using flat-plate transmission lines as Blumlein PFL's at Sandia dating back to 1963 (ie.; SPASTIC, HARP, PROTO), this device eliminates the balun impedance-matching section that currently exists for the source-to-antenna connection.

The EMBL modulator has been tested to full voltage and repetition rate into a dummy resistive load. Although still under development and construction, EMBL is expected to produce several times the power of SNIPER.

ANTENNAS

For testing and research purposes, three antennas have been developed with each antenna having unique characteristics tailored for specific applications. The SNIPER TEM horn antennas are operated in a traveling wave, endfire (TEM) configuration for broad bandwidth, low dispersion and good directivity. The SNIPER pulser is connected to the antenna through standard RG-220 coaxial cable. The coaxial cable provides flexibility in mounting of the antenna. The antenna is fed by an adapter which goes from the coaxial geometry to the flat plate configuration. The design of this adapter is critical to maintain the 50-ohm impedance while supporting high electrical field stresses in the air environment. Figure 5 shows this adapter connected to a TEM horn (with the SNIPER tank assembly in the background). The horn flares in both transverse directions. The plates of the horn are cut longitudinally to prevent transverse current flow and suppress higher order TE modes. Each plate ended abruptly at the aperture. The large horn had 5 foot wide, 8 foot long high voltage and ground plates. This antenna was then modified to make the a second horn by installing flared sections on the output. Radiated electric fields from this antenna are approximately 4-kV/m at 100 meters using the SNIPER source.

Figure 5. TEM Horn Antenna

The parabolic dish antenna was designed and built primarily for the purpose of accentuating the higher-frequency components of the SNIPER radiated field. The dish is 8-ft in diameter, with an F/D ratio of 0.3125. The feed antenna for the dish was a small TEM horn with the same kind of coax-zipper transition as was used with the large TEM horn. The parabolic dish radiates an impulse that is rich in the higher frequencies while at the same time it acts to disperse the lower frequencies spatially. This antenna is also fed by an RG-220 coaxial cable through a coaxial-to-flat plate adapter.

CONCLUSIONS

SNIPER has a very extensive field operation history. Overall, SNIPER has proven itself to be extremely reliable. It has performed day-in and day-out for months with minimal maintenance and down time and has not limited the pace of experimental test programs. A new source, EMBL (EnantioMorphic BLumlein), has been designed and constructed and is now being brought up to its design parameters. It is expected that EMBL will produce several times the radiated electric field of SNIPER.

PRECURSOR OF AN ULTRA-WIDEBAND RADAR SYSTEM*

Andrew S. Podgorski

National Research Council of Canada
Ottawa, Ontario, Canada
K1A 0R6

ABSTRACT

A precursor of an ultra-wideband mono-pulse radar system is presented. The design of the radar system is based on a new TEM-Horn antenna and a picosecond HV generator developed recently by the author. The purpose of the new radar design was to verify the ultra-wideband mono-pulse radar concept and to establish the ultimate design criteria for the ultra-wideband radar system.

INTRODUCTION

Since 1984 the author has worked towards the development of a composite electromagnetic threat concept that would combine the effects of lightning, nuclear, and microwave threats. The composite threat concept resulted from a need for unified protection against individual or a combination of many electromagnetic threats.

The introduction of the composite threat concept [1] based on the intensity (power) of an electromagnetic (EM) field that is localized in space and time rather than a continuous wave (CW) signal energy was controversial. However, the author's long-time pursuit of the composite threat concept led to the development of ultra-fast picosecond high-voltage generators, wideband antennas [2,3], and a picosecond single pulse measuring system [4]. The development of the picosecond generators and wideband antennas resulted in the assembly of the high-power ultra-wideband radar system described in this publication.

* NRC 35035

ULTRA-WIDEBAND RADAR SYSTEM

A new 100-ps high-voltage generator, when combined with a new wideband TEM-Horn antenna, forms a low-cost mono-pulse radar transmitter system covering the frequency spectrum extending form 180 MHz to 1.75 GHz. A 40-ps measuring system based on a Tektronics TEK 7250 digitizer, when connected into a 25-ps receiving antenna, forms a wideband mono-pulse radar receiver covering the frequency range extending from 375 MHz to 12 GHz.

The transmitter and receiver systems, when assembled as presented in Fig. 1, resulted in a wideband high-power radar system covering the frequency range extending from 375 MHz to 1.75 GHz. The wideband radar system was assembled and installed on a ground plane having dimensions of 4 m × 8 m. In order to limit the radiation from the scattered radar beam to minimum, a metallic wall covered with absorbing material was installed at one end of the ground plane. The time synchronization of the receiver was obtained by supplying the digitizer with the trigger signal directly from the transmitting antenna.

A New Wideband TEM-Horn Antenna

The new radar system is based on the recently developed wideband TEM-Horn antenna [2,3]. The antenna launches an electromagnetic wave from a high-frequency coaxial feed line into an expanding rectangular horn containing a plate conductor forming an asymmetrical parallel line within the horn. The parallel line extends beyond the horn by means of a forwardly extended conducting plate section that functions as a radiating element. The conducting plate establishes a large radiating volume in space between itself and the ground plane extending forward from a lower surface of the horn. Two modes of propagation occurring in the horn give complete coverage of the relevant frequency spectrum extending from a few megahertz to tens of gigahertz. The high-voltage capabilities of the new wideband TEM-Horn antennas allowed their use as radiators in high-power radar system.

Two antennas, transmitting and receiving, were built for this project. Each of the antennas had different radiated beam width and different antenna aperture. A transmitting antenna was designed to have a beam width of ±16° while the receiving

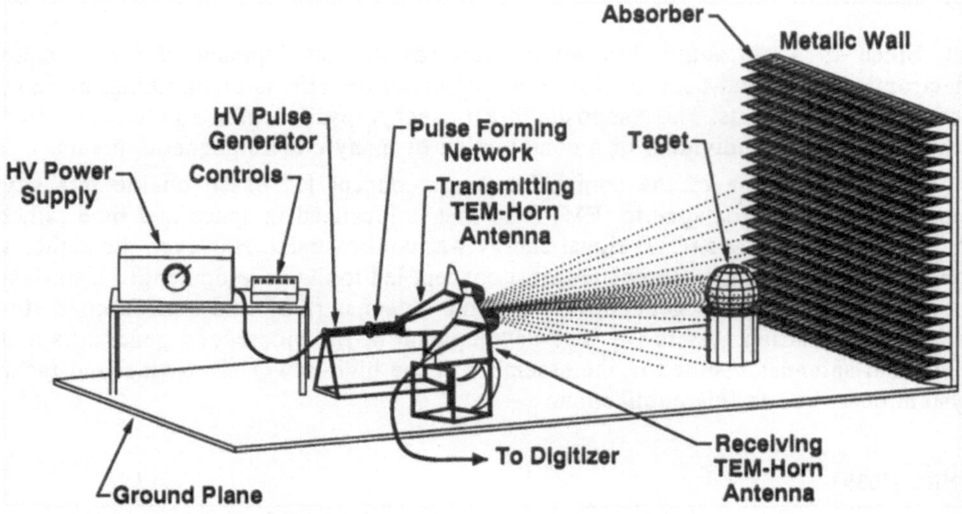

Figure 1 System for evaluation of radar performance.

antenna beam width was limited to only ±8°. The transmitting antenna horn aperture of 100 cm × 66 cm allowed transmission of pulses having rise time of 200 ps and duration of 1.75 ns. The corresponding CW bandwidth of the transmitting antenna was contained between 180 MHz and 1.75 GHz. The receiving antenna with a horn aperture of 48 cm × 32 cm allowed receiving pulses having rise times of 25 ps and durations of 0.85 ns. The corresponding CW bandwidth of the receiving antenna was between 375 MHz and 12 GHz.

The measurements assessing the performance of the antennas under pulse conditions were conducted in the set-up presented in Fig. 2. The measurement of beams from transmitting and receiving antennas reveals very high system directivity. Lack of wideband ground-free pulse measuring sensors operating at frequencies extending into the gigahertz frequency range resulted in the adaption of a ground based measuring system. Two electric field sensors were adapted for this experiment. The E-field sensor usable bandwidth was contained between 2 MHz and 1.3 GHz, while the E-dot sensor usable bandwidth was contained between 100 MHz and 10 GHz. In order to use the whole usable bandwidth of the sensors they had to be mounted directly on a large ground plane, as shown in Fig. 2. The positioning of the sensors on the ground plane allowed for accurate wideband measurements of electric field. However, such positioning, which required tilting of the antenna system with respect to the ground, resulted in the measurement of only the vertical component of the electric field.

The time domain measurements of the transmitting and the receiving antennas were made using a Tektronics TEK 7250 digitizer. The waveforms of the radiated electric field measured at different horizontal angles of the transmitting antenna system are presented in Fig. 3. Figure 3 indicates that the transmitted pulse is undistorted only in the axial direction of the antenna (0° horizontal angle). A pulse radiated in all other directions loses its high-frequency components; this can be observed as a decrease of rise time of the waveforms measured at horizontal angles of ±25° and ±35°. For angular directions of ±25° and ±35° the rise time of the waveform decreases to 2 ns, indicating therefore a total lose of bandwidth with respect to the 200-ps rise time measured at the 0° angle. Figure 3 shows that a further increase of the horizontal angle up to ±45° results in the further decrease of the electric field amplitude.

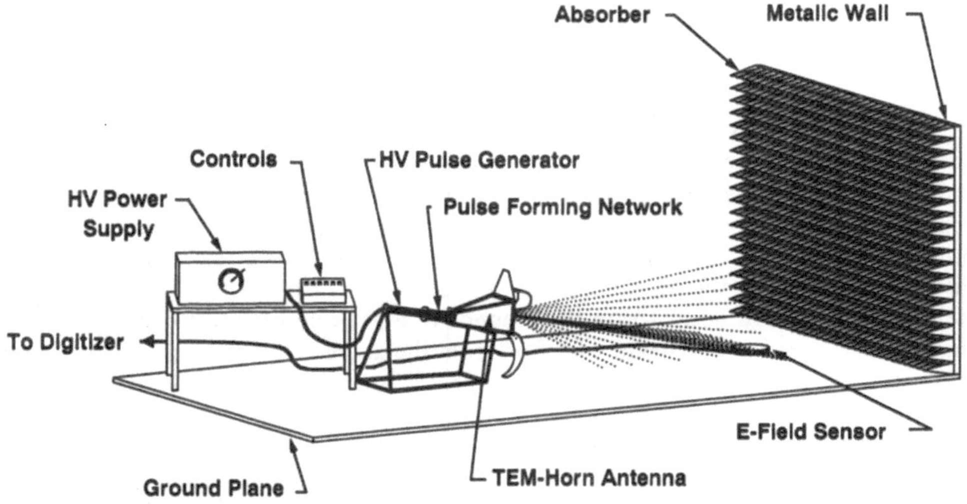

Figure 2 System for evaluation of antenna performance under pulse conditions.

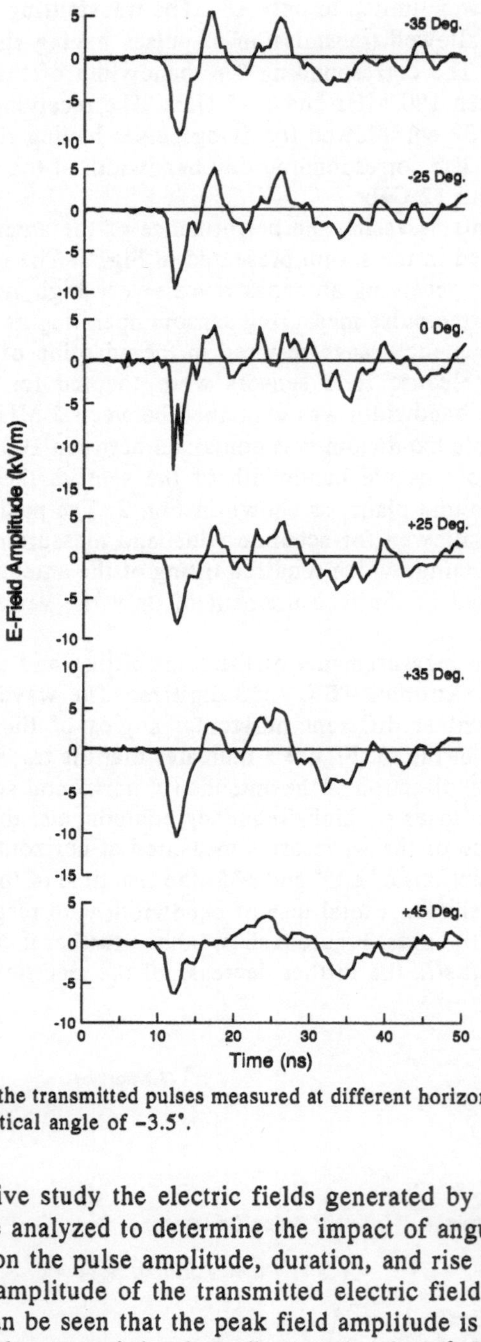

Figure 3 Waveforms of the transmitted pulses measured at different horizontal angles of the antenna system and a constant vertical angle of −3.5°.

In a comprehensive study the electric fields generated by the radar transmitter system of Fig. 2 were analyzed to determine the impact of angular resolution of the transmitting antenna on the pulse amplitude, duration, and rise time. Figure 4 represents the pulse peak amplitude of the transmitted electric field as a function of the horizontal angle. It can be seen that the peak field amplitude is fairly constant in all directions defined by the transmitting horn flare angle of ±16°. The measured electric field peak amplitude of 12 kV/m corresponds to a peak power level of 0.4 MW m^{-2} that was measured at a distance of 3.5 m from the antenna. Such a high peak power level resulted from a peak voltage of only 50 kV produced by the generator.

It is interesting to note that the duration of the transmitted pulse of approximately 2 ns is unaffected by the variation of the horizontal and the vertical angles. One must attribute this to the fact that the pulse duration depends on the low-frequency behaviour of the transmitting antenna and therefore depends primarily on the antenna radiating surface dimension.

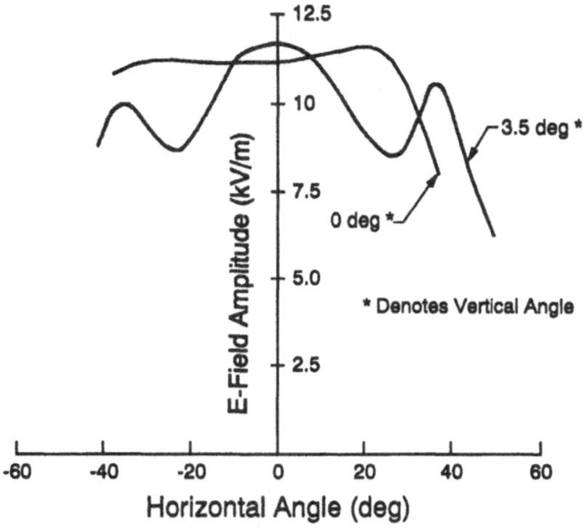

Figure 4 E-field peak amplitude of the transmitted pulse in function of the horizontal angle of the antenna system.

Figure 5 displays the measured effects of radiation angle on the rise time of the radiated electric field. A quick look at Figs. 4 and 5 shows that, although the angular resolution of the amplitude of the received signal does not change for different radiation angles, the rise time and therefore the system bandwidth change substantially. The drastic change of the rise time and the resulting change of the system bandwidth present a new opportunity for using the angular resolution of the wideband antemna to improve the antenna time domain directivity. Figure 5 reveals that the bandwidth of the signal radiated from the transmitting antenna decreases to half of its peak value at an angle of ±8°. One can postulate that the excellent bandwidth discrimination of the new TEM-Horn antenna can substantially improve the angular resolution of the radar system or, for that matter, of any wideband communication system.

High-speed Marx Generator

A 200-kV 100-ps Marx generator was built and tested by the author in the past. The generator allowed low-cost wideband testing of electromagnetic protection [5].

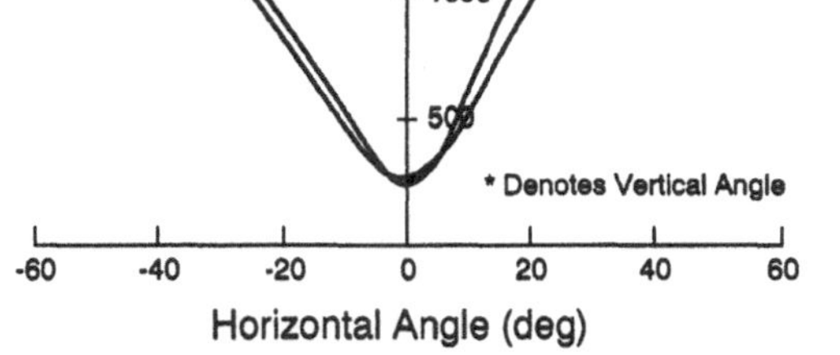

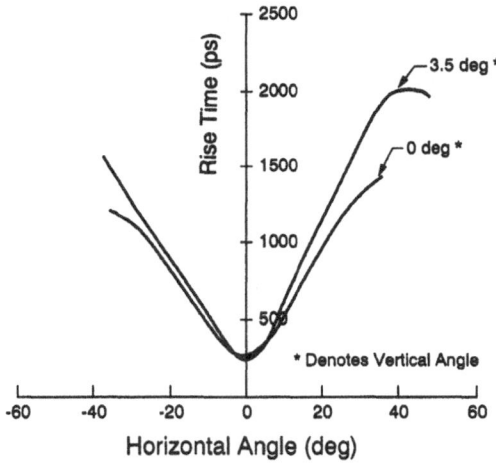

Figure 5 Rise time of the transmitted pulse in function of the horizontal angle of the antenna system.

The same generator was used in this project as a low-cost pulse source for the new radar system. The high-voltage capabilities of the picosecond pulse generators permitted study of mono-pulse radar systems at megawatt power levels.

The high-speed generator was developed as a result of previously undertaken numerical modelling of ultra-fast lightning discharges [6]. The study resulted in a corona model being incorporated into a previously developed 3-D model of lightning and helped in a better understanding of the fast, picosecond discharges.

The design of the new high-speed generator uses a commercial 100-kV 2-ns Marx generator manufactured by Veradyne Corporation. A specially designed fast sphere gap switch and peaking network assures that the system operates with rise times as fast as 100 ps. The specially designed sphere gap switch operates at atmospheric pressure and it does not require the use of special gases normally needed for fast, sub-nanosecond operation.

The development of the new high-speed generator was possible only because of the development of very fast pulse measuring systems [4]. The response of the impulse measuring systems developed by the author has been increased to 40 ps and a new signal processing package permitting reduction of measurement error and expansion of analog bandwidth of the digital measurements has been developed.

RADAR SYSTEM VERIFICATION

The radar system verification was done by measuring reflections from two targets. The first target consisted of a thin metal rectangular plate having dimensions of 1 m × 2 m. The second target was simulated using a 20-cm metallic sphere.

The radar system measurements were performed in a laboratory located in a small metallic building. The use of such a building dictated the implementation of a testing method (see Fig. 5) that involves the use of an electromagnetic absorber needed to prevent blinding of the receiver due to the signal scattered from walls. However, the use of absorbers did not prevent blinding of the receiver by signals arriving directly from the transmitter. The small size of the building prevented the physical extension of the propagation path between the transmitter, target, and the receiver to a distance of 20 m required to prevent blinding. The 7-m distance used during the experiment resulted in only a 20 ns time delay between the fronts of the generated pulse arriving directly at the receiver from the transmitter and the front of the pulse arriving at the receiver from the target. Therefore, the two signals, the scattered signal from the target and the long tail (100 ns) of the pulse arriving directly from the transmitter, had to be measured simultaneously. Because of the high cost involved, no attempt was made to shorten the duration of the 30-ns pulse provided by the generator in order to shorten the tail of the radiated pulse.

The blinding effect was mostly visible during the measurements of the reflection from the smaller target — the 20-cm metallic sphere. The reflection from the large 1 m × 2 m metallic plate was strong enough to be measured on the top of a 1000-V pulse coupling directly from the transmitter. Figure 6 shows a reflection from a metallic plate measured by the radar receiver. The same figure also displays a received signal measured without the target present in front of the radar. The comparison of the two signals shows that the signal from the target, having an amplitude of 200 V, is easily visible on the top of the 1000-V pulse tail arriving directly from the transmitter antenna. A large amplitude shift of the two signals shown in Fig. 6 is related to an instability of the generator voltage. The large amplitude difference of the two signals was easily alleviated by averaging 10 separately measured responses. Figure 7 shows the effects of averaging on the level of amplitude shift between the

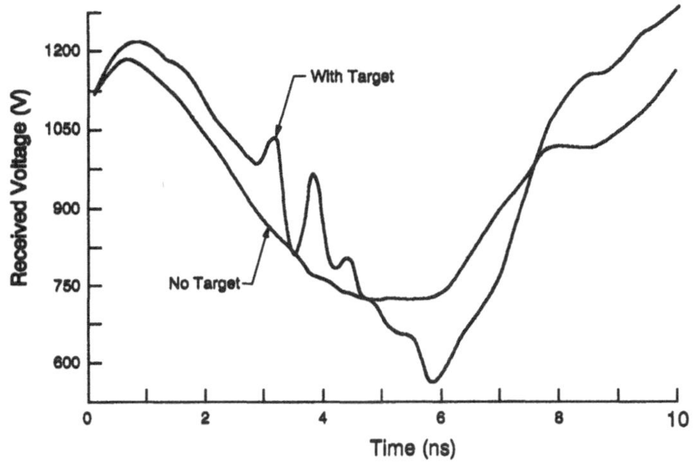

Figure 6 Waveforms of single pulses measured at the radar receiver with a thin metallic target 1 m × 2 m installed.and removed.

two signals of Fig. 6. Figure 7 was created by averaging 10 individually measured signals without the target and an additional 10 individually measured signals with the target present. The averaging of each of the 10 signals produced the response that can be seen in Fig. 7. The averaging of the 10 individually measured signals removed the generator signal level instability and allowed an accurate estimation of the reflection from the target. The blinding effect of the direct coupling of the tail of the long pulse into the receiver was easily removed by subtracting the two signals presented in Fig. 7. The result of this subtraction is presented in Fig. 8. In order to show the accuracy of the measurements, Fig. 8 presents for comparison a second measurement of an averaged signal reflected from the same target. The small difference between the two signals indicates the high measurement accuracy of the radar system.

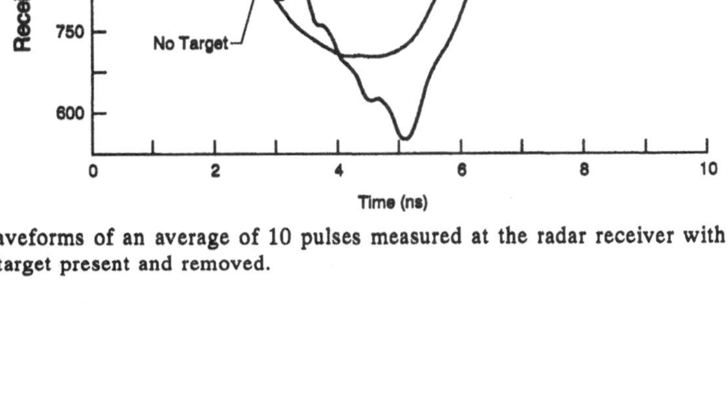

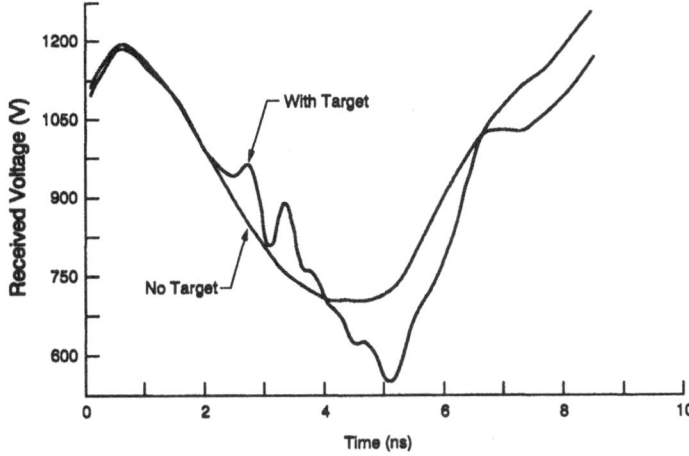

Figure 7 Waveforms of an average of 10 pulses measured at the radar receiver with a 1 m × 2 m thin metallic target present and removed.

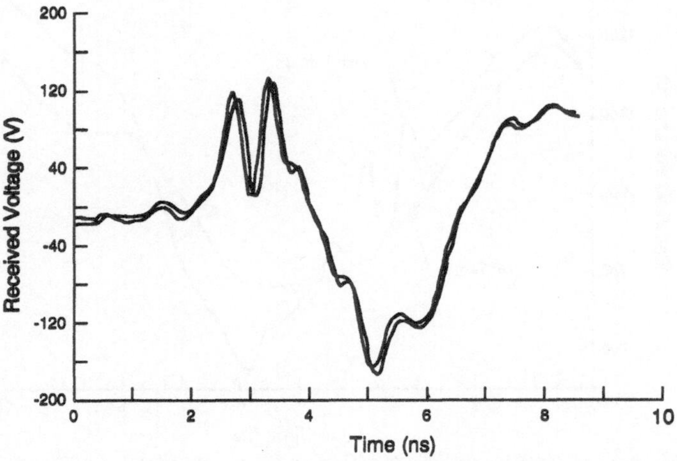

Figure 8 Waveforms of a net reflection from a target. The signal coming directly from the transmitter has been removed by subtracting the waveforms of Fig. 7. For comparison a second waveform identical to the first has been measured and calculated in order to allow the estimation of radar system error.

CONCLUSIONS

By developing new wideband antennas and high-power fast-rise-time generators it was possible to assembly a high-power wideband radar system.

A wideband radar system operating in the frequency range extending over one decade of frequencies (180 MHz to 1.75 GHz) has been build and tested. One of the aims of this project was to assess the reliability of the new radar concept and define the system limits. It has been concluded during the course of this project that it will be possible in the future to build a wideband narrow beam radar system extending over many decades of frequencies and operating at terawatt peak power levels.

REFERENCES

1. Podgorski, A.S. "Composite electromagnetic field pulse threat." *Proceedings of the 1990 IEEE International Symposium on Electromagnetic Compatibility*, Washington, DC. August 21–23, 1990. pp. 224–227. NRC 31665.
2. Podgorski, A.S., and Gibson, G. "Broadband antennas and electromagnetic field simulator." Canadian and International Patent application.
3. Podgorski, A.S., and Gibson, G. "New broadband gigahertz field simulator." *Proceedings of 1992 the IEEE International Symposium on Electromagnetic Compatibility*, Anaheim, CA. August 17–21, 1992. pp. 435–437. NRC 33203.
4. Podgorski, A.S., Dunn, J., and Yeo, R. "Study of picosecond rise time in human- generated ESD." *Proceedings of the 1991 IEEE International Symposium on Electromagnetic Compatibility*, Cherry Hill, NJ. August 12–16, 1991. pp. 263– 264. NRC 31802.
5. Podgorski, A.S., and Campagna, P. "Report on a new, broadband HERO testing method." *Proceedings of the EMC Technology EXPO-92 Symposium, Reston, VA*. May 19–22, 1992. pp. 135–139. NRC 33197.
6. Podgorski, A.S. "Three-dimensional time domain model of lightning including corona effects." *Proceedings of the International Aerospace and Ground Conference on Lightning and Static Electricity*, Cocoa Beach, FL. April 16–17, 1991. pp.75.1–75.7. NRC 31801.

MULTI-GHz BANDPASS, HIGH-REPETITION RATE SINGLE CHANNEL MOBILE DIAGNOSTIC SYSTEM FOR ULTRA-WIDEBAND APPLICATIONS

Lynn M. Miner[1] and Donald E. Voss[2]

[1]Phillips Laboratory
Kirtland AFB, New Mexico 87117

[2]Voss Scientific
Albuquerque, New Mexico 87108

ABSTRACT

Characterizing radiated UWB signals poses unique challenges due to requirements for: (1) multi-GHz bandpass recording of the signal's leading edge; (2) GHz-bandpass recording of long record lengths (10s-100s of ns); and (3) determining shot-to-shot reproducibility at rep-rates exceeding 10 kHz. The System Verification Apparatus (SVA) is a novel diagnostic system which can measure 60-ps rise-time signals on a single-shot basis, while monitoring pulse-to-pulse variation. The fully-integrated SVA includes a broadband sensor, signal and trigger conditioning electronics, multiple parallel digitizers with deep local storage, and automated software for acquiring, archiving, and analyzing waveform data with rapid (secs-minute) turnaround time. The instruments are housed in a portable 100-dB shielded aluminum enclosure. The SVA utilizes a 6-GHz bandpass free-field D-dot sensor to measure the incident electric field. Three separate digitizers together meet the requirements of high bandwidth, long record length, and high repetition rate. A 6-GHz bandpass scan converter digitizer captures the leading edge (few ns) of the radiated signal. 1-GHz and 600 MHz bandwidth solid-state digitizers supporting long record lengths (> 2 μs) record the balance of the signal, which typically contains negligible content above 1 GHz. These solid-state digitizers can store > 900 waveforms locally at rep-rates exceeding 65 Hz and 100 kHz, respectively. Data management and instrument control use an 80486-based PC, operating in a user-friendly Windows environment. All waveform and system configuration data are automatically stored in a built-in database. A fiber-optic link, up to 2 km long, provides electromagnetic isolation of the computer.

INTRODUCTION

A broad range of research devices generate ultra-wideband (UWB) electromagnetic signals, loosely defined here as those with a frequency spectrum extending into the microwave region and spanning 1 octave or more. While these signals can be produced by single shot devices, they also often are generated at repetition rates of a kHz or higher. The high frequency content of UWB signals usually results from an electromagnetic signature with fractional-ns risetime. The broad bandwidth is due partially to the lower frequency features which follow the initial fast-rise portion, and can extend to 10s-100s of nanosecond pulse lengths. Characterizing UWB signals encountered in typical research applications poses new challenges for diagnostic systems due to simultaneous requirements for: (1) multi-

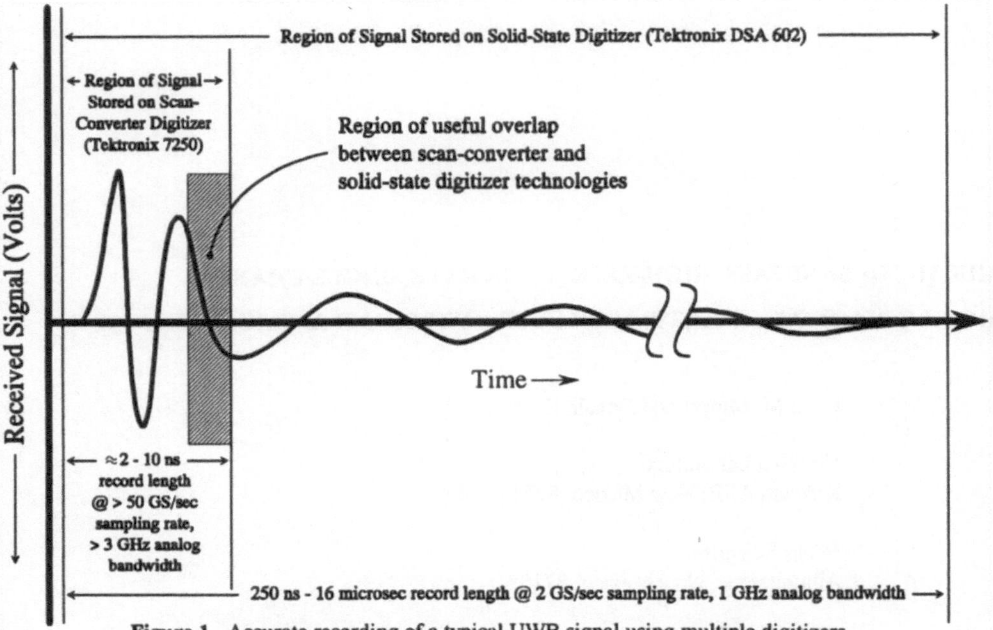

Figure 1. Accurate recording of a typical UWB signal using multiple digitizers.

GHz bandpass recording of the UWB signal's leading edge; (2) high bandwidth, though typically sub-GHz bandwidth, recording of the long record length ($\gtrsim$100 ns) portion of the signal associated with other than the leading edge; and (3) determining shot-to-shot reproducibility for repetitive signals. Note that the ultra-wideband system described here is similar in many ways to one designed for narrowband use (Voss et al., 1992).

INSTRUMENTATION AND HARDWARE

At this time, no single instrument can simultaneously meet the requirements for tens to hundreds of nanosecond record length and multi-GHz bandwidth. However, a combination of commercial digitizer technologies can. An example of how the technique can be applied to a typical UWB waveform is shown schematically in Figure 1. Because the typical UWB waveform has a fast leading edge, often with sub-nanosecond risetime, a multi-GHz analog bandwidth, single-shot capable digitizer is necessary to record it. The only commercially available digitizers with bandwidths exceeding 2 GHz use scan-converter technology combined with a traveling-wave e-beam deflection system[1]. In the SVA apparatus, we utilize the 6-GHz bandpass Tektronix 7250 for recording the leading edge of the waveform. This scan converter digitizer has an intrinsic risetime for a step function source of 60 psec, and can store 30 waveforms locally at multi-Hz repetition rates. Its main limitation, like all electron beam based digitizers, is a very short record length. For example, to capture a signal with a 200 psec risetime, the maximum record length can be no longer than 10 nsec, possibly only 5 nsec. Fortunately, for the vast majority of UWB-type electrical signals, essentially all of the frequency content above 1 GHz is due to waveform variation in the first few nsec of the pulse. Thus, a Tek 7250 can be used to effectively capture that portion of the spectrum above 1 GHz.

As shown in Figure 1, the 7250 captures only a small portion of the overall waveform. However, that portion which it does capture is recorded with good fidelity due to its intrinsic 6-GHz single-shot bandwidth. Operating in parallel with the Tek 7250, the entire waveform can also be recorded by a 1-GHz analog bandwidth solid-state digitizer, the Tektronix DSA602. All but the first

[1]Model 7250 (6-GHz bandwidth) or Model SCD5000 (4.5-GHz bandwidth) from Tektronix Inc., Beaverton, Oregon 97077

few ns of the signal will be recorded with excellent fidelity in this way, because the slow time variations of the latter portions of the signal yield sub-GHz (i.e. in-band) frequency content. Thus, by splitting the signal to be diagnosed prior to the 7250, and routing it to a separate solid-state digitizer with an extremely long record length (i.e., microseconds for the DSA602) and partially overlapping time domains (note the cross-hatched region of Fig. 1 which is recorded by both digitizers), it is straightforward to time-tie the resulting stored waveforms after the fact with automated software.

Because advances in solid-state digitizers allow hundreds of shots to be temporarily stored locally at the digitizer, the SVA can preserve its multi-GHz single-shot bandpass even for bursts of many separate shots. The Tek DSA602 can rearm at rates exceeding 50 Hz, and the Tektronix RTD720 digitizer, which has a somewhat lower analog bandwidth of 500 MHz, can rearm at rates exceeding 100 kHz. Similarly, the 7250 scan-converter digitizer can store 30 shots locally, albeit at a slower, 10 Hz, repetition rate. These 3 instruments, all operated in parallel on the same signal, are used in the SVA to provide maximum possible bandwidth under a wide variety of rep-rate conditions, within the limitations imposed by the hardware's local storage and rep-rate capabilities.

The SVA is a fully integrated system for measuring radiated UWB RF signals. A 6-GHz bandpass free-field ACD D-dot sensor[2] with an associated 10-GHz bandwidth balun[3] provide a time-differentiated replica of the incident electric field. The sensor output drives a low loss cable routed to the recording instruments, which are housed in a 100-dB RF-shielded designed for field use. The 3 digitizers, trigger and signal conditioning electronics, and a fiber-optic transceiver unit are all mounted in the shock-isolated enclosure shown schematically in Figure 2.

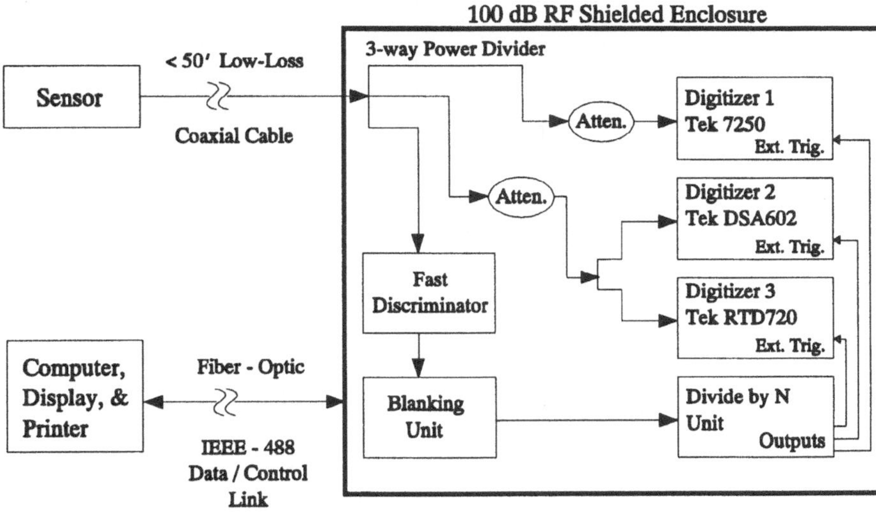

Figure 2. Schematic of SVA key hardware units.

DATA ACQUISITION, ARCHIVAL, ANALYSIS, AND CONTROL (DA^{3}C) SYSTEM

The SVA software was custom designed to meet the needs of managing UWB signals while operating in a field-based environment. The DA^{3}C system used is composed of three primary modules (acquisition/control, archival, and analysis), an automatic component-compensation module, and an on-line, hypertext-based, context-sensitive, help system. Due to how the software has been implemented in the multitasking environment, these modules can all be executing at (effectively) the same time, and relevant information is transferred between them automatically. The consistent user-interface,

[2]Model ACD-11A(R), EG&G Washington Analytical, or equivalent from Prodyn Inc., Albuquerque, NM 87107

[3]Model BIB-100G, Prodyn Inc., Albuquerque, New Mexico 87107

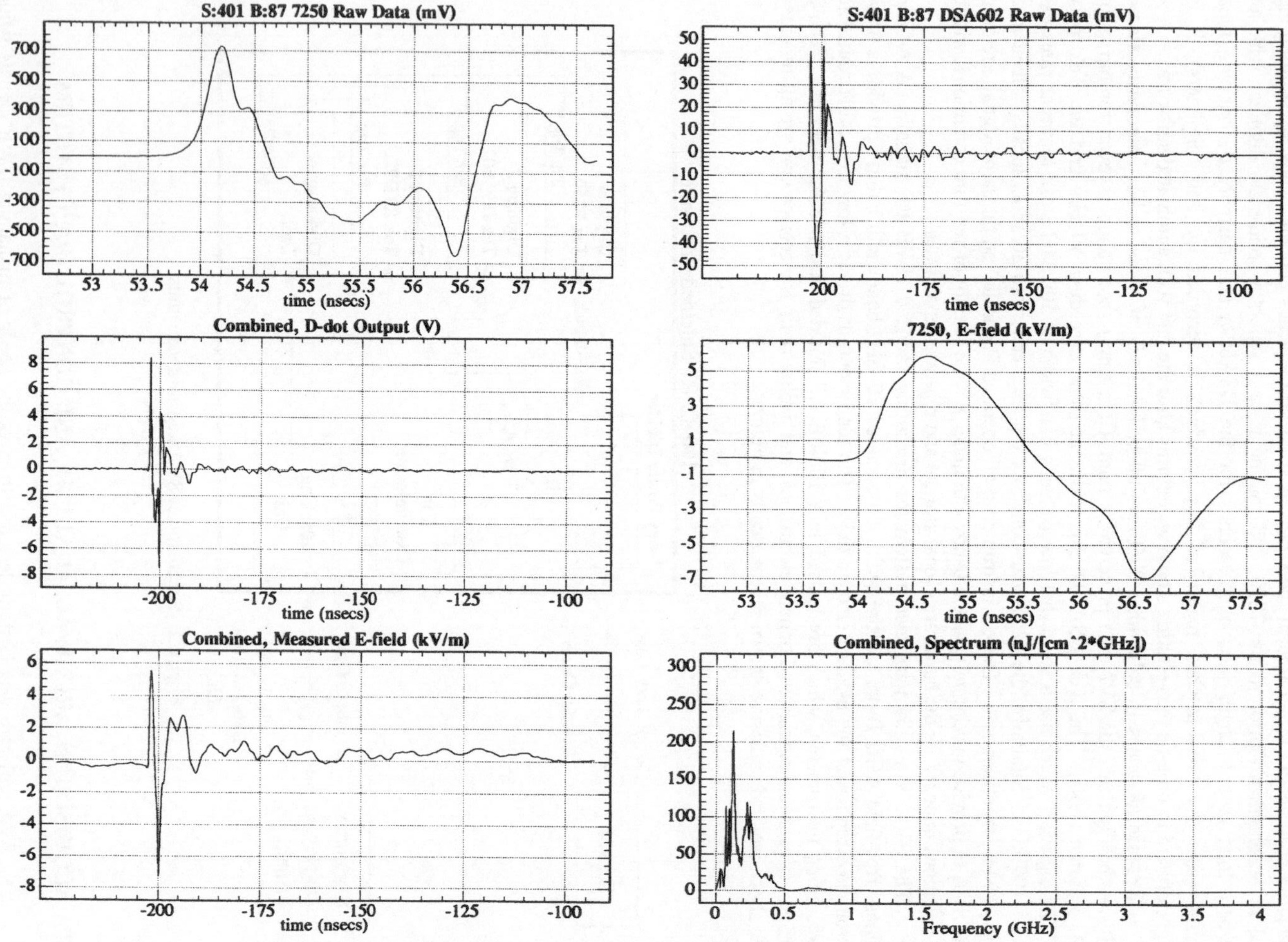

Figure 3. Typical waveforms generated in the automated data reduction process.

elimination of software incompatibilities, and greater ease of code maintainability are other advantages. A more detailed description of the DA^3C concept is available elsewhere (Voss Scientific / Tektronix, 1991). The built-in database system provides tools for rapidly selecting and retrieving specific information by specifying a variety of Boolean-combinable search criteria. Data are archived and displayed on an 80486-based PC which can be located up to 2 km from the shielded diagnostic enclosure. A fiber-optic-based data link provides electromagnetic isolation of the data recording computer from any spurious currents that would otherwise be generated on conducting lines. System turnaround is very rapid. For example, all data channels from a 10 shot burst can be automatically retrieved, reduced to meaningful electric field vs. time data, archived on hard disk, displayed to a super VGA monitor, and output from a laser printer in under 2 minutes. The rapid turnaround is particularly helpful in field tests where workers are required to quantitatively monitor electromagnetic field levels on a near-real-time basis. Typical data are shown in Figure 3.

REFERENCES

1. Voss, D.E, Koslover, R.A, Hoy, J.M., Jerome, C.B, Lovesee, A.L., Preston, T.J., and Slemp, D.A. , "Phillips Laboratory Narrowband, High Rep-Rate Mobile HPM Diagnostic System," these proceedings.

2. Fully-Integrated Multichannel Systems for Single-Shot and Repetitively Pulsed Waveform Applications, Voss Scientific / Tektronix Application Note 34W-8016, Tektronix Inc., Beaverton Oregon, May, 1991.

AN IMPULSE RADIO COMMUNICATIONS SYSTEM

Paul Withington, II

Pulson Communications
Suite 500, 8280 Greensboro Dr.
McLean, Virginia 22102

Larry W. Fullerton

Time Domain Systems, Inc.
Suite 3, 4825 University Square
Huntsville, Alabama 35816

OVERVIEW

Pulson Communications has built a prototype impulse radio. The transmitter has an average power of 450 microwatts as measured out of the antenna. This prototype, transmitting high qualiity audio, was tested to ranges in excess of 7 kilometers. The potential range of this type of system at this power level may exceed 20 kilometers. The prototype had a center frequency of 675 MHz and an approximately equal bandwidth. We used a small wideband omni-directional antenna. This is significant because impulse radios:

- Can share spectrum without affecting conventionial radio transmissions and allow thousands of impulse transmitters to operate simultaneously in a very confined area without disrupting other radio systems.
- Would appear to be adaptable for type acceptance under the rules of Part 15.
- Consume substantially less power than existing conventional radios.
- Cost less than other sophisticated radio designs, especially spread spectrum systems.

While Pulson's laboratory hardware has been optimized for short range, there is no theoretical reason it could not be used for longer range communications, e.g., satellite-to-ground communications or interplanetary probes.

IMPULSE RADIO TECHNOLOGY BASICS

Pulson's impulse transmitters emit ultra-short "Gaussian" monocycle pulses with a tightly controlled average pulse-to-pulse interval. Pulson has been working with pulse widths of between 0.5 and 1.5 nanoseconds and pulse-to-pulse intervals of between one hundred nanoseconds and one microsecond.

The system uses pulse position modulation, with the actual pulse-to-pulse intervals being varied on a pulse-by-pulse basis by two components: an information component and a pseudo-random code component. Unlike spread spectrum systems, the pseudo-random code is not necessary for energy spreading (because the impulses themselves are inherently wideband), but rather for channelization, energy smoothing in the frequency domain, and jamming resistance.

A Pulse

The most basic element of Pulson's impulse radio technology is the practical implementation of a Gaussian monocycle. A Gaussain monocycle is the first derivative of the Gaussian function. Figure 1 shows the monocycle in both the time and frequency domains. (Pulson's impulses are not gated sine waves, which have significantly different and much less desirable characteristics.)

Ultra-Wideband, Short-Pulse Electromagnetics
Edited by H. Bertoni *et al.*, Plenum Press, 1993

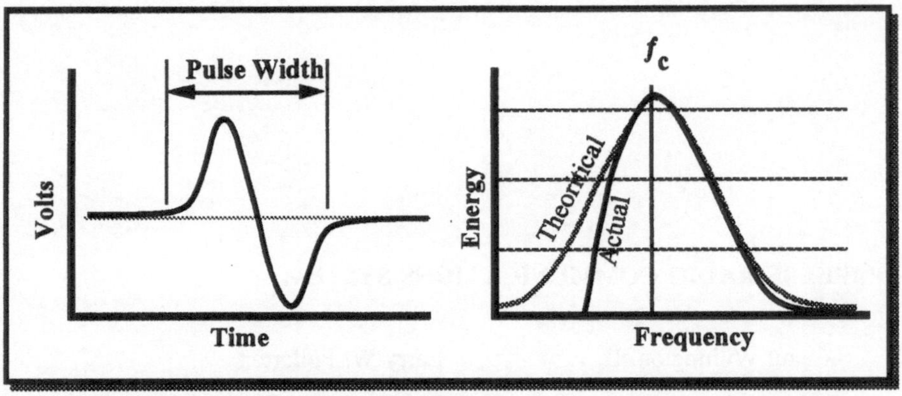

Figure 1. Gaussian monocycle in both time and frequency domains

The Gaussian monocycle waveform is naturally a wide bandwidth signal, with the center frequency and the bandwidth completely dependent upon the pulse's width. In the time domain, the Gaussian monocycle is described mathematically by:

$$V(t) = Ate^{-\left(\frac{t}{\tau}\right)^2} \tag{1}$$

Where:

A is the peak amplitude of the pulse.
t is time.
τ (tau) is a decay constant.

The carrier frequency is then:

$$f_c = \frac{1}{\sqrt{2}\pi\tau} \tag{2}$$

And the half-power bandwidth is:

$$\Delta f = \frac{0.2714}{\tau} \tag{3}$$

From equations 2 and 3, one can see that both the carrier frequency and bandwidth are specified in terms of the decay constant tau. Since tau also defines the pulse width, then the pulse width specifies both the center frequency and bandwidth.

In practice, Pulson has determined that the center frequency of a pulse is approximately the reciprocal of the pulse's length. Thus, a "1 ns" pulse has a center frequency of approximately 1 gigahertz.

A Pulse Train

Impulse systems use pulse trains, not single pulses, for communications. Pulson is conducting laboratory tests with pulse repetition frequencies of between 1 and 10 megapulses per second (mpps).

Figure 2 contains an illustration of a 1 mpps system with 1 ns pulses. In the frequency domain, this highly regular pulse train produces energy spikes (comb lines) at one megahertz intervals; thus, the already low peak power is spread among over the comb-lines. This pulse train carries no information and, because of the regularity of the energy spikes, might interfere with conventional radio systems.

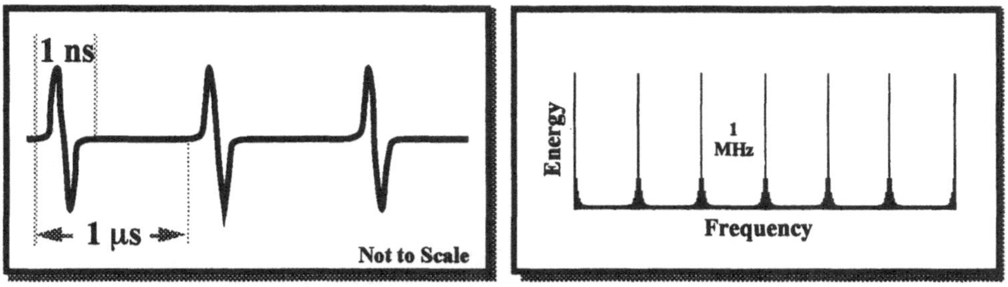

Figure 2. An Gaussian monocycle pulse train in both time and frequency domains.

Impulse systems have very low duty cycles and an average power significantly lower than its peak power. In the example in Figure 2, the impulse transmitter operates 0.1% of the time, i.e., 1 ns per microsecond (μs).

Modulation

Additional processing is needed to modulate the pulse train and to smooth the energy distribution in the frequency domain so that it does not interfere with conventional radio systems.

Amplitude and frequency/phase modulation are unsuitable for this particular form of impulse communications; the only suitable choice is pulse position modulation. As illustrated in Figure 4, a modulating signal changes the pulse repetition interval (PRI) in proportion to the modulation.

If the modulating signal were to have three levels, the first level might shift the generation of the pulse forward in time from the nominal by ∂ picoseconds; the second level might not shift the pulse position in time from the nominal at all; and the third level might delay the pulse by ∂ picoseconds. This would be a digital modulation scheme. Analog modulation would allow deviations between PRI-∂ and PRI+∂.

In the frequency domain, pulse position modulation distributes the energy over more frequencies. Thus in Figure 3, for an individual pulse in the sequence, there are three possible positions (+∂, 0, -∂). There are, then, five possible intervals between two pulses (Table 1).

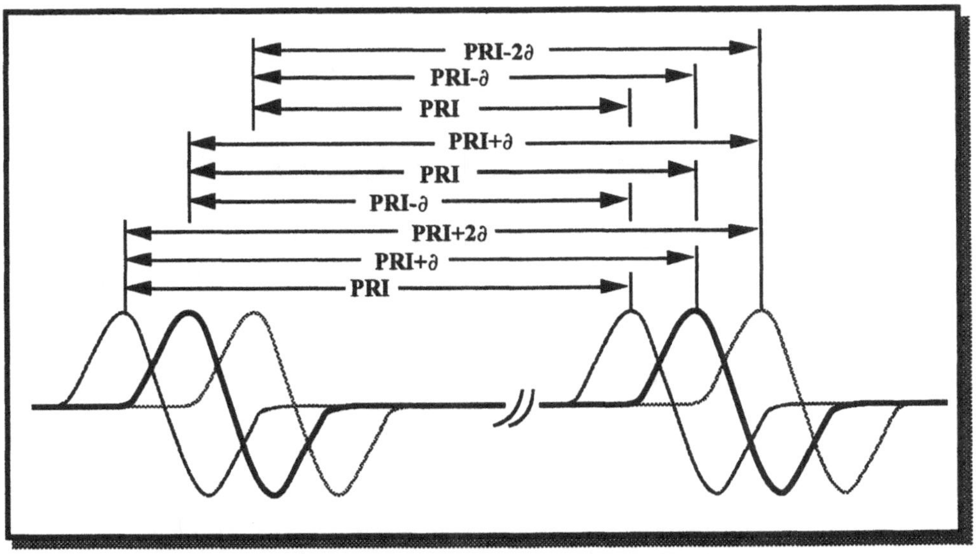

Figure 3. Pulse position modulation.

Table 1. Pulse repetition intervals in a three position modulation scheme

Preceding Pulse's Position	Pulse's Position		
	-∂	0	+∂
-∂	PRI	PRI-∂	PRI+2∂
0	PRI-∂	PRI	PRI+∂
+∂	PRI-2∂	PRI+∂	PRI

Pulse position modulation tends to smooth the energy distribution in the frequency domain For example, in the case of a 1 mpps system if the modulation dither (∂) were ±100 picoseconds, the PRI is 1,000,000 Hertz and the additional frequency components are: 999,800.04 Hertz, 999,900.01 Hertz, 1,000,100.01 Hertz, and 1,000,200.04 Hertz. Transmitted energy is now distributed among more spikes (comb lines) in the frequency domain. If the total transmitted energy remains constant, the energy in each frequency spike decreases as the number of possible pulse positions increases.

Coding for energy smoothing and channelization

With a maximum modulation dither of ±25% of the pulse width the smoothing is negligible. If, however, the dither were much greater the smoothing effect would be greater. In a system with a 1 ns pulse width, the optimal smoothing dither is 10 nanoseconds, because the lowest significant frequency component of a 1 ns Gaussian monocycle is 100 MHz; 10 nanoseconds = $^1/_{(100\ \mathrm{MHz})}$.

Pulson achieves optimal smoothing by applying to each pulse a pseudo-random noise (PN) dither with a much larger magnitude than the modulation dither. Figure 4 when compared to Figure 2 shows the impact of using a 256 position code relative to an uncoded signal.

PN dithering also creates a method for specifying channels. In an uncoded system, differentiating between separate transmitters would be very hard. PN codes create communications channels, if they are relatively orthogonal, i.e., there is low correlation between the different codes being used.

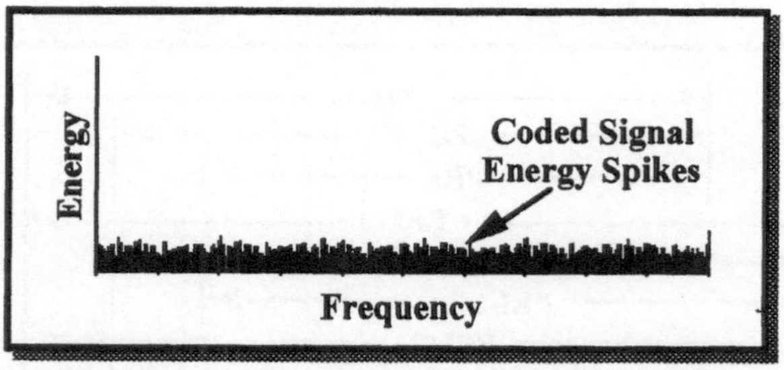

Figure 4. The impact of pseudo-random dither on energy distribution in the frequency domain

Reception and demodulation

Clearly, if there were a large number of impulse radio users within a confined area, there might be mutual interference. And while the PN coding minimizes that interference, as the number of users rises, the probability of an individual pulse from one user's sequence being received simultaneously with a pulse from another user's sequence increases. Fortunately, Pulson's implementation of an impulse radio does not depend on receiving every pulse; it is a correlating, synchronous receiving process that uses a statistical sampling of many pulses to recover the transmitted information.

Pulson receivers typically integrate over 1000 or more pulses to yield the demodulated output. The optimal number of pulses over which the receiver integrates is dependent on a number of variables. This process reduces noise induced uncertainties about the exact arrival time of individual pulses.

Jam Resistance

Besides channelization and energy smoothing, the pseudo-random coding also makes impulse radio highly resistant to jamming from all radio communications systems, including other impulse radio transmitters. This is critical as any other signals within the band occupied by an impulse signal act as jammers to the impulse radio. Since there are no unallocated 1+ gigahertz bands available for impulse systems, they must share spectrum with other conventional and impulse radios without being adversely affected. The PN code helps impulse systems discriminate between the intended impulse transmission and transmissions from others (Figure 5).

In Figure 5, a narrowband sinusoidal signal overlays the impulse signal. The input to the cross correlator would include that narrowband signal as well as the received ultrawide band impulse signal. Without pseudo-random coding, the cross correlator would sample the jamming signal with such regularity that the sinusoidal signals could cause significant interference to the impulse radio receiver.

Figure 6 illustrates when the transmitted impulse signal is encoded with the pseudo-random dither (and the receiver synchronized with that identical pseudo-random dither) it samples the jamming signal randomly. Integrating over many pulses would then negate the impact of jamming.

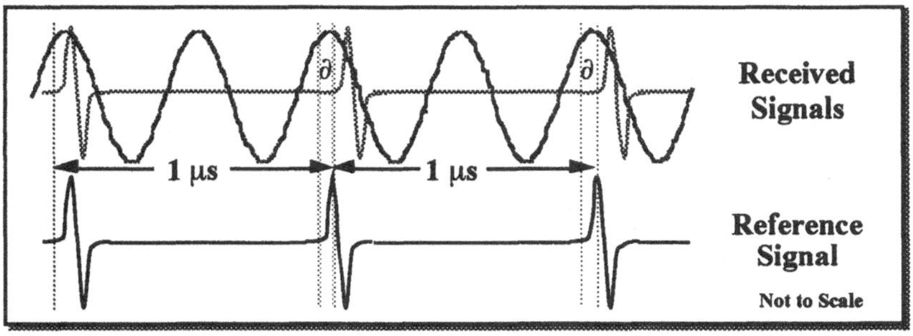

Figure 5. Cross correlation in the presence of jamming and no pseudo-random coding.

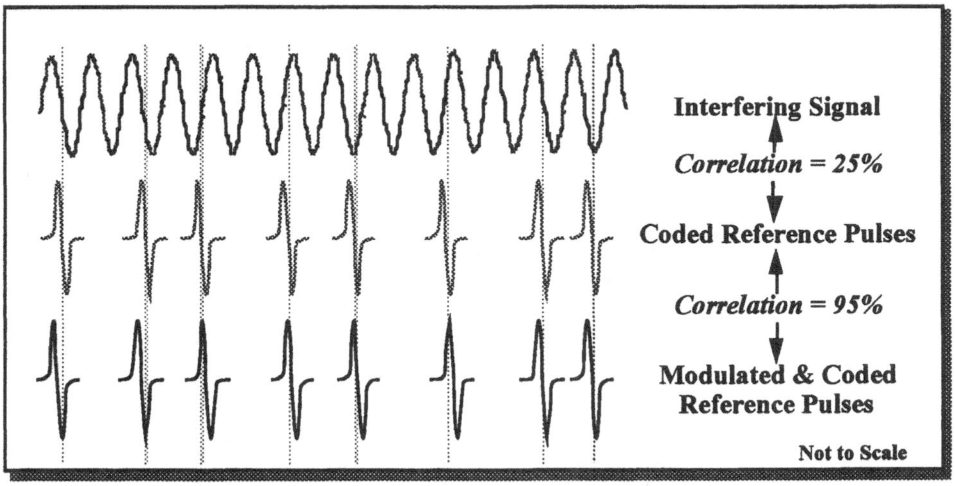

Figure 6. Cross correlation in the presence of jamming with PN coding.

In statistical terms, the pseudo-randomization in time of the receive process creates a stream of randomly distributed values with a mean of zero (for jamming signals). All that is necessary to eliminate the impact of jammers is to sample over enough pulses, i.e., integrate over a sufficiently large number of pulses, to drive the impact of the jamming signals to zero.

Processing Gain

Impulse radio is jam resistant because of its huge processing gain. A definition of processing gain, which quantifies the decrease in channel interference when wideband communications is used, is the ratio of the bandwidth of the signal to the bandwidth of the information signal. For example, a direct sequence spread spectrum system with a 10 kHz information bandwidth and a 16 MHz channel bandwidth yields a processing gain of 1600 or 32 dB. However, far greater processing gains are achieved with impulse systems where for the same 10 kHz information bandwidth and a 2 GHz channel bandwidth the processing gain is 200,000 or 53 dB.

Receiver Selectivity

The receiver used for impulse communications is designed according to a radically different paradigm than that used for conventional communications. Whereas conventional communications employ frequency conversion by use of a local oscillator (heterodyne receiving), impulse communications employs front-end correlation techniques without frequency down-conversion (homodyne receiving). Due to correlation conversion, the receiver achieves high signal-to-noise levels over the length of a pulse sequence or train. This increase in signal-to-noise by correlation conversion is known as "receiver gain".

PERFORMANCE IN THE REAL WORLD

Pulson's Test System

In an advanced impulse radio personal communications prototype transmitter built by Pulson the average transmitted power is 450 microwatts. The center frequency is 1.9 GHz and smoothed by a pseudo-random code with at least 256 positions.

Range

Table 2 compares the power transmitted by a cellular portable, a cordless telephone, and an impulse transmitter. Given that cellular's average power is 942 times the impulse transmitter's and that the cordless telephone's is 158 times greater, one might expect the impulse transmitter to have very limited range. In fact, even with numerous powerful jammers, the impulse signal can be received at far greater distances than a cordless phone and may have a range comparable to a portable cellular telephone.

In simulations using attenuators a 450 microwatt prototype has achieved a range in excess of 9 kilometers.

Table 2. Comparison of transmitted powers.

Radio Type	Average Power
Cellular Portable	424 milliwatts
Cordless Telephone	71 milliwatts
Impulseradio	450 μwatts

Multipath and Propagation

Multipath fading, the bane of sinusoidal systems, will be less of a problem for impulse systems than for conventional radio systems. In an impulse system in order for there to be multipath effects, very particular and rare conditions must persist. Either:

- The path length traveled by the multipath pulse must be less than the pulse width times the speed of light. (For a one nanosecond pulse, that equals 0.3 meters or about 1 foot, i.e., [1 ns] x [300,000,000 meters/second].) (See Figure 7, in the case where the pulse traveling "Path 1" arrives one half a pulse width after the direct path pulse.); or
- The multipath pulse travels a distance that equals the interval of time between pulses times the speed of light might interfere times an integral number with the next pulse. (For a 1 megapulse per second system that would be equal to traveling an extra 300, 600, 900, etc. meters.) However, because each individual pulse is subject to the pseudo-random dither, these pulses are decorrelated.

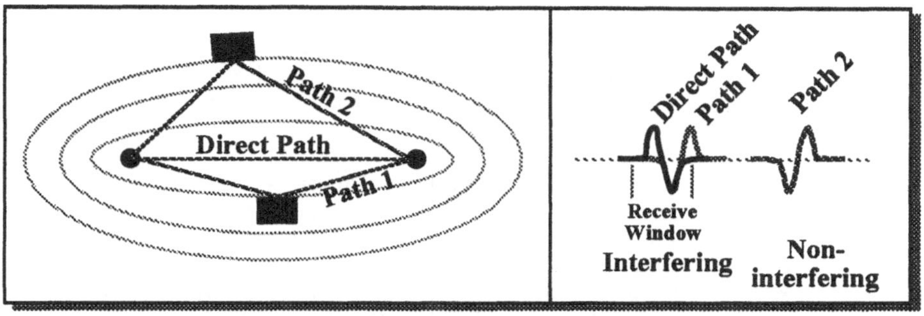

Figure 7. Multipath in an impulse system

Pulses traveling between these intervals do not cause self-interference (in Figure 7, this is illustrated by the pulse traveling Path 2).

Pulses traveling grazing paths, as illustrated in Figure 7 by the narrowest ellipsoid, are the ones that create multipath effects. As illustrated in the Figure 8, if the multipath pulse travels one half width of a pulse width further it increases the power level of the received signal (the phase of the multipath pulse will be inverted by the reflecting surface). If the pulse travels less than one half a pulse width further it will create destructive interference. For a 1 ns pulse, destructive interference will occur if the multipath pulse travels between 0 and 15 cm.

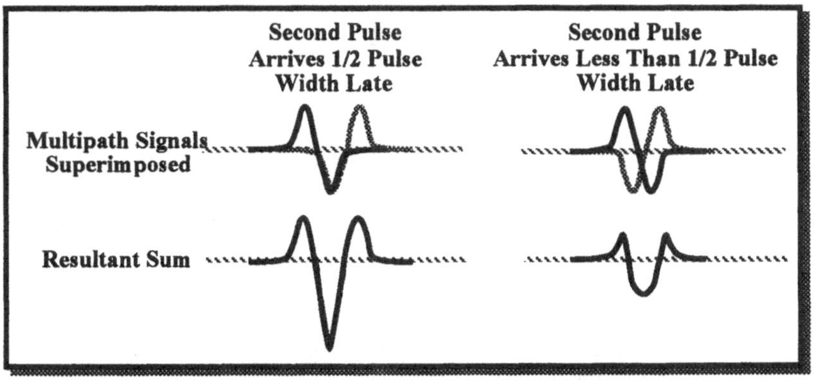

Figure 8. Interaction of a direct path pulse and a grazing path reflection in an impulse system.

Rayleigh fading, so noticeable in cellular communications, is a continuous wave phenomenon, not an impulse communications phenomenon.

Range tests of impulse systems (including impulse radar tests for the military) suggest that multipath will not present any major problems in actual operation. Additionally, Pulson plans to move to shorter pulse widths, which will further reduce the probability of destructive interference (because the reflected path length required for destructive interference will be shortened).

CONSTRUCTING AN IMPULSE RADIO SYSTEM

Pulson Communications has expended significant resources to develop its impulse communications systems and determined that while all components must be very carefully designed, certain components determine the ultimate feasibility of producing impulse communication systems.

Time base — Because this system uses picosecond deviations of the pulse position, an accurate clock is an absolutely necessity. Just a few years ago sufficiently accurate clocks were too expensive for consumer products; this is no longer true.

Antenna — Most antennae are designed to operate over very narrow bandwidths. Pulson Communications has a small, effective broad band antenna. The dimensions for an 1.9 GHz center frequency Pulson system would be about 5 cm x 7 cm x 0.7 cm.

PN codes — Codes must be orthogonal with respect to other codes as well as to themselves, i.e., there must be neither cross- nor auto- correlation. Additionally, the codes must smooth the energy distribution effectively and allow fast signal locking. There has been extensive research in the area of coding, which shows that there are huge numbers of codes which should have the proper qualities required by Pulson's impulse system.

Pulson's impulse radios are much simpler to build than equivalently sophisticated conventional radios. The receiver is more complicated than the transmitter, since it includes a control loop to attain and maintain synchronization with the transmitter. But a Pulson receiver is a homodyne receiver, i.e., it converts directly from radio transmissions to baseband. Thus, the only wideband analog components of the circuitry are the antenna, pre-amplifiers, cross-correlator, and the reference pulse generator.

Patents have been issued or applied for worldwide on its impulse systems and technology.

RADAR APPLICATIONS

Time Domain Systems has tested several prototype radar systems. Developments include:

- Optimum correlation receivers.
- Time delay multi-element antennae.
- Interferometric beam forming.

System tests have included real time detection and tracking objects.

ACKNOWLEDGMENTS

Pulson Communications wishes note that Larry W. Fullerton, Time Domain Systems, Inc., invented the impulse radio system described in this paper. Pulson has been funding these efforts. Additionally, Terrence W. Barrett, Ph.D., Barrett Science and Engineering, and Jeremy K. Raines, Ph.D., P.E., Raines Engineering, have each contributed significantly to the development of this document.

ULTRAWIDEBAND CLUSTERED-CAVITY™ KLYSTRON

John G. Siambis and Robert S. Symons

Litton Electron Devices
San Carlos, CA 94070

I. INTRODUCTION

The generation of ultrawideband electromagnetic signals at high powers has been at the frontiers of research and development for a long time in conventional microwave tube technology. More recently, a lot of publicity and attention has been focused on non-conventional approaches to generate ultrawideband, short pulse signals via high speed photoconductive bulk avalanche semiconductor switching and spark gap switching. The emerging capabilities in conventional and non-conventional RF sources are anticipated to lead to novel applications in radar, communications and material interactions.

Ultrawideband is a relatively new term in the microwave tube radar/communications community. The conventional way to describe bandwidth δf is as a percentage of center frequency, f_o. This leads to the following three bandwidth definitions:

Narrowband: $\delta f / f_o < 1\%$

Broadband: $1\% < \delta f / f_o < 25\%$

Ultrawideband: $\delta f / f_o > 25\%$

The non-conventional, novel sources of ultrawideband signals require significant engineering development for integration and deployment into proposed novel commercial and military radar and communications systems. In contrast, conventional microwave tubes have a very well established technology base and require modest technology upgrades to perform in the ultrawideband, high power regime.

In this paper we shall discuss the ultrawideband clustered-cavity™ klystron, which was proposed in the early 1980s by R. S. Symons[1]. At that time, an experimental program was initiated at Litton Electron Devices to apply this principle in order to upgrade the bandwidth performance of existing klystron tubes in radar installations by doubling their bandwidth[2]. Successful completion of this program demonstrated doubling of band-

Ultra-Wideband, Short-Pulse Electromagnetics
Edited by H. Bertoni *et al.*, Plenum Press, 1993

width from 6.5% to 12.8% in an interaction circuit length of only 70 cm, for the L-5792 klystron, maintaining the original tube's outside envelope[3]. Current studies are focusing on designs for high power klystrons with bandwidths in the range of 30% to 40% by replacing individual intermediate cavities with triplets, in longer klystrons[4].

In Section II, the basic principles of operation of the one- intermediate cavity klystron, Fig. 1, the staggered cavity klystron, Fig. 2, and the clustered cavity klystron, Fig. 3, will be presented, discussed and compared. Particular emphasis will be placed on the elements of the respective klystron system that impact bandwidth performance. In Section III, the clustered-cavity klystron™ will be discussed in detail. In Section IV conclusions will be given.

II. KLYSTRON BANDWIDTH PERFORMANCE

The key circuit element responsible for bandwidth performance for the klystron systems of Figs 1-3 is the coaxial or reentrant cavity surrounding the beam in the drift tube, in the following ordered sequence of importance and criticality: Intermediate or Gain cavities, Output cavity, Input cavity. Each cavity is described by the quality factor Q, defined by

$$\frac{1}{Q} = \frac{Power\ Loss}{2\pi f_o\ x\ (Energy\ Stored)} = \frac{P_L}{2\pi f_o E_o} = \frac{\delta f}{f_o} \tag{1}$$

where f_o is the center frequency and δf is the half-power bandwidth for the subject cavity. The several component parts for the Q of the cavities in Figs 1-3 are given by

$$Q^{-1} = Q_u^{-1} + Q_r^{-1} + Q_e^{-1} + Q_b^{-1} \tag{2}$$

where Q_u is the unloaded, cold, copper walled cavity Q, typically $Q \simeq 1{,}000$. The Q_r stands for the power loss inside the cavity to resistive films deposited on the cavity walls or to resistors and resistive posts placed inside the cavity in order to reduce the Q in a controlled way, thereby increasing the bandwidth δf. Resistive loading is usually applied to the intermediate or gain cavities and reduces the Q of the cavity to $Q_r \simeq 100$, or less. The Q_e stands for power flows to or from external, to the cavity, sources or loads. Specifically, in the RF input and output cavities, respectively, Q_e is given by

$$Q_e^{-1} = Q_{in}^{-1} = P_{in} / (2\pi f_o E_o)\ ;\ Q_e^{-1} = Q_{out}^{-1} = P_{out} / (2\pi f_o E_o) \tag{3}$$

where P_{in} is the external input power and P_{out} is the power output to the load. Typically, $Q_{in} \simeq 100$ and $Q_{out} \simeq 20$. The Q_b stands for the loading of the cavity by the modulated beam interacting with the cavity at the cavity gap, given by[5,6]

$$Q_b^{-1} = G_b(R / Q)\ ;\ R / Q = \sqrt{L / C} \tag{4}$$

$$G_b = (G_o /2)\ M\ [M - \cos(\theta/2)]\ ;\ G_o = I_o / V_o \tag{5}$$

$$B_b = G_b \cot(\theta/2)\ ;\ M = \sin(\theta/2) / (\theta/2) \tag{6}$$

$$\theta = \omega l_g / u_o \tag{7}$$

where G_b is the beam loading RF susceptance, G_o is the dc beam conductance given by the ratio of dc beam current I_o over the dc beam voltage V_o. The beam-cavity coupling

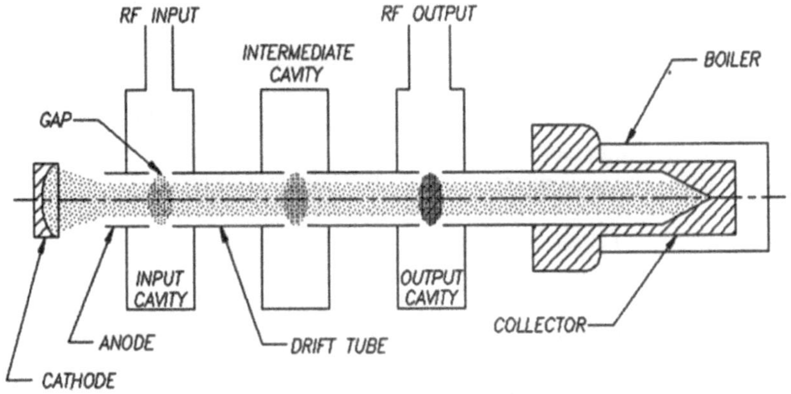

Figure 1. Basic elements of the klystron

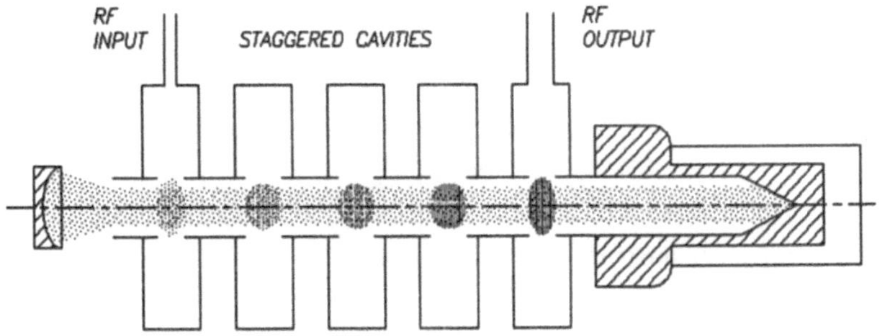

Figure 2. Basic elements of the stagger tuned klystron

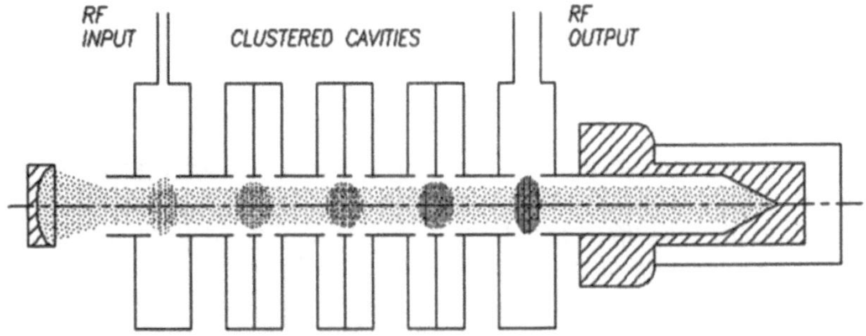

Figure 3. Basic elements of clustered cavity klystron

coefficient M, given in Eq. (6), as a function of the electron's dc transit angle θ, in turn defined by Eq. (7) as a function of the angular frequency ω, the gap length l_g and the dc beam velocity u_o. The ratio of cavity shunt resistance R to cavity Q in Eq. (4) is a purely geometric factor for the unloaded cavity as shown in the second Eq. (4). Finally, the several types of cavity quality factors identified in Eq. 2 can be represented by their equivalent circuit, as shown in Fig. 4. The total response of each cavity will be the sum of the specific elements that comprise it. Figure 4 illustrates that the total impedance of the cavity is represented by a real equivalent shunt resistance only at the center frequency, f_o. For other frequencies the magnitude of the Q follows the Lorentz line shape and the equivalent cavity impedance has real and imaginary parts.

The operation of the basic klystron of Fig. 1 is as follows. A thermionic cathode generates an electron beam of current I_o and voltage V_o that drifts to the gap of the input cavity. The RF signal to be amplified by the klystron amplifier is fed into the input cavity and sets up a voltage across the input cavity gap that modulates the energy and velocity of the beam electrons as they drift under the gap. The beam electrons are confined in axial motion by an externally applied strong, uniform axial magnetic field. The velocity modulated beam exits the input cavity gap and as it proceeds to the intermediate cavity gap, it converts the velocity modulation of the beam into density modulation or electron bunching on the beam. If the intermediate or gain cavity gap is placed on the first maximum current bunching location, l_o, given by

$$\omega_{pr} l_o \,/\, u_o = k_p \, l_o = \pi/2 \tag{8}$$

where ω_{pr} is the beam relativistic plasma frequency, then the gain cavity is strongly excited by the density bunch current modulation on the beam. The resulting gain in kinetic beam voltage across the gap in given to lower order by the gain G,

$$G = M^2 R \,/\, Z_o \tag{9}$$

where R is the equivalent shunt resistance of the cavity and Z_o is the characteristic impedance of the beam modulation in the drift tube, given by[7]

$$Z_o = (\omega_{pr} / \omega) \, [(u_o^2 M_o \, \gamma^3) \,/\, I_o \, q] \tag{10}$$

where M_o, q and γ are the rest mass, charge and relativistic mass factor respectively for the electrons of the beam. When the density bunch modulated beam reaches the output gap, satisfying the condition of Eq. (8), it feeds energy to the output cavity, which in turn drives the RF output to the load. The frequency bandwidth of the simple klystron if Fig. 1 is determined by the superposition, on a frequency scale, of the gain/bandwidth performance of the three cavities. In general, such klystrons have bandwidth in the narrowband or lower broadband range.

Systematic attempts to increase the bandwidth of the basic klystron led to the concept of the stagger-tuned klystron, shown in Fig. 2. A sequence of intermediate cavities is introduced, each tuned at a different center frequency, thereby broadening the spectrum. Resistive loading of the intermediate cavities also broadens the bandwidth of the so loaded cavity. The concept of stagger-tuned cavities has lead to bandwidth improvements of the order of 10-15%. Unfortunately, as more intermediate cavities are added the possibilities for phase cancellations between the various space-charge wave excited on the beam increase.

Further, significant success in obtaining broadband and ultrawideband

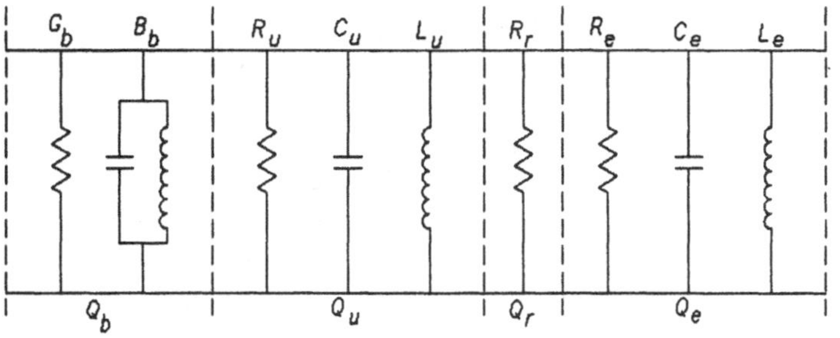

Figure 4. Equivalent circuit representation of several cavity impedances including off-resonance components

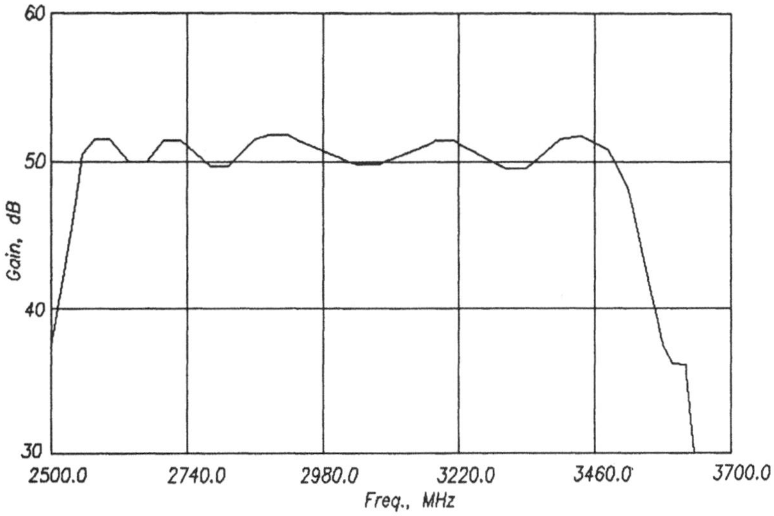

Figure 5. Calculated gain versus frequency characteristics for a 50MW clustered-cavity klystron with 6 intermediate cavity triplets

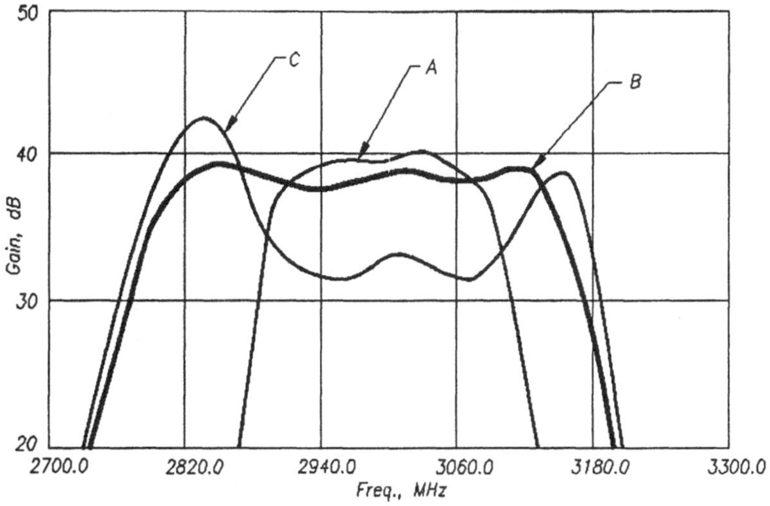

Figure 6. Calculated gains of Megawatt klystrons of the same length with various intermediate cavity arrangements

performance from the conventional klystron tube technology has been achieved via the Clustered-Cavity™ Klystron concept shown in Fig. 3. In the Clustered-Cavity™ Klystron, the individual intermediate cavities of a stagger-tuned multicavity klystron are replaced by pairs or triplets. Each cavity in the pair or triplet has, in addition, resistive loading in the cavity, such that its Q is one-half or one-third respectively of the Q of the single cavity they replace. The spacing of the gaps of the individual cavities in the multiplet is as close as possible. The spacing from the center of a multiplet to the center of the next multiplet is about one-quarter of a space-charge wavelength, as given by Eq. 8. A simple way of explaining the improved bandwidth is to realize that the two or three cavities in the multiplet are tuned to essentially the same frequency and are excited by essentially the same RF current so that they act on the beam in unison to form a new bunch, in the same way a single cavity might, but over two or three times the bandwidth. The actual physics is more complex and in particular, when cavity spacings in a klystron differ from the one-quarter of space-charge wavelength of Eq. 8, the velocity modulation beam current becomes important and must be accounted along with the main density bunch component of the RF current. These velocity modulation effects have in the past been routinely neglected in analytical and computational evaluations of broadband klystron designs. Symons and Vaughan[8] have recently developed an analytical and computational method that includes the velocity modulation effects and results in very much improved agreement between theory/computational predictions and experimental results. This work will be discussed briefly in the next section. We conclude this section by referring to Fig. 5 which illustrates a design for an ultrawideband clustered-cavity klystron, with bandwidth in excess of 30%.

III. CLUSTERED CAVITY KLYSTRON

In order to design and analytically/computationally evaluate and predict the performance of a clustered cavity klystron, an analysis of the interaction of the beam with a long sequence of cavities is required. The quarter wavelength condition of Eq. (8) is not generally satisfied and both beam density and beam velocity modulation coexist on the beam. In addition the sources for these beam density and velocity modulations are the gaps of each cavity the beam traverses. The phases with which these modulations are initiated on the beam at each gap are different and careful account has to be kept of all the amplitudes and phases as the beam interacts with successive cavity gaps. This has been accomplished in a recent analytical work by Symons and Vaughan[8]. The analytical results have been programmed in a computational model and code called SSKLY 26, which has extensively been exercised to generate data like the ones shown in Figs.5 and 6. The formulation utilized to keep track of the gap interactions is based on that introduced by Kreuchen, Auld and Dixon[6] and is given by

$$I_m = \Sigma_{n=1}^{m-1} g_{mn} V_n \qquad (11)$$

where I_m is the RF current at the mth gap due to the voltages V_n at all preceding gaps and the gap-drift space transconductances g_{mn}. The algorithm for carrying out the calculation is straight forward, but the actual expressions for the transconductances become complicated as the number of cavities increase. As an illustrative example, Fig 6a shows the calculated response of a 6 cavity stagger-tuned klystron in which the drift lengths are all about 60° of plasma phase. Figure 6b shows the calculated response when the four intermediate cavities are replaced by four intermediate-cavity pairs with periodic spacings of 40 and 20 degrees of plasma phase. Q's of the intermediate cavities for figure 6b are half of those for figures 6a and the fractional-detunings of the stagger tuning pattern are doubled. Figure 6c shows another calculated response for a klystron of the same length as those represented in figures 6a and 6b, but now with the 8 intermediate cavities used in the calculation of figure 6b spaced equally along the beam. One can conclude that zeros have moved into the band and have reduced the gain in the middle of the band about 10dB. Clearly, the curve of figure 6b is more desirable.

IV. CONCLUSIONS

The recent analytical, computational and experimental exploration of the clustered cavity klystron has ushered conventional tube technology into the ultrawideband region of bandwidth with modest incremental improvements and modifications of existing tube technology. Further increments in bandwidth appear feasible for longer tubes with additional clustered cavities. The full exploration of the concept and the deliniation of its limitations await to be investigated. As the bandwidth capabilities and power gain of the intermediate clustered cavity chain is increased, it is anticipated that the RF output cavity and its window will have to be improved via coupled cavity concepts capable of efficiently coupling out of the tube the larger bandwidth at higher powers.

V. ACKNOWLEDGEMENTS

The authors are indebted to Rodney Vaughan for many useful discussions.

REFERENCES

1. R.S. Symons, "Broadband Klystron Cavity Arrangement," U.S. Patent #4,800,322, Jan 24, 1989.
2. R.S. Symons, B. Arfin, R.E. Boesenberg, P.E. Ferguson, M. Kirshner and J.R.M. Vaughan, "An Experimental Clustered-Cavity™ Klystron," Proceedings (Technical Digest) of the International Electron Devices Meeting, Washington DC, December 1987, IEDM 87:153 (1987).
3. R.S. Symons, "The Development of an Ultrawideband Klystron," May 1991, to be published.
4. R.S. Symons, "Scaling Laws and Power Limits for Klystrons," Proceedings (Technical Digest) of the International Electron Devices Meeting, Los Angeles, CA, December 7-10, 1986, IEDM 86:156 (1986).
5. R.S. Symons and J.R.M. Vaughan, "Modification of Klystron Beam Loading By Initial Velocity Modulation of the Beam," Proceedings (Technical Digest) of the International Electron Devices Meeting, San Francisco, CA, December 1990, IEDM 90:885 (1990).
6. K.H. Kreuchen, B.A. Auld and N.E. Dixon, "A Study of the Broadband Frequency Response of the Multicavity Klystron Amplifier," J. Electronics 2:529 (1957).
7. J.G. Siambis and M. Friedman, "Klystron Instability in the Coaxial Autoaccelerator," Particle Accelerators 8:217 (1978).
8. R.S. Symons and J.R.M. Vaughan, "The Theory of the Clustered-Cavity™ Klystron" to be submitted for publication in IEEE Transactions on Plasma Science.

GENERATION OF A FREQUENCY CHIRPED PULSE USING PHASE VELOCITY TRANSITIONS IN A RAPIDLY CREATED PLASMA

S. P. Kuo, A. Ren and J. Huang

Weber Research Institute
Polytechnic University
Farmingdale, NY 11735

ABSTRACT

Conversion of a monochromatic source microwave into a frequency up-shifted and chirped pulse has been demonstrated experimentally. The experiment uses a high voltage (~100KV) DC discharge to generate a dense plasma suddenly between two parallel plates to interact with the passing through source microwave. A signal with a very broad frequency spectrum is recorded. The spectrum in the peak region has a relative flat and asymmetric distribution over 500MHz bandwidth manifesting a 2 ns chirped pulse. The central frequency (~6.4GHz) of the pulse is up-shifted from the frequency (~4.7GHz) of the source wave by about 40%. Moreover, frequency components which are up-shifted as high as 80% are also recorded. A theoretical model is developed to simulate the experiment. The numerical result is in good agreement with the experimental observation.

INTRODUCTION

In using very high-power microwave pulses for the application of radar and directed energy system, it is crucial to identify the effect of self-generated plasma on the propagation of the pulses. It has been shown that the pulse self-generated plasma can indeed radically modify wave propagation[1-6]; in particular, its role on up shifting the carrier frequency of the pulse has received considerable attention because such a "frequency auto-conversion process" keeps the self-generated plasma to be underdense to the pulse and unable to cause cutoff reflection[7-9]. Physically, this frequency up-shifting phenomenon can be explained by the result that the phase velocity of the wave increases with the plasma density. Hence, the tail part of the pulse propagates faster than the leading part of the pulse in the self-generated plasma whose density has an increasing profile from the leading edge to the tail end of the pulse, and the pulse is squeezed toward its front as time evolves[9]. Consequently, the wave frequency is up-shifted.

Similarly, such a phase velocity transition mechanism for frequency shift can also be applied to the situation that the plasma is created suddenly by the separate (from the wave) means and uniformly throughout a volume of several wavelength in each dimension. It is

Ultra-Wideband, Short-Pulse Electromagnetics
Edited by H. Bertoni *et al.*, Plenum Press, 1993

noted that the wavelength $\lambda_0=2\pi/k_0$ of the wave does not change after the sudden increase of the plasma density. This is because the wave experiences only temporal variation of the plasma. However, the plasma causes a sudden reduction on the index of refraction $\sqrt{\varepsilon_r}$ of the background medium and forces the wave to propagate with a larger phase velocity. Thus, the new wave has to oscillate with higher angular frequency $\omega=(\omega_0^2+\omega_p^2)^{1/2}$ at the subsequent time, in order to satisfy the dispersion relation $\varepsilon_r=k_0^2c^2/\omega^2$, where ω_0 and ω_p are the initial frequency of the wave and the plasma frequency, respectively.

Since the plasma can be created by the separate means, the source wave does not have to be high power and the amount of frequency shift and bandwidth may be varied by adjusting the characteristics of the plasma. It suggests that a plasma device be devised to tune the frequency of existing source of microwave or to convert a CW source wave into a wide band pulse train, where the period of the pulse train and the length of each pulse are determined by the repetition rate of the plasma discharge and the length of the plasma volume respectively. Such a wide band pulse train has a potential application as the radar jamming source.

A comprehensive theory and numerical simulation on this wave frequency shift by interacting with a suddenly created uniform plasma were recently presented by Wilks et al.[10]. Several experiments were then followed. One of them conducted by our group[11] was performed by using two crossed microwave pulses for plasma generation in their intersection region inside a Plexiglas chamber of 2ft cube. The frequency spectrum of the same pulses at the exit side of the chamber were examined. The results showed that the central frequencies of the transmitted pulses were up-shifted. The percentage of frequency shift was found to be consistent with the theory of Wilks et al.[10].

Other demonstrations of frequency up-shifting of microwave pulses by rapid plasma creation were given by the experiments of Joshi et al.[12] and Rader and Alexeff[13]. The experiment of Joshi et al was carried out in a cavity in which the plasma was created rapidly by ionizing the background azulene vapor with a laser pulse. The frequency of the incident RF wave at 33.3GHz was up-shifted by 5% with greater than 10% of efficiency. Signals with much higher up-shifted frequencies were also observed, but their generation mechanism was later confirmed experimentally by Savage et al.[14] to be associated with the relativistic Doppler effect. The experiment of Rader and Alexeff[13] was performed in a long cylindrical tube in which the plasma was generated by DC discharge and the source wave of 2.68GHz was passed and reflected from the other end. A 2% up-shift from the source wave frequency was reported.

The purpose of the present work is to explore the possibility of converting a CW source wave into a periodic pulse by passing the source wave through a repetitively generated fast growing plasma. The new pulse shall have a up-shifted and chirped carrier frequency. The plasma in the experiment is generated by a fast DC discharge of the gas between two conducting plates. The experiment is described in the next section, in which the experimental result is also reported. A theoretical model is then presented in the following section, where the numerical result of the model is also reported and compared with the experimental result. In the last section, the implication of the experimental result is discussed and the present work is also summarized.

EXPERIMENT AND RESULTS

The experiment is conducted by using a DC discharge to generate a fast growing dense plasma for frequency shifting purpose. The electrodes for the electrical discharge consist of a pair of parallel conducting plates placed inside a Plexiglas vacuum chamber.

The electrode pair is also used to guide the incident wave. A Marx bank facility rated at 240KV and 13KJ maximum energy is used for required fast generation of a dense plasma. In the present experiment, the Marx bank is regulated by an internal impedance of 15Ω and only charged to about 100KV. The chamber is filled with dry air to about 1 torr pressure which is near the highest pressure allowed before getting into constricted arc discharge. The separation between the plates is about 15 cm leading to an open circuit average field strength of 6.6KV/cm which is much higher than the breakdown threshold field of 40V/cm. After the gas breakdown caused by the applied high voltage, a dense plasma with reasonably uniform density distribution between the two parallel plates is generated and the voltage across the electrodes drops considerably to slowdown the ionization so that a broadband frequency shift can be achieved. A schematic of the complete experimental setup is presented in Fig.1., in which the dimensions of the electrodes and the vacuum chamber are also indicated.

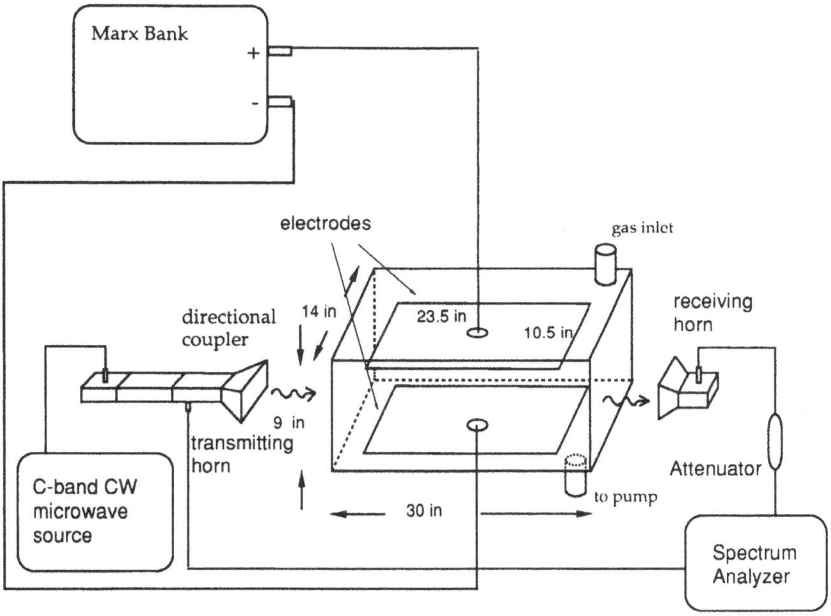

Fig. 1. Schematic of the experimental setup.

A CW source wave of 4.7GHz is injected into the chamber to propagate along the parallel plate waveguide. Without plasma generation the spectrum of the signal received at the other end of the chamber by spectrum analyzer is a single spike at 4.7GHz. On the other hand, whenever the discharge takes place, the suddenly created plasma modifies the characteristics of the interacting wave and different spectral components appear. The density of the generated plasma is measured indirectly in terms of the enhanced airglow whose relationship with the actual plasma density is calibrated by the airglow coresponding to microwave cutoff density and measured in a separate experimental setup[4]. Shown in Fig.2a is the spatial distribution of the peak intensity of the airglow emitting from an 1μs microwave pulse-induced plasma whose peak near the entrance of the pulse adjusted by the microwave power to be about $10^{11}cm^{-3}$, the cutoff density of the 3.27 GHz microwave. The temporal evolution of the intensity of the airglow of the DC discharge plasma is shown in

Fig.2b which is compared with Fig.2a and indicates that the peak plasma density is about 8 times the cutoff density of the source microwave of 4.7GHz. In both systems, background air pressure are set at 1 torr. Since the plasma density is much higher than the cutoff density of the incident wave, the incident wave is blocked by the plasma from propagation through the chamber and only the portion of the wave inside the chamber prior to the discharge can

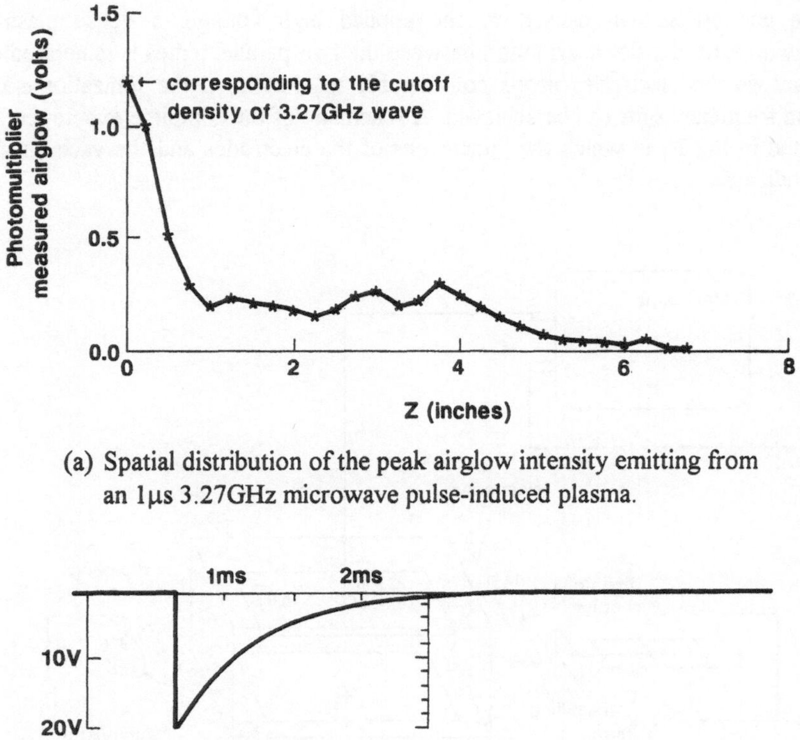

(a) Spatial distribution of the peak airglow intensity emitting from an 1μs 3.27GHz microwave pulse-induced plasma.

(b) Temporal evolution of the airglow intensity from the DC discharge plasma.

Fig. 2. Airglow intensities of the plasmas.

still propagate through the chamber. Thus, the transmitted signal is expected to be a pulse of about 2ns duration, i.e., the transit time of the source wave propagating through the chamber. Since the plasma is generated rapidly, it is also expected that the central frequency of the spectrum of the pulse is up-shifted. Moreover, since the plasma density increases with finite growth rate, the pulse is expected to experience a time dependent frequency up-shift. Including all of the expected features, a frequency up-shifted and chirped 2ns pulse is expected to be generated by the interaction. Such a pulse is indeed generated by the experimental setup and demonstrated by its spectral distribution shown in Fig.3. This spectrum is recorded by using the time sampling method. The spectrum analyzer measures randomly in one sweep a single frequency component of the output signal in each discharge. Each measured component represents a sampling value of the frequency spectrum of the transmitting signal at a given frequency. The spectrum of Fig.3 is constructed from the data of more than 100 consective discharges. The peak line of 4.7GHz is the spectrum of the incident CW signal. The remaining lines are the sampling lines of the frequency spectrum of the chirped pulse. As seen, it is a broadband spectrum whose central frequency is up-shifted away from the original line of 4.7GHz. It is noted that the spectral distribution around 6.4GHz has a bandwidth about 500MHz and is asymmetric. It represents a 2ns chirped pulse whose pulse length is consistent with the transit time of the source wave propagating through the chamber as discussed before. The appearance of the other lines in Fig.3 is believed to be attributed to the non-stationary growth rate and decay rate of the plasma.

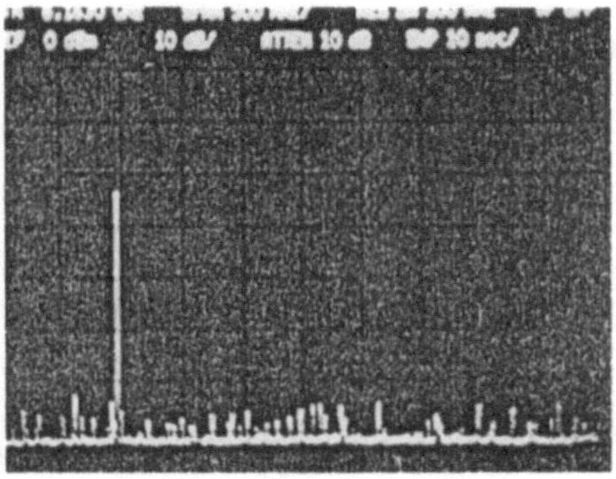

Fig. 3. Sampling spectrum of the output signal. The peak line at 4.7GHz is the spectrum of the source microwave. The horizontal scale is 500MHz/div. and centered at 6.1630GHz. The vertical scale is 10dB/div. and has a maximum of 0dBm.

THEORETICAL MODEL

For achieving a better understanding of the experimental results, a theoretical model of the experimental system is developed for numerical simulation. It involves a CW wave propagation through a fast growing plasma slab which is located between z=0 and L. a schematic of the problem is illustrated in Fig.4. The time variation of the plasma density in

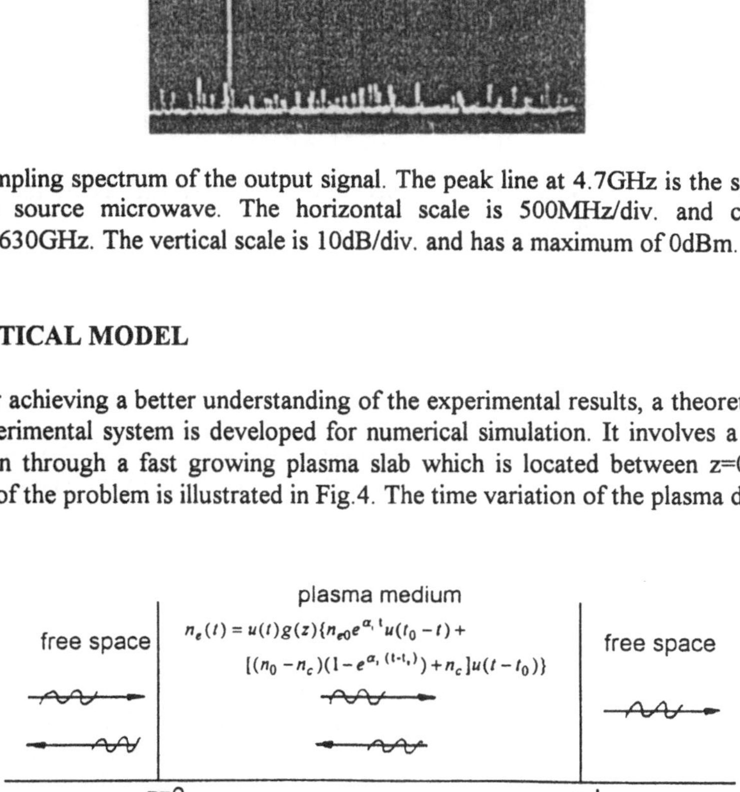

Fig. 4. Initial and boundary value problem.

the slab is modeled as $n_e(t) = u(t)g(z)\{n_{e0}e^{\alpha_1 t}u(t_0 - t) + [(n_0 - n_c)(1 - e^{\alpha_1 (t-t_1)}) + n_c]u(t - t_0)\}$, where u(t) is an unit step function and g(z)=1 for 0<z<L and 0 for z at elsewhere; n_{e0} is the initial background density, α_1=<v_i> is the average ionization frequency within the time interval $(0,t_0)$, $t_0=1/\alpha_1 \ln(n_c/n_{e0})$, n_c is the cutoff density for the incident source wave, and n_0 is the maximum electron density of the gaseous discharge. The relevant equations are the electron momentum equation

$$\frac{\partial}{\partial t} n_e \vec{v}_e = -(e/m) n_e \vec{E}$$

the Maxwell's equations

$$\nabla \times \vec{E} = -\frac{1}{c}\frac{\partial}{\partial t}\vec{B} \text{ and } \nabla \times \vec{B} = (4\pi/c)\vec{J} + \frac{1}{c}\frac{\partial}{\partial t}\vec{E}$$

and the constitutive relation

$$\vec{J} = -en_e\vec{v}_e$$

These equations can be combined into a wave equation describing the wave propagation through the plasma slab:

$$\left[\frac{\partial^2}{\partial z^2} - \frac{1}{c^2}\frac{\partial^2}{\partial t^2} - \omega_p{}^2/c^2\right]\vec{E}(z,t) = 0 \tag{1}$$

where $\omega_p{}^2 = 4\pi n_e e^2/m$

Before the plasma discharge (i.e. $t<0$), a CW source wave is propagating through the space between the two electrode plates. The wave fields are assumed to be $\vec{E} = \hat{x}E_0 e^{-i(\omega_0 t - k_0 z)}$ and $\vec{B} = \hat{y}B_0 e^{-i(\omega_0 t - k_0 z)}$, where $E_0=B_0$ and $\omega_0=k_0c$. Thus, the two initial conditions of (1) are obtained to be

$$E(z,t=0^+)=E(z,t=0^-)=E_0 e^{ik_0 z}$$

$$\frac{\partial}{\partial t}E(z,t=0^+)=\frac{\partial}{\partial t}E(z,t=0^-)=-i\omega_0 E_0 e^{ik_0 z} \tag{2}$$

Thus, the experimental system is described by the wave equation (1) subject to the initial conditions (2) together with the continuity (boundary) conditions at z=0: $E(0^-,t) = E(0^+,t)$ and $\frac{\partial}{\partial z}E(0^-,t) = \frac{\partial}{\partial z}E(0^+,t)$, and z=L: $E(L^-,t) = E(L^+,t)$ and $\frac{\partial}{\partial z}E(L^-,t) = \frac{\partial}{\partial z}E(L^+,t)$.

Eq.(1) is solved numerically, where α is treated as a variable parameter whose value is chosen to fit the numerical result with the experiment. Presented in Fig.5 is the numerical result of the spectrum of the transmitted signal. The spike at $f/f_0=1$ is the spectrum of the incident source wave. In the calculation $\omega_{p0}{}^2=8\omega_0{}^2$ and $\alpha_1=2.36\times 10^{10}\text{sec}^{-1}$ for $t<0.68$ ns and $\alpha_2=2.95\times 10^7\text{sec}^{-1}$ for $t>0.68$ ns are used. The value of $\omega_{pe}{}^2$ is based on the measurement and a modeling of α by a two step function is based on the understanding that the discharging voltage provided by the Marx bank decreases in time. The initial α_1 value for $t<0.68$ ns used in the calculation is chosen to agree with the average ionization rate of the dry air during the beginning 0.68 ns time interval. The Maximum discharge current of the Marx bank regulated by its internal impedance is taken into account to determine the time of 0.68 ns. The second α_2 value for $t>0.68$ ns is chosen to be a variable parameter. The resemblance between the spectrum of Fig.5 and the experimentally measured spectrum of Fig.3 suggests that a vital model be developed and the frequency shift mechanism be validated. The model can be used to instruct the future experiments.

SUMMARY AND DISCUSSION

Initial value problem concerning wave propagation through a rapidly created plasma has been studied experimentally. The preliminary result of the experiment was presented in a separate publication[15]. Plasma generation instantaneously over a relatively (to the wave length of the source microwave) large volume is via a DC discharge between two parallel plates. The rapidly growing plasmas force the frequencies of the passing source wave to up-shift. The purpose of the experiments is to use this frequency up-shift phenomenon to convert a CW wave into a periodic pulse featured with up-shifted and chirped carrier frequency. The spectral result of Fig.3 shows that a frequency up-shift of the source microwave as large as 80% is achieved. The monochromatic source wave is converted into a frequency up-shifted and chirped short pulse after interacting with the plasma of each gaseous discharge. The central frequency of the generated pulse is about 6.4GHz which is

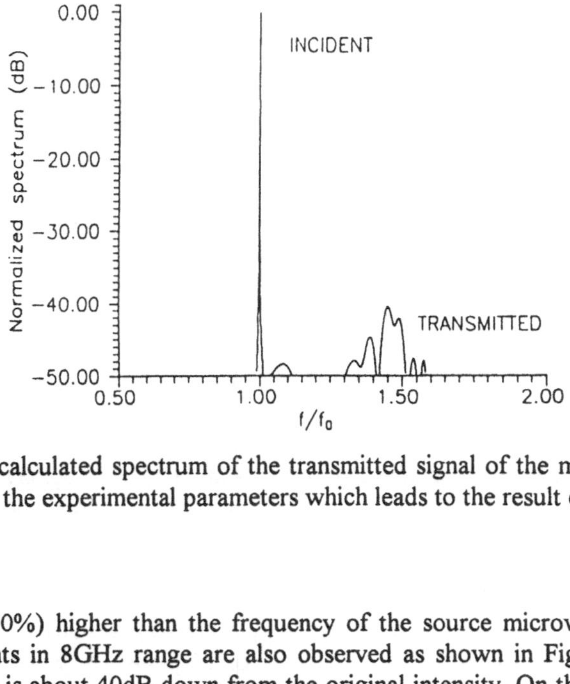

Fig. 5. Numerically calculated spectrum of the transmitted signal of the modeling problem (Fig.4) using the experimental parameters which leads to the result of Fig.3.

1.7GHz (or about 40%) higher than the frequency of the source microwave. Moreover, frequency components in 8GHz range are also observed as shown in Fig.3. The spectral intensity of the pulse is about 40dB down from the original intensity. On the other hand, its spectral bandwidth is broadened from the original 100KHz range to a width of about 500MHz. Thus the conversion efficiency in terms of generating new wave of up-shifted frequency (not the conversion efficiency from a CW wave into a pulse) is estimated to be about 50%. A theoretical model is also developed to simulate the experiment numerically. The numerically calculated spectrum of the transmitting signal as shown in Fig.5 has almost reproduced all the features of the experimentally measured spectrum of Fig.3. This theoretical model will be used to instruct how to upgrade the experiment. One possible application of present work is to develop a source of high power periodic microwave pulse. This is done by using periodic fast gaseous discharge to chop a high power CW microwave into a train of frequency up-shifted and chirped short pulses. The repetition rate and duty cycle of the periodic pulse depend on the capability of the electrical charging system and the plasma decay time respectively.

ACKNOWLEDGMENTS

We would like to thank Raytheon ESD-LI for providing the test equipment for the experiments. This work was supported by the Air Force Office of Scientific Research of the Air Force system command, Grant No. AFOSR-91-0002.

REFERENCES

1. J. H. Yee, R. A. Alverez, D. J. Mayhall, D. P. Byrne, and J. DeGroot, "Theory of Intense Electromagnetic Pulse Propagation through the atmosphere", Phys. Fluids 29, 1238 (1986).
2. M. J. Mulbrandon, J. Chen, P. J. Palmadesso, C. A. Sullivan, and A. M. Ali, "A Numerical Solution of the Boltzmann Equation for High-Power Short Pulse Microwave Breakdown in Nitrogen", Phys. Fluids B. 1, 2507 (1989).
3. S. P. Kuo and Y. S. Zhang, "Bragg Scattering of EM Waves by Microwave Produced Plasma Layers", Phys. Fluids B 2, pp.667-673 (1990).
4. S. P. Kuo, Y. S. Zhang and Paul Kossey, "Propagation of High Power Microwave Pulses in air Breakdown Environment", J. Appl. Phys. 67, pp.2762-2766 (1990).
5. S. P. Kuo and Y. S. Zhang, "A Theoretical Model for Intense Micowave Pulse Propagation in an Air Breakdown Environment", Phys. Fluids B 3, 2906 (1991).
6. S. P. Kuo, Y. S. Zhang, M. C. Lee, Paul Kossey, and Robert J. Baker, "Laboratory Chamber Experiments Explcring the Potential Use of Artificial Ionized Layers of Gas as a Bragg Reflector for Over-the-horizon Signals", Radio Science, in press.
7. V.B.Gildenburg, V.A.Krupnov, and V.E.Semenov, "Frequency Autoconversion and Reflectionless Propagation of a Powerful Electromagnetic Pulse in an ionizing Media", Pis'ma Zh. Teleh. hiz.14, 1695 (1988).
8. S. P. Kuo, Y. S. Zhang, and A. Ren, "Observation of Frequency Up-conversion in the Propagation of a high Power Microwave Pulse in a Self-generated Plasma", Phys. Lett. A 150, pp.92-96 (1990).
9. S. P. Kuo and A. Ren, "Frequency up-conversion of a high-power microwave Pulse Propagating in a Self-generated Plasma", J. Appl. Phys. 71, pp.5376-5380 (1992).
10. S. C. Wilks, J. M. Dawson, and W. B. Mori, "Frequency Up-conversion of Electromagnetic Radiation with Use of an Overdense Plasma", Phys. Rev. Lett., Vol.61, No.3, pp.337-340 (1988).
11. S. P. Kuo, "Frequency Up-conversion of Microwave Pulse in a Rapidly Growing Plasma", Phys. Rev. Lett., Vol.65, No.8, pp.1000-1003 (1990).
12. C. Joshi, C. E. Clayton, K. Marsh, D. B. Hopkins, A. Sessler, and D. Whittum, "Demonstration of the Frequency Up-shifting of Microwave Radiation by Rapid Plasma", IEEE, Trans. Plasma Sci., Vol.18, No.5, pp.814-818 (1990).
13. M. Rader and I. Alexeff, "Microwave Frequency Shifting Using Phase Velocity Transitions in a Plasma", submitted to IEEE Trans. Plasma Sci.
14. R.L. Savage, Jr., C. Joshi, and W. B. Mori, "Frequency up-conversion of Electromagnetic Radiation upon Transmission into an Ionization Front", Phys. Rev. Lett., Vol.68, pp946-949 (1992).
15. S. P. Kuo and A. Ren, IEEE Trans. Plasma Sci., accepted for publication.

ANTENNAS AND ARRAYS

IMPULSE RADIATING ANTENNAS

Carl E. Baum
Phillips Laboratory
Kirtland AFB, NM 87117

Everett G. Farr
Farr Research
Albuquerque, NM 87123

ABSTRACT

A number of applications require radiation of a short pulse of electromagnetic energy out to large distances. These applications include target discrimination in a cluttered environment (e.g., looking over the ocean), aircraft identification by taking a "TDR" of its major scattering centers, and target location through foliage. The Impulse Radiating Antenna (IRA) has generated widespread interest for its ability to radiate a broadband pulse. The purpose of this paper is to summarize recent work on Impulse Radiating Antennas.

INTRODUCTION

Impulse Radiating Antennas (IRAs) are meant to satisfy the need for a radiating device for one class of transient radars. This class of radar attempts to solve several problems, such as target identification by scattering centers on aircraft, target discrimination in a highly cluttered environment such as when looking over the ocean, and foliage penetration. We provide here the basic theory of Impulse Radiating Antennas, and apply that theory to some practical examples.

BASIC THEORY

Impulse radiating antennas consist of a conical TEM feed that attaches to a reflector antenna. The load at the attachment point is chosen so that at low frequencies the antenna behaves as a $\vec{p} \times \vec{m}$ dipole with a cardioid radiation pattern[1,2]. An example of a IRA is shown in Figure 1.

Before providing specific designs for IRA's, we first describe a simple theory on which the devices are based. Consider a planar aperture as shown in Figure 2, whose tangential electric field is described by[3]

$$\tilde{\vec{E}}_t(\vec{r},s) = e^{-\gamma \vec{1}_o \cdot \vec{r}'} E_o \tilde{f}(s) \vec{g}(\vec{r}') \tag{2.1}$$

where the usual convention of primed coordinates referring to the source and unprimed coordinates referring to the observation point has been used. This is just a plane-wave distribution with a uniform time dependence of $f(t)$ in the time domain. The spatial distribution in the aperture is handled by $\vec{g}(\vec{r}')$, and the direction of maximum radiation is $\vec{1}_o$. It is shown by Baum[3] that the radiated field is just

$$\tilde{\vec{E}}_f(\vec{r},s) = e^{-\gamma r} \frac{E_o A}{2\pi r c} s \tilde{f}(s) \tilde{\vec{F}}_a(\vec{1}_r,s) \tag{2.2}$$

where the aperture function is

$$\tilde{\tilde{F}}_a(\vec{1}_r,s) = \frac{1}{A}\left[(\vec{1}_z\cdot\vec{1}_r)\tilde{1} - \vec{1}_z\vec{1}_r\right]\cdot\int_S e^{\gamma[\vec{1}_r-\vec{1}_o]\cdot\vec{r}'}\,\vec{g}(\vec{r}')\,dS' \tag{2.3}$$

and the following definitions apply

$$A = \text{Aperture area}, \qquad \tilde{1} = \vec{1}_x\vec{1}_x + \vec{1}_y\vec{1}_y + \vec{1}_z\vec{1}_z$$

$$\tilde{1}_z = \vec{1}_x\vec{1}_x + \vec{1}_y\vec{1}_y, \qquad \tilde{1}_r = \tilde{1} - \vec{1}_r\vec{1}_r \tag{2.4}$$

If we now specialize the above equations to a uniform plane wave in the aperture, then

$$\vec{g}(\vec{r}') = \begin{cases} \vec{1}_y & \text{on } S \\ 0 & \text{off } S \end{cases} \tag{2.5}$$

Furthermore, if we specify that the observation point is on boresight $(\vec{1}_r = \vec{1}_o)$, then the above equations simplify considerably to

$$\tilde{\tilde{F}}_a(\vec{1}_r,s) = \vec{1}_y\ , \qquad \tilde{\vec{E}}_f(\vec{r},s) = e^{-\gamma r}\,\frac{E_o A}{2\pi rc}\,s\,f(s)\,\vec{1}_y \tag{2.6}$$

Thus, if the aperture field has a step-function time dependence, then the radiated field on boresight is

$$\tilde{\vec{E}}_f(\vec{r},t) = \frac{E_o A}{2\pi rc}\,\delta(t-r/c)\,\vec{1}_y \tag{2.7}$$

This delta function requires some interpretation.

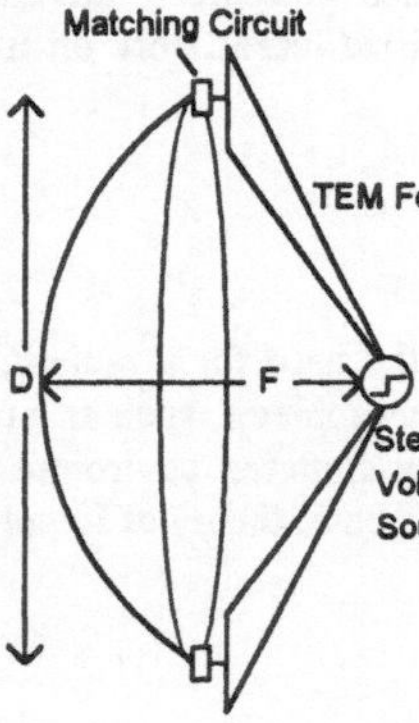

Figure 1. An IRA.

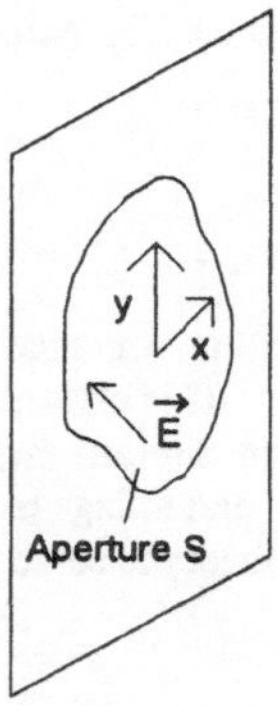

Figure 2. The aperture field.

The problem we face is that on boresight, the energy density is infinite, for bounded aperture fields. Since this is clearly nonphysical, greater care must be taken with the aperture integral in the boresight direction. It has been shown that a more careful aperture field integration leads to[3]

$$\vec{E}(z\vec{1}_z) = \vec{1}_z E_o u(t-z/c), \qquad 0 < t-z/c < \Delta t \tag{2.8}$$

where Δt is the clear time required to first see the edge of the aperture. The above is a step function lasting for a time Δt. For a circular aperture of radius a, $\Delta t \approx a^2/(2rc)$, so the integral of the electric field for the duration of this step function is just

$$\int \vec{E}(z\vec{1}_z,t)\,dt = \vec{1}_y\,\frac{E_o a^2}{2rc} \tag{2.9}$$

This provides the same integral as that generated by integrating the delta function of (2.7).

In order to account for the unusual behavior on boresight, we define an approximate delta function $\delta_a(r,t)$, which has the following properties. Its area is unity, its peak magnitude is proportional to r, and its width is proportional to $1/r$. As $r \to \infty$, this becomes a true delta function, and $\delta_a(r,t)/r$ has a peak which is constant with r. Thus, $\delta_a(r,t)$ should be used instead of $\delta(t)$ where required to maintain finite energy.

The Aperture-Field Integral

The integral of the aperture field was carried out in the previous section for the trivial case of a circular aperture excited uniformly with a step function. In this section we generalize the above results for more arbitrary aperture distributions. These results will provide an effective height that characterizes the radiation on boresight for an aperture.

Consider an aperture field generated by the transmission line formed by two parallel wires. A diagram of this is shown in Figure 3. As we saw in the previous section, the radiated electric field on boresight is

$$E_f(\vec{r},t) = \frac{\delta_a(t)}{2\pi c r}\int_{S_a} \vec{E}_t(x,y)\,dS' \tag{3.1}$$

where $\delta_a(t)$ is the approximate Dirac delta function as described in the previous section. The integral can be carried out for several geometries using contour integration in the complex plane. Consider the aperture distribution generated by two parallel wires as shown in Figure 3. If there is a voltage V between the two conductors, and the two conductors form a transmission line of characteristic impedance $f_g Z_o$, then let us define an aperture height such that[4]

$$\frac{V}{f_g}\vec{h}_a = \int_{S_a} \vec{E}_t(x,y)\,dS' \,, \qquad Z_o = \sqrt{\mu_o / \varepsilon_o} \tag{3.2}$$

This provides a simple way of expressing the far field

$$\vec{E}_f(\vec{r},t) = \frac{V}{2\pi c f_g r}\vec{h}_a\,\delta_a(t) \tag{3.3}$$

One now needs to perform the integration in (3.2) in order to find the aperture height.

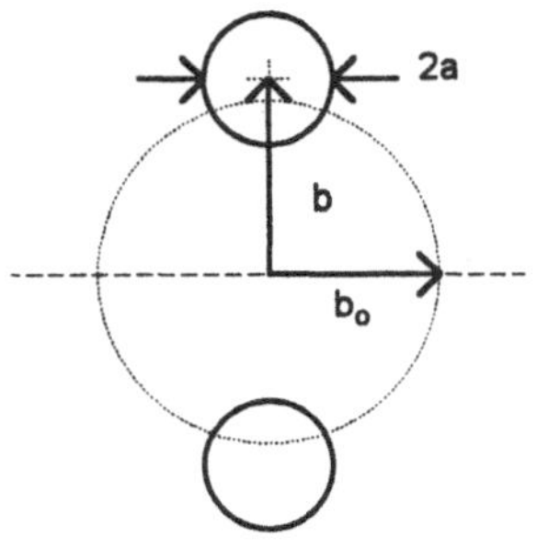

Figure 3. Transmission line for forming the aperture field.

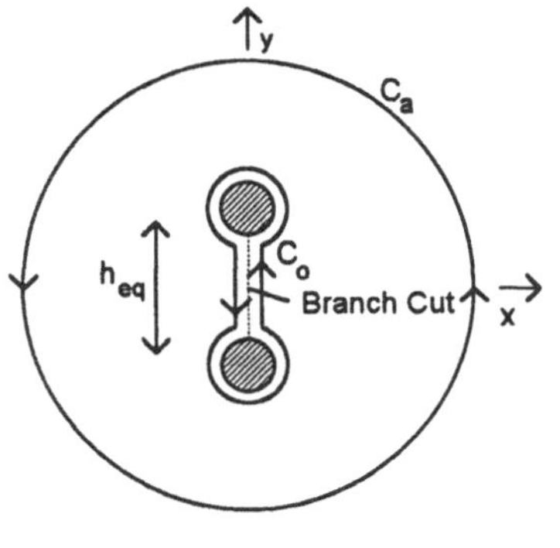

Figure 4. The contour to be integrated.

In order to perform the integration, we convert the vector electric field to a single complex function in the complex plane (Figure 4). The aperture electric field can be expressed as the two-dimensional gradient of a complex potential function. Thus,

$$\zeta = x - jy \equiv \text{complex coordinate} \,, \qquad w(\zeta) = u(\zeta) - jv(\zeta) \equiv \text{complex potential}$$

$$f_g = \Delta u / \Delta v \,, \qquad E(\zeta) = E_x(\zeta) - jE_y(\zeta) = -\frac{V}{\Delta u}\frac{dw(\zeta)}{d\zeta} \tag{3.4}$$

We now cast the integral into a complex surface integral as follows

$$\bar{\Upsilon} = \Upsilon_x \bar{1}_x + \Upsilon_y \bar{1}_y = \int_{S_a} \bar{E}(x,y)\, dS' \tag{3.5}$$

$$\Upsilon = \Upsilon_x - j\Upsilon_y = \int_{S_a} E(\zeta)\, dS' = -\frac{V}{\Delta u} \int_{S_a} \frac{dw(\zeta)}{d\zeta}\, dS' \tag{3.6}$$

By Green's theorem, the surface integral of a complex function can be recast as a contour integral[4], so

$$h_a = h_{a_x}(\zeta) - jh_{a_y}(\zeta) = \frac{f_g}{V}\Upsilon = \frac{1}{\Delta v} \int_{S_a} \frac{dw(\zeta)}{d\zeta}\, dS' = \frac{j}{2\Delta v} \oint_{C_a} w\, d\zeta^*$$

$$= \frac{j}{\Delta v} \oint_{C_a} u\, d\zeta^* = -\frac{j}{\Delta v} \oint_{C_a} v\, d\zeta^* \tag{3.7}$$

Note the branch cut in Figure 4 which makes v (the magnetic potential) single-valued. Therefore, the contribution from C_o is strictly subtracted, but gives zero in some of the above formulas. We now have a simple form for calculating the effective height of an antenna. For the two-wire problem, the complex potential function is obtained from a standard text such as Smythe[5], so

$$w(\zeta) = 2j\ \text{arccot}(\zeta / a) \tag{3.8}$$

This can be substituted into (3.7) for various aperture shapes, so effective heights can be determined simply. If the feed wires are thin, we find the aperture integral for various aperture shapes to be

$$h_a / h_{eq} = \begin{cases} 1/2 & \text{circular aperture} \\ 0.55 & \text{square aperture} \\ 1 & \text{strip aperture} \end{cases} \tag{3.9}$$

where $h_{eq} = 2b_o$ is the equivalent height (spacing) of two parallel conductors. Furthermore, if the wires are of any arbitrary shape, then the result for an infinite aperture is

$$h_a / h_{eq} = 1/2 \tag{3.10}$$

Thus, the radiated field on boresight for the usual case of a circular aperture is

$$E_r(r,t) = \frac{V}{r} \frac{h_{eq}}{4\pi c f_g} \delta_a(r,t) \tag{3.11}$$

where the approximate delta function is defined in the previous section.

Note that the above contour integral was calculated with the assumption that the feed is long. This is an unnecessary constraint, as shown by Farr and Baum[6], Appendix A. It is shown there that a reflection off a paraboloidal reflector produces a flat phase front with the same aperture field distribution as an infinitely long cylindrical transmission line (TEM mode), and is exactly expressed by the usual stereographic transformation (with minus sign, reflector transformation).

A SIMPLE MODEL OF THE IRA

In the previous section we described the impulsive portion of the radiation from the IRA on boresight. Now, we develop a simple expression for the entire waveform [6].

Consider what happens when a step voltage drives the IRA shown in Figure 1. There is a prepulse for a time $2F/c$, where F is the focal length of the reflector. This is due to the direct radiation of the currents on the feed arms. The shape of this prepulse is a step function, similar to the driving voltage. In the last two sections we saw that the impulsive portion of the radiated field is an approximate delta function, so let us consider a radiated waveform shown in Figure 5. The area of the impulse is known from Sections 2 and 3. Furthermore, it is possible to calculate the direct radiation from a conical feed by using various stereographic projections. If the area under the prepulse is equal to the area under the impulse, then the tail portion of the waveform can be made small with proper tuning of the matching circuit.

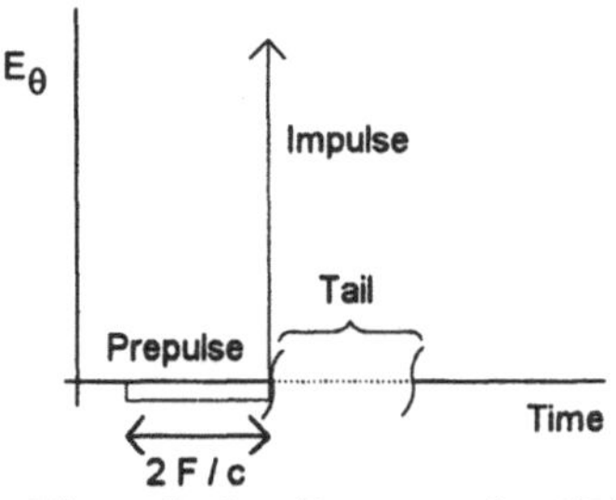

Figure 5. Step Response of an IRA.

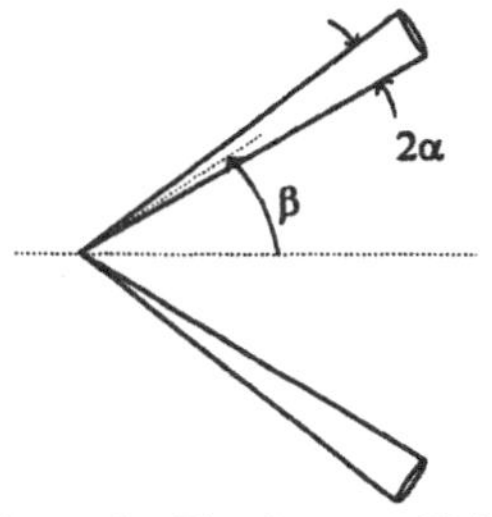

Figure 6. Circular cone IRA feed.

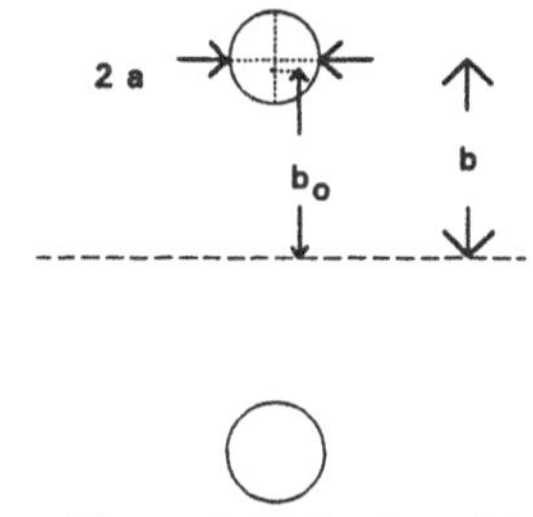

Figure 7. Projection of the conical feed onto a plane.

We calculate now the magnitude of the prepulse. It is simplest to calculate this for the geometry of two circular cones, as shown in Figure 6. If one were interested in a feed consisting of flat plates (either facing or coplanar), then for high impedances (small α) the results for the circular cones describe the plate geometries of angular width 4α as well. For lower impedances, more exact expressions are available[6].

It is now necessary to project the spherical geometry of the circular cones onto a planar surface. In order to do so, we invoke the usual stereographic projection[5]. Thus, the polar coordinates in the projection plane are

$$x = 2F\cos(\phi)\tan(\theta/2), \qquad y = 2F\sin(\phi)\tan(\theta/2) \tag{4.1}$$

The projection of the spherical cone generates a cylindrical structure whose cross-section is two circles with half height b and radius a such that

$$b = \frac{2F\sin(\beta)}{\cos(\alpha)+\cos(\beta)}, \qquad a = \frac{2F\sin(\alpha)}{\cos(\alpha)+\cos(\beta)} \tag{4.2}$$

where a and b are as shown in Figure 7. It is simple to find the electric field at the center of the projected structure[7]. Thus, we find the backward radiated field (forward as far as our antenna is concerned) to be

$$E_\theta = -\frac{V}{r_o}\frac{\cos(\alpha)-\cos(\beta)}{\pi f_g \tanh(\pi f_g)\sin(\beta)} \tag{4.3}$$

This is just what we need for calculating the ratio of the prepulse area to the impulse area. The impulse area is found by integrating (3.10) with respect to time. The prepulse area is found by multiplying (4.3) by $2F/c$, the round trip transit time of the feed. The ratio of the prepulse area to the impulse area is found to be

$$\left|\frac{A_p}{A_i}\right| = \frac{4(F/D)\left[\cos(\alpha)-\cos(\beta)\right]}{\tanh(\pi f_g)\sin(\beta)} \tag{4.4}$$

If this ratio is approximately equal to unity, then the two areas are equal. A plot of the difference of this ratio from unity is shown in Figure 8. For the types of parameters in which one is typically interested ($F/D = 0.4$, $Z_{feed} = 400\ \Omega$), the areas are equal to better than 1%. Thus, a reasonable expression for the on-boresight radiated field is

$$E(r,t) = \frac{V_o}{r}\frac{D}{4\pi c f_g}\left[\frac{c}{2F}\left[-u(t)+u(t-2F/c)\right]+\delta_a(t-2F/c)\right] \tag{4.5}$$

Of course, this expression makes use of a number of assumptions, including the use of an ideal matching circuit at the boundary between the feed and reflector. A second assumption is that the aperture blockage is small (valid for thin feed projections). Nevertheless, it is very helpful that such a simple result is available for such a complex structure.

A number of design features are still to be considered in order to minimize the tail on the IRA. A design for the matching circuit is still under development. This design requires a calculation of the electric and magnetic moments of the IRA. Once these are known, the resistance of the matching circuit at low frequencies can be calculated, which will balance the two dipole moments. A second area of further work will be to calculate the field scattered from the TEM feed and the reflector edges using diffraction theory. Both of these issues are currently being addressed.

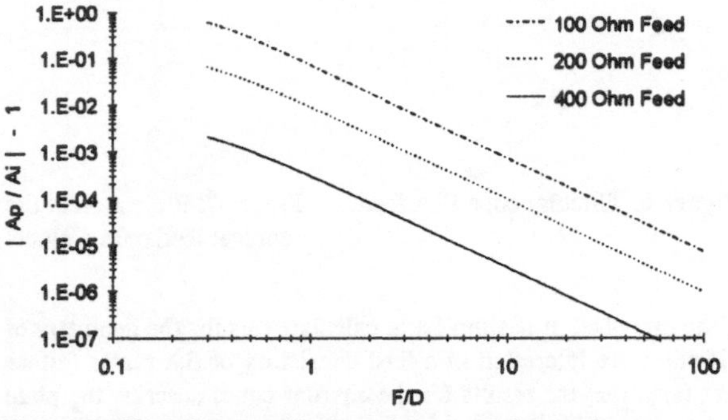

Figure 8. Error between the impulse area and prepulse area.

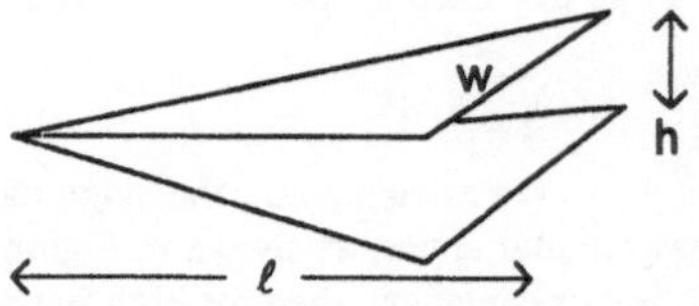

Figure 9. A TEM Horn.

TEM HORN

Another antenna sometimes used for radiating fast transients is the TEM horn (Figure 9). It is simplest to provide an analysis of a TEM horn by starting with the low frequencies. A low-frequency model of the TEM horn can be generated by using a simple open-circuit transmission-line model[8]. The model is generated by breaking the transmission line into a cascade of differential segments. The current and charge are known on each of these segments as a function of time, so they can be considered a cascade of small electric and magnetic dipoles. The fields radiated by each of these dipoles can be summed on boresight, leading to a low frequency model of

$$E(r,t) = -\frac{V_o}{r}\frac{h}{4\pi c f_g}\left[\delta_a(t) + \frac{c}{2\ell}\left[-u(t) + u(t - 2\ell/c)\right]\right] \tag{5.1}$$

where f_g is the ratio of the horn impedance to the impedance of free space, h is the height of the horn at its aperture, ℓ is the length of the horn, and V_o is the magnitude of the voltage step launched onto the horn.

The above model is incorrect at high frequencies, because the early part of the step response is actually a step function. This is merely the $1/r$ extrapolation of the field at the aperture. In order to correct for this, it is necessary to flatten the top of the δ-function in the above equation, and broaden its width, while maintaining the same area[8]. The effective width (in seconds) of this flattened δ-function is just the area of the δ-function as determined by the low-frequency model, divided by the peak magnitude[4]. In the time domain, the step response has a form shown in Figure 10. This is roughly equivalent in the frequency domain to running the ideal step response through a low-pass filter of the form

$$G(\omega) = \frac{1}{1 + j\omega/\omega_2}, \qquad \omega_2 = \frac{4\pi\ell c f_g}{h^2} \tag{5.2}$$

Although the this filter function must be considered only approximate, it has been demonstrated that it has the correct mid-frequency and high-frequency behavior. The only ambiguity is its behavior near $\omega=\omega_2$.

The above results are valid only under the assumption of a matched load at the apex of the horn. If the source is not matched to the characteristic impedance of the horn, then there are multiple reflections within the TEM horn, and the above description is no longer valid. Furthermore, there is an assumption that there is no loss of signal as it passes through the matching circuit. This is therefore an idealized model, since normally there would be a factor of two loss in voltage as it passed through a matching circuit. Some suggestions for such matching networks are provided by Farr and Baum[8].

There are a number of variations on the above rather simple design. In particular, one might consider using a lens in front of the TEM horn, in order to provide additional focusing[9]. This has the effect of sharpening the "broadened" delta function back to its original form. In doing so, this converts the antenna into a lens IRA[8]. Although this arrangement can considerably reduce the high-frequency rolloff of a TEM horn, it should be pointed out that lenses can be very heavy.

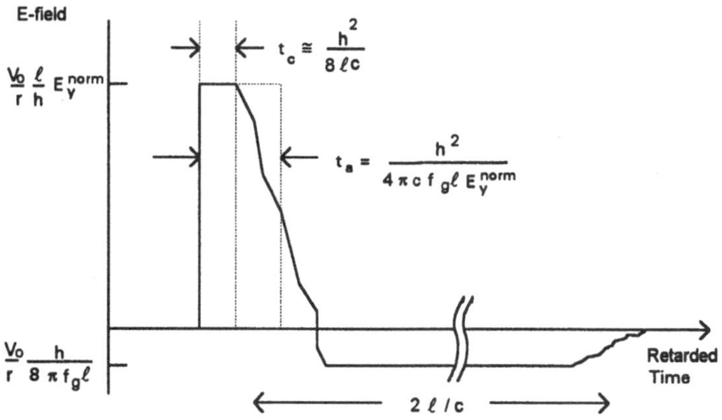

Figure 10. Step Response of a TEM horn.

AN EXAMPLE: TEM HORN AND TWO SIZES OF IRA

We compare now the performance of a TEM horn to two IRAs. One IRA (medium size) has a diameter equal to the length of the TEM horn, while the other IRA (small size) has the same aperture area as the TEM horn, as shown in Figure 11. The TEM horn has an impedance of ~116 Ω, while both IRA's have feed impedances of 400 Ω and *F/D* ratios of 0.4. The TEM horn is driven with 200 kV peak voltage, while the IRA's are driven with 371 kV, providing the same power to both antennas. Note that various baluns are required for both cases, which may require conversion of single-ended signal to differential. The driving voltage is a double exponential with a 200 ps. 10-90% risetime and a long decay time.

The frequency domain step response of the three antennas at 1 km on boresight is shown in Figure 12. The response to the driving voltage at 1 km is shown in the time domain in Figure 13. The response of both IRAs is better than that of the TEM horn everywhere except at very low frequencies for the small IRA. The high-frequency rolloff of the TEM horn is clear from these results.

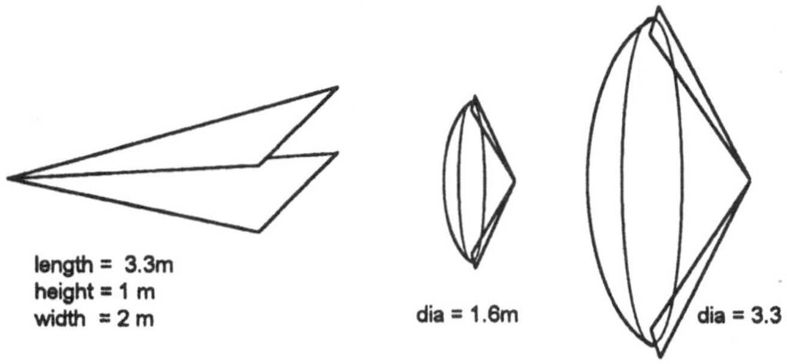

Figure 11. TEM horn and two sizes of IRA for comparison. The center IRA (small) has the same aperture area as the TEM horn, while the right IRA (medium) has a diameter equal to the length of the TEM horn.

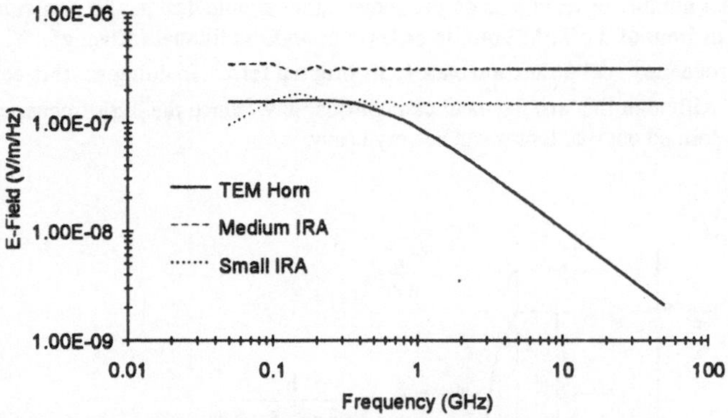

Figure 12. Step responses of the TEM horn and two IRA's at 1 km.

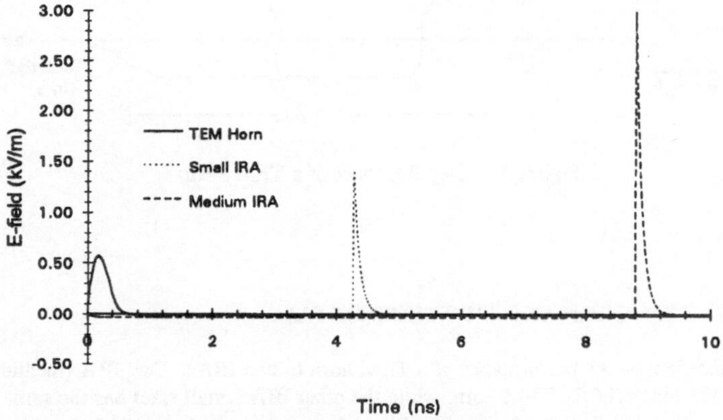

Figure 13. Time domain response of the three antennas at 1 km.

CONCLUSION

We have described in detail the behavior of Impulse Radiating Antennas in a number of different ways. In addition, we have compared the response of IRAs to TEM horns. We find that IRAs are best for radiating fast risetime pulses because their step response does not roll off at the high end.

It is appropriate to note a number of caveats in the analysis presented here. We have assumed an ideal matching circuit in the IRA, as well as an idealized feed circuit in the TEM horn. We have made no attempt to design the matching circuits or baluns, (although some ideas on baluns for IRAs are provided by Baum[10]). Furthermore, the high-voltage properties of the two antennas have been ignored. Nevertheless, we can draw some important conclusions from the above data. The data suggest that an IRA is a promising antenna for radiating fast transient pulses. When comparing antennas of similar size (largest dimension), the IRA has a considerable advantage in performance. When comparing antennas of similar aperture area, the IRA still has clear advantages at the high end, since it does not suffer the high-frequency rolloff problem of the TEM horn. Finally, the IRA has some desirable directionality at low frequencies.

REFERENCES

1. E. G. Farr and J. S. Hofstra, An Incident Field Sensor for EMP Measurements, Sensor and Simulation Note 319, November 6 1989.
2. C. E. Baum, General Properties of Antennas, Sensor and Simulation Note 330, July 23 1991.

3. C. E. Baum, Radiation of Impulse-Like Transient Fields, Sensor and Simulation Note 321, November 25 1989.
4. C. E. Baum, Aperture Efficiencies for IRAs, Sensor and Simulation Note 328, June 24 1991, and 1992 APS/URSI/NEM Symposium Proceedings, July 20-24, 1992, Chicago.
5. W. R. Smythe, *Static and Dynamic Electricity*, Third Ed., Hemisphere, 1989, p. 76 and p. 460.
6. E. G. Farr and C. E. Baum, Prepulse Associated with the TEM Feed of an Impulse Radiating Antenna, Sensor and Simulation Note 337, March 1992.
7. C. E. Baum, Impedances and Field Distributions for Symmetrical Two Wire and Four Wire Transmission Line Simulators, Sensor and Simulation Note 27, October 10 1966.
8. E. G. Farr and C. E. Baum, A Simple Model of Small-Angle TEM horns, Sensor and Simulation Note 340, May 1992.
9. C. E. Baum and A. P. Stone, *Transient Lens Synthesis*, Hemisphere, 1991.
10. C. E. Baum, Configurations of TEM Feed for an IRA, Sensor and Simulation Note 327, April 27, 1991.

ACCURATE MODELING OF ANTENNAS FOR RADIATING SHORT PULSES, FDTD ANALYSIS AND EXPERIMENTAL MEASUREMENTS

James G. Maloney* and Glenn S. Smith

School of Electrical Engineering
Georgia Institute of Technology
Atlanta, GA 30332

INTRODUCTION

Antennas used to radiate short pulses often require different design rules than those that are used to radiate essentially time-harmonic signals. If one considers only the performance in the time-domain, i.e., time varying waveforms, then the following are reasonable criteria for the performance of the antenna: (i) The radiated pulse should be a faithful reproduction of the excitation; i.e., there should be little pulse distortion on radiation. (ii) The reflected signal at the input of the antenna should be small. For practical systems, the peak amplitude of the reflected signal typically is required to be 40 dB below that of the incident signal. (iii) The amplitude of the radiated signal in the desired direction should be as large as possible.

In the past, empirical approaches were often used to design an antenna that satisfied these criteria [1]. An exception is the resistively loaded dipole/monopole where an analytical approach was used to obtain a profile for the resistive loading that produced an outward traveling wave on the antenna, i.e., greatly reduced the reflections from the open ends [2],[3]. Resistive loading has also been used to reduce the reflections in other antennas, e.g., TEM horns [4].

The finite-difference time-domain (FDTD) method is a very flexible numerical approach that can be used to treat a variety of electromagnetic problems in the time domain. It is well suited to the analysis and design of antennas for radiating short pulses; however, several advances had to be made before the method could be applied to this problem, these include: an accurate formulation of the antenna problem suitable for FDTD solution [5], radiation boundary conditions to effectively truncate the space surrounding the antenna [6], efficient transforms for obtaining far-field information from the calculated near field [5],[7], and efficient methods for including resistive materials in the FDTD method (good conductors via surface impedance concepts and

*Now with the Georgia Tech Research Institute, Signature Technology Laboratory.

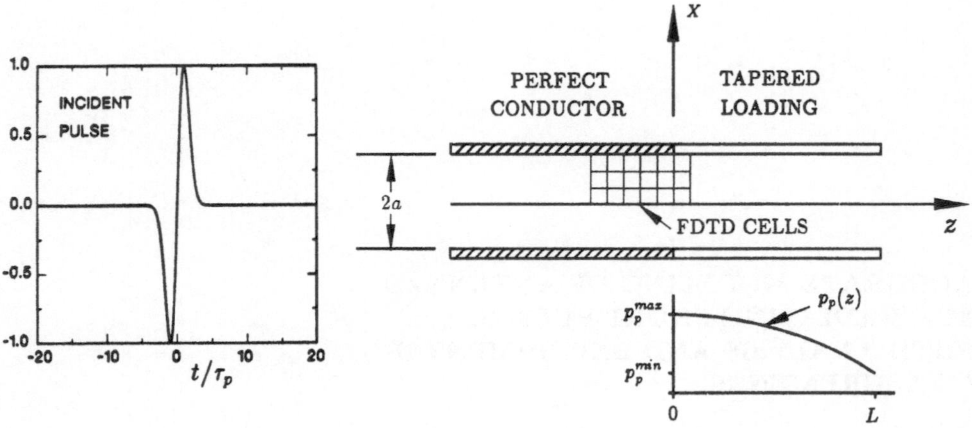

Figure 1. Geometry for the parallel-plate radiator.

electrically-thin sheets via sub-cell models) [8],[9].

In this paper, we will illustrate the use of the FDTD method with two antennas designed for the radiation of short pulses. The first is a simple, two-dimensional geometry, an open-ended parallel-plate waveguide, while the second is a three-dimensional, rotationally symmetric geometry, a conical monopole fed through an image plane by a coaxial transmission line. Both antennas are "optimized" according to the criteria stated above by adjusting geometrical parameters and including resistive loading that varies continuously with position along the antenna. The predicted performance for the conical monopole antenna is compared with experimental measurements; this verifies the optimization and demonstrates the practicality of the design.

DISCUSSION OF EXAMPLES

Parallel-Plate Radiator

Fig. 1 shows the geometry for the parallel-plate radiator. The excitation for this example is a TEM mode in the perfectly conducting section of the waveguide whose electric field, shown on the left of Fig. 1, is a differentiated Gaussian pulse with characteristic time τ_p. The waveguide is resistively loaded over a section of length L at the end, so that the loss tangent varies quadratically with position z (see the inset in Fig. 1 which shows the loss tangent p_p at the peak frequency of the pulse). In this section of the waveguide, the plates are assumed to be several skin depths thick at all significant frequencies in the pulse, so that they can be modeled by a surface impedance boundary condition [8],[10].

The three parameters, p_p^{max}, p_p^{min}, and L/a that determine the quadratic profile were varied in a parametric study, and values were selected that optimize the performance of the antenna according to the three criteria stated earlier ($p_p^{max} = 31.6$, $p_p^{min} = 10.0$, $L/a = 10.0$). Fig. 2 compares results for the optimized, resistively loaded radiator with those for a similar radiator with perfectly conducting plates; both the radiated pulse in the direction of the z axis and the reflected signal in the waveguide are shown. Notice that the optimization has improved the shape of the radiated pulse. It now closely resembles the excitation, the differentiated Gaussian pulse, as required by criterion (i). Also notice that the optimization has significantly reduced the reflection within the waveguide (by a factor of about 5), as required by

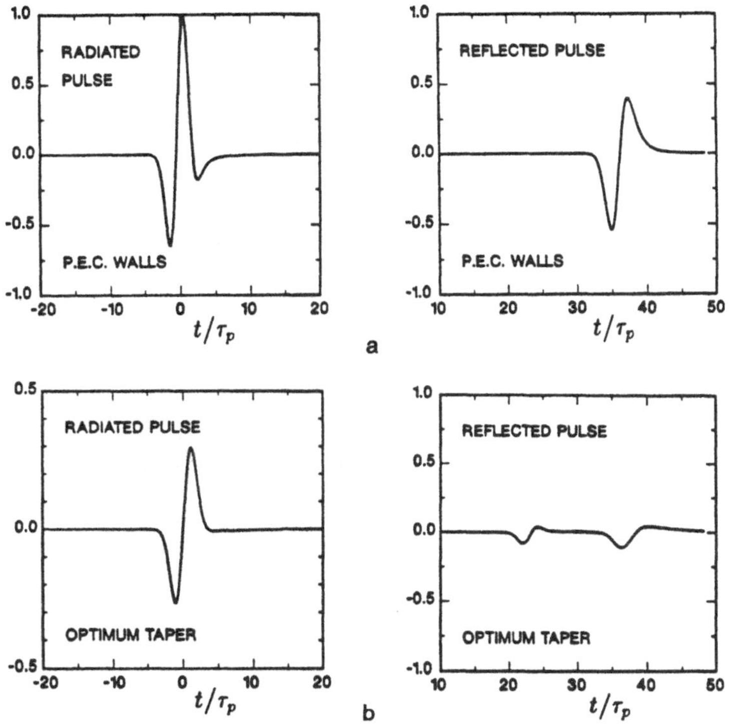

Figure 2. Radiated and reflected pulses for the parallel-plate radiator. (a) Perfectly conducting antenna. (b) Resistively loaded antenna.

criterion (ii). The resistive loading, however, has reduced the peak amplitude of the radiated pulse by about a factor of 3.

The magnitude of the electric field, $|\vec{\mathcal{E}}|$, in the space interior to and surrounding the optimized antenna is displayed using a gray scale in Fig. 3. The plot on the left is for a time when the incident pulse has not reached the end of the waveguide, while the plot on the right is for a time after the pulse has been radiated from the end of the waveguide. Graphs of the surface charge density on the inside and outside of the top plate of the waveguide are below each plot. The six graphs surrounding the plot on the right show the temporal variation of the radiated (far-zone) electric field.

Conical Monopole

The previous example illustrates the use of the FDTD method in the optimization of a radiator for short pulses; however, it is for an ideal, two-dimensional structure. We will now consider a more practical radiator, the resistively loaded, conical monopole antenna shown in Fig. 4, a rotationally symmetric, three-dimensional structure.

The conical monopole is fed through an image plane by a coaxial transmission line. The electrically thin sheet forming the side of the cone has a resistance per square, R, that varies continuously with position, l/h_r (see the inset in Fig. 4). This produces an internal impedance per unit length for the conical surface that is proportional to the reciprocal of the distance from the end of the cone $(1 - l/h_r)$. The thin sheet is efficiently modeled in the FDTD method using a technique described in reference [9].

The first step in the optimization was to reduce the reflections from the drive point

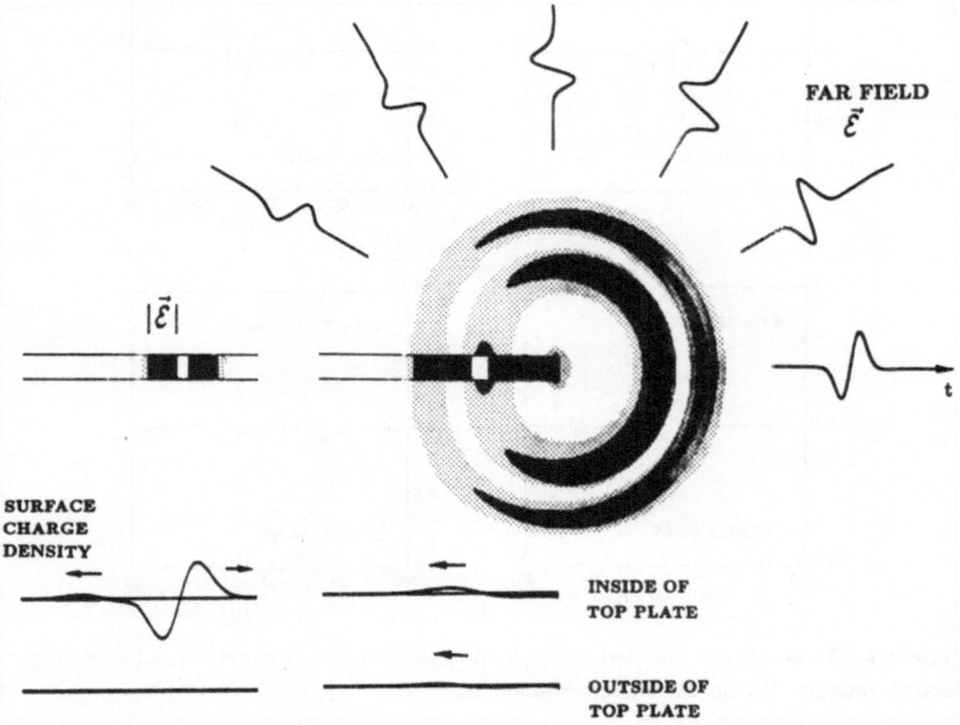

Figure 3. Radiation from the optimized, resistively loaded, parallel-plate radiator.

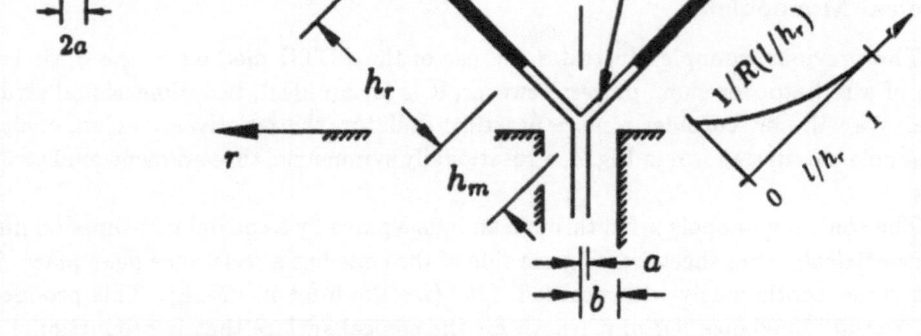

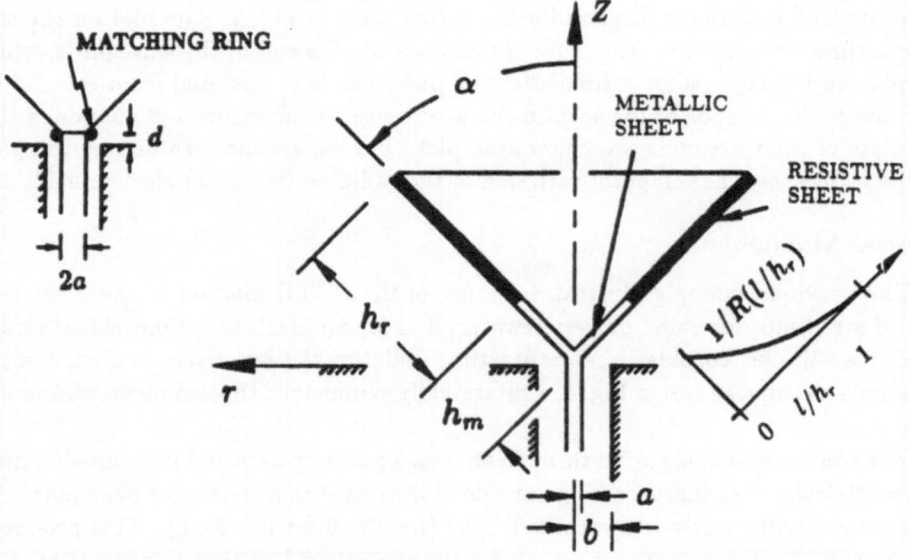

Figure 4. Geometry for the conical monopole antenna.

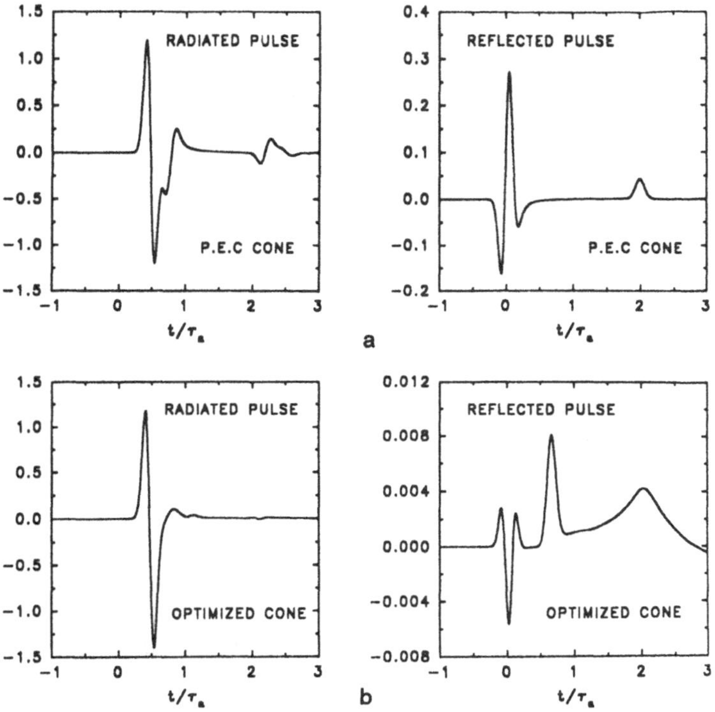

Figure 5. Radiated and reflected pulses for the conical monopole antenna. (a) Perfectly conducting cone. (b) Optimized, resistively loaded cone.

of the antenna: The half angle of the cone ($\alpha = 47^o$) was chosen to make the characteristic impedance of the conical and coaxial transmission lines the same ($b/a = 2.30$, $Z_c = 50\ \Omega$). The height of the conical apex above the image plane was adjusted ($d/a = 1/\tan\alpha$), and a small, metallic, matching ring was added at the junction (see the inset in Fig. 4). Next, the reflection from the end of the cone was reduced by adjusting the parameters for the resistive loading ($R(0) = 25\ \Omega$, $h_r/h_m = 2.0$).

Fig. 5 compares results for the optimized, resistively loaded cone with those for a similar perfectly conducting cone. The excitation is a TEM mode in the coaxial line whose electric field is again a differentiated Gaussian pulse with characteristic time τ_p. Both the radiated pulse at broadside ($\theta = 90^o$) and the reflected signal in the coaxial line are shown as functions of the normalized time, t/τ_a. Here, $\tau_a = (h_m + h_r)/c$ is the time required for light to travel the slant height of the cone, and $\tau_p/\tau_a = 6.43 \times 10^{-2}$. Notice that the optimization has improved the performance of the antenna: The radiated pulse now resembles the excitation (differentiated Gaussian pulse), as required by criterion (i); the reflected voltage within the coaxial line has been significantly reduced (peak amplitude reduced by 30 dB), as required by criterion (ii). Also notice that the amplitude of the radiated pulse has not been reduced by the inclusion of resistive material in the optimized antenna (it is essentially the same for both the optimized and perfectly conducting cones); thus, criterion (iii) is satisfied.

The magnitude of the electric field, $|\vec{\mathcal{E}}|$, in the space surrounding these antennas is displayed using a gray scale in Fig. 6 for three times. The top half of each plot is for the optimized cone, while the bottom half is for the perfectly conducting cone. It is interesting to compare the fields for these two antennas. In Fig. 6(a) a wavefront has moved out along the cone; for the optimized cone, this wavefront has "leaked"

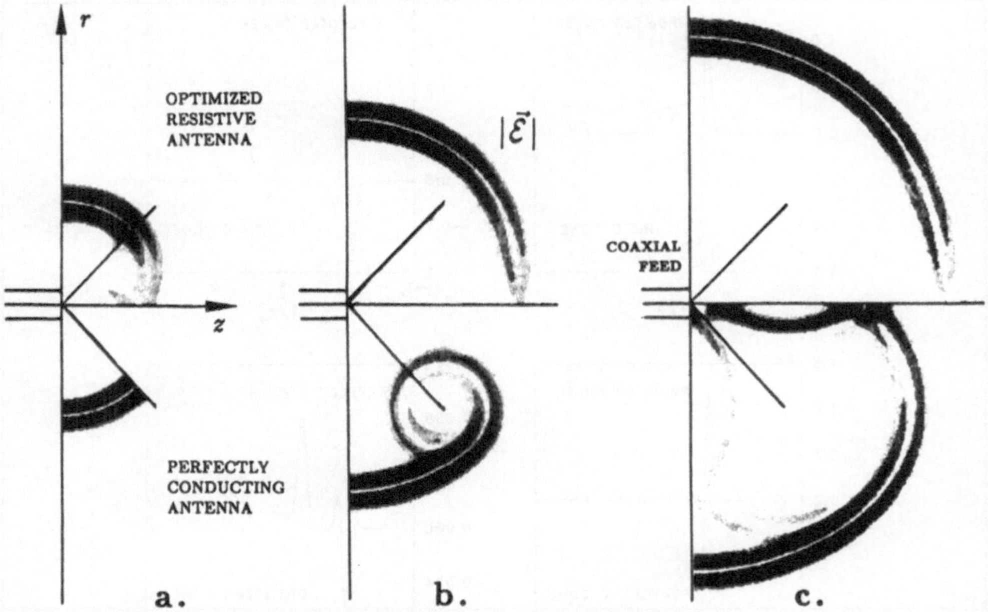

Figure 6. Radiation from conical monopole antennas. The top half of each plot is for the optimized, resistively loaded antenna while the bottom half is for the perfectly conducting antenna.

through the resistive sheet to appear in the interior of the hollow cone; while for the perfectly conducting cone, this wavefront has been confined to the space between the cone and the image plane. In Fig. 6(b) the primary wavefront has "slipped" off the end of the optimized cone without producing any substantial additional radiation, while a reflection has occurred at the end of the perfectly conducting cone producing a secondary torroidal wavefront. In Fig. 6(c) the primary wavefront has been "cleanly" radiated from the optimized cone, while multiple reflections have occurred on the perfectly conducting cone producing several additional wavefronts. These results show that the optimized, resistive cone "works" because it allows the primary wavefront to continually "leak" through the side of the cone, thereby producing no significant reflection when it encounters the end.

EXPERIMENTAL RESULTS

The flexibility in the FDTD method allows the modeling of the fine structural details of experimental antennas; thus, it is reasonable to expect the theoretical calculation made with the FDTD method to be in excellent agreement with experimental measurements. This agreement has been demonstrated in a series of publications which began with results for metallic, cylindrical and conical monopoles [5], followed by results for cylindrical monopoles with uniform resistive loading [9] and tapered resistive loading [3]. In the last case, the resistive loading in the experimental model was obtained by soldering end-to-end a group of discrete, high-frequency, precision resistors. The flexibility of the FDTD method allowed the geometrical and electrical details of each resistor to be included in the analysis, and excellent agreement was obtained with the measured results. Most recently, results were obtained for conical monopole antennas with uniform and tapered resistive loadings [11],[12], and representative examples from these results will be discussed next.

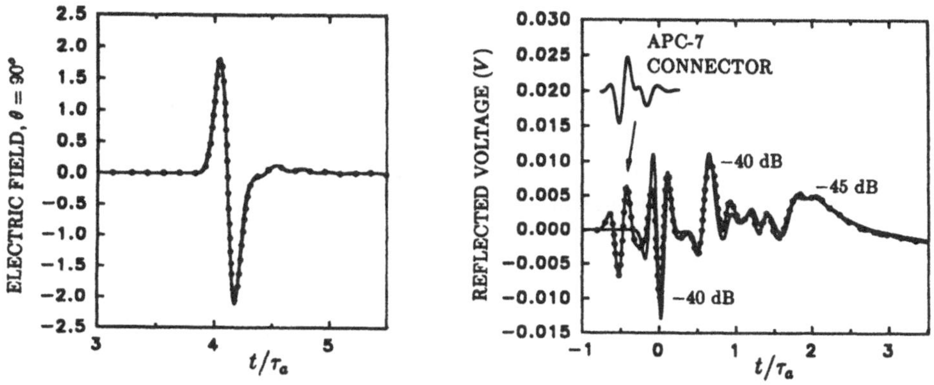

Figure 7. Comparison of the FDTD theoretical (dots) and the measured (solid line) results for the optimized, resistively loaded conical monopole antenna.

Fig. 7 compares the measured (dots) and FDTD (solid) results for an optimized, resistively loaded, conical monopole antenna. This antenna is similar to the one discussed above, except that the resistive section of the cone did not have a continuously varying resistance per square. Instead, it was formed from six individual sections, each made from a sheet of material (carbon-black filled, conductive, polycarbonate film) with a fixed value of resistance per square. In other words, a "discretized" profile was used instead of a "continuous" profile. The details of the discretization were included in the FDTD analysis.

In Fig. 7 the measured electric field at broadside ($\theta = 90^o$) is seen to be in excellent agreement with the theoretical results. The measured reflected voltage in the coaxial line is also in excellent agreement with the theoretical results, and all measured reflections are below the desired level, −40 dB relative to the peak value of the incident pulse. Several reflections can be identified in this waveform: at $t/\tau_a \approx 0.0$ the reflection from the drive point, at $t/\tau_a \approx 0.67$ the reflection from the beginning of the resistive section of the cone, for $0.8 \leq t/\tau_a \leq 1.8$ reflections from the discontinuities in resistance of the discretized profile, and finally at $t/\tau_a \approx 2$ the reflection from the end of the cone. Notice that for $t/\tau_a \leq -0.3$, an additional reflection occurs in the measurements where none appears in the theory. This reflection is due to the connectors (APC-7) which join the antenna to the network analyzer; thus it is not directly associated with the antenna. For the purpose of comparison, the measured reflection from a connector pair is also shown in the figure. Also, notice that the internal reflections for the antenna are now comparable to those from the pair of precision connectors. The small discrepancies that remain between the measured and FDTD reflected voltage are most likely the result of imperfections in the construction and approximations in the characterization of the experimental model.

CONCLUSION

Although the FDTD method has been discussed in the literature for over 25 years, it has only recently been applied to radiating structures, i.e, antennas; the first results being published around 1989 [5],[13]. Since then, it has been applied to a variety of antennas [3],[9],[11],[14],[15], and when care is taken to make the theoretical model an accurate representation of the experimental model, excellent agreement between theory and experiment is always obtained.

The method is currently being used in an interactive mode to design antennas for radiating short pulses. In this approach, graphics, like the gray scale plots shown in Figs. 3 and 6 or similar color movies, are used to identify regions on the antenna that need improvement, such as those that produce reflections. Modifications are then introduced and the plots re-made and examined. The procedure is repeated until the desired performance is obtained. This approach was used in optimizing the conical monopole antenna described above.

ACKNOWLEDGMENT

This work was supported in part by the Joint Services Electronics Program under contracts DAAL03-87-K-0059, DAAL-03-90-C-0004 and DAAL-03-89-G-0078

REFERENCES

1. R. J. Wohlers, "The GWIA, an Extremely Wide Bandwidth Low-Dispersion Antenna," Calspan Corp., Buffalo, NY, 1971.
2. T. T. Wu and R. W. P. King, "The Cylindrical Antenna with Nonreflecting Resistive Loading," *IEEE Trans. Antennas Propagat.*, vol. AP-13, pp. 369-373, May 1965.
3. J. G. Maloney and G. S. Smith, "The Role of Resistance in Broadband, Pulse-Distortionless Antennas," *Proc. 1991 IEEE AP-S Int. Symp.*, London, Ontario, June 1991, vol. 2, pp. 707-710; also "A Study of Transient Radiation from the Wu-King Resistive Monopole - FDTD Analysis and Experimental Measurements," submitted for publication to *IEEE Trans. Antennas Propagat.*
4. S. Evans and F. N. Kong, "TEM Horn Antenna: Input Reflection Characteristics in Transmission," *IEE Proc.*, vol. 130, pt. H, pp. 403-409, Oct. 1983.
5. J. G. Maloney, G. S. Smith, and W. R. Scott, Jr., "Accurate Computation of the Radiation from Simple Antennas Using the Finite-Difference Time-Domain Method," *Proc. 1989 IEEE AP-S Int. Symp.*, San Jose, CA, June 1989, pp. 42-45; also *IEEE Trans. Antennas Propagat.*, vol. 38, no. 7, pp. 1059-1068, July 1990.
6. J. Fang, "Time Domain Finite Difference Computation for Maxwell's Equations," Ph.D. Dissertation, University of California at Berkeley, Nov. 1989.
7. K. S. Yee, D. Ingham, and K. Shlager, "Time-Domain Extrapolation to the Far Field Based on FDTD Calculations," *IEEE Trans. Antennas Propagat.*, vol. 39, pp. 410-413, Mar. 1991.
8. J. G. Maloney and G. S. Smith, "The Use of Surface Impedance Concepts in the Finite-Difference Time-Domain Method," *IEEE Trans. Antennas Propagat.*, vol. 40, pp. 38-48, Jan. 1992.
9. J. G. Maloney and G. S. Smith, "The Efficient Modeling of Thin Material Sheets in the Finite-Difference Time-Domain (FDTD) Method," *IEEE Trans. Antennas Propagat.*, vol. 40, pp. 323-330, Mar. 1992.
10. J. G. Maloney and G. S. Smith, "Optimization of Pulse Radiation from a Simple Antenna using Resistive Loading," *Microwave and Optical Technology Letters*, vol. 5, pp. 299-303, June 1992.
11. J. G. Maloney and G. S. Smith, "Optimization of a Resistively Loaded Conical Antenna for Pulse Radiation," *Proc. 1992 IEEE AP-S Int. Symp.*, Chicago, IL, vol. 4, July 1992, pp. 1968-1971; also "Optimization of a Conical Antenna for Pulse Radiation: An Efficient Design Using Resistive Loading," submitted for publication to *IEEE Trans. Antennas Propagat.*
12. J. G. Maloney, "Analysis and Synthesis of Transient Antennas using the Finite-Difference Time-Domain (FDTD) Method," Ph.D. Dissertation, Georgia Institute of Technology, Atlanta, GA, 1992.
13. A. Reineix and B. Jecko, "Analysis of Microstrip Patch Antennas Using the Finite-Difference Time-Domain Method," *IEEE Trans. Antennas Propagat.*, vol. 37, pp. 1361-1369, Nov. 1989.
14. D. S. Katz, M. J. Piket-May, A. Taflove, and K. R. Umashankar, "FDTD Analysis of Electromagnetic Wave Radiation from Systems Containing Horn Antennas," *IEEE Trans. Antennas Propagat.*, vol. 39 , pp. 1203-1212, Aug. 1991.
15. P. A. Tirkas and C. A. Balanis, "Finite-Difference Time-Domain Method for Antenna Radiation," *IEEE Trans. Antennas Propagat.*, vol. 40, pp. 334-340, Mar. 1992.

WIDE-BANDWIDTH RADIATION FROM ARRAYS OF ENDFIRE TAPERED SLOT ANTENNAS

Daniel H. Schaubert

Electrical & Computer Engineering
University of Massachusetts
Amherst, MA 01003

INTRODUCTION

Endfire tapered slot antennas, also known as Vivaldi or notch antennas, have been demonstrated to be effective radiators of wide-bandwidth signals [1]-[4]. The antennas can be used singly or in arrays, and they can be designed to operate at microwave or millimeter-wave frequencies. Although many types of "frequency-independent" antennas have been developed, endfire slot antennas are unique in their potential for use in wide scanning arrays where element spacings less than one-half wavelength at the highest frequency are required.

Endfire tapered slot antennas are frequently fabricated by etching a flared slot pattern in the metallization of a microwave substrate as indicated in Figure 1, but self-supporting metallic fins without any dielectric substrate are used also. The narrow slotline at the antenna's feed can be connected to a variety of transmission lines. Figure 1 illustrates the use of a strip transmission line that couples to the slotline via a wide-band transition.

When used as single elements, the polarization and spatial distribution of the signal (radiation pattern) are important characteristics. Experimental and calculated results showing how these characteristics depend upon antenna parameters such as length and flare angle are presented for linearly tapered slot antennas. Cross-polarization levels in the diagonal plane of the antennas are quite high and are illustrated. When used in an array to achieve more directive radiation or electronic beam pointing, active element patterns are an important characteristic. Several examples will be presented to illustrate that good scan performance can be achieved in arrays with element spacings less than one-half wavelength at the upper frequency. Examples that produce a scan blindness are also presented.

SINGLE ELEMENT RADIATION CHARACTERISTICS

Endfire tapered slot antennas display different radiation characteristics depending upon their size. When used as single elements, the antennas often are designed to produce a reasonably directive beam, which is achieved by making the antenna length L (see Fig. 1) to be 3-10 wavelengths and the aperture height A to be 1-3 wavelengths. The antenna then operates somewhat like a traveling wave radiator and produces an endfire beam with directivity that increases slowly with frequency. Empirical studies have provided design criteria for antennas that can be used in this way [5]-[6].

When the length of the tapered slot antenna is less than approximately one wavelength and its aperature height is less than approximately one-third wavelength, the antenna radiates a broad beam, which becomes somewhat omnidirectional as the antenna size decreases further. In this regime, the antenna is useful for wide-band, wide-scan phased arrays. For antenna dimensions between the wide-beam case and the traveling wave case, the antenna passes through a resonance region where the continuous transition from one case to the other is occasionally disrupted by resonance effects. Asija [7] has studied this region by using a

Ultra-Wideband, Short-Pulse Electromagnetics
Edited by H. Bertoni *et al.*, Plenum Press, 1993

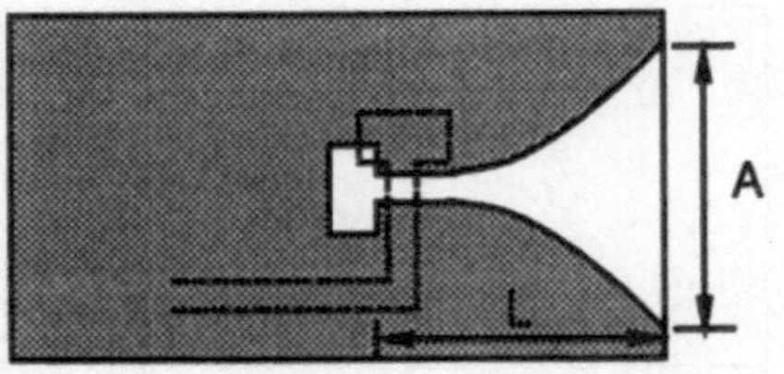

Figure 1. Endfire tapered slot antenna with strip feed etched from metallization on micorwave substrate.

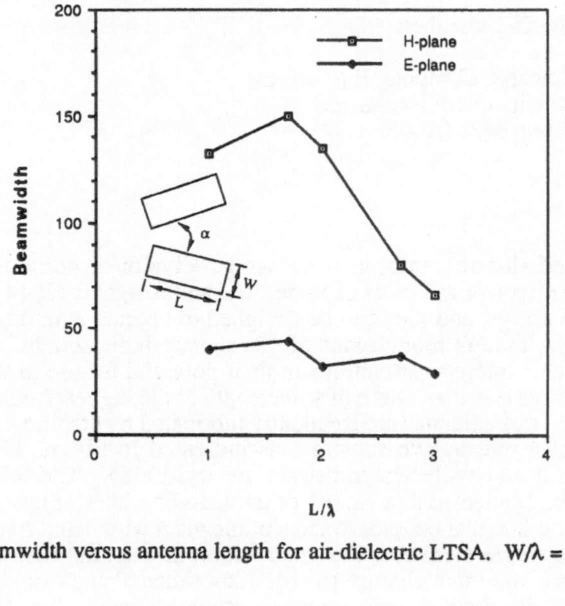

Figure 2. Beamwidth versus antenna length for air-dielectric LTSA. $W/\lambda = 0.33$, $\alpha = 10^{o}$.

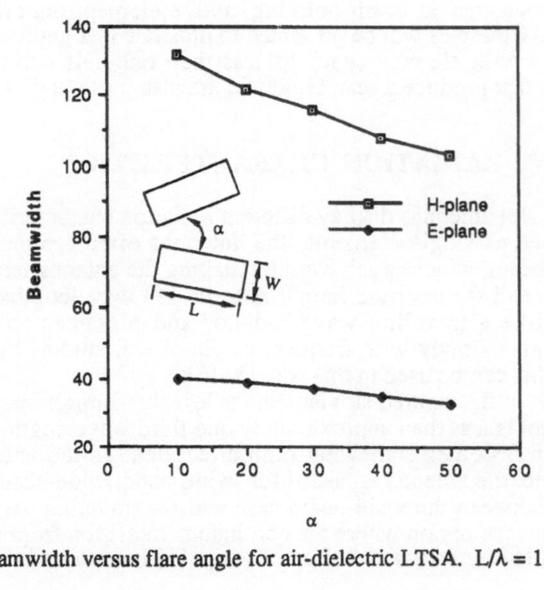

Figure 3. Beamwidth versus flare angle for air-dielectric LTSA. $L/\lambda = 1.0$, $W/\lambda = 0.33$.

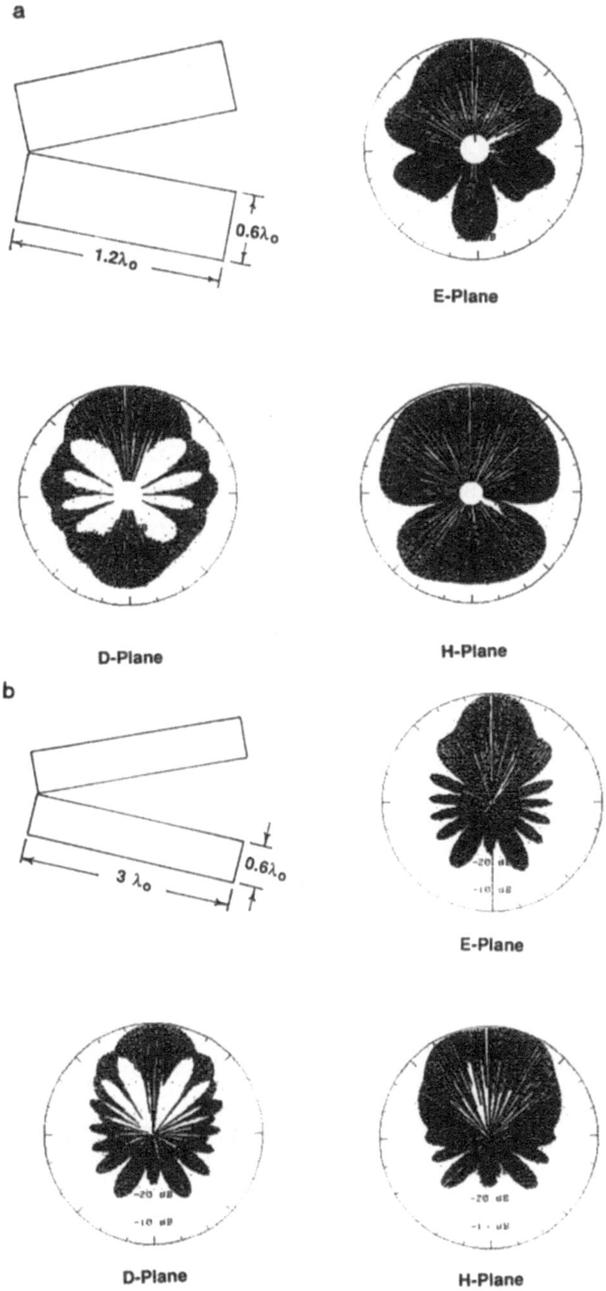

Figure 4. Spin-linear radiation patterns of simplified air-dielectric LTSAs. (a) 1.2-λ_0 antenna. (b) 3.0-λ_0 antenna.

moment method analysis of thin conducting fins forming a linearly tapered slot antenna (LTSA) with air dielectric (no substrate). His results for beamwidth variations with antenna length and flare angle are shown in Figures 2 and 3.

One of the least desirable aspects of endfire tapered slot antennas is the high levels of cross-polarization that they radiate in the diagonal plane. Measured spin-linear radiation patterns for two simplified antennas are shown in Figure 4. The longer antenna produces a narrower beamwidth, but the polarization characteristics of the antennas are similar. Specifically, the antennas are linearly polarized in the principal planes, but in the diagonal plane, the axial ratio decreases rapidly at scan angles away from boresight. The axial ratio is typically about 10 dB at the angle corresponding to -3 dB of the copolarized signal. A feature

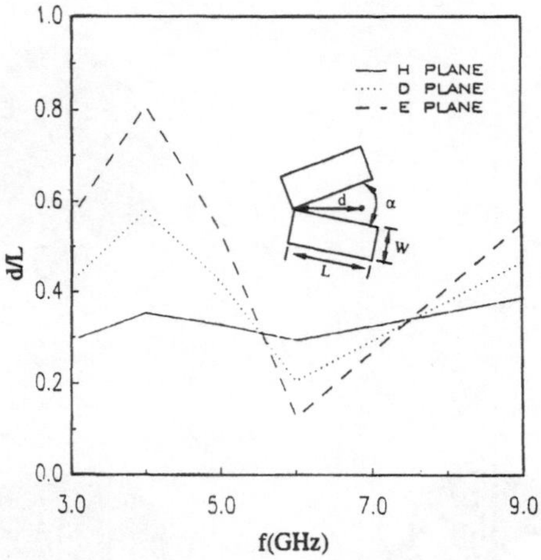

Figure 5. Phase center location of typical air-dielectric, simple LTSA. L = 6 cm, W = 3 cm, $\alpha = 14^{\circ}$.

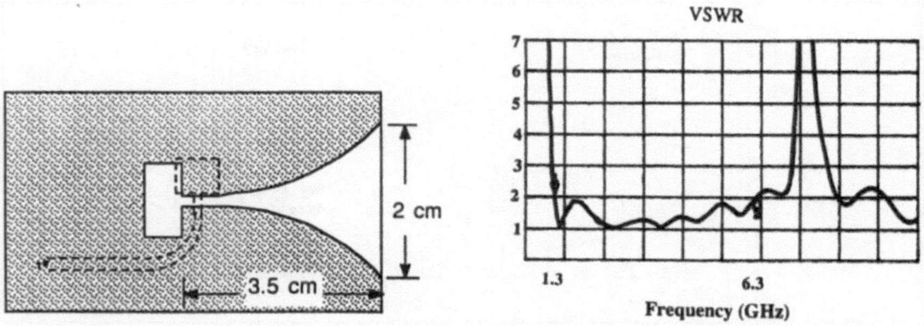

Figure 6. VSWR of Vivaldi notch element with stripline feed.

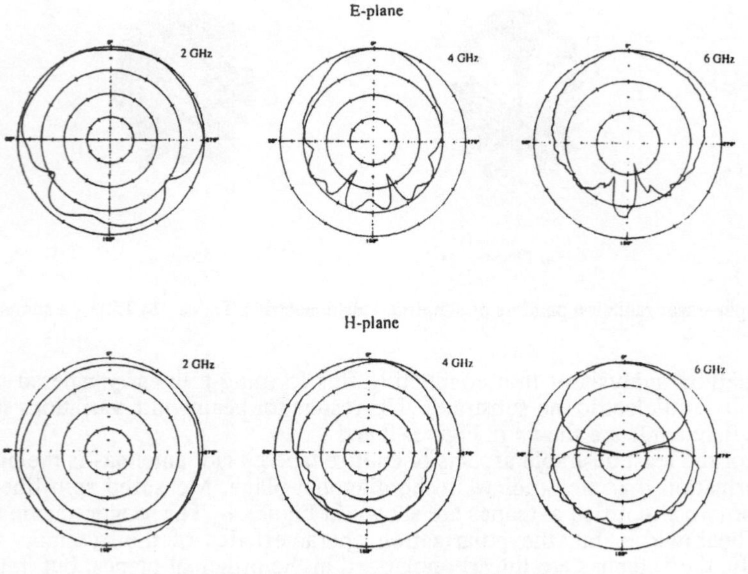

Figure 7. Radiation patterns of antenna in Figure 6.

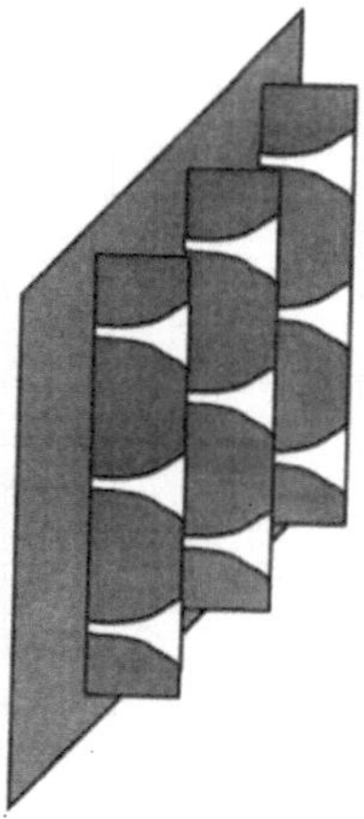

Figure 8. Portions of infinite array of Vivaldi antennas.

closely related to the cross-polarization in the diagonal plane is the location of the antenna's phase center. Figure 5 is taken from [7] and it shows that the phase centers for three scan planes are located at different distances along the antenna's axis for most frequencies. In addition to the high cross-polarized radiation associated with such differences in phase center location, this particular antenna would not be a very good wide-band feed for a reflector because it would result in significant defocussing in the E-plane at frequencies where the phase center is far from its nominal location.

One of the most desirable aspects of endfire tapered slot antennas is their ability to radiate most of the power delivered to them even though they are electrically small. The antenna in Figure 6 is well matched at 1.5 GHz where it is only 0.17 wavelengths long and 0.10 wavelengths high at the aperture. The radiation patterns of the antenna are reasonably well behaved over a 3:1 frequency band, except for a narrowing of the E-plane beamwidth in the vicinity of 4 GHz (Fig. 7).

RADIATION FROM INFINITE ARRAYS

As an initial step toward understanding the performance of large arrays of endfire tapered slot antennas, infinite arrays are being analyzed by using a method of moments technique [8]. A portion of an infinite array is depicted in Figure 8. The analysis is general enough to treat dipole and monopole arrays and the results have been compared to published results for these limiting cases. Also, the analysis of LTSA arrays has agreed well with waveguide simulator results for H-plane scan. However, most of the cases presented in this paper cannot be validated with simulations, so independent verification has not yet been obtained. Nonetheless, several of the observed phenomena agree, at least qualitatively, with results that have been described by other researchers for large scanning arrays so the results are believed to be correct.

The radiation characteristics of the tapered slot arrays have been evaluated by computing the radiated field per unit cell of the infinite array. This approach, as opposed to simply computing the active reflection coefficient, provides information about the polarization of the radiated field. The first sets of data show active element gain at broadside versus frequency for three element shapes. Each case is normalized to its maximum value. The antennas being considered are depicted in Figure 9 and the results are shown in Figure 10. Antennas 2 and 3 display an anomaly that appears to be a blind spot at broadside.

At frequencies not associated with scan anomalies, the antennas generally display a well behaved active element pattern with half-power beamwidths between 45 and 60 degrees. A few typical results are shown in Figures 11 and 12. These figures also contain data about the polarization of the radiated field. As is the case for single elements, the field is linearly polarized in the principal planes, but in the diagonal plane the axial ratio decreases rapidly as the beam scans away from broadside. The axial ratio is of the order of 10 dB at the angle

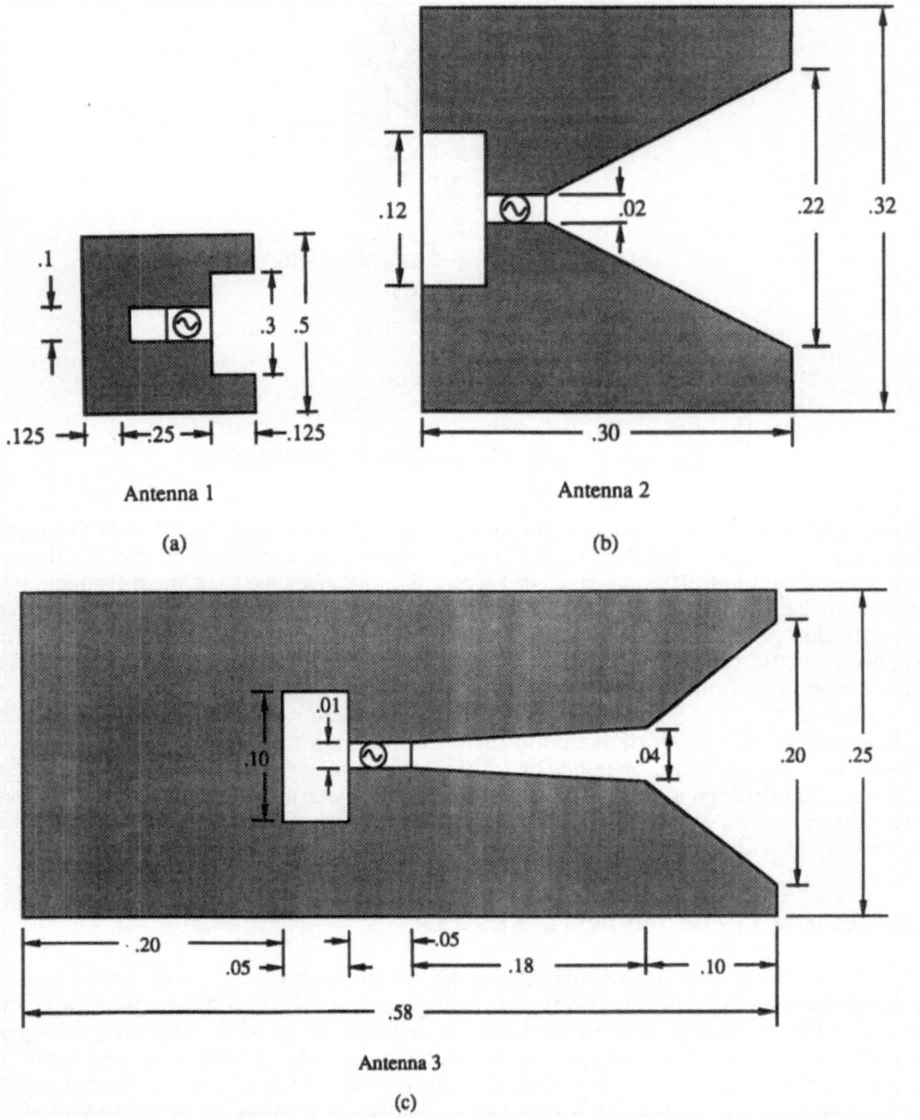

Figure 9. Unit cell of arrays that are evaluated. All dimensions are in meters. (a) Antenna 1, a = b = 0.5; (b) Antenna 2, a = b = 0.32; (c) Antenna 3, a = b = 0.25.

corresponding to -3 dB of the "copolarized" component, which is approximately aligned with the direction of the metallic fins (vertical in Fig. 8).

A radiation pattern displaying scan blindness is shown in Figure 13. The frequency of this pattern is very close to that for which the blindness occurs at broadside (see Fig. 10), so it appears that these anomalies are confined to narrow frequency bands within the operating range of the antennas.

SUMMARY

Endfire tapered slot antennas can be used singly or as elements in phased arrays. The antennas exhibit stable radiation characteristics over bandwidths of one octave or more except for isolated anomalies that occur in large arrays. The results of computer analysis of infinite arrays are being used in an attempt to identify the relationship between these anomalies and the physical parameters of the array. It is hoped that this will lead to effective methods for designing arrays that are free of scan blindness problems.

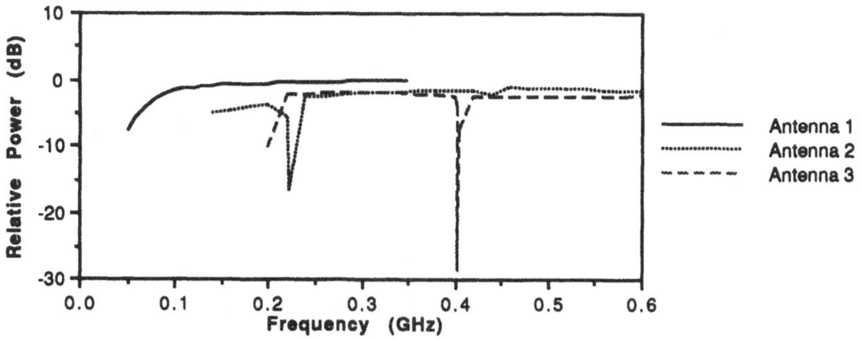

Figure 10. Relative gain in brodside direction for antennas in Figure 9. Curves for antennas 2 and 3 are offset 1 and 2 dB, respectively, for clarity.

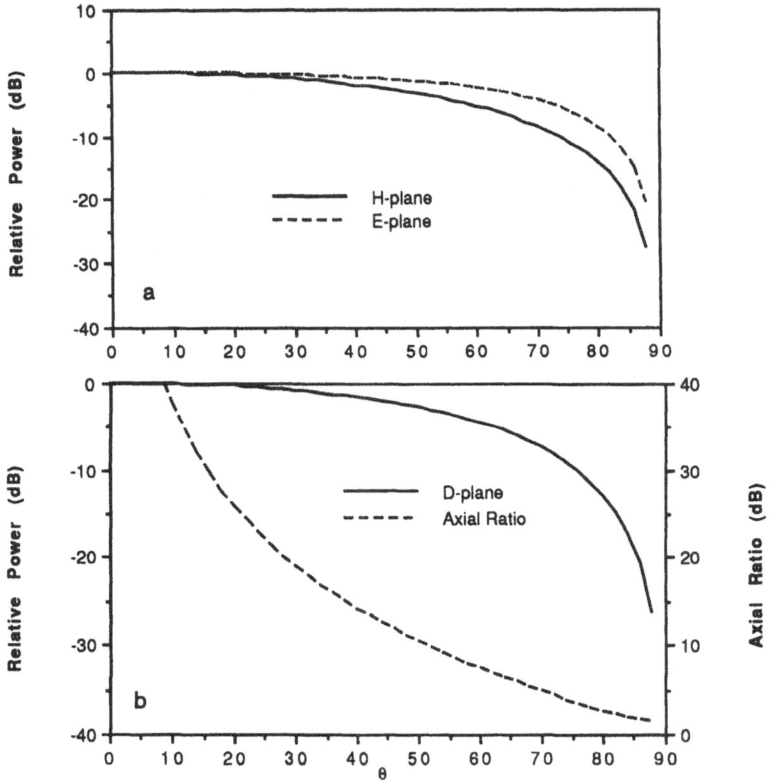

Figure 11. Active element radiation patterns of antenna 1 at 0.20 GHz, Z_g = 100 ohms. (a) E- and H-plane. (b) D-plane and axial ratio in D-plane.

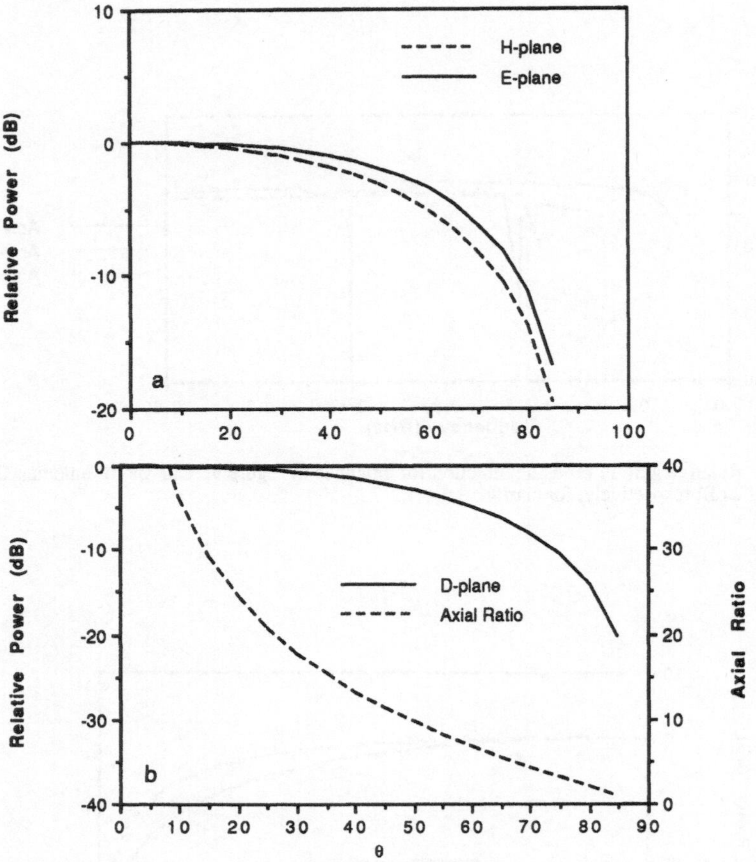

Figure 12. Active element pattern of antenna 2 at 0.3516 GHz, Z_g = 100 ohms. (a) E- and H-plane. (b) D-plane and axial ratio in D-plane.

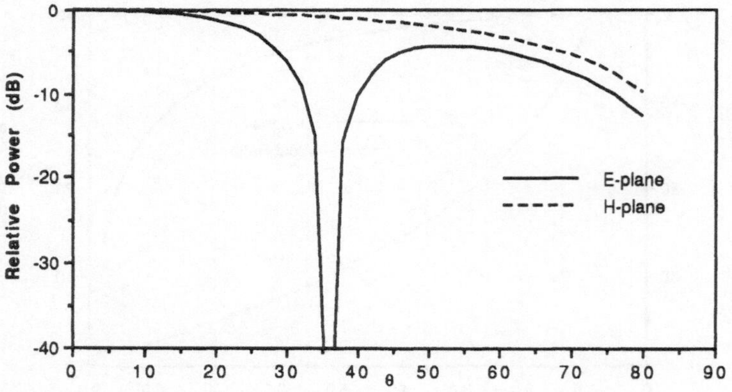

Figure 13. Principal plane active element patterns of antenna 2 at 0.448 GHz.

ACKNOWLEDGEMENTS

This work was supported in part by the U.S. Army Research Office, contract number DAAL03-92-G-0295, and by the University of Massachusetts.

REFERENCES

1. L.R. Lewis, M. Fasset and J. Hunt, "A broadband stripline array element", *Digest of 1974 IEEE Symp. Ant. Prop.*, 335-337, Atlanta, GA.
2. P.J. Gibson, "The Vivaldi aerial", *Digest of the 9th Eur. Microw. Conf.*, 120-124, Brighton, UK, 1979.
3. E.L. Kollberg, J. Johansson, T. Thungren, T.L. Korzeniowski and K.S. Yngvesson, "New results on tapered slot endfire antennas on dielectric substrates", *8th Int'l. Conf. Infrared & mmwaves*, Miami, FL, 12-17, Dec. 1983.
4. D.H. Schaubert, "Endfire slotline antennas", *JINA'90 Digest*, pp. 253-265, November 1990.
5. K.S. Yngvesson, D.H. Schaubert, T.L. Korzeniowski, E.L. Kollberg, T. Thungren and J. Johansson, "Endfire tapered slot antennas on dielectric substrates", *IEEE Trans. Ant. Prop.*, AP-33, 1392-1400, December 1985.
6. K.S. Yngvesson, T.L. Korzeniowski, Y.S. Kim, E.L. Kollberg and J. Johansson, "The tapered slot antenna – a new integrated element for millimeter wave applications", *IEEE Trans. Microw. Th. Tech.*, MTT-37, 365-274, February 1989.
7. A. Asija, "Endfire tapered slot antennas and their polarization characteristics", M.S. Thesis, Univ. of Massachusetts, September 1989.
8. M.E. Cooley, D.H. Schaubert, N.E. Buris and E.A. Urbanik, "Radiation and scattering analysis of infinite arrays of endfire slot antennas with a ground plane", *IEEE Trans. Ant. Prop.*, AP-39, 1615-1625, November 1991.

THE FAR FIELD SYNCHRONIZATION OF UWB SOURCES BY CLOSED LOOP TECHNIQUES

Gerald F. Ross

Anro Engineering, Inc.
450 Bedford Street
Lexington, MA 02173-1520

1. INTRODUCTION

It is shown in paper how the generation of very short pulse bursts of high microwave energy can be achieved by synchronizing a number of low-cost sources. The peak power radiated in the far field, at boresight, is given by N^2 times the effective radiated pulse (erp) of a single element. Thus, if a source has an erp of 1 kW, then a cluster of ten such generators will radiate 100 kW. Since the signal radiated by an individual source can be just a few rf cycles of microwave energy, very high speed circuitry must be employed to achieve synchronization. The technique described in this paper is presented in Section 2. It maintains synchronization circumventing thermal drift by closed loop feedback techniques. Short-term jitter is reduced to below the resolution of the sampling oscilloscope (e.g., less than 20 ps) by over-triggering the solid state generator.

In the example, described in Section 3, each source generated in the far field has an erp of 1 kW at S-band; the root-mean-squared (rms) pulse duration of the signal was about 1 ns at pulse repetition frequency (prf) as high as 35 kHz. Four sources were actually synchronized, producing a peak erp of 16 kW. Ten such sources will soon be connected generating up to 100 kW. The work is supported under a current U. S. Navy Small Business Innovation Research (SBIR) program. [1] Our conclusions are presented in Section 4.

2. THE SYNCHRONIZATION OF MICROWAVE SOURCES

ANRO Engineering, Inc. (ANRO) developed, under an internal research and development (IR&D) program, a 13-stage Marx generator which produces a 1250 V, 3 ns baseband pulse at a prf up to 50 kHz. [2] a Marx generator essentially charges capacitors from a low voltage source in parallel and then discharges them in series, creating a high voltage pulse into a load. [3] In this case, the load is a microwave cavity, namely: an antenna. The pulse rise time, which is nominally about 1.0 ns, is sharpened to 70 ps by an avalanche diode. Avalanche transistors are used as the primary switching element to discharge the capacitors in series. The particular cavity/antenna selected for the system was a dipole tuned for 2.6-2.8 GHz placed above a flat ground plane. The radiated signal is shown in Figure 1. It has a nominal center frequency of 2.6 GHz and an rms pulse duration of 1 ns. Since there are only 3 rf cycles present in the waveform, it is difficult to imagine how one would synchronize an array of these sources to gain the N^2 power gain advantage, where N is the number of ganged sources.

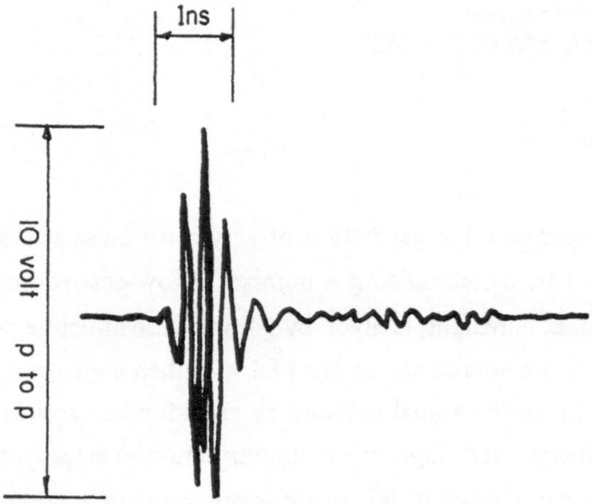

Figure 1. S-band pulse packet radiated by each element.

The basic synchronization scheme is shown in Figure 2. [4] An S-band CW local oscillator 2.3 GHz oscillator is used as the synchronizing "clock" for the four sources as well as the eventual sources in the ten-element array. A small sample of the transmitter signal is sampled by the "sniffer" at each element and mixed with the CW reference. A bandpass filter (BPF) is used to select only the lower sideband signal. *The LO frequency is selected such that the resulting beat frequency signal, because of the duration of the transmission, is approximately a plus or minus half-cycle of baseband energy, whose amplitude depends upon the phase difference between CW oscillator and pulse output.* The baseband pulse (plus or minus) is peak detected and stretched. This *dc* voltage is used to control the delay of the avalanche transistor output with respect to the trigger pulse. A functional relationship of the *dc* voltage between base and emitter of the avalanche transistor and the time delay of the output pulse is shown in Figure 3. When the loop is closed, the gain in the loop (about 30 dB) locks the microwave output of the CW S-band reference. Although the gain in the loop can be increased to make the long-term drift arbitrarily small, the time constant of the loop must be made long to prevent oscillation (e.g., 100 milliseconds and depends upon anticipated thermal drift).

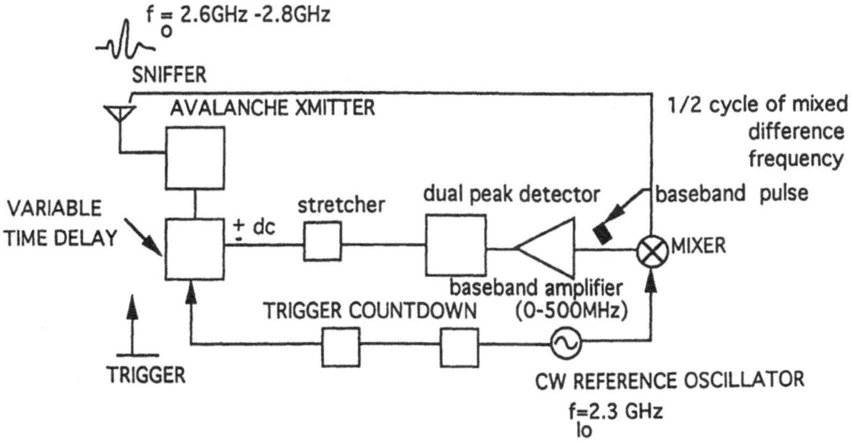

Figure 2. Basic synchronization scheme to accommodate long-term drift.

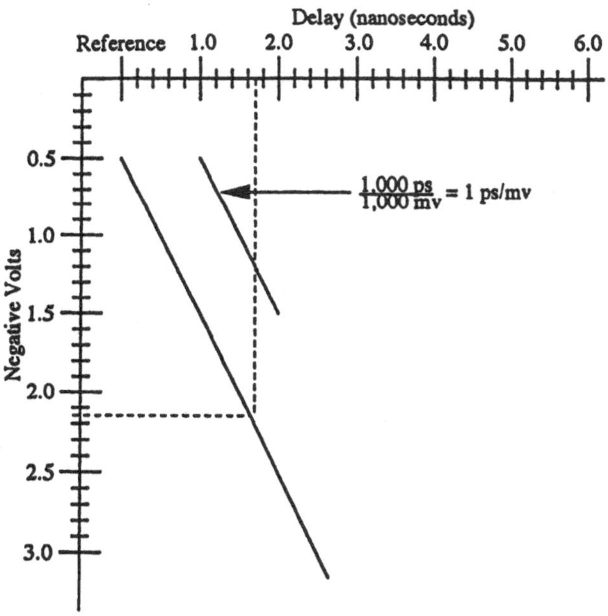

Figure 3. Time delay vs. V_{bo} for steering application (1 ps/mV).

The relationship between time delay vs. voltage for the type RS-3500 avalanche transistor was found by laboratory experiment. The properties found by examining hundreds of these transistors were remarkably similar.

The feedback loop created for temperature stabilization can be mathematically modeled. Consider the electrical equivalent circuit of the loop as shown in Figure 4.

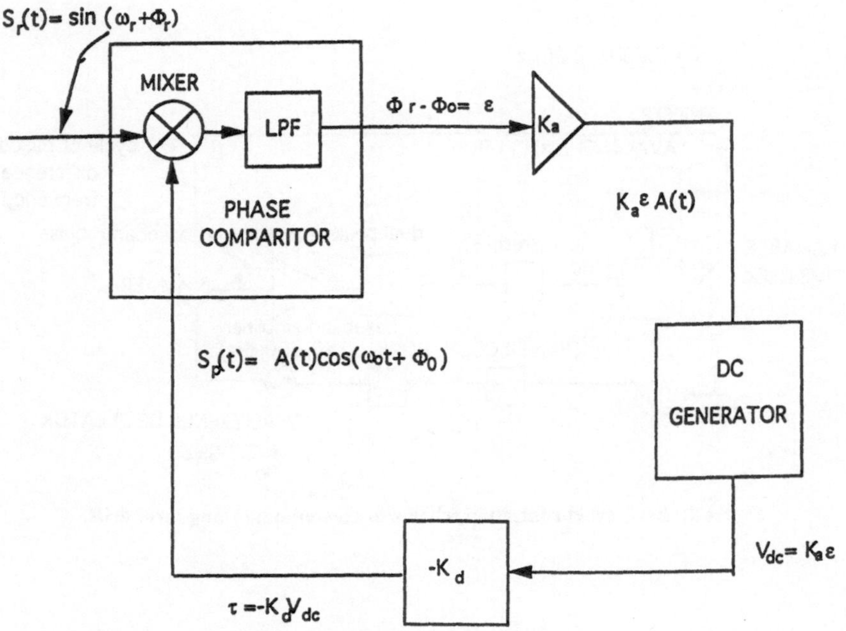

Figure 4. Equivalent circuit of the synchronization loop.

If the CW reference is expressed as

$$S_r(t) = \sin(\omega_t + \Phi_r), \tag{1}$$

then the element waveform is given by

$$S_p(t) = A(t)\cos(\omega_0 t + \Phi_0), \tag{2}$$

where *A(t)* is the pulse shape and $\Phi_0 = \omega_0\tau$. The error angle output of the phase comparator is given by

$$\Phi_r - \Phi_0 = \varepsilon. \tag{3}$$

The output of the comparator is amplified to produce a *baseband* voltage waveform of the form

$$V(t) = K_a \varepsilon A(t), \tag{4}$$

which is peak detected and stretched to produce the DC voltage

$$V_{dc} = K_a \varepsilon. \tag{5}$$

This DC voltage is applied to the emitter base junction of the avalanche transistor to implement a transmitted pulse delay given by

$$\tau = -K_d V_{dc}. \tag{6}$$

The closed loop equations are obtained by substituting (5) and (6) into (7), as follows:

$$\Phi_0 = \omega_0\tau = -\omega_0[-K_d K_a \varepsilon] = \omega_0 K_d K_a[\Phi_r - \Phi_0]. \tag{7}$$

Let

$$\omega_0 K_d K_a = G, \tag{8}$$

then

$\Phi_0 = G\Phi_r - G\Phi_0,$

or, rewriting, $\Phi_0[1 + G] = G\Phi_r$, and substituting into (8) one obtains

$$\Phi_0 = \frac{G}{[1+G]}\Phi_r, \text{ and } \Phi_r - \Phi_0 = \varepsilon = \Phi_r\left[1 - \frac{G}{[1+G]}\right], \tag{9}$$

which shows, as expected, that the phase error decreases as the loop gain, G, increases (e.g., $\varepsilon \to 0$ as $G \to \infty$).

3. SYNCHRONIZATION OF A FOUR-ELEMENT ARRAY

To demonstrate feasibility of the concept, it was decided to monitor the radiated field in our indoor laboratory five feet away. This saved considerable time, logistically, but created a differential time delay problem that was not apparent initially. For example, consider the array of four elements spaced d feet apart, as shown in Figure 5, when a receiving antenna probe is located at a distance R. Then from geometry

$$\tan\theta = \frac{3d}{2R}, \tag{10}$$

and for $R >> d$,

$$\tan\theta \sim \theta = \frac{3d}{2R}. \tag{11}$$

Similarly,

$$\sin\theta = \frac{2\Delta}{3d}, \tag{12}$$

where Δ is the differential delay between the array center and the end element, and from (11),

$$\frac{3d}{2R} = \frac{\Delta}{3d_2}. \tag{13}$$

Solving for R,

$$R = \frac{9d^2}{4\Delta}. \tag{14}$$

For $\Delta = 0.01$ feet (corresponding to a time delay error of 10 ps) and $d = 1$ foot, we require a range, R, to be equal to 225 feet. Since only 5 feet is conveniently available, it was necessary to mount the elements on the arc of a circle.

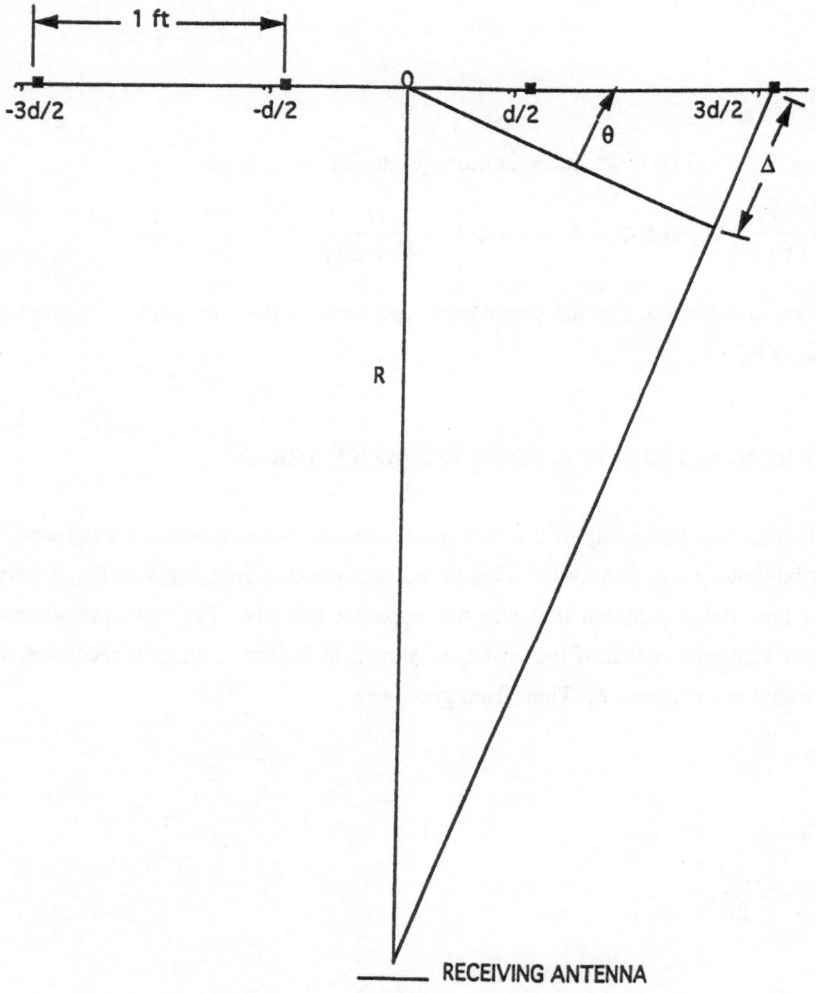

Figure 5. Differential time delay geometry.

Accordingly, a 1' × 8' × 4' pine board was purchased and a five-foot radius was established and the transmitters mounted via a bracket such that the front of each unit was parallel to the tangent of the arc, as shown in Figure 6. Each element was synchronized to the master oscillator and the elements mounted on a wood board separated one foot apart. Numerous ground loops and circuitry coupling problems existed initially. It took several weeks to implement proper shielding before it was possible to begin testing.

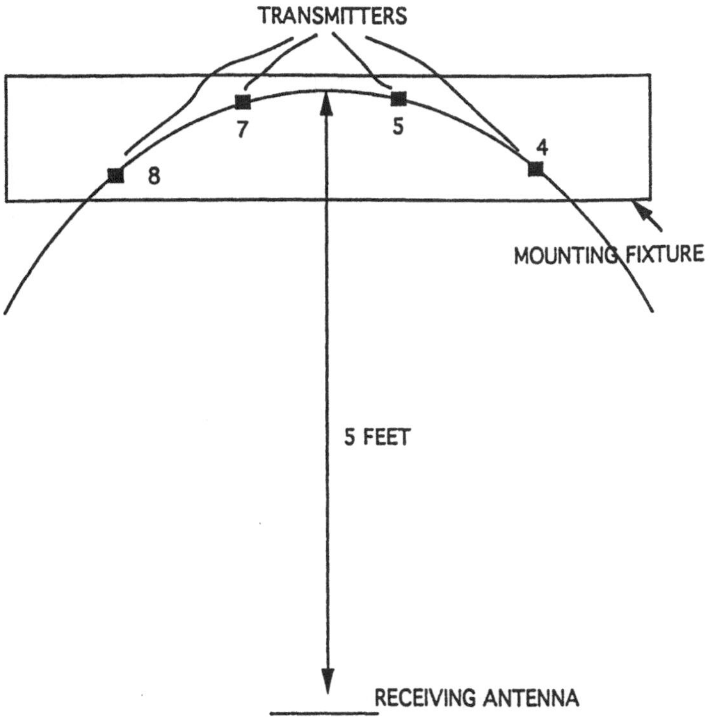

Figure 6. Four-element phased array test configuration.

Four transmitters were each set to 230 volts, E_{cc}. A dipole probe mounted in a corner reflector and balsa wood fixture served as the receiver. With E_{cc} removed from all but one of the transmitters, the peak-to-peak (pp) output voltage received by the probe was recorded, as shown in Table 1, on a sampling oscilloscope.

Table 1. Transmitter Array Test Results

Transmitter #	Vpp Received Volts
4	6.0
5	4.6
7	4.0
8	6.4

Then all the transmitters were activated. The peak voltage was measured to be 20 Vpp+, thus proving virtually perfect synchronization. It is very difficult to discern the peak voltage with more than a five percent accuracy on the oscilloscope. The shape of the pulse at boresight was virtually undistorted.

The time scale on the scope was expanded to 100 ps/division in search as time jitter. If the actual summed waveform jitters, then the waveform is not shown in the scope as a fine line, but rather as a band. The observable jitter appeared to be less than 20 ps. The jitter present in the scope itself is in the same region so that it is difficult to quantitatively determine the actual time jitter. But since the peak value of the received signal was virtually equal to the arithmetic sum of the individual sources, one would have to assume that the jitter was negligible.

In a separate measurement, the peak power of a similar transmitter was measured one meter from its source by the same type of receiving element; the Vpp voltage was 10 volts. Therefore, the peak power received at a 50 ohm receive antenna load is given by

$$P_p = \frac{E^2}{50} = \frac{\left(\frac{5}{\sqrt{2}}\right)^2}{50} = \frac{1}{4}. \tag{15}$$

The effective power radiated over a sphere by the transmitter is given by

$$\frac{P \cdot G_T}{4\pi R^2}, \tag{16}$$

where G_T is the gain of the dipole radiating element. The power received by the dipole antenna having a gain of G_R is obtained by multiplying (16) by the effective area of the antenna, viz:

$$A_{eff} = \frac{\lambda_0^2 G_R}{4\pi}, \tag{17}$$

so that the equation for determining peak power is given by:

$$\frac{P \cdot G_T}{4\pi R^2} \cdot \frac{\lambda_0^2 G_R}{4\pi} = P_p = \frac{1}{4}. \tag{18}$$

Solving for the effective radiated power (erp) = $P \cdot G_T$, we have

$$erp = P \cdot G_T = \frac{P_p (4\pi)^2 R^2}{\lambda_0^2 G_R} = 1.37 \text{ kW} \tag{19}$$

for a 1 meter distance and the following parameters:

R^2 = 1;

P_p = 1/4;

λ_0 = 1.44×10^2 m^2 for $f_0 = 2.5$ GHz;

G_R = 2 for a dipole in a corner reflector.

Note: To determine the peak power at the load, $\frac{5}{\sqrt{2}}$ was assumed to be the rms voltage; that is, the pulse was fully present for all three cycles. In fact, this should be derated by determining the rms value of the transmitted waveform. The answer appears to be closer to $\frac{(5\sqrt{3})^2}{50}$. This reduces (19) by a factor of 1.37 x 3/4 = 1.03 kW.

The approximate average power for a 10 kHz prf is given by:

$$\text{Average Power} = 1.03 \times 10^3 \cdot \frac{\text{pulse duration}}{\text{pulse repetition period}} = \frac{10^{-9}}{100 \times 10^{-6}} \sim 10^{-2} \text{ watts.}$$

Thus, by increasing the voltage in the load by a factor of 4, we have increased the peak power at boresight by 16; the effective radiated power of the four-element array is about 16 kW because the voltages are *coherently* added in the far field. The result is a dramatic increase in peak radiated power.

4. CONCLUSIONS

It has been shown that individual lower power solid state sources can be ganged by simple feedback techniques to produce an erp proportional to the square of the number of sources used. Feedback techniques can conveniently reduce the effects of thermal drift; drift is reduced, essentially, directly by the loop gain. The maximum gain depends on the permissible thermal drift. For the example given, a loop gain of 30 dB was acceptable.

The advantages of achieving higher erp by a multiplicity of low-cost sources with, essentially, baseband hardware must be compared to the use of higher cost laser activated switches and then adjunct power supplies and circuitry. One clear advantage here is the ability to achieve pulse repetition frequencies up to 35 kHz.

REFERENCES

[1]U. S. Navy Contract No. N00024-90-C-4535, Phase II, SBIR Program, U. S. Naval Sea Systems Command, Washington, DC; expires April 1993.

[2]G. F. Ross, R. M. Mara, K. W. Robbins, "Short Pulse Microwave Source with a High PRF and Low Power Drain", U. S. Patent Applied For.

[3]E. Miller, Editor, *Time Domain Measurements in Electromagnetics*, pp. 100-101, Van Nostrand Reinhold, NY, 1986.

[4]G. F. Ross, R. M. Mara, "Synchronization of Very Short Pulse Microwave Signals for Array Applications", U. S. Patent #5,084,706, issued January 28, 1992.

POLARIZATION DIVERSE ULTRA-WIDEBAND ANTENNA TECHNOLOGY

Michael C. Wicks[1] and Paul Antonik[2]

[1]Rome Laboratory
26 Electronic Parkway
Griffiss AFB, NY 13441

[2]Kaman Sciences Corporation
258 Genesee Street
Utica, NY 13502

INTRODUCTION

This paper presents an ultra-wideband (UWB) antenna design that is capable of multiple polarizations. The multiple element design provides polarization diversity while maintaining desirable phase and standing wave characteristics.

The prospect of replacing multiple single purpose antennas on a modern military platform with a single antenna capable of wideband and multi-function operation is of significant interest. Considering space availability, maintenance simplification and aerodynamics, antenna count is now carefully considered in the planning of each new aircraft and space vehicle and in any modification of existing equipment. The broadband radiation capability of this antenna together with its polarization and phase characteristics suggest the possibility of its service in such applications. This antenna can also be of use in other environments such as satellite communications for both the orbital vehicle and the earthbound receptor functions. In the latter, the earthbound satellite receptor use of this antenna can be supplemented with a parabolic dish or other reflector arrangement.

The difficulty encountered in simultaneously achieving minimum phase dispersion with desirable spatial patterns in a single wideband antenna is demonstrated by the commonly accepted compromises which lead to the use of the log periodic or cavity backed spiral antennas. Each of these antennas can achieve bandwidths exceeding a decade while also providing respectable spatial patterns and radiation efficiency. However, these antennas are known to have undesirable voltage standing wave ratios, with values in the range of 2 to 1 or greater. These antennas also exhibit severe time or phase dispersion of a transmitted or received signal. Such antennas, if fed with a very short radio frequency pulse of less than several cycles duration, provide a radiated electromagnetic waveform which contains severe phase and time dispersion effects, distorting the waveform. The combination of non-dispersion and broad frequency band characteristics is particularly unusual but is required for use in a spread spectrum signal environment and for other broadband applications.

The antenna described in this paper provides notably improved operating characteristics, and is capable of transmitting and receiving polarization diverse signals with high fidelity, phase dispersionless operation over a wide frequency band. It operates in a non-resonant mode, providing a low input voltage standing wave ratio, and therefore, also a low radar cross section. This antenna is suitable for mounting in the nose cone of an aircraft or missile, and is capable of use as a radar tracking antenna, an electronic support measure or electronic intelligence antenna. One other important application of this design is for a high precision ultra wide band dual polarized antenna for use in calibrated laboratory and anechoic chamber measurements. The polarization diverse UWB antenna is shown in Figure 1.

BACKGROUND

The TEM horn[1,2] is a two-conductor, end-fire, traveling-wave structure suitable for wideband applications. In the typical design, two parallel plates are flared at one end in such a way as to provide a gradual impedance transition to free space from that of the parallel plate transmission line which feeds it. With proper design, a low VSWR can be obtained.

A variant of the TEM horn, and one which is more closely akin to the antenna discussed in this paper, is the short axial length horn[3,4]. It has open sides like a TEM horn, but it is loaded with a double ridge, resulting in a frequency independent gain over bandwidths as large as 18:1. The logarithmic ridges resemble in shape the inner surfaces of the airfoil structures for the antenna in this study.

The TEM and short axial length horns are two-conductor structures and are thus limited in the types of patterns which they can produce. The antenna discussed in this paper is a novel extension of the short axial length horn idea to incorporate more than two conductors and provide, in effect, an array of TEM horns in a limited space centered about a common axis. Each airfoil element can be thought of as working against its image in the conical section, similar to a short axial length horn. The combination of the four elements, then, is like having four axial length horns located on the same axis, each emitting its own polarization and phase. A variety of patterns can be produced by different phases of excitation of the four airfoil elements, as discussed above. Since the antenna has many of the features of the short axial length horn, one would expect it to be broadband.

DESCRIPTION OF THE ANTENNA

Figure 1 shows an overall perspective view of the Orthogonally Polarized Quadraphase Electromagnetic Radiator[5]. This antenna consists of a truncated cone shaped ground plane surrounded symmetrically by four radiating elements. Also shown in Figure 1 is a physical support structure which may be used to retain the radiating elements in fixed positions with respect to the conical ground plane. The radiating element shown in Figure 2 may be described as resembling the cross-section of an airfoil member. The element shape includes a rounded leading edge, a somewhat flattened "lower surface", a curved "upper surface", and a tapered trailing edge region. This airfoil-like shape provides for the desired surge impedance and radiating characteristics. The set of four radiating elements may also be described as the cross-sectional elements of a horn antenna.

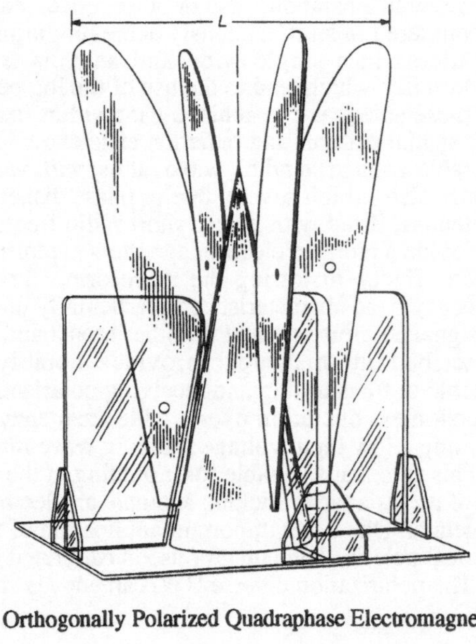

Figure 1. Orthogonally Polarized Quadraphase Electromagnetic Radiator

A profile view of the antenna is presented in Figure 2. Three of the four coaxial cable fittings can be seen in this view. These fittings serve both as a portion of the physical structure of the antenna and as terminating fixtures for coaxial cable transmission line elements used in coupling electrical signals through the ground plane and to the radiating elements. Therefore, no balun is required.

Twenty-two chord lines are also illustrated in Figure 2. The dimensions of each of these chord lines are given in Table 1, in inches. Also note that the distance between points A and H is 0.25 inches, for a 0.125 inch thick radiating element.

A procedure for the empirical selection of radiating element size, shape, and spacing parameters using a time domain reflectometer or similar instrument and the concept of surge impedance is shown in Figure 3. Generally, this fabrication procedure assumes the presence of an initial cut or try at the antenna, which may be arrived at from the designers previous experience and from conventional continuous wave antenna theory together with a consideration of space allocation and shape configuration.. With this initial cut of the antenna, time domain reflectometer measurements are made through a length of coaxial transmission line selected to achieve an impedance match with the signal source or receiver.

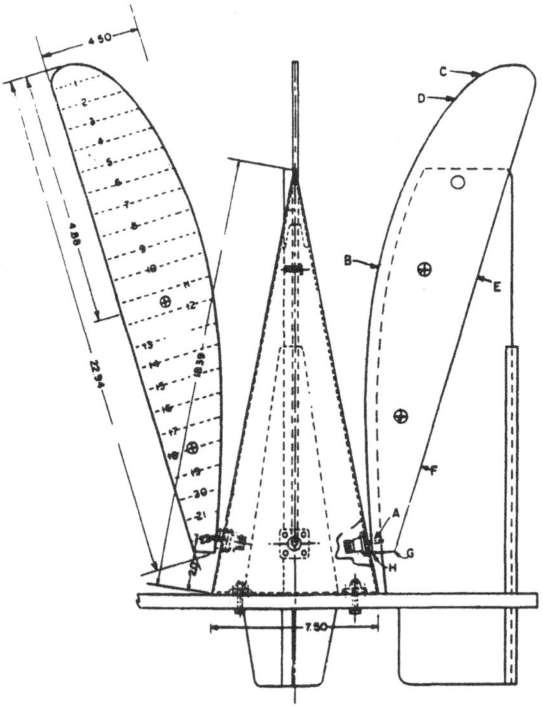

Figure 2. Profile view of the polarization diverse UWB antenna.

The feed region of the radiating element, identified with the letter A in Figure 2, is preferably arranged to having a surge impedance of 50 ohms for a well matched coupling with common coaxial cable. The configuration of the feed region of the radiating elements can be approximated theoretically by regarding the spacing between the radiating element and the ground plane element as the slot portion of a slot radiator as described in Johnson and Jasik[6]. The Johnson and Jasik text is helpful in the initial configuration of the radiating element and its spacing.

Theoretical consideration and the initial cut of an antenna also utilize the conceptual dual of a single radiating element antenna. According to the dual concept, when a radiating

Table 1. Radiating element chord line dimensions for microwave band antenna with 22.94 inches overall length.

Chord Line Number	Chord Length Dimension
1	2.44
2	3.00
3	3.38
4	3.70
5	3.90
6	4.18
7	4.34
8	4.46
9	4.50
10	4.48
11	4.42
12	4.30
13	4.12
14	3.90
15	3.60
16	3.38
17	3.10
18	2.98
19	2.54
20	2.10
21	1.72
22	1.34

element is located above a metal ground plane, the dual of this element appears below the ground plane and image theory provides a tool for analyzing the resulting properties. Removal or alternately shrinking of the ground plane cone in the antenna until only two radiating elements remain is included in an analysis of this type. Transmission line slot theory may then be applied to theses remaining two elements and their spacing. The slot width may be presumed to open exponentially from the throat to the mouth regions of the elements with the radiating element end opposite the feed point considered as a constant radius arc. A slot radiator of this type has a transverse electromagnetic mode (TEM) of propagation.

As indicated in the second and third blocks of Figure 3, the initial cut antenna may be refined through the use of a time domain reflectometer. Desirable surge impedance characteristics of the antenna elements are shown in Figure 4. The surge impedance at the element feed point, which is also identified as point A in Figure 4, is 50 ohms. Commencing with this feed point impedance, a smoothly increasing value of surge impedance progressing from the feed point through the throat, mid region, and leading edge region is desired.

As indicated above, the radius of the element at the leading edge or open end of the horn shape in Figure 2, lies between a too small radius value in which excessive slope and unwanted energy feedback to the input region of the radiating element horn occurs and too large of a radius value in which the physical size of the antenna becomes excessive. The radiating element backside configuration, that is, the geometry of the element along the points designated as D, E, and F in Figure 2, is not critical with respect to antenna electrical characteristics and may be cut as a straight line or otherwise arranged according to structural or other considerations. A long length for the antenna, together with the relatively slow change of surge impedance, is desirable in order to realize a low voltage standing wave ratio characteristic for the antenna.

A low voltage standing wave ratio is desirable not only for its usual benefits of minimizing power reflected back to the transmitter and maximally coupling radio frequency energy into the antenna and to free space, but also to obtain a reduced radar cross section for the antenna. A low radar cross-section is clearly desirable for military uses of antennas as may be surmised from the current interest in stealth aircraft.

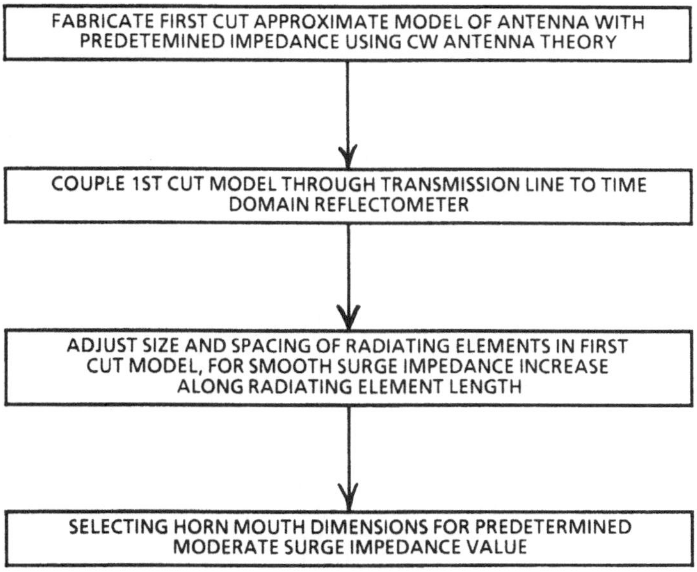

Figure 3. A procedure for the empirical selection of antenna parameters

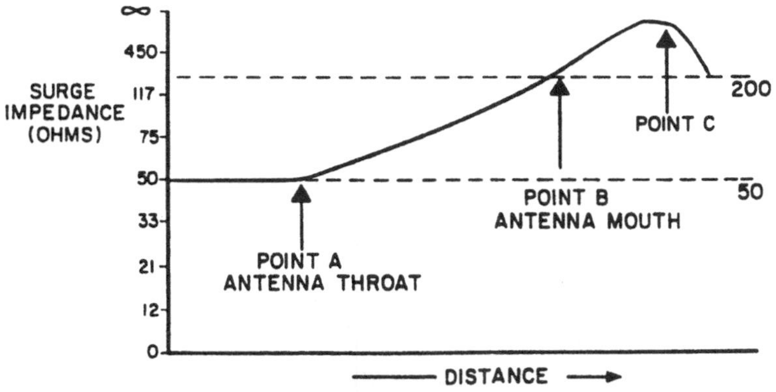

Figure 4. Desirable surge impedance characteristics of the radiating element.

The polarization and electromagnetic field patterns achieved with the antenna are variable in accordance with the relative phasing of the radiating elements. These elements are fed with coaxial cable, the grounded conductor of which is connected to the conical ground plane. The center conductors of the element feeding four different coaxial cables are connected to the insulated center conductors of the fittings. The distal end of these coaxial cable transmission lines may be connected to a variety of energy source (or sink) associated phasing equipment, such as 180 degree hybrid couplers or Magic Tees or the 45 degree and 90 degree phasing apparatus described below.

In Figure 5, a 180 degree hybrid coupler or a broadband magic T network is used to couple between a radio frequency source and the transmission lines feeding two opposing elements. The radiating elements are fed in anti-phase, that is, 180 degrees out of phase, by applying a signal to the difference port of the coupler and terminating the sum port in a matched load. With this coupling arrangement, the electric field between opposite radiating elements extends from one element to the opposite. The field pattern resulting from anti-phase connection of antenna elements is illustrated in Figure 6 when measured at 3 GHz.

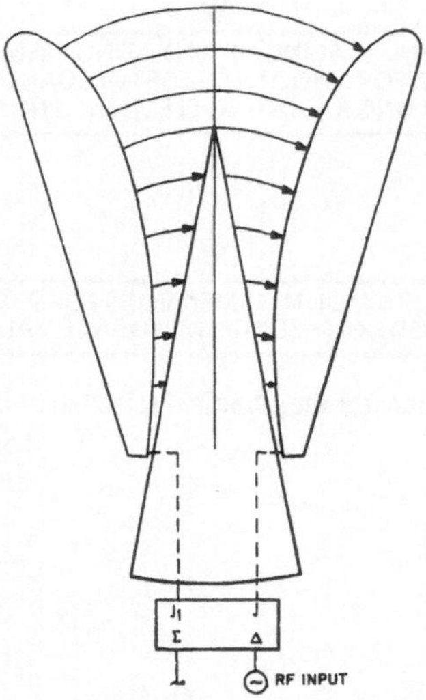

Figure 5. Anti-phase excitation for achieving sum pattern.

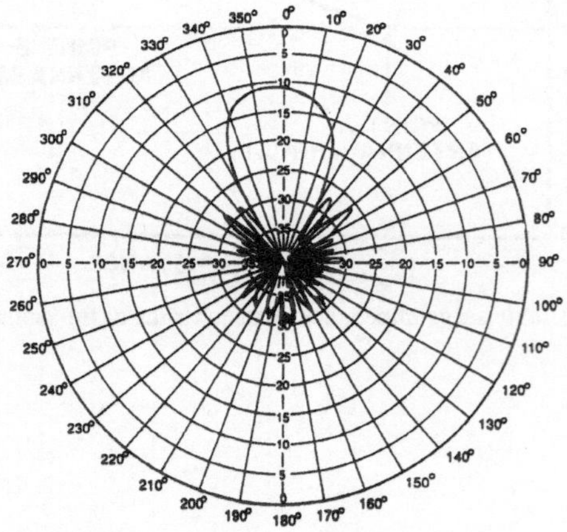

Figure 6. Measured sum pattern at 3 GHz.

The antenna pattern shape resulting from a short axial length horn somewhat resembles that shown in Figure 6 for anti-phase excitation of two opposing airfoil shaped elements. Indeed, if one looks at the E-field pattern for anti-phase excitation shown in Figure 5 for two opposed radiating elements, the cone is superfluous; the same pattern could be obtained by driving the two elements directly without the cone, although the cone helps to provide the desired impedance match to the feed line.

When the radio frequency input signal to the antenna is applied to a sum port of the 180 degree hybrid coupler or broadband magic T as shown in Figure 7, the resulting antenna element electric field extends from both radiating elements to the conical ground plane. The field pattern for the antenna measured at 8 GHz is illustrated in Figure 8. A good monopulse null and low antenna sidelobes were obtained across the microwave band with this arrangement. In the Figure 9 feed arrangement, one of the networks is fed with a phase adjustable signal from the primary network in order to control the antenna element phase relationships and the resulting antenna radiation. With the use of variable phase shifting elements, signals of any polarization can be radiated from this antenna. Typical values of phase shift and the resulting polarization are listed below in Table 2

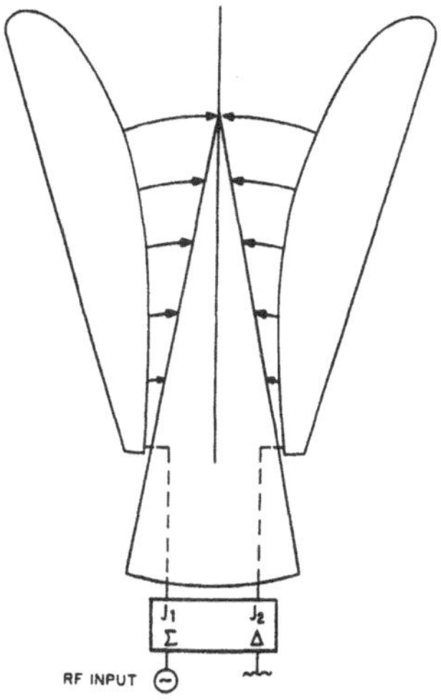

Figure 7. In-phase excitation for achieving difference pattern

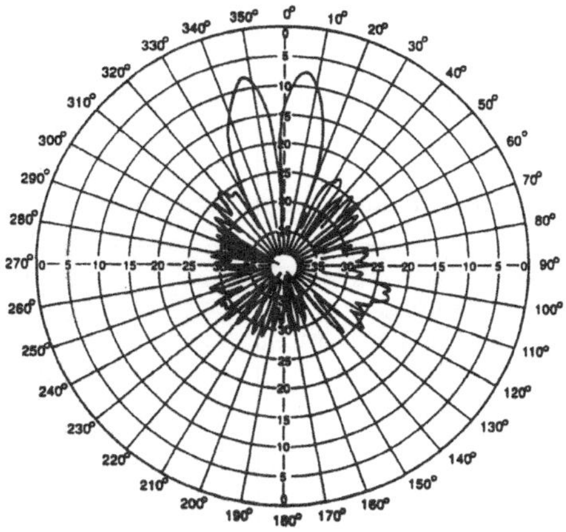

Figure 8. Measured difference pattern at 8 GHz.

Antenna dimensions as shown in Table 3 provide good sum or main beam patterns and difference or monopulse patterns as well as desirable circular and elliptical polarization performance and desirable VSWR performance. The "opening" of the mouth of the antenna (dimension "L" in Figure 1) is specified as 16 inches, or about 41 cm. For a TEM horn the mouth opening should be at least one-half wavelength for the horn to radiate efficiently. From this dimension, one can estimate, therefore, that the lower limit of the operating band for the antenna should be about 360 MHz. This antenna provides good performance over the frequency range of 0.5 to 18 GHz.

Table 2. Polarizations for an antenna with a variable phase shifter..

Value of Phase Shift	Radiation Polarization
0 Deg	+45 Deg Linear
45 Deg	Elliptical CW
90 Deg	Circular CW
135 Deg	Elliptical CW
180 Deg	-45 Deg Linear
225 Deg	Elliptical CCW
270 Deg	Circular CCW
315 Deg	Elliptical C
360 Deg	+45 Deg Linear

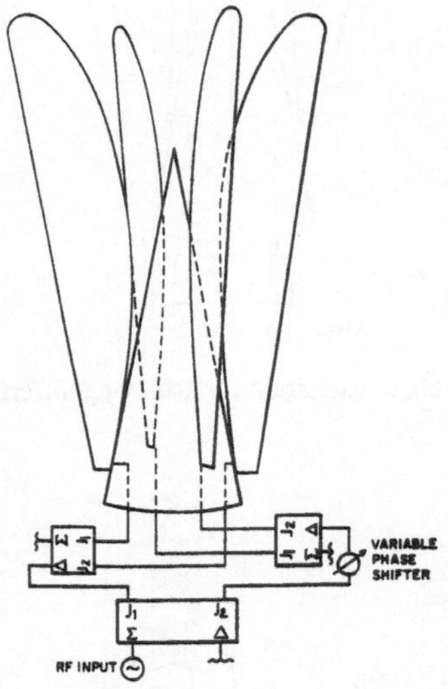

Figure 9. Feed arrangement with adjustable phase shifter for achieving multiple polarizations.

Table 3. Dimensions for an antenna operating in the 0.5 to 18 GHz band.

Radiating Element Length:	23 inches
Mouth Opening:	16 inches
Radiating Element Thickness:	0.1 inches
Height:	25 inches
Cone Half Angle:	12.5 degrees

An antenna constructed according to the Table 3 parameters provides the measured performance found in Table 4.

Table 4. Measured performance for an antenna with the dimensions of Table 3.

Frequency:	4 GHz	6 GHz	8 GHz
Gain:	20.3 dB	24.1 dB	25.4 dB
Beamwidth:	19 Deg	12 Deg	10.5 Deg
VSWR:	1.09:1	1.10:1	1.11:1

Other operating frequency ranges can be achieved by scaling the stated dimensions according to wavelength. For example, to increase the lower frequency performance limit from 0.5 GHz to 1.0 GHz, antenna dimensions should be one-half the values given in Table 3.

Figures 10 and 11 compare the distortion performance of the orthogonally polarized UWB antenna with that of a commercially available broadband antenna. Each of the antennas are impressed with a short duration pulse of radio frequency energy, the pulse being 0.2 nanoseconds long. Clearly, the ringing and distortion of the commercial antenna as shown in Figure 11 indicates significantly poorer signal fidelity than does the pattern for the orthogonally polarized antenna as shown in Figure 10. The response of Figure 10 is, of course, desirable for use with a high resolution radar since the large instantaneous bandwidth of the applied short duration pulse is radiated and received without incurring measurable distortion. The dispersive characteristics shown in Figure 11 preclude the use of such commercial antennas with large instantaneous bandwidth waveforms.

In a similar manner, Figures 12 and 13 show the response of a typical circularly polarized broadband antenna to a rotating linearly polarized source in an anechoic test chamber. The response of the typical antenna in Figure 13 is clearly secondary to the response of the present antenna as shown in Figure 12. The response of the present antenna is constant to within limitations of the measuring equipment and indicates desirable antenna performance.

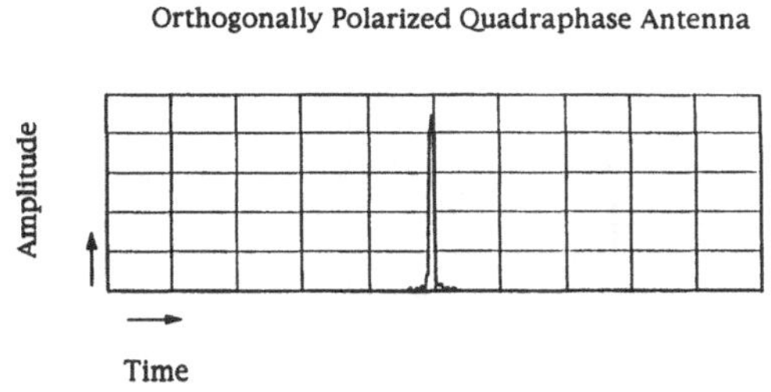

Figure 10. Dispersion of the Orthogonally Polarized Quadraphase Electromagnetic Radiator.

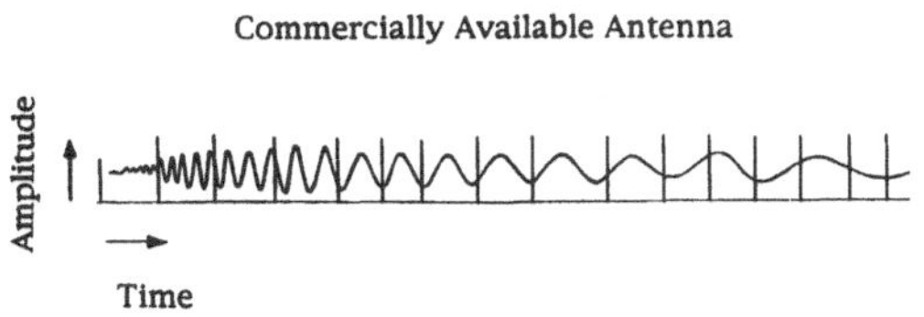

Figure 11. Dispersion of a typical broadband commercial antenna.

The orthogonally polarized antenna has also been found to have desirable collimation between the horizontally and vertically polarized beams over the indicated operating frequency range. Many dual polarized antennas have beam collimation problems, which arise when the horizontally polarized beam points in a different direction than the vertically polarized beam, and may wander about with respect to each other, even over moderate frequency ranges. Examples of this performance have been reported in the literature, especially with respect to military equipment antennas. Measurements made on this antenna show that it has overcome this problem as a result of the described antenna structure and feed arrangement.

The present antenna also provides a small monopulse angle tracking error, a characteristic which does not change appreciably with frequency over the entire indicated microwave operating frequency band. This characteristic is used in target detection and tracking. Low angle tracking or pointing direction errors are desirable for present and future radars in which small target tracking capability is needed.

SUMMARY

This paper has described a design for a polarization diverse, ultra-wideband antenna. The antenna possesses a number of desirable physical and electrical characteristics including:

- Broadband capability
- Minimal phase dispersion
- Low input voltage standing wave ratio
- A non-resonant structure
- Simple and low cost construction
- Polarization diversity
- Desirable phase comparison monopulse response for tracking

The orthogonally polarized quadraphase antenna has broad potential applications. These include use in UWB systems, replacing multiple narrowband antennas with a single wideband antenna, and for use as a standard in calibrated laboratory or field measurements.

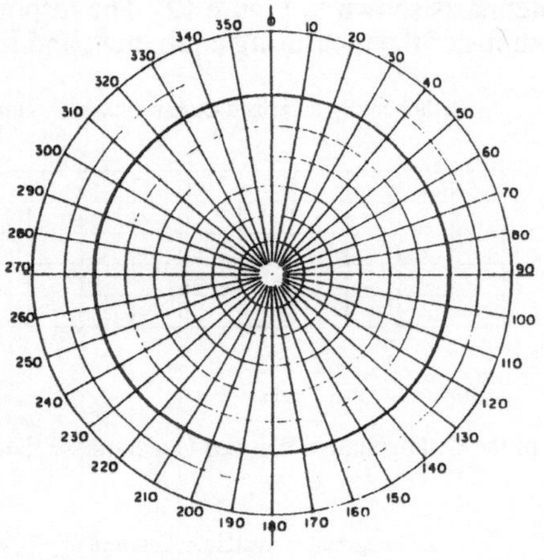

Figure 12. Polarization response of the Orthogonally Polarized Quadraphase Electromagnetic Radiator

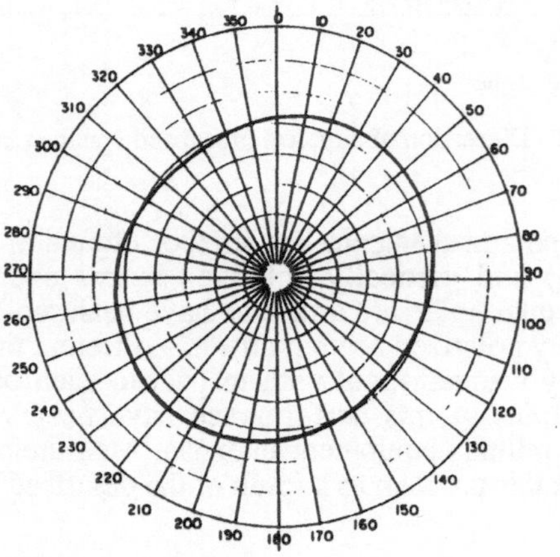

Figure 13. Polarization response of a typical broadband commercial antenna.

ACKNOWLEDGMENTS

The authors would like to thank the following individuals for their reviews, analyses, inputs, and technical advice: Dr. K.S.H. Lee, Dr. F.C. Yang, Dr. G.T. Capraro, and Mr. A. Drozd, all of Kaman Sciences Corporation; Dr. R. Smith, formerly of Kaman; and Mr. K. Siarkiewicz and Mr. T. Blocher, of Rome Laboratory.

REFERENCES

1. S. Evans and F.N. Kong, "Gain and Effective Area for Impulse Antennas", Third International Conference on Antennas and Propagation ICAP 83 , 12-15 April 1983, Part 1 Antennas, IEEE (U.K.), pp. 421-424.

2. M. Kanda, "Transients in a Resistively Loaded Linear Antenna Compared With Those in a Conical Antenna and a TEM Horn", IEEE Trans. Vol AP-28, Jan. 1980, pp. 132-136

3. G. Ross and D. Lamensdorf, "Balanced Radiator System", U.S. Patent No. 3,659,203, 25 April 1972.

4. J.L. Kerr, "Short Axial Length Broadband Horns", IEEE Trans. Vol. AP-21, Sept. 1973, pp.710-714.

5. M.C. Wicks and P. Van Etten, "Orthogonally Polarized Quadraphase Electromagnetic Radiator", U.S. Patent No. 5,068,671, 26 November 1991.

6. R.C. Johnson and H. Jasik, "Antenna Engineering Handbook", 2nd Edition, McGraw-Hill, 1984.

WIDEBAND CIRCULARLY POLARIZED APERTURE-COUPLED MICROSTRIP ANTENNAS

S.D. Targonski, D.M. Pozar

ECE Department
University of Massachusetts at Amherst
Amherst, MA 01003

I. INTRODUCTION

The aperture coupled microstrip antenna[1] has several advantages over transmission line or probe fed patch antennas. Separate substrates can be used for the feed circuit and the antenna element to isolate spurious feed radiation from the antenna element by use of a common ground plane, and to allow more space for the feed network. The input impedance is easily controlled by the size and position of the aperture, and any excess reactance caused by the coupling aperture can be removed through the use of a tuning stub. The aperture coupled configuration also exhibits very low cross-polarization levels, making it well-suited to circularly polarized antenna designs.

A common technique to produce circular polarization is to excite two orthogonal linearly polarized elements with a 90 degree phase difference. This method can be utilized in the aperture coupled microstrip antenna either by using two off-center coupling apertures,[2] or through the use of a crossed slot.[3] The use of two off-center apertures results in an inherent asymmetry in the antenna which produces higher cross-polarization levels, thereby deteriorating the level of circular polarization purity that can be achieved. This effect is especially apparent when a thicker antenna substrate is used to increase bandwidth. In addition, the offset slot configuration limits the slot length to less than half the patch dimension, which in turn limits the substrate thickness over which impedance matching can be obtained, and hence the maximum bandwidth. The crossed slot retains symmetry, and therefore is capable of producing circularly polarized radiation with very good polarization purity. It also permits the use of slot lengths greater than half the patch width, which is critical to achieving adequate coupling when thick antenna substrates are used for wide bandwidths.

The aperture coupled microstrip antenna can generally be impedance matched with an aperture that is well below resonant size, thus limiting the level of back radiation to about -20dB relative to the main lobe. But a microstrip antenna coupled in this fashion is only capable of about 5% bandwidth, due to the fact that the small coupling aperture limits the antenna substrate thickness that may be used. By using a thick antenna substrate with a low dielectric constant a bandwidth of 20% to 25% can be achieved.[4,5] However, because of the thick antenna substrate, a larger slot size is needed to obtain the necessary coupling to impedance match the antenna, resulting in a higher level of back radiation. As an alternative, the required coupling can also be achieved through the use of a "dogbone" aperture,[6] which can provide increased coupling compared to a rectangular slot of the same length.

In this paper the steps taken in the design of three separate wideband circularly polarized aperture coupled microstrip antennas are discussed, and the relative advantages and disadvantages of each design are compared. Details of a series feed design and two parallel feed designs will be presented. The series design is capable of 15% bandwidth for an axial ratio of less than 3dB and return loss of better than -10dB, with a gain greater than 7dB over this range. The parallel feed designs display even better performance, exhibiting impedance and axial ratio bandwidths of 22% and 25%, respectively, and an axial ratio of better than 2dB.

II. DESIGN OF THE WIDEBAND LINEARLY POLARIZED PROTOTYPE ELEMENT

Using the geometry of Figure 1 with a single aperture, a dual-fed linearly polarized element was constructed. The patch was 17mm square, the slot size was 17x0.8mm, and the superstrate was 1.6mm thick with ε_r=2.2. The foam layer was 5mm thick with ε_r=1.07, and the feed substrate was 0.8mm thick with ε_r=2.2. The 50Ω feed lines were offset 5.2mm from the center of the slot. The thick foam antenna substrate provides wide bandwidth, and a large aperture with a resonant frequency that is reasonably close to that of the patch element is used, creating a bandwidth (SWR < 2:1) of 34% centered at 5.23GHz. The large bandwidth of this antenna can be explained by the fact that the aperture acts as a second resonator in combination with the patch element. This creates an effect similar to the stacked patch antenna, which uses two patches of slightly different resonant frequency to achieve a larger bandwidth. However, the fact that the resonant frequency of the aperture is close to that of the patch causes it to radiate a high level of back radiation. For this large bandwidth configuration, the level of back radiation is approximately -10dB.

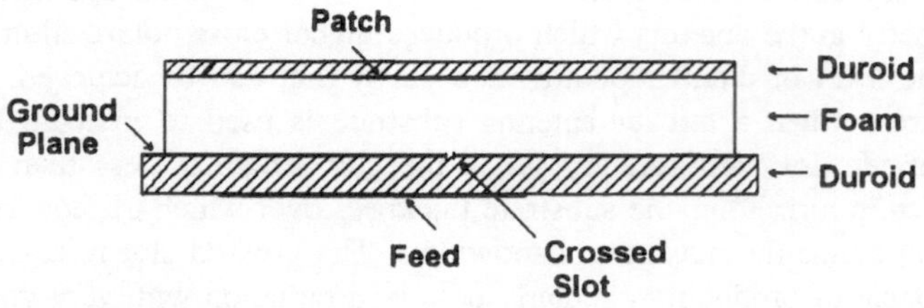

Figure 1. Cross-sectional view of the aperture-coupled microstrip antenna.

A balanced feed configuration as used in the linearly polarized element described above is difficult to implement on each arm of the crossed slot. In Tsao et al.,[3] a microstrip crossover was used in the design, but this technique creates potential fabrication problems. This problem can be overcome by implementing a 180 degree phase shift in the feed design, and using a series or parallel feed for the two arms of the crossed slot, as discussed below.

III. THE SERIES FEED CONFIGURATION

The series feed configuration is shown in Figure 2a and its equivalent circuit in Figure 2c. Each port is fed 180 degrees out of phase in order to account for the phase reversal caused by the oppositely directed feedlines at the two feed points for each aperture. A quarter-wavelength section of transmission line is placed between each aperture to create the 90 degree phase difference required for circular polarization. In order to achieve an excitation of the two orthogonal apertures that is equal in amplitude with a 90 degree phase difference, the effective impedance must be matched to 50Ω. Any impedance mismatch present will create a standing wave on the quarter-wave section of line between the orthogonal apertures and produce phase and amplitude errors, resulting in an increased axial ratio.

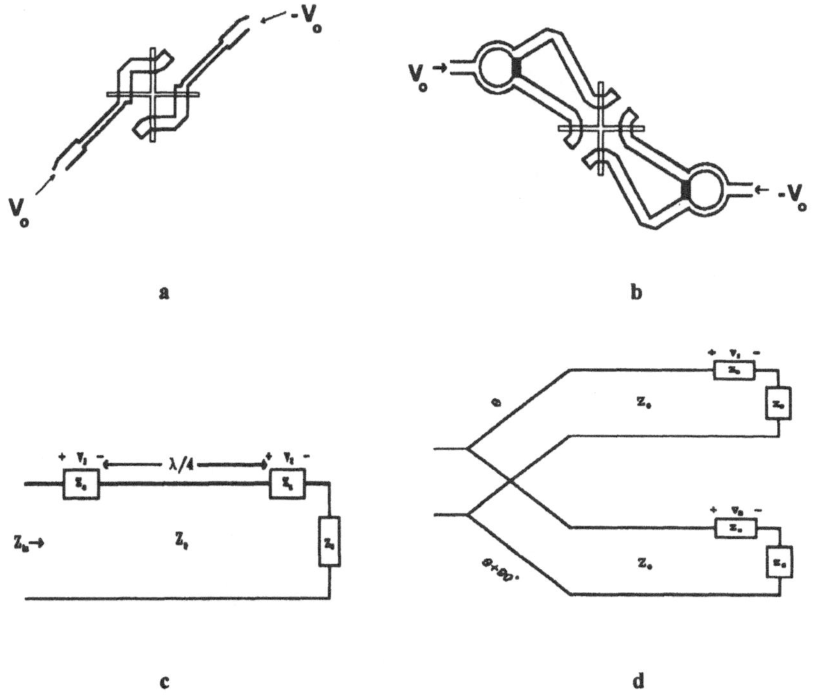

Figure 2. a) Configuration of the series-fed crossed slot. b) Configuration of the parallel-fed crossed slot. c) Equivalent circuit for one of the feed lines in a). d) Equivalent circuit for one of the feed lines in b).

In the equivalent circuit, the series impedance Z_a is equal to the effective impedance minus the impedance of the tuning stub. Because of the lack of a tuning stub on the first aperture, a large reactance will be present in the input impedance referenced at the first aperture. This reactance may be removed by implementing a short section of transmission line before the first aperture. The resulting input impedance can then be easily matched to 50Ω with a quarter-wave transformer.

A series fed antenna was constructed from the linearly polarized prototype design. The effective impedance, Z_{eff}, was nominally matched to 50Ω, thereby creating a VSWR of less than 2:1 on the quarter-wave section of line between the apertures. Since the VSWR was small a good axial ratio was obtained. A measured spinning linear far-field pattern for the series fed element is shown in Figure 3. Also, the measured gain and axial ratio of this element are plotted versus frequency in Figure 4. The axial ratio at 5.1 GHz is less than 1dB over a wide angle; the size of the ground plane in this measurement was 61x45.75cm. An axial ratio of less than 3dB was obtained over a 12% bandwidth. The bandwidth for 10dB return loss was in excess of 30%, leaving the axial ratio as the limiting factor.

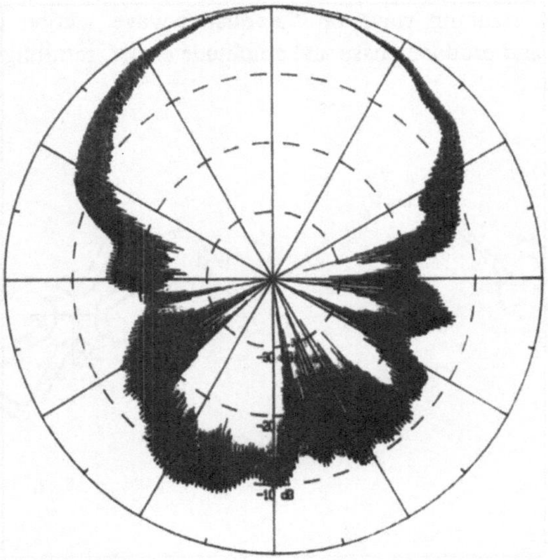

Figure 3. Measured spinning linear pattern of the series fed microstrip antenna at 5.1 GHz.

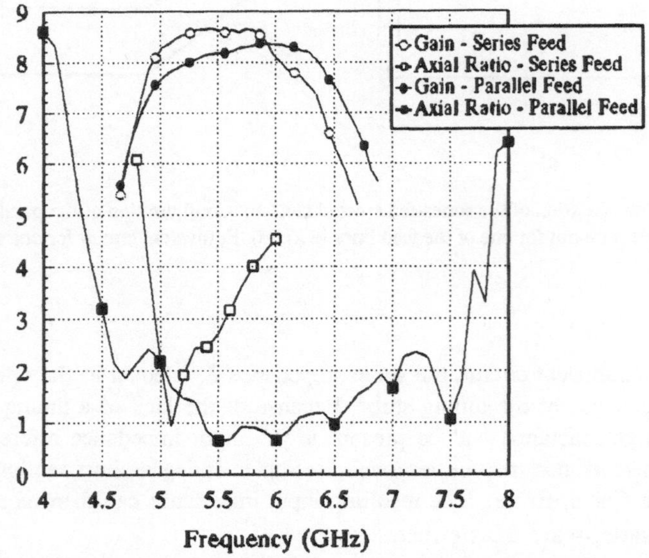

Figure 4. Measured gain and axial ratio for the series fed and parallel fed microstrip antennas.

IV. THE PARALLEL FEED CONFIGURATION

The parallel feed structure is shown in Figure 2b along with its equivalent circuit in Figure 2d. Here the two orthogonal apertures are fed through a Wilkinson power divider, with one arm of the output feedlines a quarter-wavelength longer than the other in order to produce a 90 degree phase shift.

A circularly polarized antenna was constructed using the parallel feed, based on the linearly polarized element of Section II. The pattern at 5.6GHz is essentially the same as the pattern shown in Figure 3 since the same linearly polarized element was used in the design. Notice again that the axial ratio is quite good over a wide angle. Figure 4 shows the axial ratio and gain of the parallel fed antenna versus frequency. Note that the axial ratio characteristics versus frequency are much improved over the series feed, due to the fact that amplitude error between the two orthogonal polarizations has been eliminated. The gain of the antenna then becomes the limiting factor in the bandwidth. The bandwidth for the gain, defined as being within 1dB of the maximum gain over the band, is 29%, which represents a substantial improvement over the bandwidth of the series feed.

V. DISCUSSION

For a simple uniform linear or planar array, a single 180 degree hybrid can be used for the input to an entire feed network for several elements. For a phased array, however, each element must be fed separately. This would require a ring hybrid for each element in the array, which most likely would not be feasible due to space limitations. As an alternative in this case, a half-wavelength section of transmission line can be used to produce the 180 degree phase shift needed to drive both ports.

The half-wavelength section of transmission line will produce a 180 degree phase difference only at the center frequency, and a phase error in the excitation of each aperture will result over the band. A phase error in the excitation of each aperture excites a mode in the orthogonal aperture. Due to the symmetry of the crossed slot, however, the effects of this excitation on the axial ratio will cancel and no decrease in circular polarization purity will result. Experimental results performed on the wide bandwidth parallel fed CP antenna presented in Section IV have shown that a phase error of as much as 25 degrees produced no noticeable effects on the axial ratio, pattern, or gain of the element over the entire bandwidth, thereby showing that a half-wavelength section of transmission line used between the two ports can operate over a wide bandwidth. This type of arrangement will not work on the series feed configuration, however, since the feed network lacks the symmetry of the parallel configuration.

Two novel geometries for feeding a wideband circularly polarized aperture coupled microstrip element have been presented, using either a series or a parallel feed. The parallel feed configuration provides better axial ratio and bandwidth characteristics, but is more complicated to design and fabricate because of the Wilkinson power dividers needed for isolation between the output ports. If a very large array of these elements is being constructed and a bandwidth in excess of 12% is not needed, the series feed design will greatly improve the ease of fabrication.

Acknowledgment: The authors would like to thank Professor Robert W. Jackson for his input and help in the design of the parallel feed structure. We also would like to thank Mr. Baha Pelin at RohmTech for the donation of several samples of ROHACELL foam which were used in the antenna.

REFERENCES

1. D.M. Pozar, Microstrip antenna aperture-coupled to a microstripline, *Electronics Letters*, Vol. 21, pp. 49-50, January 1985.

2. A. Adrian and D.H. Schaubert, Dual aperture-coupled microstrip antenna for dual or circular polarization, *Electronics Letters*, Vol. 23, pp. 1226-1228, November 1987.

3. C.H. Tsao, Y.M. Hwang, F. Killburg, and F. Dietrich, Aperture-coupled patch antennas with wide-bandwidth and dual polarization capabilities, *IEEE Antennas and Propagation Symposium Digest*, pp. 936-939, 1988.

4. J.-F. Zurcher, The SSFIP: a global concept for high-performance broadband planar antennas, *Electronics Letters*, Vol. 24, pp. 1433-1435, November 1988.

5. F. Croq, A. Papiernik, and P. Brachat, Wideband aperture-coupled microstrip subarray, *IEEE Antennas and Propagation Symposium Digest*, pp. 1128-1131, 1990.

6. D.M. Pozar and S.D. Targonski, Improved coupling for aperture-coupled microstrip antennas, *Electronics Letters*, Vol. 27, pp. 1129-1131, June 1991

LOW CROSS-POLAR, SHORT-PULSE RADIATION CHARACTERISTICS OF PRINTED ANTENNAS COVERED BY A POLARIZATION GRATING

Atanu Mohanty, and Nirod K. Das

Weber Research Institute
Polytechnic University
Route 110, Farmingdale, NY 11735

1 INTRODUCTION

Printed antennas are attractive elements for broadband/short-pulse radiation [1,2], and are particularly suited for integration with microwave and optoelectronic circuits. However, it can often be difficult to obtain low cross-polar radiation from such printed antennas over an ultra-broad bandwidth and/or a wide beam angle. In an integrated circuit environment, it is particularly difficult to avoid the unwanted, often significant, levels of cross-polar radiation from the antenna feed circuitry and other integrated components. Also, other than along the principal (E- or H-) planes, cross-polarization components can be introduced in off-broadside directions, if the antenna substrate is electrically thick. It is important to avoid or suppress such unwanted cross-polarization components for applications in radar and communications.

Strip gratings printed on a thin substrate have been commonly used for the suppression of cross-polarization. Though, in the past, the use of printed gratings have been limited only to narrow-bandwidth operations, they can be equally effective under ultra-wideband or short-pulse conditions [3]. This paper will present a rigorous study of the polarization filtering characteristics of printed gratings, and their effects on the short-pulse radiation from printed antennas. A full-wave analysis of a printed antenna element loaded on top by a polarization grating will be described, with representative results on the polarization characteristics. The analysis generates the short-pulse responses via a broadband frequency synthesis, and assumes an infinite periodic grating neglecting any truncation effects. The effects of the antenna and grating substrates, and the interaction between the antenna element and the grating, are rigorously accounted for using multilayer spectral-domain Green's functions [4] and a Floquet mode decomposition of the grating currents.

Ultra-Wideband, Short-Pulse Electromagnetics
Edited by H. Bertoni *et al.*, Plenum Press, 1993

2 ANALYSIS

Figure 1 shows the geometry of a printed antenna element covered by a periodic strip grating. Standard geometries of printed antennas on single or multiple dielectric substrates can be analyzed using a spectral domain moment method approach [5,6,7]. However, when an additional grating layer is included, the antenna can not be directly analyzed using spectral-domain Green's functions, because, unlike a dielectric layer, the grating introduces nonuniformity in the transverse plane. In order to account for the grating effects, first in section 2.1 we analyze the grating with the dielectric substrates, excited by a 'spectral current sheet' placed on the plane of the antenna. Under this current-sheet excitation, an infinite number of plane wave modes(the Floquet modes) are used to describe the total field. In section 2.2, the current on the antenna is expressed as a spectral decomposition of phased current sheets, and the contribution due to each spectral current sheet is obtained repeating the procedure of section 2.1. The total field of the antenna including the grating is then computed via a spectral integration. The far field radiation is deduced using a stationary phase evaluation method. Note that in the absence of the grating only a particular spectral component of the antenna current, with $k_x = k_0 \sin\theta \cos\phi$ and $k_y = k_0 \sin\theta \sin\phi$ (see Fig.1), contributes to the far field. But when the grating is introduced, it may also diffract some of the evanescent fields of the antenna into the far field. However, if the separation between the antenna and the grating is large, the evanescent coupling from the antenna to the grating is small and may be neglected for far-field computation. This results in considerable computational simplification. We have presented representative results for a short dipole antenna in section 3.

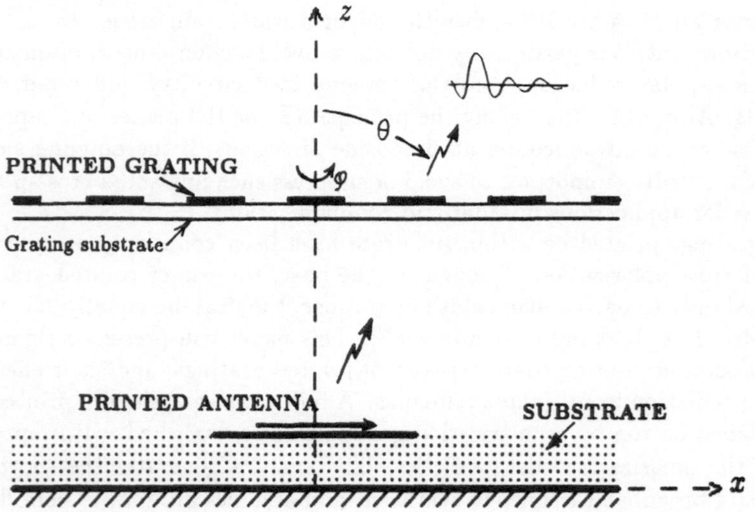

Figure 1. Periodic grating to suppress the cross-polarized radiation from a printed antenna for broadband/short-pulse applications. The printed grating may be placed directly on top, or properly spaced above, the antenna surface.

2.1 Current Sheet Excitation of the Grating

When a spectral current sheet, $\vec{J}_s(k_x, k_y) = \hat{s}e^{-j(k_{sx}x+k_{sy}y)}$, is placed on a given plane in a layered structure, the Fourier transform of the induced currents on the infinite grating, $\tilde{\vec{J}}_g(\vec{k}) = \tilde{\vec{J}}_g(k_x, k_y)$, can be expressed in terms of basis current transforms, $\tilde{\vec{J}}_m(k_x)$, on each strip, using a Floquet mode decomposition:

$$\tilde{\vec{J}}_g(\vec{k}) = \sum_{m=1}^{N} a_m \tilde{\vec{J}}_{mg} = \frac{(2\pi)^2}{D} \sum_{m=1}^{N} a_m \tilde{\vec{J}}_m(k_x) \sum_{\mu=-\infty}^{\infty} \delta(k_x - k_{sx} - \mu\frac{2\pi}{D})\delta(k_y - k_{sy}) \quad (1)$$

The transverse (x) variation of currents on each strip are expanded using a complete set of basis functions, $\vec{J}_m(x)$, as described in the appendix.

Now, the transform of the electric field due to the grating is obtained from the transform of the grating current via the spectral domain dyadic Green's function [4,8]

$$\tilde{\vec{E}}_{mg}(\vec{k}) = \underline{\underline{\tilde{G}_g}}(\vec{k}) \cdot \tilde{\vec{J}}_{mg}(\vec{k}). \quad (2)$$

Similarly, the spectrum of the incident field, $\tilde{\vec{E}}_s(\vec{k})$, produced by the source sheet, $\vec{J}_s(k_x, k_y)$, is given by:

$$\tilde{\vec{E}}_s(\vec{k}) = (2\pi)^2 \underline{\underline{\tilde{G}_s}}(\vec{k}) \cdot \hat{s}\delta(k_x - k_{sx})\delta(k_y - k_{sy}). \quad (3)$$

The dyadic Green's functions $\underline{\underline{\tilde{G}_g}}$ and $\underline{\underline{\tilde{G}_s}}$ referred to above, correspond to the source currents on the grating, and the source currents on the source sheet, respectively.

The N unknown basis amplitudes (a_m's) of the current expansion on the grating can now be solved using a Galerkin testing procedure:

$$\sum_{m=1}^{N} a_m \left(\int_{-\infty}^{\infty} \tilde{\vec{J}}_n^* \cdot \tilde{\vec{E}}_{mg}\, dk_x \right) = \sum_{m=1}^{N} a_m \left(\int_{-\infty}^{\infty} \tilde{\vec{J}}_n^* \cdot \underline{\underline{\tilde{G}_g}} \cdot \tilde{\vec{J}}_{mg}\, dk_x \right)$$
$$= -\int_{-\infty}^{\infty} \tilde{\vec{J}}_n^* \cdot \tilde{\vec{E}}_s\, dk_x \;; n = 1, \ldots, N \quad (4)$$

$$\sum_{m=1}^{N} a_m \left(\sum_{\mu=-\infty}^{\infty} \tilde{\vec{J}}_n^*(k_{sx} + \mu\frac{2\pi}{D}) \cdot \underline{\underline{\tilde{G}_g}}(k_{sx} + \mu\frac{2\pi}{D}, k_{sy}) \cdot \tilde{\vec{J}}_m(k_{sx} + \mu\frac{2\pi}{D}) \right)$$
$$= -D\tilde{\vec{J}}_n^*(k_{sx}) \cdot \underline{\underline{\tilde{G}_s}}(\vec{k}_s) \cdot \hat{s} \quad (5)$$

Once the unknown current on the grating is determined, the field can be determined via suitable Green's functions. Clearly from (5), the total field produced by the current sheet, $\vec{J}_s(k_x, k_y) = \hat{s}e^{-j(k_{sx}x+k_{sy}y)}$, in the presence of the grating, include infinite discrete spectral values at $k_x = k_{sx} + \frac{2\pi}{D}\mu, k_y = k_{sy}; \mu = 0, \pm1, \pm2, \ldots$, out of which only a few modes could contribute to the far field. For close spacing of the grating, $D < \lambda/2$, however, only one mode radiates, all others being evanescent.

2.2 Radiation Pattern of Antenna Including Grating

A known current distribution on the antenna, $\vec{J}_a(x, y)$ can be decomposed using its Fourier transform, $\tilde{\vec{J}}_a(k_x, k_y)$:

$$\vec{J}_a(x,y) = \frac{1}{(2\pi)^2} \int_{-\infty}^{\infty} \int_{-\infty}^{\infty} \tilde{\vec{J}}_a(k_x, k_y) e^{-jk_x x} e^{-jk_y y}\, dk_x\, dk_y \tag{6}$$

The total field due to the antenna current including grating can now be expressed as a superposition of the fields produced by the spectral sheets, $\tilde{\vec{J}}_a(k_x, k_y)e^{-jk_x x}e^{-jk_y y}$, obtained in the last section, via a Fourier spectral integral. The radiated far field, $\vec{E}(r, \theta, \phi)$, can then be obtained from the total field, $\vec{E}_t$, on the top surface of the layered structure, via a stationary phase evaluation. This allows significant computational simplification:

$$\vec{E}(r,\theta,\phi) = \frac{jk_0 \cos\theta}{2\pi r} e^{-jk_0 r} \tilde{\vec{E}}_t(k_0 \sin\theta \cos\phi, k_0 \sin\theta \sin\phi) \tag{7}$$

Note that the far-field expression in (7) requires the evaluation of $\tilde{\vec{E}}_t$ only at a single spectral point, $k_x = k_0 \sin\theta \cos\phi = k_a, k_y = k_0 \sin\theta \sin\phi = k_b$. This is, in general, contributed by the antenna current spectrum at $k_x = k_a + \mu\frac{2\pi}{D}, \mu = 0, \pm1, \pm2, \pm3, \ldots, k_y = k_b$, if the evanescent coupling of the antenna to the grating is rigorously included. However, as mentioned before, when the antenna and the grating are electrically far apart, the evanescent coupling can be practically neglected, and the far field can be obtained from the antenna current transform at $k_x = k_a, k_y = k_b$, alone.

The above discussion assumes a known current distribution on the antenna. For an arbitrary antenna, however, the current can be expanded using a basis set. The unknown basis coefficients can be solved using a spectral-domain Galerkin testing method [5]. As discussed before, the spectral-fields produced by a source basis function need to be repeatedly evaluted using the procedure of section 2.1 for each spectral component of the source current. Therefore, testing of this source field on the test basis function will involve the spectral integration of an infinite spectral series. As mentioned, if the evanescent coupling between the antenna and the grating can be neglected, the above procedure can be greatly simplified.

In the present paper, we specifically concentrate on the general broadband/short-pulse polarization characteristics of printed gratings, and their fundamental effects on printed antenna radiation. So we assume a dipole antenna for simplicity. Time domain performance is obtained via Fourier synthesis.

3 RESULTS

A short dipole antenna is excited by a current pulse with a spectrum:

$$F(\omega) = \frac{(\omega t_i)^4}{12} e^{-|\omega| t_i} \tag{8}$$

Figure 2a and 2b show the spectrum and the corresponding pulse for $1/t_i = 30GHz$, respectively. The radiation characteristics in the direction $\theta = 45°, \phi = 45°$ are presented in Figs. 3-5. Referring to Fig.1, if there were no dielectric substrates, the radiation from the dipole without the grating on top would be purely TM_x. When one or more dielectric layers are introduced, the radiation is still TM_x along the principal planes, but some TE_x radiation may be present in off-broadside directions along non-principal planes. Therefore, in the absence of the grating the TM_x and TE_x components are considered the co-polarization and cross-polarization components, respectively.

To achieve an effective suppression of the cross-polarized radiation, the grating strips must be placed perpendicular(y) to the dipole. Therefore, the TE_y field components of the antenna almost pass through, whereas the TM_y fields are strongly reflected by the grating. Thus, when the grating is introduced, the far-field radiation is predominantly TE_y, and hence the TE_y radiation is justifiably considered the co-polarized component when the grating is present.

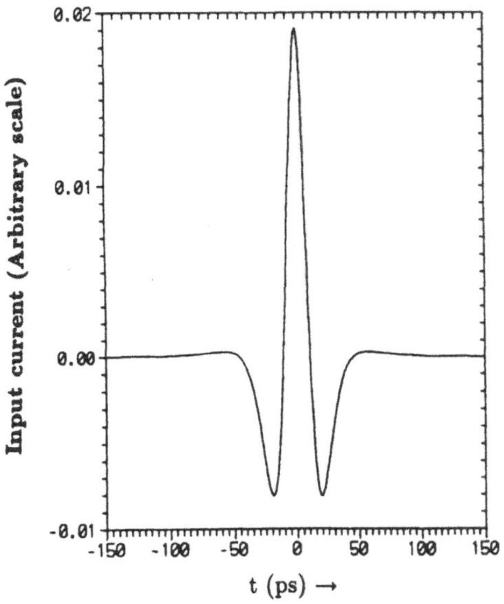

Figure 2a. Input pulse.

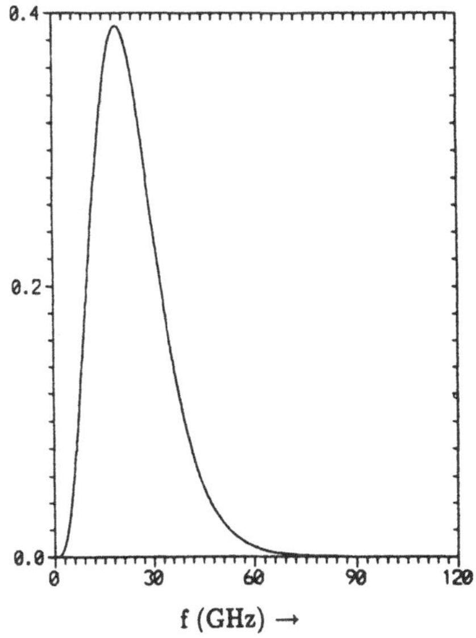

Figure 2b. Spectrum of input pulse.

Figure 3 shows the radiation from a dipole printed on a dielectric substrate, without a grating layer, at an off-broadside direction along the diagonal plane. It can be seen, there is a significant level of cross-polarized radiation from the antenna due to the effect of the dielectric substrate. Figure 4 shows the corresponding results including a grating layer on top. It can be seen that the level of cross-polarization has now been significantly reduced. The cross-polarized radiation is only 14.11 dB below the co-polarized component without the grating, but it significantly drops down to 29.44 dB below the co-polarized component when the grating is placed. Better improvements can be expected using a grating with narrower strip-widths and smaller strip-spacings.

As discussed, the TM_y (see Fig.1) component of radiation from the antenna is strongly reflected from the grating layer. These reflected TM_y fields again get reflected back by the groundplane below the antenna, possibly establishing a resonance condition for an appropriate separation between the groundplane and the grating. This resonance will result in a large magnitude of the TM_y fields. However, through every reflection, the dielectric substrates of the antenna or the grating convert a part of the TM_y fields to TE_y fields, which escapes through the grating to far-field radiation. The external radiation, therefore, exhibits a strong oscillation in the time domain. The frequency of this oscillation, f_R, will be different in different angles of observation, θ, related to the electrical distance, H, between the grating and the groundplane as: $H \cos\theta = c/(2f_R)$,

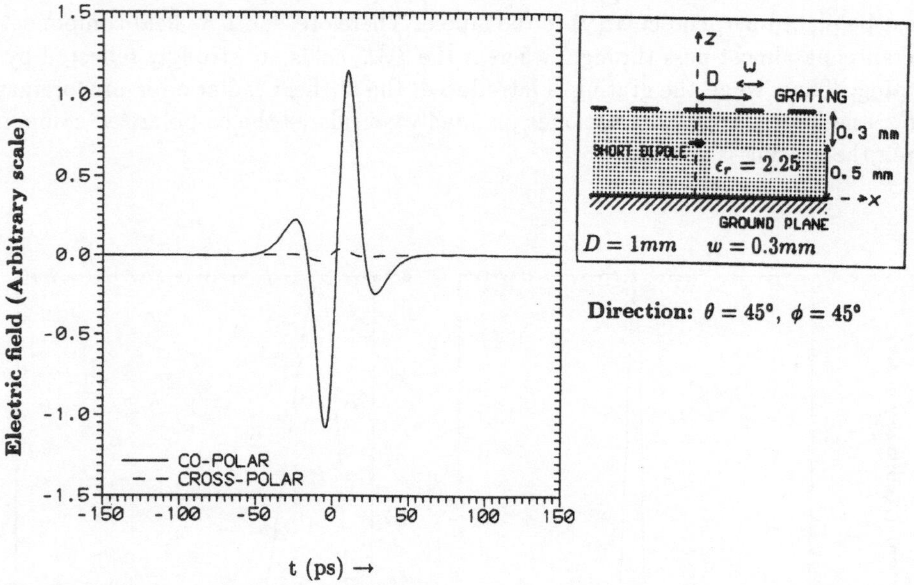

Figure 3. Dielectric layers introduce cross-polarization in directions away from the principal planes.

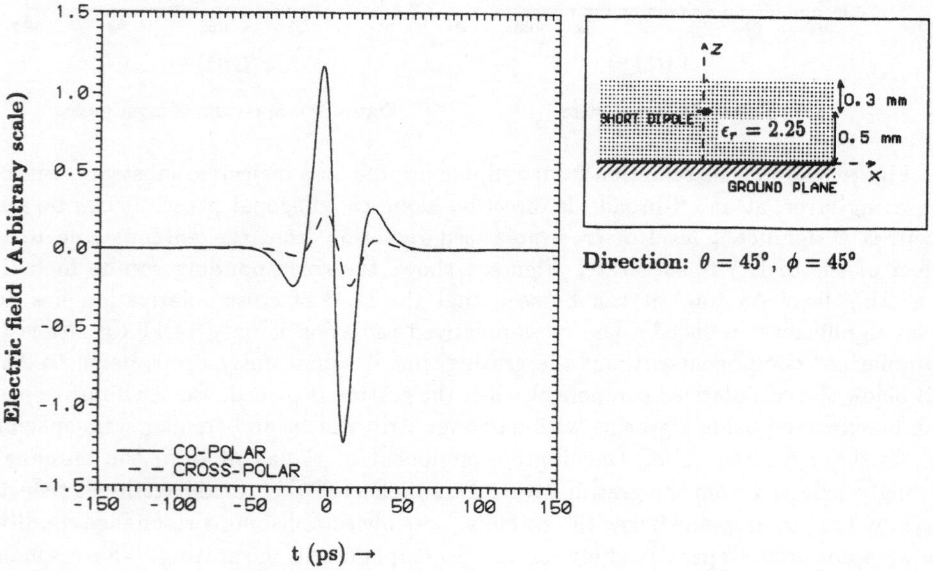

Figure 4. Cross-polarization suppression by a grating. The ratio of the energy of the co-polarized component to that of the cross-polarized component is improved by 15.33 dB compared to the case without the grating.

where c = the velocity of light in free space. Equivalently in the frequency domain, if the antenna is excited by the frequency, f_R, a strong radiation will be observed along the corresponding angle, θ. Figure 5 clearly shows this interesting resonance effect. The time period of oscillation of the radiated pulse agrees with the above calculations.

4 CONCLUSION

Significant levels of cross-polarized radiation can be introduced if the dielectric substrate of a printed antenna is electrically thick. A strip grating can be effectively used for the suppression of the cross-polarized radiation for short-pulse/ultra-wideband applications. The strip width and spacing of the grating, the spacing between the antenna

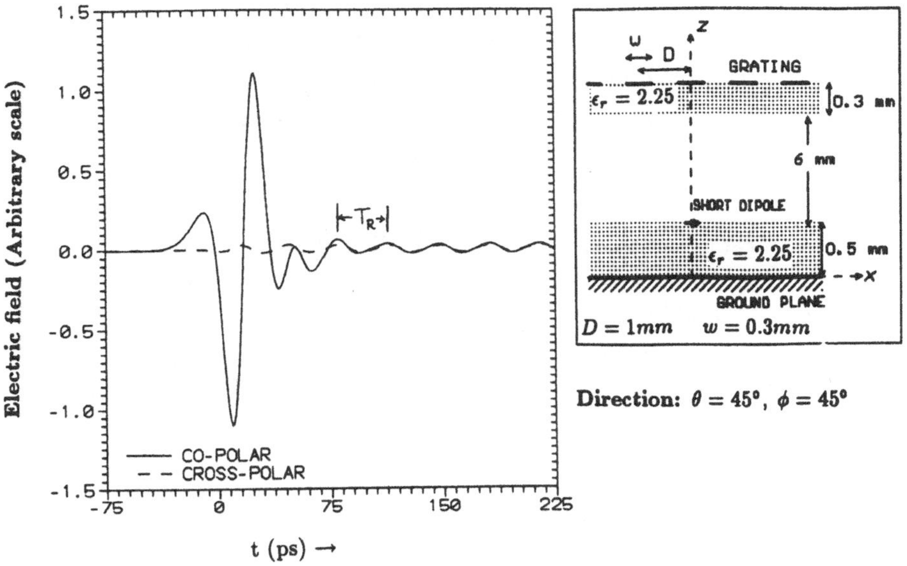

Figure 5. Pulse ringing caused by grating resonance. Here, ringing period $T_R \approx 33ps$. Calculated value of T_R assuming a thin substrate is $33.94ps$.

and the grating, and the substrate thicknesses, are the important design parameters that will determine the level of cross-polarization suppression, the pulse distortion, and the possible ringing effects due to grating resonance. Properly spacing the grating above the antenna may avoid the unwanted grating resonance.

5 APPENDIX

A suitable choice for the basis functions is:

$$\vec{J}_m(x) = \begin{cases} \hat{y} f_{yp}(x) & p = \frac{m-1}{2}, m = 1,3,5,\ldots; \\ \hat{x} f_{xp}(x) & p = \frac{m}{2}, m = 2,4,6,\ldots; \end{cases} \tag{9}$$

where, $f_{yp}(x)$ and $f_{xp}(x)$ are zero for $|x| > w/2$, and for $|x| < w/2$:

$$f_{yp}(x) = \frac{1}{\pi\sqrt{(\frac{w}{2})^2 - x^2}} \begin{cases} cos(p\pi x/w) & p \text{ even}; \\ sin(p\pi x/w) & p \text{ odd}; \end{cases} \tag{9a}$$

and

$$f_{xp}(x) = \frac{1}{w} \begin{cases} sin(p\pi x/w) & p \text{ even}; \\ cos(p\pi x/w) & p \text{ odd}. \end{cases} \tag{9b}$$

For convenience, the above functions are expressed choosing the center of the strips as reference.

References

[1] A. P. Defonzo and C. R. Lutz. Optoelectronic Transmission amd Reception of Ultrashort Electrical Pulses. *Applied Physics Letters*, 51:212–214, 1987.

[2] G. Arjavalingam, Y. Pastol, J. M. Halbout, and W. M. Robertson. Optoelectronically Pulsed Antennas: Characterization and Applications. *IEEE Antennas and Propagation Society Magazine*, 33(1):7–11, February 1991.

[3] G. Arjavalingam, Y. Pastol, L. W. Epp, and R. Mittra. Characterization of Quasi-Optical Filters with Picosecond Transient Radiation. *IEEE Transactions on Antennas and Propagation*, 40(1):63–66, January 1992.

[4] N.K. Das and D.M. Pozar. A Generalized Spectral-Domain Green's Function for Multilayer Dielectric Substrates with Applications to Multilayer Transmission Lines. *IEEE Transactions on Microwave Theory and Techniques*, MTT-35(3):326–335, March 1987.

[5] D. M. Pozar. Input Impedance and Mutual Coupling of Rectangular Microstrip Antennas. *IEEE Transactions on Antennas and Propagation*, AP-30(6):1991–1996, November 1982.

[6] N. K. Das and D. M. Pozar. Multiport Scattering Analysis of Multilayered Printed Antennas Fed by Multiple Feed Ports, Part I: Theory; Part II: Applications. *IEEE Transactions on Antennas and Propagation*, AP-40(5):469–491, May 1992.

[7] N. K. Das and D. M. Pozar. A Generalized CAD Model for Printed Antennas and Arrays with Arbitrary Multilayer Geometries. In L. Safai, editor, *Computer Physics Communication, Thematic Issue on Computational Electromagnetics*, North-Holland Publications, 1991.

[8] N. K. Das. *A Study of Multilayered Printed Antenna Structures.* PhD thesis, Department of Electrical and Computer Engineering, University of Massachusetts, Amherst, September 1987.

REFLECTOR ANTENNAS FOR ULTRAWIDEBAND USAGE

P R Foster

Microwave and Antenna Systems, UK

INTRODUCTION

The provision of high gain antennas for ultrawideband systems is limited not only by the electromagnetic problems but also by the mechanical aspects such as the volume required. Single antenna elements such as TEM horns can provide gains of 20.0 dBi at the highest operating frequencies but any additional gain requires the use of other techniques. The simplest techniques are to use an array of elements or to use a reflector antenna. This paper deals with the design of reflector antennas.

The transforming effect of the reflector surface is such that a constant aperture feed results in a constant gain reflector while a constant gain feed results in a constant aperture reflector. However this elegant postulate is somewhat degraded in practice by the presence of blockage, spillover and diffraction from the rim in a reflector antenna. Diffraction is particularly impor- tant because it is frequency sensitive and therefore the treatment of UWB reflector antennas is best carried out in the frequency domain.

The interactions between the feed and the reflector and the need to keep the antenna dimensions to reasonable values mean that the available bandwidth is limited to less than 15:1, and the lowest frequency is likely to be restricted and an examination of reflector performance at frequencies below 1 GHz is necessary.

TRANSFORMATION CAUSED BY A REFLECTOR SURFACE

The reflector provides a Fourier Transform of the illumination on the aperture reflector [1]. The aperture distribution is based on the feed radiation pattern which is based on the feed aperture distribution. This amounts to a double Fourier transform. The final radiation pattern from the reflector is therefore a version of the feed aperture distribution scaled in angle. This is only true when a sufficient angular extent of the feed radiation pattern is captured by the reflector surface, say, at least down to the first null.

Low gain antenna elements suitable for use as UWB feeds divide into two classes,

Ultra-Wideband, Short-Pulse Electromagnetics
Edited by H. Bertoni *et al.*, Plenum Press, 1993

those which provide a constant beamwidth and gain with frequency (loaded dipoles and frequency independent antennas) and those which provide a constant aperture which results in gain increasing with increasing frequency. The first class provides a constant aperture illumination for the reflector which results in a variable increasing with frequency and the second provides an aperture illumination on the reflector which decreases with increasing frequency. This results in constant gain.

In practise, blockage, diffraction and other effects will distort this simple picture.

CHOICE OF CONFIGURATION

The antenna elements suitable for use with a UWB system are limited in number [2]. In order to demonstrate the effects on reflector performance, one of each class has been examined in detail, a loaded dipole for the constant gain feed and a TEM horn for the constant aperture feed.

Reflectors with a shaped surface are not appropriate for UWB usage as they are frequency sensitive. The options (Figure 1) are

A prime focus reflector. Blockage will be caused by the feed and its support.

B axisymmetric dual reflector, e.g, Cassegrain. Apart from blockage problems, the secondary reflector should be a few wavelengths in diameter to avoid excessive edge diffraction and generation of crosspolarised radiation.

C single offset reflector which will avoid blockage

D dual offset reflector. Again the secondary reflector should be a few wavelengths across to avoid excessive diffraction effects.

In all four cases, the highest frequency of operation will be determined by the surface accuracy and the lowest frequency by the dimensions of the reflector (Figure 2).

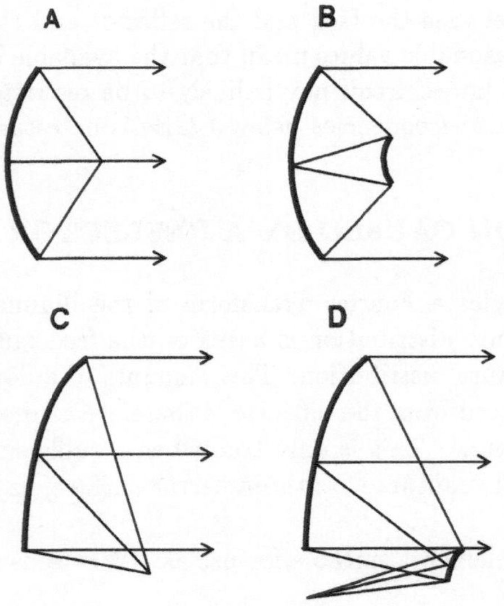

Figure 1: Schematic geometry of reflector types

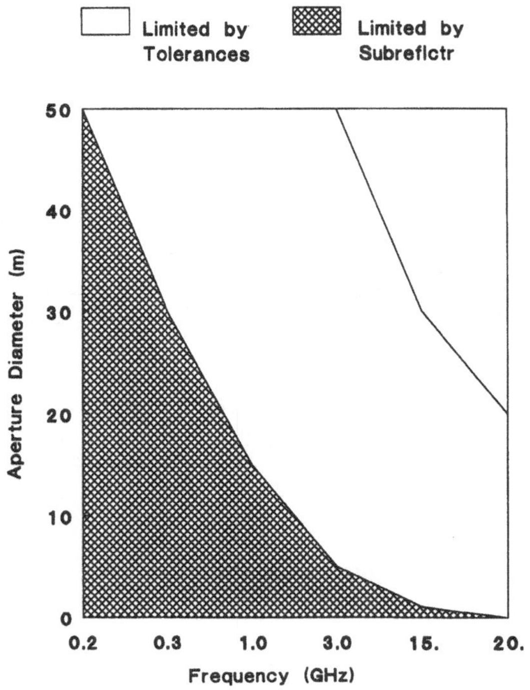

Figure 2: Frequency limits for a Cassegrain reflector

REFLECTOR PERFORMANCE WITH TEM HORN

Prime Focus Reflector

Due to mechanical constraints on the feed support, an F/D of about 0.45 is reasonable and gives an included angle of 116.0 degrees. The dimensions for the aperture of a TEM horn can be set by allowing the lowest frequency to illuminate the reflector edge at a -3.0 dB level. The aperture is 600 mm square at 250.0 MHz. The length is a function of highest frequency if an optimum horn is to be used. The length can be expressed as $0.5D^2\frac{F_h}{F_l}\lambda_l$ and is based on the constraint that a maximum phase error of $\pi/2$ takes place across the horn aperture at the highest frequency, F_h. D is the number of wavelengths, λ_l, in the aperture at the lowest frequency, F_h. This length is excessive if F_h is taken as 5.0 GHz and a value of 2.5 GHz was adopted, giving a length of 1.5 metres for the horn.

Computations of the reflector radiation patterns between 200 MHz and 2.5 GHz were carried out for aperture diameters of 5 and 50 metres. The peak gain is shown in Figure 3. The gain is not constant at the lowest frequencies because all the power from the horn is effectively captured by the reflector only at frequencies greater than 500 MHz. Because a smaller portion of the reflector is used at the higher frequencies, the percentage blockage increases with frequency. Since the two reflector geometries and the TEM feed horn are identical, the larger reflector is less affected by this problem. The radiation patterns are badly deformed above 1.0 GHz for the 5.0 m reflector with high sidelobes due to blockage (Figure 4). The patterns for the 50.0 m reflector are not deformed up to 5.0 GHz and shows the low sidelobes typical of this configuration when blockage is low. There is some distortion due to the phase error across the horn

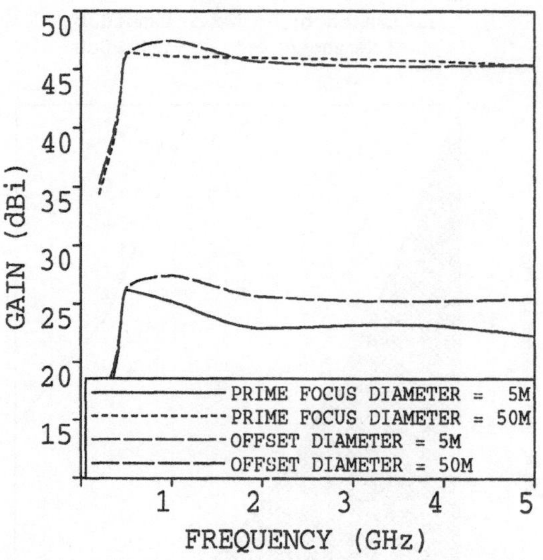

Figure 3: Peak gain vs frequency for a reflector with a TEM horn as feed

aperture at the highest frequencies. Crosspolarisation levels are determined by the levels generated by the TEM horn.

The mechanical problem of supporting such a large horn at the prime focus has not been addressed here but is recognised as a severe drawback to this configuration since the support mechanism will add to the blockage.

The problem of blockage will also affect an axisymmetric dual reflector used with a TEM horn. In addition, since a dual reflector will have a longer effective focal length, the TEM horn will have a larger aperture which will result in a longer horn or an increased phase error across the aperture at the highest frequencies. The TEM horn can be more easily supported at the secondary focus of a dual reflector system. The crosspolarised response will be large at the lowest frequencies since the subreflector will be only a few wavelengths in diameter

Offset Reflector

The offset reflector has certain constraints on geometry if high crosspolarisation is to be avoided [3]. A geometry with an offset angle of 45.0 degrees and a included angle of 70.0 degrees was adopted to obtain sufficient clearance past the TEM Horn and to provide a minimum focal length. The horn aperture must either be larger or the lowest frequency must be raised. At 250.0 MHz, the aperture will be 1000.0 mm and the length will be 1.5 m for a optimum length at 1.08 GHz. Computations were carried out for aperture diameters of 5.0 and 50.0 metres (Figure 3). An increase in gain over the prime focus reflector for a diameter of 5.0 metres is due to the removal of blockage. Crosspolarisation is present at a peak level of -19 dB at 250 MHz and falling to -26 dB at 5 GHz since the effective included angle decreases with increasing frequency.

Spillover And Diffraction

Spillover occurs when radiation from the feed is not intercepted by the reflector. Because a decreasing amount of the reflector is used as the frequency increases, spillover

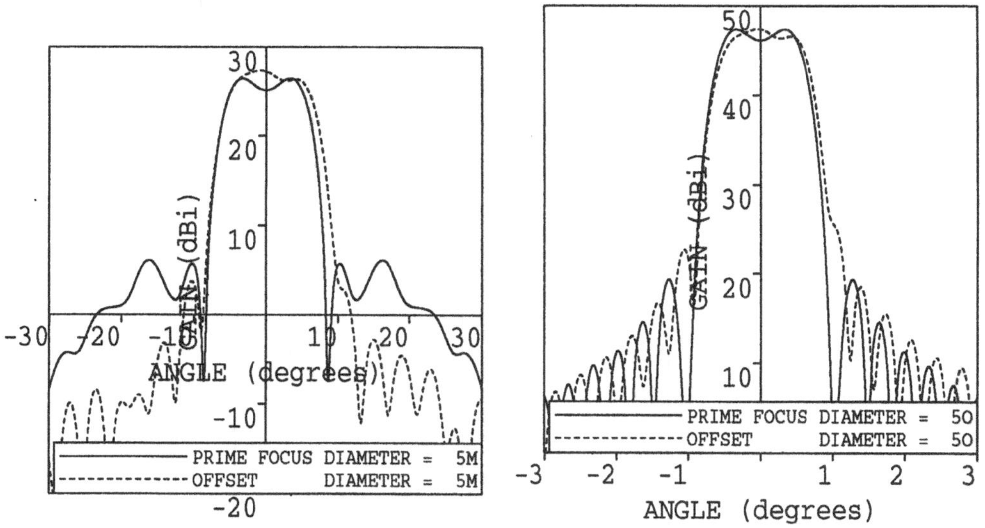

Figure 4: Radiation patterns with a TEM horn at 1 GHz

only occurs at the lowest frequencies and, in a single reflector, will occur in the rearwards direction. If a dual reflector system is used, the spillover will be in the forward direction and will provide energy in advance of the pulse from the main reflector.

Diffraction will occur from the edge of a reflector. When diffraction occurs from the main reflector rim, the time of travel will be that of the pulse from the aperture. When diffraction takes place at a subreflector, the time of travel will be less than that of the pulse from the aperture. As with spillover, the level of signal reaching the edges of the reflector and subreflector (if present) decreases rapidly with frequency and the diffracted signal will be low. In the examples of Figure 3, the edge illumination is below -18.0 dB for frequencies above 1 GHz.

REFLECTOR PERFORMANCE WITH LOADED DIPOLE

Prime Focus Reflector

The radiation pattern of a loaded dipole is that of a halfwave dipole at all frequencies. A short focal length single reflector must be used although it may be either an axisymmetric or an offset reflector. Blockage with a loaded dipole will be small although a support structure will be needed.

The major problem with a loaded dipole is the forward radiation which will produce a prepulse. The only way of suppressing this forward radiation would be to use a RAM-loaded plate behind the dipole which would have to be at least one wavelength in diameter to provide any protection for the lowest frequencies. Use of a conducting plate would make the feed frequency sensitive. A RAM loaded plate would cause blockage as severe as the blockage of a TEM horn.

Radiation patterns have been computed for a prime focus reflector with a diameter of 5 and 50 m and an F/D ratio of 0.45 for comparison with the results using a TEM horn. As expected, the gain increases and the beamwidth decreases with frequency (Figures 5 and 6). A shorter F/D ratio could be used but pulse behaviour is likely to degrade.

Off boresight, the highest frequencies will provide less gain and the risetime of the pulse will reduce. There is also an effect due to the radiation from both ends of the dipole which will produce a single pulse in the reflector aperture in the central ray but a double pulse having increasing separation towards the edge of the aperture. This is worsened by the use of a short F/D ratio.

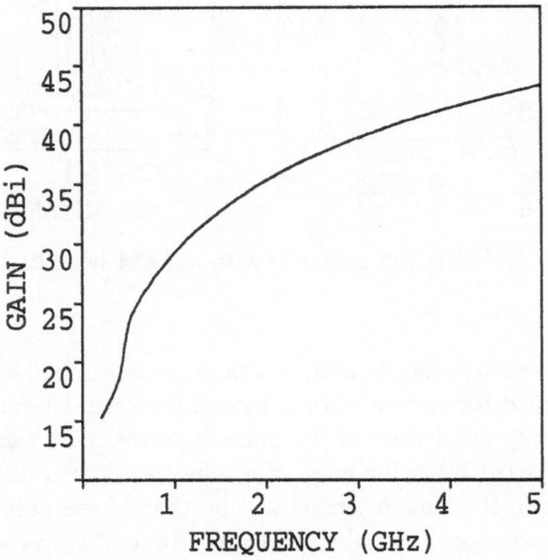

Figure 5: Peak gain vs frequency for a prime focus reflector with a loaded dipole as feed

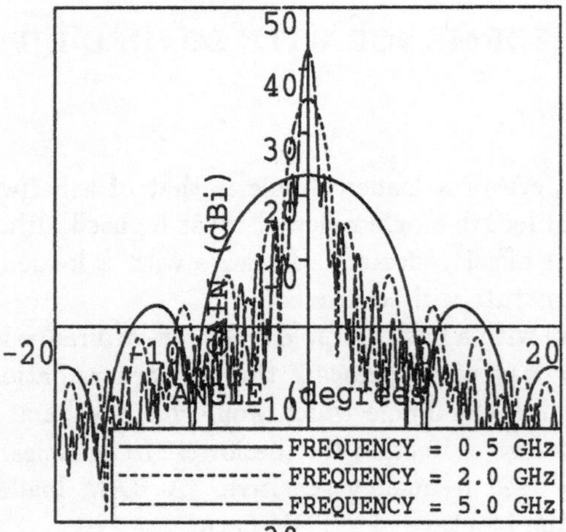

Figure 6: Radiation patterns of a prime focus reflector (F/D = 0.45 and aperture diameter = 5.0 metres)

CONCLUSIONS

The degradation caused by blockage forces the use of an offset reflector when a TEM horn is used unless the aperture diameter of the reflector is greater than 15 metres. The bandwidth is restricted to 10:1 for constant gain from the reflector. The lower limit to frequency is set by the breakdown of the feed aperture transform while the upper frequency is by the difficulties of making a high frequency TEM horn. The use of a single offset reflector with a TEM horn provides the least degradation to pulse transmission.

A loaded dipole may be used with an axisymmetric reflector, provided no attempt is made to remove the dipole back radiation. However the pulse is much more distorted when a loaded dipole is used due to the offaxis pulse behaviour of the dipole itself. A reflector used with a loaded dipole is a constant aperture antenna, at least, near boresight and, as such, is suitable for a receive antenna.

Dual reflector systems are not appropriate for UWB usage because of the problems caused by forward spillover past the subreflector and the difficulty of providing a narrow beam of a less than 20.0 degrees at the lowest frequencies. A narrow beam requires a large aperture in the TEM horn and the length of a TEM horn increases with D^2 where D is the side dimensions of the horn aperture. The length is increased still further if an optimum balun is used since the balun will have a length of about 0.5m.

REFERENCES

[1] J. S. Ajioka and J. L. McFarland, "Beam-forming Feeds", Chapter 19 in "Antenna Handbook", Editors Y. T. Lo and S. W. Lee, Van Nostrand Reinhold Company, New York, 1988

[2] P. R. Foster, "Performance of Ultrawideband Antennas", SPIE Conference on Ultrawideband Radar, Los Angeles, January,1992

[3] W. V. R. Rusch et al, "Quasi-optical Antenna Design", Chapter 3 in "The Handbook of Antenna Design", editors, A. W. Rudge et al, Peter Peregrinus, 1982.

TRANSIENT LENSES FOR TRANSMISSION SYSTEMS AND ANTENNAS

Carl E. Baum and Alexander P. Stone

Phillips Laboratory
Kirtland AFB, New Mexico 87117

Department of Mathematics and Statistics
University of New Mexico
Albuquerque, NM 87131

THE LENS DESIGN PROBLEM

In the design of waveguide transitions we desire to transmit a TEM wave, ideally with no reflection or distortion, from one transmission line to another. Such waveguide transition regions are usually referred to as EM lenses or more specifically, transient lenses. This goal is accomplished by specifying the lens geometry and constitutive parameters (i.e., the shape and medium of the EM lens). The physical properties of these lenses, given by the permeability μ and the permittivity ϵ, may be a function of position, but we assume that these properties are frequency independent. The conductivity of the medium is taken to be zero, and cross sectional dimensions are large compared to the wavelengths at the high frequencies of interest. This is in contrast to a lens such as the Luneburg lens or Maxwell fish eye lens, both of which are based on a geometric optics approximation. The need for a low dispersion system also argues for TEM guiding structures. Since we may also wish to change the direction of propagation of a wave being transmitted from one region to another, we must also allow for distortion introduced at bends. While in many cases we can obtain exact solutions to the lens design problem, approximations are generally involved in the practical realizations of most EM lenses. For example, one may have to cut off lenses that are theoretically infinite in extent. Moreover, frequency independence may not be realized exactly. The exact solutions to design problems are usually obtained by one of two basic approaches [1]. The first method is a differential-geometric approach, while the second method is a differential-impedance-matching and transit-time-conservation approach.

TWO APPROACHES TO EXACT LENS DESIGN

Differential Geometric-Scaling

The basic idea is the creation of a class of electromagnetic problems, each having a complicated geometry and medium, which are equivalent under a differential-geometric scaling to an electromagnetic problem having a simpler geometry and medium. The latter problem we term the formal problem while the former problem is the lens (or real-world) problem. Solutions to Maxwell's equations can then be used to specify EM lenses for transitioning TEM waves, without distortion or reflection, between certain types of transmission lines. Thus we consider an orthogonal curvilinear coordinate system (u_1, u_2, u_3) with a line element

$$(d\ell)^2 = h_1^2(du_1)^2 + h_2^2(du_2)^2 + h_3^2(du_3)^2 \tag{1}$$

where the scale factors h_i are given by

$$h_i^2 = \left(\frac{\partial x}{\partial u_i}\right)^2 + \left(\frac{\partial y}{\partial u_i}\right)^2 + \left(\frac{\partial z}{\partial u_i}\right)^2 \tag{2}$$

with (x, y, z) rectangular Cartesian coordinates. The Maxwell equations and constitutive relations are

$$\nabla \times \vec{E} = -\frac{\partial \vec{B}}{\partial t}\ , \nabla \times \vec{H} = \vec{J} + \frac{\partial \vec{D}}{\partial t}\ , \nabla \cdot \vec{D} = \rho\ , \nabla \cdot \vec{B} = 0$$

$$\vec{D} = (\epsilon_{i,j}) \cdot \vec{E},\ \vec{B} = (\mu_{i,j}) \cdot \vec{H}\ . \tag{3}$$

In our situation $\rho = 0$ and $\vec{J} = 0$, except on boundaries. We have assumed that $(\epsilon_{i,j})$ and $(\mu_{i,j})$ are real constant matrices, independent of frequency; they may, however, be functions of position. We introduce matrices $(\alpha_{i,j})$ and $(\beta_{i,j})$, dependent on the scale factors h_i, and obtain

$$(\alpha_{i,j}) = (1_{i,j} h_i), (\beta_{i,j}) = (1_{i,j} H/h_i)\ , \quad H = h_1, h_2, h_3$$

$$\vec{E}' = (\alpha_{i,j}) \cdot \vec{E}\ ,\ \vec{H}' = (\alpha_{i,j}) \cdot \vec{H}\ ,\ \vec{B}' = (\beta_{i,j}) \cdot \vec{B}\ ,\ \vec{D}' = (\beta_{i,j}) \cdot \vec{D} \tag{4}$$

where the primed parameters are the formal ones. The formal Maxwell equations are obtained by defining

$$(\epsilon'_{i,j}) = (\beta_{i,j}) \cdot (\epsilon_{i,j}) \cdot (\alpha_{i,j})^{-1}\ ,\ (\mu'_{i,j}) = (\beta_{i,j}) \cdot (\mu_{i,j}) \cdot (\alpha_{i,j})^{-1} \tag{5}$$

and replacing all quantities by primed quantities. The restriction to inhomogeneous isotropic media will restrict the form of the matrices to diagonal forms

$$(\epsilon_{i,j}) = \epsilon(1_{i,j})\ ,\ (\mu_{i,j}) = \mu(1_{i,j}) \cdot (\epsilon'_{i,j}) = \epsilon(\beta_{i,j}) \cdot (\alpha_{i,j})^{-1}\ ,\ (\mu'_{i,j}) = \mu(\beta_{i,j}) \cdot (\alpha_{i,j})^{-1}\ . \tag{6}$$

Examples of this technique include the design of converging, diverging, and bending lenses (based on bispherical, toroidal, and cylindrical coordinate systems, respectively), for transitioning TEM waves between conical and/for cylindrical transmission lines [1]. Special two dimensional lenses, based on conformal transformations, are also found in which one is required to vary only scalar ϵ or μ, but not both [1]. Perfectly conducting boundaries are arranged to follow the appropriate coordinates in the orthogonal curvilinear system for which tangential $\vec{E}$ is zero. Figure 1 shows an inhomogeneous diverging lens, based on toroidal coordinates, with cylindrical and conical transmission lines.

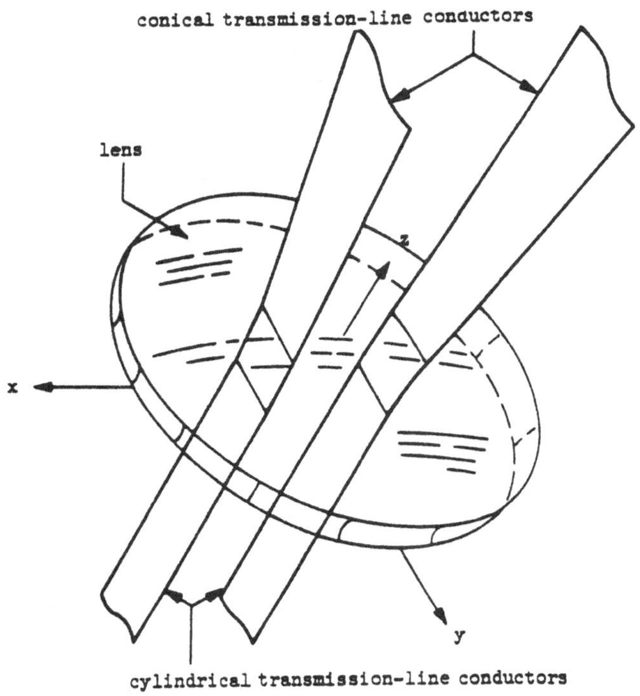

Figure 1. Toroidal Lens with Cylindrical and Conical Transmission Lines

A recent simple example of an exact uniform isotropic lens, used in bending the direction of propagation of a TEM wave between two wide parallel plates, is a wedge dielectric lens [3] shown in Figure 2. The bend angle, ψ_b is a function only of the relative permittivity, ϵ_r, of the lens. It is given by

$$\psi_b = 4\arctan(\sqrt{\epsilon_r}) - \pi \ , \ \ \epsilon_r = \tan^2\left(\frac{\psi_b + \pi}{4}\right) \ , \ \ \psi_1 = \frac{\pi + \psi_b}{4} \ , \ \ \psi_2 = \frac{\pi - \psi_b}{4} \tag{7}$$

Hence, if we wish to bend a waveguide by $\psi_b = \pi/4$, then $\psi_1 = 5\pi/16 \simeq 56.3°, \psi_2 = 3\pi/16 \simeq 33.8°$, and $\epsilon_r = 2.24$ which is the dielectric constant that approximates that of transformer oil or polyethylene. We observe that the Brewster condition, $\psi_1 + \psi_2 = \pi/2$, (no reflection at interfaces) is satisfied. We assume that the plate spacing h is small compared to the width w. As $w \to \infty$ the lens is exact in the sense that the various TEM waves are matched. However, in practical applications a finite width will introduce imperfections.

IMPEDANCE-MATCHING AND TRANSIT-TIME CONSERVATION

An alternative approach to exact transient lens design is one which might be termed a differential-impedance-matching and transit-time conservation method. We must first differentially match impedances at all waveguide and lens boundaries so that a TEM wave may be transmitted from one region to another without reflection. Secondly, in order that a wave be transmitted undistorted, a plane or spherical TEM wave in one region should go into one such wave in another region and consequently the travel times for waves following different paths should be conserved. These two conditions usually give rise to a system of nonlinear differential equations whose solutions will determine the lens geometry and medium. In [6] we have a simple example to explain this technique for bending TEM waves by dividing the space between two infinitely wide parallel perfectly conducting plates by the insertion of perfectly conducting sheets. A transient broad-band lens then can be constructed by bending these plates (Fig. 3). The wavefront in the lens is taken as a symmetry plane and the conducting sheets are bent so that the lens boundaries form equal angles ψ_w on both sides of one such

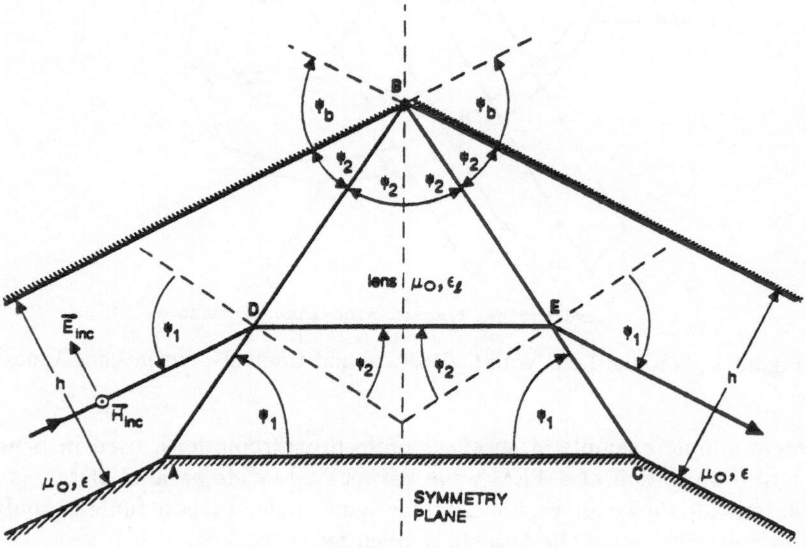

Figure 2. Wedge Lens in Parallel-Plate Geometry

wavefront. The spacing Δ, and hence, the admittance per unit width, is maintained between adjacent plates, and thus there is no reflection at the symmetry plane (in the transmission-line approximation if the wavelength λ is much less than Δ). We note in Fig. 3 that the total angle of bend, ψ_B, is $2(\psi_1 - \psi_w)$. This angle can be strictly in terms of ψ_1 or ψ_w and when the permeabilities and permittivities are the same we find that

$$\psi_b = 2[\psi_1 - \arctan\left(\frac{\sin(\psi_1)\cos(\psi_1)}{1+\sin^2(\psi_1)}\right)], \quad \psi_b = 4\psi_1^3 + 0(\psi_1^5) \text{ for } \psi_1 \to 0\,. \tag{8}$$

This lens represents still another way of controlling TEM propagation in a way that allows λ to be much less than the parallel plate spacing in the original waveguide. Since the plates cannot be of infinite width there are fringe fields at the edges of the plates which are not accommodated by the two-dimensional lens structure. The lens will intercept some fraction of the energy in the incident TEM wave, approximately that between the plates. If this fraction is near 1, good lens performance is expected. While this kind of lens can have various materials between the conducting sheets, one application may lie in evacuated transmission systems.

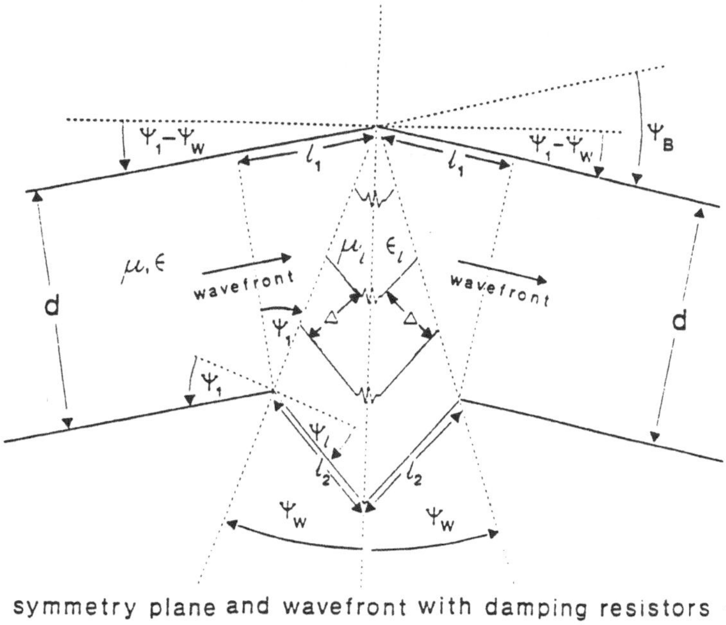

Figure 3. Bending Lens

This general approach to lens design, which is described in detail in [1], has been used to construct a number of examples of exact anisotropic lenses. In these examples the permittivity is in general also a function of position (inhomogeneous) while the closely spaced metal sheets make the medium anisotropic. These examples include a transition region between two cylindrical waveguides of different size (Fig. 4), a circular conical lens feeding a circular conical antenna, and a lens for transitioning plane waves between media of different permittivities.

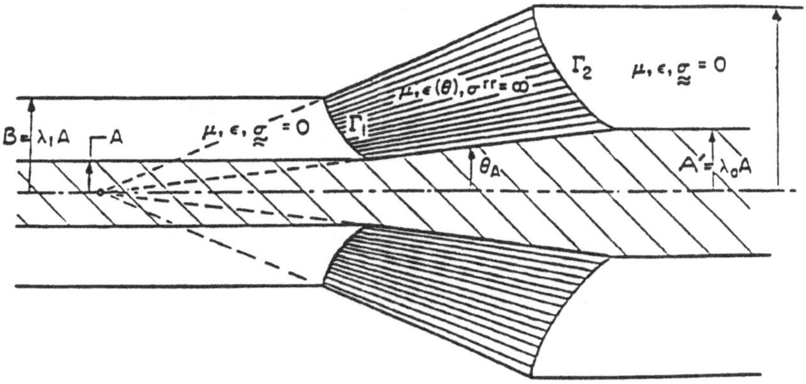

Figure 4. Geometry of Fixed μ, Variable ε, Anisotropic Lens Matching Circular Coaxial Lines of Different Size

APPROXIMATE LENSES

If the equal time requirement is maintained, but the impedance matching condition relaxed from a differential one (along every ray) to a global one in which the transmission line impedance is kept constant on both sides of the lens (as well as inside the lens if possible), then one may develop a class of approximate lenses. What results is an average kind of impedance matching, allowing some small mismatch along ray paths, particularly at lens boundaries. By carefully choosing lens parameters the high-frequency properties can sometimes be made to have small degradation. A specific example [2] of this approximate procedure is that of a prolate spheroidal lens feeding a circular coax (Fig. 5). The transition region is a concentric uniform isotropic dielectric lens feeding a circular coaxial line, specified so that a wave launched at an apex propagates through the lens region onto the coaxial line with minimum reflection and distortion. The equal-transit-time condition dictates that the lens shape is a prolate spheroid whose equation is

$$\left(\frac{\Psi_b}{\ell}\right)^2 + 2\left(1 - \frac{1}{\sqrt{\epsilon_r}}\right)\frac{z_b}{\ell} + \left(1 - \frac{1}{\epsilon_r}\right)\left(\frac{z_b}{\ell}\right)^2 = 0\,. \tag{9}$$

A macroscopic impedance matching condition, allowing a uniform isotropic lens, can then be imposed and one finds that the lens shape is determined by ϵ_r and the characteristic impedance of the coaxial line. This condition is given by

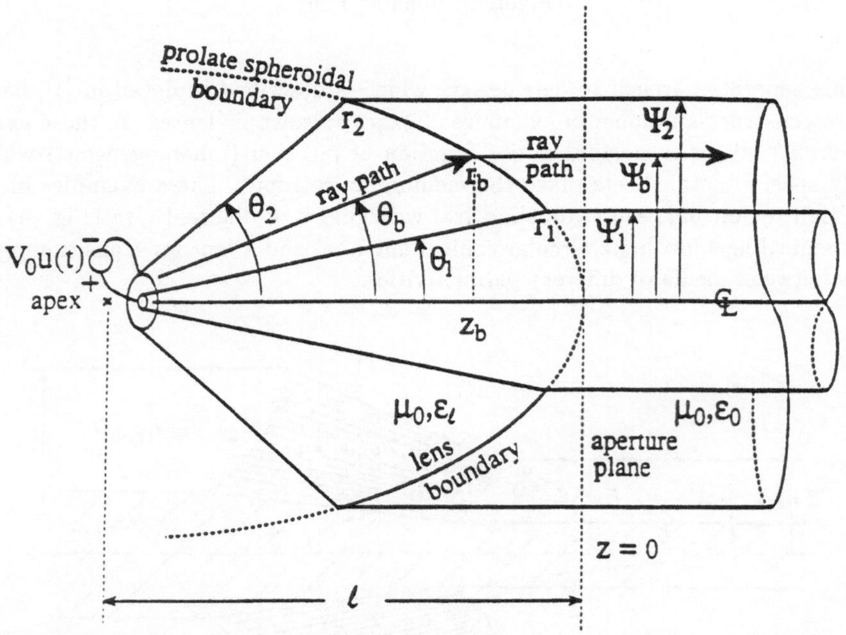

Example for $Z_c = 50\Omega$ (free space medium) with $\varepsilon_r = 2.26$ (transformer oil or polyethylene)

Figure 5. Prolate Spheroidal Lens with Circular Conical Transmission Line Feeding Circular Coax

$$\frac{Z_0}{2\pi}\ln\left(\frac{\Psi_2}{\Psi_1}\right) = \frac{Z_0}{2\pi\sqrt{\epsilon_r}}\ln\left[\frac{\tan\left(\frac{\theta_2}{2}\right)}{\tan\left(\frac{\theta_1}{2}\right)}\right] \tag{10}$$

A calculation then leads to expressions for the lens parameters θ_1 and θ_2 as a function of ϵ_r and the characteristic impedance. We may also define a high-frequency transfer function T_V which compares an initial voltage associated with the TEM made on the coax to a step rising voltage $V_0 u(t)$ in the conical section. For the general case this is

$$T_V = \frac{V(t_{a+})}{V_0} = \frac{\int_{S_a} \vec{E}(\vec{r}_s, t_{a+}) \cdot \vec{e}_0(\vec{r}_s) dS}{\int_{S_a} \vec{e}_0(\vec{r}_s) \cdot \vec{e}_0(\vec{r}_s) ds} \tag{11}$$

where $\vec{e}_0(\vec{r}_s) \equiv$ electric-field distribution on S_a of TEM mode, and $\vec{E}(\vec{r}_s, T_{a+}) \equiv$ initial electric field on S_a for $\vec{e}_0$ normalized to give 1 volt between conductors. S_a is the effective aperture plane (perpendicular to the z-axis of a general cylindrical transmission line), and $V(t_{a+})$ is the initial voltage on the coax at time $t = t_{a+}$ just after the wave initial rise. For our special coaxial geometry this is

$$T_V = \frac{1}{V_0}\int_{\Psi_1}^{\Psi_2} E_\Psi(\Psi, t_{a+}) d\Psi \, . \tag{12}$$

Since the conical and cylindrical sections have the same characteristic impedance, $|T_V|$ is bounded by 1 We find that T_V may be expressed through an integral as a function of ϵ_r and the coax impedance. Moreover, over a useful range of coax impedance and lens permittivity the lens is quite efficient. The impedance condition assures perfect low-frequency performance (no reflection) and the high frequency/early-time performance, as measured by T_V, is also quite good. For example, when $\epsilon_r = 2.26$ (the approximate dielectric constant for polyethylene or transformer oil) T_V varies from .999 to .977 as the characteristic impedance varies from 1 ohm to 50 ohms. The quantity T_V^2, which is also quite close to 1, represents the power in the TEM mode on the coax. the quantity $(1 - T_V^2)$ represents the power in the reflections at the lens boundary plus all the higher order modes on the coax.

The examples of such approximate lenses need to be expanded to include those for TEM waveguides more complicated than circular coaxes. Another interesting geometry is the balanced two-wire structure such as might be used for the feed structure launching a spherical TEM wave into a paraboloidal reflector [7], (as in the IRA (impulse radiating antenna)). As one approaches the conical apex, the electric field strength can get quite large so that various dielectric insulation techniques, as illustrated in Fig. 6, are needed. Where the dielectric has a relative permittivity significantly greater than 1.0, the dielectric medium can be thought of as a lens which needs to be shaped (along with the conductors) to minimize the pulse distortion in this region.

CONCLUDING REMARKS

These transient lenses have application to impulse radars, EMP simulators, pulse-power equipment, and in general devices in which it is important to transport electromagnetic waves in a low-loss dispersionless way over a broad band of frequencies, including (at the high frequency end) radian wavelengths shorter than the cross-section dimensions of the transmission and/or radiating systems. Transmission systems may include coaxial wave-guides (cables), strip lines, etc. In this case they may need to be bent or changed in some way, and transitions (lenses) constructed for matching between the various parts. Lenses have often been used in directional antennas for straightening out the phase front over a large aperture to give a highly directive beam. Using the present concepts this is extended to use with broadband pulses [4] [5]. A similar application is for the launching of fast-rising high-amplitude TEM waves in electromagnetic-pulse (EMP) simulators. In expanding the wave from a relatively smaller feed line to a large aperture a conical TEM structure is appropriate and transient lenses are also useful for establishing the spherical TEM wave on such structures.

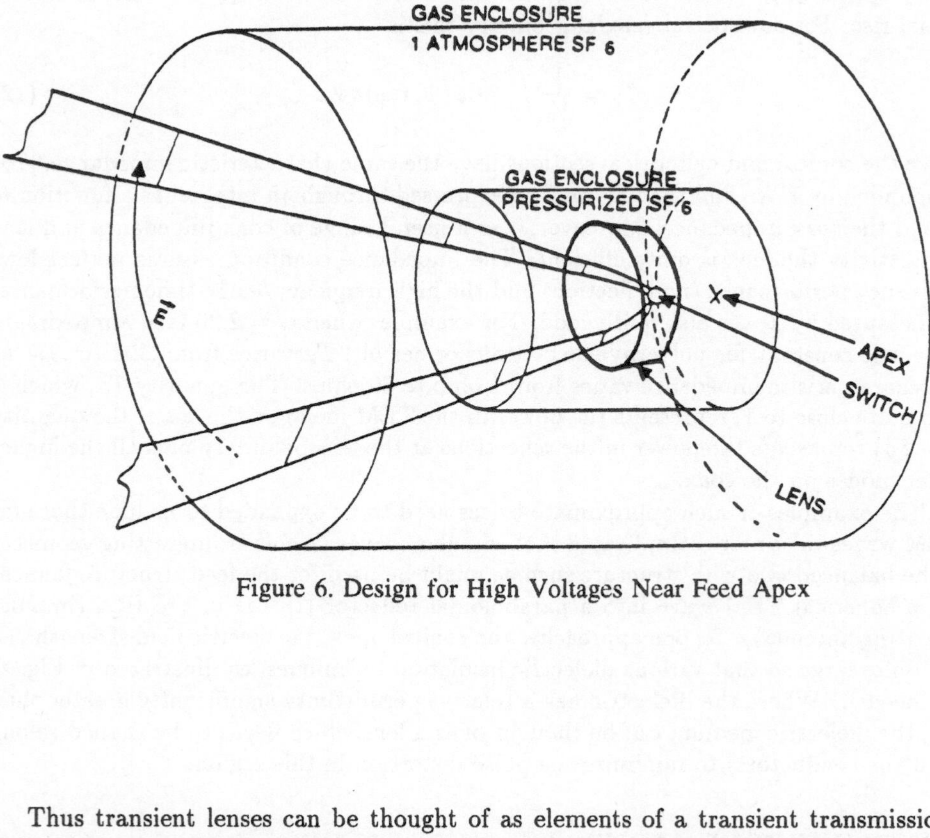

Figure 6. Design for High Voltages Near Feed Apex

Thus transient lenses can be thought of as elements of a transient transmission/ radiation system. They transition waves between various parts of the system, receiving the wave from one part and reshaping it for transmission into the next part with a minimum of pulse distortion. Such elements are essential if one is to transport pulses with significant frequencies so high that corresponding radian wavelengths are less than cross section dimensions of the transmission/radiation system.

REFERENCES

1. C. E. Baum and A. P. Stone, *Transient Lens Synthesis: Differential Geometry in Electromagnetic Theory*, Hemisphere Publishing Corp., 1991.
2. C. E. Baum, J. J. Sadler, and A. P. Stone, "A Prolate Spheroidal Uniform Isotropic Dielectric Lens Feeding a Circular Coax," Sensor and Simulation Note 335, December 1991.
3. C. E. Baum, "Wedge Dielectric Lenses for TEM Waves Between Parallel Plates," Sensor and Simulation Note 332, September 1991.
4. C. E. Baum, "Aperture Efficiencies for IRAs," Sensor and Simulation Note 328, June 1991.
5. C. E. Baum and E. G. Farr, "A Simple Model of Small-Angle TEM Horns," Sensor and Simulation Note 340, May 1992.
6. C. E. Baum, "Arrays of Parallel Conducting Sheets for Two-Dimensional E-Plane Bending Lenses," Sensor and Simulation Note 341, April 1992.
7. C. E. Baum, "Configurations of TEM Feed for an IRA," Sensor and Simulation Note 327, April 1991.

PULSE PROPAGATION AND GUIDANCE

ASYMPTOTIC DESCRIPTION OF ELECTROMAGNETIC PULSE PROPAGATION IN A LINEAR DISPERSIVE MEDIUM

Kurt E. Oughstun, Judith E.K. Laurens, and Constantinos M. Balictsis

Department of Computer Science and Electrical Engineering
University of Vermont
Burlington, Vermont 05405

INTRODUCTION

The asymptotic description of linear dispersive pulse propagation phenomena in lossy dielectrics is critically dependent upon the dominant contribution (i.e. the contribution with the least exponential decay) to the propagated field behavior at a given space-time point. This asymptotic description is based upon the exact integral representation that describes plane wave pulse propagation through a linear dispersive medium occupying the half space $z > 0$ and is given by

$$A(z,t) = \frac{1}{2\pi}\int_C \tilde{f}(\omega)\exp\{\frac{z}{c}\phi(\omega,\theta)\}d\omega \tag{1}$$

for all $z \geq 0$, where

$$\tilde{f}(\omega) = \int_{-\infty}^{\infty} f(t)e^{i\omega t}dt \tag{2}$$

is the temporal Fourier spectrum of the initial pulse $f(t) = A(0,t)$ at the plane $z = 0$. Here $A(z,t)$ represents any scalar component of the electric field, magnetic field, or Hertz vector whose spectral amplitude $A(z,\omega)$ satisfies the dispersive Helmholtz equation

$$(\nabla^2 + \tilde{k}^2(\omega))A(z,\omega) = 0 \tag{3}$$

throughout the half-space $z > 0$. The complex wave number $\tilde{k}(\omega) = \beta(\omega) + i\alpha(\omega)$ appearing here is given by

$$\tilde{k}(\omega) = \frac{\omega}{c}n(\omega) \ , \tag{4}$$

where c denotes the speed of light in vacuum. The complex index of refraction $n(\omega) = n_r(\omega) + in_i(\omega)$ of the dispersive dielectric occupying the half-space $z > 0$ is given by the appropriate branch of the square root relationship

$$n(\omega) = (\mu\epsilon(\omega))^{1/2} \ , \tag{5}$$

where μ is the relative magnetic permeability and where $\epsilon(\omega) = \epsilon_r(\omega) + i\epsilon_i(\omega)$ is the relative complex-valued dielectric permittivity of the medium.

If $f(t) = 0$ for $t < 0$, then Eq.(1) is taken to be a Laplace representation in which the contour of integration C is the line $\omega = \omega' + ia$ with a being a fixed positive constant that is greater than the abscissa of absolute convergence[1] for the function $f(t)$ and where $\omega' = \Re\{\omega\}$ varies from negative to positive infinity. The complex phase function $\phi(\omega, \theta)$ appearing in the integrand of Eq.(1) is given by

$$\phi(\omega, \theta) = i\frac{c}{z}(\tilde{k}(\omega)z - \omega t) = i\omega(n(\omega) - \theta) \ , \tag{6}$$

where $\theta = ct/z$ is a dimensionless parameter that characterizes any particular space-time point (z, t) in the plane wave field.

Any given dielectric material is both dispersive and absorptive with regard to the propagation of electromagnetic waves. As a consequence, both the phase velocity $v_p(\omega) = \omega/\beta(\omega)$ and the attenuation coefficient $\alpha(\omega)$ of a time-harmonic wave in a dispersive medium are functions of the angular frequency of oscillation ω of the wave. In addition, the physical requirement of causality[2] intimately relates the frequency dispersion of the phase velocity to that of the attenuation coefficient via the Hilbert transform pair relationship between the real and imaginary parts of the dielectric permittivity. These two interrelated properties of the medium have a profound influence upon the propagation characteristics of transient fields, such as electromagnetic pulses, as they propagate through the material.

The exact integral representation given in Eq.(1) describes the propagation of any given transient electromagnetic field in terms of a superposition of time-harmonic wave components with different angular frequencies. Because both the phase velocities and attenuation coefficients of these spectral components are functions of the angular frequency ω in a dielectric, then both the relative phases and amplitudes of the different spectral components change in a continuous fashion as the transient field propagates through the dielectric. These changes, in turn, can result in profound changes in both the envelope structure and intensity of the pulse as well as in its instantaneous oscillation frequency, all of which evolve in a continuous manner as the pulse propagates through the dispersive medium. For ultrashort pulses, this evolution may be separated into two disjoint regimes. In the first, or immature dispersion regime, the propagation distance is smaller than a single absorption depth of the medium (at the appropriate frequency) and the properties of the pulse rapidly change in a complicated manner. Eventually, as the propagation distance increases above the absorption depth, the field evolution settles into the second, or mature dispersion regime[3], and remains there for the remainder of the propagation, where the pulse now evolves in a simple, systematic manner.

The behavior of the propagated field throughout the mature dispersion regime is given by the asymptotic expansion of the exact integral representation (1) for large z. Because the medium is absorptive, and hence $n(\omega)$ is complex-valued, the saddle point method[4,5] must be used to obtain the required asymptotic approximation. The classic theory of signal propagation in a Lorentz model dielectric using the asymptotic method of steepest descents was originally developed by Sommerfeld[6] and Brillouin[7,8] in 1918 and was partially modified by Baerwald[9] in 1930. This classic theory has recently been corrected[10,11] with independent numerical verification[12–14] and expanded upon[15–19] using modern asymptotic theory.

For a given dispersive medium, the dominant contribution (i.e. the contribution with the least exponential decay) to the asymptotic approximation of the integral representation (1) of the field behavior at a fixed space-time value θ is completely determined by the saddle points of the complex phase function $\phi(\omega, \theta)$ that is characteristic of the dispersive medium as well as by the initial pulse spectrum $\tilde{f}(\omega)$. If the initial rise and/or fall time of the input pulse is less than the medium relaxation time, then the appropriate form of the complex phase function is that given in Eq.(6). The saddle point dynamics then depend only upon the dispersive properties of the medium. As a consequence, the asymptotic theory requires a causal, analytic model of the frequency dispersion of the medium. The construction of such a classical model for triply distilled H_2O that is reasonably accurate for all $\omega\epsilon[0, \infty)$ is considered in the second Section of this paper. However, as the initial rise and/or fall time of the input pulse increases above the medium relaxation time (in the appropriate frequency domain), it is found that the complex phase function $\phi(\omega, \theta)$ must be modified so as to include the initial rise/fall-time parameter. The resultant saddle point dynamics then depend not only upon the medium dispersion but also upon the input pulse properties. This is considered in the third Section of this paper for the case of an input gaussian envelope pulse as the input pulse width is allowed to vary from the ultrashort (sub-femtosecond) limit up to the quasimonochromatic regime.

FREQUENCY DISPERSION MODEL OF THE DIELECTRIC PERMITTIVITY OF TRIPLY-DISTILLED LIQUID WATER AND ITS THEORETICAL IMPULSE RESPONSE

Experimental data of the real frequency dispersion of the real index of refraction $n_r(\omega) = \Re\{n(\omega)\}$ and of the absorption coefficient

$$\alpha(\omega) = \frac{\omega}{c} n_i(\omega) \ , \tag{7}$$

where $n_i(\omega) = \Im\{n(\omega)\}$, of triply-distilled liquid water is illustrated in Fig.1. Here $\Re\{\cdot\}$ denotes the real part and $\Im\{\cdot\}$ denotes the imaginary part of the quantity appearing in the brackets. Although this experimental data may be sufficient for some numerical models of electromagnetic pulse propagation, it is insufficient for the asymptotic description of such phenomena. The asymptotic theory requires that $n(\omega) = n_r(\omega) + in_i(\omega)$ be known throughout the complex ω-plane, and physical reality requires that the real and imaginary parts of $n(\omega)$ satisfy the Kramers-Krönig relations[2] or, equivalently, that the real and imaginary parts of $\epsilon(\omega)$ form a Hilbert transform pair. The latter condition restricts the class of functions that may be used to construct a model of $n(\omega)$ that yields the measured real frequency dependence depicted in Fig.1. Furthermore, since a correct description of the electromagnetic energy flow in a given causally dispersive medium requires[20] that the dynamical model of the medium response to the externally applied electromagnetic field be known, it is inappropriate to simply construct an arbitrary causal model of $\epsilon(\omega)$ that has as many variable parameters as necessary to fit a given set of experimental data if that model has no physical basis.

Analytic, Frequency Dispersion Model of the Dielectric Permittivity of Triply-Distilled Water

The low frequency dependence of $n_r(\omega)$ that is depicted in Fig. 1 for $0 \leq \omega \lesssim 10^{13} hz$ is characteristic of rotational polarization in dielectrics, while the anomalous dispersion phenomena observed for $\omega > 10^{13} hz$ is characteristic of resonance polarization in dielectrics. Rotational polarization effects may be adequately described by the Rocard-Powles model[21], which is a first-order correction to the classic Debye model[22,23] of polar media, while resonance polarization phenomena may be described by the classical Lorentz oscillator model[24,25]. Both of these physical models result in causal models for the dielectric permittivity.

Linear combination of the analytic expression for the dielectric permittivity given by the Rocard-Powles extension of the Debye model for two relaxation times with the analytic expression for the dielectric permittivity given by the Lorentz oscillator model for four distinct

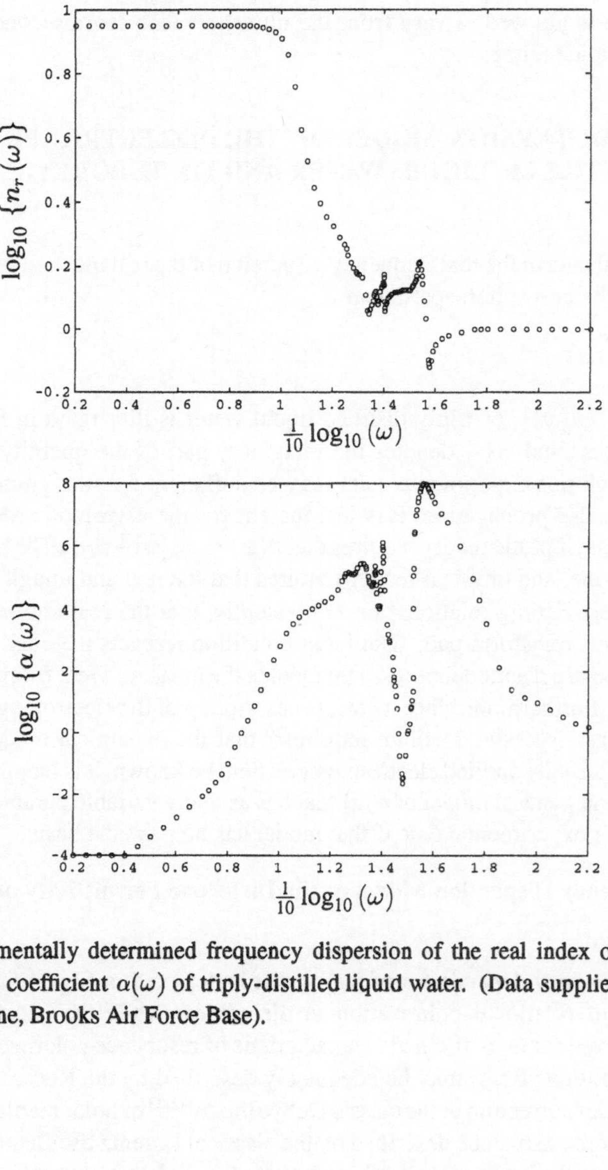

Figure 1. Experimentally determined frequency dispersion of the real index of refraction $n_r(\omega)$ and the absoprtion coefficient $\alpha(\omega)$ of triply-distilled liquid water. (Data supplied by the School of Aerospace Medicine, Brooks Air Force Base).

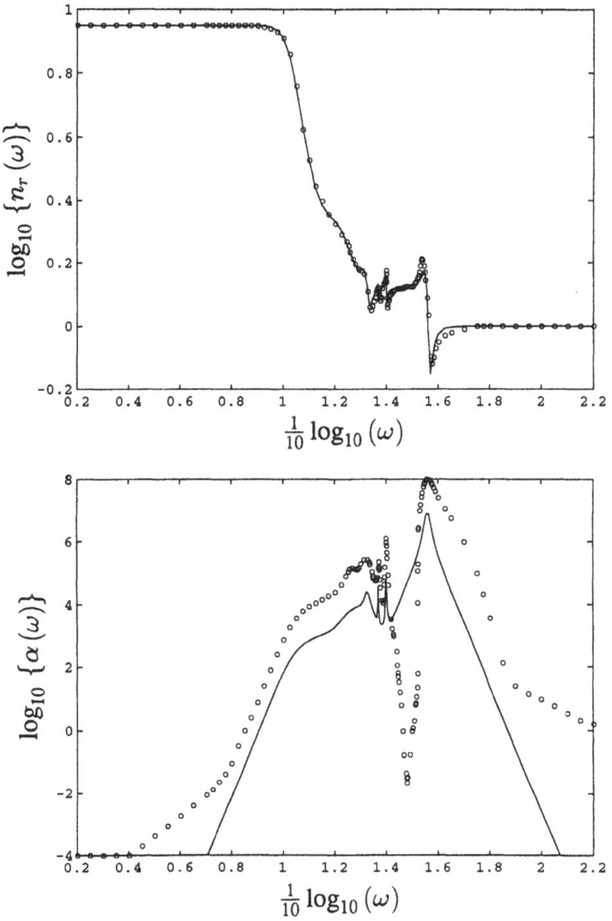

Figure 2. Comparison of the theoretical curves (solid curves) with the experimental data (o o o) for the real index of refraction (upper diagram) and the absorption coefficient (lower diagram) of triply-distilled liquid water.

resonance lines results in the expression

$$\begin{aligned}\epsilon(\omega) = \epsilon_\infty &+ \frac{a_0}{(1-i\omega\tau_0)(1-i\omega\tau_{f0})} + \frac{a_2}{(1-i\omega\tau_2)(1-i\omega\tau_{f2})} \\ &- \frac{b_{11}^2}{\omega^2-\omega_{11}^2+i2\delta_{11}\omega} - \frac{b_{13}^2}{\omega^2-\omega_{13}^2+i2\delta_{13}\omega} \\ &- \frac{b_{15}^2}{\omega^2-\omega_{15}^2+i2\delta_{15}\omega} - \frac{b_{17}^2}{\omega^2-\omega_{17}^2+i2\delta_{17}\omega} ,\end{aligned} \tag{8}$$

which is a causal model that is appropriate for the observed frequency dispersion of triply-distilled liquid water. Here $\epsilon(\omega)$ denotes the relative dielectric permittivity where a_0 and a_2 are phenomenological parameters, τ_0 and τ_2 are relaxation times and τ_{f0} and τ_{f2} are friction times of the relaxation model. Each Lorentz oscillator line is denoted by the index j and is described by the natural resonance frequency ω_j, the phenomenological damping constant δ_j, and the square of the plasma frequency $b_j^2 = 4\pi N_j e^2/m$, where e is the charge and m the mass of the electron, and where N_j denotes the number density of the j-type oscillators.

Substitution of Eq. (8) into Eq.(5) with $\mu = 1$ then yields the appropriate causal model for the real and imaginary parts of the complex index of refraction for water. The parameters appearing in this model were then determined by fitting the model for the real index of refraction $n_r(\omega)$ to the experimental data, depicted in the upper part of Fig.1, using a simple root mean square curve fitting technique based on the relationship

$$rmsnr = \left[\frac{1}{N}\sum_N (n_{r,calc} - n_{r,expr})^2\right]^{\frac{1}{2}} \quad . \tag{9}$$

Here N denotes the total number of data points employed, $n_{r,expr}$ is the experimental data value for the real index of refraction at a given value of ω, and $n_{r,calc}$ is the value of the real index of refraction that is calculated from Eqs.(5) and (8) at that same value of ω. Theoretical results require that the minimization of the fit variable $rmsnr$ simultaneously maximize the friction times τ_{f0} and τ_{f2} in order to give the best correction of the Debye plateau problem[20]. Following the completion of this minimization for the relaxation time parameters over the frequency domain $10^2 hz \leq \omega \lesssim 10^{13} hz$, optimization of the Lorentz parameters b_j, ω_j, and δ_j as well as ϵ_∞ is then performed in a sequential manner over the appropriate frequency domain as each individual Lorentz line is included in the optimized composite model. Reoptimization for the entire composite model is then performed, first for each of the relaxation parameter sets individually, followed by each of the Lorentz lines. This reoptimization procedure is repeated until convergence to a stable value of the fit variable $rmsnr$ is achieved; this occurred after three iterations for the example considered here. The final set of model parameters for the dielectric permittivity model given in Eq.(8) for triply-distilled liquid water is given in Table 1. Comparisons of the theoretical (solid curves) real index of refraction $n_r(\omega)$ and the absorption coefficient $\alpha(\omega) = \omega n_i(\omega)/c$ with their corresponding experimental values are depicted in the upper and lower diagrams, respectively, of Fig.2 over the entire frequency domain of interest. Because the model parameter values given in Table 1 were optimized for the real index of refraction data, the theoretical curve is seen to provide an accurate fit to the experimental data everywhere except in the neighborhood of the ultraviolet line $\omega_{17} = 3.7 \times 10^{15} hz$. The resultant fit of the theoretical absorption coefficient $\alpha(\omega)$ to the experimental data is quite good (considering that these data were not used in the parameter fitting procedure) except in the visible region of the spectrum about $\omega \sim 10^{15} hz$. Notice that the solid curves in Fig.2 rigorously satisfy the Kramers-Krönig relations and thus that the experimental data clearly does not. The experimental absorption data then appears to be too large overall and is in error when either $\omega \lesssim 10^8 hz$ or $\omega \gtrsim 10^{18} hz$. On the other hand, the theoretical model needs to be modified with regard to its behavior from the visible

Table 1. Optimized parameter values for the dielectric permittivity model given in Eq.(8) for triply-distilled H_2O.

Relaxation Parameters (Microwave Region)		
$a_0 = 74.1$		$a_2 = 2.90$
$\tau_0 = 5.30 \times 10^{-11}\, sec$		$\tau_2 = 3.80 \times 10^{-13}\, sec$
$\tau_{f0} = 3.10 \times 10^{-13}\, sec$		$\tau_{f2} = 5.40 \times 10^{-14}\, sec$
Resonance Parameters (Infrared Region)		
$\omega_{11} = 1.80 \times 10^{13}\, sec^{-1}$	$\omega_{13} = 4.90 \times 10^{13}\, sec^{-1}$	$\omega_{15} = 1.00 \times 10^{14}\, sec^{-1}$
$b_{11} = 1.20 \times 10^{13}\, sec^{-1}$	$b_{13} = 6.80 \times 10^{12}\, sec^{-1}$	$b_{15} = 2.00 \times 10^{13}\, sec^{-1}$
$\delta_{11} = 4.30 \times 10^{12}\, sec^{-1}$	$\delta_{13} = 8.40 \times 10^{11}\, sec^{-1}$	$\delta_{15} = 2.80 \times 10^{12}\, sec^{-1}$
Resonance Parameters (Ultraviolet Region)		
	$\omega_{17} = 3.70 \times 10^{15}\, sec^{-1}$	
	$b_{17} = 3.20 \times 10^{15}\, sec^{-1}$	
	$\delta_{17} = 8.00 \times 10^{14}\, sec^{-1}$	
Resonance Parameters (Very High Frequency Region)		
	$\epsilon_\infty = 1$	

($\omega \sim 10^{14} hz$) to the ultraviolet ($\omega \sim 10^{16} hz$) region of the electromagnetic spectrum. Since half of the window in the absorption data is lost due to the Rocard-Powles model of the low frequency behavior ($\omega \lesssim 10^{13} hz$) while the remainder is lost due to the Lorentz model of the resonance behavior at $\omega_{17} = 3.7 \times 10^{15} hz$, these two aspects of the overall model of $\epsilon(\omega)$ given in Eq.(8) should be addressed first in any future research. Finally, the theoretical frequency dependence of the real and imaginary parts of the complex-valued dielectric permittivity $\epsilon(\omega)$ given by Eq.(8) for triply-distilled liquid water is depicted in Fig.3. These curves rigorously maintain the Hilbert transform pair relationship that exists between $\epsilon_r(\omega)$ and $\epsilon_i(\omega)$.

Impulse Response of the Model Dielectric Permittivity of Triply-Distilled Water

The propagated field due to an input delta function pulse $f(t) = \delta(t)$ is given by the integral representation [cf. Eq.(1) with $\tilde{f}(\omega) = 1$]

$$A(z,t) = \frac{1}{2\pi} \int_C \exp\{\frac{z}{c}\phi(\omega,\theta)\} d\omega \tag{10}$$

for all $z \geq 0$. The propagated field behavior in the mature dispersion regime then depends solely upon the frequency dispersion of the complex index of refraction of the medium through the dynamics of the dominant saddle points of $\phi(\omega,\theta)$. The propagated field over several θ domains at the fixed propagation distance $z = 3 \times 10^{-5} m$ is depicted in the upper diagrams of Figures 4 and 5 while the corresponding instantaneous frequency of oscillation of this numerically determined field evolution is depicted in the lower diagram of each figure when $\epsilon(\omega)$ is given by Eq.(8). Since the field evolution identically vanishes for all $\theta < 1$, only the behavior for $\theta \geq 1$ is depicted. The Sommerfeld precursor dominates the propagated field evolution for $1 \leq \theta < \theta_{SB}$, where $\theta_{SB} \approx 1.19$; this field structure is completely determined by the dynamical evolution of the distant saddle points of $\phi(\omega,\theta)$. The Brillouin

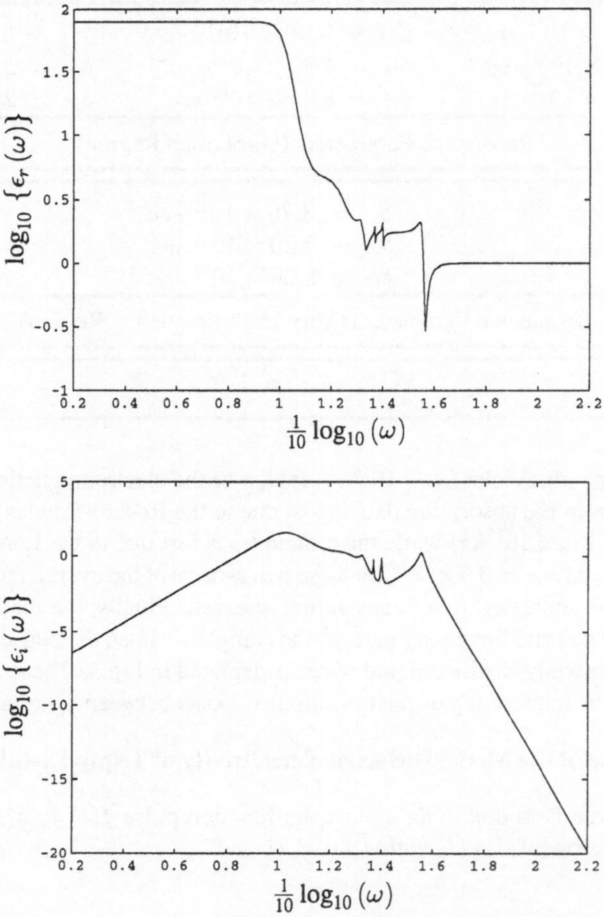

Figure 3. Theoretical curves for the real (upper diagram) and imaginary (lower diagram) parts of the complex-valued dielectric permittivity $\epsilon(\omega)$ given by Eq.(8) for triply-distilled liquid water using two relaxation times and four resonance lines.

precursor then dominates the propagated field evolution for $\theta > \theta_{SB}$; the initial portion of this field structure is determined by the dynamical evolution of the near saddle points of $\phi(\omega, \theta)$. The overall structure of these two precursor fields is similar to that obtained previously for either a single[10,11] or a double[15] resonance Lorentz medium. However, the detailed medium model considered here produces several unique features in the propagated field evolution following the initial portion of the Brillouin precursor that have not been observed in previous studies. The most prominent of these is the observed quenching of the Brillouin precursor field amplitude following its initial peak. This quenching phenomena is due to the presence of interference between two nearly equal but distinct frequency components in the propagated field structure, as is clearly evident in the evolution of the instantaneous frequency of oscillation depicted in Fig.5. These two frequency components are due to two sets of saddle points that have nearly the same attenuation and evolve in neighboring regions of the complex ω-plane.

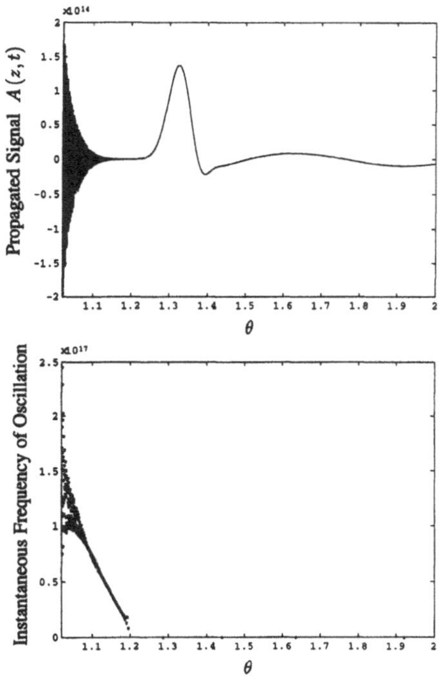

Figure 4. Impulse response of the model dielectric permittivity of triply-distilled water for small values of $\theta \geq 1$.

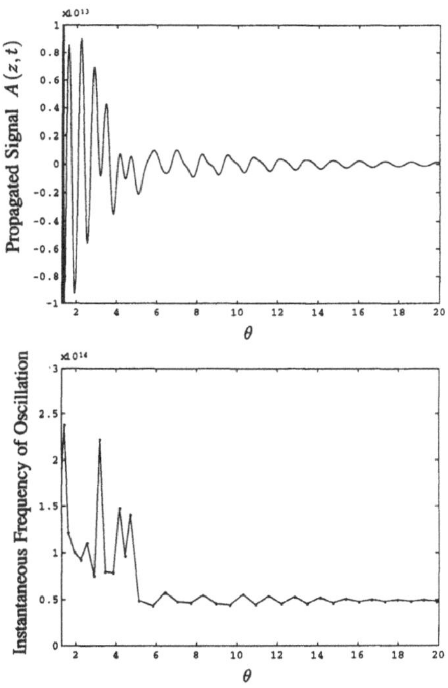

Figure 5. Impulse response of the model dielectric permittivity of triply-distilled water for intermediate values of θ.

ASYMPTOTIC DESCRIPTION OF GAUSSIAN PULSE PROPAGATION IN A SINGLE RESONANCE LORENTZ MODEL DIELECTRIC

For an input unit amplitude gaussian-modulated harmonic signal of fixed but otherwise arbitrary carrier frequency ω_c, the initial time behavior of the pulse at the plane $z = 0$ is given as

$$f(t) = u(t)\sin(\omega_c t + \psi) \ , \tag{11}$$

where the constant phase term ψ appearing in this expression is taken as $\psi = 0$ for a sine wave or as $\psi = \pi/2$ for a cosine wave. The initial pulse envelope is given by

$$u(t) \ = \ \exp[-\left(\frac{t - t_0}{T}\right)^2] \ , \tag{12}$$

where $2T$ is the input pulse width that is centered at the time $t = t_0$ at the plane $z = 0$. In contrast to previously published research that has considered an input delta function pulse[10-12], an input unit step-function modulated harmonic signal[6-13], or an input rectangular envelope modulated harmonic signal of arbitrary initial pulse width[16], the initial field for an input gaussian envelope pulse does not identically vanish for $t < 0$ at the plane $z = 0$. Nevertheless, one can always choose t_0 sufficiently large (at least as large as a few initial pulse widths $2T$) such that the important characteristics of the propagated field[17,18] occur when $\theta \geq 1$ and the field amplitude is negligible for all $\theta < 1$.

The Ultrashort Pulse Regime

Upon substitution of Eq.(11) with (12) into Eqs.(1)-(2), there results the standard integral representation of the propagated field due to an input unit amplitude gaussian envelope modulated harmonic signal[17-19]

$$A(z,t) = \frac{1}{2\pi}\Re\left\{ i\int_C \tilde{U}(\omega-\omega_c)\exp[\frac{z}{c}\phi(\omega,\theta')]d\omega\right\}, \tag{13}$$

for all $z \geq 0$, where

$$\tilde{U}(\omega-\omega_c) = e^{-i\psi}\pi^{1/2}Te^{-i\omega_c t_0}\exp[-\frac{T^2}{4}(\omega-\omega_c)^2] \tag{14}$$

is, apart from the factor $e^{-i\psi}$, the input pulse envelope spectrum. The compex phase function $\phi(\omega,\theta')$ appearing in the integrand of Eq.(13) is given by Eq.(6) with $\theta = ct/z$ replaced by

$$\theta' = \theta - \frac{c}{z}t_0 = \frac{c}{z}(t-t_0)\,. \tag{15}$$

Asymptotic analysis of the integral representation (13) of the propagated field in the mature dispersion regime requires that a specific model of the complex index of refraction be used. Physical reality requires that this medium model be causal. For simplicity, the present analysis will be restricted to a single resonance Lorentz model dielectric with complex index of refraction

$$n(\omega) = \left(1 - \frac{b^2}{\omega^2-\omega_0^2+2i\delta\omega}\right)^{1/2}, \tag{16}$$

which satisfies the Kramers-Krönig relations[2].

In order to determine the propagated field one always has the option of numerically evaluating the integral representation given in Eq.(13). Although such an approach may be performed with some (usually unspecified) degree of accuracy using a variety of numerical techniques, it does not show explicitly the dependence of the propagated field on the various input field and/or medium parameters. Alternatively, one has the option of approximately evaluating the integral representation (13) for the propagated field using mathematically well-defined asymptotic expansion techniques. Although this approach is difficult to implement, it does yield analytical expressions for the propagated field which explicitly display its dependence on the various input field and/or medium parameters. This then leads to physical insight concerning the dynamical evolution of ultrashort pulses in a given dispersive medium.

In recent research[19] concerning gaussian pulse propagation both numerical experiments and theoretical asymptotic approaches have been employed to describe the propagated field. Specifically, two independent numerical algorithms for the evaluation of the integral representation (13) have been constructed and implemented. The first is the so-called Hosono code[12] which performs a numerical inversion of Laplace-transform type integrals such as that given by the integral representation (13) for the propagated field. The other independent numerical approach is the so-called asymptotic code which is a numerical implementation of the more physically appealing method of steepest descents.[15]. The results of these two independent numerical procedures are found to be in excellent agreement with each other and therefore provide a secure standard with which to compare the description provided by the theoretical asymptotic analysis.

The first step in the theoretical analysis is a straightforward application of the asymptotic method of Olver[5], which is an extension of the saddle point method with less stringent requirements on the deformation of the original path of integration through the appropriate relevant dominant saddle points.[10]. The asymptotic representation of the field is then found as[17–19]

$$A(z,t) \sim A_S(z,t) + A_B(z,t) \tag{17}$$

as $z \to \infty$ for all $\theta' > 1$. The propagated field may then be represented as the superposition of a generalized Sommerfeld and Brillouin precursor. The Sommerfeld precursor results from the asymptotic expansion of the integral representation (13) about the distant saddle points of $\phi(\omega, \theta')$, while the Brillouin precursor results from the asymptotic expansion about the near saddle points of $\phi(\omega, \theta')$. The resultant analytical nonuniform theory code, which is a numerical implementation of the analytic expressions for the propagated field, is found to be in good qualitative but not quantitative agreement with the results from the two numerical experiments. Furthermore, this asymptotic representation is nonuniform about the two space-time points $\theta' = 1$ and $\theta' = \theta_1$, where[10]

$$\theta_1 \cong \theta_0 + \frac{2\delta^2 b^2}{\theta_0 \omega_0^2 (3\omega_0^2 - 4\delta^2)} , \tag{18}$$

$$\theta_0 = n(0) = (1 + \frac{b^2}{\omega_0^2})^{1/2} . \tag{19}$$

For Brillouin's choice of the medium parameters ($\omega_0 = 4 \times 10^{16} hz, \delta = 0.28 \times 10^{16} hz, b^2 = 20 \times 10^{32} hz^2$), which are used throughout this section, one obtains $\theta_0 = 1.5$ and $\theta_1 \cong 1.501$.

The origin of the observed discrepancies between this asymptotic representation of the theoretical description and the numerical experiments has been traced to the use of the approximate analytic expressions for the saddle point locations (derived in Ref. 10) and the values of the derivatives of the complex phase function at them. Indeed, when these approximations are removed from this asymptotic representation, which results in the so-called numerical nonuniform asymptotic theory code, the calculated behavior of the propagated field is found to be in very good agreement with the two numerical experiments. As in the previous asymptotic representation, this approach is nonuniform at both $\theta' = 1$ and $\theta' = \theta_1$. Modern uniform asymptotic expansion techniques[11,16,19] may then be employed to yield an asymptotic representation of the propagated field that is uniformly valid for all $\theta' \geq 1$. It should be noted that some approximations have been retained in this last step of our theoretical development in order that the resulting analytic expressions for the propagated field remain analytically tractable. It is then expected that the predictions of the so-called uniform theory code, which numerically implements these expressions, will be in good agreement with the two numerical experiments, although not as good as the predictions of the numerical nonuniform theory code, but much better than the predictions of the analytical nonuniform theory.

A complete comparison of the results of the two numerical experiments with the theoretical predictions from the three asymptotic codes is illustrated in Fig.6 for an input unit amplitude gaussian-modulated harmonic signal with carrier frequency $\omega_c = 5.75 \times 10^{16}/\text{sec}$, which is close to the upper end of the absorption band, and an input pulse width $2T = 0.2 fsec$,

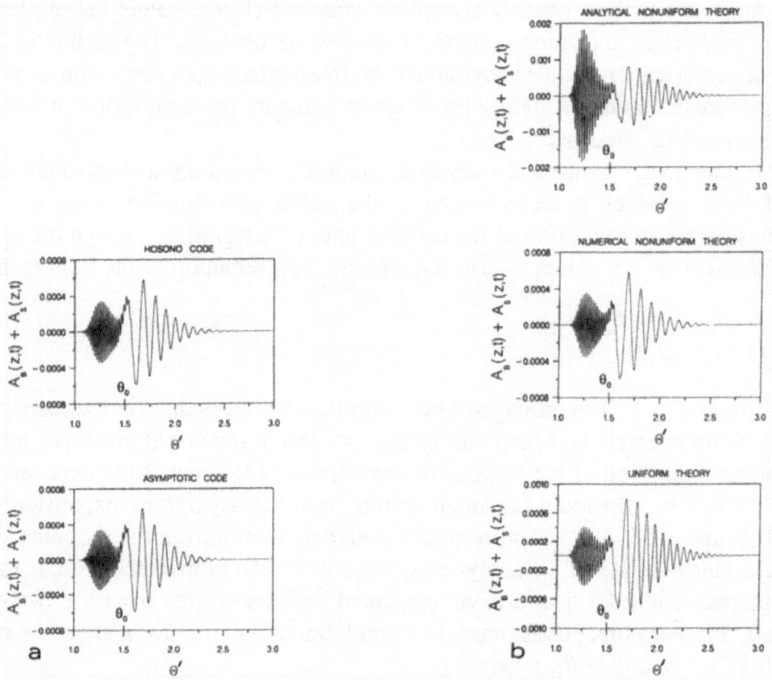

Figure 6. The dynamical field evolution due to an input gaussian-modulated cosine field with initial pulse width $2T = 0.2fsec$ and carrier frequency $\omega_c = 5.75 \times 10^{16}hz$, which is near the upper edge of the absorption band of the single resonance Lorentz medium. The propagation distance is fixed at $z = 83.92z_d$, where z_d is the e^{-1} absorption depth of the dielectric medium at ω_c. Part (a) of the figure depicts the results of the two numerical experiments, while part (b) depicts the results of the three asymptotic approximations.

at the propagation distance $z = 83.92z_d = 1.0\mu m$ inside the dielectric medium, where z_d is the absorption depth of the medium at the carrier frequency ω_c. For each of these calculations the center of the input pulse was chosen at $t_0 = 10T$.

It is clearly evident in Fig.6(a) that the results of the two numerical experiments are in excellent agreement. By comparison, the prediction of the analytical nonuniform theory shown at the top of Fig.6(b), is seen to be in good qualitative but not quantitative agreement with the two numerical experiments. This is primarily due to the use of the approximate analytic expressions[10] for both the near and distant saddle point locations and the values of the derivatives of the complex phase function at them. When these approximations are removed, the resultant numerical nonuniform theory whose predicted field behavior is shown in the middle of Fig. 6(b), is found to be in very good agreement with the two numerical experiments except at the points $\theta' = 1$ and $\theta' = \theta_1 \cong 1.501$ where both the analytical and numerical nonuniform asymptotic theories break down. In order to overcome this nonuniform behavior, the uniform asymptotic theory has been invoked along with some of the approximations for the saddle point locations. As illustrated at the bottom of Fig. 6(b), the predictions of the propagated field by the uniform theory are seen to be in good agreement with the two numerical experiments.

The break-up of the input gaussian envelope pulse into a generalized Sommerfeld precursor field, which is due to the high frequency components of the input pulse spectrum, and a generalized Brillouin precursor field, which is due to the low frequency components of the spectrum, is clearly evident in Fig. 6. It is found[17–19] that the generalized Sommerfeld precursor decays exponentially with increasing propagation distance, whereas the peak of the generalized Brillouin precursor is attenuated only as $z^{-1/2}$. Therefore, as the propagation distance increases the propagated field will be dominated by the generalized Brillouin precursor field. Finally, the instantaneous frequency of oscillation of the generalized Sommerfeld precursor approaches the upper end of the absorption band from above as θ' increases, while that for the generalized Brillouin precursor approaches the lower end of the absorption band from below as θ' increases.

When the initial pulse width $2T$ of the input gaussian-modulated harmonic field is decreased toward zero, the input pulse approaches the limiting case of an input δ-function pulse and the agreement between the two numerical experiments and each of the theoretical representations considered here becomes excellent. Such is not the case, however, as the initial pulse width is increased and the input pulse approaches the quasimonochromatic or slowly-varying envelope regime from below.

Transition to the Quasimonochromatic Pulse Regime

As the initial pulse width of the input gaussian modulated harmonic field is increased and allowed to approach the quasimonochromatic regime from below, the asymptotic representation described thus far is found to increasingly deviate from the predictions of the two numerical experiments. This limitation in the applicability of the classical asymptotic analysis to ultrashort gaussian pulse propagation necessitates that the asymptotic description be modified in such a manner that the resultant asymptotic representation is uniformly valid with respect to both the space-time parameter θ' and the initial pulse width $2T$ of the input pulse. To that end, the classical exact integral representation for the propagated field that is due to an input unit amplitude gaussian-modulated harmonic field, as given in Eq. (13)-(14), is rewritten as

$$A(z,t) = \frac{1}{2\pi}\Re\left\{ i \int_{ia-\infty}^{ia+\infty} \tilde{U}_M(\omega-\omega_c) e^{\frac{z}{c}\Phi_M(\omega,\theta')}\, d\omega \right\} , \tag{20}$$

for all $z \geq 0$, where the modified spectral amplitude is now given by

$$\tilde{U}_M(\omega-\omega_c) = e^{-i\psi}\pi^{1/2} T e^{-i\omega_c t_0} , \tag{21}$$

and the modified complex phase function is now given by

$$\Phi_M(\omega,\theta') = i\omega[n(\omega)-\theta'] - \frac{cT^2}{4z}(\omega-\omega_c)^2 , \tag{22}$$

with

$$\theta' = \theta - \frac{ct_0}{z} = \frac{c}{z}(t-t_0) \tag{23}$$

as before. With this form of the integral representation, the saddle point locations are now determined from the modified complex phase function $\Phi_M(\omega,\theta')$ given in Eq. (22). The saddle point locations now depend not only upon the medium dispersion, as displayed by the first term on the right hand side of Eq. (22), but also upon the pulse width and carrier frequency of the input field, as displayed by the second term on the right hand side of Eq. (22). For an input ultrashort pulse, the first term on the right-hand side of Eq. (22) dominates the second term and the modified analysis reduces to the classical description presented in the previous subsection. On the other hand, for broader input pulses that may be classified in the quasimonochromatic regime, the second term appearing on the right hand side of Eq. (22) completely dominates the first term and the asymptotic expansion of the integral representation (20) is then taken about the carrier frequency ω_c. The asymptotic approximation of the modified integral representation (20) then reduces to the well-known slowly-varying envelope approximation [26–30] as the input pulse width increases.

The asymptotic analysis of the modified integral representation for the propagated field given in Eqs. (20)-(22) begins with an analysis of the topography of the real part $X_M(\omega, \theta')$ of the modified complex phase function $\Phi_M(\omega, \theta')$ in the complex ω-plane. The exact saddle point equation for the location of the saddle points of $\Phi_M(\omega, \theta')$ in the complex ω-plane, given by

$$\frac{d\Phi_M(\omega, \theta')}{d\omega} = 0 \ , \tag{24}$$

then becomes, for a single resonance Lorentz medium,

$$\left[\frac{\omega^2 - \omega_1^2 + i2\delta\omega}{\omega^2 - \omega_0^2 + i2\delta\omega}\right] + \left[\frac{\omega b^2(\omega + i\delta)}{(\omega^2 - \omega_0^2 + i2\delta\omega)^2}\right] = \left[\theta' - \frac{icT^2}{2z}(\omega - \omega_c)\right]\left[\frac{\omega^2 - \omega_1^2 + i2\delta\omega}{\omega^2 - \omega_0^2 + i2\delta\omega}\right]^{1/2} , \tag{25}$$

where $\omega_1^2 = \omega_0^2 + b^2$. Upon squaring this equation in order to eliminate the square root expression appearing in it, thereby doubling the number of possible solutions, one obtains the exact expression

$$\begin{aligned} &[\omega^2 - \omega_1^2 \ + \ i2\delta\omega]^2[\omega^2 - \omega_0^2 \ + \ i2\delta\omega]^2 \\ &+ \ [\omega^2 b^4(\omega + i\delta)^2] \\ &+ \ 2[\omega^2 - \omega_0^2 + i2\delta\omega][\omega^2 - \omega_1^2 + i2\delta\omega][\omega b^2(\omega + i\delta)] \\ &- \ [\theta' - \frac{icT^2}{2z}(\omega - \omega_c)]^2[\omega^2 - \omega_1^2 + i2\delta\omega][\omega^2 - \omega_0^2 + i2\delta\omega]^3 = 0 \end{aligned} \tag{26}$$

The left hand side of Eq. (26) is a 10th-order polynomial in the complex frequency ω so that it then has 10 complex-valued roots. The exact saddle point equation (25) must then have 5 complex-valued roots. As a consequence, 5 saddle points now enter the asymptotic description.

It is very difficult, if not impossible, to obtain analytic solutions of the exact saddle point equation (26), so that one has to resort to numerical techniques in order to determine the exact saddle point locations. Fortunately, the limiting behavior of the exact saddle point locations as $\theta' \to +\infty$ can easily be found analytically. To that end, consider the exact saddle point equation (25) which may be rewritten as

$$n(\omega) + \left[\frac{\omega b^2(\omega + i\delta)}{(\omega^2 - \omega_0^2 + i2\delta\omega)^2 n(\omega)}\right] \ + \ \frac{icT^2(\omega - \omega_c)}{2z} = \theta' \, . \tag{27}$$

Hence, when $\theta' \to +\infty$ then either $\omega^2 - \omega_0^2 + i2\delta\omega = 0$, which yields the exact roots

$$\omega_{1,4} \ = \ -i\delta \pm \sqrt{\omega_0^2 - \delta^2} \, , \tag{28}$$

or $n(\omega) = 0$, which yields the exact roots

$$\omega_{2,5} \ = - \ \ i\delta \pm \sqrt{\omega_1^2 - \delta^2} \quad . \tag{29}$$

Hence, four of the five saddle points of the modified complex phase function $\Phi_M(\omega, \theta')$ approach the four branch points $\omega_\pm = \pm\sqrt{\omega_0^2 - \delta^2} - i\delta, \omega'_\pm = \pm\sqrt{\omega_1^2 - \delta^2} - i\delta$ of $n(\omega)$ in the limit as $\theta' \to \infty$.

The results of a numerical solution of the exact saddle point equation (25) are illustrated in Fig. 7. The five saddle points of the modified complex phase function are numbered in

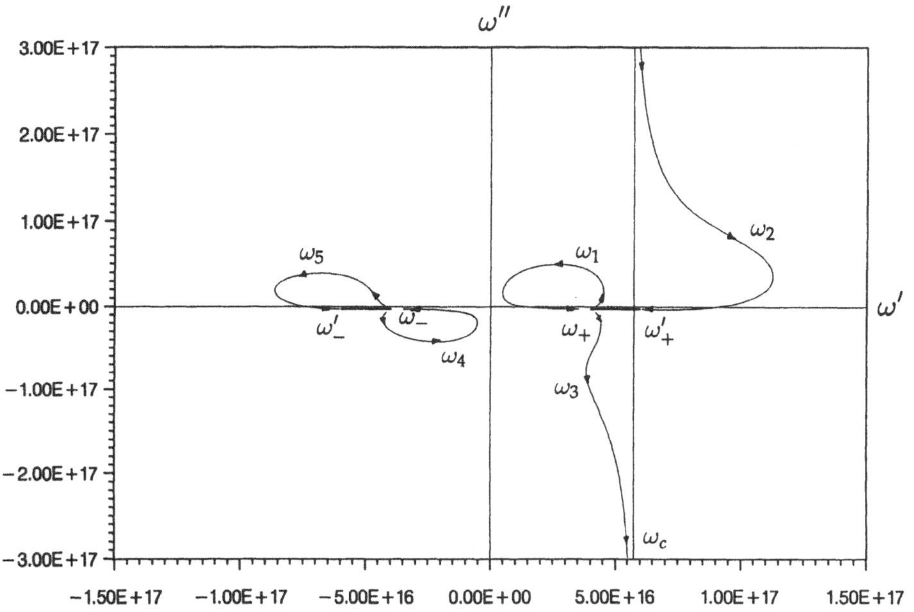

Figure 7. Dynamical motion of the saddle points of the modified complex phase function $\Phi_M(\omega, \theta')$ of a single resonance Lorentz medium for an input gaussian envelope pulse with $2T = 0.2fsec, t_0 = 15T, \omega_c = 5.75 \times 10^{16}hz$ at a propagation distance $z = 1\mu m$. The arrow on each path indicates the direction of motion of that saddle point as θ' increases over the domain $[-10.0, +10.0]$.

accordance with the limiting behavior predicted in Eqs. (28) and (29). The arrows in the figure indicate the direction of movement of each saddle point for increasing values of θ' in the range $\theta\epsilon[-10.0, +10.0]$. Four of the five saddle points are clearly seen to approach the branch points $\omega_\pm, \omega'_\pm$ of $n(\omega)$, and hence of the modified complex phase function $\Phi_M(\omega, \theta')$as $\theta' \to +\infty$. The saddle point ω_2 is seen to approach the line $\omega = \omega_c$ asymptotically in the upper half plane as $\theta' \to -\infty$, while the saddle point ω_3 approaches the line $\omega = \omega_c$ asymptotically in the lower half-plane in the opposite limit as $\theta' \to +\infty$.

A comparison of the predictions of the propagated field due to an input gaussian envelope pulse with initial pulse width $2T = 0.2fsec$ centered at $t_0 = 15T$ with the carrier frequency $\omega_c = 5.75 \times 10^{16}hz$ at a propagation distance of $z = 1 \times 10^{-6}m$ is illustrated in Fig. 8. The propagated field evolution predicted by the numerical experiment is illustrated in part (a) of the figure. The propagated field evolution described by the nonuniform asymptotic approximation of the unmodified integral representation[(13)-(14)], illustrated in part (b) of the figure, is in very good agreement with the result of the numerical experiment except in small neighborhoods about the points $\theta' = 1$ and $\theta' = \theta_1 \cong 1.501$. The propagated field evolution described by the asymptotic approximation of the modified integral representation (20)-(22) illustrated in part (c) of the figure, is virtually indistinguishable from the numerical experiment for all θ'. Because of the modified saddle point dynamics, this description is uniformly valid for all θ'.

This level of accuracy in the asymptotic description that is obtained from the modified integral representation given in Eqs. (20)-(22) remains intact as the initial pulse width $2T$ is broadened into the quasimonochromatic regime. In that limit, the dominant saddle point of the modified phase function $\Phi_M(\omega, \theta')$ moves into the vicinity of the point $\omega = \omega_c$ along the real frequency axis about which the initial pulse spectrum is peaked. The asymptotic approximation of the propagated field then reduces to the familiar result obtained in the slowly-varying envelope approximation.

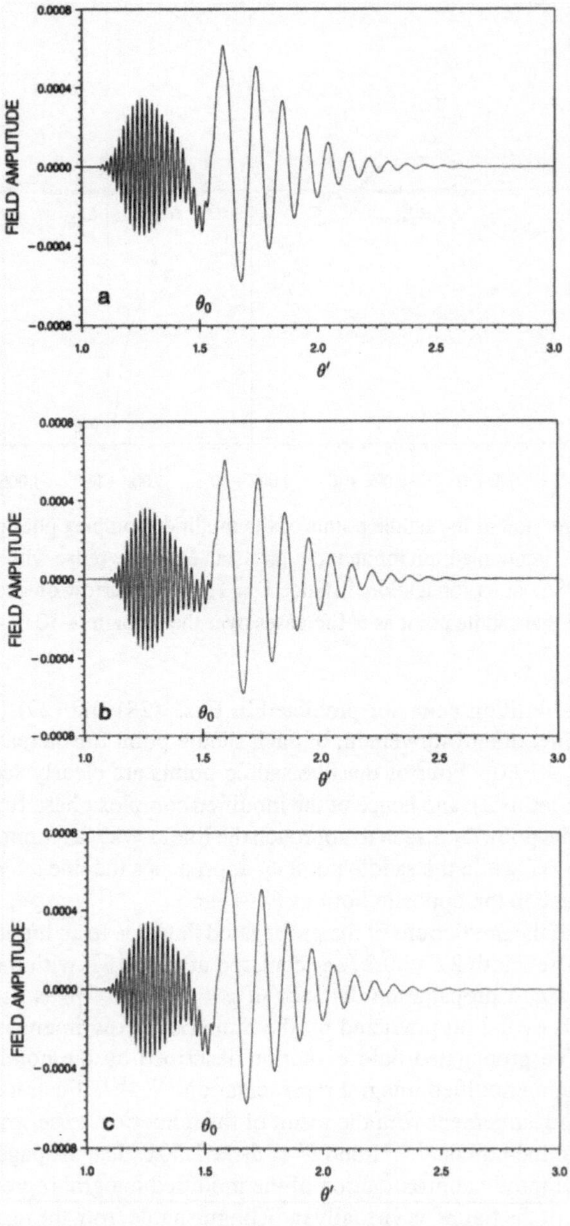

Figure 8. Dynamical evolution of the propagated field due to an input gaussian-modulated cosine wave pulse with initial pulse width $2T = 0.2 fsec$ and carrier frequency $\omega_c = 5.75 \times 10^{16} hz$ at a propagation distance of $z = 1 \times 10^{-6} m$ in a single resonance Lorentz medium. The field evolution depicted in part (a) is from the numerical experiment, part (b) is from the nonuniform asymptotic theory with the unmodified phase function $\phi(\omega, \theta)$, and part (c) is from the asymptotic theory with the modified phase function $\Phi_M(\omega, \theta)$.

CONCLUSIONS

The asymptotic description of dispersive pulse propagation phenomena has been shown here to be critically dependent upon the dominant contribution to the propagated field behavior at a given space-time point. For a given dispersive medium, the dominant contribution to the field behavior is completely determined by the dynamics of the saddle points of the complex phase function $\phi(\omega, \theta)$ of the dispersive medium, the initial pulse envelope spectrum, and the input pulse carrier frequency. For initial rise and/or fall times of the input pulse envelope that are typically less than the medium relaxation time, the appropriate form of the complex phase function is $\phi(\omega, \theta) = i\omega[n(\omega) - \theta]$, where $n(\omega)$ is the complex index of refraction of the medium and $\theta = ct/z$ is a dimensionless space-time parameter. The saddle point dynamics then depend only upon the dispersive properties of the medium. However, as the initial rise and/or fall time of the input pulse increases above the medium relaxation time, it has been shown that the complex phase function must be modified so as to include the rise/fall time parameter. The saddle point dynamics then depend not only upon the medium dispersion but also upon the input pulse properties. In either case, it is critical that the model of the dispersive medium be causal in order that a close connection with physical reality be maintained.

ACKNOWLEDGMENT

The research presented in this paper was supported, in part, by the United States Air Force Office of Scientific Research under Grant No. F49620-92-J-0206.

REFERENCES

1. J.A. Stratton. "Electromagnetic Theory," McGraw-Hill, New York, (1941), pp. 333-340.
2. H.M. Nussenzveig. "Causality and Dispersion Relations," Academic, New York (1972), ch. 1.
3. G.R. Baldock and T. Bridgeman. "Mathematical Theory of Wave Motion," Halsted, New York (1981), ch. 5.
4. E.T. Capson. "Asymptotic Expansions," Cambridge University Press, London (1971), ch. 7-8.
5. F.W.J. Olver, Why steepest descents?, *SIAM Rev.* 12:228 (1970).
6. A. Sommerfeld, Über die fortpflanzung des lichtes in disperdierenden medien, *Ann. Phys.* 44:177(1914).
7. L. Brillouin, Über die fortpflanzung des lichtes in disperdierenden medien, *Ann. Phys.* 44:203 (1914).
8. L. Brillouin. "Wave Propagation and Group Velocity," Academic, New York (1960).
9. H. Baerwald, Über die fortpflanzung von signalen in disperdierenden medien, *Ann. Phys.* 7:731 (1930).
10. K.E. Oughstun and G.C. Sherman, Propagation of electromagnetic pulses in a linear dispersive medium with absorption (the Lorentz medium), *J. Opt. Soc. Am., B* 5:817 (1988).
11. K.E. Oughstun and G.C. Sherman, Uniform asymptotic description of electromagnetic pulse propagation in a linear dispersive medium with absorption (the Lorentz medium), *J. Opt. Soc. Am. A* 6:1394 (1989).
12. P. Wyns, D.P. Foty, and K.E. Oughstun, Numerical analysis of the precursor fields in linear dispersive pulse propagation, *J. Opt. Soc. Am. A* 6:1421 (1989).
13. K.E. Oughstun, P. Wyns, and D. Foty, Numerical determination of the signal velocity in dispersive pulse propagation, *J. Opt. Soc. Am. A* 6:1430 (1989).

14. R.M. Joseph, S.C. Hagness, and A. Taflove, Direct time integration of Maxwell's equations in linear dispersive media with absorption for scattering and propagation of femtosecond electromagnetic pulses, *Opt. Lett.* 16:1412 (1991).
15. S. Shen and K.E. Oughstun, Dispersive pulse propagation in a double resonance Lorentz medium, *J. Opt. Soc. Am. B* 6:948 (1989).
16. K.E. Oughstun and G.C. Sherman, Uniform asymptotic descritpion of ultrashort rectangular optical pulse propagation in a linear, causally dispersive medium, *Phys. Rev. A* 41:6090 (1990).
17. K.E. Oughstun and J.E.K. Laurens, Asymptotic description of ultrashort electromagnetic pulse propagation in a linear, causally dispersive medium, *Radio Sci.* 26:245 (1991).
18. K.E. Oughstun, Pulse propagation in a linear, causally dispersive medium, *Proc. IEEE* 79:1379 (1991).
19. C.M. Balictsis and K.E. Oughstun, Uniform asymptotic description of ultrashort gaussian pulse propagation in a causal, dispersive dielectric, *Phys. Rev. A* (submitted).
20. Yu.S. Barash and V.L. Ginzburg, Expressions for the energy density and evolved heat in the electrodynamics of a dispersive and absorptive medium, *Sov. Phys. Usp.* 19:263 (1976).
21. J. McConnell. "Rotational Brownian Motion and Dielectric Theory," Academic, London (1980).
22. P. Debye. "Polar Molecules," Dover, New York (1929).
23. C.J.F. Böttcher and A. Bordewijk. "Theory of Electric Polarization," Vol. II, Elsevier, Amsterdam (1980).
24. H.A. Lorentz. "The Theory of Electrons," 2nd. ed., Dover, New York (1952).
25. L. Rosenfeld. "The Theory of Electrons," Dover, New York (1965).
26. G. Garrett and D. McCumber, Propagation of a gaussian light pulse through an anomalous dispersion medium, *Phys. Rev. A* 1:305 (1970).
27. J. Jones, On the Propagation of a Pulse through a Dispersive Medium, *Amer. J. Phys.* 42:43 (1974).
28. D. Anderson and J. Askne, Wave packets in strongly dispersive media, *Proc. IEEE* 62:1518 (1974).
29. D. Anderson, J. Askne, and M. Lisak, Wave packets in an absorptive and strongly dispersive medium, *Phys. Rev. A* 12:1546 (1975).
30. D. Anderson and M. Lisak, Analytic study of pulse broadening in dispersive optical fibers, *Phys. Rev. A* 35:184 (1987).

WAVEPACKET SOLUTIONS OF THE TIME-DEPENDENT WAVE EQUATION IN HOMOGENEOUS AND IN HOMOGENEOUS MEDIA

Ehud Heyman[1]

Department of Electrical Engineering – Physical Electronics
Tel-Aviv University
Tel-Aviv 69978, Israel

1. INTRODUCTION

Pulsed beam (PB) are highly localized space-time wavepackets solutions of the time-dependent wave equation that propagate along ray trajectories. Because they have these properties, PB may be useful in various applications including modeling of highly focused energy transfer, ultra wide band Radars and local interrogation of the propagation environment. Several classes of wavepacket solutions of the homogeneous wave-equation in *free space* have been introduced recently. They include the focus wave mode (FWM),[1–4] the "bullets,"[5,6] and the complex source pulsed beam (CSPB)[7–9] solutions. The FWM-type solutions are slowly diffracting wavepackets. Their special structure is a consequence of a balance between forward and backward propagating waves.[4] Removing the backward propagating part that cannot be excited by an antenna causes a diffraction of the remaining wavepacket at about the classical Fresnel distance. The "bullets" are wavepackets whose far zone radiation patterns vanish identically outside a specified cone.[5,6] Their near zone structure, however, does not have a well confined wavepacket shape.[6] The CSPB are exact solutions that are modeled by radiation from a pulsed source located at a complex coordinate point. Physically, they are generated by radiation from real time-dependent source distributions of *finite support* so that the complex source model is just a mathematical trick to derive simple field solutions. These wavepackets stay collimated up to a certain distance from the sources and thereafter spread along a constant diffraction angle and decay like r^{-1}. Except for the far zone spreading, the wavepacket structure remains essentially unchanged all the way from the sources to the far zone. Other important features of the CSPB will be discussed in Sec. 2.7.

The discussion above illustrates the different propagation characteristics of each class of wavepacket solutions. Here we are concerned with establishing some *general characteristics* for the wavepacket solutions. The only requirement is that the solution will remain localized in space-time. We derive an approximate form of the time-dependent wave equation that controls the wavepacket within a moving space-time window, and then construct the general solutions of this "wavepacket equation." This is done first for free space (Sec. 2) and latter on for a smoothly varying medium (Sec. 3). The paraxial structure of these wavepackets resembles that of the *globally exact* CSPB, but they have a more general form that admits wavepacket astigmatism and medium inhomogeneities. An important feature in these solutions is that, except for a

[1] *Acknowledgement:* This work has been supported by the US Air force System Command, Rome Laboratory, under Contract No. F19628-91-C-0113. Part of the work has been performed while the author was a Visiting Professor at Chuo University, Tokyo, Japan

far field spreading, the wavepacket shapes remain essentially unchanged along the propagation path. This establishes these new CSPB-type solutions as the most general *eigen-wavepacket* solutions of the time-dependent wave equation. The PB solutions introduced here can also be described by an ultra wide spectrum of time-harmonic Gaussian beams that have a frequency independent collimation distance (Sec. 2.6). This implies that their beamwidth is proportional to $\omega^{-1/2}$, a fact that has important implications for the design of PB aperture antennas. However, direct treatment in the time domain is preferable, both numerically and physically, over a representation as an ultra wide band superposition of time-harmonic solutions.

2. PULSED BEAMS IN A UNIFORM MEDIUM

2.1 The Wavepacket Equation

We are interested in pulsed beam (PB) solutions $u(\mathbf{r},t)$ of the time-dependent wave equation

$$(\partial_{x_1}^2 + \partial_{x_2}^2 + \partial_z^2 - v^{-2}\partial_t^2)u(\mathbf{r},t) = 0 \tag{1}$$

in a medium with uniform wave velocity v. Referring to Fig. 1, it is assumed that the PB propagates along the z-axis in the coordinate frame $\mathbf{r} = (\mathbf{x}, z)$, $\mathbf{x} = (x_1, x_2)$. From reasons which will be clarified soon (see (7)) we utilize the analytic signal representation of $u(t)$, defined by the analytic inverse Fourier transform

$$\overset{+}{u}(t) = \int_0^\infty d\omega\, e^{-i\omega t}\hat{u}(\omega), \qquad \operatorname{Im} t \le 0 \tag{2}$$

where $\hat{u}(\omega)$ is the frequency spectrum of $u(t)$. Clearly, integral (2) defines an analytic function in the lower half of the complex t-plane. The real t limit of $\overset{+}{u}$ yields the real function u via $\overset{+}{u}(t) = u(t) + i\mathcal{H}u(t)$ where $\mathcal{H}$ is the Hilbert transform. Thus if $\overset{+}{u}(\mathbf{r},t)$ is an analytic solution of the wave equation, then both

$$u_R(\mathbf{r},t) \equiv \operatorname{Re}\overset{+}{u}(\mathbf{r},t) \qquad \text{and} \qquad u_I(\mathbf{r},t) \equiv \operatorname{Im}\overset{+}{u}(\mathbf{r},t) = \mathcal{H}u_R \tag{3a,b}$$

are real solutions. We usually consider only u_R since any linear combination of u_R and u_I may be obtained by multiplying $\overset{+}{u}$ by an appropriate complex constant and taking the real part.

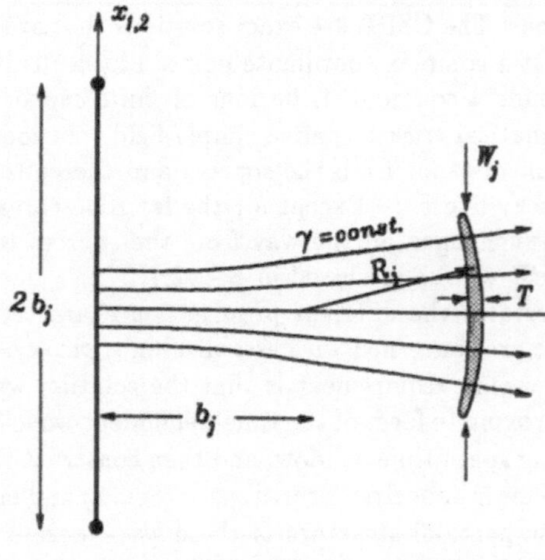

Fig. 1 PB (shaded) in a homogeneous medium. The Figure depicts a cross sectional cut in the principal plane (x_j, z). T, W_j, R_j and b_j denote the pulse length, beam width, wavefront curvature and collimation distance, respectively. Note that $W_j \gg T$ (see (20a). The lines $\gamma = const.$ are the propagation lines. The Heavy line represents the rigorous source distribution for the *globally exact* complex source pulsed beam (Sec. 2.7).

Since the PB is localized in space-time, we express $\overset{+}{u}$ in a moving coordinate frame

$$\overset{+}{u}(\mathbf{r},t) = \overset{+}{U}(\mathbf{r},\tau), \qquad \tau = t - z/v. \tag{4}$$

It is assumed next that

$$|\partial_z \overset{+}{U}| \ll |v^{-1}\partial_\tau \overset{+}{U}| \tag{5}$$

so that $(\partial_z^2 - v^{-2}\partial_t^2)u \simeq (-2v^{-1}\partial_z\partial_\tau)U$ and eq. (1) reduces to the "wavepacket equation"

$$(\partial_{x_1}^2 + \partial_{x_2}^2 - 2v^{-1}\partial_z\partial_\tau)\overset{+}{U} = 0. \tag{6}$$

2.2 Pulsed Beam Solutions

We seek a solution to (6) in the form

$$\overset{+}{U} = A(z)\overset{+}{f}\left[\tau - \tfrac{1}{2}(\mathbf{x},\mathbf{\Gamma}(z)\mathbf{x})\right] \tag{7}$$

where $\overset{+}{f}$ is an *arbitrary* time-pulse with a typical length T, $\mathbf{\Gamma}$ is a 2×2 complex symmetrical matrix, $(\mathbf{x},\mathbf{y})$ denotes a scalar product and $(\mathbf{x},\mathbf{\Gamma}(z)\mathbf{x}) = \Gamma_{11}x_1^2 + 2\Gamma_{12}x_1x_2 + \Gamma_{22}x_2^2$ is a quadratic form. To guarantee confinement of $\overset{+}{U}$ near the z-axis, $\mathrm{Im}\mathbf{\Gamma}$ must be positive definite. This and other properties of (7) will be discussed latter on.

Eq. (7) is a solution of (6) for *any* f if

$$v\mathbf{\Gamma}^2 + \mathbf{\Gamma}' = 0 \quad \text{and} \qquad A\,\mathrm{trace}\,\mathbf{\Gamma} + 2v^{-1}A' = 0 \tag{8a,b}$$

where a prime denotes a derivative with respect to the argument. To solve (8a) we set $\mathbf{\Gamma} = \mathbf{Q}^{-1}$ and obtain $\mathbf{Q}' = v\mathbf{I}$ where $\mathbf{I}$ is the unit matrix, hence

$$\mathbf{\Gamma}(z) = \left[\mathbf{\Gamma}^{-1}(0) + zv\mathbf{I}\right]^{-1}. \tag{9}$$

Thus if the initial condition matrix $\mathrm{Im}\mathbf{\Gamma}(0)$ is positive definite then $\mathrm{Im}\mathbf{\Gamma}(0)$ is also positive definite as required in (7). Finally we use the general relation

$$\ln'\det\mathbf{Q} = \mathrm{trace}(\mathbf{Q}'\mathbf{Q}^{-1}), \qquad \ln' f = f'/f, \tag{10}$$

which applies to any $\mathbf{Q}$ and in particular here with $\mathbf{Q}' = v\mathbf{I}$, to replace $\mathrm{trace}\,\mathbf{\Gamma}$ in (8b) by $v^{-1}\ln'\det\mathbf{Q}$. Eq. (8b) then yields $A(z) = (\det\mathbf{Q}(z))^{-1/2}$. The PB solution of (1) is therefore given by

$$\overset{+}{u}(\mathbf{r},t) = \sqrt{\det\mathbf{\Gamma}(z)/\det\mathbf{\Gamma}(0)}\,\overset{+}{f}\left[t - z/v - \tfrac{1}{2}(\mathbf{x},\mathbf{\Gamma}\mathbf{x})\right]. \tag{11}$$

The solution in (11) has the characteristics of a pulsed beam (PB). Axial confinement along the beam axis is due to the pulsed behavior of $\overset{+}{f}$ while transverse confinement is due to the fact that $\mathbf{\Gamma}$ is complex and due to the property of analytic signals which generally decay as the imaginary part of their argument becomes more negative (see (2)). Since $\mathrm{Im}\mathbf{\Gamma}$ is positive definite, the argument of $\overset{+}{f}$ in (11) has a negative imaginary part which increases quadratically away from the axis. The waveform of (11) is therefore strongest on the beam axis and weaken away from the axis. The beamwidth is determined by $\mathrm{Im}\mathbf{\Gamma}$ and by the decay rate of $\overset{+}{f}$ in the lower half of the complex t-plane. The latter depends, typically, on the frequency content in $\overset{+}{f}$; the higher the frequency content, the faster the decay and the narrower the beam (see examples in Sec. 2.4).

To clarify the physical characteristics of the PB we may consider a rotated transverse coordinates system $\mathbf{x}$ in which $\mathbf{\Gamma}$ is diagonal. In the general case, however, the two real symmetrical matrices $\mathrm{Re}\mathbf{\Gamma}$ and $\mathrm{Im}\mathbf{\Gamma}$ cannot be diagonalized simultaneously. In this case the major axes of the wavefront curvature and of the wavepacket amplitude are not aligned. We therefore consider first the simpler case where ReΓ and ImΓ have the same major axes. The more general case will be discussed in Sec. 2.5.

2.3 Special Case: Isoaxial Pulsed Beam

If the principal axes of ReΓ and of ImΓ coincide for $z = 0$ then, from (9), they coincide for all z. Assuming, without loss of generality that the principal coordinate frame is $\mathbf{x}$, than in that frame $\mathbf{\Gamma}$ has the form

$$\mathbf{\Gamma} = \text{diag}\{1/vq_j(z)\}, \quad q_j(z) = a_j - ib_j + z, \quad b_j > 0 \tag{12}$$

where the complex constants $a_j - ib_j$, $j = 1, 2$, are found from $\mathbf{\Gamma}(0)$. Eq. (12) becomes

$$\overset{+}{u}(\mathbf{r},t) = \sqrt{\frac{q_1(0)\,q_2(0}{q_1(z)\,q_2(z)}}\;\overset{+}{f}\left[t - \frac{z}{v} - \frac{x_1^2}{2vq_1(z)} - \frac{x_2^2}{2vq_2(z)}\right]. \tag{13}$$

To quantify the properties of $\overset{+}{u}$ we separate $q_j(z)$ to real and imaginary parts via

$$1/q_j = 1/R_j + i/I_j \tag{14}$$

hence

$$R_j(z) = (z + a_j) + b_j^2/(z + a_j)\,, \qquad I_j(z) = b_j[1 + (z + a_j)^2/b_j^2] > 0. \tag{14a,b}$$

Eq. (13) now has the form

$$\overset{+}{u}(\mathbf{r},t) = (A_R + iA_I)\overset{+}{f}\;[t - \bar{\tau}(z,\mathbf{x}) - i\gamma(z,\mathbf{x})] \tag{15}$$

where we expressed the amplitude term in (13) in the form $A = A_R + iA_I$, and

$$\bar{\tau}(z,\mathbf{x}) = v^{-1}[z + x_1^2/2R_1 + x_2^2/2R_2]\,, \qquad \gamma(z,\mathbf{x}) = v^{-1}[x_1^2/2I_1 + x_2^2/2I_2]. \tag{15a,b}$$

Clearly $\bar{\tau}$ defines the paraxial propagation delay hence R_j are the wavepacket radii of curvature. The imaginary part of the argument of $\overset{+}{f}$ controls the transverse envelope decay of $\overset{+}{u}$ since the level of the waveform decays as γ increases away from the beam axis (Fig. 1). In a plane $z = const.$, the amplitude contour lines (lines of constant γ) are ellipses. In the plane (x_j, z), they are described by the condition $x_j^2/I_j(z) = const.$, hence the waist occurs at $z = -a_j$ where I_j has a minimum. Near the waist, for $|z + a_j| \ll b_j$, $I_j \simeq b_j$ hence the PB stays collimated. For $|z + a_j| \gg b_j$, on the other hand, $I_j \simeq (z + a_j)^2/b_j$ hence for $z \to \infty$ the propagation paths $\gamma = const.$ satisfy $x_j/z = const.$ and the PB opens up along a constant diffraction angle Θ_j. The discussion above identifies b_j as the collimation (Rayleigh) length of the PB in the (x_j, z) plane. In fact, a key feature in this PB solution is that all its frequency components have the same collimation distance (see discussion in Sec. 2.6).

To understand the structure of the *real* PB fields we introduce the real waveform $f_\gamma(t)$ via (see (2))

$$\overset{+}{f}(t - i\gamma) \equiv f_\gamma(t) + i\mathcal{H}f_\gamma(t). \tag{16}$$

From (15), the real field solutions u_R can be express in the form

$$u_R = \{A_R - A_I\mathcal{H}\}\; f_\gamma\,[t - \bar{\tau}]\,. \tag{17}$$

Since the real waveforms f_γ decay as γ grows, u_R is strongest on the beam axis (where $\gamma = 0$ and $f_\gamma \equiv f$) and decays as γ increases away from the axis. Furthermore, the waveforms in (17) are gradually Hilbert transformed along the propagation paths (defined by $\gamma = const.$) as the balance between A_R and A_I changes. It is instructive to consider here the special case of a stigmatic PB with $a_1 = a_2 \equiv a$ and $b_1 = b_2 \equiv b$. Here $A(z) = [a - ib]/[a - ib + z]$ so that $A_R = [a(a + z) + b^2]/[(a + z)^2 + b^2]$ and $A_I = -bz/[(a + z)^2 + b^2]$. Substituting into (17) one finds that the waveforms of u_R are partially Hilbert transformed from f_γ in the $z = 0$ plane to $z^{-1}\{a + b\mathcal{H}\}f_\gamma$ as $z \to \infty$. Another simple example is $b_1 \neq b_2$ but $a_1 = a_2 = 0$ (i.e., the waists in both principal directions are at $z = 0$). Here the waveforms change from f_γ in the $z = 0$ plane to $z^{-1}\sqrt{b_1 b_2}\,\mathcal{H}f_\gamma$ as $z \to \infty$.

2.4 Specific Signal: Analytic Delta (Rayleigh Pulse)

A simple example for a pulse shape is given by

$$\overset{+}{f}(t) = \overset{+}{\delta}(t - iT) \equiv (\pi i)^{-1}(t - iT)^{-1}, \qquad T > 0 \tag{18}$$

where $\overset{+}{\delta}$ is the analytic δ function (i.e., $\overset{+}{\delta}(t - iT) \to \delta(t) + \mathcal{P}/\pi i t$ as $T \to 0$ with $\mathcal{P}$ denoting Cauchy's principal value). The real field solutions are given by (17) with

$$f_\gamma(t) = \mathrm{Re}\overset{+}{\delta}(t - iT - i\gamma) = \pi^{-1}(T + \gamma)\,/\,(t^2 + (T + \gamma)^2) \tag{19a}$$

$$\mathcal{H}f_\gamma(t) = \mathrm{Im}\overset{+}{\delta}(t - iT - i\gamma) = -\pi^{-1}t\,/\,(t^2 + (T + \gamma)^2). \tag{19b}$$

The 3db pulse-width and the peak value in (19a) are $2(T+\gamma)$ and $\pi^{-1}(T+\gamma)^{-1}$, respectively; the waveform is shortest and strongest on the beam axis ($\gamma = 0$) and it decays as γ increases away from the axis. For γ satisfying $(T + \gamma) = \sqrt{2}T$ the peak is 3db weaker than the axial peak. Solving for γ and using (14b) and (15b) yields the 3db beamwidth in the principal direction x_j

$$W_j(z) = W_{0j}\sqrt{1 + (z + a_j)^2/b_j^2}, \qquad W_{0j} = K_0\sqrt{vTb_j}, \qquad K_0 = 2\sqrt{2(\sqrt{2} - 1)} \simeq 1.8 \tag{20a}$$

with W_{0j} being the beamwidth at the waist. As $z \to \infty$, W_j opens up along the diffraction angle

$$\Theta_j = K_0\sqrt{vT/b_j} \tag{20b}$$

Finally it is important to note that the waist beamwidth W_{0j}, the 3db collimation length b_j and the axial 3db pulse-length $T_0 = 2T$ are related by what may be termed "the time-dependent Rayleigh limit"

$$b_j = K_1 W_{0j}^2/vT_0, \qquad K_1 = [4(\sqrt{2} - 1)]^{-1} \simeq 0.6. \tag{21}$$

Representative plots for stigmatic PBs are given in Figs. 2 and 3 of Ref. 8. The figures show in fact the *globally exact* complex source pulsed beam field but, except for the discontinuity in the source plane $z = 0$ which is attributable to the complex source model, they also apply for the present paraxial solution (see Sec. 2.7). The parameters in Ref. 8 are translated to the present generalized representation by setting $a_j = 0$, $b_j = b$ and $T = \beta/v$. One observes that the normalized pulse-length $vT/b = 0.0005$ in Fig. 2 there is ten times shorter than in Fig. 3, giving a stronger and narrower PB (cf. (19) and (20a)). (The reader should disregard Fig. 2(a) where the 3D graphics failed to sample the very short pulse maximum so that the distribution shown is inaccurate).

The frequency spectrum of the analytic delta pulse is $\hat{f}(\omega) = \exp(-\omega T)$. The low frequency components have a negligible effect on the wavepacket, but they generate a weak tail at $t \to \infty$. Analytic pulses with no (or weak) low frequency components are the high frequency analytic delta and the analytic modulated Gaussian (Ref. 8 eqs. (35) and (38), respectively). We shall not discuss these solutions here. However, it should be pointed out that within the wavepacket region their physical characteristics are similar to those in (20)-(21).

2.5 The General Astigmatic Case

In the general case, the two real symmetrical matrices $\mathrm{Re}\Gamma$ and $\mathrm{Im}\Gamma$ cannot be diagonalized simultaneously as in (12). There are, therefore, two principal coordinate frames. In the first frame $\mathrm{Re}\Gamma$ is diagonalized to $\mathrm{diag}\{v^{-1}R_j^{-1}\}$. Expressing the paraxial delay $\bar{\tau}$ in that frame yields an expression similar to (15a), hence R_j are the wavefront radii of curvature in that frame and, in general, $v\text{Re}\boldsymbol{\Gamma}$ is the curvature matrix. In the other coordinate frame $\text{Im}\boldsymbol{\Gamma}$ is diagonalized to $\text{diag}\{v^{-1}I_j^{-1}\}$ with $I_j > 0$ because $\text{Im}\boldsymbol{\Gamma}$ is positive definite. In this frame, γ has the form in (15b) hence the wavepacket has an elliptic cross section. The behavior of the parameters R_j and I_j as a function of z is complicated but they may readily be calculated from the initial condition $\boldsymbol{\Gamma}(0)$ via (9). In particular one finds that the principal coordinate frames rotate as a function of z. Furthermore, $I_j \sim O(z^2)$ as $z \to \infty$ hence the PB spreads within a diffraction cone. The discussion above identifies $\overset{+}{u}$ as an astigmatic wavepacket whose curvature and amplitude axes rotate along the propagation axis.

2.6 Relation to Time-Harmonic Gaussian Beam

A time-harmonic Gaussian beam that propagates along the z-axis in a uniform medium with wave velocity v can be express in the form

$$\hat{u}(\mathbf{r},t) = \sqrt{\det\boldsymbol{\Gamma}(z)/\det\boldsymbol{\Gamma}(0)}\, \exp[ik(z + \tfrac{1}{2}(\mathbf{x},\boldsymbol{\Gamma}\mathbf{x})], \tag{22}$$

where $k = \omega/v$ and an over caret denotes time-harmonic field constituents with an assumed $\exp(-i\omega t)$ time-dependence. $\boldsymbol{\Gamma}(z)$ is the symmetrical complex matrix with a positive definite $\text{Im}\boldsymbol{\Gamma}$ in (9). From the discussion above one finds that $\text{Re}\boldsymbol{\Gamma}$ and $\text{Im}\boldsymbol{\Gamma}$ describe, respectively, the phase front curvature and the Gaussian envelope of the beam. In the general case, (22) is an astigmatic beam whose phasefront and amplitude axes are not aligned as discussed in Sec. 2.5. In the special case when $\boldsymbol{\Gamma}$ can be diagonalized as in (12), $\hat{u}$ is an isoaxial astigmatic Gaussian beam. Its waist in the principal planes (x_j, z) occurs at $z = -a_j$ and its collimation (or Rayleigh) distances is b_j. The beamwidth and diffraction angles are given by

$$\hat{W}_j(z;\omega) = 2\sqrt{b_j/k}\sqrt{1 + (z + a_j)^2/b_j^2}, \qquad \hat{\Theta}_j(\omega) = 2/\sqrt{kb_j}. \tag{23}$$

The PB (11) is obtained now if (22) is multiplied by the frequency spectrum $\hat{f}(\omega)$ and then inverted to the time domain via (2). It is assumed in this process that the matrix $\boldsymbol{\Gamma}$ in (22) is *frequency independent.* This implies that all frequency components in the PB (11) are Gaussian beams with the same waist planes and collimation distances. At a given z, their phase front curvatures are frequency independent but their widths are proportional to $\omega^{-1/2}$ as implied by (23). These observations are important for synthesizing source distributions (say in the $z = 0$ plane) that generate the PB fields.

2.7 Relation to the Complex Source Pulsed Beam

The complex source pulsed beams (CSPB) are field solutions generated by sources located at complex space-time points. In the real coordinate space, they furnish exact real PB solutions of the time dependent wave equation. Their physical characteristics, namely the location of the waist, the direction, the collimation distance and the space-time size of the wavepacket are determined by the complex extension of the source coordinates. Physically, the CSPB are generated by *real* time-dependent source distributions of finite support, hence the complex source model is just a mathematical trick to find the desired source distribution and to generate the field solutions.

In Ref. 8 we considered a complex source pulsed beams that propagate along the z-axis with waist at the $z = 0$ plane. The paraxial form in eq. (24) there is a special case of (13) with $q_1 = q_2 = z - ib$. The present wavepacket are therefore CSPB-type solutions. As discussed in Secs. 2.3 and 2.5 they have, however, a more general form that admits wavepacket astigmatism and medium inhomogeneity.

For practical applications we are mainly interested with the PB field in the paraxial region, which is described by the techniques considered in this paper. The *globally exact* complex source model, however, is an important analytical tool: By substituting complex source coordinates into the time-dependent Green's function of an environment one may generate exact solutions for PB interaction with that environment. For certain canonical configurations one may actually obtain closed form solutions for the PB response, from which it is possible to extract local scattering models that may be extended to non-canonical configuration.[10,11] For non-critical incidence conditions these models are usually simple and may be obtained by the techniques considered in this paper, but for critical incidence conditions (e.g., near the critical angle of total reflections or near a vertex of a wedge) they may be more complicated and require careful analysis of the exact solutions. Basic canonical PB scattering problems have been explored along these lines.[10,11]

The CSPB may also be used as basis functions to construct solutions for the time-dependent wave equation. Several alternative schemes for PB expansion of pulsed radiation from localized or extended source distributions have been formulated in Refs. 12-15. The advantages of the PB approach over the conventional plane-wave or Green's function approaches are the spectral compactization achieved and the use of local basis function that may be tracked locally in complicated environment.

3. PULSED BEAMS IN AN INHOMOGENEOUS MEDIUM

3.1 Beam Coordinate System

Here we consider PB solutions of the time-dependent wave equation (1) in a smoothly inhomogeneous medium with $v = v(\mathbf{r})$. Let Σ be the PB propagation path (it will be shown in (32) that Σ is a ray trajectory). Point on Σ are denoted as $\mathbf{r}_0(\sigma)$ with σ being the arc length along Σ. Points $\mathbf{r}$ near Σ are usually expressed by the coordinates (σ, n, n_b) where n and n_b denote length along the normal and the binormal of Σ at $\mathbf{r}_0(\sigma)$, respectively. However, due to possible torsion of Σ these coordinates are, in general, nonorthogonal. This may easily be verified by using the Fernet equations $\mathbf{r}_0' = \hat{\mathbf{t}}$, $\hat{\mathbf{t}}' = K\hat{\mathbf{n}}$, $\hat{\mathbf{n}}' = -K\hat{\mathbf{t}} + \kappa\hat{\mathbf{n}}_b$, $\hat{\mathbf{n}}_b' = -\kappa\hat{\mathbf{n}}$ and $\hat{\mathbf{n}}_b = \hat{\mathbf{t}} \times \hat{\mathbf{n}}$ where $\hat{\mathbf{t}}$, $\hat{\mathbf{n}}$ and $\hat{\mathbf{n}}_b$ are unit vectors along the tangent, the normal and the binormal, K and κ are the curvature and torsion and the prime denotes derivative with respect to σ. An orthogonal system (σ, x_1, x_2) may be constructed by rotating the coordinates via

$$x_1 = n\cos\vartheta - n_b \sin\vartheta\,, \qquad x_2 = n\sin\vartheta + n_b\cos\vartheta \tag{24}$$

where the rotation angle $\vartheta(\sigma)$ satisfies

$$\vartheta'(\sigma) = \kappa(\sigma) \tag{24a}$$

In this system $d\mathbf{r} = \hat{\mathbf{x}}_1 dx_1 + \hat{\mathbf{x}}_2 dx_2 + \hat{\mathbf{t}} h_\sigma d\sigma$ with the Lamé coefficient

$$h_\sigma = 1 - K(x_1\cos\vartheta + x_2\sin\vartheta) = 1 - Kn. \tag{25}$$

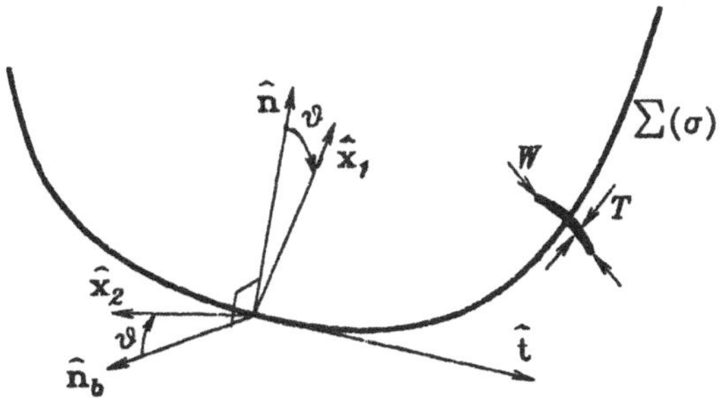

Fig. 2 Pulsed beam propagation in an inhomogeneous medium.

3.2 The Wavepacket Equation

In the coordinate system $(\sigma, \mathbf{x})$, $\mathbf{x} = (x_1, x_2)$, the wave equation is given by (1) with

$$\nabla^2 = h_\sigma^{-1}\{\partial_\sigma h_\sigma^{-1}\partial_\sigma + \sum_{j=1,2} \partial_j h_\sigma \partial_j\} \tag{26}$$

where $\partial_j \equiv \partial/\partial x_j$. To derive the wavepacket equation we make the following three assumptions:

a) Moving coordinate frame: We express $u(\mathbf{r}, t)$ in the form

$$u(\mathbf{r}, t) = U(\mathbf{r}, \tau), \qquad \tau = t - \int^\sigma \frac{d\sigma'}{v_0(\sigma')} \tag{27}$$

where $v_0(\sigma) \equiv v\,|_\Sigma$. The wave equation now yields

$$\{h_\sigma^{-1}(\partial_\sigma^2 - 2v_0^{-1}\partial_\sigma\partial_\tau + v_0^{-2}v_0'\partial_\tau + v_0^{-2}\partial_\tau^2) + (h_\sigma^{-1})'(\partial_\sigma - v_0^{-1}\partial_\tau)$$
$$+ \textstyle\sum_j \partial_j h_\sigma \partial_j - h_\sigma v^{-2}\partial_\tau^2\}U \;=\; 0. \tag{28}$$

b) Space-time window: We look for PB solutions of (28) which are characterized by the short pulse length T. From (20a) they are expected to be within the space-time window

$$\tau \sim O(T), \qquad x_j \sim O(T^{1/2}), \tag{29}$$

hence $\partial_\sigma U \sim O(1)$, $\partial_\tau U \sim O(T^{-1})$ and $\partial_j U \sim O(T^{-1/2})$. Next we substitute these constraints into (28) keeping dominant terms of order T^{-1} and lower (i.e., neglecting terms of order $T^{-1/2}$ and higher). We therefore expand $v(\mathbf{r})$ to second order near Σ in the form $v(\mathbf{r}) \simeq v_0(\sigma) + (\mathbf{v}_1, \mathbf{x}) + \frac{1}{2}(\mathbf{x}, \mathbf{V}_2\mathbf{x})$ where the 2-vector $\mathbf{v}_1(\sigma)$ and the 2×2 symmetrical matrix $\mathbf{V}_2(\sigma)$ are given by $v_{1_j} = \partial_j v\,|_\Sigma$ and $V_{2_{ij}} = \partial_i\partial_j v\,|_\Sigma$. We also expand to second order $h_\sigma^{-1} \simeq 1 + Kn + (Kn)^2$ and $v^{-2} = v_0^{-2}[1 - 2v_0^{-1}(\mathbf{v}_1, \mathbf{x}) - v_0^{-1}(\mathbf{x}, \mathbf{V}_2\mathbf{x}) + 3v_0^{-2}(\mathbf{v}_1, \mathbf{x})^2]$. Substituting into (28) and keeping terms of order T^{-1} and lower we obtain

$$\{(-2v_0^{-1}\partial_\sigma + v_0^{-2}v_0')\partial_\tau + \textstyle\sum_j \partial_j^2 +$$
$$v_0^{-2}[2Kn + (Kn)^2 + 2v_0^{-1}(1 - Kn)(\mathbf{v}_1, \mathbf{x}) + v_0^{-1}(\mathbf{x}, \mathbf{V}_2\mathbf{x}) - 3v_0^{-2}(\mathbf{v}_1, \mathbf{x})^2]\partial_\tau^2\}U = 0. \tag{30}$$

c) Σ is a ray: We assume now that Σ is a ray. It is described, therefore, by the eikonal equation that can be expressed in the form

$$K\hat{\mathbf{n}} = -\nabla_\perp \ln v(\mathbf{r}) \tag{31}$$

where $\nabla_\perp$ denotes transverse gradient with respect to Σ. From (31), points $\mathbf{r} = (\sigma, \mathbf{x})$ near Σ satisfy $Kn + v_0^{-1}(\mathbf{v}_1, \mathbf{x}) = 0$ so that (30) simplifies to

$$\{(-2v_0^{-1}\partial_\sigma + v_0^{-2}v_0')\partial_\tau + \sum_j \partial_j^2 + v_0^{-3}(\mathbf{x}, \mathbf{V}_2\mathbf{x})\,\partial_\tau^2\}U = 0 \tag{32}$$

(note that the terms in (30) that have been eliminated in (32) due to the fact that Σ is a ray are $O(T^{-3/2})$ and are otherwise dominant). Finally (32) may be further simplified into the "wavepacket equation"

$$\{-2v_0^{-1}\partial_\sigma\partial_\tau + \sum_j \partial_j^2 + v_0^{-3}(\mathbf{x}, \mathbf{V}_2\mathbf{x})\,\partial_\tau^2\}V = 0, \qquad U(\mathbf{x}, \sigma, \tau) = \sqrt{v_0(\sigma)}\,V(\mathbf{x}, \sigma, \tau) \tag{33}$$

3.3 Pulsed Beam Solutions

We express the solution of (33) in the form

$$\overset{+}{V} = A(\sigma)\overset{+}{f}\,[\tau - \tfrac{1}{2}(\mathbf{x}, \Gamma(\sigma)\mathbf{x})] \tag{34}$$

where, as before, we utilize analytic signal representation that admits arguments with negative imaginary parts (see (2)) and $\mathbf{\Gamma}$ is a 2×2 complex symmetrical matrix with a positive definite imaginary part. Following the discussions in Secs. 2 expression (34) describes a PB that propagates along Σ. This expression is a solution of (33) if (cf. (8))

$$\mathbf{\Gamma}' + v_0\mathbf{\Gamma}^2 + v_0^{-2}\mathbf{V}_2 = 0 \quad \text{and} \quad A\,\mathrm{trace}\,\mathbf{\Gamma} + 2v_0^{-1}A' = 0. \tag{35a, b}$$

Setting

$$\mathbf{\Gamma} = \mathbf{P}\mathbf{Q}^{-1} \tag{36}$$

the matrix Riccati equation (35a) reduces to the coupled linear equations

$$\mathbf{Q}' = v_0\mathbf{P}, \quad \mathbf{P}' = -v_0^{-2}\mathbf{V}_2\mathbf{Q} \tag{37}$$

with the initial conditions (see (36)) $\mathbf{Q}(0) = \mathbf{\Gamma}^{-1}(0)$, $\mathbf{P}(0) = \mathbf{I}$. Furthermore, by using the general matrix relation (10) one may replace in (35b) $\mathrm{trace}\,\mathbf{\Gamma} = v_0^{-1}\ln'\det\mathbf{Q}$ and obtain the solution $A(\sigma) = (\det\mathbf{Q}(\sigma))^{1/2}$. The PB solution is given therefore by

$$\overset{+}{u}(\mathbf{r},t) = \sqrt{\frac{v_0(\sigma)}{v_0(0)}}\sqrt{\frac{\det\mathbf{Q}(0)}{\det\mathbf{Q}(\sigma)}}\,\overset{+}{f}\left[t - \int_0^\sigma \frac{d\sigma'}{v_0(\sigma')} - \tfrac{1}{2}(\mathbf{x},\mathbf{\Gamma}(z)\mathbf{x})\right]. \tag{38}$$

In this expression $\overset{+}{f}$ may be any analytic pulse (e.g., the examples in Sec. 2.4). The following general properties of the solutions of the system of equations (37) can be proved:[16]

$$\det\mathbf{Q}(\sigma) \neq 0, \qquad \mathbf{\Gamma} = \mathbf{\Gamma}^T, \qquad \mathrm{Im}\mathbf{\Gamma} = \tfrac{1}{2}(\mathbf{Q}\mathbf{Q}^\dagger)^{-1} \tag{39}$$

where $\dagger$ denotes the Hermitian conjugate. Thus, along Σ $\mathbf{\Gamma}$ stays symmetrical with a positive definite imaginary part. Accordingly $\overset{+}{u}$ is confined along Σ and is regular for all σ. Following the discussion in Secs. 2.3 and 2.5, the matrix $v_0(\sigma)\mathrm{Re}\mathbf{\Gamma}(\sigma)$ is recognized as the wavepacket curvature while the matrix $v_0(\sigma)\mathrm{Im}\mathbf{\Gamma}$ controls the transverse decay of amplitude. All the expressions in Sec. 2 regarding the PB characteristics can therefore be extended to the present case in the respective coordinate frame.

4. SUMMARY AND CONCLUSIONS

We presented PB solutions in inhomogeneous medium. Utilizing the fact that these fields are highly localized in space-time we derived the time-dependent wavepacket equation which describes the field within a moving space-time window that brackets the PB, and then construct the exact solutions of this equation. The wavepackets are expressed in terms of a rather arbitrary pulse function $\overset{+}{f}$ and the "complex curvature matrix" $\mathbf{\Gamma}$ that can be found by solving a matrix ordinary differential equation along the ray. The real part and the positive definite imaginary part of $\mathbf{\Gamma}$ describe, respectively, the curvature and the amplitude of the wavepacket. The principal axes of astigmatism of the curvature and of the amplitude are in general not aligned and may rotate along the propagation axis. Finally, the *real* PB fields are gradually Hilbert transformed along the propagation axis. Simple examples for these properties in free space are given for the special case when the principal axes of astigmatism coincide (Secs. 2.3-4). The paraxial wavepackets constructed here are related to the CSPB, but they have a more general form that admits wavepacket astigmatism and medium inhomogeneities. The *globally exact* CSPB can therefore be used to derive local models for scattering and diffraction of the paraxial PB solutions considered here.[10,11]

The motivation in this paper was also to clarify some general characteristics of space-time wavepacket solutions. This has been done by means of the wavepacket equation that describes the field within a moving space-time window. The PB fields are exact solutions of this equation. It has been shown that, except for a far field spreading, the wavepacket shape remain essentially unchanged along the propagation path. This establishes these new CSPB-type solutions as the most general *eigen-wavepacket* solutions of the time-dependent wave equation.

REFERENCES

1. J.N. Brittingham, Focus wave modes in homogeneous Maxwell's equations: Transverse electric mode, *J. Appl. Phys.* 54:1179-1189 (1983).

2. R.W. Ziolkowski, Exact solutions of the wave equation with complex locations, *J. Math. Phys.* 26:861-863 (1985).

3. I.M. Besieris, A.M. Shaarawi and R. W. Ziolkowski, A biderectional traveling plane wave representation of exact solutions of the scalar wave equation, *J. Math. Phys.* 30:1254-1269 (1989).

4. E. Heyman, B.Z. Steinberg and L.B. Felsen, Spectral analysis of focus wave modes, *J.Opt. Soc. Am. A*, 4:2081-2091 (1987).

5. H.E. Moses and R.T. Prosser, Initial conditions, sources, and currents for prescribed time-dependent acoustic and electromagnetic fields in three dimensions. Part I: The inverse initial value problem. Acoustic and Electromagnetic "bullets", expanding waves, and imploding waves, *IEEE Trans. Antennas Propagat.* 34:188-196 (1986).

6. H.E. Moses and R.T. Prosser, Acoustic and electromagnetic bullets. New exact solutions of the acoustic and Maxwell's equation, *SIAM J. Appl. Math.*, 50:1325-1340 (1990).

7. E. Heyman and B.Z. Steinberg, A spectral analysis of complex source pulsed beams, *J. Opt. Soc. Am. A* 4:473-480 (1987).

8. E. Heyman and L.B. Felsen, Complex source pulsed beam fields, *J. Opt. Soc. Am. A* 6:806-817 (1989).

9. E. Heyman, B.Z. Steinberg and R. Ianconescu, Electromagnetic complex source pulsed beam fields, *IEEE Trans. Antennas Propagat.* 38:957-963 (1990).

10. E. Heyman and R. Ianconescu, Pulsed beam reflection and transmissions at a dielectric interface. Part I: Two dimensional fields, *IEEE Trans. Antennas Propagat.* 38:1791-1800 (1990).

11. R. Ianconescu and E. Heyman, Pulsed beam diffraction by a perfectly conducting wedge: Exact solution and approximate local models, *IEEE Trans. Antennas Propagat.* (submitted).

12. E. Heyman, Complex source pulsed beam expansion of transient radiation, *Wave Motion* 11:337-349 (1989).

13. B.Z. Steinberg, E. Heyman and L.B. Felsen, Phase space beam summation for time dependent radiation from large apertures: Continuous parameterization, *J. Opt. Soc. Am. A* 8:943-958 (1991).

14. B.Z. Steinberg and E. Heyman, Phase space beam summation for time dependent radiation from large apertures: Discretized parameterization, *J. Opt. Soc. Am. A* 8:959-966 (1991).

15. E. Heyman and I. Beracha, Complex multiple beam and Hermite Pulsed beam: A novel expansion scheme for time dependent radiation from well collimated apertures, *J. Opt. Soc. A* 9:October (1992).

16. V.M. Babich and V.S. Buldyrev, *Asymptotic Methods in Shortwave Diffraction Problems - The Standard Problem Method*, Nauka, Moscow (1972), Chap. 9.

ULTRA-WIDE BANDWIDTH, MULTI-DERIVATIVE ELECTROMAGNETIC SYSTEMS

Richard W. Ziolkowski

Department of Electrical and Computer Engineering
The University of Arizona
Tucson, AZ 85721
(602) 621-6173

1. INTRODUCTION

The characteristics of the beams generated by ultra-wide bandwidth (UWB) electromagnetic systems are central to their practical applications, in particular to radar and electromagnetic remote sensing systems. These characteristics include the rate of beam divergence, the beam intensity, and the energy efficiency. Analytical bounds on the characteristics of beams generated by an arbitrary pulse-driven array have been derived and supported with numerical calculations[1,2] and experiments[3]. These bounds extend the meaning of near-field distances or diffraction lengths to the situation where the array driving functions can be broad-bandwidth signals and the complete transmitter-receiver system can involve multiple time derivatives. The bounds on the beam characteristics associated with these pulse-driven multiple-derivative (PDMD) electromagnetic systems are defined in terms of effective frequencies that characterize the frequency moments of the set of time signals used to drive the transmitting array. It is found that the higher order coherence properties of the beams associated with these PDMD electromagnetic systems are degraded more slowly by diffraction than the corresponding lower-order characteristics. The higher-order derivatives simply emphasize the higher frequencies present in the finite time signals driving the array.

For transmitting and receiving array systems which consist of elements that are not large in comparison to the shortest wavelength of significance contained in the signals driving them, the output signals of such systems are multiple time derivatives of the input driving functions. For instance, an electromagnetic array whose elements are essentially capacitive will impose a time derivative on a signal as it leaves an element of the array. The resulting signal then acquires another time derivative as it propagates into the far field of that element. If the signal is received by a matched element, the receiver time differentiates the propagated signal again, a result consistent with reciprocity[4,5]. Consequently, the resulting received signal has three time derivatives imposed on the input signal. The combined transmitting and receiving array then constitutes a PDMD electromagnetic beam system, and the beam associated with it has the noted extended near-field characteristics. The bounds simply define the extent of these near-field enhancements.

These extended near-field effects are also obtainable for remote sensing purposes with a one time derivative transmitting and receiving array system, i.e., one that simply reproduces the input signal on transmit and receive. By properly designing the input signal set, one can maintain the coherence of the higher order moments of the beam and can then impose the time derivatives by signal processing upon reception. Thus, one can mimic a PDMD transmitter-receiver system with intelligent signal processing and obtain its desirable effects. These effects are confirmed by the results given in Ref. 6.

An electrically short (non-resonant) dipole transmitter-receiver system is chosen below to

Ultra-Wideband, Short-Pulse Electromagnetics
Edited by H. Bertoni *et al.*, Plenum Press, 1993

illustrate the results because it is one of the simplest examples of a system whose output signals are related to the input driving functions by several time derivatives. Since the diffraction process affects the higher order moments of the spectra more slowly than lower orders, these higher-derivative systems have more intensity and energy available at targets within these extended diffraction lengths. Moreover, if the driving signals have a high degree of correlation, these systems also exhibit enhanced transverse localization properties. Because of the different time derivatives involved, different portions of the frequency spectra of those signals will control the associated beam characteristics. Thus, a set of UWB signals and an electromagnetic transmitter-receiver system can be designed to realize enhancements of a particular beam parameter. The localized wave (LW) solutions discussed in Refs. 1-3 simply provide an immediate access to this situation. The actual transmitter/receiver system configuration then becomes extremely important if enhanced performances are desired, and it must be tailored to the particular beam parameter of interest. A preliminary analysis of the potential applicability of a localized wave PDMD system to radar and to other remote sensing systems is given. In particular, it includes the effects of having a target within the extended near-field region of the localized wave PDMD system on its overall output power, signal-to-noise ratio and resolution. Preliminary comparisons between a localized wave PDMD array system and the corresponding conventional continuous wave system will illustrate these results.

2. PULSED-BEAM SYSTEM MODEL AND ITS OPERATING CHARACTERISTICS

Simple dipole array models of the PDMD transmitting and receiving systems are sufficient to illustrate the major points. The transmitting system consists of a planar array of N independently addressible, center-fed, linear, non-resonant dipole elements and the requisite electronics; the receiving system is either a matched single dipole element or a matched array of linear dipole elements and the requisite electronics. The j-th element in the array is assumed to be excited with the ultra-wide bandwidth voltage signal $f_j(t)$. The associated frequency spectra will be labeled $F_j(\omega)$. The total physical area A of the array is $A = \sum_{j=1}^{N} A_j$. Every dipole in the array is assumed to have the same orientation relative to the z-axis (beam propagation axis). Mutual coupling effects between the transmitting dipoles are small assuming the signals are chosen so that the distance cT associated with their time record T is smaller than the separation distance between elements. If the input field energy is the energy distributed over the array, the n-th element receives $\mathcal{E}_n^{in}$ joules: $\mathcal{E}_n^{in} = C_T^{-1} \int_{-\infty}^{\infty} |f_n(t)|^2\, dt$, where C_T is a real positive constant, independent of frequency in the frequency band of interest, that has the units of impedance and, hence, the total array-weighted input energy is then given simply as: $\mathcal{E}_{in} = A^{-1} \sum_{n=1}^{N} A_n\, \mathcal{E}_n^{in}$. This form of the input energy is chosen to account for the amount of the array "real estate" dedicated to each of the driving functions.

For the unit pair of electrically small transmitting and receiving dipoles assumed here, the measured voltage signal $g(\vec{r},t)$ at the receiving dipole located at the observation point $\vec{r}$ is proportional to three time derivatives of the voltage signal $f_i(\vec{r}_i,t)$ driving the i-th transmitting dipole[2]. Therefore, the total signal at the receiving dipole is a linear superposition of three time derivatives of the pulses driving each transmitting dipole in the array: $g_{meas}(\vec{r},t) = C_j \sum_{n=1}^{N} \partial_t^3 f_n(\vec{r}_n, t - R_n/c)/[2\pi\, c\, R_n]$, where the distance between the n-th transmitting dipole and the receiver is $R_n = |\vec{r} - \vec{r}_n|$. The factor C_j is also a positive real constant, independent of frequency in the frequency band of interest. Its specific value is model dependent; it is related to the areas and the impedances of the PDMD transmitting and receiving dipoles and has units that makes the time integral of the absolute square of $g(\vec{r},t)$ an energy quantity. Then the measured field energy at the observation point $\vec{r}$ is simply $\mathcal{E}_{meas}(\vec{r}) = \int_{-\infty}^{\infty} |g(\vec{r},t)|^2\, dt$.

Bounds on the performance characteristics of beams generated by driving these model arrays with spatial distributions of UWB time signals have been derived[1,2]. The ratio of the measured energy to the input energy defines the measured energy efficiency: $\Gamma_{meas}^{enrg} = \mathcal{E}_{meas}/\mathcal{E}_{in}$. It is advantageous to renormalize the efficiency $\Gamma_{j,enrg}^{meas}$ by the term $(C_j^2/C_{in}^2)\omega_{rad}^{2(j-1)}$. This represents a weighting of the time derivatives, which result from the transmitting and receiving arrays, by the physical constants associated with those arrays and by the effective frequency associated with their conversion of the input energy to the radiated field energy or reciprocally, from the propagated to the measured field energy. In particular, the "effective" frequency of the radiated field ω_{rad} is given by the expression

$$\omega_{rad}^2 \stackrel{\text{def}}{=} \frac{\sum_{n=1}^N A_n \int_{-\infty}^{\infty} dt\, |\partial_t f_n(\vec{r}_n,t)|^2}{\sum_{n=1}^N A_n \int_{-\infty}^{\infty} dt\, |f_n(\vec{r}_n,t)|^2} = \frac{\sum_{n=1}^N A_n \int_{-\infty}^{\infty} d\omega\, \omega^2\, |F_n(\vec{r}_n,\omega)|^2}{\sum_{n=1}^N A_n \int_{-\infty}^{\infty} d\omega\, |F_n(\vec{r}_n,\omega)|^2} . \tag{1}$$

Introducing the analogus effective frequency of the measured field:

$$(\omega_j^{meas})^{2(j-1)} \stackrel{\text{def}}{=} \frac{\sum_{n=1}^N A_n \int_{-\infty}^{\infty} dt\, |\partial_t^j f_n(\vec{r}_n,t)|^2}{\sum_{n=1}^N A_n \int_{-\infty}^{\infty} dt\, |\partial_t\, f_n(\vec{r}_n,t)|^2} = \frac{\sum_{n=1}^N A_n \int_{-\infty}^{\infty} d\omega\, \omega^{2j}\, |F_n(\vec{r}_n,\omega)|^2}{\sum_{n=1}^N A_n \int_{-\infty}^{\infty} d\omega\, \omega^2\, |F_n(\vec{r}_n,\omega)|^2} , \tag{2}$$

the bound on the normalized measured energy efficiency of an UWB beam associated with a $j, j \geq 1$, time derivative system can then be written as

$$\tilde{\Gamma}_{j,enrg}^{meas} \stackrel{\text{def}}{=} \frac{\Gamma_{j,enrg}^{meas}}{(C_j^2/C_{in}^2)\,\omega_{rad}^{2(j-1)}} \leq \left(\frac{\omega_j^{meas}}{\omega_{rad}}\right)^{2(j-1)} \left[\frac{L_{rad}}{z}\right]^2 \equiv \left(\frac{L_{j,enrg}^{meas}}{z}\right)^2 , \tag{3}$$

where the energy Rayleigh distance of the field radiated by a $j = 1$ system is

$$L_{rad} = A\,\omega_{rad}/(2\pi c) = A/\lambda_{rad} , \tag{4}$$

and of the measured field

$$L_{j,enrg}^{meas} = [\omega_j^{meas}/\omega_{rad}]^{(j-1)}\, L_{rad} . \tag{5}$$

Note that these effective frequencies are quantities that characterize *by a single frequency value* all of the broad-bandwidth components contained in all of the signals involved in the radiation and measurement processes. The effective frequency ω_{rad} accounts for the spectral energies launched into the medium in our model. The measured frequency ω_j^{meas} accounts for the spectral energies measured in the receiver of our model.

In a CW or narrow-band case, $\omega_j^{meas,CW} \equiv \omega_{rad}^{CW} = \omega_{CW}$ and the energy and intensity diffraction lengths associated with the CW measured field are identical for all j:

$$L_{j,enrg}^{meas,CW} = L_{j,int}^{meas,CW} = L_{rad}^{CW} = A/\lambda_{CW} , \tag{6}$$

If each of the radiating elements have the same area $A_n = A_0$ so that $A = N\,A_0$, Eq. (3) gives

$$\tilde{\Gamma}_{j,enrg}^{CW} \leq (L_{rad}^{CW}/z)^2 = N^2\,(A_0/\lambda_{CW}\,z)^2 , \tag{7}$$

i.e., the CW energy efficiency is bounded by the field resulting from a coherent superposition of all of the radiating elements. In contrast, the diffraction length $L_{j,enrg}^{meas,UWB}$ for a UWB system can be *designed* for a PDMD system to be substantially greater than L_{rad} by a suitable choice of the UWB signals driving the array, e.g., the LW signals. In particular, for a one time derivative system, Eq. (5) indicates that $L_{1,enrg}^{meas,UWB} = L_{rad}$. In contrast, for a three time derivative system it gives

$$L_{3,enrg}^{meas,UWB} = (\omega_3^{meas}/\omega_{rad})^2\, L_{rad} , \tag{8}$$

Since the higher order moments of a ultra-wide bandwidth spectrum can be designed to occur higher in frequency than the lower order ones, a higher derivative system provides an extended near-field region; and, hence, the energy efficicency of the beam will be maintained in that region and will be enhanced beyond it. This can be described by saying that diffraction affects the higher order coherence properties of a beam more slowly than its lower orders.

In a similar manner, introducing the term

$$\Upsilon_j^{meas} \stackrel{\text{def}}{=} \frac{\max_t \sum_{n=1}^N A_n \left|\partial_t^j f_n(\vec{r}_n,t)\right|^2}{\sum_{n=1}^N A_n \int_{-\infty}^{\infty} dt \left|\partial_t^j f_n(\vec{r}_n,t)\right|^2} , \tag{9}$$

where the operator $\max_t$ takes the maximum in time, one can obtain the bound on the efficiency of the maximum field intensity for a j time derivative system normalized by the array weighted input energy and renormalized by the system factors. It leads to the diffraction length associated with the measured maximum beam intensity:

$$L_{j,int}^{meas} = \left(\frac{\omega_j^{meas}}{\omega_{rad}}\right)^{(j-1)} \left[\frac{\Upsilon_j^{meas}}{\omega_{rad}}\right]^{1/2} L_{rad} = \left(\frac{\Upsilon_j^{meas}}{\omega_{rad}}\right)^{1/2} L_{j,enrg}^{meas}, \tag{10}$$

which is generally different from the one associated with the measured beam energy. The term Υ_j^{meas} is simply the ratio of the maximum of a function to its time-averaged value. Since the LW driving signals all have broad bandwidths, one can achieve a very large instantaneous value of the square of the signals with a small average value over a time period comparable to the CW case. The entire bandwidth controls the behavior of the measured beam intensity rather than, as in the energy efficiency case, a particular frequency value. Thus one can design the LW driving functions to obtain $L_{j,int}^{meas,UWB} \gg L_{rad}^{CW}$.

The rate of divergence of the beam generated by a pulse-driven array can be obtained as well. It is measured essentially as the radius at which its energy profile has decreased to half its maximum value in the plane $z = const$ away from the aperture divided by the distance from the array. The CW "Antenna Theorem" has been extended to a general UWB system[2] and yields this quantity. In particular, the product of the source area and the far-field beam angular spread (solid angle) of the measured beam energy is on the order of the measured wavelength squared:

$$A \times \left[\pi \left(\theta_{enrg}^{meas}\right)^2\right] \sim \lambda_{meas}^2 . \tag{11}$$

Because one can control *by design* the correlation properties of the constituent time signals as well as the effective frequencies (i.e., the relative arrival times and the amounts of the various frequency components), the energy and intensity profiles of the beam associated with a PDMD system can be made more localized than the corresponding CW beam. This leads one to the energy beam width enhancement:

$$\theta_{enrg}^{meas} \sim \left[\lambda_{meas}/\lambda_{rad}\right] \theta_{enrg}^{rad} = \left[\omega_{meas}/\omega_{rad}\right]^{-1} \theta_{enrg}^{rad} . \tag{12}$$

It also must be remembered that the intensity and the energy profiles of a beam behave differently in the general UWB case. They are controlled by different, but related properties of the spectra of the input field. This difference between the measured intensity and energy beam profiles has been observed experimentally[3] and numerically[1-3,6]. In general the maximum intensity (in time) beam width θ_{int}^{meas} will be narrower than the energy beam width $\theta_{enrg}^{meas} = \lambda_{meas}/\sqrt{\pi A}$ given by (11), i.e.,

$$\theta_{int}^{meas} \leq \theta_{enrg}^{meas}, \tag{13}$$

because the maximum intensity in time represents an instantaneous rather than an average property of the beam. The equality occurs for any CW case. The resulting narrower intensity profiles of an PDMD system may have significant additional implications for several potential applications, including radar and other remote sensing systems.

We note that these bounds can be derived either directly in the time domain as was done in Refs. 1 and 2 or from the standard Fresnel-Fraunhofer formulae as follows. Consider the temporal Fourier transform $G_{rad}(x, y, z, \omega = kc)$ of the radiated field $g_{rad}(\vec{r}, t)$ which is produced by driving the aperture with the field $f(\vec{r}', t)$ in the plane $z' = 0$, which has the spatial-frequency Fourier transform $F(k_x, k_y, z' = 0, \omega = kc)$. It is given by the expression

$$G_{rad}(x, y, z, \omega = kc) \approx \frac{-i}{\lambda z} e^{ikz} e^{ik(x^2+y^2)/2z} F\left(\frac{kx}{z}, \frac{ky}{z}, 0, \omega = kc\right). \tag{14}$$

We notice that the inverse Fourier transform of this expression represents a time derivative of the driving field - the time derivative that results from the transition of the propagating field into the Fresnel region. Introducing $j-1$ more time derivatives for a j-derivative transmitter-receiver system, one then finds

$$G_{meas}(x,y,z,\omega=kc) = (-i\,\omega)^{j-1}\,G_{rad}(x,y,z,\omega=kc)$$
$$\approx (-i\,\omega)^{j}\,\frac{e^{ikz}}{2\pi c\,z}\,e^{ik(x^2+y^2)/2z}\,F\Big(\frac{kx}{z},\frac{ky}{z},0,\omega=kc\Big)\,. \qquad (15)$$

Now consider the energy of the signal received at the observation point. It is given by the expression

$$\int_{-\infty}^{\infty} dt\,|g_{meas}(\vec{r},t)|^2 = \int_{-\infty}^{\infty} d\omega\,|G_{meas}(x,y,z,\omega=kc)|^2$$
$$\approx \Big(\frac{1}{2\pi c\,z}\Big)^2 \int_{-\infty}^{\infty} d\omega\,\omega^{2j}\,\Big|F\Big(\frac{kx}{z},\frac{ky}{z},0,\omega\Big)\Big|^2$$
$$\leq \Big(\frac{1}{2\pi c\,z}\Big)^2 \omega_{meas}^{2(j-1)} \int_{-\infty}^{\infty} d\omega\,\omega^{2}\,\Big|F\Big(\frac{kx}{z},\frac{ky}{z},0,\omega\Big)\Big|^2$$
$$\leq \Big(\frac{1}{2\pi c\,z}\Big)^2 \omega_{meas}^{2(j-1)}\,\omega_{rad}^2\,A\,\mathcal{E}_{in} = \Big(\frac{A}{\lambda_{rad}\,z}\Big)^2 \omega_{meas}^{2(j-1)}\,\frac{\mathcal{E}_{in}}{A}\,. \qquad (16)$$

Treating the energy received at a point as the fluence $\mathcal{F}_{meas}$ there, i.e., the average energy per unit area A, and the quantity $\mathcal{E}_{in}/A$ as the corresponding input fluence $\mathcal{F}_{in}$, one obtains immediately

$$\frac{\mathcal{F}_{meas}}{\omega_{rad}^{2(j-1)}\,\mathcal{F}_{in}} \leq \Big(\frac{A}{\lambda_{rad}\,z}\Big)^2 \Big[\frac{\omega_{meas}}{\omega_{rad}}\Big]^{2(j-1)} \equiv \Big(\frac{L_{j,enrg}^{meas}}{z}\Big)^2\,, \qquad (17)$$

which is equivalent to the bound given by Eq. (3). It is then more apparent from traditional arguments that the energy beam width is naturally $\theta_{meas} = \lambda_{meas}/\sqrt{\pi A}$, as indicated by Eq(11).

3. MULTI-DERIVATIVE LW RADAR AND REMOTE SENSING SYSTEMS

The potential applications of PDMD electromagnetic systems include radar and remote sensing. To provide some motivation for studying these PDMD systems in more detail, we discuss below some potential improvements they provide over a conventional radar or synthetic aperture radar (SAR) remote sensing system. The emphasis is placed on the LW technology so that, for example, a LW radar refers to a PDMD transmitter-receiver system having LW solutions as the driving functions. This choice is made simply because, as noted above, the LW solutions provide an immediate access to the multi-derivative, extended-near-field enhancements. Similar or even improved results could be obtained with other appropriately-designed classes of UWB signal sets.

3a. Output Power for a LW Pulse-Driven Multi-Derivative Radar

Let the Fourier transform of the cross-section of the target be $\tilde{\sigma}_{target}(\omega)$. The overall response of a matched, j time derivative transmitter-receiver, LW pulse-driven radar system can be estimated as $\mathcal{F}_{int}^{-1}$, the time integral of the inverse Fourier transform, applied to an energy form of the radar equation:

$$\mathcal{E}_{out} \sim \mathcal{F}_{int}^{-1}\Big[\frac{\lambda^2}{4\pi}\,G_R(\omega)\,\frac{1}{4\pi r^2}\,\tilde{\sigma}_{target}(\omega)\,\frac{1}{4\pi r^2}\,G_T(\omega)\,\tilde{\mathcal{E}}_{in}(\omega)\Big]$$
$$\leq \Bigg(\frac{L_{j,enrg}^{meas,LW}}{r}\Bigg)^2 \frac{\sigma_{target}^{max}}{4\pi r^2}\,\mathcal{E}_{in}\,. \qquad (18)$$

Consequently, the received average power (energy/time record length) for this LW PDMD radar in comparison to the equivalent CW radar is

$$\frac{P_{out}^{LW}}{P_{out}^{CW}} \approx \frac{\mathcal{E}_{out}^{LW}/T}{\mathcal{E}_{out}^{CW}/T} \sim \Bigg(\frac{L_{j,enrg}^{meas,LW}}{L_{rad}}\Bigg)^2 = \Big(\frac{\omega_j^{meas}}{\omega_{CW}}\Big)^{2(j-1)}\,. \qquad (19)$$

Thus, by design, one can improve the radar response through the received energy or average power with a LW PDMD system. A factor of 10 improvement of the Rayleigh distance ratio $L_{j,enrg}^{meas,LW}/L_{rad}$ has been achieved numerically[1,2] and experimentally[3] for a $j = 3$ time derivative system. As discussed below, this would imply more than two orders of magnitude improvement in the S/N ratio might be achievable with a LW PDMD radar.

Note that the maximum intensity values are fairly constant in the extended near field so that if $r < L_{j,enrg}^{meas,LW}$, Eq. (18) becomes

$$\mathcal{E}_{out,sphere}^{nearfield} \leq \left(\frac{\sigma_{target}^{max}}{4\pi\, r^2}\right) \mathcal{E}_{in} \,, \tag{20}$$

which indicates that the usual r^{-4} radar equation behavior can be overcome over these extended near-field distances. In fact, Eq. (17) represents what would happen if the target were spherical-like; i.e., the beam is constant to the target and then the scattered field behaves like a spherical wave. The expression would be further improved when scattering from a plate by another factor of r^2:

$$\mathcal{E}_{out,plate}^{nearfield} \leq \sigma_{target}^{max}\, \mathcal{E}_{in} \,, \tag{21}$$

This means that in the extended near field the response should vary as $(2r)^{-\epsilon}$ when scattering from a plate and as $r^{-(2+\epsilon)}$ when scattering from a sphere, where $0 \leq \epsilon \ll 1$. As above, the distances over which these effects are maintained are determined by the multi-derivative diffraction lengths. Initial ultrasound experiments by Kent Lewis at Lawrence Livermore National Laboratory have verified that the PDMD beams exhibit this behavior.

3b. SNR for a LW Pulse-Driven Multi-Derivative Radar

Assuming that the receiver is limited only by uncorrelated noise, it is readily shown that the signal-to-noise ratio (SNR) of the LW PDMD radar system is approximately related to the SNR of a conventional SAR system as:

$$(S/N)_{LW} \sim \left(\frac{L_{j,enrg}^{meas,LW}}{L_{rad}}\right)^2 (S/N)_{CW} = \left(\frac{\omega_j^{meas}}{\omega_{CW}}\right)^{2(j-1)} (S/N)_{CW} \,. \tag{22}$$

A more detailed, but conservative calculation comparing a chirped SAR system having a pulse compression ratio of T/τ with a LW PDMD radar has been made[7]. It assumes comparable uncorrelated receiver noise, time record lengths, bandwidths, target response, etc.; but, because of the potential technical difficulties, assumes multiple pulse averaging only in the conventional SAR case and none in the LW PDMD radar case. The analysis gives a comparison of the SNRs for these systems:

$$\frac{(S/N)_{LW}}{(S/N)_{CW}} \sim \frac{T}{\tau} \tag{23}$$

It follows that *for a given SAR application, the amount of pulse compression necessary is a measure of the potential improvement available with a LWR.* Since pulse-compression ratios are typically 100's to 1000's, Eq. (23) makes it evident that a LW radar offers a tremendous performance advantage, which, in all likelihood, can be improved further through the use of intelligent processing of the scattered LW radar signal. In fact, the (S/N) improvements indicated by (22) could be obtained, which is directly related to extended near field properties of the LW radar system. It is initially surprising that the SAR's pulse compression ratio, a quantity which one normally attempts to maximize, should be a measure of the LW radar's superiority. This result can be explained by noting that the bandwidths and total energies in the SAR and LW radar are fixed and equal. Hence, to increase the SAR's pulse compression ratio while holding the bandwidth and transmitted energy constant is to decrease its radiated power, which decreases its SNR. This result does not mean that the SAR's performance will not be improved by increasing the compression ratio; rather it indicates that one can do that much better expending similar resources on the corresponding LW PDMD radar system.

3c. Along-Track Resolution for a LW Pulse-Driven Multi-Derivative Synthetic Aperture Radar

We wish to consider a synthetic aperture LW PDMD radar, termed here a SALWR. At each location of the transceiver array, the appropriate LW driving signals are transmitted and the scattered signals collected. Thus, in comparison to a conventional SAR, there may be different signals at different positions of the transceiver system.

For a conventional SAR, the (cross-track) range resolution δ_{CT} is controlled by the bandwidth of the driving pulses. Since the LW signals are UWB solutions, the range resolutions of a SAR and a SALWR can be made comparable by forcing their absolute bandwidths to be similar. In contrast the minimum along-track resolution δ_{AT} is controlled by the carrier frequency and the antenna size. It is obtained[8] by accounting for the two-way synthetic aperture pattern. For a conventional SAR it is given by the expression:

$$\delta_{AT}^{SAR} = \frac{\lambda_{rad}}{2L_{SAR}} R \,, \tag{24}$$

where λ_{rad} is the wavelength of the carrier, R is the target range, and L_{SAR} is the length of the synthetic aperture. Let the length of the physical transceiver aperture be D. Taking the synthetic aperture length L_{SAR} to be the same as $L_{range} = (\lambda_{rad}/D)\, R$, the length L_{range} corresponding to the beamwidth size at the distance R from the transceiver's antenna, the resolution becomes

$$\delta_{AT}^{SAR} = \frac{\lambda_{rad}}{2(\lambda_{rad}/D)\, R} R \equiv \frac{D}{2} \,, \tag{25}$$

half the size of the physical aperture and independent of the carrier wavelength.

Analogously, the corresponding SALWR has the along-track resolution at R:

$$\delta_{AT}^{LW} = \frac{\lambda_{meas}}{2L_{SAR}} R \,, \tag{26}$$

determined by the effective frequency of the measured field. Given the same size aperture as the conventional SAR, one then obtains an enhanced resolution:

$$\delta_{AT}^{LW} = \left[\frac{\lambda_{meas}}{\lambda_{rad}}\right] \delta_{AT}^{SAR} \equiv \left[\frac{\lambda_{meas}}{\lambda_{rad}}\right] \frac{D}{2} \,, \tag{27}$$

the enhancement, as one might expect, being proportional to the ratio of the effective wavelengths associated with the measured and the radiated fields. On the other hand, this result also indicates that the same resolution (25) provided by the conventional SAR can be maintained by a SALWR with a smaller aperture length, i.e., if the synthetic aperture length is taken to be

$$L_{SAR}^{LW} = \frac{\lambda_{meas}}{\lambda_{rad}} L_{SAR} \,, \tag{28}$$

the along-track resolution is again $D/2$. Moreover, because the diffraction length of the SALWR would be $(\lambda_{meas}/\lambda_{rad})^{-2}$ times larger than the conventional SAR's diffraction length, these resolution enhancement effects can be obtained with the SALWR from a much larger target range. The desired effect again is by design for the intended application.

Finally, in contrast to a narrow-bandwidth SAR, side-lobe suppression is more readily accomplished with a UWB system such as a SALWR. By properly designing the driving functions, one can optimize the beam shape at the target range[9]. Thus resolution could also be improved without a corresponding increase in the sidelobe levels.

Note that the assumption that $L_{range} = L_{SAR}$ implies that *even the conventional SAR is taking data in the near-field of the synthetic aperture*, i.e., the beamwidth at R is the same size as the aperture. As shown recently in Ref. 10, this near-field construct gives one a great deal of freedom to design the LW solutions for a particular application and realize the concomitant extended near-field enhancements. In fact any LW solution can be approximately recovered from a finite array within this extended near-field region. For instance, one can design a pencil-beam with a large depth-of-focus. Such a beam has many remote sensing applications.

Some of these enhancements have recently been realized experimentally for ultrasound remote sensing and imaging by Lu and Greenleaf at the MAYO Clinic[11].

3d. LW PDMD Radar and Remote Sensing System Summary

In summary, a LW PDMD radar or remote sensing system offers:

- High effective gains in intensity and energy
- High cross-range resolution with low sidelobe levels
- High range resolution (large absolute bandwidths)
- High S/N ratios (large time-bandwidth products)
- Pencil beams with large depth-of-focus

In addition, the extended diffraction lengths and narrower beam widths associated with the LW systems should permit one to employ smaller sources to perform equivalent system requirements. This would be a significant advantage, for example, when there are concerns about the system size. Of course, this is not the complete story and further refinements of the above calculations must be accomplished to justify their practicality. For instance, these pencil beams will require nonconventional strategies for their use. A LW PDMD radar system could be "piggybacked" with a conventional radar: broad coverage with a CW high power microwave system for detection and a pencil-beam interrogating LW PDMD system for localization and identification. A superior detection, localization, and identification radar system might be obtained in this manner.

4. CONCLUSIONS

There are many potential radar and remote sensing applications of LW beams associated with electromagnetic PDMD systems. Increased depths of foci, broad bandwidths, and narrower beams have been demonstrated with specially designed LW PDMD arrays. Preliminary simulations and experiments for high resolution imaging with LW beams have also been very favorable. The ultrasound medical imaging area is already applying these LW concepts with success. It is hoped that this success will carry-over to electromagnetic applications, including radar and remote sensing systems.

5. REFERENCES

1. R. W. Ziolkowski, "Localized wave physics and engineering", *Phys. Rev. A* 44(6): 3960-3984 (1991).

2. R. W. Ziolkowski, "Properties of electromagnetic beams generated by ultra-wide bandwidth pulse-driven arrays", *IEEE Trans. Antennas and Propagat.* 40(8): 888-905 (1992).

3. R. W. Ziolkowski and D. K. Lewis, "Verification of the localized-wave transmission effect", *J. Appl. Phys.* 68: 6083-6086 (1990).

4. M. Kanda, "Transients in a resistively loaded linear antenna compared with those in a conical antenna and TEM horn", *IEEE Trans. Antennas and Propagat.* AP-28: 132-136 (1980).

5. M. Kanda, "Time domain sensors for radiated impulsive measurements", *IEEE. Trans. Antennas and Propagat.* AP-31: 438-444 (1983).

6. R. W. Ziolkowski and J. Judkins, "Properties of Ultra-Wide Bandwidth Pulsed Gaussian Beams", to appear in *J. Opt. Soc. Am. A*, (1992).

7. Brian A. Baertlein, Ballena Systems Corporation, 1150 Ballena Blvd., Suite 210, Alameda, CA 94501, private communications.

8. J. P. Fitch, 1988, Synthetic Aperture Radar, Springer-Verlag, New York.

9. J. E. Hernandez, R. W. Ziolkowski, and S. R. Parker, "Synthesis of the driving functions of an array for propagating localized wave energy", *J. Acoust. Soc. Am.* 92(1): 550-562, 1992.

10. R. W. Ziolkowski, I. M. Besieris, and A. M. Shaarawi, "Aperture realizations of exact solutions to homogeneous wave equations", to appear in *J. Opt. Soc. Am. A*, (1992).

11. J.-Y. Lu and J. F. Greenleaf, "Experimental verification of nondiffracting X waves", *IEEE Trans. Ultrason., Ferroelec., Freq. Contr.* 39(3): 441-446, 1992.

ULTRASHORT PULSE RESPONSE IN NONLINEAR DISPERSIVE MEDIA

Richard Albanese, John Penn, Richard Medina

Radiation Analysis Branch
Armstrong Laboratory
Brooks Air Force Base
San Antonio, Texas 78235

INTRODUCTION

Development of fast electromagnetic switches, power sources and antennas is resulting in the generation of short, high-energy electromagnetic pulses.[1,2] Understanding how such pulses propagate through living tissue is of interest to basic biology and of particular interest to the field of occupational medicine which is concerned to establish safe exposure levels for humans.

Biological tissue is a complex composite with strong dispersive features.[3] Rapid rise time electromagnetic pulses cause a variety of transient structures when propagating through such media. Sommerfeld and Brillouin seem to have first studied these transients while addressing the question of the speed of light in dispersive media.[4,5] Many years later, Pleshko and Palocz produced experimental evidence of the so-called Sommerfeld (high-frequency) and Brillouin (low frequency) precursors.[6] More recently, Oughstun and Sherman with colleagues have produced a rich sequence of articles elaborating, extending and refining the asymptotic analyses of Sommerfeld and Brillouin.[7,8] Our own interest has been to study the development of precursors with reference to non-asymptotic regions in a dispersive medium. Additionally, since our application is biological and medical, we have included interfaces in our analyses, and have emphasized microwave tissue properties. We have emphasized use of Fourier Series as the computational tool.[9] A more recent development is the introduction of finite difference time domain methods.[10,11]

In this paper we inquire into the influence on transient formation of weak nonlinearity in the dispersive medium. A perturbation analysis is presented. We have studied trapezoidal modulation of a sinusoidal field as the incident signal. This is described in the following section. After fixing the incident field, the perturbation analysis is given, which is in turn followed by numerical results and discussion.

THE INCIDENT FIELD

We consider a field orthogonally incident on a dispersive half-space. Our incident field is created by trapezoidal modulation of a continuous sinusoid.

Ultra-Wideband, Short-Pulse Electromagnetics
Edited by H. Bertoni *et al.*, Plenum Press, 1993

For 0<(t-z/c)<a,

$$E_{inc}(z,t) = (1/a)(t-z/c)\sin[\omega(t - z/c)]$$

for a<(t-z/c)<(τ-a),

$$E_{inc}=\sin[\omega(t-z/c)]$$

for (τ-a)<(t-z/c)<τ,

$$E_{inc} = \{1-(1/a)[t-z/c-(\tau-a)]\}\sin[\omega(t-z/c)]$$

and, for τ<(t-z/c)<L, $E_{inc} = 0$. See Figure 1 for a representative example of this incident signal. $E_{inc}(z,t)$ is assumed periodic with period L.

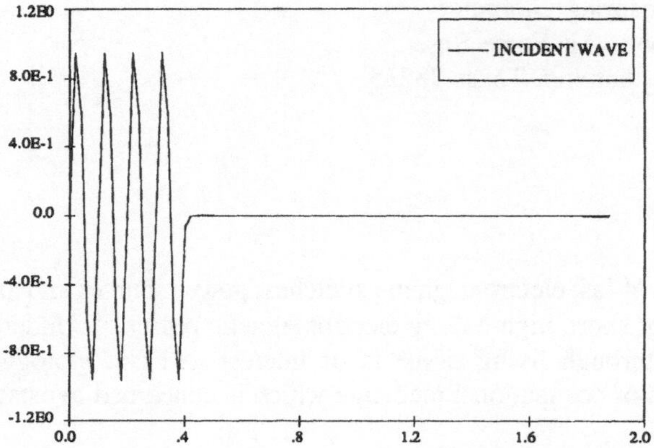

Figure 1. A representative incident signal. The x axis is in nanoseconds, the y axis is in volts/meter.

The Fourier Series of this incident signal is

$$E_{inc}(z,t) = \sum_{n=1}^{\infty} (A_n^2 + B_n^2)^{1/2} \sin[\lambda_n(t - z/c) + \phi_n]$$

for which

$$\begin{aligned} aLA_n = &\{\sin[(\omega+\lambda_n)a] + \sin[(\omega+\lambda_n)(\tau-a)] \\ &-\sin[(\omega+\lambda_n)\tau]\}/(\omega+\lambda_n)^2 + \\ &\{\sin[(\omega-\lambda_n)a] + \sin[(\omega-\lambda_n)(\tau-a)] \\ &-\sin[(\omega-\lambda_n)\tau]\}/(\omega-\lambda_n)^2 \end{aligned}$$

$$\begin{aligned} aLB_n = &\{1-\cos[(\omega+\lambda_n)a] - \cos[(\omega+\lambda_n)(\tau-a)] \\ &+\cos[(\omega+\lambda_n)\tau]\}/(\omega+\lambda_n)^2 + \\ &\{-1+\cos[(\omega-\lambda_n)a] + \cos[(\omega-\lambda_n)(\tau-a)] \\ &-\cos[(\omega-\lambda_n)\tau]\}/(\omega-\lambda_n)^2 \end{aligned}$$

where $\lambda_n = 2\pi n/L$ and $\tan \phi_n = A_n/B_n$.

THE PERTURBATION ANALYSIS

We use Maxwell's equations in the following form for our problem:

$$\nabla \times \vec{E} = -\partial \vec{B}/\partial t$$
$$\nabla \times \vec{H} = \partial \vec{D}/\partial t$$
$$\nabla \cdot \vec{D} = 0$$
$$\nabla \cdot \vec{H} = 0$$

Thus our medium is idealized in that we are neglecting conduction currents and we are assuming there is no free charge ($\rho = 0$). Further, we take $\vec{B} = \mu_o \vec{H}$.

Assuming E_{inc} is a magnitude in the x direction orthogonal to the z-axis of propagation, direct algebraic manipulation of the above equations yields

$$\frac{\partial^2 E_x(z,t)}{\partial z^2} - \mu_o \epsilon_o \frac{\partial^2 E_x(z,t)}{\partial t^2} = \mu_o \frac{\partial^2 P_x(z,t)}{\partial t^2} \qquad (1)$$

for the field within the medium.

To complete the problem we need a connection between $E_x(z,t)$ and $P_x(z,t)$, and for this we choose the following nonlinear relation:

$$E_x(z,t) = \Gamma \partial P_x(z,t)/\partial t + \omega_o^2 P_x(z,t) + s\, P_x(z,t)^3 \qquad (2)$$

In this nonlinear relationship s is a small parameter. We chose this nonlinear relationship because of our interest in biological tissue. The linear portion of the constitutive relation is simply the Debye model which, when a D.C. conductivity term is added, is quite useful in describing the dispersiveness of water.[3,9] The cubic non-linear term is chosen since most tissues are not expected to have an axis of symmetry.[12]

Inserting equation 2 into 1, the fundamental equation of our analysis is obtained.

$$\left[\left(\frac{\partial^2}{\partial z^2} - \mu_o \epsilon_o \frac{\partial^2}{\partial t^2}\right)\left(\Gamma \frac{\partial}{\partial t} + \omega_o^2\right) - \mu_o \frac{\partial^2}{\partial t^2}\right] P_x(z,t) = - s\left(\frac{\partial^2}{\partial z^2} - \mu_o \epsilon_o \frac{\partial^2}{\partial t^2}\right) P_x(z,t)^3 \qquad (3)$$

The first step in our perturbation solution is to solve the linear propagation problem for $E_x^{lin}(z,t)$ and $P_x^{lin}(z,t)$. $E_x^{lin}(z,t)$ is found in reference 9:

$$E_x^{lin}(z,t) = 2\sum_{n=1}^{\infty} \frac{(A_n^2 + B_n^2)^{1/2} \exp(-\alpha_n z) \sin(\lambda_n t - \beta_n z + \phi_n + \theta_n)}{[(1 + \beta_n c/\lambda_n)^2 + (\alpha_n c/\lambda_n)^2]^{1/2}}$$

In this equation A_n, B_n and ϕ_n are as defined above. The terms α_n and β_n are the tissue frequency dependent attenuation and phase constants respectively given by:

$$\alpha_n = \lambda_n (\mu_o \epsilon_n/2)^{1/2} \{-1 + [1 + (\sigma_n/\lambda_n \epsilon_n)^2]^{1/2}\}^{1/2}$$

$$\beta_n = \lambda_n (\mu_o \epsilon_n/2)^{1/2} \{+ 1 + [1 + (\sigma_n/\lambda_n \epsilon_n)^2]^{1/2}\}^{1/2}$$

In these equations, ϵ_n and σ_n are the tissues's frequency dependent dielectric constant and conductivity respectively. These terms are calculated from the linearized version of equation 2 (s = 0). That is:

$$\epsilon_n = \epsilon_o + \omega_o^2/(\omega_o^4 + \Gamma^2\lambda_n^2)$$

$$\sigma_n = \Gamma\lambda_n^2/(\omega_o^4 + \Gamma^2\lambda_n^2)$$

Finally, the phase angle θ_n is specified by

$$\tan\theta_n = \alpha_n/[(\lambda_n/C) + \beta_n]$$

$P_x^{lin}(z,t)$ is calculated from $E_x^{lin}(z,t)$ using the linearized version of equation 2 again. We find

$$P_x^{lin}(z,t) = 2\sum_{n=1}^{\infty} \frac{C_n \exp(-\alpha_n z)\sin(\lambda_n t - \beta_n z + \phi_n + \theta_n - \psi_n)}{\Gamma(q^2 + \lambda_n^2)^{1/2}}$$

In this last equation

$$C_n = \frac{(A_n^2 + B_n^2)^{1/2}}{[(1 + \beta_n C/\lambda_n)^2 + (\alpha_n C/\lambda_n)^2]^{1/2}}$$

$$q = \omega_0^2/\Gamma$$

and

$$\tan\psi_n = \lambda_n/q$$

To compute the polarization perturbation, $P_x^{lin}(z,t)$ must be cubed and inserted into equation 3. We find:

$$\begin{aligned} P_x^{lin}(z,t)^3 = -\frac{2}{\Gamma^3}\sum_{all\ i,j,k} B_{ijk}\exp(-r_{ijk}z)\{&\sin[(\lambda_i+\lambda_j+\lambda_k)t-(\xi_i+\xi_j+\xi_k)] \\ &+\sin[(\lambda_i-\lambda_j-\lambda_k)t-(\xi_i-\xi_j-\xi_k)] \\ &-\sin[(\lambda_i-\lambda_j+\lambda_k)t-(\xi_i-\xi_j+\xi_k)] \\ &+\sin[(\lambda_i+\lambda_j-\lambda_k)t-(\xi_i+\xi_j-\xi_k)]\} \end{aligned}$$

In this equation for $P_x^{lin}(z,t)^3$ we have used

$$B_{ijk} = C_i C_j C_k/[(q^2+\lambda_i^2)(q^2+\lambda_j^2)(q^2+\lambda_k^2)]^{1/2}$$

$$r_{ijk} = \alpha_i + \alpha_j + \alpha_k$$

and

$$\xi_n = \xi_n(z) = \beta_n z - \phi_n - \theta_n + \psi_n$$

In response to a signal of the form $\exp[i(\lambda t - kz)]$, equation 3 provides a solution

Cexp[i(λt-kz)] where k is a complex propagation constant B-iA, and

$$C=-s/(R+iI)$$
$$R=\omega_0^2+\mu_0\lambda^2 V_1/(V_1^2+V_2^2)$$
$$I=\lambda\Gamma-\mu_0\lambda^2 V_2/(V_1^2+V_2^2)$$
$$V_1=A^2-B^2+\mu_0\epsilon_0\lambda^2$$
$$V_2=2AB$$

Thus, the polarization perturbation $P_x^{(1)}(z,t)$ is:

$$P_x^{(1)}(z,t)=\frac{2}{\Gamma^3}\sum_{all\ i,j,k} B_{ijk}\exp(-I_{ijk}z)\ \bullet$$
$$\{E^{+++}\sin[(\lambda_i+\lambda_j+\lambda_k)t-(\xi_i+\xi_j+\xi_k)-\Delta_{ijk}^{+++}]$$
$$+E^{+--}\sin[(\lambda_i-\lambda_j-\lambda_k)t-(\xi_i-\xi_j-\xi_k)-\Delta_{ijk}^{+--}]$$
$$-E^{+-+}\sin[(\lambda_i-\lambda_j+\lambda_k)t-(\xi_i-\xi_j+\xi_k)-\Delta_{ijk}^{+-+}]$$
$$+E^{++-}\sin[(\lambda_i+\lambda_j-\lambda_k)t-(\xi_i+\xi_j-\xi_k)-\Delta_{ijk}^{++-}]\}$$

where each $E^{a,b,c}$ and $\Delta_{ijk}^{a,b,c}$ coefficient has the form

$$E^{a,b,c}=(R_{a,b,c}^2+I_{a,b,c}^2)^{-1/2}$$

$$\tan\Delta_{ijk}^{a,b,c}=I_{a,b,c}/R_{a,b,c}$$

and a,b, and c are plus or minus signs depending on what sum of frequencies is involved. $P_x^{(1)}(z,t)$ is the first term in the perturbation series. This series converges due to dissipation in the medium[13].

We are now in a position to calculate the E field for weak non-linear response. We have

$$P_x(z,t) = P_x^{lin}(z,t)+sP_x^{(1)}(z,t)$$

This equation is inserted into 2 to obtain, to first order in s,

$$E_x(z,t) = E_x^{lin}(z,t) + s[P_x^{lin}(z,t)]^3 + s(\Gamma\frac{d}{dt} + \omega_0^2)P_x^{(1)}(z,t)$$

Results are shown in the next section.

NUMERICAL RESULTS

We have studied a non-linear medium given by equation 2. For numerical purposes Γ and ω_o^2 were chosen to be those of distilled water[3,9]. Thus we use

$$\omega_o^2 = 1/(72.7\epsilon_o) = 1.55x10^9$$

and

$$\Gamma = 8.1\ x\ 10^{-12}\ \omega_o^2 = 0.0126$$

Figure 1 shows a typical trapezoidally modulated incident signal with fundamental frequency 10^{10} Hz. Using the chosen coefficients, nonlinear effects were first noted when the pulse incident voltage was 3 X 10^{11} Volts/meter. The nonlinear response at one centimeter within the water is shown in Figure 2, where the linear response is also shown for

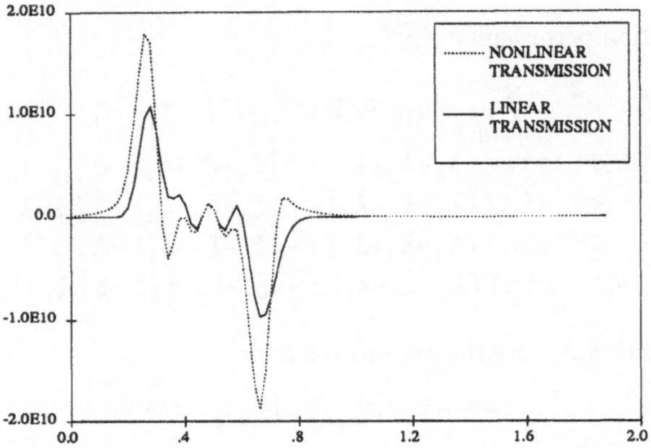

Figure 2. Linear (solid line) and nonlinear (dashed line) responses to the incident wave of figure 1 with peak at $3x10^{11}$ volts/meter.

comparison. Nonlinearity seems to be broadening and heightening the transient fields while effacing the fundamental frequency.

In the future we hope to extend this work to other media. Greater departures from linearity can be accommodated by calculating higher order perturbations. This is difficult to do using analytical expressions as given in this paper, but is much simpler using numerical methods such as the FFT. In fact, we believe a very high order perturbation expansion can be done numerically. With respect to pulses with long rise times, we hope to pursue an analysis of the relationship of our perturbation approach to slow envelope approximations.

REFERENCES

1. V.L. Granatstein and I. Alexeff editors. "High Power Microwave Sources," Artech House, Boston (1987).
2. A. Guenther, M. Kristiansen, and T. Martin editors. "Opening Switches," Plenum Publishing Corporation, New York (1987).
3. Personal communication with Professor E. H. Grant. See also E. H. Grant, R. J. Sheppard, and G.P. South. "Dielectric Behaviour of Biological Molecules in Solution," Clarendon, Oxford (1978).
4. A. Sommerfeld, Uber die Fortpflanzung des Lichtes in disperdierenden Medien, Ann Phys. 44:177 (1914).
5. L. Brillouin, Uber die Fortpflanzung des Lichtes in disperdierenden Medien, Ann Phys. 44:203 (1914).
6. P. Pleshko and I. Palocz, Experimental observation of the Sommerfeld and Brillouin precursors in the microwave domain, Phys. Rev. Letters 22:1201 (1969).
7. K.E. Oughstun and G.C. Sherman, Propagation of electromagnetic pulses in a linear dispersive medium with absorption [the Lorentz medium], J. Opt. Soc. Amer. B, 5:817 (1988).

8. K.E. Oughstun, Pulse propagation in a linear, causally dispersive medium, Proc. IEEE, 79:1379 (1991).
9. R. Albanese, J. Penn, and R. Medina, Short-rise-time microwave pulse propagation through dispersive biological media, J. Opt. Soc. Amer. A, 6:1441 (1989).
10. R. Luebbers, F. P Hunsberger, K. S. Kunz, R. B. Standler and M. Schneider, A frequency-dependent finite-difference time-domain formulation for dispersive materials, IEEE Trans. Electromagnetic Compat., 32:222 (1990).
11. T. Kashiwa, N. Yoshida, and I. Fukai, A treatment by the finite-difference time-domain method of the dispersive characteristics associated with orientation polarization, Trans. IEICE. E73:1326 (1990).
12. Y. R. Shen. "The Principles of Nonlinear Optics," John Wiley & Sons, New York, 1984.
13. E. A. Coddington and N. Levinson. "Theory of Ordinary Differential Equations," McGraw Hill Book Company, New York (1955).

MODULATION AND NOISE IN SOLITON PULSE TRAINS

John M. Arnold

Department of Electronics & Electrical Engineering
University of Glasgow
Glasgow G12 8LT
Scotland

INTRODUCTION

The propagation of ultra-short electromagnetic pulses in physical media other than vacuum is inevitably limited by dispersion. In certain circumstances, however, nonlinear behaviour in the medium can be exploited to propagate the pulse in the form of a soliton, which exhibits no dispersive spreading and can therefore propagate over large distances without change of shape. The generation of solitons is now a well-established technology in the medium of silica optical fibres, and it is only a matter of time before optical solitons will be demonstrated in a compact device such as a semiconductor integrated optical waveguide. Phenomena ascribed to solitons have also been observed at microwave frequencies in experiments on magnetostatic spin waves guided in thin films of ferromagnetic materials. In fact, nonlinear phenomena are inseparable from ultrashort pulse phenomena, since to maintain the energy of a pulse as the pulse width is reduced requires the peak intensity to increase in inverse proportion to its duration; in many communications applications the energy per pulse must exceed a certain value to achieve a prescribed signal-to-noise ratio, and the high intensities thereby implied for ultrashort pulses may well drive a nonlinear response in the material in which the pulse propagates. The soliton makes a virtue out of a necessity, since it uses the nonlinearity to offset the natural linear dispersion.

In considering the application of solitons to information-carrying or information-retrieval systems, the question naturally arises as to the stability of these pulses, since the soliton is an intrinsically nonlinear phenomenon. Although the individual soliton is itself a highly stable entity, trains of solitons carrying information in their parameters (such as arrival time or phase) exhibit complex dynamics which may render difficult the extraction of information which may have been coded into these parameters on transmission; in addition, noise is inevitably present and interacts with the solitons in a more complex way than would be expected in purely linear propagation. Because of the imminent application of soliton technology to ultra-short pulse (1-10 ps) optical communications [1,2], it has become necessary to develop simple approximate theories which describe these interactions using tractable mathematics. This paper describes some applications of dynamical stability theory to nonlinear ultra-short pulse phenomena.

LINEARISED STABILITY THEORY

Optical nonlinear phenomena are described by the Nonlinear Schrodinger Equation (NSE) [3]

$$i\partial_x\phi + \frac{1}{2}\partial_t^{\,2}\phi + |\phi|^2\phi = 0 \tag{1}$$

which governs the evolution of the complex wave envelope ϕ as a function of normalised distance x and time t; the time t is measured in a local time-frame moving along at the linear group velocity of the optical wave at its prescribed carrier frequency Ω. The stationary CW solution of (1) is given by

$$\phi = (\eta/\sqrt{2})\exp(i\eta^2 x/2) \tag{2}$$

for some constant η representing the amplitude. The (stationary) soliton solution of (1) is given by

$$\phi = \eta \exp(i\eta^2 x/2)\operatorname{sech}\eta t \tag{3}$$

for some constant η representing both the amplitude and the inverse pulse width of the soliton. It is possible, by an appropriate rescaling of x, t and ϕ, to take $\eta = 1$ in these formulae without loss of generality, due to a scaling symmetry of (1).

In practical applications such as communication systems one is interested in the behaviour of periodic pulse trains satisfying $\phi(t+T) = \phi(t)$, or more generally quasi-periodic pulse trains for which $\phi(t+T) = e^{i\alpha}\phi(t)$. Such solutions can be found in terms of elliptic functions from (1) if it is assumed that $\phi = \Phi\exp(i\eta^2 x/2)$, and that Φ is independent of distance x, in which case Φ is said to be stationary. Making this substitution in (1) gives for the function Φ

$$i\partial_x\Phi + \frac{1}{2}\left(\partial_t^{\,2} - \eta^2\right)\Phi + |\Phi|^2\Phi = 0 \tag{4}$$

For $\partial_x\Phi = 0$ and $\eta T >> 1$ (period much larger than the pulse width) these solutions look like periodically repeated solitons, with an intersoliton phase of α.

One is then interested in the stability of these stationary pulse solutions. It is well-known that the CW solution (2) is subject to the Benjamin-Feir instability [4-7], in which small perturbations added to the basic form of (2) initially may grow exponentially with propagation distance; it is also well-known [7] that the soliton solutions (3) are stable with respect to small additive perturbations, which disperse away from the soliton with increasing propagation distance. The stability of arbitrary stationary solutions of (4) can be studied by representing the function Φ by $\Phi = \Phi_0+\varepsilon$, where Φ_0 is a stationary solution of (4) ($\partial_x\Phi_0 = 0$) and ε is a small complex perturbation. The equation satisfied by ε is then

$$i\partial_x\varepsilon + D\varepsilon + \Phi_0^2\varepsilon^* = 0 \tag{5}$$

where D is the differential operator

$$D = \frac{1}{2}\left(\partial_t^2 - \eta^2\right) + 2\left|\Phi_0\right|^2. \tag{6}$$

The analysis of stability proceeds by postulating a suitable form for ε. If Φ_0 is x-independent, then the differential equation (5) has x-independent coefficients, and must possess solutions of the form

$$\varepsilon = \tilde{\varepsilon}_+ e^{i\lambda x} + \tilde{\varepsilon}_-^* e^{-i\lambda^* x} \tag{7}$$

which separates (5) into two equations

$$\lambda\tilde{\varepsilon}_+ = D\tilde{\varepsilon}_+ + \Phi_0^2\tilde{\varepsilon}_- \tag{8a}$$

$$-\lambda\tilde{\varepsilon}_- = D\tilde{\varepsilon}_- + \Phi_0^{*2}\tilde{\varepsilon}_+ \tag{8b}$$

Equations (8) are coupled ordinary differential equations for the amplitudes, with an eigenvalue λ, which must be solved subject to certain boundary conditions on the amplitudes. In the case where Φ_0 is the CW solution $\Phi_0 = \eta/\sqrt{2}$, the eigenfunctions are Fourier modes with t-dependence $\exp(-i\omega t)$, and the eigenvalues are given by

$$\lambda^2 = \frac{1}{16}\omega^2(\omega^2 - 2\eta^2) \tag{9}$$

When $\omega^2 < 2\eta^2$, then λ is imaginary; the branch on which λ is negative imaginary implies exponential growth of the perturbation, according to (7). This describes the Benjamin-Feir (or modulation) instability, in which the CW solution is unstable with respect to a band of frequency components $-\sqrt{2}\eta < \omega < \sqrt{2}\eta$ in the perturbing modulation.

Similar methods can be used to determine the stability of the stationary quasi-periodic soliton train solutions; however, the mathematics is considerably more difficult than the simple example shown above [8]. The end result, which we do not derive here, is that

$$\lambda^2 = -8\eta^4 e^{-\eta T} \cos\alpha\,(1-\cos\beta) + o(e^{-\eta T}) \tag{10}$$

where T is the period of the stationary solution, α is the intersoliton phase and β indexes a particular Floquet mode of the perturbation ε. The last term indicates that the error diminishes with $T\to\infty$ faster than the dominant term.

QUASI-PARTICLE APPROXIMATIONS

A much more direct method of arriving at (10) can be obtained by the quasi-particle approximation [9,10]. In this approximation we assume that the soliton train is represented at any propagation distance x by

$$\Phi = \sum_k e^{ik\alpha} \operatorname{sech}(t - kT - q_k) \tag{12}$$

where α is the mean intersoliton phase, T is the mean intersoliton time, and q_k is the time shift of the k'th soliton. We have set the amplitudes of the solitons all equal to 1 on account of the scaling symmetry of (1), and we assume that $\alpha = 0$ or π only. Under these conditions it has been shown by Gordon [11] that (12) is a good approximation for the evolution of two solitons ($k = 1,2$) with the displacements q_k dependent on the propagated distance x, provided that the separation between the solitons satisfies $T+q_2-q_1 \gg 1$. The evolution of the q_k obeys the differential equation

$$d_x^2\, q_k = -4(-1)^k e^{-T} e^{-(q_2-q_1)} \cos\alpha \tag{13}$$

which is similar to Newton's law for particles of unit mass placed at q_k, with an interaction force given by the right-hand side of (13). For more than two particles, one would expect these forces to add linearly, giving

$$d_x^2 q_k = 4\cos\alpha\, e^{-T}\left\{e^{-(q_{k+1}-q_k)} - e^{-(q_k-q_{k-1})}\right\} \tag{14}$$

Equation (14) with $\alpha = \pi$ is the Toda lattice equation of integrable dynamical systems theory.

Equation (14) can be linearised for $q_k \ll 1$, with the result

$$d_x^2 q_k = 4\cos\alpha\, e^{-T}\{2q_k - (q_{k-1} + q_{k+1})\} \tag{15}$$

Eq. (15) is separable in the form

$$q_k = e^{i\lambda x} e^{ik\beta} \tag{16}$$

and the spatial propagation coefficient satisfies

$$\lambda^2 = -8\, e^{-T} (1-\cos\beta) \cos\alpha \tag{17}$$

Eq.(17) agrees completely with the more rigorous result (10) when it is remembered that $\eta = 1$ and $\alpha = 0$ or π in the quasi-particle derivation. This result demonstrates that the pulse train is stable for $\cos\alpha \leq 0$ ($\alpha = \pi$) and unstable for $\cos\alpha > 0$ ($\alpha = 0$). The stability length L is defined as the reciprocal of the maximum value of $|\lambda|$

$$L = (1/4)\, e^{T/2} \tag{18}$$

and forms a useful scale for the calibration of the range over which weak perturbations of soliton position evolve.

SYNTHESIS OF STABLE DISPERSIVE PULSE PATTERNS

In the stable case ($\alpha = \pi$) arbitrary distributions of position of the solitons can be synthesised by superposition of elementary Fourier components (16). Thus the most general distribution is

$$q_k(x) = \frac{1}{2\pi} \int_{-\pi}^{\pi} \{Q(\beta)\cos(\lambda x) + V(\beta)\sin(\lambda x)\} e^{ik\beta} d\beta \tag{19}$$

By choosing the functions Q and V in (19) arbitrary solutions can be synthesised; some examples follow below.

Dispersion of the displacement of a single soliton

Suppose that at $x = 0$ all but one of the solitons are at their equilibrium positions $q_k(0) = Q_0\delta_{k0}$ and all the solitons are initially at rest, so that $d_x q_k(0) = 0$. Then from (19) we find $Q(\beta) = Q_0$ and $V(\beta) = 0$. The integral in (19) can be evaluated exactly in this case, with the result that

$$q_k(x) = Q_0 J_{2k}(x/L). \tag{20}$$

As x increases from 0 the displacement of the $k = 0$ soliton disperses symmetrically onto neighbouring solitons, which themselves acquire nonzero displacements.

Intersoliton interference in pulse-position modulation (PPM)

If at x = 0 all the solitons are given a random distribution of initial displacements $q_k(0)$ the soliton interaction spreads the displacements onto neighbouring pulses; this gives rise to an intersoliton interference whose variance in the stable case is

$$\sigma_e^2 = 2\sigma_q^2 \sum_{k \neq 0} J_{2k}^2(x / L) \tag{21}$$

where σ_q^2 is the variance of the signal modulation $q_k(0)$, assumed equal on all solitons. For small x/L (21) can be approximated by $\sigma_e^2 = \sigma_q^2(x/L)^4/32$. In the unstable case the J-type Bessel functions in (21) are replaced by I-type modified Bessel functions.

Gordon-Haus noise in an interacting pulse train

The phenomenon of Gordon-Haus noise [12] occurs in the theory of amplified fibre links. In its simplest form it can be regarded as the uniformly-distributed generation of noise at all points along the propagation path. The component of noise in-phase with the solitons has little effect on the propagation. The component in-quadrature with the solitons causes velocity fluctuations of the solitons which can accumulate over large distances, causing pulse jitter. If at x = x' the solitons are subjected to additive perturbations of velocity $p_k = d_x q_k$ of variance σ_p^2, and the total perturbation integrated over all $0 < x' < x$, representing uniform distribution of the noise throughout the propagation path, the resultant variance of the positions in time of the solitons at x in the stable case is obtained from (19) as

$$\sigma_v^2 = \langle q_k^2(x) \rangle = L_1^{-1}\sigma_p^2 \sum_{k \in Z} \int_0^x \left\{ \int_{x'}^x J_{2k}(x'' / L)dx'' \right\}^2 dx' \tag{22}$$

where L_1 is the normalised mean spacing between noise sources along the path. When x/L is small (22) reduces to $\sigma_v^2 = \sigma_p^2 x^3/3L_1$. This approximation agrees with the Gordon-Haus formula [12], but the exact result (22) deviates significantly from this when the range x is larger than the stability length L.

SUMMARY AND CONCLUSIONS

At the present stage of development the quasi-particle theory is rather heuristic, being based almost exclusively on an extension by analogy of an analytical result of Gordon for two interacting solitons which was derived systematically from Inverse Scattering Theory. For the limited range of parameters used here ($\alpha = 0$ or π) the method gives agreement with more rigorous derivations using other methods. In particular the linear stability exponent λ is predicted, and the limiting form of the Gordon-Haus formula is obtained correctly. There have been more systematic attempts to derive the method from a full Lagrangian formulation of the NLE [13], but the case of general α is quite complicated, since the amplitudes can no longer be assumed all equal and so normalised away as we were able to do here. There are distinct advantages in this method, however. The equation (14) derived for the quasi-particle interactions is the Toda lattice equation in the stable case $\alpha = \pi$. Like the NSE, this equation is integrable by the ISM. Preservation of the integrability is a very strong property which should eventually be exploitable in a systematic method of reduction of the dynamical degrees of freedom from the full NSE wavefunction Φ to the reduced description in terms of a few quasi-particle coordinates.

REFERENCES

1. Mollenauer, L. F., Evangelides, S. G. and Haus, H. A., Long distance soliton propagation using lumped amplifiers and dispersion shifted fiber, *J. Light. Tech.*, **9**, 194-197 (1991)
2. Mollenauer, L. F., Nyman, B. M., Neubelt, M. J., Raybon, G. and Evangelides, S. G.: 'Demonstration of soliton transmission at 2.4 Gbit/s over 12000 km', *Elect. Lett.*, **27**, 178-179 (1991)
3. Agrawal, G., *Nonlinear Fibre Optics*, Academic Press (1989)
4. Benjamin, T. B., The stability of solitary waves, *Proc. Roy. Soc.*, **A328**, 153-158 (1972)
5. Benjamin, T. B. and Feir, J. E., The disintegration of water wave trains in deep water, *J. Fluid Mech.*, **27**, 417-430 (1967)
6. Hasegawa, A., Generation of a train of soliton pulses by induced modulation instability in optical fibres, *Opt. Lett.*, **9**, 288-290 (1984)
7. Zakharov, V. E. and Rubenchik, A. M., Instability of waveguides and solitons in nonlinear media, *Sov. Phys. JETP*, **38**, 494-500 (1974)
8. Arnold, J. M., Stability theory for pulse train solutions of the nonlinear Schrodinger equation, submitted to *IMA Jour. App. Math.* (1992)
9. Arnold, J. M., Qualitative dynamics of modulated soliton pulse trains, OSA Topical meeting Nonlinear Guided Waves, Cambridge, England (1991)
10. Arnold, J. M., Soliton pulse analogue communication, accepted for publication in *IEE Proc. (J)* (1992)
11. Gordon, J. P., Interaction forces among optical fibre solitons, *Opt. Lett.*, **8**, 596-598 (1983)
12. Gordon, J. P. and Haus, H. A., Random walk of coherently amplified solitons in optical fibre transmission, *Opt. Lett.*, **11**, 665-667 (1986)
13. Anderson, D. and Lisak, M., Bandwidth limits due to mutual pulse interaction in optical soliton communication systems, *Opt. Lett.*, **11**, 174-176 (1986)

DEEPER-PENETRATING WAVES IN LOSSY MEDIA

Liyou L. Li, Leonard S. Taylor and Hainan Dong

Electrical Engineering Department
University of Maryland, College Park, MD 20742

INTRODUCTION

Hyperthermia treatment of cancerous tissue is a well-established modality to improve the effectiveness of ionizing radiation in the treatment of cancer. Electromagnetic(plane) waves can penetrate the body and heat tissue in the interior, but a principal problem in the use of electromagnetic energy for hyperthermia treatment of cancer is that at the short wavelengths which are useful for localizing the energy deposition volume, the penetration depth of electromagnetic plane waves into the tissue is very short, of the order of a centimeter, so that it is not possible to heat deep-lying tumors without overheating the surface, even using multiple radiators to superimpose the fields in the tumor. Employing frequencies at the low end of the microwave band and "skin coolers" increases the effective depth of localized heating, but only to about four to five centimeters. The only available technique to successfully overcome this problem has been the use of invasive radiators introduced into the body. Recently, however, other types of electromagnetic radiation modes have been studied which appears to offer solutions to this difficult and important problem without requiring either interstitial or intraluminal insertion of the radiating elements into the body. Surface waves are one example; these waves cannot be expressed as an angular spectrum of plane waves, unless the spectrum is extended to include complex angles of propagation. The decay rate of these waves in absorbing media is not the same as that of the usual plane waves; the skin depth for these modes can be much larger[1]. It has also been established [2] that antenna near fields are another wave type which may possess enhanced penetration depths.

The "electromagnetic missile" is an electromagnetic beam wave generated by short current pulses. In this mode of propagation, introduced by T. T. Wu[3-5], electromagnetic energy in the form of short impulses, rather than a modulated carrier, propagates in a directed beam, rather than a spherically spreading wave, and the power density in the beam is maintained over long distances with relatively small decay. The properties of the EM missile beam wave have been confirmed both theoretically and experimentally in air. We have been able to demonstrate theoretically that in lossy media, the rate of decay can be much less than that for ordinary wave types. If the theoretical promise of this technique can realized, it may be possible to direct time-multiplexed beams to selectively heat deep-lying tumors, in much the same way that ionizing radiation beams are now employed.

In our analysis and in the numerical computation the effects of both absorption and dispersion on the EM missile beam wave taken into account. Because the numerical computation requires a specific model and because our interest was centered on medical applications, the computation was carried out using a medium whose dielectric and conductive properties were based on the published measurements of the properties of high water content biological materials (muscle and organ tissue) which is highly absorbing for electromagnetic radiation at high frequencies. Thus in these cases, the wave field of the radiator is ordinarily significantly attenuated even within the near field, and it is not possible to use Fresnel diffraction theory.

In our study, a specific magnetic current source was used as the radiator and four different excitation pulse forms were considered. The numerical results show that this type of wave applicator will greatly enhance the penetration of electromagnetic energy into lossy biological materials. A preliminary experiment undertaken to verify the possibility of deeper-penetrating waves has also provided preliminary verification that through a suitable choice of pulse shape a deeper penetrating wave can be generated.

FORMULATION OF THE PROBLEM

Consider a magnetic current source on the circular aperture of an perfect conducting plane at z=0, shown in Fig. 1. A semi-infinite lossy dielectric is in $z > 0$. Following [3], the constraint condition is that the total radiated energy is finite. Thus if $P(\omega)$ be the integrated radiated energy per unit frequency at ω; then we require $\int_0^\infty P(\omega)\, d\omega < \infty$. The equivalent magnetic current distribution is the result of the aperture field at z=0

$$\mathbf{E}_s(\mathbf{r}, t) = \begin{cases} \mathbf{e}_x \; E_0(x, y, 0) f(t), & \text{if} \quad \rho \leq a; \\ 0 \quad , & \text{if} \quad \rho > 0. \end{cases} \tag{1}$$

Let the $E(\mathbf{r}, \omega)$ be the Fourier transform of $E(\mathbf{r}, t)$, then, for the distribution (1), at z=0

$$\mathbf{E}_s(\mathbf{r}, \omega) = \begin{cases} \mathbf{e}_x E_0(x, y, 0) \tilde{f}(\omega), & \text{if} \;\; \rho \leq a; \\ 0 \quad , & \text{if} \;\; \rho > 0. \end{cases} \tag{2}$$

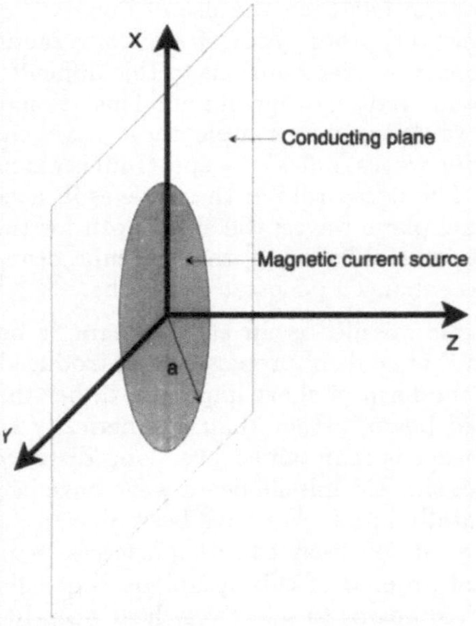

Fig. 1. Geometry of the theoretical model.

Thus the constraint condition can be written as $\int_0^\infty |\tilde{f}(\omega)|^2 \; d\omega < \infty$. The condition limits the pulse shape of the excitation source. The quantity of central interest is the magnitude of the Poynting vector $\mathbf{E} \times \mathbf{H}$, or its time integral

$$F(\varphi, \mathbf{R}) = \int_{-\infty}^{\infty} dt \int_S d\,S \; |\mathbf{E}(\mathbf{R}, t) \times \mathbf{H}^*(\mathbf{R}, t) \; . \; \hat{\mathbf{n}}|. \tag{3}$$

Here S is a bounded, oriented surface of finite area away from the antenna, and $F(\varphi, \mathbf{R})$ is the energy transmitted through S at $\mathbf{R}$ in the direction $\hat{\mathbf{n}}$. Our goal is to have the quantity $F(\varphi, \mathbf{R})$ decays as slowly as possible in the lossy dielectric. As discussed in [3], equation (3) is also expressed as

$$F(\varphi, \mathbf{R}) = \frac{\pi}{2} \int_0^{\infty} d\omega \int_S dS |\mathbf{E}(\mathbf{R}, \omega) \; \times \mathbf{H}^*(\mathbf{R}, \omega) \; . \; \hat{\mathbf{n}}|. \tag{4}$$

where $\mathbf{E}(\mathbf{R}, \omega), \mathbf{H}(\mathbf{R}, \omega)$ are the temporal Fourier transforms. From (2) the magnetic current source, $\mathbf{m}_s$ is

$$\mathbf{m}_s = \mathbf{E}_s \times \mathbf{n}(\mathbf{r}, \omega) = \begin{cases} -\mathbf{e}_y E_0(x, y, 0)\tilde{f}(\omega), & \text{if } \; \rho \leq a; \\ 0 \;\; , & \text{if } \; \rho > 0. \end{cases} \tag{5}$$

Using where $k^2 = \omega^2 \mu\epsilon$ and Here $r = \sqrt{(x - x\prime)^2 + (y - y')^2 + z^2}$ and s is the aperture. We consider the EM power into the lossy material along the z-axis, i.e. on x=0, y=0, and we restrict our analysis to the case where E_0 on the aperture is a constant to obtain $\mathbf{A}_m$ as Setting S equal to zero and putting the of expressions of E and H fields into the expression (4) of $F(\varphi, \mathbf{R})$ we obtain the formula for the electromagnetic power radiated from the source into the dielectric

$$F(0,0,z) = \int_0^\infty |\tilde{f}(\omega)|^2 \frac{E_0^2 c}{4} \sqrt{\frac{\epsilon_0}{\mu_0}} \{[e^{jk_0\sqrt{\epsilon_r}z} - \frac{z}{\sqrt{a^2+z^2}} e^{jk_0\sqrt{\epsilon}\sqrt{a^2+z^2}}]$$

$$[\frac{e^{-jk_0\sqrt{\epsilon_r^*}}\sqrt{a^2+z^2}}{2k_0}(-k_0\sqrt{\epsilon_r^*}(\frac{a^2+2z^2}{a^2+z^2} + \frac{a^2}{(a^2+z^2)^{3/2}}) + \sqrt{\epsilon^*} e^{-jk_0\sqrt{\epsilon_r^*}z}]\} |dk_0 \tag{6}$$

where $k_0 = \frac{\omega}{c}$, $c = \sqrt{\mu_0 \epsilon_0}$, ϵ_r is the relative complex permittivity of the dielectric outside the source, and $\tilde{f}(\omega)$ is a function of k_0. It can be seen that $\tilde{f}(\omega)$ will contribute to the value of integral in (6). In other words, if we choose different $\tilde{f}(\omega)$ satisfying the constraint condition , we will obtain different electromagnetic power in the lossy dielectric at the same depth: The rate of decay of electromagnetic power into lossy materials depends on the choice of the pulse shape $f(t)$.

SLOW DECREASE OF EM ENERGY IN THE FAR-FIELD ZONE

For any finite frequency ω, the far field is described by a field pattern $\mathbf{f}(\widehat{\mathbf{N}})$, viz.,

$$\mathbf{E}(\mathbf{R}) \propto \widehat{\mathbf{N}} \times \mathbf{f}(\widehat{\mathbf{N}}) R^{-1} e^{ikR}, \qquad \mathbf{H}(\mathbf{R}) \propto \widehat{\mathbf{N}} \times \mathbf{E}(\mathbf{R}) \tag{7}$$

where $k^2 = \omega^2 \mu\epsilon$ and k is a complex number in lossy materials. Thus electromagnetic energy in lossy materials decays exponentially and is proportional to $e^{-Im(k)R}/R^2$. From equation (6) it can be understood that if we choose the excitation function $\tilde{f}(\omega)$ carefully we will possibly be able to eliminate the exponential decaying after the integral. It is almost impossible to carry out the integral in (4.2.12) analytically.

Thus we let $z >> a$ and $(a^2+z^2)^{3/2} \approx z^3(1+3a/z)$, and we consider a nondispersive lossy dielectric outside the antenna. Then we obtain

$$F(0,0,z) \approx K \int_0^\infty |\tilde{f}(\omega)|^2 dk_0 e^{-2k_0 Re(\sqrt{\epsilon_r})z}(1-e^{jk_0\sqrt{\epsilon_r}a})(1-e^{-jk_0\sqrt{\epsilon_r^*}a}) \quad (8)$$

where K is a constant. If we choose $\tilde{f}^2(\omega) = \sqrt{(\omega/\omega_0)^2 + 0.01}e^{-2(\omega/\omega_0)}$, when $z >> a + c/\omega_0$ and $n_i z >> n_r a$, F(0,0,z) finally becomes

$$F(0,0,z) \propto (\frac{5a}{z+c/\omega_0})^{-\frac{3}{2}}. \quad (9)$$

F(0,0,z) represents the electromagnetic energy in the nondispersive lossy material. It is noticed that electromagnetic energy in the lossy material is no longer exponentially decaying, nor does it decay proportional to R^{-2}, which is the decay rate for waves in the far-field zone in free space.

In order to verify (9), a numerical calculation of relative electromagnetic power in the nondispersive lossy dielectric was carried out for a=6.0 cm, $\epsilon_r = (66, 40)$ and $\tilde{f}(\omega)$ in (2) where ω_0 = 0.99 Ghz (Fig. 2). From Fig. 2 it can be seen that the relative electromagnetic power for the transient excitation decays much more slowly in the lossy dielectric than for a sinusoidal excitation at ω = 0.99 Ghz which is the "center" frequency for the transient excitation we used in (9), shown in Fig. 3. Thus the electromagnetic missile effect can be obtained in a lossy dielectric in the far zone.

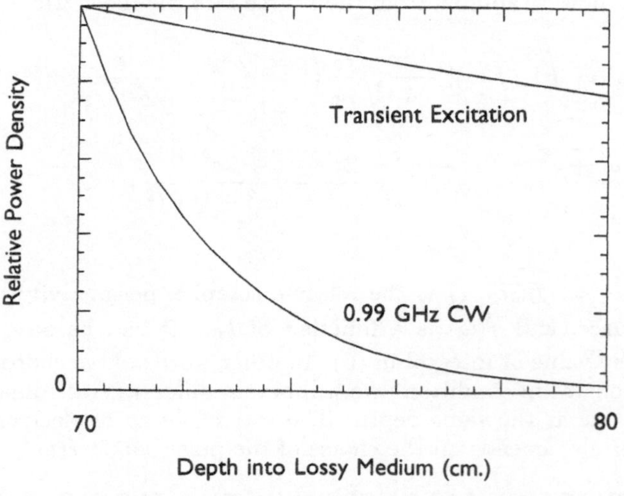

Fig. 2 Power density in far-field zone of a nondispersive lossy medium.

Because of the analytic complications of (6) in the near-field zone we have carried out a numerical calculation using a Fortran program. The programs are provided in an appendix. First, we again considered a nondispersive lossy dielectric with $\epsilon_r = (60, 40)$ as before.

In order to see how the wave penetration into lossy dielectric depends on the choice of pulse shapes we have considered four different pulse shapes in our numerical calculations: $\tilde{f}_1(\omega) = (\frac{\omega}{\omega_0})^{1/2}\, e^{-4(\frac{\omega}{\omega_0})^2}$, $\tilde{f}_2(\omega) = (\frac{\omega}{\omega_0})\, e^{-(\frac{\omega}{\omega_0})^2}$, $\tilde{f}_3(\omega) = (\frac{\omega}{\omega_0})^{1/2}\, (e^{\frac{\omega}{\omega_0}} +$

$e^{-\frac{\omega}{\omega_0}})^{-1}$, $\tilde{f}_4(\omega) = \delta(\omega - \omega_0)$. The numerical results are shown in Fig. 4a. It can be noticed that the choice of the pulse shape f(t) may affect the penetration of electromagnetic energy into lossy material.

Because the assumption of a nondispersive lossy dielectric is not realistic for tissue we have also taken into consideration the case of a dispersive lossy dielectric. Since our purpose is to consider the application of this method to hyperthermia treatment, the dispersive dielectric properties were chosen from the reported dielectric permittivity and conductivity of various canine tumor and normal tissues [6]. The frequency response range for the dielectric constants is available from 10 Mhz to 18 Ghz, which is sufficient for our calculation for the chosen transient excitation, $\tilde{f}_1(\omega)$ shown in Fig. 3. The relative electromagnetic powers in the far-field zone of the dis-

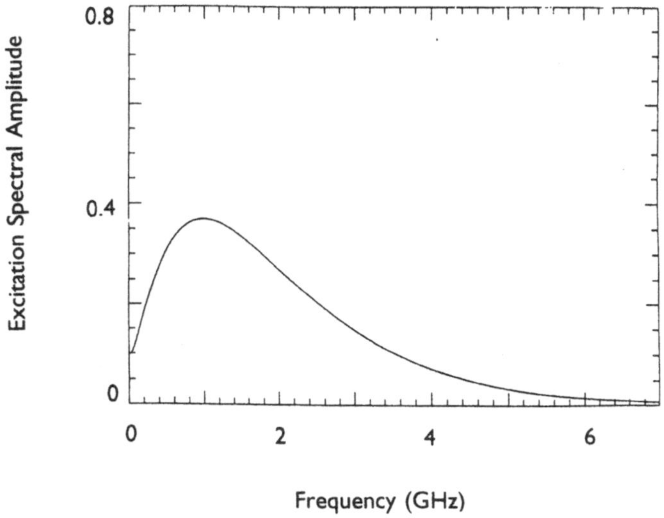

Fig. 3 The frequency spectrum of the transient excitation.

persive lossy dielectric are calculated and shown in Fig. 4 respectively. It can be seen that the penetration of electromagnetic energy into the dispersive lossy material can be greatly enhanced by using the proper transient excitation. In the far-field zone of the material the decaying of the transient excited wave is surprisingly slower than either sinusoidal waves operating at f=0.915 Ghz and f=0.25 Ghz which is the "center" frequency for $\tilde{f}_1(\omega)$.

As stated before, the concept of the Electromagnetic Missile in free space can be understood that the existance of the Electromagnetic Missile depends upon the high frequency components in the electromagnetic spectrum pulse because for short distances in the Fresnel zone, i.e., when $\frac{ka^2}{2r} >> 1$, the variation of the energy is different from r^{-2} [3], [5]. But in lossy materials, the situation is much more complicated. Does the slow decrease of electromagnetic energy in lossy materials depends only upon the low frequencies in the spectrum? In order to answer this question we compare relative electromagnetic powers for pulses with different low and high frequency spectrums.

We choose $pulse_0$ with $f_0(\omega/\omega_0)$ (shown in Fig. 3), $pulse_1$ with $f_1(\omega/\omega_0)$ (shown in Fig. 5) to be the same as $pulse_0$ but with the lower frequency part from "center frequency" f_c=0.99 GHz cut out of the spectrum of $pulse_0$. $pulse_2$ with $f_2(\omega/\omega_0)$

(shown in Fig. 6) is produced by cutting the lower frequency part from f_c. In Fig. 7 we compare the relative EM powers of $pulse_1$ and $pulse_2$ with that of the $pulse_0$ in the near-field zone of the same dispersive lossy dielectric. From Fig. 7 it is clear that $pulse_1$ with lower frequency components is more penetrating than $pulse_2$ which consists of higher frequency components, but $pulse_0$ is more penetrating than $pulse_1$ or $pulse_2$. Thus it is apparent that the slow decrease of electromagnetic energy in lossy materials depends not only on the presence of the low frequency spectrum of the pulse but also relates to the high frequency spectrum as well. In other words, for a given lossy dielectric we can choose the time dependence (or frequency spectrum) for a pulse so that it will decay more slowly in the material than expected from simple considerations of the Fresnel zone dependency.

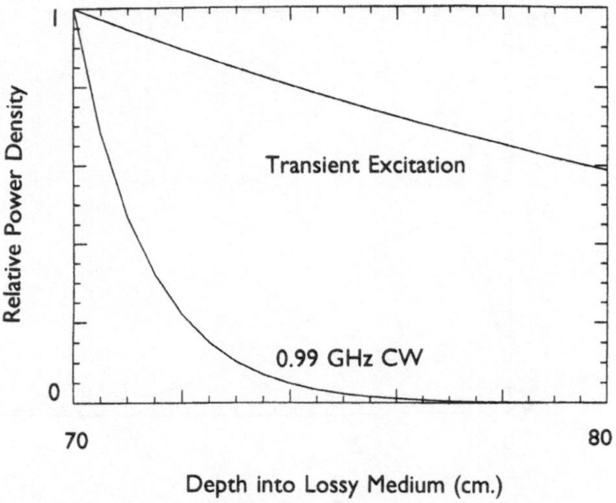

Fig. 4 Power density in the far-field zone of a dispersive lossy medium.

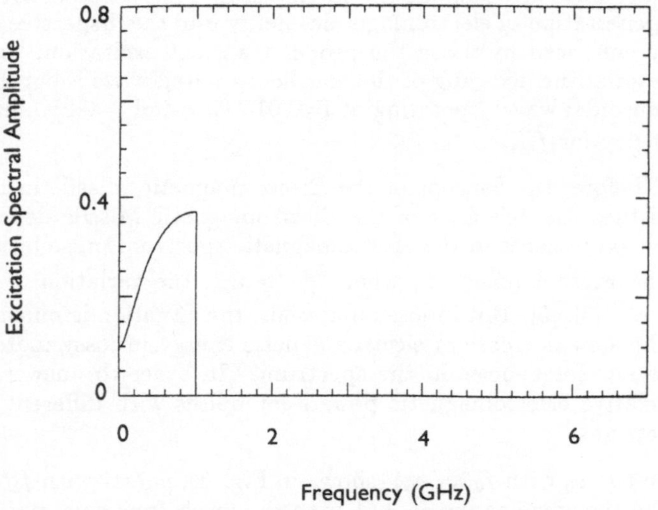

Fig. 5 The lower frequency part of the transient excitation.

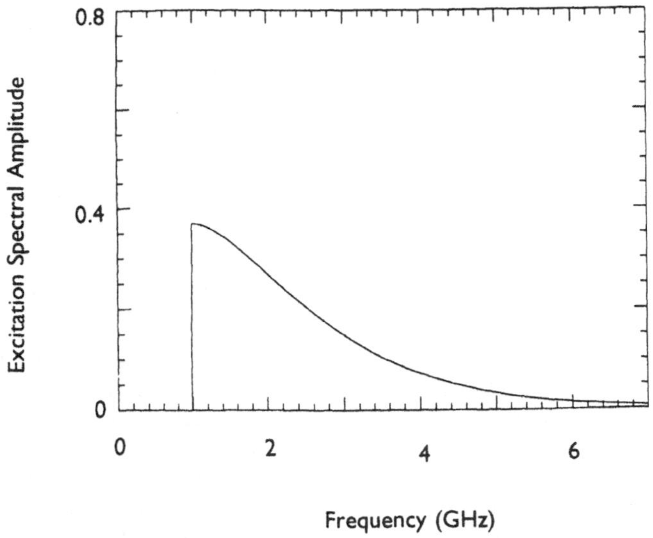

Fig. 6 The higher frequency part of the transient excitation.

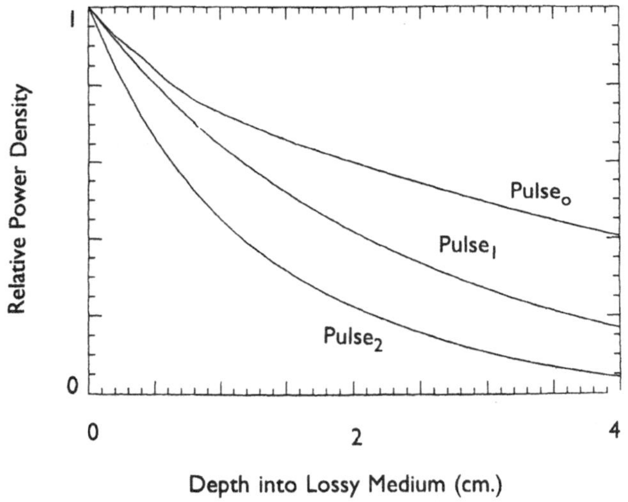

Fig. 7 Power density in the near-field zone of a dispersive lossy medium.

EXPERIMENTAL RESULTS AND DISCUSSION

Our results have shown that by a suitable choice of time dependence for a pulse, electromagnetic energy transmitted into lossy material can decay very slowly. Although the pulse shape $\tilde{f}_1(\omega)$ used in the calculation may not be optimal, it does answer on the theoretical level the question about whether an applicator which produces deeper-penetrating wave is possible. The dielectric properties of the lossy biological material we chose in the calculation is close to that of human body tissue, and we conclude that it may be possible to bypass the limitation of present applicators used for deep-heating cancer treatment.

An experiment to verify the possibility of deeper-penetrating wave generator would include: (1) Measurement of wave penetrations in a lossy dielectric as a function of pulse shape. (2) Comparison of pulse penetration into the lossy material with that of sinosoidal waves. (3) Comparison of the penetration of pulses in the near-field and far-field zones of the launcher and of the effects of antenna configuration on the decay of pulses in lossy material.

We have collaberated with the Institute of Remote Sensing in Beijing in the design of an experiment to verify the possibility of deeper-penetrating waves. The experiment is still in progress. We describe preliminary results. The setup for the experiment is shown in Fig. 10. The lossy material used is 3% saline solution. The water tank is 50 cm long, 35 cm wide and 10 cm high. The transmitter and receiver are the same broadband antennas. The receiver is attached to the bottom of the water tank and the transmitter can be elevated in different heights of the water. The pulses used in the experiment are close to Gaussian in shape and have of pulse widths 0.65 ns, 1.0 ns and 1.2 ns respectively, with the same 100 kHz repetition frequency. The measurements are for distances of 4 cm, 5 cm and 6 cm between the transmitter and receiver.

Because the pulses with different pulse widths were generated with different peak voltages, it is necessary to normalize the peak voltages relative to the voltage at 4 cm depth in order to compare the decay rates of the three pulses. The decay curve at depths from 4 cm to 6 cm were also measured for a continuous wave with frequency

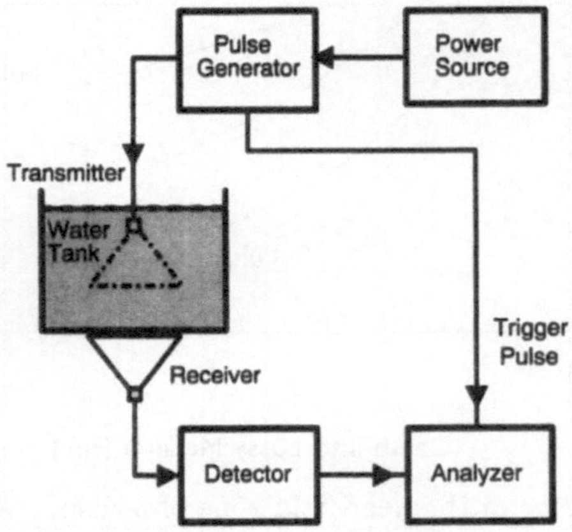

Fig.8 Experimental setup.

450 MHz. The experimental results are shown in Fig. 11. The results verify the claim that changing the shape of a pulse can result in a change of its decay rate in a lossy material. In Fig. 11 the decay of the pulse with a 1 ns pulse width in saline solution is clearly smaller than that for the pulses with 0.65 ns and 1.2 ns widths, (and smaller than that for a continuous wave with frequency 450 MHz). In order to compare experimental results with theoretical ones we have calculated the decay rates for the pulses used in the experiment. The dispersion relationship for a 3% saline solution

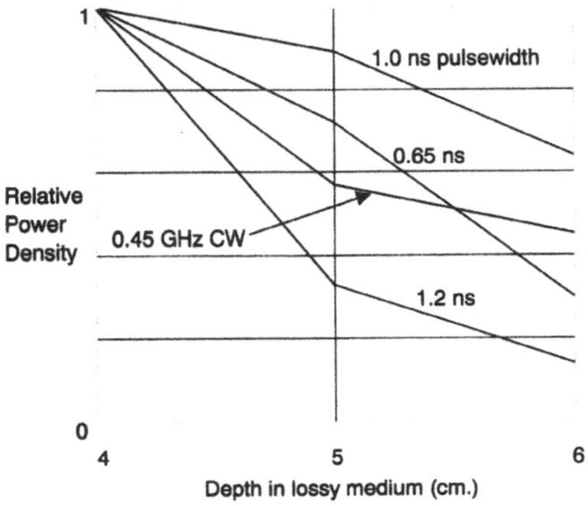

Fig. 9 Experimental results of the decay for different pulses.

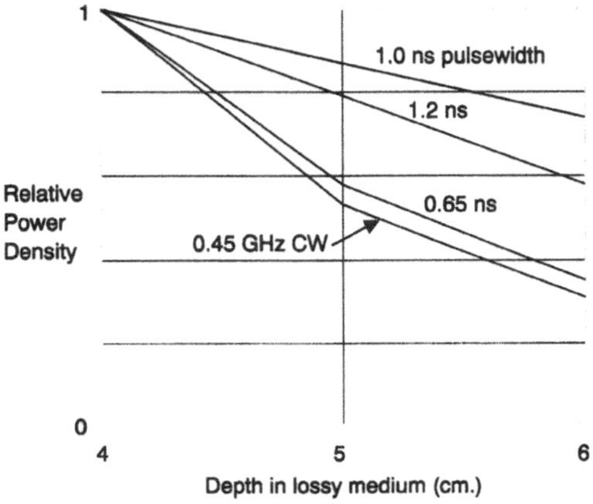

Fig. 10 Theorectical results of the decay for different pulses.

was not available and we used the dispersive dielectric in our previous calculation. The theoretical results for decay rates of the pulses with 0.65 ns, 1.0 ns and 1.2 ns pulse widths are shown in Fig. 12. It can be seen that the theoretical results are similar to the experimental ones except that the relative EM power density for the pulse with 1.2 ns pulse width is higher than that for the continuous wave at 6 cm in lossy material. It can be concluded from preliminary experimental and theoretical results that using transient excitation with a proper pulse shape one may significantly enhance the penetration depth into the near-field zone of lossy materials.

REFERENCE

[1] L. S. Taylor,"Penetrating electromagnetic wave applicators", IEEE Trans. AP-32, 1138 (1984).

[2] J. B. Andersen, "Focusing in lossy media", Radio Science, 19, 1195-1198(1984).

[3] T. T. Wu, "Electromagnetic missiles", J. Appl. Phys.,57, 2370-2373(1984).

[4] T. T. Wu, W. P. King and H. Shen, "Spherical lens as a launcher of electromagnetic missiles", J. Appl. Phys., 62, 4036-4039, (1987).

[5] H-M Shen, et al,"The properties of the electromagnetic missiles", J. Appl. Phys., 66, 4025 (1989)

[6] H. P. Schman, "Interaction of microwave and RF radiation with biological systems", J. Micro. Power,16, 107-119(1981).

ON-AXIS FIELDS FROM A CIRCULAR UNIFORM SURFACE CURRENT

D. J. Blejer,[1] R. C. Wittmann,[2] and A. D. Yaghjian[3]

[1]MIT Lincoln Laboratory, Lexington, MA 02173
[2]NIST, Boulder, CO 80303
[3]RL/ERCT, Hanscom AFB, MA 01731

INTRODUCTION AND SUMMARY

Exact closed-form expressions are derived for the on-axis electric and magnetic fields of a circular aperture excited by a uniform surface current with arbitrary time dependence. (Of course, corresponding expressions hold for a uniform magnetic or electric field exciting the circular aperture.) Necessary and sufficient conditions on the current are given to overcome the usual $1/z^2$ far-field energy dependence.

A single pulse of surface current initiated at time $t = 0$ excites two similar field pulses (waveforms) of opposite sign traveling down the center axis (z axis) normal to the circular current disk of radius a. The leading pulse begins at time $t = 0$, travels with the speed of light, and represents the direct radiation from the center of the disk. The trailing pulse of opposite sign begins at $t = a/c$, travels faster than the speed of light c, and represents interference between the waves radiated, in effect, by the rim of the current disk. In other words, the energy associated with the trailing pulse crosses the z axis at an angle, and thus the waveform along the z axis moves faster than c. The speed of the trailing pulse reduces asymptotically to the speed of light as it travels a distance $z \gg a$ from the disk. It approaches but never quite reaches the position of the leading pulse as the distance of travel z gets larger and larger. This confirms the general result, obtainable from Maxwell's equations, that "superluminary" waveform speeds result from near-field interference and cannot be maintained by a source of finite extent as the propagation distance becomes much larger than the size of the effective source region. And, of course, no signal reaches an observer in less time (after the sources are turned on) than the time interval equal to the minimum distance, from the source to the observer, divided by c.

The two pulses maintain their amplitude as they travel to infinity. Nevertheless, since they approach each other, and become equal and opposite as they travel to infinity, their sum decays with the familiar $1/z$ dependence multiplied by the derivative of the current pulse. Emphatically, however, this is true only if the first time derivative of the surface current exists and is finite. (Herein we say that a time derivative "exists" even if it has the value of plus or minus infinity.) If, for example, the surface current is discontinuous at points in time, in particular, if it has a step function in time, the two opposite pulses subtract as z gets indefinitely large to produce a total pulse whose amplitude remains constant with z but whose pulse length decreases as $1/z$.

In the direction transverse to the z axis, this pulse maintains a fixed fractional-power waist diameter on the order of the diameter of the current disk as $z \to \infty$, and thus the energy in the pulse of length $1/z$ also decays as $1/z$ (rather than the usual $1/z^2$). Thus, in principle, a pulse can be excited that overcomes the $1/z^2$ energy dependence in this single direction of propagation, but under very stringent conditions. It can be excited only by a current pulse whose secant slope is unbounded, that is, whose first time derivative, if it exists, is infinite. In other words, the current generator must produce a pulse with a zero instantaneous rise time. Such a current generator does not exist.

Also, in principle, a pulse can be sent to infinity with finite or even infinite energy through a finite area; for example, if the surface current had a delta function time dependence. However, if the surface current radiates a finite energy, the energy in the pulse through a finite area will always approach zero as it travels to infinity. This confirms the general result proven by Wu[1] that the energy in a finite transverse area of any pulse radiated by a finite-size distribution of finite-energy sources must approach zero as the propagation distance becomes large.

TIME-HARMONIC FIELDS

Consider the circular disk of uniform, linearly polarized, time-harmonic surface electric current, $\mathbf{K}_\omega = K_\omega \hat{\mathbf{x}}$, within the radius a as shown in Figure 1.

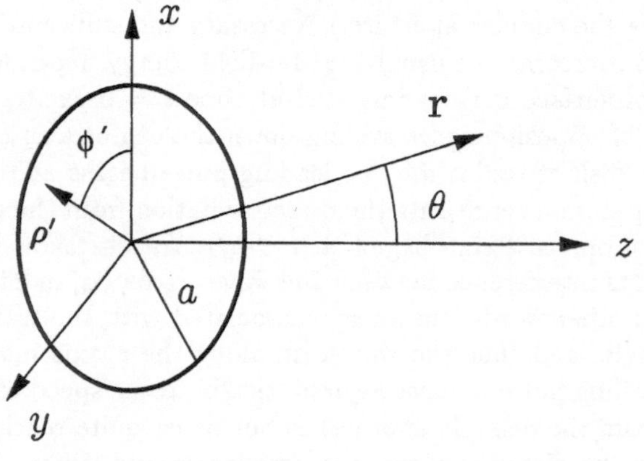

Figure 1. Circular disk of uniform surface current.

The time-harmonic magnetic and electric fields for $z > 0$ can be expressed in terms of the vector potential as

$$\mathbf{H}_\omega(\mathbf{r}) = \nabla \times \mathbf{A}_\omega(\mathbf{r})/\mu_0 \tag{1}$$

$$\mathbf{E}_\omega(\mathbf{r}) = -\nabla \times \mathbf{H}_\omega(\mathbf{r})/(i\omega\epsilon_0) \tag{2}$$

where the vector potential is given in terms of the uniform surface current by

$$\mathbf{A}_\omega(\mathbf{r}) = \frac{\mu_0 K_\omega \hat{\mathbf{x}}}{4\pi} \int_{disk} \frac{e^{ik|\mathbf{r}-\boldsymbol{\rho}'|}}{|\mathbf{r}-\boldsymbol{\rho}'|} dS' \tag{3}$$

and the $\exp(-i\omega t)$ time-harmonic dependence has been suppressed.

Taking the curl of (3) shows that the x component of the magnetic field is zero

$$H_{\omega x}(\mathbf{r}) = 0 \tag{4}$$

and that the y and z components are given by

$$H_{\omega y}(\mathbf{r}) = \frac{K_\omega}{4\pi}\frac{\partial}{\partial z}\int_0^a\int_0^{2\pi}\frac{e^{ik\sqrt{r^2+\rho'^2-2r\rho'\sin\theta\cos\phi'}}}{\sqrt{r^2+\rho'^2-2r\rho'\sin\theta\cos\phi'}}\rho' d\phi' d\rho' \tag{5}$$

$$H_{\omega z}(\mathbf{r}) = -\frac{K_\omega}{4\pi}\frac{\partial}{\partial y}\int_0^a\int_0^{2\pi}\frac{e^{ik\sqrt{r^2+\rho'^2-2r\rho'\sin\theta\cos\phi'}}}{\sqrt{r^2+\rho'^2-2r\rho'\sin\theta\cos\phi'}}\rho' d\phi' d\rho' \tag{6}$$

where the surface integral in (3) has been expressed with polar coordinates in (5) and (6). Along the z axis ($\theta = 0$), (6) shows that the z component of the magnetic field is zero

$$H_{\omega z}(z) = 0 \tag{7}$$

and the integrand of (5) is a perfect differential so that the y component of the magnetic field can be written in closed form as

$$H_{\omega y}(z) = \frac{-K_\omega}{2ik}\frac{\partial}{\partial z}\left[e^{ikz} - e^{ik\sqrt{z^2+a^2}}\right] \tag{8a}$$

or

$$H_{\omega y}(z) = \frac{-K_\omega}{2}\left[e^{ikz} - \frac{z}{\sqrt{z^2+a^2}}e^{ik\sqrt{z^2+a^2}}\right] . \tag{8b}$$

To obtain a closed-form expression for the electric field along the z axis, separate the curl in (2) into transverse (xy) and longitudinal (z) components, that is

$$\mathbf{E}_\omega = \frac{-1}{i\omega\epsilon_0}\left[\nabla_{xy}\times\mathbf{H}_{\omega xy} + \hat{\mathbf{z}}\times\frac{\partial\mathbf{H}_{\omega xy}}{\partial z} - \hat{\mathbf{z}}\times\nabla_{xy}H_{\omega z}\right] . \tag{9}$$

From (4) and (5) we see that the first term on the right side of (9), and thus the z component of the electric field, is zero along the z axis

$$E_{\omega z}(z) = 0 . \tag{10}$$

The second term of (9) can be evaluated directly from (8). Thus, we are left with evaluating $\nabla_{xy}H_{\omega z}$ in order to find the x and y components of the electric field along the z axis.

The gradient of $H_{\omega z}$ can be written from (6) in the form

$$\nabla_{xy}H_{\omega z} = -\frac{K_\omega}{4\pi}\nabla_{xy}\frac{\partial}{\partial y}\int_0^a\int_0^{2\pi}\frac{e^{ik\sqrt{r^2+\rho'^2-2r\rho'\sin\theta\cos\phi'}}}{\sqrt{r^2+\rho'^2-2r\rho'\sin\theta\cos\phi'}}\rho' d\phi' d\rho' . \tag{11}$$

Because $\frac{\partial^2}{\partial y\partial x}$ of the integral in (11) is zero along the z axis, the x component of $\nabla_{xy}H_{\omega z}$ is zero along the z axis. This implies from (9) that the y component of the electric field along the z axis is zero

$$E_{\omega y}(z) = 0 \tag{12}$$

and that (11) can be written along the z axis as

$$\nabla_{xy}H_{\omega z} = -\frac{K_\omega}{4\pi}\hat{\mathbf{y}}\frac{\partial^2}{\partial y^2}\int_0^a\int_0^{2\pi}\frac{e^{ik\sqrt{r^2+\rho'^2-2r\rho'\sin\theta\cos\phi'}}}{\sqrt{r^2+\rho'^2-2r\rho'\sin\theta\cos\phi'}}\rho' d\phi' d\rho' . \tag{13}$$

Since $\frac{\partial^2}{\partial y^2} = \frac{\partial^2}{\partial x^2}$ along the z axis in (13), and the integral in (13) obeys the scalar wave equation

$$\left[\frac{\partial^2}{\partial x^2} + \frac{\partial^2}{\partial y^2} + \frac{\partial^2}{\partial z^2} + k^2\right] \int\int = 0 \tag{14}$$

(13) can be recast in the simpler form

$$\nabla_{xy} H_{\omega z} = \frac{K_\omega}{8\pi}\hat{\mathbf{y}} \left(k^2 + \frac{\partial^2}{\partial z^2}\right) \int_0^a \int_0^{2\pi} \frac{e^{ik\sqrt{z^2+\rho'^2}}}{\sqrt{z^2+\rho'^2}} \rho' d\phi' d\rho' \tag{15}$$

which integrates to

$$\nabla_{xy} H_{\omega z} = \frac{K_\omega}{4ik}\hat{\mathbf{y}} \left(k^2 + \frac{\partial^2}{\partial z^2}\right) \left(e^{ik\sqrt{z^2+a^2}} - e^{ikz}\right) . \tag{16}$$

Substitution of $H_{\omega y}(z)$ from (8) and $\nabla_{xy} H_{\omega z}$ from (16) into (9) yields the following closed-form expression for $E_{\omega x}$ along the z axis

$$E_{\omega x}(z) = -\frac{Z_0 K_\omega}{2}\hat{\mathbf{x}} \left[e^{ikz} - \frac{1}{2}\left(1 - \frac{1}{k^2}\frac{\partial^2}{\partial z^2}\right) e^{ik\sqrt{z^2+a^2}}\right] \tag{17}$$

which becomes, upon taking the two derivatives with respect to z and rearranging terms

$$E_{\omega x}(z) = -\frac{Z_0 K_\omega}{2}\hat{\mathbf{x}} \left\{e^{ikz} - \left[\frac{z^2 + a^2/2}{z^2+a^2} + \frac{1}{2ik}\frac{a^2}{(z^2+a^2)^{3/2}}\right] e^{ik\sqrt{z^2+a^2}}\right\} . \tag{18}$$

($Z_0 = \sqrt{\mu_0/\epsilon_0}$ is the impedance of free space.)

In summary, the time-harmonic magnetic and electric fields along the z axis of the circular disk of current can be written from (4,7,8) and (10,12,18), respectively, as

$$\mathbf{H}_\omega(z) = -\frac{1}{2}\hat{\mathbf{z}} \times \mathbf{K}_\omega \left[e^{ikz} - \frac{z}{\sqrt{z^2+a^2}} e^{ik\sqrt{z^2+a^2}}\right] \tag{19}$$

$$\mathbf{E}_\omega(z) = -\frac{Z_0}{2}\mathbf{K}_\omega \left\{e^{ikz} - \left[\frac{z^2 + a^2/2}{z^2+a^2} + \frac{1}{2ik}\frac{a^2}{(z^2+a^2)^{3/2}}\right] e^{ik\sqrt{z^2+a^2}}\right\} . \tag{20}$$

Note from (19) and (20) that as $z \to \infty$, the electric and magnetic fields obey the usual far-field relation

$$Z_0 \mathbf{H}_\omega = \hat{\mathbf{z}} \times \mathbf{E}_\omega , \quad z \to \infty . \tag{21}$$

Also, for a *finite* frequency ω, the expressions on the right sides of (19) and (20) can be expanded asymptotically to show that the electric and magnetic fields behave as $1/z$ for $z \to \infty$, or more precisely, for z greater than the "Rayleigh distance"

$$z \gtrsim 4a^2/\lambda . \tag{22}$$

The expression (19) for the magnetic field can be found in a number of references.[2,3] For the acoustic piston radiator, the on-axis expression corresponding to (19) can be found in numerous early references, reference 4 being the earliest of which we are aware. As far as we know, (17) and (20) were first derived by R. C. Wittmann[5] from a linear operator approach to calculating fields.

ARBITRARY TIME-DEPENDENT FIELDS

Having obtained the time-harmonic expressions (19) and (20) for the on-axis magnetic and electric fields, it is a simple matter of taking the Fourier transform of (19) and (20) to obtain the magnetic and electric fields for a circular uniform surface current with arbitrary time dependence. Specifically, (19) and (20) transform to the time domain as

$$\mathbf{H}(z,t) = -\frac{1}{2}\hat{\mathbf{z}} \times \left[\mathbf{K}\left(t - \frac{z}{c}\right) - \frac{z}{\sqrt{z^2+a^2}}\mathbf{K}\left(t - \frac{\sqrt{z^2+a^2}}{c}\right)\right] \tag{23}$$

$$\mathbf{E}(z,t) = -\frac{Z_0}{2}\left[\mathbf{K}\left(t - \frac{z}{c}\right) - \frac{z^2+a^2/2}{z^2+a^2}\mathbf{K}\left(t - \frac{\sqrt{z^2+a^2}}{c}\right)\right.$$
$$\left. + \frac{1}{2}\frac{ca^2}{(z^2+a^2)^{3/2}}\int_{-\infty}^{t-\frac{\sqrt{z^2+a^2}}{c}} \mathbf{K}(t')dt'\right] \tag{24}$$

where, of course

$$\mathbf{K}(t) = \int_{-\infty}^{\infty} \mathbf{K}_\omega e^{-i\omega t} d\omega \,. \tag{25}$$

Equation (23) was first derived, as far as we know, by Blejer.[6] Equation (24) follows immediately from (20), which, as mentioned above, was obtained by Wittmann.[5] Closed-form solutions for the impulse response of the acoustic circular piston radiator can be integrated on-axis to obtain expressions analogous to (23) and (24).[7] The same procedure could be applied to the circular current disk to get the on-axis expressions (23) and (24). Also, the far-field step response of circular and rectangular apertures have been derived by Hill.[8]

The first or leading pulse in (23) or (24) emanates from the center of the current disk. The second or trailing pulse in (23) or (24) emanates effectively from the rim of the current disk and lags the leading pulse by the difference in time for light to travel to the observation point on the z axis from the rim and center of the disk. The trailing pulse travels faster than the speed of light and eventually approaches, but never quite catches the leading pulse. The "superluminary" speed of the trailing pulse is a near-field phenomenon caused by the interference of the waves radiated effectively by the rim of the current disk, and crossing the z axis at an angle. As one can show in general from Maxwell's equations, the speed of the pulse approaches the speed of light when the propagation distance becomes much greater than the effective size of the source distribution.

Once the pulses have traveled a distance along the z axis much greater than the radius of the disk, (23) and (24) simplify to

$$\mathbf{H}(z,t) \sim -\frac{1}{2}\hat{\mathbf{z}} \times \left[\mathbf{K}\left(t - \frac{z}{c}\right) - \mathbf{K}\left(t - \frac{z}{c} - \frac{a^2}{2cz}\right)\right], \quad z \gg a \tag{26}$$

$$\mathbf{E}(z,t) \sim -\frac{Z_0}{2}\left[\mathbf{K}\left(t - \frac{z}{c}\right) - \mathbf{K}\left(t - \frac{z}{c} - \frac{a^2}{2cz}\right)\right], \quad z \gg a\,. \tag{27}$$

If the first time derivative of the current exists and is finite, then by definition of the derivative, (26) and (27) reduce to

$$\mathbf{H}(z,t) \sim -\frac{a^2}{4cz}\hat{\mathbf{z}} \times \frac{\partial}{\partial t}\mathbf{K}\left(t^- - \frac{z}{c}\right), \quad z \gg a \tag{28}$$

$$\mathbf{E}(z,t) \sim -\frac{Z_0 a^2}{4cz}\frac{\partial}{\partial t}\mathbf{K}\left(t^- - \frac{z}{c}\right), \quad z \gg a \tag{29}$$

provided z is large enough that the time derivative of the current does not change by a significant fraction during the time interval $a^2/(2cz)$. (The superscript "$-$" on t^- in (28) and (29) indicates that the time derivative is defined from the left in accordance with (26) and (27). This left-side definition of the time derivative allows $\partial \mathbf{K}/\partial t$ to have finite jump discontinuities.[9]) The latter proviso can be expressed as

$$a^2/(2cz) \ll t_{min} \tag{30}$$

where t_{min} is the minimum time interval in which the time derivative of the current changes by a significant fraction in the neighborhood of t. When the current pulse has a finite frequency bandwidth (or at least an effectively finite frequency bandwidth) of maximum frequency ω_{max}, or minimum wavelength λ_{min}, t_{min} can be expressed as

$$t_{min} = 2\pi/\omega_{max} = \lambda_{min}/c \tag{31}$$

and (30) becomes

$$z \gg a^2/(2\lambda_{min}) \tag{32}$$

which is essentially the same as the Rayleigh criterion (22) for the minimum wavelength in the frequency spectrum of the finite bandwidth pulse.

Probably the most interesting results for the on-axis fields are obtained from (26) and (27) when the time derivative of the current pulse does not exist or is infinite at points in time, and thus (28) and (29) are no longer valid. In that case (26) and (27), for x-directed current $\mathbf{K}(t) = K(t)\hat{\mathbf{x}}$, can be written in the form

$$\mathbf{H}(z,t) \sim -\frac{a^2}{4cz}\hat{\mathbf{y}}S(t-\frac{z}{c},\frac{a^2}{2cz})\,, \quad z \gg a \tag{33}$$

$$\mathbf{E}(z,t) \sim -\frac{Z_0 a^2}{4cz}\hat{\mathbf{x}}S(t-\frac{z}{c},\frac{a^2}{2cz})\,, \quad z \gg a \tag{34}$$

where $S(t,\Delta t)$ is the slope of the secant line[9] between the points on the current pulse at times t and $t-\Delta t$:

$$S(t,\Delta t) = \frac{K(t)-K(t-\Delta t)}{\Delta t}\,. \tag{35}$$

If the secant slope $S(t,\frac{a^2}{2cz})$ is bounded as $z \to \infty$, the far fields decay as $1/z$. If the secant slope is unbounded as $z \to \infty$, the far fields decay more slowly than $1/z$; that is, there exists an electromagnetic "missile".[1] When the secant slope has a limit, it equals the derivative of the current. That is,

$$\lim_{z\to\infty} S(t,\frac{a^2}{2cz}) = \frac{\partial K(t^-)}{\partial t} \tag{36}$$

when the limit in (36) exists. When the limit of the secant slope in (36) is $+\infty$ or $-\infty$, the derivative is said to be infinite,[9] and the fields decay more slowly than $1/z$ as $z \to \infty$. (The current pulse with time dependence given by $\sqrt{|t|}$ has an infinite derivative at $t = 0$. The current functions, $t\sin(\ln|t|)$ and $t\sin(1/t)$, are examples of continuous current functions that have bounded and unbounded secant slopes, respectively, at the time $t = 0$ where the limit in (36) and thus the derivative does not exist.)

From the relationship

$$\left|\frac{\partial K(t)}{\partial t}\right| = \left|\int_{-\infty}^{\infty} -i\omega K_\omega e^{-i\omega t} d\omega\right| \le \int_{-\infty}^{\infty} |\omega K_\omega| d\omega \tag{37}$$

one sees that the time derivative cannot be infinite, and thus no electromagnetic "missile" can be excited if the spectrum of the current is finite and its magnitude $|K_\omega|$

decays as $|\omega|^{-(2+\alpha)}$ ($\alpha > 0$) or faster as $|\omega| \to \infty$. (Also, it can be shown that no electromagnetic "missile" will be excited if K_ω decays as ω^{-2} as $|\omega| \to \infty$.) Conversely, if $|K_\omega|$ behaves as $|\omega|^{-(2-\alpha)}$ as $|\omega| \to \infty$, a finite function of frequency K_ω can be found that will produce an electromagnetic "missile".

Consider the example of a surface current with time variation given by a step function. Then the electric and magnetic fields in (26) and (27) are given by a rectangular pulse of constant amplitude independent of the large propagation distance z. The length of this constant-amplitude pulse is given by $a^2/(2z)$, which decreases as $1/z$ with the distance z to the pulse. In other words, a time step function in the surface current generates a far-field rectangular pulse along the z axis that does not decay in amplitude as $z \to \infty$, but decays in length like $1/z$; see Figure 2.

The total energy in this step-function electromagnetic "missile" also decays as $1/z$. To show this we need to determine how far the rectangular pulse maintains its amplitude in the directions transverse (xy) to its direction of propagation z. This can be done directly by looking at the radiation integral (3) a fixed distance away from the z axis. However, the same results are obtained from a simple argument involving the effective bandwidth of the rectangular pulse: The rectangular pulse propagates to infinity with an effective minimum wavelength λ_z (in its frequency spectrum) on the order of its pulse length $a^2/(2z)$. (Note that $\lambda_z \to 0$ as $z \to \infty$.) In the sidelobe region of the circular aperture radiator given by $\theta \gtrsim \lambda_z/a$, one expects (what (3) can be used to prove) that the anomalous on-axis "missile" behavior reverts to the usual $1/r$ far-field amplitude dependence. Thus the anomalous"missile" behavior is confined predominantly to a transverse distance from the z axis on the order of

$$R = O\left(\frac{z\lambda_z}{a}\right) = O\left(\frac{z}{a}\frac{a^2}{2z}\right) = O(a)\,. \tag{38}$$

That is, the transverse dimensions of the on-axis rectangular pulse of constant amplitude are on the order of the size of the circular current disk. Integrating the energy density $(\epsilon_0 E^2 + \mu_0 H^2)/2$ over this constant amplitude pulse with constant fractional-power transverse width, and with length equal to $a^2/(2z)$, produces a total energy in the "missile" that decays as $1/z$.

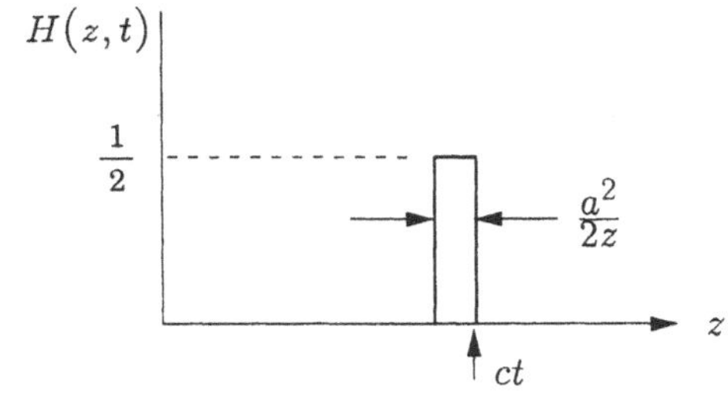

Figure 2. Magnetic field vs large z (at fixed time t) from a surface current with unit step function time dependence.

In brief, this step-function electromagnetic "missile" is a pulse, satisfying Maxwell's equations, that avoids spreading in the direction transverse to propagation by continuously decreasing its pulse length while increasing the average value of its frequency spectrum. Its total energy content vanishes as it propagates, and thus its contribution to the total radiated energy in the far field is zero.

In general, Wu[1] has shown that a current source of finite energy in a finite region of space cannot radiate a "missile" of finite (nonzero) energy to an indefinitely large distance from the source. (The argument goes as follows. Because a source pulse of finite bandwidth cannot generate a "missile", any "missile" must maintain itself on the energy in the increasingly higher frequencies. If the total energy radiated by the sources is finite, the energy in the frequency spectrum at frequencies higher than any finite value must approach zero. Incidentally, as a consequence of this result, no photon-like energy packet is a causal solution to Maxwell's equations with finite energy sources in a finite region of space.) However, if one allows current sources to radiate infinite energy, such as a current source with delta function time dependence, (23) and (24) show immediately that two pulses with infinite energy content are radiated to an indefinitely large distance in the direction of the z axis.

Although these pulses that propagate to infinity with amplitude and energy content that decay slower than $1/z$ and $1/z^2$, respectively, are valid solutions to Maxwell's equations, it is unlikely that such fields could be produced in practice because they require an input current pulse with a secant slope that is unbounded. Conceptually, this requirement is met by current that has a discontinuity or an infinite first derivative in time; that is, by a current generator with zero instantaneous rise time (reciprocal of the secant slope). Realistically, no current generator meets this specification.

REFERENCES

1. T.T. Wu, Electromagnetic missiles, *J. Appl. Phys.* 57:2370 (1985).
2. R.C. Rudduck and C.J. Chen, New plane wave spectrum formulations for the near fields of circular and strip apertures, *IEEE Trans. Antennas and Propagation* AP-24:438 (1976).
3. A.D. Yaghjian, Efficient computation of antenna coupling and fields within the near-field region, *IEEE Trans. Antennas and Propagation* AP-30:113 (1982).
4. H. Backhaus and F. Trendelenburg, Vibration of circular membranes, *Z. Tech. Phys.* 7:630 (1926).
5. R.C. Wittmann, Spherical-wave expansions of the electromagnetic fields of a uniformly excited circular aperture, *URSI Digest*, Boulder, CO, p. 53 (1992); also R.C. Wittmann and A.D. Yaghjian, Spherical-wave expansions of piston-radiator fields, *JASA* 90:1647 (1991).
6. D.J. Blejer, On-axis transient fields from a uniform circular distribution of electric or magnetic current, *Proceedings of PIERS*, Boston, MA, p. 363 (1989).
7. M. Greenspan, Piston radiator: some extensions of the theory, *JASA* 65:608 (1979).
8. D.A. Hill, Far-field transient response of an antenna from near-field data, *NBSIR 86-3063*, Boulder, CO (1986).
9. J.M.H. Olmsted. "Intermediate Analysis," Appleton-Century-Crofts, New York (1956), ch. 4.

TIME DOMAIN CHARACTERIZATION OF ACTIVE MICROWAVE CIRCUITS — THE FDTD DIAKOPTICS METHOD

Tian-Wei Huang, Bijan Houshmand, and Tatsuo Itoh

Electrical Engineering Department
University of California at Los Angeles
Los Angeles, California 90024

ABSTRACT

A finite-difference time-domain (FDTD) Diakoptics method is developed in this paper. A two overlapped cells boundary is proposed for implementation of this method. The numerical dispersion error of FDTD in finding the impulse response is circumvented by a deconvolution method with a band-limited excitation. The transient responses of several linear and nonlinear active circuits are computed and compared with simulated results from Microwave SPICE (MWSPICE).

INTRODUCTION

To analyze an ultra-wideband system, the time-domain methods are often more suitable than the frequency-domain methods. For a linear passive system, the wideband system response can be obtained from a wideband time-domain input signal. Currently, time-domain methods, such as the transmission line matrix (TLM), the spacial network (SNW), and the finite-difference time-domain (FDTD), can combine the lumped circuits analysis into their field calculation. These methods can include the nonlinear or active elements directly into their calculation to get a more accurate result.

These time-domain methods require a fine spacial discretization which results in a large computer memory storage, in order to maintain the accuracy at high frequencies. For a large and complicated circuit, this memory requirement of the whole circuit is often not practical. One possible solution for this memory requirement problem is to divide a whole circuit into several segments which can be calculated individually, including the interaction among segments. The time-domain Diakoptics idea which was first introduced in TLM [1]-[3], provides such segmentation algorithm. This paper presents the implementation of the time-domain Diakoptics method in FDTD.

This paper is organized as follows. Diakoptics method as it applies to a general linear system is described first. The application of the method to the FDTD algorithm is illustrated next. The computational aspects and numerical results are then demonstrated.

THE DIAKOPTICS METHOD

The Diakoptics idea originates from the linear system theory. If the impulse response h(t) of a one-port linear passive system is available, then the system output Y(n) can be determined from the convolution operation between h(t) and the system input X(n). This relationship allows the replacement of the whole one-port linear passive system by its impulse response h(t). This idea is illustrated in Fig. 1.

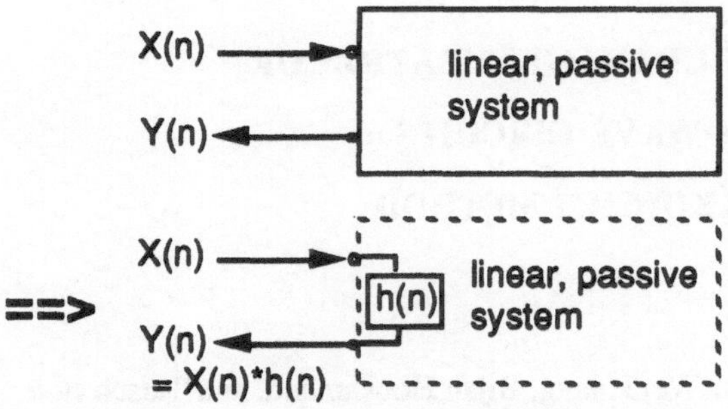

Figure 1. Linear system theory: a linear passive system can be replaced by its impulse response h(t).

The impulse response idea can be used in the microwave circuit and field simulation. A whole circuit is divided into a linear passive region and a nonlinear active region, which are connected through a multi-port boundary as shown in Fig. 2(a). A linear passive region can be replaced by an impulse response matrix [g], Fig. 2(b). The response of this linear passive region is calculated by Eq. (1) which relates the output of port m at time k, $Y_m(k)$, to the input signal of all ports $X_n(k')$, n=1..N, k'=0..k. Eq.(1) is the convolution operation for a multi-port system and [g] is similar to a time-domain Green's function [4].

$$Y_m(k)=\sum_{n=1}^{N}\sum_{k'=0}^{k} g(m,n,k-k')\, X_n(k') \tag{1}$$

Where

g(m,n,k') : the impulse response of port m at t =k' due to the unit excitation of port n at t = 0.

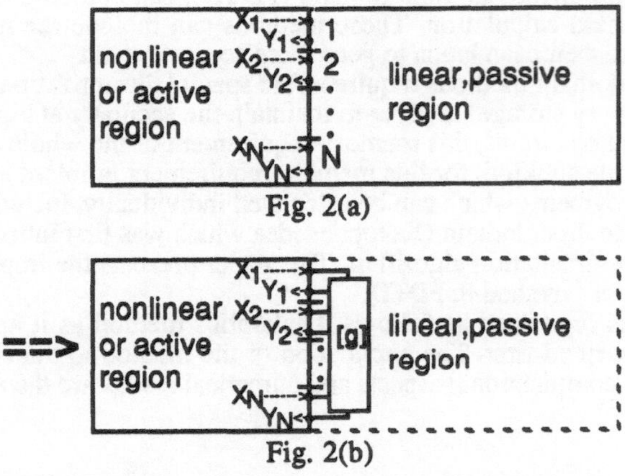

Figure 2. A multi-port linear passive region is replace by its impulse response matrix [g].

THE FDTD DIAKOPTICS METHOD

The FDTD Diakoptics method can be illustrated by the following example, that is, an infinitely long parallel-plate transmission line loaded with lumped elements. It is assumed that the distance between two plates is small in terms of wavelength; therefore, no field variation occurs in the perpendicular direction to the plates. The fields only vary along the propagation direction. For a comparison purpose, such a circuit can be analyzed by a conventional time-domain circuit simulator, like SPICE.

This structure is divided into three sections in Fig. 3 : one center section which contains a nonlinear active region and two outer uniform transmission lines as a linear passive region. Since the field variables in each FDTD cell represent the total field, the outgoing and incoming waves cannot be defined within one FDTD cell. On the segmentation boundary, two overlapped cells are used to represent the input and output of one port. During the time iterations of FDTD Diakoptics method, the fields of a boundary cell can be calculated from the convolution operation, as shown in Fig. 4.

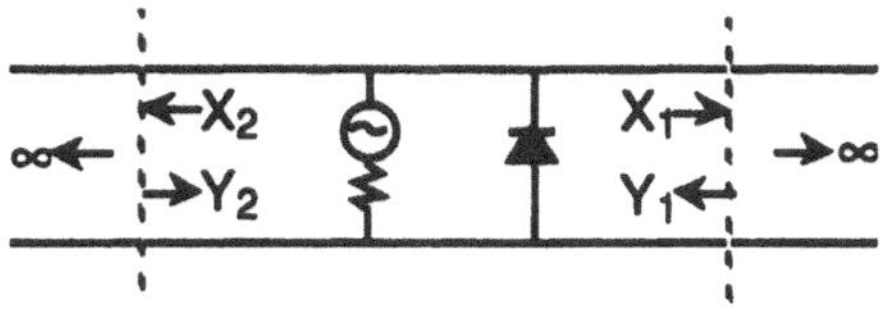

Figure 3. An infinitely long parallel-plate transmission line loaded with lumped elements.

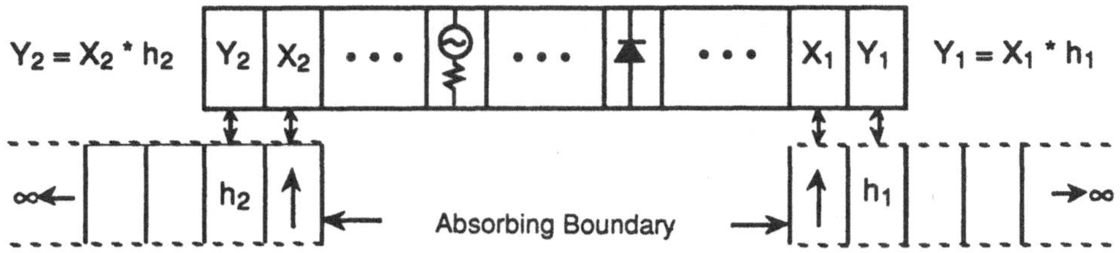

Figure 4. Two overlapped cells boundary to implement the FDTD Diakoptics method.

The crucial step to implement the FDTD Diakoptics is the evaluation of the impulse response of passive segments. For a circuit shown in Fig.5, the impulse response of a one-directional infinitely long transmission line can be calculated analytically. In this case, the impulse response, observed on the next cell is a time-delayed impulse. The time delay depends on the cell size, and the time step used for the FDTD computation. Using the analytical impulse response in the FDTD Diakoptics method, a load current of this linear circuit is plotted in Fig. 5, and compared with simulated results from MWSPICE, which demonstrates an excellent agreement.

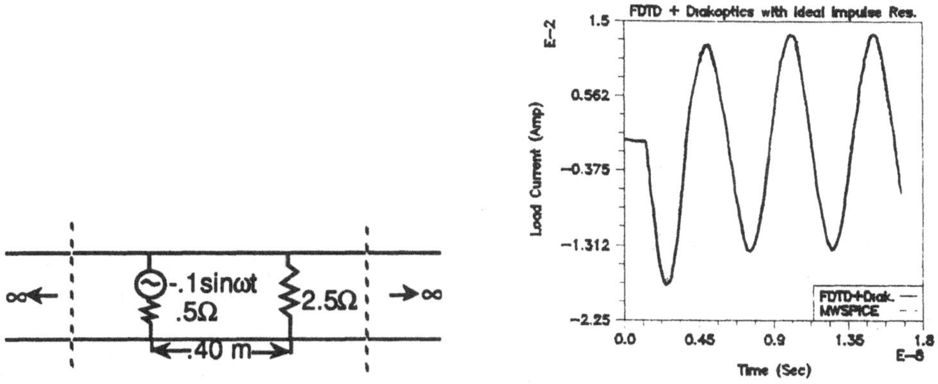

Figure 5. A linear circuit with two infinitely long transmission lines is simulated by the FDTD Diakoptics method and with an ideal impulse response (f = 200 MHz, Cell Size = 1 cm x 1cm)

COMPUTATION OF THE IMPULSE RESPONSE

For most circuits, it is not practical to find their impulse response analytically. Instead of analytical methods, some numerical methods, like FDTD, can be used to find the time-domain impulse response. In this paper, an FDTD-based method is developed to produce the time-domain impulse response numerically.

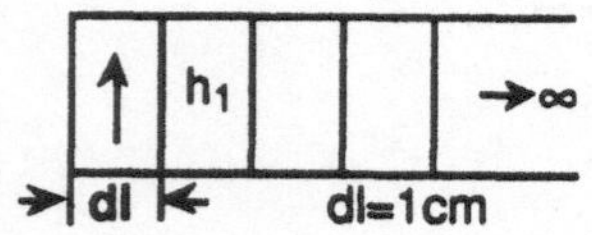

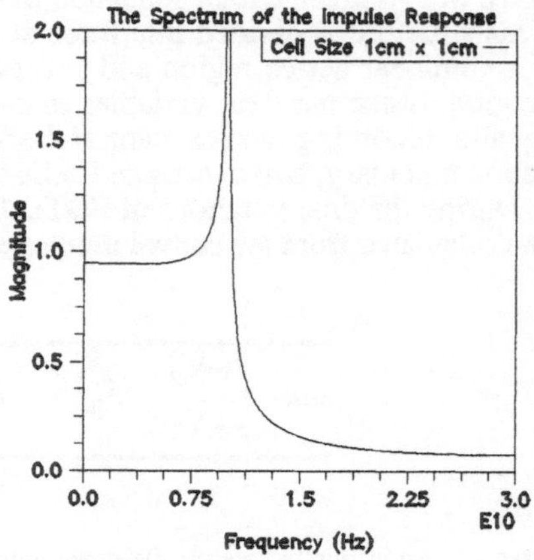

Figure 6. The spectrum of the impulse response of a one-directional uniform transmission line by FDTD simulation (Cell Size 1cm x 1cm, Cut-off Frequency at 10 GHz)

Direct computation of the impulse response is numerically unstable, due to the numerical dispersion [5] in FDTD, which originates from the spacial discretization of the computational space. The numerical dispersion introduces a cut-off for high frequency components of the propagating signal. Near the cut-off frequency, the slow-traveled components of the input signal will accumulate on the structure and behave as undamped oscillation. This oscillation resembles the existence of a pole in the system transfer function, as shown in Fig. 6. One possible solution to this stability problem is using a band-limited excitation, which has a frequency band below the cut-off frequency of FDTD. After the excitation is launched, the impulse response h(n) can be generated from the band-limited excitation X(n) and band-limited response Y(n) by using a recursive deconvolution formula Eq.(2) [6].

$$h(0) = \frac{Y(0)}{X(0)} \qquad n = 0$$

$$h(n) = \frac{Y(n) - \sum_{k=0}^{n-1} h(k) \cdot X(n-k)}{X(0)} \qquad n \geq 1 \qquad (2)$$

NUMERICAL RESULTS

The following three numerical results are based on using band-limited excitation and deconvolution in FDTD to find out the impulse response which is used in the FDTD Diakoptics method. The load currents, calculated from the FDTD Diakoptics method, in three different circuits are plotted in Figs. 7, 8, 9. All these three results have good agreement with simulated results from MWSPICE.

Fig. 7 shows a delayed sinusoidal load current on a resistor. This delay is due the signal propagating time from source to load. The first cycle is the transient response and the current gradually reaches the steady state after the second cycle.

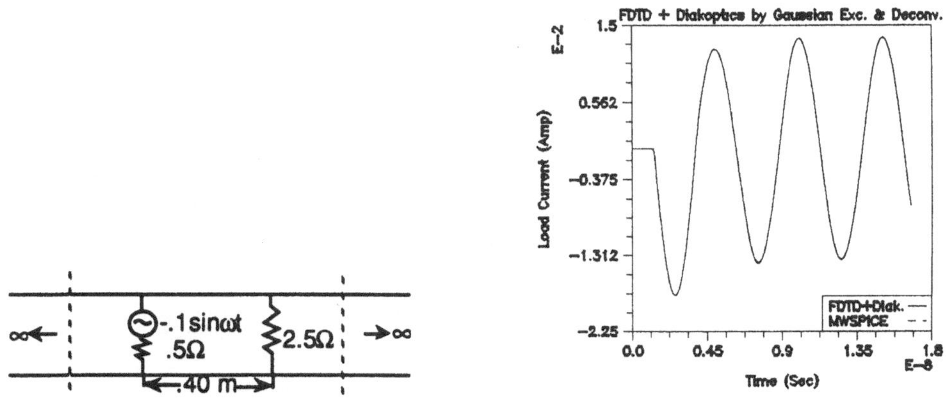

Figure 7. A linear circuit with two infinitely long transmission lines is simulated by the FDTD Diakoptics method and with an deconvolved impulse response (f = 200 MHz, Cell Size = 1 cm x 1cm)

Fig. 8 shows a similar circuit as Fig. 7 except using a finite-length transmission line to replace the infinitely long transmission line in Fig. 7. This replacement means a different impulse response should be used in the FDTD Diakoptics method. During that first half cycle, due to the signal propagation delay, the PEC walls in the transmission line have no effect on the current. On the second half cycle, the reflection from the PEC walls result in a large current amplitude.

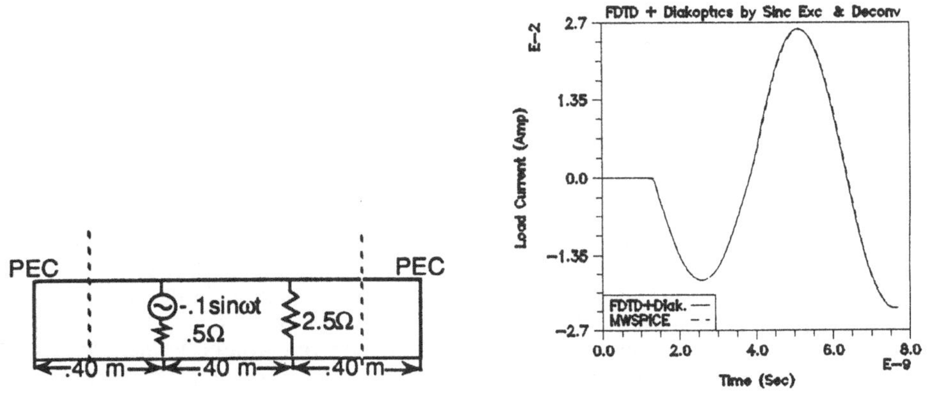

Figure 8. A linear circuit with two shorted transmission lines is simulated by the FDTD Diakoptics method and with an deconvolved impulse response (f = 200 MHz, Cell Size = 1 cm x 1cm)

Fig. 9 illustraes the capability of analyzing nonlinear active circuits in the FDTD Diakoptics method. The I-V characteristic of the active device is represented by a third order polynomial. The dot line in Fig. 9 is the load voltage of the nonlinear active device, and the solid line represent the output current of this device. Due to the third order polynomial in the I-V curve, the frequency of the output current is tripled. This circuit is regraded as a frequency tripler.

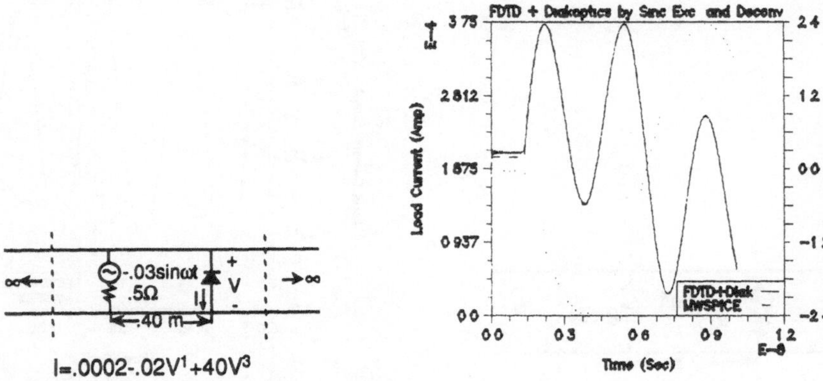

Figure 9. A nonlinear circuit with two infinitely long transmission lines is simulated by the FDTD Diakoptics method and with an deconvolved impulse response (f = 100 MHz, Cell Size = 1 cm x 1cm)

CONCLUSION

The FDTD Diakoptics method is used to simulate linear and nonlinear active circuits, which shows good agreement with simulated results from the MWSPICE. Using this method, a large linear passive region can be replaced by an impulse response matrix of a multi-port boundary. If the number of ports is small and the replaced linear region is large, then the computer storage is reduced. This method also can reduce the CPU time for device optimization problems, where the active device is optimized over a range of parameters. In such problems, the field simulation of the passive segments are performed only once and the iteration is confined to the active segment. For the future development, the cascaded Diakoptics method and 3-D circuits simulation are considered.

ACKNOWLEDGMENTS

This work was supported by the Office of Naval Research N00014-91-J-1651 and the Joint Services Electronics Program F49620-92-C-0055.

REFERENCES

[1] P. B. Johns, and K. Akhtarzad, "The use of time domain diakoptics in time discrete models of fields," *Int. J. Num. Methods Eng.*, vol. 17, pp. 1-14, 1981.

[2] P. B. Johns, and K. Akhtarzad, "Time domain approximations in the solution of fields by time domain diakoptics," *Int. J. Num. Methods Eng.*, vol. 18, pp. 1861-1373, 1982.

[3] W. J. R. Hoefer, "The discrete time-domain Green's function or Johns matrix - a new powerful concept in transmission line modelling (TLM)," *Int. J. Num. Modelling.*, vol. 2, pp. 215-225, 1989.

[4] C. W. Kuo, and T. Itoh, "A new time-domain transverse resonance method in solving guided wave problems," *Int. J. Num. Modelling.*, vol. 3, pp. 229-234, 1990.

[5] A. Taflove, and K. R. Umashankar, "Advanced numerical modeling of microwave penetration and coupling for complex structures," Final Rept. No. UCRL-15960, Contract 6599805, Lawrence Livermore Nat. Lab., 1987.

[6] J. G. Proakis, and D. G. Manolakis, *Introduction to Digital Signal Processing*. New York: Macmillan, 1988.

WIDE-BAND REDUCTION OF COUPLING AND PULSE DISTORTION IN MULTI-CONDUCTOR INTEGRATED CIRCUIT LINES

James P. K. Gilb and Constantine A. Balanis

Telecommunications Research Center
Department of Electrical Engineering
Arizona State University
Tempe, AZ 85287-7206

INTRODUCTION

Ultrafast clock speeds and short rise times of digital signals give them a very large bandwidth. With clock speeds in the Gigabit/second range and rise times on the order of 10 picoseconds, these signals will have significant frequency components in the tens to hundreds of GHz. This wide frequency range requires a full-wave analysis, rather than a a quasi-static one, to accurately predict the distortion and coupling of high-speed digital signals. High speed circuits also have small inter-line spacing to increase packing density, which in turn increases coupling distortion and crosstalk. Successful design of these high-speed circuits requires accurate prediction of pulse distortion and coupling, as well as a technique to reduce coupling over a very wide band of frequencies.

Many different techniques have been used to reduce coupling in high-speed digital circuits. One approach is to place an extra line that is grounded at both ends between adjacent lines[1,2]. Although this method can provide some reduction in crosstalk, it increases fabrication costs, increases inter-line spacing, and requires via holes. In addition, while the direct crosstalk is reduced, the return voltage can increase the near-end crosstalk. Another technique that can be used is substrate compensation which reduces crosstalk is through the choice of the electrical parameters of a multi-layer substrate[3]. This technique allows very small inter-line spacings, less than one conductor width, while reducing coupling over a very wide range of frequency. For the case on two adjacent conductors, the crosstalk and coupling distortion can be essentially eliminated. While it is not always possible to completely eliminate crosstalk for multi-conductor cases, it is usually possible to obtain a significant reduction in the level of crosstalk through substrate compensation.

Ultra-Wideband, Short-Pulse Electromagnetics
Edited by H. Bertoni *et al.*, Plenum Press, 1993

In this paper, distortion and coupling of wide-band signals on integrated circuit lines is analyzed using a full-wave approach to obtain frequency-domain data, and a Fourier transform approach to obtain the time-domain results. The effect of additional lines on the distortion and crosstalk of multi-line structures is examined, showing how crosstalk can be significantly reduced for a wide range of conductor configurations. Variation of crosstalk and distortion for different pulse widths is also investigated to illustrate the effect of dispersion for multi-conductor lines.

ANALYSIS

The Spectral Domain Approach (SDA)[3,4] is used to compute the frequency-domain parameters used in the time-domain analysis. Chebyshev polynomials modified by the edge condition are used as expansion functions[4]. In this paper four expansion functions are used for J_z and three for J_x for each of the conductors. The geometry for a multi-layer, multi-conductor structure is shown in Fig. 1. Only one conductor is shown, with

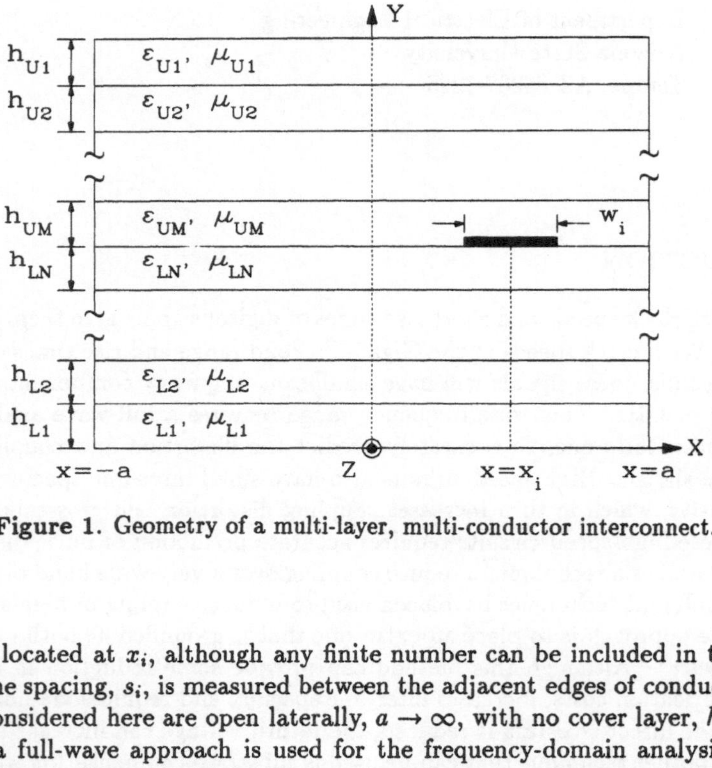

Figure 1. Geometry of a multi-layer, multi-conductor interconnect.

a width w_i located at x_i, although any finite number can be included in the analysis. The interline spacing, s_i, is measured between the adjacent edges of conductors. All of the cases considered here are open laterally, $a \rightarrow \infty$, with no cover layer, $h_{U1} \rightarrow \infty$.

Since a full-wave approach is used for the frequency-domain analysis, the most direct method for computing the time-domain response is to use a modal analysis[4]. For an N conductor system, it is necessary to compute N modal propagation constants and an N by N matrix that specifies the relative magnitudes of the current densities on each of the N conductors for each of the N modes. Since the propagation constants and the current densities of the modes vary as a function of frequency, they must be computed over a wide spectrum to accurately predict pulse distortion and coupling. The time-domain response is then computed with and inverse Fourier transform of the frequency domain data[4].

RESULTS

For all of the following results, a ramped pulse is used with a time domain response given by:

$$v(t) = \begin{cases} 0 & \text{for } |t| > \tau + \tau_r \\ 1 - \dfrac{|t| - \tau}{\tau_r} & \text{for } \tau \le |t| \le \tau + \tau_r \\ 1 & \text{for } |t| \le \tau \end{cases}$$

where τ_r is the rise time and τ is the pulse duration. Only symmetric conductor configurations are considered and the lines are numbered consecutively from one side to the other, i.e. for the four conductor case, line 1 and line 4 are on the outside, with lines 2 and 3 in the center.

To design a substrate-compensated, low-coupling structure, the substrates materials and heights need to be chosen such that the maximum level of crosstalk on the adjacent lines is minimized. For the single layer case, the lines closest to the excited line have the highest level of crosstalk. However, for the multi-layer case, the crosstalk level on the closest lines may be less than on those further away. The maximum crosstalk on adjacent lines as a function of substrate height ratio, $h_{L1}/(h_{L2} + h_{L1})$ is shown in Fig. 2 for two, three and four symmetric microstrip lines and in Fig. 3 for two and five symmetric microstrip lines. The maximum crosstalk is calculated using the quasi-static values of the phase constants and the current matrix for a ramped pulse with $\tau = 100$ picoseconds and $\tau_r = 10$ picoseconds at a distance of 25 mm. Only the results for unique excitations are shown, for example, in the four conductor case exciting line 4 is equivalent to exciting line 1, since the structure is symmetric.

For the two-conductor case, shown in both Fig. 2 and Fig. 3, there are two values of the height ratio which eliminates the crosstalk on the adjacent line. Likewise, for the three-conductor case with the middle line (line 2) being excited, it is also possible to essentially eliminate the crosstalk on both of the adjacent lines. For this case, the behavior of the maximum crosstalk level is very close to the two-conductor case, except that the highest level of crosstalk is only about 0.36 compared with 0.5 for the two line case. These cases are similar because there is at most one conductor on either side of the excited line.

While it was possible to essentially eliminate the crosstalk for the 2 line case and the three line case with line 2 excited, for the four and five conductor cases, and the three conductor case with line 1 excited, there is a limit to the reduction of the crosstalk. The reason for this is that while the crosstalk to the nearest lines is greatly reduced, the crosstalk to lines farther away is not reduced as much and so the maximum crosstalk of all of the lines is larger than the crosstalk level on the nearest line. Thus, while it is usually possible to essentially eliminate the crosstalk to the nearest lines in a multi-conductor case, the overall crosstalk reduction is limited by the crosstalk to lines farther away.

Just as the 2-line case was similar to the 3-line case with line 2 excited, the crosstalk levels for 3 lines with line 1 excited and 4 lines with line 2 excited show very similar behavior, especially in the low-coupling regions. These cases are similar because there is at most 2 conductors on either side of the excited conductor. Thus it would be expected that the case of 4 lines with line 1 excited will give results similar to the following cases; 1) 5 lines, line 2 excited, 2) 6 lines, line 3 excited, and 3) 7 lines, line 4 (the middle line) excited, since these cases have at most three conductors on either side of the excited line. Indeed, the 5 line case with line 2 excited in Fig. 3 is similar to the 4 line case with line 1 excited in Fig. 2. Since additional lines, i.e. more than four

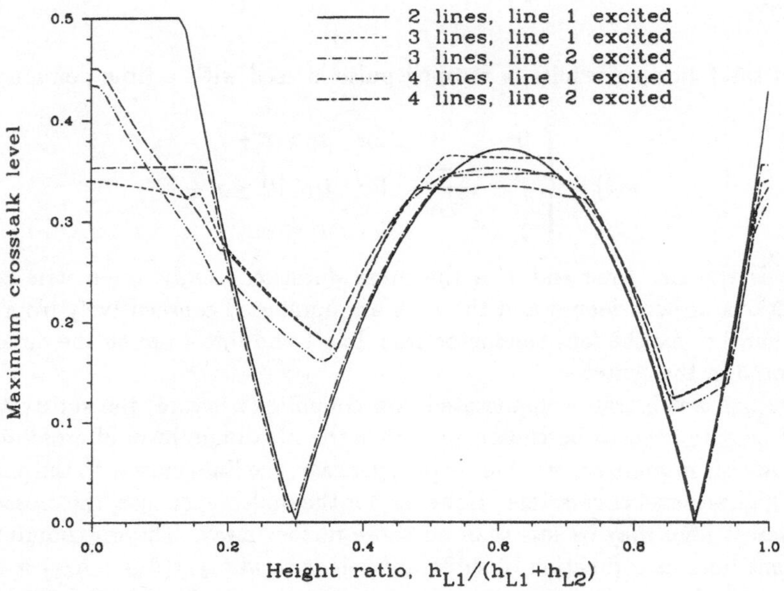

Figure 2. Maximum crosstalk on a two-layer substrate for two, three, and four signal conductors as a function of the substrate height ratio, $h_{L1}/(h_{L1}+h_{L2})$ ($w = s = 0.6$ mm, $\epsilon_{rL1} = 2.94$, $\epsilon_{rL2} = 10.8$, $\epsilon_{rU1} = 1.0$, $h_{L1} + h_{L2} = 0.635$ mm, $h_{U1} \to \infty$).

on either side of the excited line, would have little effect on the crosstalk and coupling distortion, it is expected that the low-coupling regions shown in Figs. 2 and 3 would have reduced coupling for any number of lines of these widths and spacings on this substrate.

In the low coupling regions of Figs. 2 and 3, the crosstalk to the nearest lines is drastically reduced, even essentially eliminated, but the crosstalk on lines farther away is not reduced by as much and so these lines have the highest level of crosstalk. This effect can be seen in Fig. 4 where the maximum crosstalk as a function of the substrate height ratio on each of the individual lines is shown for the five conductor case with the outside line (line 1) being excited. The crosstalk level on the nearest line, line 2, is greatly reduced, from 0.42 to 0.04, for two different values of the substrate height ratio, as with the two conductor case. However, near those two values, the crosstalk on line 3 is much higher, and so the maximum crosstalk on all of the lines is not reduced as much. Furthermore, for height ratios between 0.35 and 0.85 the maximum crosstalk level occurs on line 4, two lines removed from the excited line. Thus for multi-conductor, multi-layer structures the highest crosstalk level is not necessarily on the nearest conductor. However, while it may not be possible to essentially eliminate the crosstalk on a multi-line structure with substrate compensation, it is still possible to get a substantial reduction in the crosstalk using substrate compensation. For the five line case, the maximum crosstalk level on the other lines can be a reduced by up to a factor of four.

Results for distortion and coupling of a ramped pulse on a four-conductor line are shown in Figs. 5 and 6 where τ = 100 picoseconds and τ_r = 10 picoseconds. For this example, line 1 is excited and the responses on all four lines are shown at a

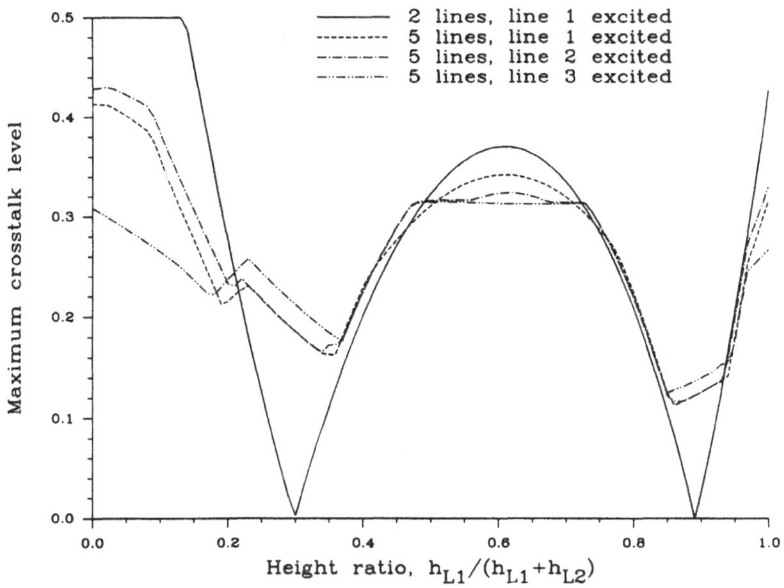

Figure 3. Maximum crosstalk on a two-layer substrate for two and five signal conductors as a function of the substrate height ratio, $h_{L1}/(h_{L1}+h_{L2})$ with the same dimensions as in Fig. 2.

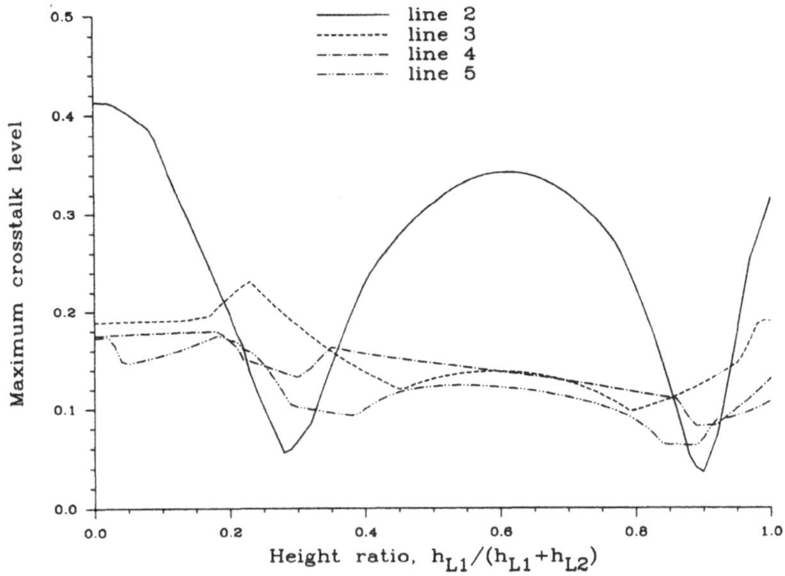

Figure 4. Maximum crosstalk on the four adjacent lines of a two-layer structure with five signal conductors as a function of the substrate height ratio $h_{L1}/(h_{L1}+h_{L2})$ with the same dimensions as in Fig. 2.

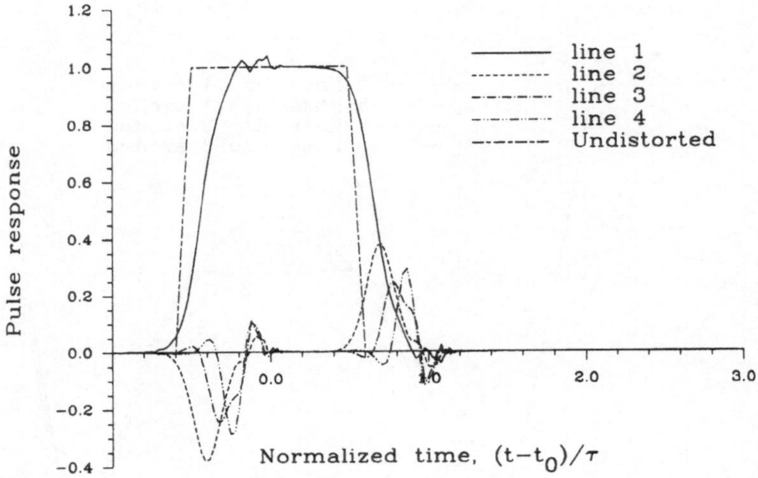

Figure 5. Pulse distortion on four symmetric conductors at a distance $l = 50$ mm with line 1 excited ($w = s = 0.6$ mm, $\epsilon_{rL1} = 2.94$, $\epsilon_{rL2} = 10.2$, $\epsilon_{rU1} = 1.0$, $h_{L1} = 0.635$ mm, $h_{L1} = 0$ mm, $h_{U1} \to \infty$).

distance of $l = 50$ mm as a function of the normalized time, $(t - t_0)/\tau$, where t is time, $t_0 = (l/c)\sqrt{\epsilon_{\text{reff(min)}}(0)}$, c is the speed of light, and $\epsilon_{\text{reff(min)}}(0)$ is the minimum effective dielectric constant of the four modes in the quasistatic region. Fig. 5 is a single-layer substrate, while Fig. 6 is a two layer substrate designed to have low coupling through substrate compensation by choosing $h_{L1}/(h_{L1} + h_{L2}) = 0.875$.

In Fig. 5, the crosstalk on the adjacent line is almost 40% of the amplitude of the initial signal, representing a very high level of crosstalk. The highest level of crosstalk occurs on the nearest line, line 2. The pulse on line 1 has also broadened in time, due to coupling distortion. On the compenstated substrate, however, the maximum crosstalk on the adjacent lines is only 15% of the amplitude of the initial signal. The maximum crosstalk occurs on line 3, which is one removed from the excited line. On the other hand, the response on line 2 is only 5% of the amplitude of the intial signal, the lowest of the three adjacent lines. Thus, this substrate configuration minimizes the coupling to the nearest lines, but only slightly decreases the coupling to lines farther away. However, the crosstalk has been substantially decreased using substrate compensation.

As the pulse duration, τ, or rise time, τ_r, decreases, the signal will suffer more degradation due to dispersion. Using the same structures as in Figs. 5 and Fig. 6, the time domain response on the four lines is shown in Figs. 7 and 8 for a ramped pulse with one-half the duration, $\tau = 50$ picoseconds, and the same rise time, $\tau_r = 10$ picoseconds. The pulse on the uncompensated substrate shows much more coupling distortion and pulse widening than the longer pulse in Fig. 5. The crosstalk on the adjacent lines is also much greater, due to the shorter pulse width. On the other hand, the pulse on the compensated substrate shows much less distortion due to coupling, and the crosstalk on the adjacent lines is much lower.

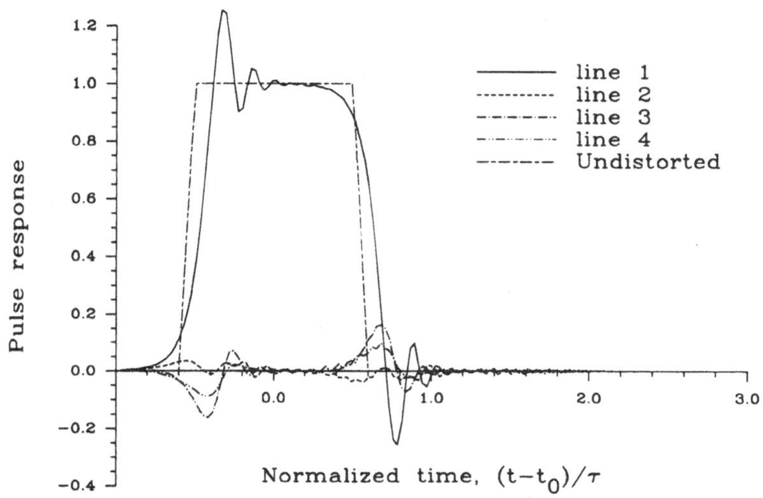

Figure 6. Pulse distortion on four symmetric conductors at a distance l = 50 mm with line 1 excited ($w = s = 0.6$ mm, $\epsilon_{rL1} = 2.94$, $\epsilon_{rL2} = 10.8$, $\epsilon_{rU1} = 1.0$, $h_{L1} = 0.55562$ mm, $h_{L1} = 0.07938$ mm, $h_{U1} \rightarrow \infty$).

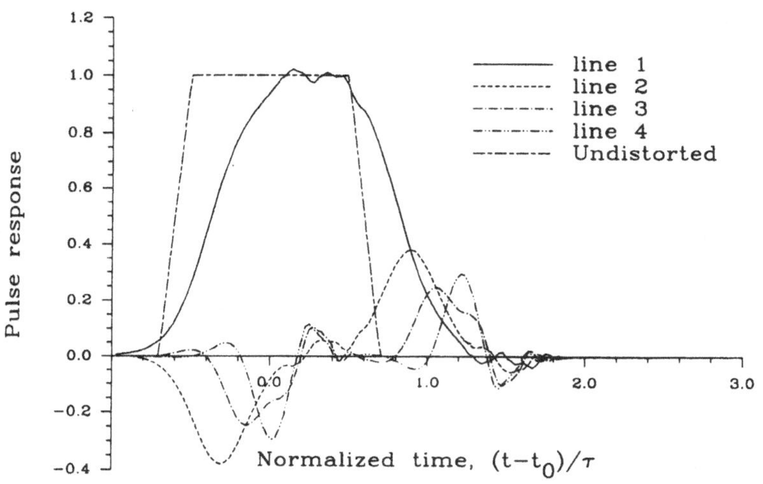

Figure 7. Pulse distortion on four symmetric conductors at a distance l = 50 mm with line 1 excited and the same dimensions as in Fig. 5.

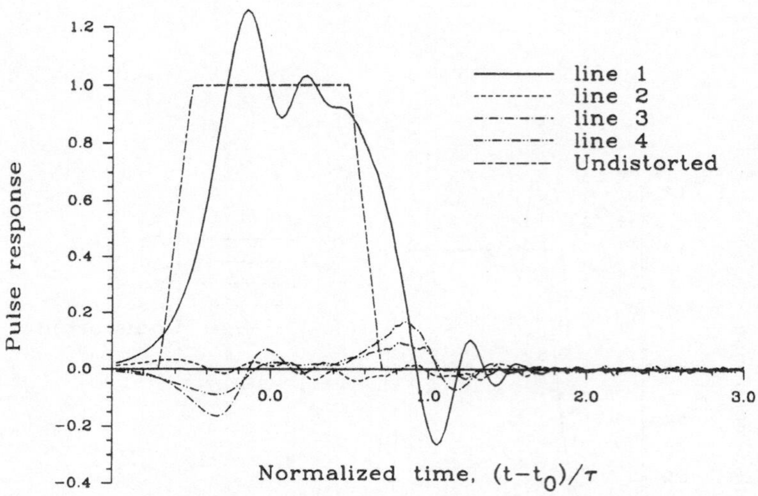

Figure 8. Pulse distortion on four symmetric conductors at a distance $l = 50$ mm with line 1 excited and the same dimenstions as in Fig. 6.

REFERENCES

1. L. Carin and K. J. Webb, Isolation effects in single- and dual-plane VLSI interconnects, *IEEE Trans. Microwave Theory Tech.*, MTT-38:396–404 (1990).

2. S. Seki and H. Hasegawa, Analysis of crosstalk in very high-speed LSI/VLSI's using a coupled multi-conductor MIS microstrip line model, *IEEE Trans. Microwave Theory Tech.*, MTT-32:1715–1720 (1984).

3. J. P. Gilb and C. A. Balanis, Pulse distortion on multilayer coupled microstrip lines, *IEEE Trans. Microwave Theory Tech.*, MTT-37:1620–1628 (1989).

4. J. P. K. Gilb and C. A. Balanis, Asymmetric, multi-conductor, low-coupling structures for high-speed, high-density digital interconnects, *IEEE Trans. Microwave Theory Tech.*, MTT-39:2100–2106 (1991).

PRESERVATION OF THE SHAPE OF ULTRA-WIDEBAND PULSES IN MMIC TRANSMISSION LINES BY USING FREQUENCY DEPENDENT SUBSTRATES

Nicolaos G. Alexopoulos[1] and Rodolfo E. Díaz[2]

[1]Department of Electrical Engineering
University of California
Los Angeles, CA 90210

[2]Hexcel APD
2500 W. Frye Rd.
Chandler, AZ 85224

ABSTRACT

A new class of transmission lines is proposed which approximately preserves the shape of time domain pulses without sacrificing the manufacturing advantages of standard MMIC transmission lines. By incorporating an inhomogeneity of specified conductivity and shape under the strip conductor of conventional microstrip, the substrate is given an apparent frequency dependent permittivity which undoes the dispersive behavior of the real part of the microstrip's effective dielectric constant. The rules for the design of the frequency dependent substrate are given and its behavior verified by using Finite Difference Time Domain calculations. Other possible applications of these transmission lines are discussed.

I- INTRODUCTION

Because of their compactness and producibility, Monolithic Microwave Integrated Circuits (MMIC) are ideally suited for microwave applications ranging from the construction of conformal antennas to the implementation of gigabit logic circuits. However, the ever increasing need for broader frequency bands of operation in these applications highlights a limitation in traditional MMIC design; namely, that MMIC printed transmission lines are inherently frequency dependent structures.

Although design approaches exist that alter the dispersive properties of these lines by the use of multiple dielectric layers (e.g. suspended microstrip or the use of superstrates) there is currently no unified method for the custom design of this frequency dependence. The analytic model of dielectric materials, developed in reference 1, provides us with such a method: It will be shown in this paper that by using customized frequency dependent substrates it is possible to alter the frequency dependence of the transmission line's propagation constant in a prescribed fashion. To the extent that the permittivity of the substrate's material can be designed, to that extent the dispersive properties of a MMIC transmission line can be controlled.

As a detailed example of this technique, it is shown that the real part of the effective permittivity of a microstrip transmission line can be rendered essentially frequency independent by the use of a substrate whose material dispersion cancels the microstrip's natural electrodynamic dispersion. The price paid is some added loss. Since the other widely used MMIC transmission lines (slotline, finline, coplanar waveguide, etc.) exhibit similar electrodynamic dispersion, the solution proposed here for microstrip, can be used for them as well.

II- PROOF OF PHYSICAL REALIZABILITY

Consider a typical microstrip geometry[2] as shown in Figure 1a. When the substrate's permittivity is frequency independent, (by permittivity is to be understood the relative permittivity unless otherwise noted), the microstrip's effective permittivity is frequency dependent, as in Figure 1b. The effective permittivity increases with frequency from a DC value of 7.8 to the asymptotic limit 11.7, the substrate's value. This approach to the value of the substrate's permittivity is characteristic of all MMIC transmission lines and it emphasizes the resemblance, at high frequencies, between the hybrid mode structure of such non-TEM lines and that of the surface waves that can propagate on the substrate. It is this asymptotic behavior that suggests the way to create an essentially dispersionless microstrip.

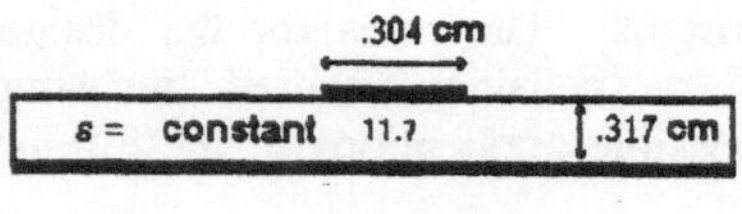

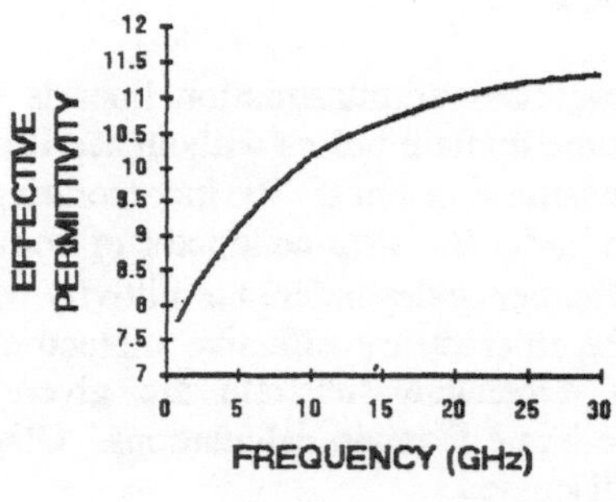

Figure 1. Typical microstrip constructed on a frequency independent substrate (a) has a frequency dependent permittivity (b).

Suppose that the substrate's permittivity is allowed to decrease with frequency so that, although at DC it is 11.7, at high frequencies it becomes 7.8. Then the microstrip's effective permittivity, which starts off as 7.8, will also end at the new asymptotic value of 7.8. Clearly, if the substrate's permittivity drops with frequency at just the right rate, the effective permittivity of the microstrip can be maintained at a constant, frequency independent, value of 7.8. The necessary profile for the substrate's permittivity as a function of frequency is very closely approximated by the frequency dependent behavior of a Debye relaxation material (Figure 2). Its equivalent circuit[1] is shown inset in the Figure.

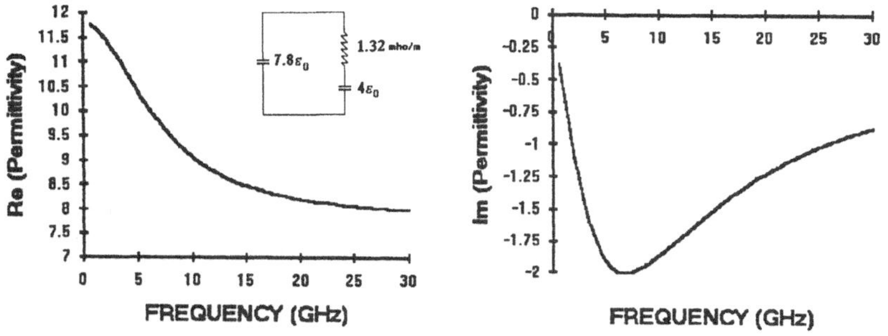

(a) (b)

Figure 2. The real part of a Debye relaxation material's permittivity drops with frequency (a). This drop can be used to compensate microstrip dispersion at the expense of the loss due to the imaginary part (b).

At DC, such a material exhibits its highest permittivity, given by the parallel sum of its two capacitances (11.8). As frequency increases, the permittivity is seen to be complex, with the real part dropping from its initial peak value and the imaginary part increasing towards a peak at the relaxation (angular) frequency given by $(RC)^{-1}$. Thereafter both real and imaginary parts drop with frequency; the first towards the value of the isolated capacitor (7.8), the second towards zero. Since Debye relaxation materials are causal, the material described by the circuit in Figure 2 is physically realizable and therefore the material needed to cancel the electrodynamic dispersion of microstrip can be constructed. The question remaining is, How?

III- THE LAYERED SLAB AS A DEBYE MATERIAL

To construct the desired Debye relaxation material we consider first the structure of Figure 3a in the presence of TEM waves. This composite material, consisting of alternating layers of conducting and insulating dielectrics, can be expected to exhibit a frequency dependent permittivity as follows: At low frequencies, the conduction current dominates the behavior of the conducting regions. As a result, the electric field is "shorted out" in the interior of those layers and it is, therefore, concentrated in the insulating regions. Since all the "voltage" that used to drop across a given volume of space now only drops across the insulating layers, a net increase of effective capacitance occurs. This is equivalent to an artificially enhanced permittivity. As the frequency increases, the displacement current becomes a significant player in the conducting and insulating regions and the "voltage" no longer drops entirely across the insulating layer. Thus the effective capacitance drops and the artificial permittivity approaches that of the original individual materials. The circuits of Figure 3b show that this structure is equivalent to a simple Debye relaxation material.

The frequency dependent permittivity thus obtained will slow down the plane wave by the usual $\varepsilon_r^{-1/2}$ factor, as long as the magnetic field of the plane wave is not perturbed. That is, it will work as long as the conductivity of the conducting layer is not "so high" that the induced eddy currents expel the magnetic field lines. In the limit of infinite conductivity, both electric and magnetic field lines are denied access to the conducting regions and so, although the permittivity is increased by the concentration of the electric field lines, the permeability is correspondingly decreased by the diamagnetic expulsion of the magnetic field lines. The net effect would then be an unchanged wave velocity.

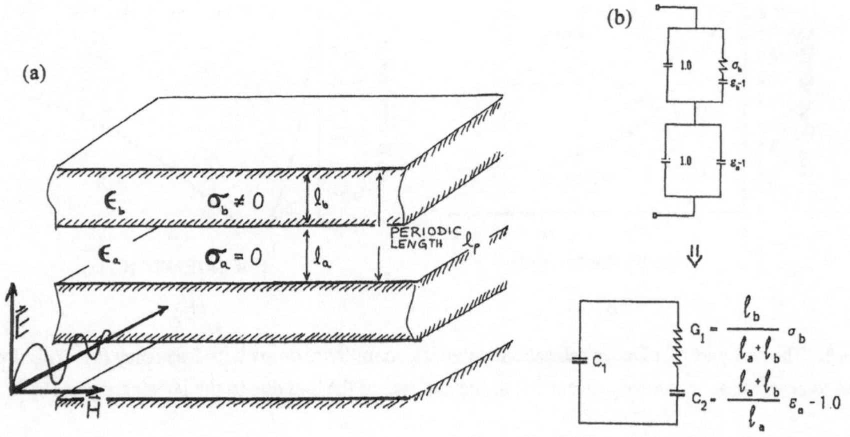

Figure 3. The layered slab composite material (a) offers an effective permittivity to plane waves that is equivalent to that of a Debye material (b).

Therefore, apart from the diamagnetic restriction on the conductivity and the obvious restriction that the periodic unit be smaller than a wavelength, the layered slab provides us with a means to construct an arbitrary Debye relaxation material. The relative thicknesses of the layers control the values of the two permittivity capacitors while the bulk conductivity of the conducting layer controls the conductivity parameter (and therefore the relaxation frequency.)

Solving the circuits of Figure 3b to obtain the Debye material of Figure 2, yields the ratio of thicknesses of the layers and the first initial guess at the required conductivity. Implementation of the layered slab in microstrip could then resemble either Figure 4a or Figure 4b. We call such a microstrip, a Debye microstrip*.

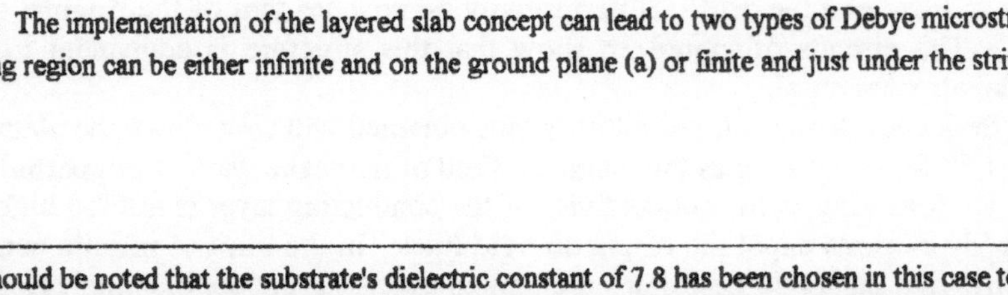

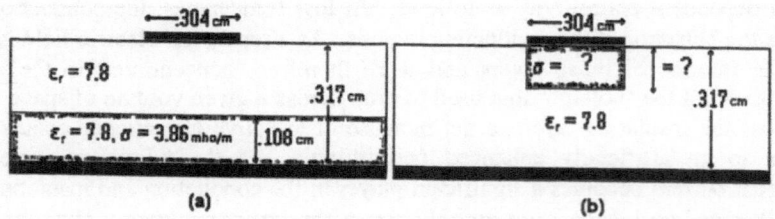

Figure 4. The implementation of the layered slab concept can lead to two types of Debye microstrip. The conducting region can be either infinite and on the ground plane (a) or finite and just under the strip (b).

---

* It should be noted that the substrate's dielectric constant of 7.8 has been chosen in this case to allow the low frequency limit of the computations to be compared with the work of previous authors (Reference 2). If instead, a base substrate dielectric constant of 11.7 had been chosen, the ratio of thicknesses would be the same as calculated here but the substrate's apparent permittivity would vary from approximately 20.0 at DC to 11.7 at high frequencies. In that case, we would be designing the microstrip's effective permittivity to remain at the frequency independent value of 11.7.

IV- ANALYSIS OF THE FIRST DEBYE MICROSTRIP

In this work we chose the configuration of Figure 4a for two reasons. First, placing the conducting layer directly on the ground plane allows it to interact with the microstrip fields that most closely resemble the original TEM plane wave assumption of the layered slab design. Second, assuming that the conducting region is infinite in the transverse direction allows the FDTD solution of the problem to be checked against a spectral domain computer code**. After this initial check, other configurations requiring the versatility of FDTD for their analysis (such as the one of Figure 4b) can be addressed with confidence. The configuration of Figure 4a is also simpler to optimize as there are only two parameters involved, the thickness of the conducting region and its conductivity. In the case of Figure 4b, by placing the region in contact with the strip, we expose it to the strong diverging edge-fields. As a result, if we wish to maintain full control of the frequency dependent permittivity, the conducting region must be allowed to be an arbitrarily shaped inhomogeneity. Its cross section becomes an additional, necessary, optimization parameter.

Even in the case of Figure 4a, since the microstrip field structure is not that of a plane wave, the final dimensions of the Debye microstrip are expected to differ from those derived from the layered slab approximation. However, the given values are close enough to the right answer to be useful as starting values for a fullwave solution of the problem. Iteration of the parameters thereafter by using either the time domain or the spectral domain computer codes leads to a conducting slab of conductivity 3.38 mhos/meter and a thickness of 0.128 cm.

Figure 5a shows the real part of the effective permittivity as calculated by the spectral domain method and the FDTD method. For comparison, the curve for the unmodified microstrip is also shown. As desired, the modified microstrip exhibits an essentially flat effective permittivity. It is of the order of 7.1 from DC through 20 GHz and thereafter it matches the unmodified microstrip's eventual rise to 7.8. The price paid for attaining this approximate frequency independence is the added loss due to the imaginary part of the permittivity of the Debye material. The imaginary part of the effective microstrip permittivity is shown in Figure 5b as obtained from both computational methods.

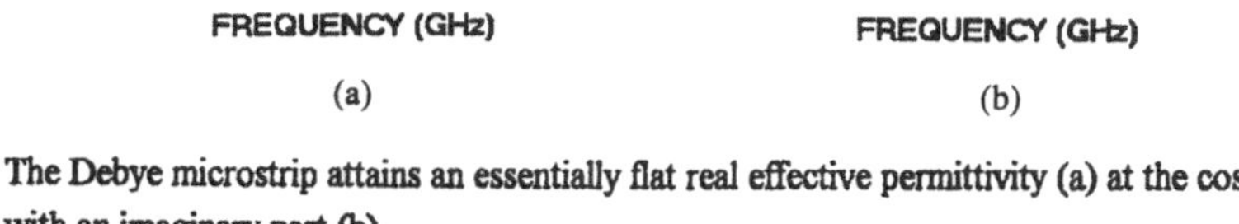

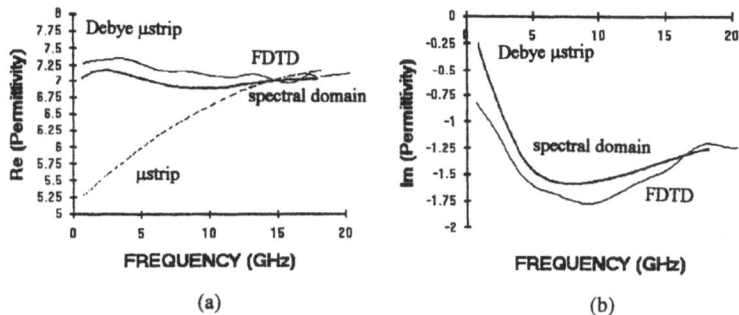

Figure 5. The Debye microstrip attains an essentially flat real effective permittivity (a) at the cost of the loss associated with an imaginary part (b).

Whereas conventional microstrip can significantly distort picosecond pulses, the Debye microstrip essentially preserves the shape of the pulse. The FDTD calculations

** The fullwave spectral domain program used in this problem was written by Dr. Nirod K. Das of the Polytechnic University in Brooklyn, NY. Dr. Das kindly provided his time and the use of his program and computing facilities for the spectral domain computations.

allow us to examine the evolution of a sharp gaussian pulse as it travels along either line. Figure 6a is the result for conventional microstrip. Figure 6b is the result for the Debye microstrip. As the Figure shows, the price paid for nearly distortionless transmission of the pulse is a drop in its amplitude.

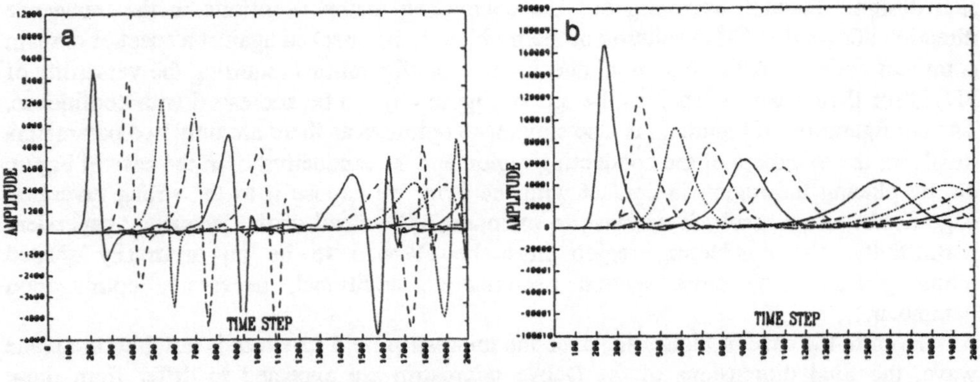

Figure 6. A sharp gaussian pulse gets significantly distorted in conventional microstrip (a). The Debye microstrip essentially preserves the pulse shape at the expense of some loss (b).

V- DISCUSSION

It has been shown that the electrodynamic dispersion of microstrip, which leads to the distortion of time domain pulses, can be undone using a dispersive substrate. A method for constructing such a substrate by imbedding a conducting inhomogeneity under the strip has been given, and its performance demonstrated by the use of fullwave computer codes. In particular, the FDTD solution shows that the resulting Debye microstrip essentially preserves the shape of a gaussian pulse at the expense of some loss.

The artificial increase in effective permittivity achieved at the low frequencies by the layered slab effect of the Debye microstrip is commonly referred to in the literature as a slow wave propagation mode. It is believed that the approach presented in this paper is general enough to encompass the design of such structures. The determination of the conducting region's thickness and conductivity by the equivalent Debye relaxation material circuit immediately gives design parameters that are good enough to start a fullwave optimization. Besides interesting pulse propagation properties, other applications can be envisioned for this new type of MMIC transmission line with modified frequency dependence. Combining the Debye microstrip with amplifying elements distributed along the transmission line, may result in net distortionless and lossless transmission of pulses for ultra-wideband antenna systems. Additionally, because the Debye microstrip has a higher effective permittivity at low frequencies than the unmodified microstrip, its propagation constant can match that of low order surface waves in the original substrate. Thus a Debye microstrip of the type suggested in Figure 4b (which only alters the substrate locally) can couple strongly to surface waves and surface wave devices constructed on the same substrate.

REFERENCES

1. R. E. Díaz, "The Analytic Continuation Method for the Analysis and Design of Dispersive materials," Ph.D. Dissertation, UCLA, Los Angeles (March 1992).
2. X. Zhang, J. Fang, K. K. Mei and Y. Liu, Calculations of the dispersive characteristics of microstrips by the Time-Domain Finite Difference method, *IEEE Trans. Microwave Theory Tech.* MTT-36:263 (1988)

MODELING OF SKIN EFFECT IN TLM

Peter Russer[1,2] and Bertram Isele[2]

[1] Ferdinand-Braun-Institut für Höchstfrequenztechnik
Rudower Chaussee 5
O-1199 Berlin, Fed. Rep. Germany

[2] Lehrstuhl für Hochfrequenztechnik
Technische Universität München
Arcisstrasse 21
8000 Munich 2, Fed. Rep. Germany

Introduction

The transmission line matrix (TLM) method [1,2,3] is an efficient method for time-domain analysis of electromagnetic fields. For the modeling of long interconnection lines within planar circuits, the conductor losses have to be considered. The discretization of thin conducting sheets in a usual TLM scheme is not applicable, because the number of nodes within the TLM mesh will increase extremely, when the penetration depths in the metallic sheet are small compared to the other geometric parameters such as line width and substrate height. On the other hand the use of large gridding factors to reduce the number of nodes will cause stability problems and the excitation of spurious modes will be inevitable. Therefore the thin conducting sheets have to be described by a seperate model enclosed in a TLM mesh modeling the environment.

A review of contributions concerned with the skin-effect is given in [6]. Time-domain models of the skin-effect have been investigated by various authors [4,5,7,8] and were applied mostly to transmission line interconnections. Lossy subregions can be modeled in TLM by connecting a lumped resistor or an infinitely long transmission line stub across each mesh node [9]. The use of lumped elements in TLM networks was first introduced by Johns et al. and has been applied to many different topics [10,11,12].

Principle of Skin Effect Surface Resistance Modeling

The investigation of structures with metallic boundaries requires a correct modeling of the skin-effect surface impedance taking also into account its $f^{\frac{1}{2}}$ dependence.

The skin effect surface impedance Z_A is given by

$$Z_A = R_A + jX_A = (1+j)\sqrt{\frac{\omega\mu}{2\sigma}} \tag{1}$$

where μ is the permeability and σ is the conductivity of the metal. The surface impedance Z_A is the impedance of a quadratic surface element of arbitrary dimension if the tangential field may be considered homogenious within the surface element. The skin effect penetration depth d is given by

$$d = \sqrt{\frac{2}{\omega\mu\sigma}} \tag{2}$$

A rectangular surface element with the side lengths Δu and Δv exhibits impedances Z_u and Z_v for the surface current components flowing in the $u-$ and $v-$direction given by

$$Z_u = \frac{\Delta u}{\Delta v} Z_A, \qquad Z_v = \frac{\Delta v}{\Delta u} Z_A \tag{3}$$

The skin effect surface impedance can be modeled with arbitrary accuracy by a lumped element ladder network consisting of cascaded sections according to Fig. (1). In the following we describe this model for a rectangular arbitrarily curved surface coordinate mesh. It is assumed that the surface area and its curvature are large compared with the skin effect penetration depths d. Under this assumption the electromagnetic field components inside the metallic conductor are parallel to the surface. Under each surface element cell the electromagnetic field is modeled by an attenuated plane electromagnetic wave propagating into the metal normally to the surface. This plane wave behavior allows to neglect the coupling of adjacent cells inside the metallic conductor. For each surface element the two impedances Z_u and Z_v have to be modeled.

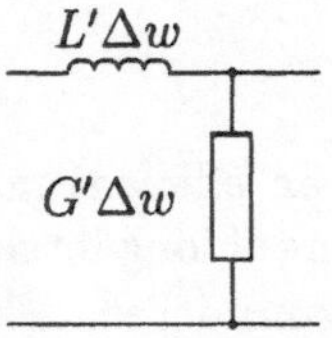

Figure 1. Equivalent circuit of a section inside the metal region.

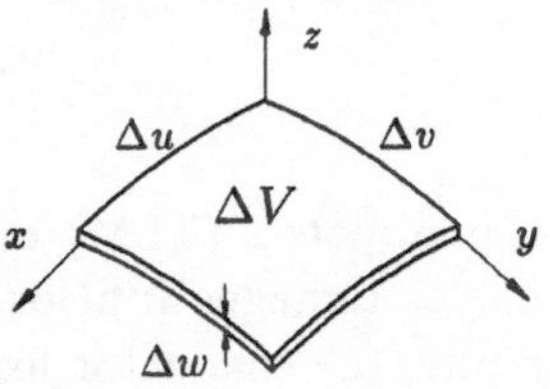

Figure 2. Volume element ΔV in the x–y plane

The conductor surface is subdivided into surface elements with side lengths Δu and Δv, respectively. The surface impedance elements Z_u and Z_v are now modeled by ladder networks. For this purpose in the w direction normal to the conductor surface a discretization with the interval length Δw is introduced. The length Δw shall be chosen small compared with the skin effect penetration depth d and may be therefore considerably smaller than Δu and Δv respectively. Modeling a volume element ΔV with the dimensions Δu, Δv, and Δw, the inductances $L'_u\Delta w$ and $L'_v\Delta w$ for the current components in $u-$ and $v-$direction are given by

$$L'_u\Delta w = \mu\frac{\Delta u \Delta w}{\Delta v}, \qquad L'_v\Delta w = \mu\frac{\Delta v \Delta w}{\Delta u} \tag{4}$$

and the parallel conductances $G'_u\Delta w$ and $G'_v\Delta w$ for both cases are given by

$$G'_u\Delta w = \sigma\frac{\Delta v \Delta w}{\Delta u}, \qquad G'_v\Delta w = \sigma\frac{\Delta u \Delta w}{\Delta v} \tag{5}$$

Cascading an infinite number of elements according to Fig. (1) yields the left side characteristic impedances

$$\tilde{Z}_u = \frac{\Delta u}{\Delta v} Z_A \sqrt{1 + 2j\left(\frac{\Delta w}{d}\right)^2}$$

$$\tilde{Z}_v = \frac{\Delta v}{\Delta u} Z_A \sqrt{1 + 2j\left(\frac{\Delta w}{d}\right)^2} \tag{6}$$

For $\Delta w \to 0$ we obtain $\tilde{Z}_{u,v} \to Z_{u,v}$. For $\Delta w \ll d_0$ we obtain from eq. (6)

$$\begin{aligned} \tilde{Z}_u &\doteq \frac{\Delta u}{\Delta v}\sqrt{\frac{\omega\mu}{2\sigma}}\left[1 + j - (1-j)\left(\frac{\Delta w}{d}\right)^2\right] \\ \tilde{Z}_v &\doteq \frac{\Delta v}{\Delta u}\sqrt{\frac{\omega\mu}{2\sigma}}\left[1 + j - (1-j)\left(\frac{\Delta w}{d}\right)^2\right] \end{aligned} \tag{7}$$

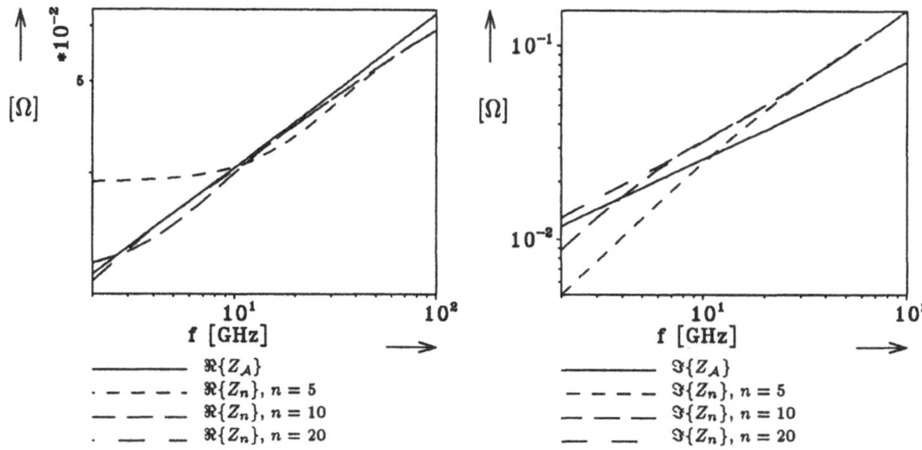

Figure 3. Surface impedance of cell with $\Delta w = d(200GHz)$

For simplicity, we only investigate cells with a quadratic shape, i.e. $\Delta v/\Delta u = 1$ in the following discussion. The complex surface impedance Z_A modelled by a ladder network with a finite number of L-R – sections is given recursively by

$$Z_i = \omega L'\Delta w + \frac{1}{G'\Delta w + \frac{1}{Z_{i-1}}} \tag{8}$$

with i ranging from 1 to n. We choose Δw to be the penetration depth d_0 at a sufficiently high frequency f_0. Therefore the ladder network, containing n sections represents a wide-band model of the skin effect. The accuracy of the model at the lower frequency bound depends on the penetration depth d_0 and the number of sections. Fig. 3 and 4 compares the frequency dependences of the real and imaginary part of surface impedances for different assumed penetration depths d_0 with results given by eq. (1). As shown, the quality of simulation will increase with the number n of L-R – sections.

Discrete State Equation Modeling of the Skin Effect

The skin-effect ladder model is included into a TLM mesh to describe a thin conducting sheet, positioned in the middle between two layers of TLM nodes. To retain symmetry inside the metal region we use a ladder network consisting of n lumped resistors connected in parallel and $n + 1$ lumped inductances connected in series.

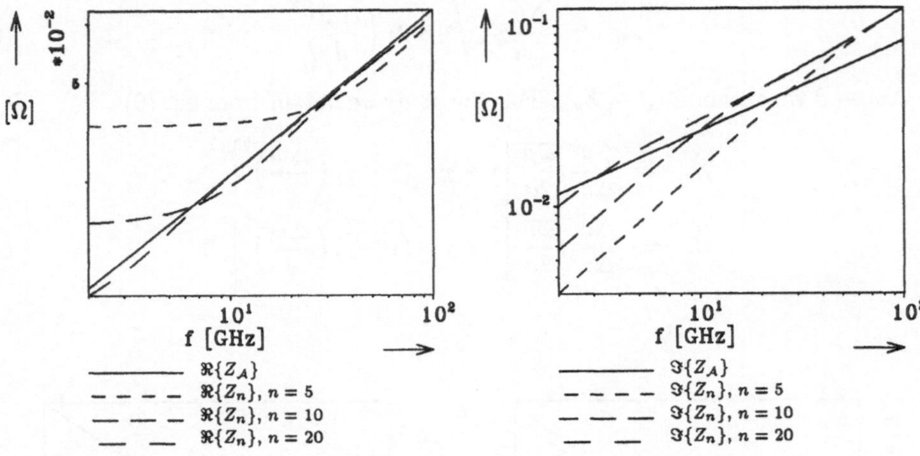

Figure 4. Surface impedance of cell with $\Delta w = d(500GHz)$

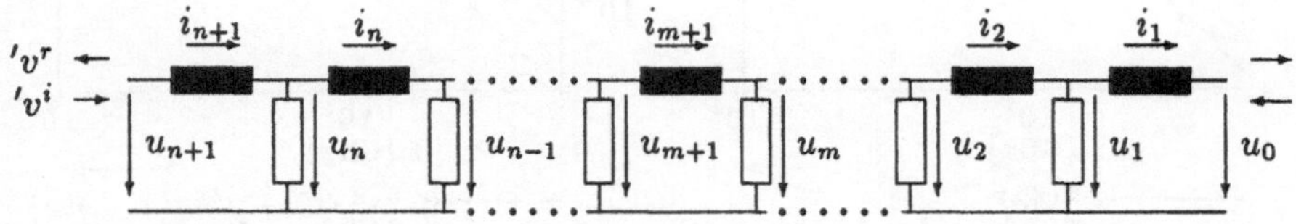

Figure 5. *L-R* – ladder network for modeling skin effect

The ladder network is described by a set of state equations. The *L-R* – sections connected to the TLM-network, are described by the following discrete time state equation

$$\frac{d}{dt}i_{n+1} \quad = \quad -\left(\frac{1}{LG} + \frac{1}{LY}\right) i_{n+1} + \frac{1}{LG}i_n + \frac{2}{L}v^i \tag{9}$$

while the state equations at the inner *L-R* – sections results in

$$\frac{d}{dt}i_{n+1} = \frac{1}{LG}i_{n+2} - \frac{2}{LG}i_{n+1} + \frac{1}{LG}i_n \tag{10}$$

where v^i is the incident voltage wave pulse on the ladder and Y is the characteristic line admittance of the interfaced TLM network. The output equation at the end of the ladder is given by

$$v^r = v^i + \frac{1}{Y}i \tag{11}$$

where i is the current through the adjacent inductance. In the simplest case all sections of the ladder network are identical. Therefore the entire ladder network is described by the following set of state equations

$$\frac{d}{dt}\begin{bmatrix} i_1 \\ i_2 \\ \cdot \\ \cdot \\ i_n \\ i_{n+1} \end{bmatrix} = \begin{bmatrix} -\left(\frac{1}{LY}+\frac{1}{LG}\right) & \frac{1}{LG} & & & & 0 \\ \frac{1}{LG} & \frac{-2}{LG} & \frac{1}{LG} & & & \\ & \cdots & \cdots & \cdots & & \\ & \cdots & \cdots & \cdots & & \\ & & & \frac{1}{LG} & \frac{-2}{LG} & \frac{1}{LG} \\ 0 & & & & \frac{1}{LG} & -\left(\frac{1}{LG}+\frac{1}{LY}\right) \end{bmatrix} \begin{bmatrix} i_1 \\ i_2 \\ \cdot \\ \cdot \\ i_n \\ i_{n+1} \end{bmatrix} + \begin{bmatrix} -\frac{2}{L}{}''v^i \\ 0 \\ 0 \\ 0 \\ 0 \\ \frac{2}{L}{}'v^i \end{bmatrix} \tag{12}$$

with the incident voltage wave pulses ${}'v^i$ and ${}''v^i$ at both ends of the ladder as shown in Fig. 5. These set of ordinary differential equations

$$\frac{d}{dt}\mathbf{I} = \mathbf{AI} + \mathbf{BV}^i \tag{13}$$

is discretized by the implicit Euler scheme (backward Euler scheme)

$$(1 - \Delta t\mathbf{A})_{k+1}\mathbf{I} =_k \mathbf{I} + \Delta t\mathbf{B}\,_k\mathbf{V}^i. \tag{14}$$

Due to the band structure of the matrices, the set of time discrete state equations is solved by tridiagonal matrix algorithms using a minimal number of operations. The equations have to solved for both polarisations seperately. Therefore the algorithm need $N = 2\,n\,n_x\,n_y$ additional memory locations for the storage of all state variables, where n is the number of L-R – sections within a surface cell, and $n_x n_y$ is the number of cells within the metallic sheet, respectively.

Numerical Example

We choose a triplate line as a simple test configuration for demonstrating the feasibility of our skin effect model. Triplate lines have symmetric cross sections and therefore the dispersion is only caused by the loss mechanism in the metal conductor. The TLM model of the line was extended with a periodic boundery condition to simulate the propagation of a Gaussian impulse a very long triplate line with the total length L_t, exceeding the length of the TLM mesh L by a factor n. The outgoing Gaussian impulse at a plane at the position x_1 of the TLM mesh is injected into a plane at the position x_n in the direction of the propagation within the mesh and vice versa.

$$_{k+1}{}''\mathbf{V}^i|_{x_1=const} = {}_k{}'\mathbf{V}^r|_{x_n=const} \tag{15}$$

$$_{k+1}{}'\mathbf{V}^r|_{x_n=const} = {}_k{}''\mathbf{V}^i|_{x_1=const} \tag{16}$$

If the initial width of the Gaussian impulse is choosen small compared to the length of the mesh L, the remaining amplitudes from former passes are negligible. The length L of the single triplate line is $16\,mm$ and the height h is $1.2\,mm$. The ratio h/w of the triplate line is 12/7, corresponding to a line impedance of 95 Ohms. We assume the ground and cover plane to be ideal conducting walls. The inner conductor is modeled by a ladder network consisting of 20 sections in normal direction.

Fig. 6 shows the spreading of a Gaussian pulse traveling along the triplate line and the dispersion properties are depicted. The results are compared with a ideal model of the triplate line using a electric wall representing the inner conductor in Fig. 7 In the case of the ideal conductor the amplitude and the waveform remains unchanged. In the case of a real conductor the pulses are attenuated and the waveform gets distorted after several passes.

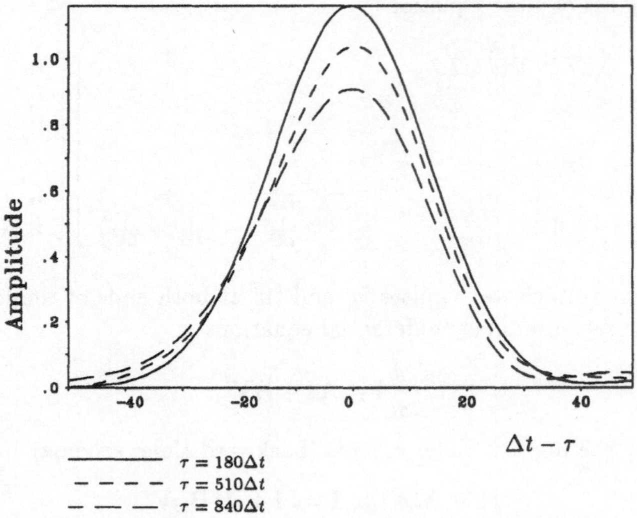

Figure 6. Spreading of Gaussian pulse after a delay τ

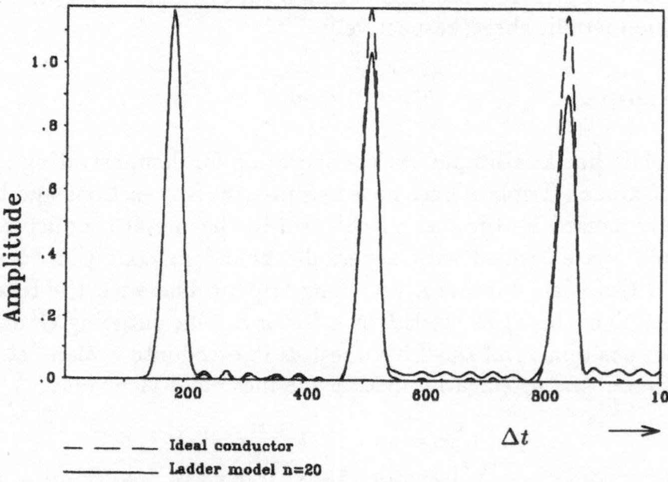

Figure 7. Propagation of Gaussian impulse traveling in a periodically extended triplate line

Conclusion

A new time domain model, describing the skin-effect in thin metallic sheets, has been introduced. The skin effect is modeled by a ladder network of series inductances and conductances. We have introduced a state variable representation for the modeling of thin dispersive metallic sheets. The model is embedded into a three-dimensional TLM algorithm and is applicable to metallic sheets of arbitrary shape and thickness, whereas the requirements on additional storage and computing time are moderate.

Acknowledgements

This work was supported by the Deutsche Forschungsgemeinschaft (DFG), Bonn.

References

1. P.B. Johns, R.L. Beurle, "Numerical Solution of 2-Dimensional Scattering Problems using a Transmission-Line Matrix," *Proc. IEE*, vol.118, no. 9, pp 1203-1208, Sept. 1971.

2. W.J.R. Hoefer, "The Transmission Line Matrix Method-Theory and Applications," *IEEE Trans. Microwave Theory Tech.*, vol. MTT-33, no.10,pp.882-893, Oct 1985.

3. W.J.R. Hoefer, "The Transmission Line Matrix (TLM) Method," Chapter 8 in "Numerical Techniques for Microwave and Millimeter Wave Passive Structures," edited by T. Itoh, New York, 1989, John Wiley & Sons, pp. 496-591.

4. A.J. Gruodis, C. W. Ho, E. F. Miersch and A.E. Ruehli, " Delay line approach for analyzing lossy transmission lines," *IBM Technical Disclosure Bulletin,* Vol.19, pp.2366–2368, 1976.

5. P.A. Brennan and A.E. Ruehli, " Time-domain skin-effect using resitors and lossless transmission lines," *IBM Technical Disclosure Bulletin,* Vol.21, pp.2362–2363, 1978.

6. L.-T. Hwang, I. Turlik, "A Review of the Skin-Effect as Applied to Thin Film Interconnections," *IEEE, Trans. Comp. Hybrids Manuf. Technol.*, Vol.15, No.1, pp.43–55, Feb. 1992.

7. C.-S. Yen, Z. Fazarinc, R.L. Wheeler, Time-Domain Skin-Effect Model for Transient Analysis of Lossy Transmission Lines," *Proc. IEEE* , Vol.70, No.7, pp.750–757, July 1982.

8. C.-C. Wang, Y.-F. Chan, " Transient Response of Lossy Transmission Lines with Arbitrary Loads," *Intl. J. of Numerical Modelling*, Vol.5, No.2, pp.111–120, May 1992.

9. S. Akhtarzad, "Analysis of Lossy Microstrip Structures and Microstrip Resonators by the TLM Method," Ph.D dissertation, University of Nottingham, England, July 1975.

10. P.B. Johns, M. O'Brien, "Use of the Transmission Line Modelling (t.l.m.) Method to Solve Nonlinear Lumped Networks", *The Radio and Electronic Engineer*, Vol.50, No.1/2, pp.59–70, Jan./Febr. 1980.

11. C.R. Brewitt-Taylor, P.B. Johns, "On the Construction and Numerical Solution of Transmission-Line and Lumped-network models of Maxwell's Equations." *Int. J. Numer. Methods Eng.*, Vol.15, pp.13-30, 1980.

12. P. Russer, P.P.M. So, W.J.R. Hoefer "Modeling of Nonlinear Active Regions in TLM," *IEEE Microwave and Guided Wave Letters*, Vol.1, No.1, pp.10–13, January 1991.

ON THE PULSE RADIATION FROM A LOADED MICROSTRIP LINE USING A SPACE-TIME DYADIC GREEN'S FUNCTION APPROACH

Renato Cicchetti

Department of Electronic Engineering
University of Rome "la Sapienza"
Via Eudossiana 18, 00184 Rome, Italy

ABSTRACT-A space-time analysis of the field radiated from microstrip lines is presented. The analysis is based on the dyadic Green's function method. The dyadic Green's function is derived by means of an appropriate field decomposition in TE and TM waves. The transient signals excited on the line and their radiated fields are computed by using IFFT numerical technique. The radiation mechanism, as well as the influence of the structure's electrical and geometrical parameters on the radiation phenomenon are analyzed. Indications to reduce the amplitude of the radiated field, or its distortion as well as the distortions of the transient signals excited on the line, are given.

I. INTRODUCTION

The emerging technologies make now possible the generation, radiation and detection of short-pulse electromagnetic signals. Short signals have been proposed for radar target identifications, high-speed digital and radio communications [1,2]. As it is well known these fields have broad frequency spectrum produced by fast time-varying signals (some picoseconds). Circuits particularly suited to operate with such signals are those based on microstrip technology. These circuits are of particular importance for the mentioned applications because they can guarantee the appropriate bandwidth.

Presently numerous methods based on a frequency domain full-wave theory have been developed to determine the electromagnetic behavior of microstrip circuits [3,4,5]. Only recently models for analyzing the radiated transient electromagnetic field from dipoles [6] and microstrip lines [7] have been presented. In [6] the field has been computed solving the Maxwell's equation in the spectral domain, while in [7] using the space-time electric dyadic Green's function method.

In this paper, the model presented in [7] is extended in order to analyze the transient response of a loaded microstrip line excited with short pulse signals. For this purpose, the transient far-field is expressed as a function of the dipole moment of the surface current excited by the generator on the metal strip. The time behavior of the line current, of the dipole moment, and of the radiated field are then compared in order to establish the influence played by the electrical and geometrical parameters of the planar structure on the line response.

II. THEORY

Fig. 1 shows a lossless microstrip transmission line, of width w and length ℓ, printed on a grounded dielectric slab of thickness d and permittivity ε_r.

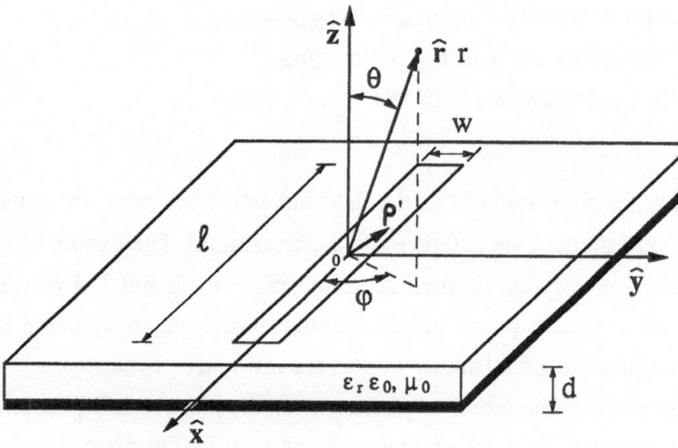

Fig. 1 Microstrip transmission line printed at the air-dielectric interface.

The field excited in the structure has been evaluated by means of the space-time electric dyadic Green's function. This function has been obtained by solving the inhomogeneous dyadic wave equation using an appropriate TE and TM (with respect to the normal at the interface) field decomposition in the space domain. It is shown that these fields form a complete set in terms of which an arbitrary field can be represented. Using this approach the field equations are scalarized and the dyadic field is expressed by means of an appropriate set of integro-differential operators that act on scalar functions (one for TE, and two for TM waves) [8]. The far-field is then derived using the saddle point technique, or the image principle, as suggested in [7]. This approximation correspond to the "high-frequency" regime which represents a valid approximation (asymptotic) of the field when the distance is large compared to the local wavelength of the radiated field, and the observation point is not near the dielectric interface. Finally, the resulting field in the time domain is obtained using one-dimensional Fourier inverse transform of the corresponding spectral field densities. The resulting field of a narrow strip line oriented along the x-axis is given by

$$e_\theta(\mathrm{r},\tau) \sim -\frac{\mu_0}{2\pi}\frac{1}{\mathrm{r}}\frac{\sqrt{\varepsilon_r - \sin^2\theta}\,\cos\theta}{\varepsilon_r\cos\theta + \sqrt{\varepsilon_r - \sin^2\theta}}\cos\phi\,.$$

$$\cdot\left(\sum_{n=0}^{\infty}(-\Gamma_v(\varepsilon_r,\theta))^n\left\{\frac{\partial}{\partial\tau}m_x[\tau - 2n\,\tau_\theta] + \right.\right.$$

$$\left.\left. - \frac{\partial}{\partial\tau}m_x[\tau - 2\,(1+n)\,\tau_\theta]\right\}\right) \tag{1}$$

$$e_\phi(\mathrm{r},\tau) \sim \frac{\mu_0}{2\pi}\frac{1}{\mathrm{r}}\frac{\cos\theta}{\cos\theta + \sqrt{\varepsilon_r - \sin^2\theta}}\sin\phi\,.$$

$$\cdot\left(\sum_{n=0}^{\infty}\Gamma_u^n(\varepsilon_r,\theta)\;\frac{\partial}{\partial\tau}m_x[\tau - 2n\,\tau_\theta] + \right.$$

$$\left. - \frac{\partial}{\partial\tau}m_x[\tau - 2\,(1+n)\,\tau_\theta]\right\}\Bigg) \tag{2}$$

where $\Gamma_v(\varepsilon_r, \theta)$, and $\Gamma_u(\varepsilon_r, \theta)$ are the Fresnel reflection coefficients, τ is the delayed time, τ_θ represents the half delay time between two consecutive reflections of the electromagnetic field radiated from the line at the air-dielectric interface. In (1)-(2) $m_x(\cdot)$ represents the dipole moment of the surface current dominant mode $j_x(\cdot)$ excited on the metal strip

$$m_x(\theta,\phi,\tau) = \int_{S'} j_x(\rho', \tau + \frac{\hat{\mathrm{r}}\cdot\rho'}{c})\,dS' \tag{3}$$

where S' is the domain of the metal strip, $\rho' = x'\,\hat{x} + y'\,\hat{y}$ is the current source vector, $\hat{r}$ is the observation unit-vector, and c is the speed velocity in vacuum. Eqs. (1)-(2), and (3) state that the field produced by the line is a superposition of the elementary field generated by the surface current elements excited on the strip.

To evaluate (3) the Maxwell distribution for the longitudinal x-component of the surface current is used. In fact, as shown in [8], the longitudinal current weakly depends on frequency and its shape is accurately described by the Maxwell distribution. So we can write

$$j_x(x,y,t) = \frac{2}{\pi w}\frac{i\,(x,t)}{\sqrt{1-\left(\frac{2y}{w}\right)^2}} \tag{4}$$

where $i\,(x, t)$ is the current in the microstrip line. The expression of the current $i\,(x, t)$ is obtained by means of an inverse Fourier transform of the corresponding spectral current.

This current is calculated solving, with the appropriate boundary condition at both ends, the distributed equivalent circuit of the quasi-TEM mode.

Since the response of the line is to be analyzed over a wide frequency band the influence of the dispersive hybrid mode of propagation cannot be neglected and the frequency dependent effective microstrip permittivity $\varepsilon_{eff}(f)$ should be taken into account. In this case, to obtain the surface current, the dipole moment, and the response of the line, it is necessary to perform a numeric inverse transform of the relative spectral expressions.

With regard $\varepsilon_{eff}(f)$ a dependence of the Getsinger type is assumed

$$\varepsilon_{eff}(f) = \varepsilon_r - \frac{\varepsilon_r - \varepsilon_{eff}(0)}{1 + \frac{f^2}{f_p^2}} \tag{5}$$

where f_p (inflection frequency), depends on the static characteristic impedance of the microstrip line, on the thickness d, and on the relative permittivity of the substrate [9].

As regards the characteristic impedance we assume the expression

$$Z_c(f) = Z_c \sqrt{\varepsilon_{eff}(f) / \varepsilon_{eff}(0)} \tag{6}$$

Eq. 6, which gives the appropriated dependance of the characteristic impedance on the frequency (V/I definition), has been obtained by the author [8] using a rigorous asymptotic development of its exact electrodynamic expression based on the dyadic Green's function method.

The propagation of the current signal $i(x, t)$ along the line is of great importance to understand the radiation mechanism. Unfortunately a closed-form analytical expression of the line current is not available. However an asymptotic expression of the current near the formation of the wavefront is obtained by expanding the spectral current density in the limit of $\omega \to \infty$, and then by applying the inverse Fourier transform. Using this technique it is easy to verify that at a given observation point x the current vanishes identically before the arrival of the initial response (wavefront) that propagates from the source generator with finite speed. The impulse response of the current near the formation of the wavefront for a matched line excited at $z = -\ell/2$ is

$$i(x,t) \sim \left\{ \delta\left(t - \frac{x+\ell/2}{v_\infty}\right) + \right.$$

$$\left. \frac{\partial}{\partial t} J_0\left[\omega_p \sqrt{\frac{\varepsilon_r - \varepsilon_{eff}(0)}{\varepsilon_r}} \sqrt{t^2 - \left(\frac{x+\ell/2}{v_\infty}\right)^2}\,\right] u\left(t - \frac{x+\ell/2}{v_\infty}\right) \right\} / \{2 Z_{c\infty}\} \tag{7}$$

where $v_\infty = c/\sqrt{\varepsilon_r}$, $Z_{c\infty} = Z_c \sqrt{\varepsilon_r}/\sqrt{\varepsilon_{eff}(0)}$, $\omega_p = 2\pi f_p$, $\delta(.)$ is the Dirac delta distribution, $J_0(.)$ is the Bessel function of order zero, and $u(\cdot)$ is the Heaviside unit step function.

Eq. (7) shows that very near the wavefront, the relevant frequencies are so high that dispersion is not effective. Dispersion becomes noticeable when the argument of the Bessel function assumes moderately large value, and the field behaves like a wave packet [10]. So, the velocity of the wavefront is independent of the geometrical line's parameters and depends only on the substrate permittivity and permeability.

III. NUMERICAL RESULTS

The dipole moment and the transient far-field, excited by the voltage generator, are compared in this section in order to analyze the different rule played by the structure's parameters on the transient line response. The parameters of the considered structure are d=1.57 mm, ℓ = 10 cm, w=1.3 mm, and $\varepsilon_r = 10$. To the mentioned parameters corresponds a characteristic impedance $Z_c(0)$ of 50 Ω, and an effective permittivity $\varepsilon_{eff}(0)$ of 6.66. The internal impedance of the voltage generator (at $z = -\ell/2$) and the line's load (at $z = \ell/2$) are assumed to be of 50 Ω. The voltage generator considered is of Gaussian type pulse

$$V_g(t) = V_0 e^{-(t-t_0)^2/t_s^2} \tag{8}$$

with $V_0 = 1$ V, $t_s = 30$ ps, and $t_0 = 0.1$ ns. The numerical results are obtained using the IFFT numerical technique (1024 points).

Fig. 2 reproduces the dipole moment for $\theta = 45°$, and $\phi = 0°$ associated with the current excited on the line by the voltage generator. We notice a distortion of the signal which is produced by the dispersive feature of the hybrid mode of propagation. Dispersive characteristic of the hybrid mode (quasi-TEM) has been considered in [6] due to a suitably combination of the TEM field with the lateral-wave field. These fields have different phase velocity. Therefore the fields combination produces dispersion. Finally, in Fig. 3 the electric field transient response (θ-component) is presented.

Comparing Figs. 2 and 3 it is easy to observe the effects of the dispersion and reflection of the current on the line, and of the radiated field at the air-dielectric interface (see Eqs. (1), and (2)), respectively. In particular the distortion increases as the decay time of the Gaussian pulse decreases, or when a wider strip is used. In fact, wider strip or substrate with high relative permittivity reduce the inflection frequency, and consequently, an increment of dispersion distortions appears. Moreover, we notice that the energy emission can be considered spitted into two different phases of emission. The first emission happens when the microstrip line is excited by the generator at the first end; the second one takes place when the propagating signal excites the load at the second end (the wavefront propagate from the generator to the load at about 1.054 ns later). In all the mentioned cases to the primary emission corresponds a secondary one which is produced by the multiple reflections of the field at the air-dielectric interface.

Taking into consideration that every line have to be designed according to specific design constrains (low-distortion of the guiding signals, low level of the radiated field, etc.), we can conclude that only an accurate choice of the microstrip's parameters (which have to be designed as function of the specific application) is necessary in order to satisfy specific design constrains.

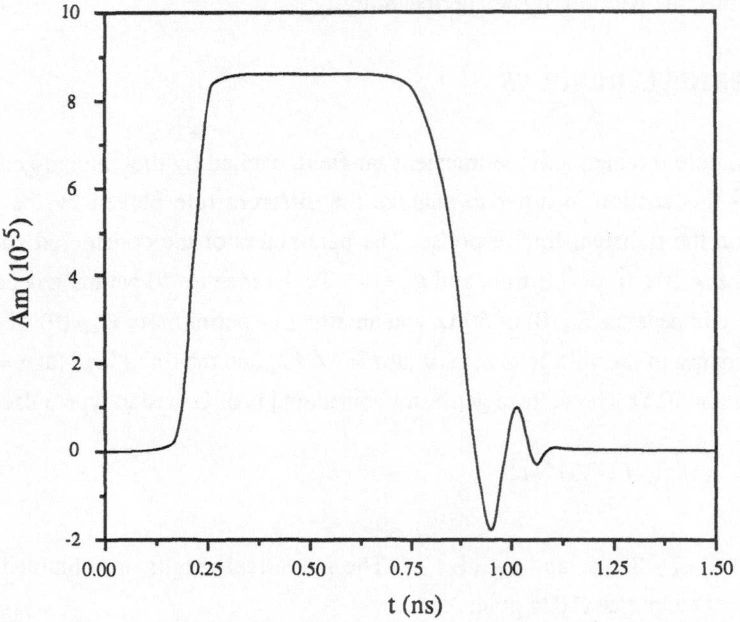

Fig. 2 Dipole moment (d = 1.57 mm, ℓ = 10 cm, w = 1,3 mm, ε_r = 10, θ = 45°, and ϕ = 0°). The ripples that appear at t ≅ 1 ns are those produced by the dispersive propagation of the hybrid mode in the microstrip line.

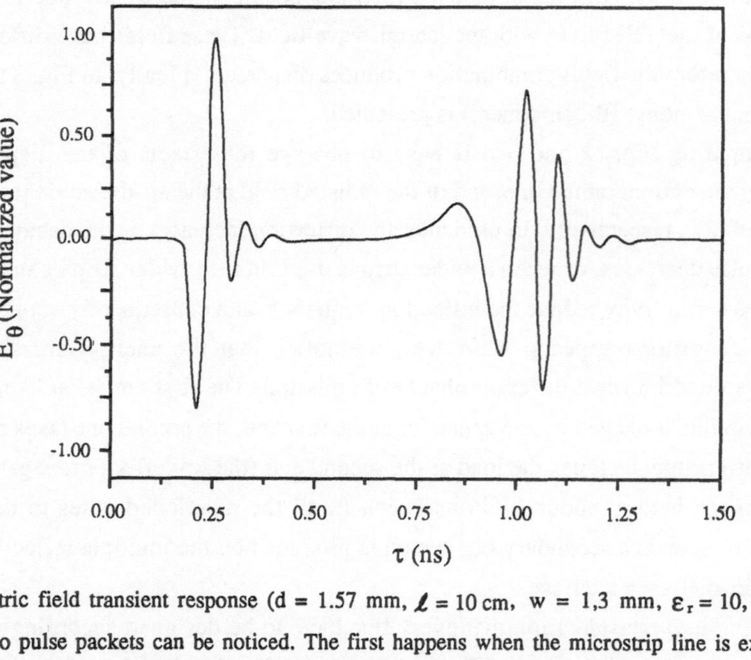

Fig. 3 Electric field transient response (d = 1.57 mm, ℓ = 10 cm, w = 1,3 mm, ε_r = 10, θ = 45°, and ϕ = 0°). Two pulse packets can be noticed. The first happens when the microstrip line is excited by the generator; the second takes place when the propagating signal excites the load at the second end.

VI. CONCLUSION

An analysis of the transient response of a loaded microstrip line excited by rapidly time varying signal has been presented. The model of analysis is based on the space-time dyadic Green's function method. The transient signals excited on the line and their radiated fields are computed by using IFFT numerical technique. The radiation mechanism, as well as the influence of the structure's electrical and geometrical parameters influencing the radiation phenomenon have been considered. Particularly, it can be shown that the response is particularly influenced by the line length, the thickness and the permittivity of the dielectric substrate. It is therefore evident that an accurate choice of these parameters is necessary in order to reduce the amplitude of the radiated field, or its distortion as well as the distortions of the transient signals excited on the microstrip line.

AKNOWLEDGEMENT

This paper is dedicate to Profs. L. B. Felsen, N. G. Alexopoulos, and T. Itoh whose researches have been used in the theoretical and analytical developments of this paper. The author is also grateful to Mr. Maurizio Fascetti for his assistance in the presentation of diagrams and figures.

REFERENCES

1. T. Leung, and C. Balanis, "Attenuation Distortion of Transient Signals in Microstrip, " *IEEE Trans. Microwave Theory Tech..* 36:765 (1988).
2. L. B. Felsen, "Very Short Pulse Scattering: Time Osservable-Based Parametrization," Proceedings of the "*Second International Conference on Electromagnetics in Aerospace Applications*", pp. 221, Turin (1991).
3. N. G. Alexopoulos, "Integrated-Circuit Structures on Anisotropic Substrates,"*IEEE Trans. Microwave Theory Tech..* 33:847 (1985).
4. R. H. Jansen, "The Spectral-Domain Approach for Microwave Integrated Circuits," *IEEE Trans. Microwave Theory Tech..* 33:1043 (1985).
5. T. Itoh, "Numerical Techniques for Microwave and Millimeter-Wave Passive Structures," New York, Chichester, Brisbane, Toronto, Singapore: John Wiley & Sons (1989).
6. R. W. King, "Lateral Electromagnetic Waves and Pulses on Open Microstrip," " *IEEE Trans. Microwave Theory Tech..* 38:38 (1990).
7. R. Cicchetti, "Transient Analysis of Radiated Field from Electric Dipoles and Microstrip Lines," *IEEE Trans. Antennas Propagat.* 39:910 (1991).
8. P. Bernardi, and R. Cicchetti, "Dyadic Green's Function for Conductor-Backed Layered Structures Excited by Arbitrary Sources," submitted for publication on *IEEE Trans. Microwave Theory Tech..*
9. W. J. Getsinger, "Microstrip dispersion model, " *IEEE Trans. Microwave Theory Tech..* 21:34 (1973).
10. L. B. Felsen, "Transient Electromagnetic Fields," in *Topics in Applied Physic*, vol 10, Berlin, Heidelberg, New York: Springer-Verlag (1976).

SCATTERING THEORY AND COMPUTATION

PHASE SPACE ISSUES IN ULTRAWIDEBAND/SHORT PULSE WAVE MODELING

Leopold B. Felsen

Department of Electrical Engineering
Polytechnic University
Six MetroTech Center
Brooklyn, NY 11201

I. INTRODUCTION

Continuing advances in the development of radiators and receivers for electromagnetic, acoustic, elastic, etc., waves, have provided the capability for reliable data gathering. Recent trends have been toward ever wider signal bandwidths because of the enlarged data base provided thereby. This has led to the serious consideration of ultrawideband/short pulse systems with high temporal-spatial resolution. To extract from received signals in the data the desired information for missions pertaining to communication, target location and identification, remote sensing, and others, one requires data processing methods that link the features (observables) in the signal to features in the environment encountered by the signal during its travel from source to receiver. The writer has termed such methods Observable-Based Parametrizations (OBP) (Fig. 1). The presentation here raises in outline relevant issues pertaining to the systematic construction of predictive OBP algorithms for forward and inverse modeling of time-dependent wave phenomena in complex environments.

The synthesizing basis elements in an OBP are "physically correct" localized wave objects in the configurational (space-time) and the spectral (wavenumber-frequency) domains, i. e. in the 8-dimensional $(x, y, z, t) - (k_x, k_y, k_z, \omega)$ phase space, where (x, y, z) are spatial coordinates, (k_x, k_y, k_z) are the corresponding spatial spectral wavenumbers, and t, ω represent time and frequency. This phase space with its various subdomains furnishes the blueprint for tracking individual OBP wave objects and combining them appropriately in a global OBP for the overall problem. The subdomains also systematize pre-processing (input signal shaping) and post-processing (output signal filtering) options. OBP rules for complex environments are learned from the OBP analysis of canonical, rigorously solvable test problems that contain those localized individual scattering mechanisms which appear interactively in the conglomerate. The strategy and methodology has been discussed in previous publications [1, 2], and OBP algorithms have been constructed and validated for a variety of test problems (see list of references in [1] and [2]). In their most versatile form, the algorithms are constructed with self-consistent combinations of ray fields, beam fields, trapped and leaky guided mode fields, resonant mode fields, Floquet modes in periodic environments, etc., both in the frequency and time domains. In the time domain (TD), the direct treatment of dispersion has recently enlarged the OBP knowledge base [3-7].

II. PHASE SPACE ARCHITECTURE

Wave-oriented data processing requires merging of two disciplines which have traditionally been pursued separately: wave radiation, propagation and diffraction on the one hand, and signal processing on the other (Fig. 1). Furthermore, it is possible to phrase

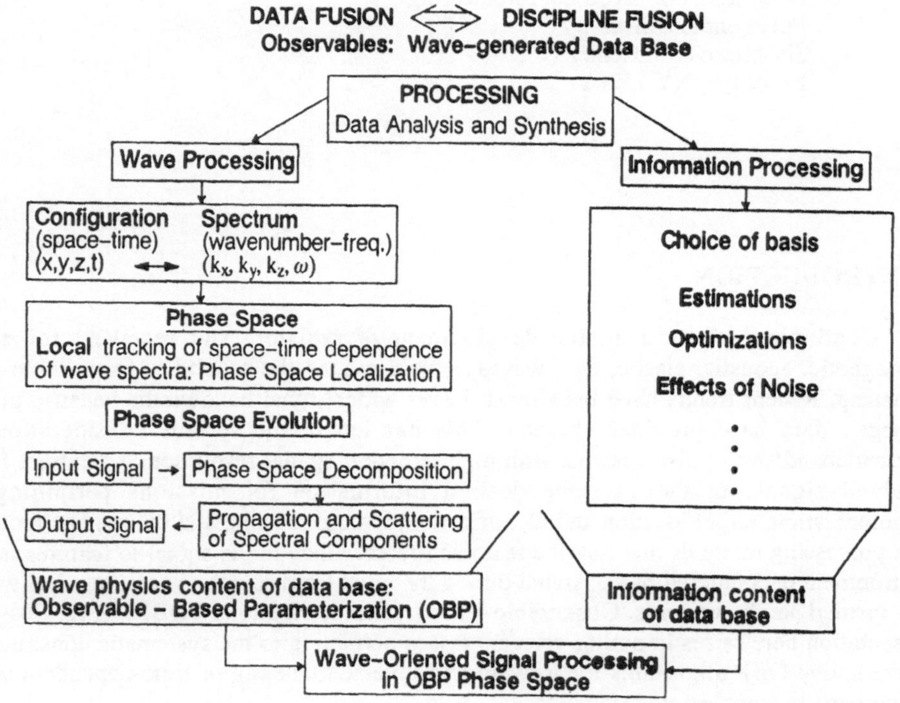

Figure 1

the phase space manipulations in a system format which is often better matched to the user community. Traditional signal processing has been confined to the time-frequency (t, ω) domain, with recent advances emphasizing alternative windowing schemes, (t, ω) transforms which enhance various measures of the time signal, and general multiresolution transforms for improved resolution and efficiency [8, 9]. These schemes can, in principle, also be applied to the spatial observables, for example, to the (x, p_x) subdomain and, more generally, to hybrid subdomains like (t, p_x) (Fig 2). However, to qualify as good OBP's for

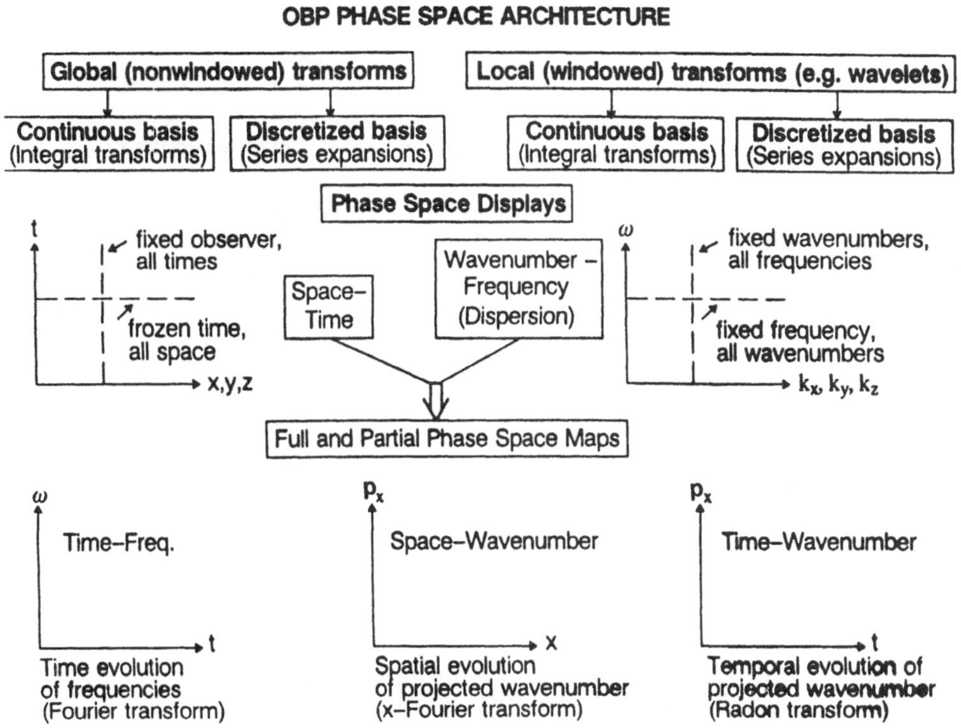

Figure 2

wave phenomena, the transform basis functions and windows should define good wave propagators which remain compact (localized) during their forward trek (direct problem) or backward trek (inverse problem) through the phase space. This favors transforms with plane-wave (i.e. Fourier) type kernels and smooth wavelet windows. Good candidates for OBP window functions are exemplified by complex-argument time-harmonic and time-dependent delta function sources; these analytic wavelets generate Gaussian-like beams (GB) and pulsed beams (PB), respectively, which are smooth collimated propagators without strong sidelobes [10]. Moreover, these basis elements form complete sets for modeling arbitrary space-time fields [11]. Although OBP adapted basis and window functions of this type may appear unsophisticated to the signal processor, they play a very useful role in wave-oriented processing.

Access to the phase space subdomains is gained by applying to the space-time data appropriate discrete or continuous, nonwindowed or windowed transforms (Fig. 2). For TD wave processing which is emphasized here, these subdomains are (x, t), (y, t), (z, t),(p_x, t), $(p_{x,y,z}, t)$ parametrizations, where $p_{x,y,z} = k_{x,y,z}/\omega$ is the normalized wavenumber (reciprocal of phase speed) and c is the frequency-independent propagation speed, are derived from (x, y, z, t) data via Radon transforms, (x, y, z, t) plots define space-time rays, (ω, t) plots track the instantaneous frequency content of the signal reaching a fixed observer at (x, y, z), etc.

The defining wave equation in nondispersive media imposes the constraint $|p| = 1/c$. Dispersion, as formalized by the relevant dispersion equation $D(\underset{\sim}{k}, \omega) = 0$, $\underset{\sim}{k} = (k_x, k_y, k_z)$, generalizes this constraint away from $k = \omega/c$ to frequency-dependent wave propagation speeds. For dispersive materials,

$\underset{\sim}{k} = k(\omega)$ or $\omega = \omega(\underset{\sim}{k})$ relates frequency ω to all spectral wavenumber components $k_{x,y,z}$. For geometries with boundaries or interfaces that give rise to multiple traveling wave interactions along a particular space coordinate - - for example, z - - the collective effect of these interactions can be incorporated into the spatial wavenumber $k_z = k_z(k_x, k_y, \omega)$ via the structural dispersion relation $D(\underset{\sim}{k_t}, \omega) = 0$, $\underset{\sim}{k_t} = (k_x, k_y)$, even when the materials in the configuration are nondispersive.

Asymptotic methods applied to nonwindowed or windowed transforms furnish in any two-dimensional subdomain (ω, ς) of the phase space the trajectory traversed by the relevant wave object. This "bare-bones" trajectory is surrounded by a domain of influence (the spectral flesh) which defines the degree of localization. The domain of influence, in turn, depends on the asymptotic procedure and/or the window function. For radiation from apertures, these aspects have been explored in previous studies [12-14], and the concepts utilized there can be generalized to the scattered fields encountered when the incident field interacts with the scattering environment. Localization achieved solely by constructive wave interference (stationary phase asymptotics) can be interpreted physically in terms of contributions arriving at the observer (ω, ς) from the first Fresnel zone surrounding the "bare-bones" trajectory. General considerations pertaining to this can be found in [15], while application to scattering from periodic arrays is treated in [16].

In any two-dimensional subdomain (ω, ς), the remaining six phase space variables are regarded as parameters. For example, the (p_x, t) subdomain tracks the temporal evolution of the x-component of the "phase slowness" (reciprocal of phase speed) or, equivalently, of the instantaneous angle of arrival projected onto $y = z =$ const. planes. Throughout this tracking, the observer is fixed at (x, y, z); the arrival directions p_y, p_z relative to the y and z axes and the frequency ω are fixed as well. Clearly, this information is incomplete because the p-behavior is strongly dependent on frequency via the (p, ω) dispersion relation. Therefore, the auxiliary (p_x, ω) and (t, ω) subdomains need to be considered together with the (p_x, t) domain for complete understanding of the x-projected time domain observables at a fixed observation point. To gain insight into the imprint produced in the various subdomains by distinct OBP wave processes, it is instructive to resort to relatively simple test configurations with well defined OBP wave objects, and to chart the phase space evolution of these objects. Before embarking on this phase space trek in Sec. IV, we shall make some observations pertaining to dispersion.

III. PARAMETRIZING DISPERSION

A. Material Dispersion

Taken at their most fundamental level, all electromagnetic (EM) wave phenomena take place in a nondispersive (locally and instantaneously reacting) background: empty space. Material media are composed of particles which respond to impinging radiation (i.e., external forces) according to the dynamical equations governing the interaction. Linear wave motion in the combined EM-dynamical system is described in terms of a multicomponent EM-dynamical field which can be organized into operator form for the impulsive-point-source excited field, the Green's function $\underset{\sim}{G}$ [17],

$$\underset{\approx}{L}(\nabla, \frac{\partial}{\partial t}; Q)\, \underset{\sim}{G}(\underset{\sim}{r}, t; \underset{\sim}{r}', t') = \underset{\sim}{1}\, \delta\, (\underset{\sim}{r} - \underset{\sim}{r}')\, \delta\, (t - t'), \underset{\sim}{r} = (x, y, z)\,. \tag{1}$$

Here, $\underset{\approx}{L}$ is a $N \times N$ matrix whose elements contain the first-order spatial and temporal partial derivatives as well as the constitutive parameters of the nondispersive backgrounds symbolized collectively by Q, $\underset{\sim}{G}$ is a column vector composed of the N EM-dynamical field variables, $\underset{\sim}{1}$ is a unit vector defining the source strengths, $\delta(\alpha)$ is the delta function and $(\underset{\sim}{r}', t')$ are the space-time source coordinates. These equations define nondispersive wave motion for the self-consistently coupled EM-dynamical field.

Dispersion (nonlocal or noninstantaneous reaction) takes place when the N-dimensionality of the overall system is reduced. For EM interaction with materials, this implies that the dynamical part is absorbed by the EM part through elimination of the dynamical field variables. This can be illustrated formally by separating the overall system (1) into explicitly coupled EM and dynamical systems, suppressing the argument of $\underset{\sim}{G}$ in (1) for convenience,

$$\underset{\approx}{L}_1 (\nabla, \frac{\partial}{\partial t}; Q_1) \underset{\sim}{G}_1 = \underset{\sim}{1}_1 \, \delta(\underset{\sim}{r} - \underset{\sim}{r}') \, \delta(t - t') + \underset{\approx}{U}_{12} \underset{\sim}{G}_2 \tag{2a}$$

$$\underset{\approx}{L}_2 (\nabla, \frac{\partial}{\partial t}; Q_2) \underset{\sim}{G}_2 = \underset{\approx}{U}_{21} \underset{\sim}{G}_1 \, . \tag{2b}$$

Here, subscripts 1 and 2 distinguish the EM (Maxwell field) and dynamical portions, respectively, with dimensionalities N_1 and $N_2 = N - N_1$. The terms $\underset{\approx}{U}_{12} \underset{\sim}{G}_2$ and $\underset{\approx}{U}_{21} \underset{\sim}{G}_1$ act as induced sources introduced into systems 1 and 2 by coupling to systems 2 and 1, respectively, with the operators $\underset{\approx}{U}_{12}$ and $\underset{\approx}{U}_{21}$ detailing the coupling strengths of the individual field constituent. Formally solving (2b) via $\underset{\sim}{G}_2 = \underset{\approx}{L}_2^{-1} \underset{\approx}{U}_{21} \underset{\sim}{G}_1$, inserting into (2a) and rearranging, one finds, suppressing the arguments of $\underset{\approx}{L}_1$ and $\underset{\approx}{L}_2$,

$$\hat{\underset{\approx}{L}}_1 \hat{\underset{\sim}{G}}_1 = \underset{\sim}{1}_1 \, \delta(\underset{\sim}{r} - \underset{\sim}{r}') \, \delta(t - t'), \hat{\underset{\approx}{L}}_1 (\nabla, \frac{\partial}{\partial t}; \hat{Q}_1) = \underset{\approx}{L}_1 - \underset{\approx}{U}_{12} \underset{\approx}{L}_2^{-1} \underset{\approx}{U}_{21} \, . \tag{3}$$

The inverse operator $\underset{\approx}{L}_2^{-1}$, through its dependence on $\nabla = (\partial/\partial x, \partial/\partial y, \partial/\partial z)$ and $\partial/\partial t$, implies that the equivalent Maxwell operator $\hat{\underset{\approx}{L}}_1$ for the EM field in (3) is integro-differential. Thus, the operator contains a portion with the constitutive parameters Q_2 of the dynamical field, which -- via the space-time integrations $\int d\underset{\sim}{r}'' \int dt''$ -- requires knowledge of the EM field $\underset{\sim}{G}_1 (\underset{\sim}{r}, t; \underset{\sim}{r}'', t'')$ prior to the observation time t and from spatial points other than the observation point $\underset{\sim}{r}$. This nonlocally reacting behavior in time and space represents temporal and spatial dispersion, respectively. The "equivalent" constitutive parameters $\hat{Q}_1$ in the Maxwell field equations therefore become dispersive convolution operators on $\underset{\sim}{G}_1$ rather than constants.

The integral operator in (3) can be eliminated on multiplying (3) by $\underset{\approx}{L}_2$, assuming commutability of $\underset{\approx}{L}_2$ and $\underset{\approx}{U}_{12}$. The resulting differential operator $\underset{\approx}{L}_2 \underset{\approx}{L}_1$ is now of higher order than the first order Maxwell operator $\underset{\approx}{L}_1$. Thus, the dynamically loaded EM field $\underset{\sim}{G}_1$ is determined by a system of partial differential equations of higher order than the nonloaded EM field.

The above considerations, restricted entirely to the configuration domain, become clarified and physically more transparent when viewed from a phase space perspective. In the spatial and temporal spectral domains $(\underset{\sim}{k}, \omega)$, reached via Fourier space-time transforms, the operators $\nabla \rightarrow ik$ and $\partial/\partial t \rightarrow -i\omega$ become algebraized since the Fourier kernels are the plane wave and time harmonic eigenbasis functions $exp(ik\cdot r)$ and $exp(-i\omega t)$ respectively. Now, the operation L_2^{-1} in (3) can be carried out explicitly as $\left[\underset{\approx}{L}_2(i\underset{\sim}{k}, -i\omega)\right]^{-1}$, thereby making the spatially and temporally dispersive equivalent constitutive parameters evident through their k and ω dependence, respectively. Spectral singularities (resonances) located at the spectral poles defined by $\underset{\approx}{L}_2(i\underset{\sim}{k}, -i\omega) = 0$ can be parametrized either as $\underset{\sim}{k}_q = \underset{\sim}{k}_q(\omega)$, $q = 1, 2, \ldots$, or as $\omega_m = \omega_m(\underset{\sim}{k})$, $m = 1, 2, 3, \ldots$.

In materials research, it has been conventional to project onto the temporal frequency (ω) domain, thereby expressing the spatial wavenumber spectrum as a function of frequency via $\underset{\sim}{k}_q = \underset{\sim}{k}_q(\omega)$ [18] with synthesizing wavefunctions $exp[i\underset{\sim}{k}_q(\omega)\cdot \underset{\sim}{r} - i\omega t]$. This is convenient when the exciting signal is narrowband or decomposed into narrowband subintervals. However, for very short pulse signals, retaining time explicitly by projecting onto spatial normalized wavenumber (p, t) subdomains, may provide a phenomenologically better parametrization than synthesis via the frequency domain. The synthesizing wave functions then are $exp[i\underset{\sim}{k}\cdot\underset{\sim}{r} - i\omega_m(\underset{\sim}{k})t]$.

B. Structural Dispersion

Although applied to materials, the discussion above shows in general terms that dispersion occurs when the dimensionality of a wave system is reduced by absorbing the deleted variables into the remaining system. The dimensionality may be reduced also due to configurational constraints imposed by boundaries. Staying entirely within the Maxwell EM field $\underset{\sim}{G}_1$ in (2a) in a medium with negligible dispersion, "equivalent source" terms can be made to appear on the right-hand side if the wave dynamics, now subject to boundary or periodicity conditions, is decomposed into multiple interactions among nondispersive wavefields propagating between reflecting boundaries or around closed surfaces. Assuming that this has been done, the hierarchy of these nondispersive interactions replaces $\underset{\approx}{U}_{12}\,\underset{\sim}{G}_2$ in (2a). If the operator $\underset{\approx}{M}$ accounts for a single nondispersive interaction (for example, via a traveling wavefront), the totality of interactions $\sum M^n$ can be expressed collectively as $\hat{M} = (1-M)^{-1}$. By eliminating individual interactions, the collective form $\hat{M}$ reduces the dimensionality by global restructuring. In the wavenumber spectral domain, the algebraized $\hat{M}$ operator generates frequency-dependent pole singularities (spatial resonances) descriptive of structural dispersion. For a more general restructuring, one retains $\bar{n}$ individual (wavefront) interactions $(0 \le n \le \bar{n} - 1)$ and treats those with $\bar{n} \le n < \infty$) collectively. This results in a hybrid nondispersive-dispersive (wavefront-resonance) algorithm

$$\sum_{n=0}^{\infty} \underset{\approx}{M}^n = \sum_{n=0}^{\bar{n}-1} \underset{\approx}{M}^n + \underset{\approx}{M}^{\bar{n}}\left(\underset{\approx}{1} - \underset{\approx}{M}\right)^{-1} \tag{4}$$

which weights the collective (resonance) portion according to the number of retained wavefronts.

The above concepts are best illustrated by examples where special symmetries clarify the wave physics and bookkeeping in the phase space. These examples involve multiple scatterings and consequent structural dispersion produced (i) by layered nondispersive configurations and (ii) by periodic arrays of scatterers. Only case (i) is discussed in detail.

IV. EXAMPLES

A. Plane Stratified Medium Composed of Nondispersive Materials [3-5]

Here, it is assumed that material dispersion can be neglected over the frequency interval covered by the spectrum of the incident signal. If the stratification occurs along the z-coordinate, the thereby constrained z-domain cannot be reduced to the spectral domain via the global infinite Fourier transform with its plane wave eigenbasis $exp(ik_z z)$. However, the transverse $\underset{\sim}{k}_t = (k_x, k_y)$ domain is still accessible via basis functions $exp(i\underset{\sim}{k}_t \cdot \underset{\sim}{\rho})$, where $\underset{\sim}{\rho} = (x, y)$.

The z-domain parametrization can be handled in two ways:

a. Assuming in each layer that waves propagate as in the infinite nondispersive medium of that layer, with reflections and transmissions at interfaces so as to satisfy the EM boundary conditions there. This leads to a hierarchy of multiple reflected and transmitted traveling wavefields, each propagating locally with a full spectrum $\underset{\sim}{k} = (k_x, k_y, k_z)$ along its particular ray trajectory. In the formal treatment in Sec. III B, each such interaction is tagged by an interaction operator $\underset{\approx}{M}$.

b. Restricting propagation and wavenumber spectra to the transverse $\underset{\sim}{\rho} \to \underset{\sim}{k}_t$ domain, and treating the spectral evolution along z collectively as the function $k_z = k_z(\underset{\sim}{k}_t, \omega)$. In the formal treatment in Sec. III B, the collective interaction operator $\hat{M}_z = \sum_{n=0} M_z^n = (1 - M_z)^{-1}$ introduces resonances into the transverse spectrum that are characterized by the dispersive constraint relation $1 - M_z \equiv D(\underset{\sim}{k}_t, \omega) = 0$. These resonances define guided modes propagating along ρ. The more general hybrid wavefront-resonance form follows directly from (4).

Phase space maps for scenarios a. and b. above are schematized in Figs. 3 and 4, respectively. Extensive captions describe the contents. Since all wave processes in Fig. 3 are nondispersive wavefront (ray) fields, frequency domain and time domain phenomena give rise to identical phase space trajectories. Therefore, no distinction has been made in Fig. 3. In Fig. 4, the synthesizing collective OBP wave objects are TD leaky modes guided along the transverse (ρ) coordinate.

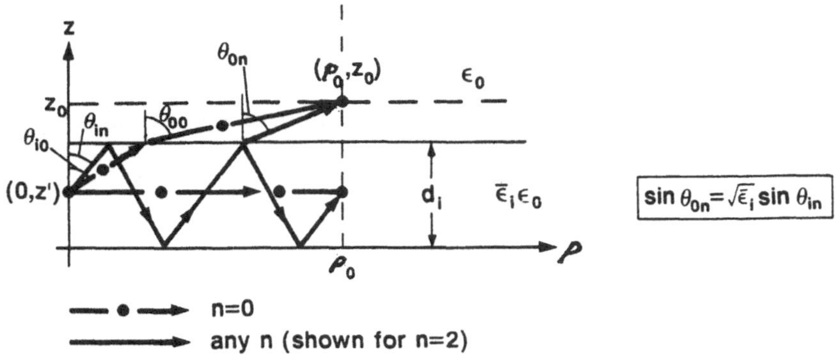

Figure 3. Subdomain phase space maps for ray parametrization of pulsed radiation from a source inside a nondispersive dielectric layer--with relative permitting ε_i -- on a ground plane. p_t is the normalized transverse wavenumber and $\theta_i = sin^{-1}(p_t/p_i)$, $\theta_0 = sin^{-1}(p_t/p_0)$ are the corresponding ray directions (angular spectra). Moreover, $p_t^2 + p_{zi}^2 = p_i^2 = p_0^2\varepsilon_i$, $p_t^2 + p_{zi}^2 = p_i^2 = p_0^2\varepsilon_i$, $c_{o,i}$ are the wave speeds in the exterior and slab regions, respectively.

3a. Physical configuration, with multiple reflected ray paths from source at $(0, z')$ to observers moving along ρ at fixed heights z. Outside observer at (ρ_0, z_0), shown in region $\rho < \rho_{c0}, \rho_{cn}$ of Fig. 3c: inside observer at (ρ_0, z'). n denotes the number of reflections at the upper boundary $z = d_i$. $\rho = (x^2 + y^2)^{1/2}$

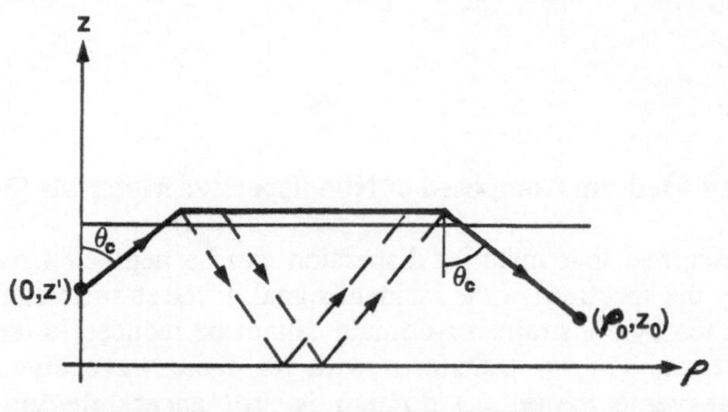

3b. Physical configuration with diffracted lateral ray field launched by a ray incident at the critical angle $\theta_c = sin^{-1}\ (p_0/p_i)$. Multiple reflected trajectories are shown dashed.

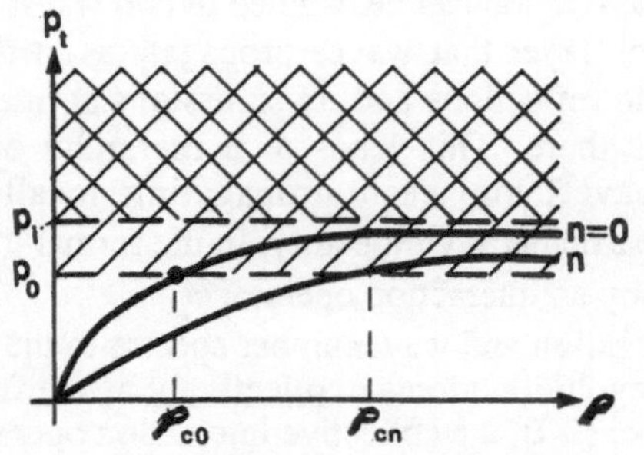

3c. $(\rho\ , p_t)$ trajectories for ray species $n = 0$ (direct ray) and arbitrary n. These trajectories track the angles of arrival, in planes z = const., as a function of lateral displacement. All rays reach an internal observer. Rays with $p_{tn} < p_0$ reach an exterior observer ($z > d_i$), while those with $p_{tn} > p_0$ do not; the latter are totally reflected internally and are evanescent (complex rays with imaging p_{zin}) on the outside. For $p_t > p_i$, all ray fields are evanescent away from the source plane $z = z'$. Note that $p_{tn} = p_i\ sin\ \theta_{in}$ in Fig. 3a. The heavy dot denotes the point spectrum and launch location of the lateral ray in Fig. 3b. ρ_{cn} = cutoff (total internal reflection) of ray species n, p_{z0} imag. //// , p_{zi} imag.

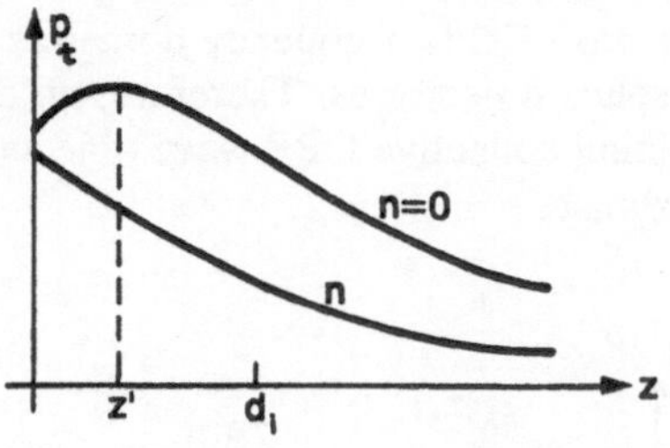

3d. (z, p_t) trajectories: arrival angles in planes ρ = const. For the direct ($n = 0$) ray, θ_{i0} reaches a maximum at $z = z'$.

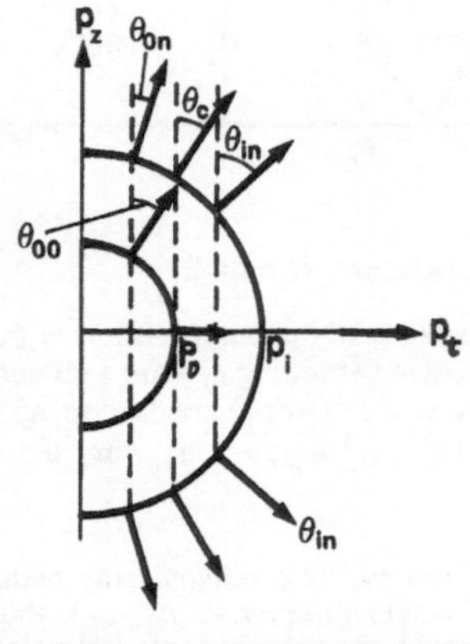

3e. (p_t, p_z) trajectories: wave slownesses (circles) and ray arrival angles (normals to circles). Transmission to the exterior occurs where p_t-line intersects the inner circle. Angles as in Fig. 3a. $p_z < 0$ are downgoing rays. $p_{tn} = p_i\ sin\ \theta_{in} = p_0\ sin\ \theta_{on}$;, $p_{zi} = (p_i^2 - p_t^2)^{1/2}$, $p_{z0} = (p_0^2 - p_t^2)^{1/2}$.

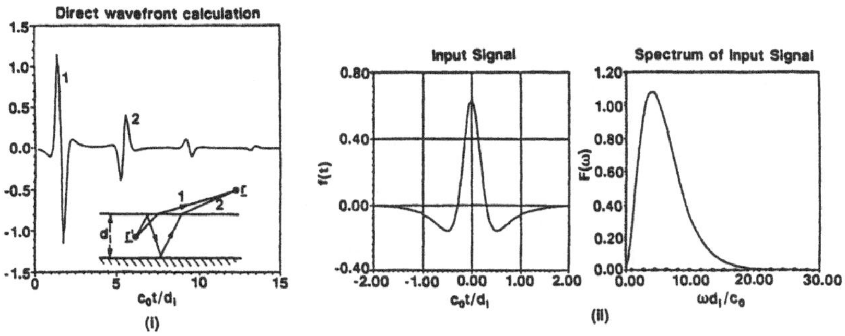

3f. (i) Multiple reflected pulsed radiated field along ray paths of Fig. 3a. (ii) Input pulse shape and frequency spectrum.

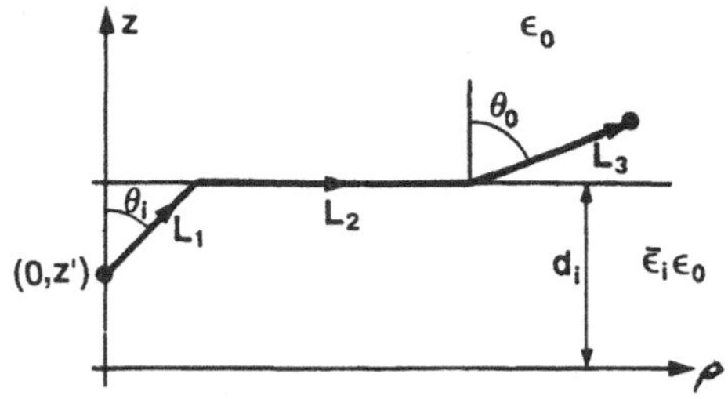

Figure 4. Subdomain phase space maps for leaky mode parametrization of pulsed radiation from a source inside a nondispersive dielectric layer--with relative permitting ε_i--on a ground plane. Leaky mode dispersion relation: $D(p_t, \omega) = 0$. Alternative solutions: a) $p_{tq} = p_{tq}(\omega)$, $q = 1, 2...$ b) $\omega_m = \omega_m(p_t)$, $m = 1, 2...$

4a. Physical configuration and typical ray-mode trajectory from source to observer. L_1 and L_3 are conventional ray paths in the nondispersive bulk media, as in Fig. 3a. L_2 is a modal ray path with $p_t = p_{tq}(\omega)$. The point spectrum p_{tq} is the wavenumber parameter for the entire ray-mode process: $p_i \sin \theta_{iq} = p_0 \sin \theta_{0q} = Re(p_{tq})$. The ray schematization is approximate and applies when $Im\, p_{tq}$ is small.

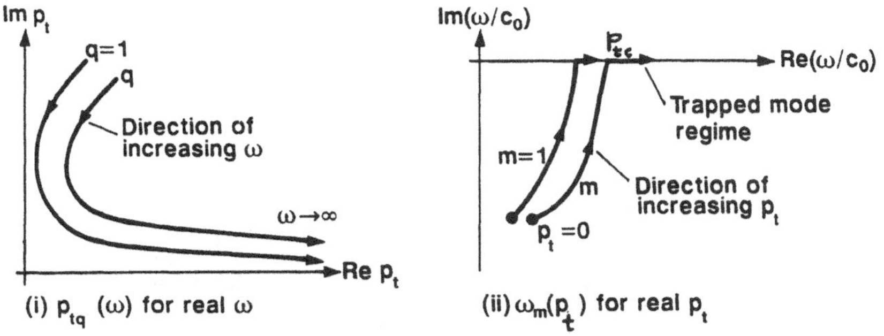

4b. Typical dispersion curves $p_{tq}(\omega)$ and $\omega_m(p_t)$. (i) $p_{tq}(\omega)$ for real ω. (ii) $\omega_m(p_t)$ for real p_t.

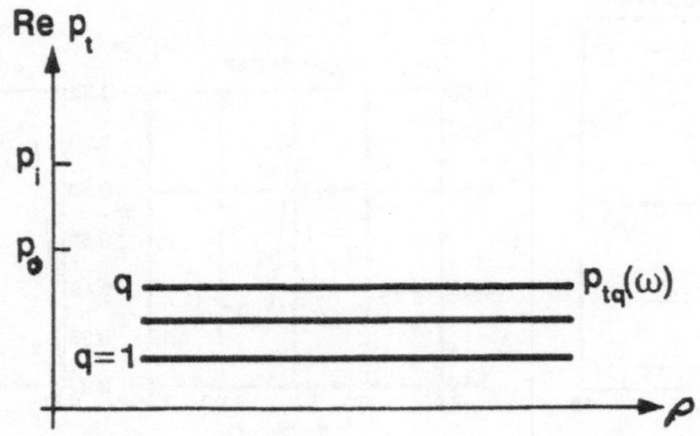

4c. (ρ, p_t) trajectories: eigenmode slownesses or eigenangles θ_{iq} and θ_{0q} for q-th leaky mode at fixed ω.

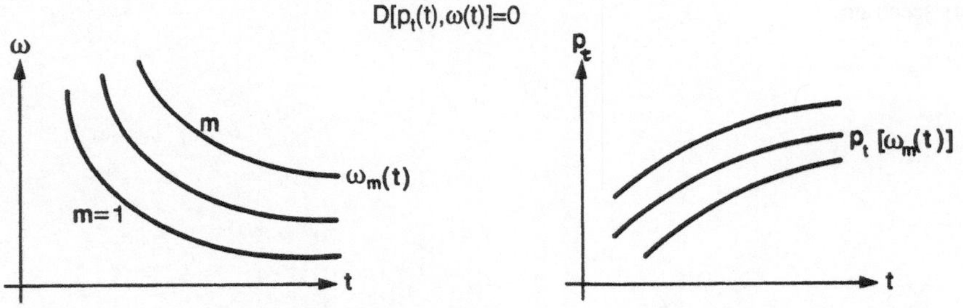

4d. (t, ω) and (t, p_t) trajectories: instantaneous frequencies and slownesses (arrival angles) at a fixed observer. The instantaneous dispersion equation provides the (ω, p_t) connection. $D[p_t(t), \omega(t)] = 0$

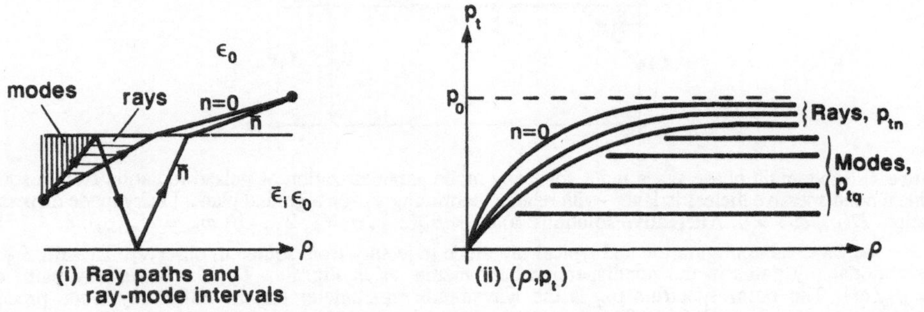

4e. Hybrid ray-mode parametrization as in (4): Retain $p_{t0} > p_{tn} > p_{t\bar{n}}$ rays (see Figs. 3a and 3c) and fill the remaining p_t interval with modes $p_{tq} < p_{t\bar{n}}$. (i) Ray paths and ray-mode intervals. (ii) (ρ, p_t).

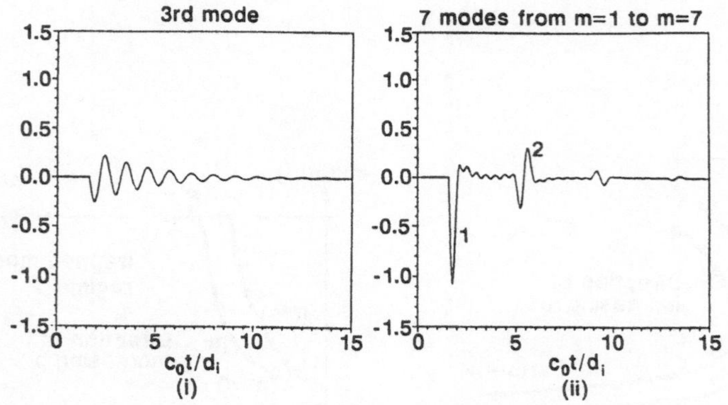

4f. TD leaky modes as in Fig. 4a generated by pulsed source in (ii) of Fig. 3f. The pulse spectrum covers 12 modes. (i) typical TD mode profile, shown for $q = 3$. (ii) buildup of multiple reflected high resolution pulses in (i) of Fig. 3f by selective mode summation, shown for $\sum_{q=1}^{7} . \sum_{q=1}^{12}$ reconstructs the signal in Fig. 3f(i) completely.

B. Finite Periodic Array of Scatterers [6, 7]

The scenario here is analogous to that for the dielectric layer in Sec. IV A: the TD scattered fields can be synthesized by a. individual tracking of multiple scattered traveling wavefronts, or b. collective treatment of the multiple interactions in terms of TD periodic structure (Bragg or Floquet) modes parametrized through wave slowness spectra p_{tq} or p_{zq}, indexed by q, in the plane of, or perpendicular to, the array, respectively. Edge effects due to truncated periodicity introduce, in addition, edge-centered waves scattered in all directions. These features have been explored in references 6 and 7 for the test configuration of a finite array of flat strips.

The phase space processing of the TD data base for scattering by the finite array has certain similarities with the processing of data for radiation from a finite phased aperture distribution (see [12], [13]) provided that the equivalent truncated aperture, which establishes the scattered field, is assumed to be composed of truncated Floquet modes. Due to space limitations, details are not shown here, but the phase space maps in [12] - [14] and those in Figs. 3 and 4 here reveal similar features. Furthermore, reference 16 in this volume deals with specific phase space processing issues for infinite periodic strip arrays; again comparison with Figs. 3 and 4 here establishes direct analogies. Included for illustration here is an example of how phase space processing of TD scattering data from a finite periodic strip array can be implemented. Figure 5 depicts the geometry, the initial high resolution TD data base, the corresponding frequency domain data obtained by conventional global Fourier transform over the temporal variable, and the (t, ω) subdomain traces obtained via windowed (short-time) Fourier transform. The latter highlights the temporal evolution of the Floquet mode spectra which resembles the leaky mode traces in Fig. 4d. For related displays, see reference 19.

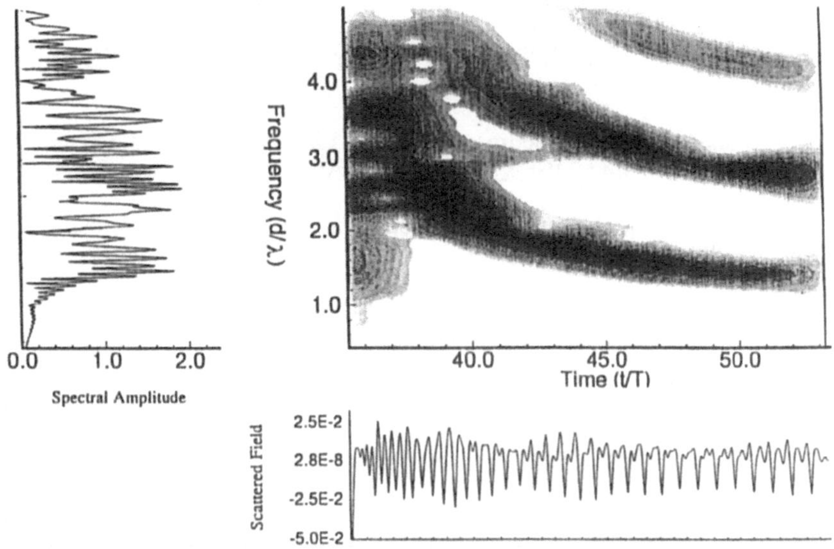

Figure 5. Time-frequency representation for scattered field due to a pulsed TE plane wave indicent normally upon a forty-strip grating. The grating period is d, the strip width is $0.06d$, and the interstrip separation is $0.94d$. The scattered fields are observed at $35.44d$ directly above the right-most edge of the array. The TD waveform (bottom) was computed using a MOM-FFT reference algorithm. The left-most figure is a Fourier transform of the entire TD waveform. The center figure was computed by performing on the TD waveform a Short-Time Fourier Transform (STFT) with a Gaussian window having $\sigma/T = 1.2$. Time is normalized to T, where $T = d/c$.

REFERENCES

1. L. B. Felsen, "Observable-Based Wave Modeling: Wave Objects, Spectra and Signal Processing," in Huygens' Principle 1690-1990: Theory and Applications," North-Holland, Amsterdam, 1992. This paper contains an extensive list of references.

2. L. B. Felsen, "Radiation and Scattering of Transient Electromagnetic Fields," Int. J. Num. Modeling: Electronic Networks, Devices and Fields, 5; 149-161, 1992. This paper contains an extensive list of references.

3. L. B. Felsen and F. Niu, "Spectral Analysis and Synthesis Options of Short Pulse Radiation from a Point Dipole in a Grounded Dielectric Layer," submitted to IEEE Trans. Antennas Propagat.

4. F. Niu and L. B. Felsen, "Time Domain Leaky Modes on Layered Media: Dispersion Characteristics and Synthesis of Pulsed Radiation," submitted to IEEE Trans. Ant. Propagat.

5. F. Niu and L. B. Felsen, "Asymptotic Analysis and Numerical Evaluation of Short Pulse Radiation From a Point Dipole in a Grounded Dielectric Layer," submitted to IEEE Trans. Antennas Propagat.

6. L. Carin and L. B. Felsen, "Time-Harmonic and Transient Scattering by Finite Periodic Flat Strip Arrays: Hybrid (Ray) - (Floquet mode) - (MOM) Algorithm and its GTD Interpretation," to be published in IEEE Trans. Ant. Propagat.

7. L. Carin, L. B. Felsen and M. McClure, "Time Domain Design-Oriented Parametrization of Truncated Periodic Strip Gratings," submitted to IEEE Microwave and Guided Wave Letters.

8. O. Rioul and M. Vetterli, "Wavelets and Signal Processing," IEEE SP Magazine, Oct. 1991.

9. F. Hlawatsch and G. F. Boudreaux-Bartels, "Linear and Quadratic Time-Frequency Signal Representation," IEEE SP Magazine, April 1992.

10. E. Heyman and L. B. Felsen, "Complex Source Pulsed Beam Fields," J. Opt. Soc. Am. A6, 806-817, 1989.

11. E. Heyman, "A General Wavepacket Solution of the Time-Dependent Wave Equation in Homogeneous and Inhomogeneous Media," this volume.

12. B. Z. Steinberg, E. Heyman and L. B. Felsen, "Phase Space Beam Summation for Time-Harmonic Radiation from Large Apertures," J. Opt. Soc. Am. A8, 41-59, 1991.

13. B. Z. Steinberg, E. Heyman and L. B. Felsen, "Phase Space Beam Summation for Time-Dependent Radiation from Large Apertures: Continuous Parametrization," J. Opt. Soc. Am. A8, 943-958, 1991.

14. B. Z. Steinberg and E. Heyman, "Phase Space Beam Summation for Radiation from Large Apertures: Discretized Parametrization," J. Opt. Soc. Am. A8, 959-966, 19.

15. Yu A. Kravtsov and Yu. I. Orlov, Geometrical Optics of Inhomogeneous Media, Springer, New York, 1991.

16. P. Borderies and L. B. Felsen, "Phase Space Analysis and Localization of Scattering by a Planar, Infinite, Weakly Resonant Array Illuminated by a Very Short Plane Pulse," this volume.

17. L. B. Felsen and N. Marcuvitz, "Radiation and Scattering of Waves," Prentice Hall, Englewood Cliffs, NJ 1973.

18. K. E. Oughston, J. E. K. Laurens and C. M. Balictsis, "Asymptotic Description of Electromagnetic Pulse Propagation in a Linear Dispersive Medium," this volume.

19. E. K. Walton and A. Moghaddar, "Analysis of Inlets and Ducts Using Time-Frequency Analysis of UWB Radar Signals," this volume.

PROGRESS IN TLM MODELING OF TIME DISCRETE FIELDS AT CONDUCTOR STRIPS, EDGES AND CORNERS

Wolfgang J.R. Hoefer

NSERC/MPR Teltech Research Chair in RF-Engineering
Department of Electrical and Computer Engineering
University of Victoria
Victoria, British Columbia, Canada V8W 3P6

INTRODUCTION

In traditional two-dimensional and three-dimensional Transmission Line Matrix (TLM models) of electromagnetic structures, conducting boundaries are introduced halfway between nodes. (For an introduction to TLM see for example a recent paper by Hoefer[1]). This approach leads to several difficulties when modeling boundaries that exhibit sharp edges or corners, such as strip and slot type structures. If, in addition, devices are to be connected to such strips, this method of boundary modeling gives very poor results. The effect of this modeling error can be observed in practically all TLM-generated results, which are systematically shifted towards lower frequencies.

The classical remedy is to make the mesh size smaller in the vicinity of the field singularities. This increases the resolution and therefore reduces the error, but leads to considerable increase in computer requirements. A highly refined mesh not only increases computer memory requirements but also imposes a finer time step, thus increasing the computation time.

The effect of mesh refinement can be observed by computing a given structure containing sharp edges with several mesh sizes and then representing the results as a function of the mesh parameter. A two-dimensional example is shown in Fig. 1. The resonant frequency of a cavity coupled to a matched waveguide through an inductive iris is computed using three different mesh densities, and then represented as a function of Δl in Fig. 2. Clearly, the results can be extrapolated and approaches the analytically exact value for $\Delta l = 0$.

An alternative technique for modeling conducting boundaries and edges, which will be described in this paper, consists of placing them directly into shunt nodes by short-circuiting them. When a given structure is modeled with the traditional and then with the alternative modeling technique, the results are affected by coarseness errors of equal magnitude but opposite sign, so that a highly accurate result can be obtained by averaging the results of two rather coarse TLM simulations with different boundary models. It is important to note that these measures for improving the modeling accuracy are essentially independent of frequency, so that their full benefit is preserved when analyzing a structure under short impulse/wideband excitation conditions.

In this paper the new boundary nodes will be described in the case of two-dimensional TLM networks, and the resulting improved accuracy will be demonstrated.

Ultra-Wideband, Short-Pulse Electromagnetics
Edited by H. Bertoni *et al.*, Plenum Press, 1993

MODIFICATION OF CORNER NODES IN 2D-TLM NETWORKS

One of the principal sources of error in the TLM analysis of structures with sharp edges and corners is the so-called coarseness error. Since it affects all strip and slot-type lines it is particularly serious in microwave and millimeter wave circuit analysis. It is due to the insufficient resolution of the edge field by the discrete TLM network. The traditional remedy for this problem is to use a finer mesh. Fig. 1 shows this approach in the case of a rectangular cavity coupled to a matched waveguide section through an inductive iris. The resonant frequency is computed with an increasingly finer mesh, and the resonant frequency is represented as a function of the mesh parameter Δl as shown in Fig. 2 (lower curve marked $p = 0$). Fitting a curve through these results and extrapolating it for $\Delta l = 0$ yields a considerably improved value for the resonant frequency. While this approach (also called Richardson approximation) is well suited for generating highly accurate results with rather coarse meshes, it is not suitable for routine computations. Refining the mesh only in the vicinity of the iris edges improves computational efficiency but, in return, introduces additional complications and computational requirements. We have recently proposed two possible solutions to this problem. The first amounts to adding a TLM branch to the node situated in front of a corner to provide a direct interaction of this node with the boundary[2]. The second is to place a node directly into the edge or corner[3]. Both approaches require the modification of the impulse scattering properties of the corner node, which represents only a minor increase in computational resources, particularly when parallel processing is used.

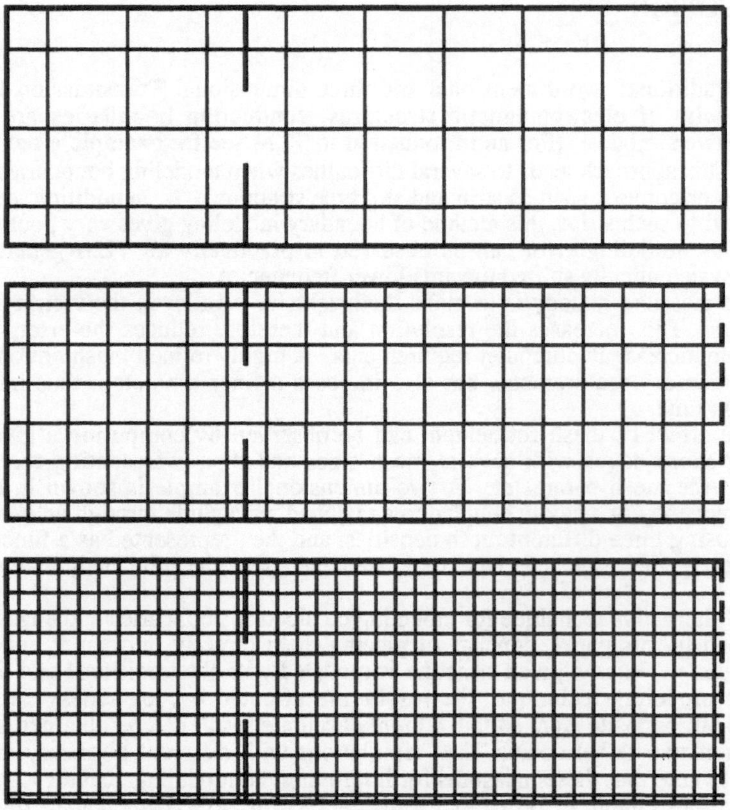

Fig. 1 Discretization of a cavity coupled to a matched waveguide through an inductive iris. The discretization is refined by a factor 2 each time.

The reason for the coarseness error is clearly seen when considering Fig. 3 The nodes situated diagonally in front of an edge are not interacting directly with the boundary but receive information about its presence only via their neighbours who have one branch connected to it. The network is thus not sufficiently "stiff" at the edge, and results obtained are always shifted towards lower frequencies.

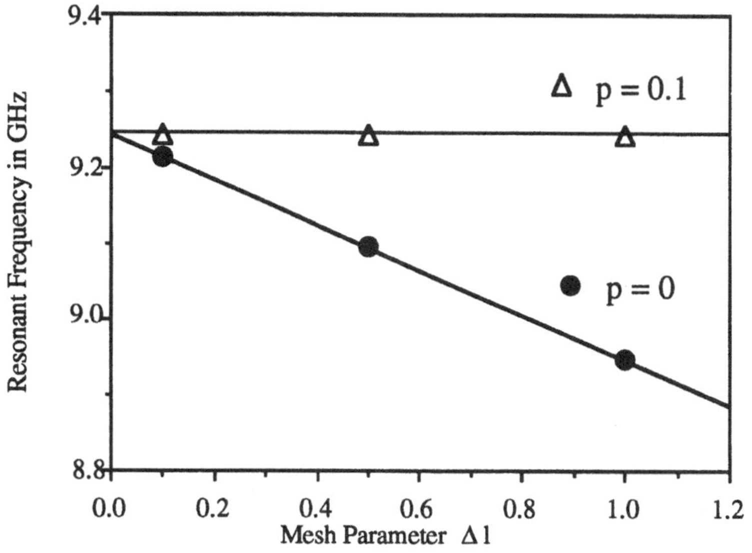

Fig. 2 Resonant frequency of a cavity computed using TLM meshes with different mesh parameters Δl. The lower curve (p = 0) shows the results obtained without compensation of the edge effect. The upper curve (p = 0.1) shows results for the same meshes with corner node compensation using a fifth stub of relative characteristic admittance 2p.

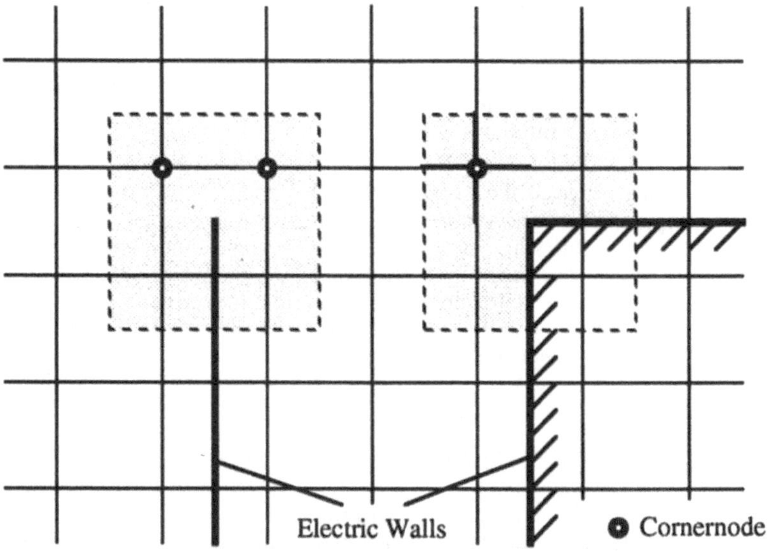

Fig. 3 Cornernodes in a 2D TLM mesh are not interacting directly with the boundaries, causing large coarseness error.

The situation is similar to that of a square lattice consisting of elastic rods which are embedded in a solid wall forming a sharp corner. The lattice is not as stiff as a continuous elastic sheet (infinitely dense lattice), and the answer here is clearly to add an additional rod between the cornernodes and the edge of the wall. The electromagnetic equivalent is to add a fifth transmission line stub to connect the cornernode directly with the conductor edge. (See Fig. 4). Since this stub is longer than the other branches by a factor $\sqrt{2}$ it is simply assumed to have a correspondingly larger propagation velocity. The effect of this corner correction is demonstrated in Fig. 2 (upper curve, marked p = 0.1). The parameter p is proportional to the fraction of power carried by the fifth branch of the corner node and is equal to half the characteristic admittance of the corner branch when normalized to the link line admittance. For p = 0.1 the frequency remains accurate even for a very coarse mesh. The best way to determine the appropriate value for p is to evaluate the resonant frequency with the coarsest discretization for several p-values and to determine which is the p-value that makes the result independent of the mesh density, yielding a horizontal line in Fig. 2. It turns out that the p-value is insensitive to frequency, so that the correction is valid over

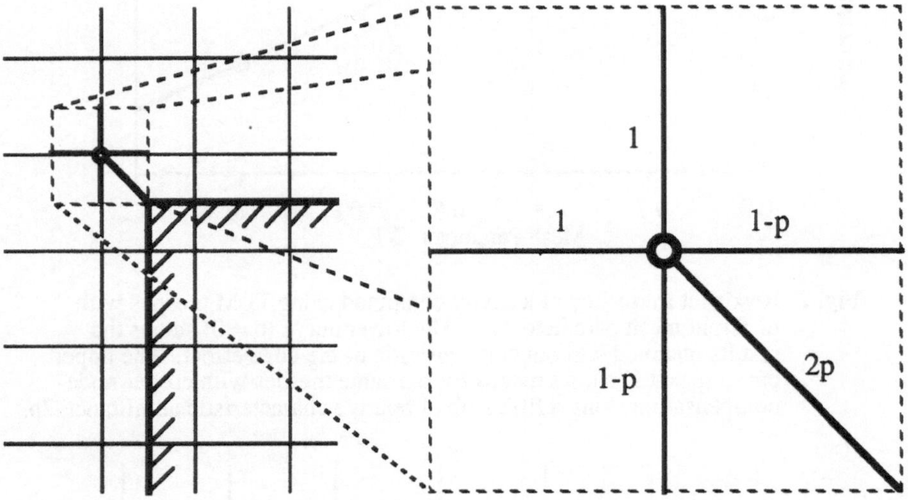

Fig. 4 Compensation of coarseness error by adding a fifth branch to the corner node. The numbers besides the node branches indicate their respective normalized characteristic admittances.

the whole frequency range of the TLM mesh ($\Delta l/\lambda \ll 1$). Note that the correction applies only to the computation of the total energy in the system and related quantities such as resonant frequency or scattering parameters. However, the field distribution in the vicinity of the edge or corner is not properly represented by the cornernode, and thus a very fine mesh is needed to properly represent the edge field. Nevertheless, in most engineering applications one is interested in the proper response of the structure as a circuit element or system rather than in the exact field distribution.

2D-TLM NETWORKS WITH BOUNDARIES ACROSS NODES

A second way to model boundaries in a TLM network is to place them across nodes. This amounts to imposing the boundary conditions not in the form of an impulse reflection coefficient in link transmission lines halfway between nodes, but rather to enforcing a zero voltage across nodes. This can easily be implemented by making the characteristic admittances of the stubs which are connected to these nodes, infinitely large (in practice this means giving them a very large value, 10^{35} for example). In this way, sharp edges or corners will coincide directly with nodes as shown in Fig. 5a. Fig. 5b shows the same resonant cavity with boundaries implemented halfway between nodes for comparison. The resonant frequencies obtained with the two boundary models are not the same. Values obtained with type (a) are too high, while those obtained with type (b) are too low, as discussed in the previous section. By refining the mesh size, one obtains progressively closer results which, finally, become identical for $\Delta l = 0$. Fig. 6 shows this tendency very clearly. In fact, the error is practically the same for both approaches and differs only in sign. It follows that a highly accurate value can be obtained by averaging both results, even if they have been obtained with a very coarse mesh.

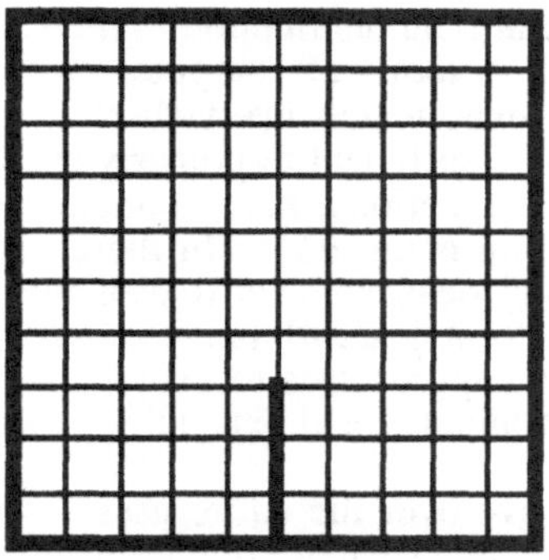

(a)

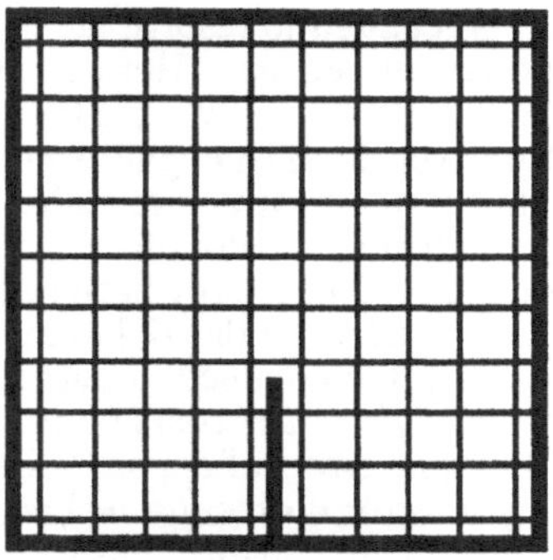

(b)

Fig. 5 Resonant cavity surrounded by conducting walls and containing a sharp edge. Walls are placed through nodes (a) and halfway between nodes (b).

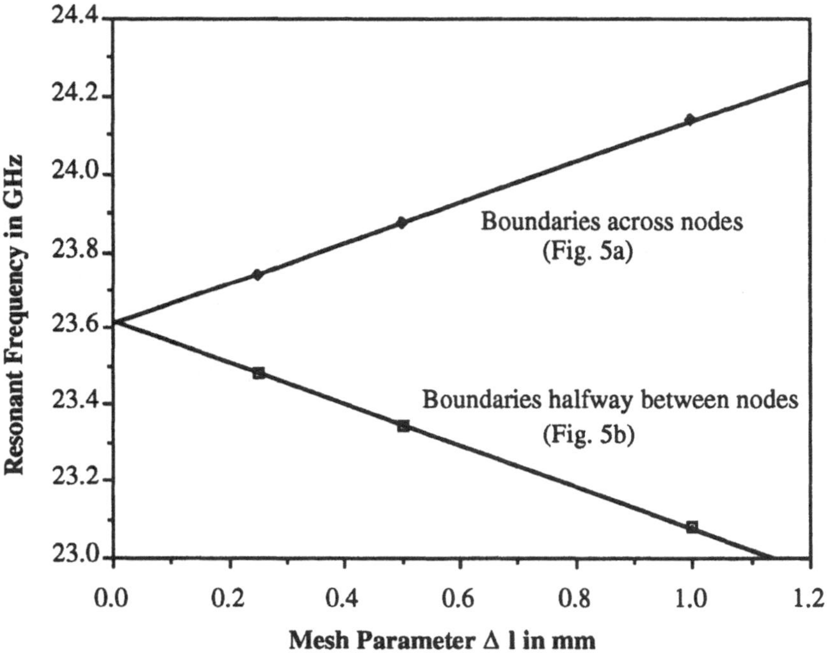

Fig. 6 Resonant frequencies of the structures of Figs. 5a and b obtained with different mesh densities. The coarseness error is reduced as Δl decreases.

Again, as in the case of the cornernode, this method does not yield accurate field values in the vicinity of the edges. However, important engineering parameters such as scattering parameters, resonant frequencies, or cutoff frequencies can be obtained in this manner with great accuracy at relatively small computational cost.

DISCUSSION AND CONCLUSION

As in any other analytical or numerical technique, the existence of field singularities represents a challenge when computing the fields and the associated circuit parameters of electromagnetic structures. When using a space discretization method such as TLM, a very fine mesh is required to resolve the highly nonuniform field distribution in the vicinity of sharp corners and edges. This can be quite costly in terms of computer memory and runtime, but it is a necessity if the fields must be known accurately everywhere.

However, in most engineering applications it is sufficient to properly evaluate the energy stored in the structure rather than the field functions, i.e. parameters such as resonant frequencies, cutoff frequencies, or scattering parameters. In these cases it is possible to correct the energy balance by incorporating localized reactive elements into the discrete field model. In the TLM case this takes the form of a fifth branch connecting the cornernodes to the conducting edge. An alternative approach is to average the results obtained with two rather coarse models (very low computational expenditure) which are affected by equal errors of opposite sign.

These error correction techniques are effective over the entire useful frequency range of the TLM network and can therefore be applied when computing impulse or transient responses. Even though these techniques have been demonstrated here for two-dimensional TLM networks, they can be extended to three-dimensional schemes as well. However, it is important to selectively set to zero only those field components that are tangential to the conductors, which requires that the characteristic admittances of all stubs connected to the 3D TLM node can be selected separately. At the present time, all corner node corrections must be introduced by hand when discretizing a structure. Further research is directed towards automatic compensation of coarseness error through automatic identification of cornernodes by the computer.

REFERENCES

1. W.J.R. Hoefer, "Huygens and the computer - a powerful alliance in numerical electromagnetics", *Proc. IEEE, Special Issue on Electromagnetics,* Vol. 79, No. 11, pp. 1459-1471, October 1991
2. U. Müller, U., P.P.M. So, and W.J.R. Hoefer, "The compensation of coarseness error in 2D TLM modeling of microwave structures", in *1992 IEEE Intl. Microwave Symp. Dig.*, Albuquerque, NM, June 1-5, 1992
3. J.S. Nielsen, W.J.R. Hoefer, "New 3D TLM condensed node structures for improved simulation of conductor strips", in *1992 IEEE Intl. Microwave Symp. Dig.*, Albuquerque, NM, June 1-5, 1992

CONTINUOUS-TIME DISCRETIZED-SPACE APPROACH TO MODELING SHORT-PULSE PROPAGATION AND SCATTERING

Anton G. Tijhuis

Laboratory of Electromagnetic Research
Department of Electrical Engineering
Delft University of Technology
P.O. Box 5031, 2600 GA Delft, The Netherlands

1. INTRODUCTION

The techniques for modeling transient electromagnetic waves in linearly reacting media can be subdivided into two categories. Local techniques, such as finite-element and finite-difference methods, are based on the discretization of Maxwell's equations in differential form. Typically, the field at any given space-time point is related to fields at neighboring points. Global techniques are based on contrast-source integral equations, and lead to representations of the scattered field in terms of physical quantities in the interior or on the surface of the scattering object. Examples are natural-mode formulations and direct solution procedures.

This contribution describes a relatively new approach of the last type. The basic idea can be formulated in two complementary ways. The first one starts from the formulation in the *time domain*. As in the marching-on-in-time method, a fixed space discretization independent of time is introduced. This reduces the pertaining integral equation to a system of time-domain differential equations of a fixed dimension. Carrying out a temporal Laplace transformation, we next obtain a matrix equation of a fixed dimension in which the complex frequency s occurs as a parameter. This equation is solved repeatedly, and the desired time-domain results are obtained by evaluating the Bromwich inversion integral with the aid of a standard FFT-operation. The second formulation starts in the *frequency domain*. As in the Method of Moments, the transformed integral equation is discretized by interpolation and subsequent analytical integration. From the behavior of the discretization error and the spectral content of the generating pulse, it is then concluded that choosing a fixed mesh size suffices to maintain, independently of frequency, a fixed absolute accuracy.

Important advantages of this approach are that dispersive media can be handled without extra computational effort, and that the accumulation of errors characteristic of time-marching methods is inherently avoided. The principal tasks in each application are the space discretization and the repeated solution of the discretized, transformed integral equation. A general mathematical formulation of the basic ideas can be found in [1,2]. In the present contribution, we avoid repeating this formulation. Instead, a historic overview is presented of the different types of scattering problems to which the approach has been applied sucessfully.

Ultra-Wideband, Short-Pulse Electromagnetics
Edited by H. Bertoni *et al.*, Plenum Press, 1993

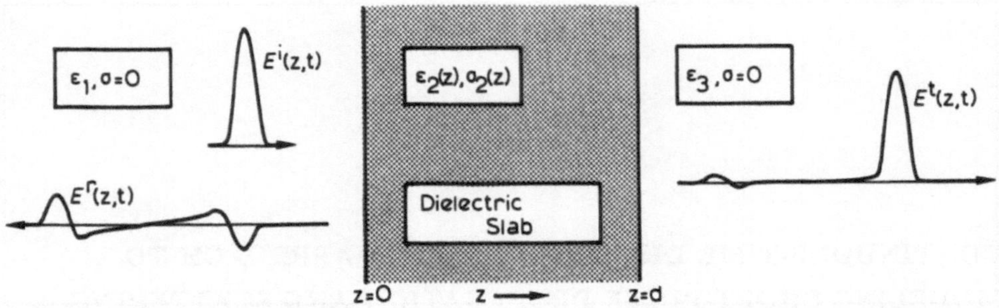

Figure 1. A pulsed electromagnetic plane wave normally incident on an inhomogeneous, lossy dielectric slab in between two homogeneous, lossless dielectric half-spaces.

2. PLANE-WAVE SCATTERING BY A DIELECTRIC SLAB

The first application involved an inhomogeneous, lossy dielectric slab as shown in Figure 1, situated in between two homogeneous, lossless dielectric half-spaces. The slab is excited by a linearly polarized electromagnetic pulse of finite duration T, normally incident from $z < 0$. The x-component of the electric-field strength in $-\infty < z \leq 0$ can be written as

$$\mathcal{E}(z,t) = \mathcal{F}(t - z/c_1) + \mathcal{E}^r(0, t + z/c_1). \tag{1}$$

In (1), $\mathcal{F}(t)$ is an integrable function that vanishes outside the interval $0 < t < T$, $\mathcal{E}^r$ denotes the reflected field, and $c_1 \stackrel{\text{def}}{=} (\epsilon_1\mu_0)^{-\frac{1}{2}}$ is the wave speed in the left half-space. For the transmitted field in $d \leq z < \infty$ we have, similarly

$$\mathcal{E}^t(z,t) = \mathcal{E}^t(d, t - (z-d)/c_3), \tag{2}$$

with $c_3 \stackrel{\text{def}}{=} (\epsilon_3\mu_0)^{-\frac{1}{2}}$.

From (1,2), we observe that it suffices determine $\mathcal{E}(z,t)$ for $0 \leq z \leq d$ and $0 \leq t < \infty$. In the context of the present discussion, we want to do this by using the contrast-source integral equation:

$$\begin{aligned}\mathcal{E}(z,t) &= \mathcal{F}(t - z/c_1) \\ &\quad - \frac{Z_1}{2}\int_0^d [\Delta\epsilon(z')\partial_t + \sigma_2(z')]\,\mathcal{E}(z', t - |z - z'|/c_1)\,dz' \\ &\quad - \frac{c_1 - c_3}{2c_3}\,\mathcal{E}(d, t - |z - d|/c_1),\end{aligned} \tag{3}$$

with $Z_1 \stackrel{\text{def}}{=} (\mu_0/\epsilon_1)^{\frac{1}{2}}$ and $\Delta\epsilon(z) \stackrel{\text{def}}{=} \epsilon_2(z) - \epsilon_1$. This equation can be derived by considering a homogeneous medium with $\epsilon(z) = \epsilon_1$ and $\sigma(z) = 0$ as background medium, and using (2) to calculate the integral over $d < z < \infty$ analytically.

In order to discretize Equation (3), we introduce the spatial grid $z_m = mh$ with $m = 0, 1, \ldots, M$ and $h = d/M$. Restricting the observation to points on this grid, we can directly approximate the space integral in (3) by a repeated trapezoidal rule with step h. This reduces (3) to

$$\begin{aligned}\tilde{\mathcal{E}}(m,t) &= \mathcal{F}(t - mh/c_1) \\ &\quad - \frac{Z_1 d}{2}\sum_{m'=0}^{M} w_{m'}[\Delta\epsilon(m'h)\partial_t + \sigma_2(m'h)]\,\tilde{\mathcal{E}}(m', t - |m - m'|h/c_1) \\ &\quad - \frac{c_1 - c_3}{2c_3}\,\tilde{\mathcal{E}}(M, t - |m - M|h/c_1),\end{aligned} \tag{4}$$

where $m = 0, 1, \dots, M$, and where $w_m = M^{-1}$ for $0 < m < M$ and $w_{0,M} = (2M)^{-1}$. Equation (4) is a linear system of equations of *fixed* dimension $M+1$ for the time signals $\{\tilde{\mathcal{E}}(m,t)\}$.

In the marching-on-in-time approach, the next step would be to discretize the time coordinate t with an interval $\Delta t \leq h/c_1$. This is the step that leads to error accumulation, and, for more complicated configurations, to instabilities. In the present method, the idea is to utilize the feature that the system (4) is linear. Therefore, we may apply a temporal Laplace transformation with $Re(s) \geq 0$. It is this treatment of the time coordinate that has lead to the designation *continuous-time discretized-space* or CTDS approach. The transformation results in the matrix equation

$$\tilde{E}(m,s) + \sum_{m'=0}^{M} C_{m'} Z^{-|m-m'|} \tilde{E}(m',s) = Z^{-m} F(s), \tag{5}$$

where $m = 0, 1, \dots, M$, $Z \stackrel{\text{def}}{=} \exp(sh/c_1)$, $F(s)$ is the Laplace transform of $\mathcal{F}(t)$, and where the dimensionless contrast parameter C_m is given by

$$C_m \stackrel{\text{def}}{=} \begin{cases} \frac{w_m}{2}\left[\frac{\Delta\epsilon(mh)}{\epsilon_1}\frac{sd}{c_1} + Z_1\sigma_2(mh)d\right] & \text{for } 0 \leq m < M, \\ \frac{w_M}{2}\left[\frac{\Delta\epsilon(d)}{\epsilon_1}\frac{sd}{c_1} + Z_1\sigma_2(d)d\right] + \frac{c_1 - c_3}{2c_3} & \text{for } m = M. \end{cases} \tag{6}$$

In this equation, the complex frequency s occurs as a parameter. Because of the special, degenerate structure of the matrix equation (5), most of its solution can be performed in closed form. The numerical effort involved amounts to carrying out a completely recursive procedure in $\mathcal{O}(M)$ operations.

Finally, the corresponding time-domain signals $\{\mathcal{E}(m,t)\}$ are obtained by determining the Bromwich inversion integral

$$\tilde{\mathcal{E}}(m,t) = (2\pi)^{-1}\exp(\beta t)\int_{-\infty}^{\infty} \exp(i\omega t)\tilde{E}(m, \beta + i\omega)\,d\omega, \tag{7}$$

with β being a real-valued, nonnegative parameter. The integral in (7) is evaluated numerically by invoking a standard FFT procedure. In this procedure, we take the frequency sampling so fine that the error in the resulting time signals is governed only by the accuracy of the space discretization leading to (4).

3. EXCITATION OF A STRAIGHT THIN-WIRE SEGMENT

For general scattering problems, a recursive solution procedure for the discretized, Laplace-transformed integral equation is not available. Therefore, we aim at solving this equation with the aid of *iterative techniques* based upon the minimization of an integrated squared error. In this section we consider a simple configuration for which this approach has been successful [2].

3.1. Formulation of the Problem

A perfectly conducting, straight thin-wire segment of length L with a circular cross-section of radius a is embedded in a homogeneous, lossless dielectric with permittivity ϵ and permeability μ (see Figure 2). A Cartesian coordinate system is introduced with the central axis of the wire located at $\{\mathbf{r} = z\mathbf{i}_z \mid 0 < z < L\}$. The wire is excited by an incident electromagnetic field $\{\mathcal{E}^i(\mathbf{r},t), \mathcal{H}^i(\mathbf{r},t)\}$, and/or driven by an impressed voltage $\mathcal{V}(t)$ across the gap $z_g - \delta < z < z_g + \delta$. The dimensions of the wire satisfy the inequality $\delta \ll a \ll L$. The aim of the computation is to determine the total current $I(z,t)$ flowing through the wire.

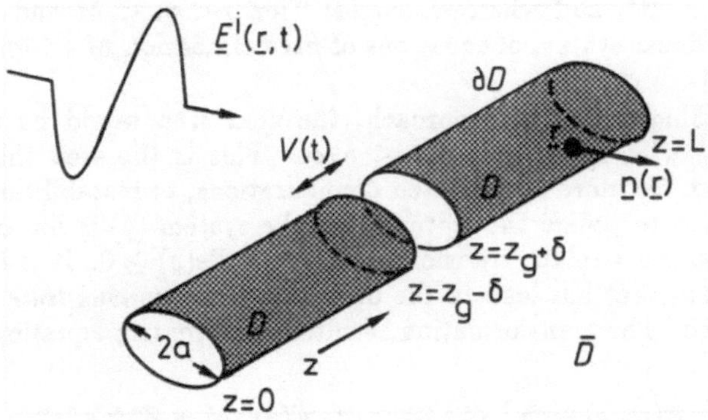

Figure 2. Transient excitation of a straight thin-wire segment by a pulsed incident field or an impressed voltage.

We consider Hallén's version of the so-called *reduced form* of the thin-wire integral equation, viz.

$$\int_0^L \frac{\mathcal{I}(z',t-R_a/c)}{4\pi R_a}\,dz' = \frac{Y}{2}\int_0^L \mathcal{E}_z^i(z'\mathbf{i}_z, t-\frac{|z-z'|}{c})\,dz' + \frac{Y}{2}\mathcal{V}(t-\frac{|z-z_g|}{c}) + \mathcal{F}_0(t-\frac{z}{c}) + \mathcal{F}_L(t-\frac{L-z}{c}), \tag{8}$$

with $0 < z < L$. In (8), we have $R_a = \sqrt{(z-z')^2+a^2}$, $c = 1/\sqrt{\epsilon\mu}$ and $Y = \sqrt{\epsilon/\mu}$. The unknown time signals $\mathcal{F}_0(t)$ and $\mathcal{F}_L(t)$ can be determined by invoking this equation for $z = 0, L$ as well, and imposing the end conditions $\mathcal{I}(0,t) = \mathcal{I}(L,t) = 0$.

We first discretize in space. The interval $0 < z < L$ is subdivided into M subintervals of equal length $h = L/M$. The grid points are taken at the boundaries of the subintervals, i.e., $z_m = mh$, with $m = 0, \dots, M$. In (8), the observation is restricted to the grid points $z = z_m$. The integral over the unknown current $\mathcal{I}(z', t-R_a/c)$ is approximated by a weighted sum over the sampled values at $z' = z_{m'}$ with $m' = 1, \dots, M-1$, through the piecewise-linear approximation

$$\mathcal{I}(z', t-\sqrt{(mh-z')^2+a^2/c}) \approx \sum_{m'=1}^{M-1} \mathcal{I}(z_{m'}, t-R_{m-m'}/c)\,\phi_{m'}(z'), \tag{9}$$

where $R_m = \sqrt{m^2h^2+a^2}$, and where $\phi_m(z)$ is the triangular expansion function

$$\phi_m(z) \stackrel{\text{def}}{=} \begin{cases} 1-|z-z_m|/h & \text{for } |z-z_m| \le h, \\ 0 & \text{otherwise.} \end{cases} \tag{10}$$

The equations with $m = 0, M$ have been included for the computation of the auxiliary signals $\mathcal{F}_0(t)$ and $\mathcal{F}_L(t)$. These approximations and a temporal Laplace transformation reduce the integral equation (8) to the system of linear equations

$$\sum_{m'=1}^{M-1} W_{m-m'} \exp(-s\frac{R_{m-m'}}{c})\,I_{m'}(s) = \frac{Y}{2}\exp(-s\frac{|z_m-z_g|}{c})\,V(s) + \frac{Y}{2}\sum_{m'=0}^{M} w_{m'} \exp(-s\frac{|z_m-z_{m'}|}{c})\,E_z^i(z_{m'}\mathbf{i}_z, s) + \exp(-s\frac{z_m}{c})\,F_0(s) + \exp(-s\frac{L-z_m}{c})\,F_L(s), \tag{11}$$

for $m = 0, \dots, M$. In the generalized trapezoidal rule on the left-hand side, the weighting coefficients are defined as

$$W_m = W_{-m} \stackrel{\text{def}}{=} \int_{z_m-h}^{z_m+h} \frac{\phi_m(z)}{\sqrt{z^2+a^2}}\, dz, \tag{12}$$

for $m = 0, 1, \ldots, M-1$. The integrals in the right-hand side of (12) are available in closed form. In the trapezoidal rule on the right-hand side, we have $w_m = h$ for $m = 1, \ldots, M-1$, and $w_0 = w_M = h/2$.

3.2. Marching on in Frequency

The discretized and transformed currents $\{I_m(s)\}$ and the transformed auxiliary signals $F_{0,L}(s)$ are determined by a dedicated version of the conjugate-gradient method. To this end, the discretized and transformed integral equation (11) is written in operator form as

$$\tilde{L}(s) \cdot \tilde{U}(s) = \tilde{F}(s). \tag{13}$$

Next, the solution of (13) is redefined as the vector that minimizes the squared error

$$\langle \tilde{L}(s) \cdot \tilde{U}(s) - \tilde{F}(s) \mid \tilde{L}(s) \cdot \tilde{U}(s) - \tilde{F}(s) \rangle, \tag{14}$$

where $\langle \tilde{F}(s) \mid \tilde{G}(s) \rangle$ is an innner product. Starting from an initial estimate $\tilde{U}^{(0)}(s)$, the minimum of this error is then searched by an iterative procedure [3].

In this method, the bulk of the computational effort is used up in the successive multiplications of the iterates with $\tilde{L}(s)$, and of the corresponding residuals with its adjoint. The efficiency of these multiplications improves considerably when the sum over m' on the left-hand side of (11) is recognized as a convolution that can be evaluated simultaneously for all relevant m with the aid of an FFT algorithm.

The key element in our implementation is the generation of the initial estimate $\tilde{U}^{(0)}(s)$. Recall that the space discretization is fixed and that the frequency step $\Delta\omega$ is relatively small. This makes it possible to *march on in frequency*, generating $\tilde{U}^{(0)}(s)$ at each frequency $s = s_n$ by extrapolating results obtained at "previous" frequencies. This is achieved as follows. As in any extrapolation scheme, the initial estimate for each new frequency s_n is expressed as a linear combination of previous "final" results:

$$\tilde{U}^{(0)}(s_n) \stackrel{\text{def}}{=} \sum_{k=1}^{K} \xi_k\, \tilde{U}(s_{n-k}). \tag{15}$$

Our special choice is to determine the expansion coefficients $\{\xi_k\}$ such that $\tilde{U}^{(0)}(s_n)$ minimizes the squared error (14). Since $\tilde{U}(s)$ is a well-behaved function of s, a good approximation of $\tilde{U}(s_n)$ can be obtained by taking a limited number of terms in the right-hand side of (15). In fact, it turns out that there is an optimum number since, for too large a K, the previous solutions $\{\tilde{U}(s_{n-k})\}$ become almost linearly dependent.

4. CIRCULAR DIELECTRIC CYLINDER

For *layered dielectric media*, the problem of solving a "general" discretized, Laplace-transformed integral equation can be avoided by applying a spatial Fourier transformation in the direction(s) where the scatterer exhibits translation symmetry, while the external incident field does not. In this section, we consider a case that leads to a discrete Fourier representation.

A radially inhomogeneous, lossy dielectric circular cylinder of radius a embedded in vacuum is excited by an electrically polarized, pulsed plane wave that is incident from the left (see Figure 3). For $a < \rho < \infty$ we have a vacuum, while for $0 \le \rho < a$ we have a dielectric with $\mu(x,y) = \mu_0$, $\epsilon(x,y) = \epsilon_2(\rho) \ge \epsilon_0$, and $\sigma(x,y) = \sigma_2(\rho) \ge 0$.

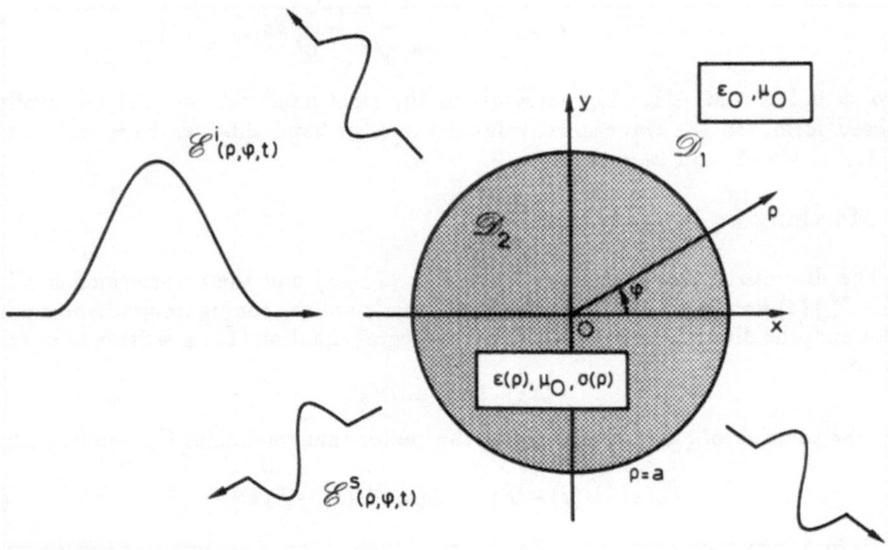

Figure 3. Scattering of an E-polarized, pulsed plane wave by a radially inhomogeneous, lossy dielectric circular cylinder.

As remarked above, the symmetry of the scattering object can be used to obtain a separation-of-variables solution. The electric-field strength can be represented in terms of the angular Fourier series

$$\mathcal{E}(\rho,\phi,t) = \lim_{L\to\infty}\left\{\sum_{\ell=-L}^{L}\exp(i\ell\phi)\,\mathcal{E}_\ell(\rho,t)\right\}. \tag{16}$$

From energy considerations, it follows that this representation converges in mean square for the scattered field outside the cylinder and for the total field inside the cylinder, uniformly in time [4,Appendix 2.5.A].

The integral equation for the spatial Fourier components $\{\mathcal{E}_\ell(\rho,t)\}$ occurring in (16) is most easily formulated in the time-Laplace domain. We end up with

$$\begin{aligned} E_\ell(\rho,s) &= A_\ell(s)\,I_\ell(s\rho/c_0) \\ &\quad -\frac{s^2}{c_0^2}\int_0^a\left[\chi(\rho')+\frac{\sigma_2(\rho')}{\epsilon_0 s}\right]\rho' I_\ell(\frac{s\rho_<}{c_0})K_\ell(\frac{s\rho_>}{c_0})E_\ell(\rho',s)\,d\rho', \end{aligned} \tag{17}$$

for $0 \le \rho \le a$. In (17), $\chi(\rho) \stackrel{\text{def}}{=} \epsilon(\rho)/\epsilon_0 - 1$ denotes the dielectric susceptibility, $\rho_< \stackrel{\text{def}}{=} \min\{\rho,\rho'\}$ and $\rho_> \stackrel{\text{def}}{=} \max\{\rho,\rho'\}$. The functions $I_\ell(z)$ and $K_\ell(z)$ are modified Bessel functions of order ℓ, and the amplitude $\{A_\ell(s)\}$ is given by

$$A_\ell(s) \stackrel{\text{def}}{=} (-1)^{\ell+1}\exp(-\frac{sa}{c_0})\,F(s), \tag{18}$$

where $F(s)$ represents the Laplace transform of the shape of the incident pulse.

Now, we introduce a *fixed* discretization for the integral over ρ' in (17), solve the resulting matrix equation repeatedly for varying s, and evaluate the Bromwich inversion integral in (7). In the space discretization, we introduce a spatial grid $\rho_m = mh$ with $m = 0,1,\ldots,M$, and $h = a/M$. Restricting the observation points to this grid, we can approximate the space integral in (17) by a repeated trapezoidal rule. The result can be written as

$$\tilde{E}(m,s) + k_m\sum_{m'=1}^{m} C_{m'}i_{m'}\tilde{E}(m',s) + i_m\sum_{m'=m+1}^{M} C_{m'}k_{m'}\tilde{E}(m',s) = i_m A_\ell(s), \tag{19}$$

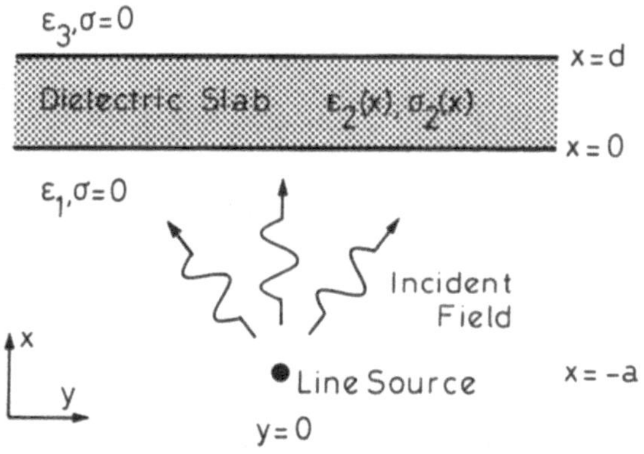

Figure 4. Transient excitation of a one-dimensionally inhomogeneous, lossy dielectric slab in between two homogeneous, lossless media by a localized, two-domensional current source.

where $m = 0, 1, \ldots, M$, and where the sums are understood to vanish when the lower index is larger than the upper one. In (19), we have

$$\begin{aligned} i_m &\overset{\text{def}}{=} I_\ell(smh/c_0) && \text{for } 0 \le m \le M, \\ k_m &\overset{\text{def}}{=} K_\ell(smh/c_0) && \text{for } 0 < m \le M, \\ C_m &\overset{\text{def}}{=} (s^2a^2/c_0^2)\, w_m \tfrac{m}{M} \left[\chi(mh) + \sigma(mh)/\epsilon_0 s\right] && \text{for } 0 < m \le M, \end{aligned} \tag{20}$$

and the weighting factors w_m are defined as $w_m = M^{-1}$ for $0 < m < M$ and $w_m = (2M)^{-1}$ for $m = 0, M$.

As in the case of the slab, we thus arrive at a linear system of equations of *fixed* dimension $M + 1$ for the transformed, discretized Fourier coefficients $\{\tilde{E}(m, s)\}$. Because of its special, degenerate form, Equation (19) can be reduced analytically to a tridiagonal matrix equation, provided that $\beta > 0$. This equation, in turn, can be solved recursively in $\mathcal{O}(M)$ operations. For details, the reader is referred to [1].

5. LINE-SOURCE EXCITATION OF A DIELECTRIC SLAB

Finally we need to discuss the application of our approach to a plane-stratified dielectric medium excited by a localized source. As our last case, we consider the most simple example of such a configuration.

We consider the two-dimensional scattering problem shown in Figure 4. The given "external" pulsed current-source distribution $\mathcal{J}_z^e(x, y, t)$ produces a transient field $\mathcal{E}_z^i(x, y, t)$ incident upon an inhomogeneous slab located in $0 < x < d$. The source is located in free space in front of the slab, i.e. in the half-space $-\infty < x < 0$. To the rear of the slab is a homogeneous, lossless dielectric half-space with coordinates $d < x < \infty$ and permittivity $\epsilon_3 = \epsilon_0\epsilon_{3r}$. It is desired to find the electric field $\mathcal{E}_z(x, y, t)$ excited in this structure. Without loss of generality, we restrict the analysis to the case where $\mathcal{J}_z^e(x, y, t)$ is a pulsed line source of arbitrary time dependence given by

$$\mathcal{J}_z^e(x, y, t) \overset{\text{def}}{=} \mathcal{I}(t)\delta(x + a)\delta(y), \tag{21}$$

where $\mathcal{I}(t)$ is the current flowing in the source.

Applying Fourier transformations in the t- and y-directions, restricting the temporal inversion to positive frequencies $0 < \omega < \infty$, and normalizing the spatial transform parameter according to $k_y = \frac{\omega}{c_0}\nu$ results in the frequency-domain Weyl representation

$$\mathcal{E}_z(x,y,t) = \frac{1}{4\pi^2}\mathrm{Re}\int_{-\infty}^{\infty}\frac{d\nu}{u_1}\int_0^{\infty} d\omega \exp\left[-i\omega\left(t-\frac{\nu y}{c_0}\right)\right] F(x,\nu,\omega), \tag{22}$$

where $u_1(\nu) \stackrel{\text{def}}{=} (\nu^2-1)^{\frac{1}{2}}$, with $\mathrm{Re}(u_1) \geq 0$ defining the "proper" Riemann sheet of the complex ν-plane. The transformed field $F(x,\nu,\omega)$ satisfies the one-dimensional integral equation

$$\begin{aligned} F(x,\nu,\omega) &= [i\omega\mu_0 I(\omega)]\exp\left(-\frac{\omega u_1}{c_0}|x+a|\right) \\ &+ \frac{iZ_0}{2u_1}\int_0^d [\sigma(x') - i\omega\epsilon_0\chi(x')]\exp\left(-\frac{\omega u_1}{c_0}|x-x'|\right)F(x',\nu,\omega)\,dx' \\ &+ \frac{u_1-u_3}{2u_1}\exp\left(-\frac{\omega u_1}{c_0}[d-x]\right)F(d,\nu,\omega), \end{aligned} \tag{23}$$

where $0 \leq x \leq d$ and $u_3(\nu) \stackrel{\text{def}}{=} (\nu^2-\epsilon_{3r})^{\frac{1}{2}}$, with $\mathrm{Re}(u_3) > 0$. This integral equation has the same structure as the frequency-domain counterpart of (3), and is discretized and solved in the same manner.

The advantage of the representation (22) is that the branch cuts due to u_1, u_3, and the collective influence of possible guided-wave modes have a fixed location in the complex ν-plane. In addition, the integrand of the ν-integral is of $\mathcal{O}(\nu^{-2})$ as $|\nu| \to \infty$, independently of x, y and t, and is an analytic function of the time coordinate in the lower half of the complex time plane. Combining these observations, we may deform the contour of integration to avoid nonuniqueness problems in the solution of (23) due to the presence of guided-wave modes, and approximate the result by a fixed combination of Gaussian quadrature rules. The transformation to the time domain proceeds as in the previous examples.

6. GENERALIZATION AND CONCLUSIONS

With the examples given in this paper, we have described the building blocks for a new way of using integral equations to model the propagation and scattering of transient electromagnetic waves. Of each example, generalizations to more complicated scattering problems have already been realized. We mention in particular the application of the marching-on-in-frequency method to the two-dimensional transient scattering by conducting and dielectric obstacles [2,5]. Applications to layered media include the excitation of a horizontally stratified dielectric slab by a point source, and the excitation of a cylindrically symmetric borehole by a line or ring source. The next challenge is to combine these ideas so that realistic configurations can be handled.

REFERENCES

1. A.G. Tijhuis, R. Wiemans and E.F. Kuester, "A hybrid method for solving time-domain integral equations in transient scattering," *J. Electro. Waves Appli.* 3:485–511 (1989).
2. A.G. Tijhuis and Z.Q. Peng, "A marching-on-in-fequency method for solving integral equations in transient electromagnetic scattering," *Proc. IEE, Part H* 138:347–355 (1991).
3. P.M. Van den Berg, "Iterative schemes based on the minimization of the error in field problems," *Electromagnetics* 5:237–262 (1985).
4. A.G. Tijhuis, "Electromagnetic Inverse Profiling: Theory and Numerical Implementation," VNU Science Press, Utrecht, The Netherlands (1987).
5. Z.Q. Peng and A.G. Tijhuis, "Transient scattering by a lossy dielectric cylinder: marching-on-in-frequency approach," *J. Electro. Waves Appli.* 6: to appear (1992).

PHASE ERROR CONTROL FOR FD-TD METHODS

Peter G. Petropoulos

Armstrong Laboratory, AL/OES
Brooks AFB, Texas 78235

INTRODUCTION

Researchers in computational electromagnetics have recently become aware of an important property[1] of finite element methods for elliptic boundary-value problems relevant to the frequency-domain Maxwell's equations. This property relates the number of points per wavelength, N_{ppw}, to frequency, ω, through a non-linear relationship. The theoretical results are for a fixed physical domain size and relate a scaled time-frequency, $k = \omega/c$, to the quantity $k\Delta = 2\pi/N_{ppw}$, where c is the speed of light in vacuum and Δ is the characteristic spatial scale of the discretization. Herein, we derive a property of the phase error for finite difference approximations to the *time-domain* Maxwell's equations that is appropriately *conjugate* to that for the conjugate, frequency-domain, problem[1].

The FD-TD[2] modelling of the interaction of ultra-wideband signals with temporally dispersive media demands faithfull propagation of high-frequency events, e.g. Sommerfeld's precursor, over short distances in computations that require thousands of time steps. Previous work[3] related N_{ppw} to electrical domain size heuristically for the FD-TD approximation to Maxwell's equations in the time domain. However, computations of scattering of a temporal pulse by a complicated structure may, or may not, involve an electrically large domain while long computations (measured in number of timesteps) over a discretized domain are almost always unavoidable. Therefore, a useful relation between a physical scale and a discretization scale would be one between total computation time and N_{ppw}. In addition, such relation must take into account the maximum tolerable phase error. The points per wavelength and the total run time are *a priori* determined for a computation. The wavelength used to determine N_{ppw} is that of the highest frequency present with significant energy in the domain.

FD-TD schemes are often run in their dispersive regime, i.e., $\Delta t \ll \Delta$, so that the solution is time-accurate. For such cases we will derive estimates for the required spatial discretization so that a predetermined phase error for the highest frequency component is achieved by the end of the timestepping. Our analysis will employ a continuous-time/discretized-space approximation to Maxwell's equations. This approximation also results from the fully discrete FD-TD schemes in the limit $\Delta t \to 0$. We examine this case since attempts to reduce errors through spatial step reduction impose a geometric rate of increase on computational resources while timestep reduction just increases computation time. In practice, a 2nd order accurate, one-step scheme is used for the time discretization and the computation time increases linearly as one applies a finer temporal sampling to improve the accuracy. However, the scheme then becomes more dispersive and a reduction of the spatial step cannot be avoided particularly for problems in which the physics themselves are of a dispersive nature. Therefore, if one is willing to work within the realm of finite difference methods that are 2nd order accurate in time, it seems logical to investigate methods that are of higher order of accuracy in space since they can achieve the same error as the usual 2nd order FD-TD scheme with larger spatial steps and less amount of timesteps.

We will show, and numerically confirm, that the relation

$$N_{ppw} \sim \alpha(\frac{P}{|e_\phi|})^{\frac{1}{s}}$$

holds for FD-TD of spatial order of accuracy s, where α depends only on the order of spatial accuracy of the scheme, P is the number of periods in time for which the computation will proceed, and e_ϕ is the phase error in radians that will be allowed to accumulate over the computation duration. P is determined from t_c and ω_*, the actual computation time and the highest frequency that is expected to be present in the calculation, respectively. It is given by $P = t_c\omega_*/2\pi$. An estimate for ω_* can be obtained by considering the initial pulse frequency content. The fraction e_ϕ is also decided *a priori*. In two and three dimensions α also depends on the angle of propagation and achieves its maximum value along directions parallel to the grid axes. The relation above then gives an upper bound on N_{ppw} in higher dimensions, and provides a justification for efforts to develop differencing schemes of higher spatial accuracy than that used in traditional FD-TD.

DISPERSION ANALYSIS

The property we seek will result from semi-discretizations of the one dimensional Maxwell's equations

$$\begin{aligned} \frac{\partial E}{\partial t} &= \frac{\partial H}{\partial x} \\ \frac{\partial H}{\partial t} &= \frac{\partial E}{\partial x}. \end{aligned} \tag{1}$$

System (1) results by scaling the electric and magnetic fields with the square root of the permittivity and permeability of free space and letting the speed of light to be one so k and ω are synonymous quantities. An exact solution corresponding to a Fourier component is the vector $\vec{M} = (E(x,t), H(x,t))^T$, where the superscript T denotes transpose, written in separable form as

$$\vec{M} = \left\{ \begin{array}{c} e(t) \\ h(t) \end{array} \right\} e^{ikx}, \tag{2}$$

and $e(t), h(t)$ are determined by inserting (2) into (1) and solving the resultant ordinary differential equations (ode) for the time dependence with initial conditions given by the constant vector $\vec{M}(x,0) = (e(0), h(0))^T e^{ikx}$. The result is

$$\vec{M} = \left\{ \begin{array}{c} e(0) \\ h(0) \end{array} \right\} e^{ik(x-t)} \tag{3}$$

and a right moving wave has been choosen by appropriatelly setting $e(0)$ and $h(0)$ in $\vec{M}(x,0)$. The analysis now proceeds by considering finite difference approximations for $\partial/\partial x$ in (1).

Representing the spatial derivatives by half-cell centered, 2nd order accurate finite differences and denoting the resulting approximate solution by $\vec{M}_{app}$, we obtain for (1)

$$\begin{aligned} \frac{dh_{app}(t)}{dt} e^{i\frac{k\Delta}{2}} &= \frac{1}{\Delta}(e^{ik\Delta} - 1)e_{app}(t) \\ \frac{de_{app}(t)}{dt} &= \frac{1}{\Delta}(e^{i\frac{k\Delta}{2}} - e^{-i\frac{k\Delta}{2}})h_{app}(t) \end{aligned} \tag{4}$$

with Δ being the cell width and assuming that $x = m\Delta$ with m being an integer. The half-cell centering is necessary since the full discretization will require a staggered space-time grid for E and H in order to achieve $O(\Delta t^2)$ accuracy with a half-step centered, one-level of storage, time differencing. Upon ellimination of $h_{app}(t)$ from (4) we get a second order ode for the time evolution of the electric field, i.e.,

$$\frac{d^2 e_{app}(t)}{dt^2} = -k^2 \frac{4}{(k\Delta)^2} \sin^2(\frac{k\Delta}{2}) e_{app}(t). \tag{5}$$

Solving we find that (using $k\Delta = 2\pi/N_{ppw}$)

$$E_{app}(x,t) = e(0)e^{ik[x - \frac{N_{ppw}}{\pi}\sin(\frac{\pi}{N_{ppw}})t]}. \tag{6}$$

The phase error is defined as the ordered difference between the exact and numerical phase, $e_\phi = \Phi_{exact} - \Phi_{numerical}$. Comparing (6) with the electric field component in (3) we find that the phase error produced by the spatial finite differencing is

$$e_\phi = -kt[1 - \frac{N_{ppw}}{\pi}\sin(\frac{\pi}{N_{ppw}})]. \tag{7}$$

Thus, e_ϕ depends *linearly* on time, and it is a lagging error. For an actual computation one would set k to be the highest wavenumber, k_*, in the problem for which we want a preset accuracy to be maintained up to the end of the timestepping. Using for t in (7) the quantity t_c, which is *a priori* determined for a calculation, and the Taylor series for the *sin* function keeping only the first term in the series for which (7) is non-vanishing, we obtain for the dependence of N_{ppw} on the allowed phase error and on the total computation time

$$N_{ppw} \sim (\frac{1}{3})^{\frac{1}{2}} \pi^{\frac{3}{2}} (\frac{P}{|e_\phi|})^{\frac{1}{2}}. \tag{8}$$

Another half-cell centered discretization for the spatial derivatives in (1) that is $O(\Delta^4)$ accurate is given by

$$\frac{\partial f}{\partial x}\Big|_{x=(m+\frac{1}{2})\Delta} = \frac{1}{\Delta}[-\frac{1}{24}f_{m+2} + \frac{9}{8}f_{m+1} - \frac{9}{8}f_m + \frac{1}{24}f_{m-1}]. \tag{9}$$

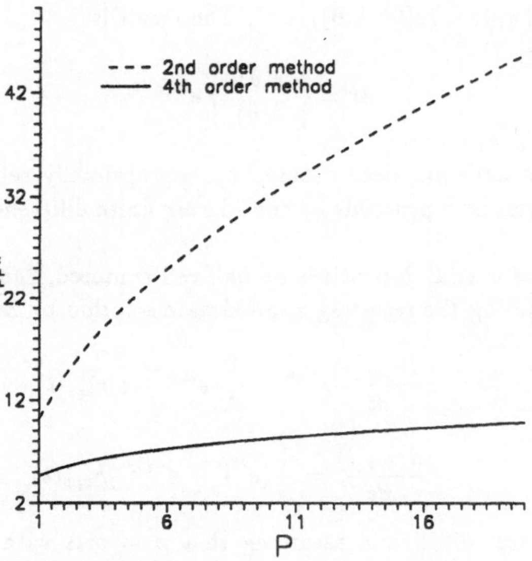

Figure 1. The dependence of N_{ppw} on P for an allowed phase error $e_\phi = 0.1$ radians.

We should note here that a similar discretization scheme was previously presented[4,5]. With (9) the set (4) becomes

$$\frac{dh_{app}(t)}{dt}e^{i\frac{k\Delta}{2}} = \frac{1}{\Delta}(-\frac{1}{24}e^{2ik\Delta} + \frac{1}{24}e^{-ik\Delta} + \frac{9}{8}e^{ik\Delta} - \frac{9}{8})e_{app}(t)$$

$$\frac{de_{app}(t)}{dt} = \frac{1}{\Delta}(-\frac{1}{24}e^{i\frac{3k\Delta}{2}} + \frac{1}{24}e^{-i\frac{3k\Delta}{2}} + \frac{9}{8}e^{i\frac{k\Delta}{2}} - \frac{9}{8}e^{-i\frac{k\Delta}{2}})h_{app}(t) \tag{10}$$

thus (5) is now

$$\frac{d^2e_{app}(t)}{dt^2} = -k^2\frac{4}{(k\Delta)^2}[\frac{9}{8}\sin(\frac{k\Delta}{2}) - \frac{1}{24}\sin(\frac{3k\Delta}{2})]^2 e_{app}(t). \tag{11}$$

Proceeding as before we obtain the phase error

$$e_\phi = -kt[1 - \frac{N_{ppw}}{\pi}(\frac{9}{8}\sin(\frac{\pi}{N_{ppw}}) - \frac{1}{24}\sin(\frac{3\pi}{N_{ppw}}))], \tag{12}$$

and as a result (8) now reads

$$N_{ppw} \sim (\frac{3}{20})^{\frac{1}{4}}\pi^{\frac{5}{4}}(\frac{P}{|e_\phi|})^{\frac{1}{4}}. \tag{13}$$

Using the same methodology, the phase error of 2nd and 4th order spatially accurate semi-discretizations for the Maxwell equations in two and three dimensions is easy to derive. For example, for two dimensional Transverse Magnetic (TM) polarization one finds the phase error of the 4th order method to be

$$\begin{aligned} e_\phi &= -|k|t[1 - \frac{N_{ppw}}{\pi}\sqrt{p_x^2 + p_y^2}] \\ p_x &= \frac{9}{8}\sin(\frac{\pi\sin\theta}{N_{ppw}}) - \frac{1}{24}\sin(\frac{3\pi\sin\theta}{N_{ppw}}) \\ p_y &= \frac{9}{8}\sin(\frac{\pi\cos\theta}{N_{ppw}}) - \frac{1}{24}\sin(\frac{3\pi\cos\theta}{N_{ppw}}) \end{aligned} \tag{14}$$

where $|k|$ is the magnitude of the wavenumber (taken as the maximum $|k_*|$), θ is the angle of propagation measured on the grid with respect to the vertical y−axis, and $|k_*|t = 2\pi P$. As seen from (14), in higher dimensions the finite difference schemes are anisotropic in addition to being dispersive. This anisotropy manifests itself through an additional dependence of the phase error on angle of propagation. However, the anisotropy is maximum for propagation parallel to the grid $y(x)$-axes where $\alpha = 0^o(90^o)$, whence $p_x(p_y) = 0$ and then (14) reduces to (12). The same holds for a two-dimensional extension of (4) and for the three dimensional case. Thus, the one dimensional estimates in (8) and (13) for N_{ppw} also apply in higher dimensions as upper bounds.

Relations (8) and (13) are the main results of this paper since they give an estimate for N_{ppw} based on *a priori* selected computation parameters. Figure 1 shows the relationship between N_{ppw} and P, the number of periods for which the computation will proceed, for an allowed phase error of 0.1 radians. The benefit of a high order spatial differencing scheme is apparent from this Figure. We should point out here that the phase error relationships derived herein hold true for the Courant-Friedrichs-Lewis number $CFL = c\Delta t/\Delta \to 0$ and this limit is relevant since that should be the case for a calculation to be time-accurate. Also, since CFL is the same for both the 2nd and 4th order methods, another deduction from Figure 1 is that the higher order methods will allow a larger time step. This is confirmed by the numerical experiments in the next section.

NUMERICAL EXPERIMENTS

The results of the dispersion analysis presented in the previous section will now be verified against numerical experiments with (1). Because our analysis does not consider the effects of the initial and boundary conditions on the interior phase error we prescribe exact initial data, $E(x, t = 0)$ and $H(x, t = -\Delta t/2)$, such that a right moving harmonic component of wavelength λ exists over the numerical grid, and periodic boundary conditions, $E(0, t) = E(L, t)$ and $H(\Delta/2, t) = H(L + \Delta/2, t)$, where L is the physical domain length, and Δt is the timestep. From now on we consider cases for which $L = \lambda$ and we will allow a maximum phase error $|e_\phi| = 0.1$ radians so that N_{ppw} is determined from Figure 1 after the decision has been made concerning the number of periods P that the schemes would be run for. We always round N_{ppw} up or down to an odd integer so that the half-period node of the initial condition is always at the center cell of the grid at $L/2$, and then that number becomes NX, the odd number of spatial cells on the grid. Thus $\Delta = L/(NX - 1)$ and then from the CFL number we obtain the time step. The total number of time steps is equal to $NMAX = P * (NX - 1)/CFL$.

The two discretization schemes considered are the traditional FD-TD scheme and the 2nd-order-time-accurate/4th-order-space-accurate FD-TD type scheme discussed in the previous section. We denote the discretized $E(x, t), H(x, t)$ by $E_i, H_{i+\frac{1}{2}}$ where $x = (i - 1) * \Delta$, and the E field is at time level n while the H field is at time level $n + \frac{1}{2}$, and i is the spatial cell index that ranges from 1 to $NX - 1$. To implement the periodic boundary conditions for the difference equations of these schemes we need to identify for all time steps $E_1 = E_{NX}$ and $H_{1+\frac{1}{2}} = H_{NX+\frac{1}{2}}$. For example, the 4th order scheme is as follows,

$$Prescribe\ H_i^{-\frac{1}{2}},\ E_i^0,\ i = 1, NX - 1$$

Time loop, $n = 0, NMAX$

Update $H-field$

$$H_1^{n+\frac{1}{2}} = H_1^{n-\frac{1}{2}} + \frac{\Delta t}{\Delta x}[\frac{9}{8}(E_2^n - E_1^n) - \frac{1}{24}(E_3^n - E_{NX-1}^n)]$$

$$H_i^{n+\frac{1}{2}} = H_i^{n-\frac{1}{2}} + \frac{\Delta t}{\Delta x}[\frac{9}{8}(E_{i+1}^n - E_i^n) - \frac{1}{24}(E_{i+2}^n - E_{i-1}^n)],\ i = 2, NX-3$$

$$H_{NX-2}^{n+\frac{1}{2}} = H_{NX-2}^{n-\frac{1}{2}} + \frac{\Delta t}{\Delta x}[\frac{9}{8}(E_{NX-1}^n - E_{NX-2}^n) - \frac{1}{24}(E_1^n - E_{NX-3}^n)]$$

$$H_{NX-1}^{n+\frac{1}{2}} = H_{NX-1}^{n-\frac{1}{2}} + \frac{\Delta t}{\Delta x}[\frac{9}{8}(E_1^n - E_{NX-1}^n) - \frac{1}{24}(E_2^n - E_{NX-2}^n)]$$

Update $E-field$

$$E_1^{n+1} = E_1^n + \frac{\Delta t}{\Delta x}[\frac{9}{8}(H_1^{n+\frac{1}{2}} - H_{NX-1}^{n+\frac{1}{2}}) - \frac{1}{24}(H_2^{n+\frac{1}{2}} - H_{NX-2}^{n+\frac{1}{2}})]$$

$$E_2^{n+1} = E_2^n + \frac{\Delta t}{\Delta x}[\frac{9}{8}(H_2^{n+\frac{1}{2}} - H_1^{n+\frac{1}{2}}) - \frac{1}{24}(H_3^{n+\frac{1}{2}} - H_{NX-1}^{n+\frac{1}{2}})]$$

$$E_i^{n+1} = E_i^n + \frac{\Delta t}{\Delta x}[\frac{9}{8}(H_i^{n+\frac{1}{2}} - H_{i-1}^{n+\frac{1}{2}}) - \frac{1}{24}(H_{i+1}^{n+\frac{1}{2}} - H_{i-2}^{n+\frac{1}{2}})],\ i = 3, NX-2$$

$$E_{NX-1}^{n+1} = E_i^n + \frac{\Delta t}{\Delta x}[\frac{9}{8}(H_{NX-1}^{n+\frac{1}{2}} - H_{NX-2}^{n+\frac{1}{2}}) - \frac{1}{24}(H_1^{n+\frac{1}{2}} - H_{NX-3}^{n+\frac{1}{2}})]$$

End time loop.

In the numerical experiments the output of the schemes after P periods should exactly reproduce the initial condition in the absence of phase error. If N_{ppw} is not choosen correctly there will be phase error and the final waveform will lag behind the initial condition by an amount $|e_\phi|$ given by (8) or (13). Using the graph of Figure 1 to determine N_{ppw} given P, we should observe a phase error of 0.1 radians at the end of $NMAX$ number of time steps.

Figures 2a and 2b show a comparison between the theoretical estimate and the computed phase error for the 2nd and 4th order schemes, respectively. The computed phase error was obtained by using linear interpolation to find the zero crossing which, for the final E field, lies in the interval $(\frac{L}{2}-\Delta, \frac{L}{2})$ (see Figures 3a) and 3b)). Then, according to the number of radians corresponding to each cell on the grid (because $L = \lambda$), the difference between the computed zero crossing and the point $x = \frac{L}{2}$, which would have been the zero crossing if there was no artificial dispersion, is converted to degrees and plotted against P. We have choosen the points per wavelength from Figure 1 so that after 10 periods the accumulated phase error would be 0.1 radians approximatelly. In the Figures we track the computed error up to the maximum number of time steps and it is seen that the linear growth and predicted final value of the phase error is confirmed. For the traditional 2nd order scheme it was necessary to have $NX = 33$, while for the 4th order scheme it was only $NX = 9$ in order to achieve the preset phase error for a $CFL = 0.025$. For $P = 10$, the 2nd order scheme was updated for 12 800 timesteps, while the 4th order scheme only for 3 200 timesteps. In this CFL limit,

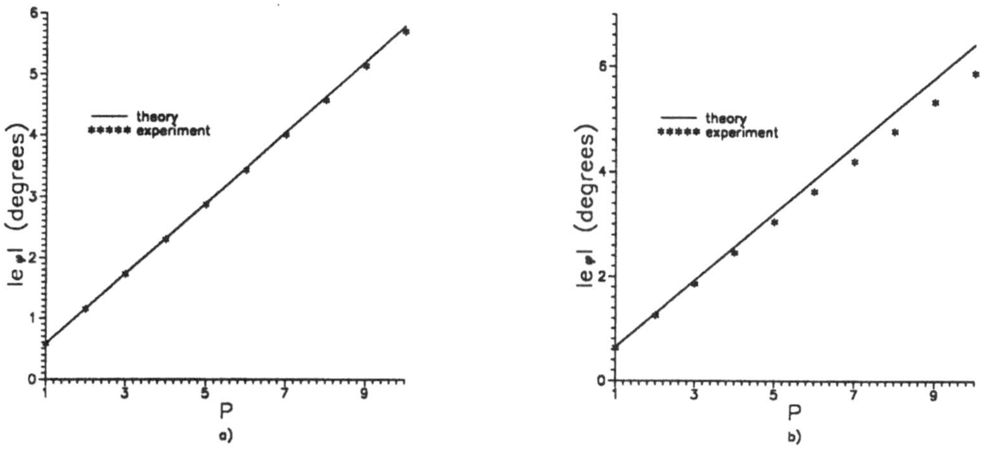

Figure 2. Computed phase error (stars) growth with the number of time period for a) the 2nd order scheme, and b) the 4th order scheme, shown with the theoretical prediction (line).

the 4th order method achieves its full theoretical accuracy. The numerical experiments also indicate a delicate sensitivity on N_{ppw} for the computed phase error of the 4th order method. From Figure 1 the optimum rounded value is $N_{ppw} = 8$ ($NX = 9$) and then the computed phase error is 5.887^o for the 4th order method. This value for the computed error is better than the predicted error and we attribute this difference to the fact that the prediction for N_{ppw} was 8.54 and not 8 which was the value employed for the experiments. Two more runs with $NX = 7$ ($N_{ppw} = 6$) and $NX = 11$ ($N_{ppw} = 10$) resulted in a computed phase error of 15.548^o and 2.520^o, respectively, for the $4th$ order method, while runs with $NX = 35$ ($N_{ppw} = 34$) and $NX = 31$ ($N_{ppw} = 30$) for the 2nd order method produced a phase error of 5.058^o and 6.463^o respectively. For the 2nd order method the optimum value from Figure 1 is $N_{ppw} = 32$ ($NX = 33$) and the computed phase error is then 5.697^o.

SUMMARY

In this paper we considered the error solely due to the spatial discretization. It was assumed that Δt was sufficient to obtain a time-accurate solution. The results of the derivations are valid for cases where the numerical computations are performed with $CFL \ll 1$ and this will always be the case with FD-TD methods of order of spatial accuracy greater than 2 if any benefit is to be gained from their use. The 2nd order FD-TD has a one dimensional non-dispersive limit ($CFL = 1$) of benefit but it is used with a $CFL \ll 1$ in order to allow a manageable number of spatial grid cells. We have shown that the number of spatial steps used can be further reduced by using a 4th order accurate spatial scheme. It should also be stressed that our results will be useful in the simulation of wave propagation in temporally dispersive media. In problems where the situation to be modeled is already dispersive, the researcher should be wary of artificial dispersion from the scheme manifesting itself in the final results as a "real" effect. In the regime of small timestep and large spatial step FD-TD schemes exhibit artificial

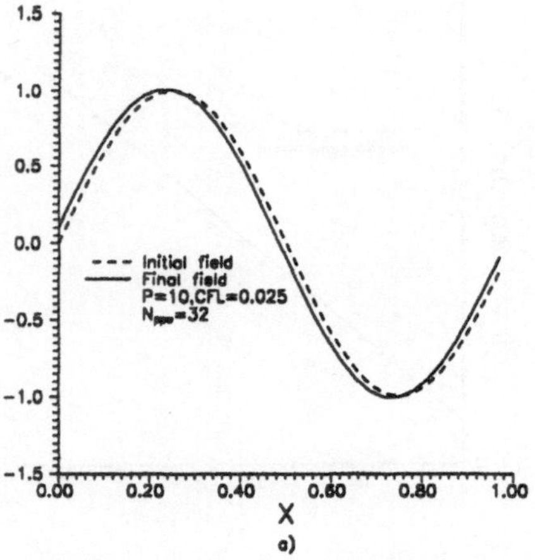

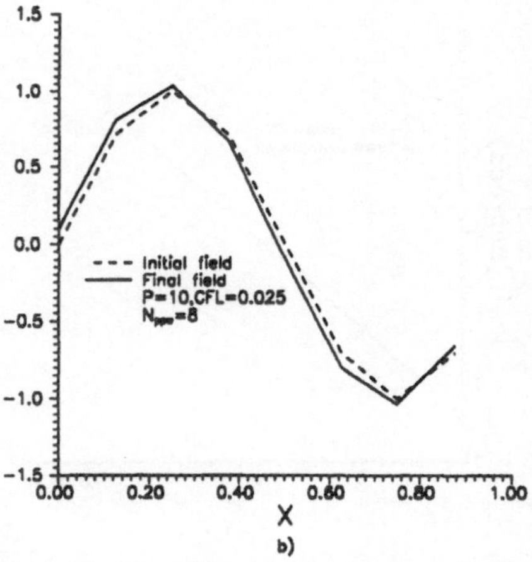

Figure 3. a) Initial and $P = 10$ field values for 2nd order scheme. b) Same as a) for 4th order scheme.

dispersion but this dispersion can be elliminated by choosing the points per wavelength as shown here so that the highest frequency events are appropriatelly sampled, e.g., the Sommerfeld precursor in a Lorentz medium or the Brillouin precursor in a Debye medium. Currently we are implementing 2-dimensional experiments for the verification of the results presented herein, and we are extending the analysis to the case where the medium is temporally dispersive.

Acknowledgements

This work was supported by the HSD Scholar Program at the USAF Armstrong Laboratory under contract F41624-92-D-4001.

REFERENCES

1. A. Bayliss, C.I. Goldstein and E. Turkel, On accuracy conditions for the numerical computation of waves, *Journal of Computational Physics,* 59:396 (1985).
2. K.S. Yee, Numerical solution of initial boundary value problems involving maxwell's equations in isotropic media, *IEEE Trans. Antennas Propagat.,* 14:302 (1966).
3. A.C. Cangellaris, Time-domain finite methods for electromagnetic wave propagation and scattering, *IEEE Trans. Magnetics,* 27:3780 (1991).
4. L. Lapidus and G.F. Pinder. "Numerical Solution of Partial Differential Equations in Science and Engineering," John Wiley, New York (1982).
5. J. Fang. "Time Domain Finite Difference Computations for Maxwell's Equations," Ph.D. Dissertation, University of California, Berkeley, CA (1989).

SIGNAL-PROCESSING APPROACH TO ROBUST TIME-DOMAIN MODELING OF ELECTROMAGNETIC FIELDS

Alfred Fettweis

Lehrstuhl für Nachrichtentechnik
Ruhr-Universität Bochum
D 4630 Bochum
Germany

Abstract

In recent publications[1-9], a new method for integrating partial differential equations describing physical systems has been presented. This method is based on simulating the actual continuous-domain system by means of a discrete-domain system, and this in such a way that the following features hold:

1. Preservation of originally existing passivity and incremental passivity, and this in such a way that these properties become available in the multidimensional (MD) sense even though they existed originally only in the one-dimensional (1-D) sense (i.e., with respect to time). As a result, one can achieve not only full stability with respect to the discretization in space and time but also full stability, and, more generally, full robustness with respect to the computational errors that are due to rounding/truncation and overflow corrections and to extraneous sources.This is possible because a multidimensional vector Liapunov function having a sufficiently simple structure is available.
2. Preservation of the exclusively local nature of the interconnections and the massive parallelism, which are inherent to all physical systems with finite propagation speed. As a result, for any given fixed time instant to be considered, the computations can be carried out simultaneously, thus fully in parallel, in all the spatial sampling points, and the computations in any of these points require previously computed results only from the immediate neighboring points.
3. Arbitrarily changing parameters as well as arbitrary boundary shapes and conditions can be tanken into account in a straightforward manner.

4. Discretization is done on the basis of the trapezoidal rule. In order to achieve recursibility (computability), the simulation may not be based on the field variables appearing in the original partial differential equations. Instead, corresponding so-called wave variables should be employed, thus variables of the type occuring in relation with the scattering-matrix formalism. This way, the mechanism involved in the physical system becomes interpretable as an incidence-to-scattering (reflection, transmission) mechanism, i.e. a mechanism exhibiting a cause-to-effect (causality) relationship. The latter in turn gives rise to computational rules that exhibit the sequential nature needed for obtaining an algorithm.

5. It appears easiest to apply the method by first representing the system by means of a multidimensional Kirchhoff circuit. From this, the desired algorithm can be derived by applying the standard procedures known from the theory of multidimensions wave digital filters[10], which has originally been developed within the context of digital signal processing. It will be discussed that the approach is applicable without difficulty to systems described by Maxwell's equations[11].

References

1. A. Fettweis, "New results in wave digital filtering", Proc. URSI Int.Symp. on Signals, Systems, and Electronics, pp. 17-23, Erlangen, Germany, Sept. 1989.
2. A. Fettweis and G. Nitsche, "Numerical integration of partial differential equations using priciples of multidimensional wave digital filters", Journal of VLSI Signal Processing, vol. 3, pp. 7-24, 1991.
3. A. Fettweis and G. Nitsche, "Transformation approach to numerically integrating PDEs by means of WDF principles", Multidimensional Systems and Signal Processing, vol. 2, pp. 127-159, May 1991.
4. A. Fettweis and G. Nitsche, "Massively parallel algorithms for numerical integration of partial differential equations", in "Algorithms and Parallel VLSI Architectures", (edited by E.F.Deprettere and A.-J. van der Veen), vol. B: Proceedings, pp. 475-484, Elsevier Science Publishers, Amsterdam, 1991.
5. A. Fettweis, "The role of passivity and losslessness in multidimensional digital signal processing - new challenges", Proc. 1991 IEEE Int. Symp. Circuits and Systems, vol. 1, pp. 112-115, Singapore, June 1991.
6. A. Fettweis, "Discrete passive modelling of viscous fluids", Proc. IEEE Int. Symp. Circuits and Systems, vol. 4, pp. 1640 - 1643, San Diego, CA, May 1992.
7. A. Fettweis, "Discrete modelling of lossles fluid dynamic systems", Arch. Elektr. Übert., vol. 46, pp. 209 - 218, July 1992.
8. A. Fettweis, "Discrete modelling of physical systems described by PDEs", Proc. 6th Eur. Conf. Signal Processing, vol. 1, pp. 55 - 62, Brussels, Belgium, Aug. 1992.

9. G. Nitsche, "Numerische Lösung partieller Differentialgleichungen mit Hilfe von Wellendigitalfiltern", Doctoral Dissertation, Ruhr-Universität Bochum, Germany, 1992.
10. A. Fettweis, "Wave digital filters: theory and practice", Proceedings IEEE, vol. 74, pp. 270-327, Feb. 1986.
11 A. Fettweis, "Multidimensional wave digital filters for discrete-time modelling of Maxwell's equations", Int. J. Numerical Modelling, vol. 5, 1992 (in print).

PHASE SPACE ANALYSIS AND LOCALIZATION OF SCATTERING BY A PLANAR, INFINITE, WEAKLY-RESONANT ARRAY ILLUMINATED BY A VERY SHORT PLANE PULSE

P. Borderies[1] and L. B. Felsen[2]

[1]Onera-Cert/Dermo
2, Avenue. E. Belin
31055 Toulouse Cédex - France
[2]Polytechnic University
NewYork - USA

1) INTRODUCTION

With the increasingly greater demands on wave channel capabilities and control for communication, remote sensing, object identification, feature extraction, etc..., predictive modeling algorithms must be parametrized so that the wave phenomena included in the model are matched as well as possible to scattering mechanisms in the environment that the signal has traversed during its progression from source to receiver. We have called such modelling "observable-based parametrization" (OBP) because it endeavours to link features (observables) in data to the relevant robust wave physics . If successful, an OBP permits a priori reconstruction of data, as well as parametric studies for forward and inverse requirements . The strategies for developing OBP 's have been reported elsewhere. They rely heavily on wave object parametrization in a configuration-spectrum (i.e, (space-time)-(wavenumber-frequency)) phase space, and on localization that defines the relevant domains of influence . Localization can be achieved by brute-force windowing (filtering), or inherently by the constructive and destructive interference among the waves in the continuum that synthesize the wave object. The latter mechanism is of special importance because it relies on the wave mechanics, i.e., the physics of signal propagation. This allows physically conditioned O.B.P.'s to be tracked forward and backward through successive encounters, and permits wave-based rather than brute-force processing of data in various subdomains of the eight-dimensional $(x,y,z,t\,;\,k_x,k_y,k_z,\omega)$ phase-space (ref.1).

Constructive and destructive interference among basis elements in a continuum is formalized by stationary phase evaluation of the synthesizing phase space integral(s). The stationary points define trajectories in the phase space, and the domain of influence around a trajectory ,established by the size of the localizing basis element "packet", is related directly to the wave object amplitude. The resulting algorithm that tracks a localized wave object through phase-space can be said to implement "phase-space ray tracing".

It has been recognized that the above localization, when applied in the configurational domain, is related to the first Fresnel zone surrounding the central ray (stationary point) trajectory from an initial reference surface to the observer (ref.2).

Ultra-Wideband, Short-Pulse Electromagnetics
Edited by H. Bertoni *et al.*, Plenum Press, 1993

This concept can be generalized to any phase-space trajectory, thereby generating Fresnel volumes around spectral, configurational or hybrid forms. Of special interest in this regard is the time-domain treatment of structurally induced dispersion, and how dispersive spectra evolve under transient conditions. For illustration, we consider the problem of short - pulse (S P) plane wave scattering by infinite, periodic arrays. Here, multiple scattering between individual elements and collective treatment via dispersive Floquet modes represent alternative approaches centered, respectively, around nondispersive and structurally dispersive parametrizations. The general concepts, which parametrize periodicity for arbitrary scattering elements, are implemented for an array of flat coplanar strips.

2) PHASE-SPACE PARAMETRIZATION STRATEGIES

A) Stationary Phase and Fresnel Localization

For synthesis of space-time dependent fields, the simplest phase space subdomain is two-dimensional (w,ξ) where w and ξ are configurational and spectral variables, respectively. Using locally adaptive basis functions $A\exp(i\Psi)$ in the spectral domain, the synthesizing integral has the generic form :

$$I_\xi(w) = \int d\xi \, A(w, \xi) \exp \{i\Psi(w,\xi) \}, \tag{1}$$

where $A(w, \xi)$ is the slowly varying spectral amplitude, and $\Psi(w, \xi)$ a rapidly varying phase. The corresponding observable is localized around the stationary (saddle) point (s) $\xi_s(w)$ defined by :

$$\Psi'(\xi_s,w) = 0 \quad , \quad ' \equiv \partial/\partial\xi \tag{2}$$

The stationary phase condition $\Psi'(\xi, w) = 0$ defines a curve in the (w, ξ) domain which can be parametrized either as $\xi = \xi(w)$ or $w = w(\xi)$. The latter form is appropriate when one performs the synthesis over the configurational variable w instead of the spectral variable ξ, with the basis functions in (1) defined accordingly, and with the outcome $I_w(\xi)$.

These dual properties of phase-space parametrizations have been discussed in the litterature (see (ref.3), for example). Asymptotic evaluation of the integral in (1) yields the familiar result (ref.4, ch.4),

$$I(w) \sim (\Delta\xi_s) \, A(\xi_s) \exp(i\Psi(w,\xi_s)) \; ; \; \xi_s = \xi_s(w) \tag{3}$$

where

$$\Delta\xi_s = [2\pi/i\Psi''(w,\xi_s)]^{1/2} \tag{3a}$$

expresses the modification of the spectral amplitude $A(\xi_s)$ due to the constructive interference among the synthesizing basis functions that surround the stationary point spectrum ξ_s.

The spectral spread $\Delta_f\xi_{|w}$, which generates (3a) at w, can be readily identified by expanding the phase function Ψ around ξ_s to the first nontrivial (quadratic) term (recalling (2)),

$$\Psi(\xi)=\Psi(\xi_s) + \Psi''(\xi_s) \, (\xi-\xi_s)^2/2 + \ldots\ldots \tag{4}$$

so that

$$\Delta_f\xi|_w = |\xi_f - \xi_s|_w = [2[\Psi(\xi)-\Psi(\xi_s)]/\Psi''(\xi_s)]^{1/2}_w \tag{4a}$$

Evidently ,

$$\Delta_f\xi|_w = \Delta\xi_s|_w \quad \text{if } |\Psi(\xi)-\Psi(\xi_s)|_w = \pi \tag{5}$$

The condition in (5) defines the spectral interval around ξ_s, over which the phase at w differs from the saddle-point phase by at most 180°. This domain of constructive interference defines the first Fresnel zone (fig 1).

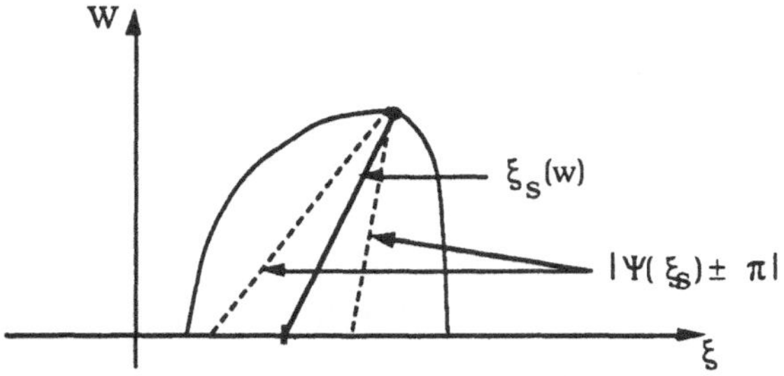

Figure 1. Fresnel localization

B) Structural Dispersion

In an environment that is spatially constrained (for example, with boundaries or interfaces along the z-coordinate), the reduced phase-space (r,t ; k_t, ω), with $k_z = k_z(k_t)$, can accommodate phenomena of structural dispersion (ref.1) defined by a dispersion relation,

$$D(k_t,\omega) = 0 \qquad , \quad k_t = (k_x,k_y) \tag{6}$$

Solving (6) in the alternative forms $k_t = k_{tn}(\omega)$, n = 1, 2,...or $\omega_m = \omega_m(k_t)$, m = 1, 2, ...leads to alternative phase-space parametrizations which have been discussed elsewhere (ref 3). Here, we follow the first option which expresses discrete spatial wavenumber (guided mode) spectra as a function of frequency ω, with subsequent ω-synthesis into the time domain. For two-dimensional, y-independent fields, where $k_z = k_z(k_x,\omega)$, the time domain is reached via (3) with ξ = ω , w = x, z, t, and

$$A(\xi)\rightarrow A_n(\omega)\, F(\omega)\ , \ \Psi(w,\xi) \rightarrow k_{xn}(\omega)x + k_{zn}(\omega)z - \omega t \tag{7}$$

For generality, the window function F(ω) incorporates constraints imposed by the spectrum of the temporal profile f(t) of the incident signal. Stationary phase evaluation of I(x,z,t) in (1) then yields the results in (2)-(4) provided that ξ and w are replaced by ω and x,z,t respectively,

$$I_n(x,z,t) \sim \Delta\omega_{fn}\, A_n(\omega_{sn})F(\omega_{sn})\, \exp\{i[k_{xn}(\omega_{sn})x+k_{zn}(\omega_{sn})z-\omega_{sn}t]\} \tag{8}$$

where $\omega_{sn} = \omega_{sn}\,(x, z, t)$.

3) APPLICATION TO PULSED PLANE WAVE SCATTERING BY A PERIODIC ARRAY

A) Instantaneous wavenumbers and frequencies

An infinite array of identical scatterers separated by a distance d along the z = 0 plane and illuminated by a plane wave exhibits structural dispersion in the form :

$$k_{xn}(\omega) = -(\omega/c)\sin(\theta_i) + 2n\pi/d\,,\ n=0,\pm1,\pm2 \tag{9}$$

$$k_{zn}(\omega) = [(\omega/c)^2 - k_{xn}^2(\omega)]^{1/2} \tag{9a}$$

which defines periodic structure (Floquet or Bragg) modes. Here, θ_i is the plane wave incidence angle (with respect to the z-axis) and c is the speed of light. Via (2) and (7) the saddle point(s) are found to be given by :

$$\omega_{sn}(t) = \frac{2n\pi c}{d}\,\frac{ct + x\sin\theta_i \pm \sin\theta_i D}{D\cos^2\theta_i}\Big|_{n\gtrless 0} \tag{10}$$

$$D = \sqrt{(ct + x\sin\theta_i)^2 - z^2\cos^2\theta_i} \tag{10a}$$

At a fixed observation point (x,z) , $\omega_{sn}(t)$ defines the instantaneous frequency at time t, which tends toward infinity for $D \to 0$, and toward the cutoff frequency

$$\omega_{cn} = 2n\pi c/[d(1\pm\sin(\theta_i)],\ n_{>}^{<}0 \tag{11}$$

as $t \to \infty$. The n-independent time $ct_0 = \pm z\cos(\theta_i) - x\sin\theta_i$, corresponding to D = 0, defines the first (causal) arrival of the scattered field (see fig. 2).

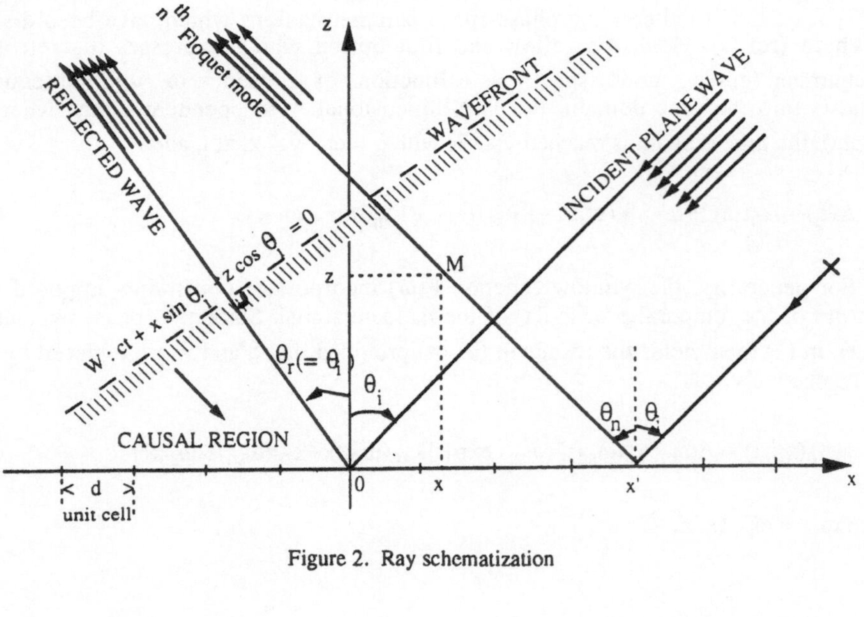

Figure 2. Ray schematization

The instantaneous spatial wavenumbers $k_{xn}(\omega_{sn})$ are obtained from (9) and (10), which yields via (9a) the compact form

$$k_{zn}[\omega_{sn}(t)] = 2n\pi D/d \tag{12}$$

for k_{zn}. When the n-dependent rectilinear wavenumbers k_{xn} or k_{zn} are converted into angular spectra θ_n via the relation

$$k_{xn} = (\omega/c) \sin \theta_n(\omega) \tag{13}$$

it is found that the instantaneous angular spectra $\theta_n[\omega_{sn}(t)]$ are independent of n :

$$\theta_{sn}(t)=\tan^{-1}\left[\frac{-\sin\theta_i\left(ct + x\sin\theta_i\right) \pm D}{z\cos^2\theta_i}\right]\Bigg|_{n \gtrless 0} \tag{14}$$

This has important implications for TD scattered field synthesis by Floquet mode superposition (ref.5 and 6). At the first arrival time t_0 and as $t \to \infty$, one obtains $\theta_{sn}(t_0) = \theta_i$, $\theta_{sn}(\infty) \to \pi/2$. The corresponding instantaneous frequencies are $\omega_{sn}(t_0) \to \infty$ and $\omega_{sn}(\infty) \to \omega_{cn}$.

A variety of phase space displays can be generated from the results in (9)-(14). A (ω_{sn}, k_{xn}, t) surface derived from (9) with $\omega \to \omega_{sn}$ depicts at a fixed observation point (x,z) the temporal evolution of ω_{sn} and k_{xsn}. A surface in $(\omega_{sn}; k_{xsn}, x$ or $z)$ space shows the spatial x or z profile of ω_{sn} and k_{xsn} at a frozen instant in time. In an (x,z,t) frame, one may parametrize the nth TD Floquet mode contribution in terms of the arrival angle $\theta_{sn}(x,z,t)$ and the corresponding frequency $\omega_{sn}(x,z,t)$; the arrival angle defines space-time ray trajectories which, when extrapolated back to the array plane z = 0, identify on the array plane the instantaneous location $x = x'_{sn}(x,z,t)$ and launch time $t = t'_{sn}(x,z,t)$ which contributes the response at (x,z,t) (fig. 3). Since the travel time of a dispersive wave process is tied to the group velocity $v_g = \nabla_k(\omega)$, the space-time ray equations, in parametric form, are given by :

$$z = v_{gzsn}\,(t - t'_{sn}) \quad , \quad x - x'_{sn} = v_{gxsn}\,(t - t'_{sn}) \tag{15}$$

where

$$v_{gzsn} = (\partial\omega/\partial k_{zsn}) = ck_{zsn}/\omega_{sn}\,,\ v_{gxsn} = (\partial\omega/\partial k_{xsn}) = ck_{xsn}/\omega_{sn} \tag{15a}$$

Solving these equations yields for the local launch time and location on the array ,

$$t'_{sn} = t - \frac{ct + x\sin\theta_i \pm D\sin\theta_i}{c\cos^2\theta_i} \tag{16a}$$

$$x'_{sn} = x - \frac{\sin\theta_i\,(ct + x\sin\theta_i\,) \pm D}{c\cos^2\theta_i} \tag{16b}$$

B) Fresnel localizations

The first Fresnel zones corresponding to the space-time dependent saddle point frequencies $\omega_{sn}(x,z,t)$ and wavenumbers $k_{xsn}(x,z,t) = k_{xn}[\omega_{sn}(x,z,t)]$ can be determined by application of (3a), (5) and (7). One finds by direct computation from (7), with (9), (9a), (10) and (10a) that

$$\Delta\omega_{fn}(x,z,t) = \sqrt{\frac{n}{d}}\,\frac{z}{\left[(ct + x\sin\theta_i)^2 - z^2\cos^2\theta_i\right]^{3/4}} \approx \sqrt{\frac{n}{d}\,\frac{z}{D^3}} \qquad (17)$$

which decays as $t^{-3/2}$ as $t \to \infty$. It may be recalled from (3) that the Fresnel interval $\Delta_f\xi$ affects the space-time behaviour of the OBP phase space object synthesized by I(w). From $\Delta\omega_{fn}$ in (17), one may determine the corresponding temporal spread Δt_{fn} (fig.3)

$$\Delta t_{fn}(t)|_{x,z} = 2\,\Delta\omega_{fn}(t)|_{x,z}\,(\partial\omega_{sn}/\partial t)^{-1} \qquad (18)$$

$$\Delta t_{fn} = \frac{1}{\pi}\sqrt{\frac{d}{n}}\,\frac{z\,D^3\cos^4\theta_i}{c\,z^2\cos^2\theta_i + c\sin^2(ct + x\sin\theta_i)\,(ct + x\sin\theta_i \pm D\sin\theta_i)} \qquad (18a)$$

The Fresnel spreads for the various phase space displays enumerated at the end of Sec. IIb can be generated from the above results . Thus , for the active region on the array that surrounds the space-time ray launch coordinates (x'_{sn}, t'_{sn}) , one finds

$$\Delta x'_{fn} = \Delta t'_{fn}\,(\partial t'_{sn}/\partial x)^{-1}|_{z=0} = \Delta t_{fn}\left(\frac{c}{D\cos^2\theta_i}\right)\left[-D\sin\theta_i \pm (ct + x\sin\theta_i)\right] \qquad (19)$$

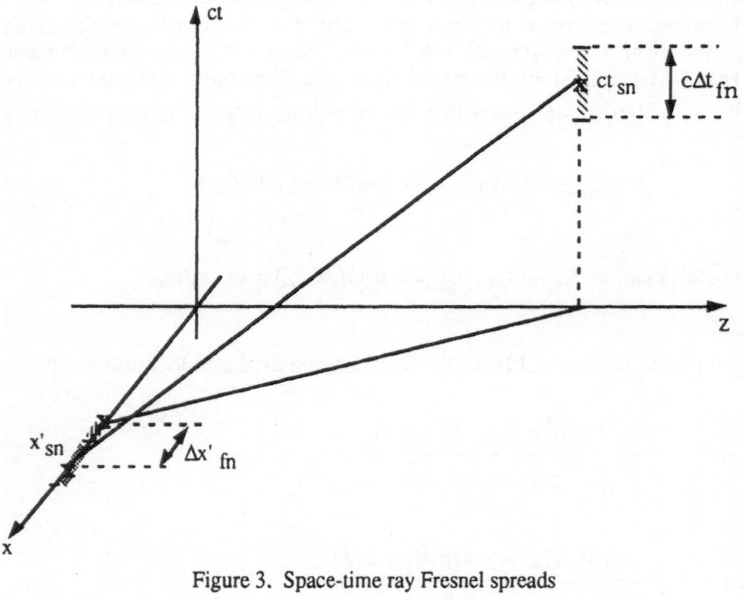

Figure 3. Space-time ray Fresnel spreads

C) Numerical example

For illustration , we consider an array with period d = 5 cm illuminated by a pulsed plane wave at normal incidence (i.e. θ_i=0). The scattered field is observed at x = 0, z=20d (Fig. 4).

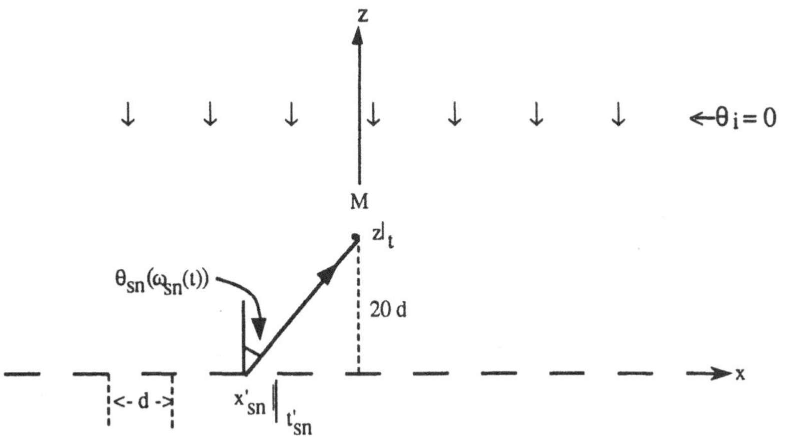

Figure 4. Physical configuration and typical TD Floquet mode space-time ray

From (9) , the specularly reflected n=0 Floquet mode is nondispersive. For the n ≠ 0 modes, the dispersion curve is a hyperbola in a (ω/c,k_{zn}) display (see Fig. 5), and the time dependent frequencies ω_{sn}(0,z,t) in (10) are located graphically by the space-time ray construction in Fig. 5 (ref.4, ch.1). The time-frequency plot for various modes is shown in Fig. 6 ; the effect of windowing introduced by the spectrum of the assumed pulse (Fig. 7) can be assessed by showing also the F(ω) profile. The temporal evolution of various Floquet modes at the observer is shown in Fig. 8. Each mode behaves typically like an oscillatory signal with varying frequency as in Fig. 6 and decaying amplitude (see(17)) ; in addition , the F(ω) profile contributes to the overall envelope for these waveforms. Dominance of the n=1 mode at very early times is clearly predictable from Fig 6. In this example , the scattering elements are specified to be flat perfectly conducting coplanar strips (Fig. 4) of 3cm width ; the excitation amplitudes of the modes depicted in Fig. 8 were evaluated for this case from the individual currents determined numerically through use of a method of moments algorithm .

Since the time-dependent scattering angle θ_{sn}(t) in (14) is the same for all mode orders n, summation over all modes within the spectral window F(ω) synthesizes the composite pulse return from the location x'_{sn}(t) on the array . The result , shown in Fig 9 , depicts an initially diffuse signal that stabilizes at late times toward well resolved pulse shapes emanating from the edges of the strips . Except for the first n = 0 nondispersive specular response , which replicates the incident pulse length, the poorly resolved follow-up signals arise because not all Floquet modes in the spectrum have been excited in this time frame. Resolution increases as more modes become active.

In the above, the pulsed return has been synthesized by TD Floquet mode superposition. Alternatively, the emergence of the various Floquet modes can be monitored by sequential tracking of multiple scatter from successive strip elements (TD-GTD). This procedure is presented in (ref.6).

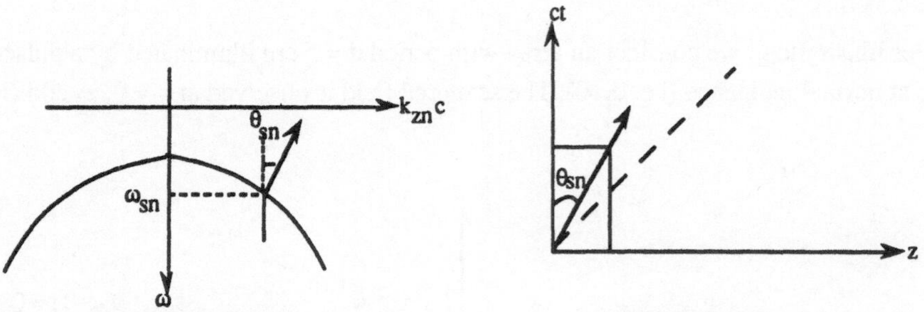

a) Dispersion relation b) Space-time ray

Figure 5. Dispersion relation and space-time rays

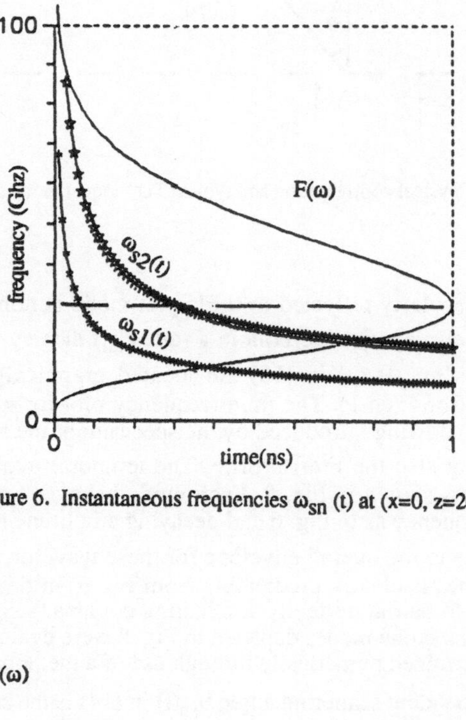

Figure 6. Instantaneous frequencies ω_{sn} (t) at (x=0, z=20d).

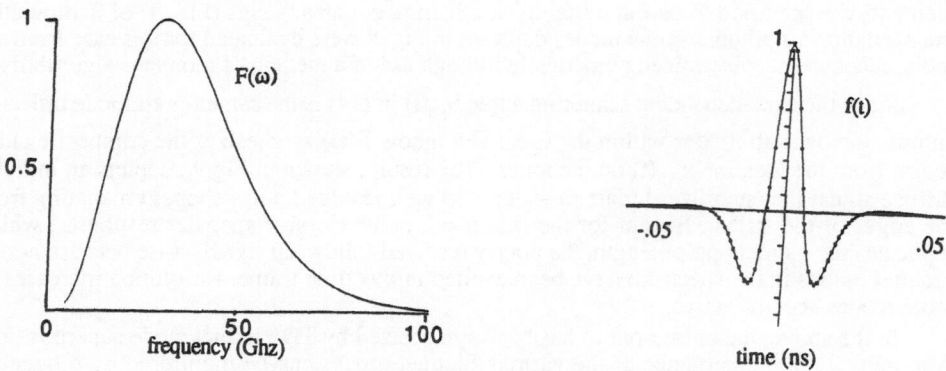

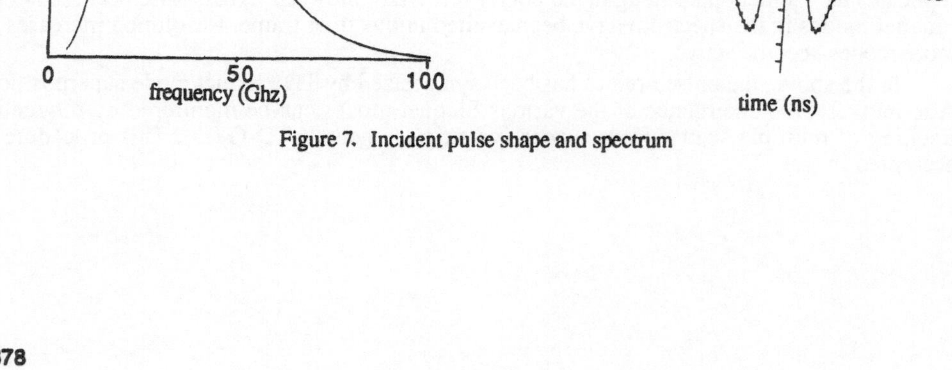

Figure 7. Incident pulse shape and spectrum

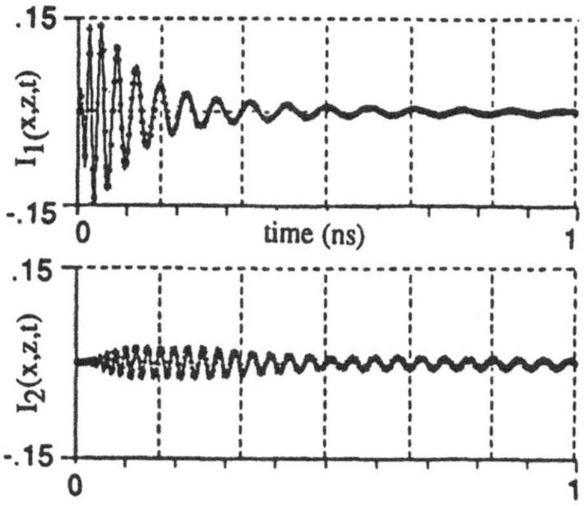

Figure 8. n = 1 and n = 2 mode amplitudes I_n at (x = 0, z = 20d)

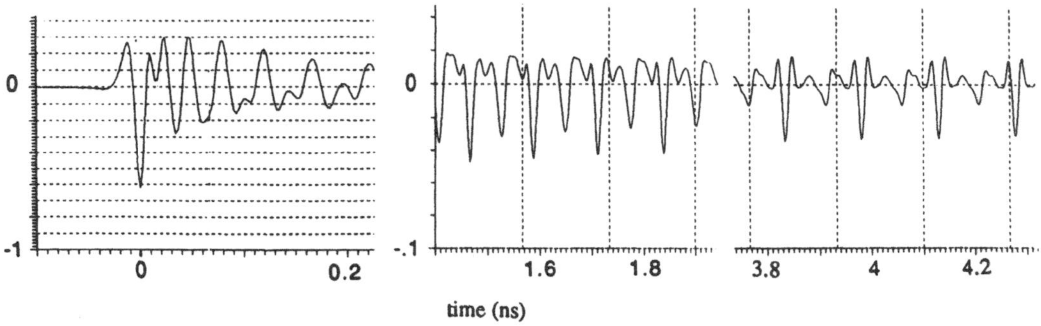

Figure 9. Total field (sum of first 18 modes) at (x=0, z=20d) in various time windows

REFERENCES

1. L.B. Felsen, "Phase Space Issues in Ultrawideband/Short Pulse Electromagnetics", this volume.
2. Yu.A. Kravtsov and Yu.I. Orlov. "Geometrical Optics of Inhomogeneous Media", Springer Series on Wave phenomena, Springer-Verlag.
3. R.W. Ziolkowski and G.A. Deschamps. "Asymptotic Evaluation of high-Frequency Fields Near a Caustic : An Introduction to Maslov's Method", Radio Sci.19, p.1001-1025,1984.
4. L.B. Felsen and N. Marcuvitz. "Radiation and Scattering of Waves", Prentice Hall, 1973.
5. L. Carin and L.B. Felsen. "Time-Harmonic and Transient Scattering by Finite Periodic Flat Strip Arrays : Hybrid (Ray) - (Floquet mode)-(MOM) Algorithm and its GTD Interpretation", to be published in IEEE Trans. Antennas Propagat.
6. P. Borderies and L.B. Felsen. "Time-Harmonic and Transient Scattering by an Infinite Periodic Flat Strip Array", in preparation.

ON THE SHARPEST POSSIBLE PULSE THAT A FINITE RADIATOR CAN GENERATE

Rodolfo E. Díaz

Hexcel APD
2500 W. Frye Rd.
Chandler, AZ 85224

ABSTRACT

Analytical and numerical techniques are used to discuss the shape of the narrowest possible pulse that a radiator of finite size can generate. Although most time domain numerical calculations use sharp Gaussian pulses to model the interaction of an ultra-wideband pulse with a target, no practical finite-sized radiator can generate such a pulse. Analytical solutions due to Heaviside for two-dimensional and three-dimensional time domain sources are compared with a frequency domain argument based on spherical mode decomposition and Finite Difference Time Domain (FDTD) computations. It is shown that the sharpest naturally occurring time domain pulse from a finite radiator is a "bipolar" pulse. The role of time domain diffraction from the edges of apertures in shaping the pulse and the nature of the time domain near field, radiating near field and far field, are discussed.

I - INTRODUCTION

The definition of a pulse used in this paper is: A waveform of finite extent containing a finite amount of energy and generated by an input of finite duration. In particular we are interested in sharp radiated pulses with an ultra-wideband spectrum for the examination of remote targets.

It is implicitly assumed in the use of such pulses that the propagated pulse reaching the target and the input pulse that generated it bear a tight enough correlation to each other that information gathered from backscattering of the propagated pulse can be used for target analysis and recognition. Therefore, in this application the shape of the propagated pulse is of paramount importance to the design and functioning of the receiving system. The shape of the pulse is also important if computational electromagnetics is to be used to assess the backscattering signatures of anticipated targets. The usual assumption is that this pulse shape can be adequately modeled by a Gaussian waveform.

It is shown in this paper that no practical finite-sized radiator can generate such a "unipolar" pulse. This is initially demonstrated in Section II by analyzing the radiated fields generated by a planar aperture in two dimensions. It is shown that the diffracted wave from the edge of the finite boundary region always makes the radiated pulse "bipolar" in the far field. Furthermore, the shape of the pulse, in general, changes with distance from the source. The effects of aperture size, pulse width and aperture excitation are discussed. In Section III the same effect is seen to occur for a three-dimensional spherical source. Finally, using spherical mode decomposition it is shown that the same practical objections against the radiation of supergain beams from a finite radiator stand also against the generation of unipolar pulses from such a source.

II- A PLANAR APERTURE IN TWO DIMENSIONS

The practical question to be answered is: Assuming that a sharp, unipolar pulse can be generated inside a transmitter system and transferred to a radiating aperture, what is the final shape of the pulse radiated into free space? The issue is one of distortion. It is well known that TEM waveguiding structures, being dispersionless, can propagate a pulse with no distortion, whereas non-TEM waveguiding structures in general distort the pulse. Is the system consisting of free space and a finite aperture a distortionless system? The answer is no. Although in the far field the radiated pulse must tend to be TEM, in the neighborhood of the source the fields are in general non-TEM. Therefore, we must expect the pulse to be distorted upon its creation.

Consider a linear aperture of length L on the two-dimensional y-z plane. (Clearly, in this case our finite energy definition of a pulse is taken to mean finite energy per unit x-length.) If an x-directed electric field is suddenly impressed on this aperture and maintained for all time ($E_x(t) = E_0\, U(t)$), then the electromagnetic discontinuity generates a wave that propagates outward at the speed of light. This wave, as illustrated in Figure 1, consists of two parts. The first, is the plane wave from the body of the aperture, propagating in a straight line perpendicular to the face of the source. The second, is a pair of negative cylindrical waves generated at the edges of the aperture. The job of these waves is to "eat up" the plane wave so that as $t \rightarrow \infty$, all that is left behind is the steady state electrostatic field due to the impressed field source.

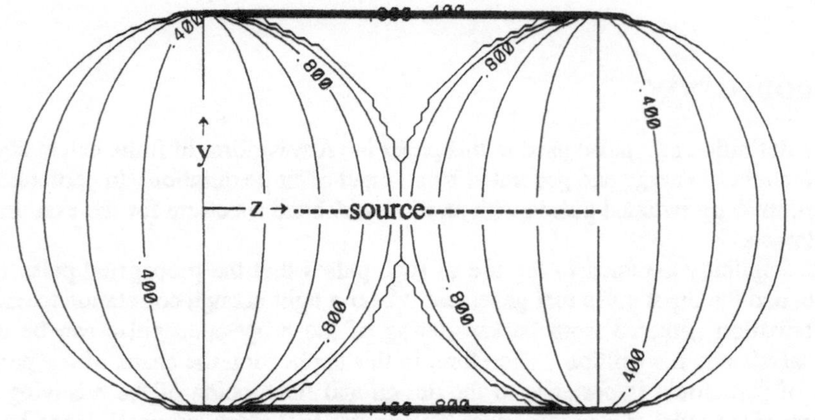

Figure 1. The pulse generated in the y-z plane by a suddenly excited E_x aperture. The contours are level curves of constant Electric field.

Heaviside[1] gives the expression for the total fields of the wave as:

$E = E_1$ between the two circles

$$E = \frac{E_1}{\pi}\left\{\frac{\pi}{2} + \arcsin\left(\frac{z}{\sqrt{c^2t^2 - y^2}}\right)\right\} \quad \text{inside the left circle} \tag{1}$$

And the right circle is obtained by symmetry. Given this expression, the most elementary finite extent pulse of finite energy content is a square pulse, produced by turning the source on at t=0 (with U(0)) and turning it off at t=dt (with -U(t-dt)). Snapshots of the propagation of such a pulse with cdt=0.085L are shown in Figure 2a. There the contour lines denote level curves of constant electric field, solid lines being positive fields and dashed lines being negative.

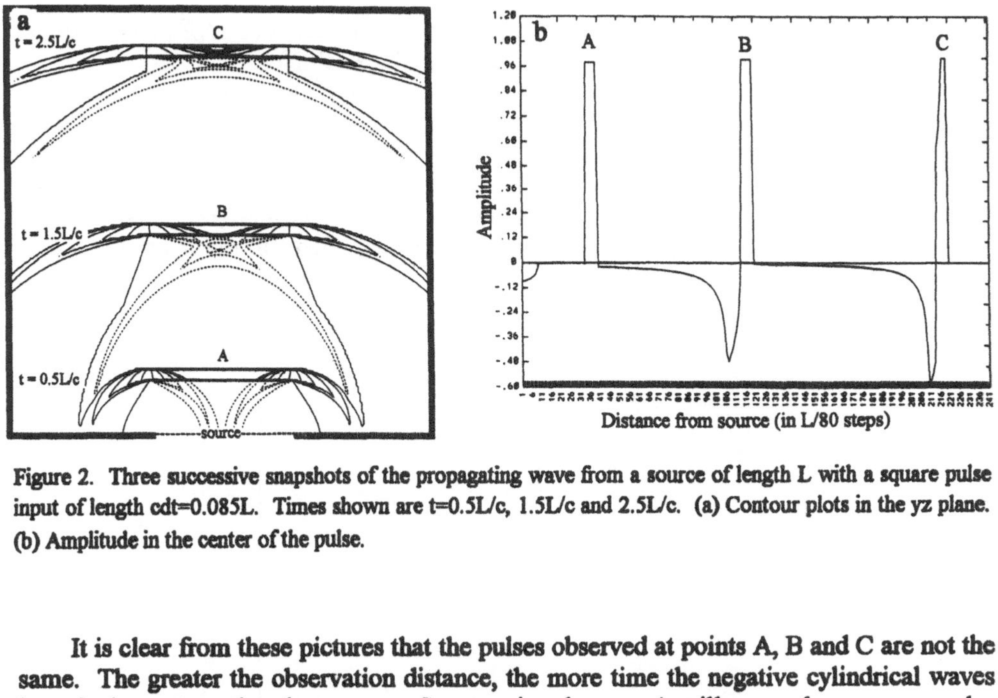

Figure 2. Three successive snapshots of the propagating wave from a source of length L with a square pulse input of length cdt=0.085L. Times shown are t=0.5L/c, 1.5L/c and 2.5L/c. (a) Contour plots in the yz plane. (b) Amplitude in the center of the pulse.

It is clear from these pictures that the pulses observed at points A, B and C are not the same. The greater the observation distance, the more time the negative cylindrical waves have had to eat up the plane wave. So, near in, observer A will see a sharp square pulse pass by, followed later by a more diffuse negative pulse. However, observer B at an intermediate distance sees the cylindrical waves arrive immediately behind the passing square pulse. And finally, far away, observer C cannot tell that there was a square pulse but rather he sees a bipolar pulse pass by. Figure 2b illustrates this situation.

A further interesting observation is the opinion that these three different observers would have about the nature of the "main pulse." Since the cylindrical waves have not caught up with the middle of the plane wave, the square pulse appears unchanged as it travels from A to B; its field strength is constant. Therefore, if observers A and B were to compare notes, they would conclude that a non-dispersing "electromagnetic bullet" had gone past them. However, when observer C and later observers compare their notes on the

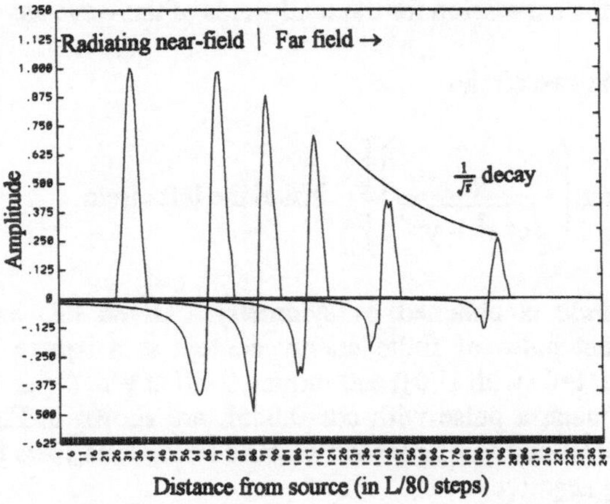

Figure 3. The evolution of the Gaussian pulse as it travels from the source plane to the far field.

bipolar pulse, they will conclude that the bipolar pulse is decaying as $1/\sqrt{r}$, as it should. Figure 3 emphasizes this point by using as input a Gaussian excitation.

As Figure 3 suggests, the disagreements between the observers is nothing more than the classic division of the radiated field into Fresnel and Fraunhofer regions. The extent of the "Fresnel" region in the time domain is exactly, $R_F = L^2/(8cdt) + cdt/2$. The sharper the pulse is ($cdt<<L$) the larger R_F is. Nevertheless, in the far field, where $R>>R_F$, the pulse is always bipolar and decaying as $1/\sqrt{r}$. The only way to produce an apparent unipolar pulse in the far field is to use the Heaviside unit step as an excitation. In that case, the pulse evolves as shown in Figure 4. But, of course, the input function is not of finite duration.

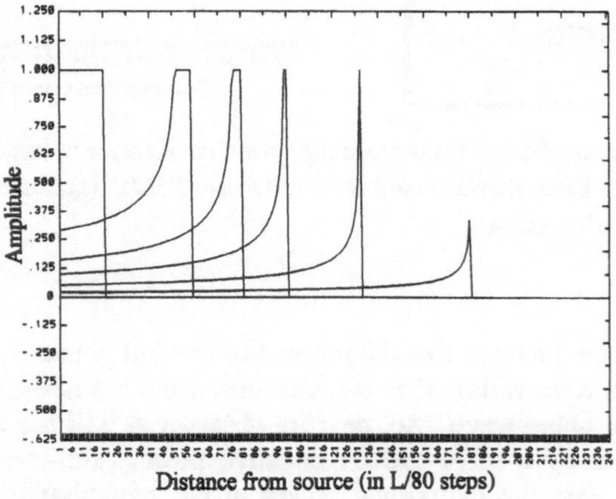

Figure 4. The evolution of the pulse due to a U(t) excitation of the linear aperture.

Therefore it is found that any pulse generated from a uniform aperture develops a negative tail as a consequence of the electromagnetic discontinuity at the edges of the aperture. Is it possible then to get rid of this tail by tapering the aperture distribution? The answer is no because in so doing we merely redistribute the diffracting sources over the whole aperture. Figure 5 shows the evolution of a Gaussian pulse radiated by a Gaussian aperture as calculated with FDTD.

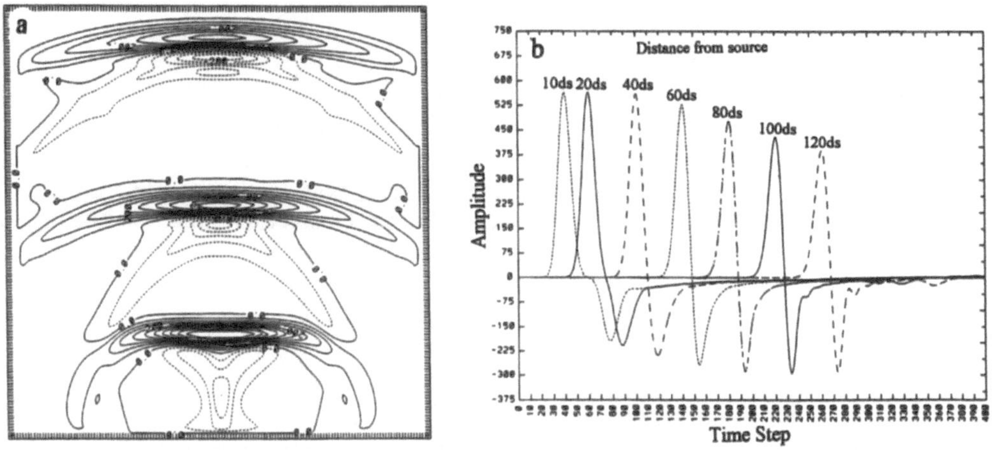

Figure 5. FDTD calculation of the propagation of a Gaussian pulse radiated by a Gaussian aperture.

All that has been accomplished here is to smooth out the field metamorphosis evidenced in Figure 2. The observed pulse shape is now almost the same at all observation positions. The price paid is a reduced Fresnel region. Therefore, the far field pulse in two dimensions from a finite radiator is always a bipolar pulse. The situation in three dimensions is analyzed in Section III.

III -THE SPHERICAL SOURCE

The most elementary spherical source is the field distribution of the first TM mode of the expansion in spherical harmonics[2]. In this mode there is an electric current flowing upwards from the south pole to the north pole of the sphere along lines of longitude. There is no azimuthal ϕ dependence, while the θ dependence goes as the first Associated Legendre polynomial $P_1^1(\cos(\theta)) = \sin(\theta)$. Therefore, this mode can be represented by a current density on the surface of the sphere of radius **a** given by $J = (I_0/2\pi a)\sin(\theta)$. Alternately, we can represent this mode by the corresponding electric dipole charge distribution where the upper hemispherical cap is negative, the lower hemispherical cap is positive and the charge density varies along the sphere as $\rho = \rho_0 \cos(\theta)$. In the electrostatic limit, such a charge distribution encases a displacement vector **D**, everywhere uniform inside the sphere and equal in value to the charge density on the sphere.

To study the fields radiated by a current impulse we can either let $I_0(t) = \delta(t)$ or we can let $D(t) = U(t)$, since then $\partial D/\partial t = \delta(t)$. The first is a conduction current impulse, the second is a displacement current impulse. They are fundamentally different. The conduction current is an infinitesimally thin shell of surface current whereas the displacement current is a volumetric current, discontinuous on the spherical surface. Heaviside[3] gives the result for the displacement current impulse. The magnetic field of the propagated wave has the form:

$$H = H_0 \sin(\theta)\left\{1 - \frac{c^2t^2 - a^2}{r^2}\right\}, \quad ct-a<r<ct+a, \text{ zero elsewhere.} \qquad (2)$$

It is a bipolar pulse positive on the leading half, negative on the trailing half and it leaves behind itself the electrostatic field of the spherical dipole. To obtain the field from the surface current shell we go to Weeks[2] and write down the frequency domain expression:

$$vH = -\frac{I_0}{2\pi j}\,\hat{J}_1'\left(\beta a\right)\,\sin(\theta)\left(\frac{1}{r}+\frac{1}{j\beta r^2}\right)e^{-j\beta r} \tag{3}$$

Where β is ω/c and $\hat{J}_1'$ is the derivative of the first order spherical Bessel function, $\hat{J}_1(x) = \frac{\sin(x)}{x} - \cos(x)$. Equation (2) can be used to calculate the radiated pulse from an arbitrary excitation by breaking up the desired input into a succession of impulses. Equation (3) can be used for the same purpose by letting $I_0(\omega)$ be the frequency spectrum of the desired input and then taking the inverse Fourier Transform. Thus, we obtain the pulses of Figure 6, where three Gaussian inputs with cdt<<a, cdt≈a and cdt>>a, at a distance of 50a are shown.

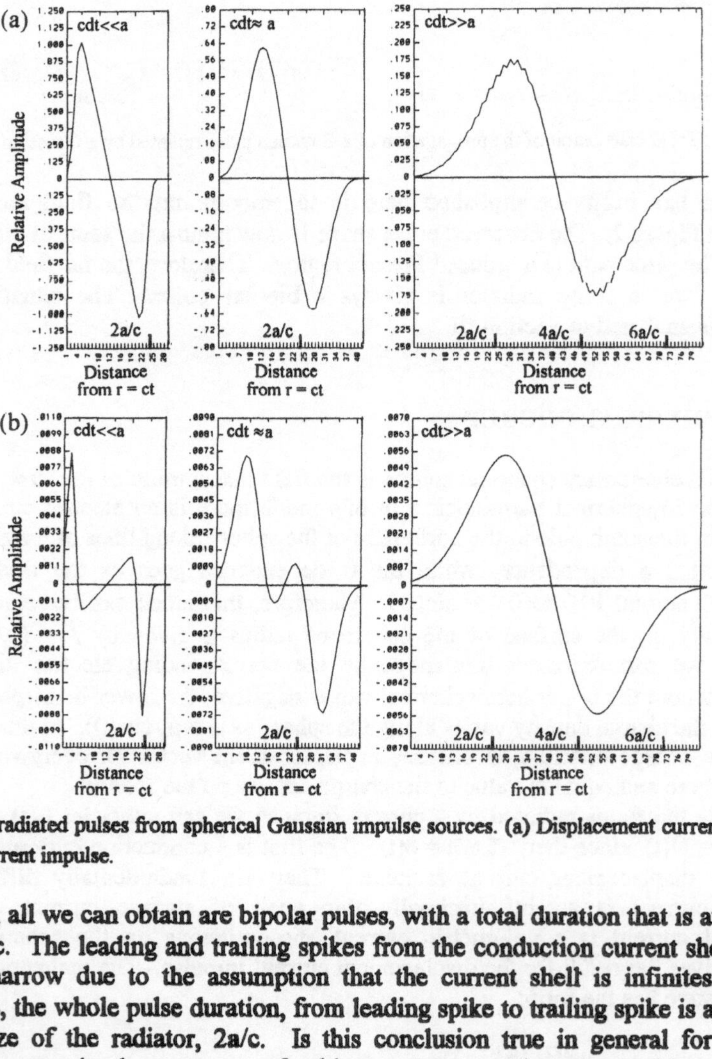

Figure 6. The radiated pulses from spherical Gaussian impulse sources. (a) Displacement current impulse. (b) Conduction current impulse.

Clearly, all we can obtain are bipolar pulses, with a total duration that is at its shortest equal to 2a/c. The leading and trailing spikes from the conduction current shell for sharp pulses are narrow due to the assumption that the current shell is infinitesimally thin. Nevertheless, the whole pulse duration, from leading spike to trailing spike is a measure of the finite size of the radiator, 2a/c. Is this conclusion true in general for all current distributions, even in the presence of arbitrary structures (antennas) able to shape the fields?

To answer this question assume that such an arbitrary current and structure combination can be contained within the sphere of radius a. Then the radiated fields can be determined by decomposing the source fields tangential to the sphere, into TE and TM spherical harmonics and then propagating these harmonics to the far field. For simplicity consider a pure TM field with no azimuthal variation. Then the radiated magnetic field is of the form:

$$H_\phi\left(\omega,r,\theta\right) = -\sum_n \frac{D_n(\omega)}{r}\hat{H}_n^{(2)}\left(\frac{\omega r}{c}\right)P_n^1(\cos(\theta)) \tag{4}$$

where $\hat{H}_n^{(2)}(x)$ is the outgoing spherical Hankel function whose limit for large x is $j^{n+1}e^{-jx}$.

The coefficients $D_n(\omega)$ are determined by matching this radiated wave to the source spectrum at the radius a. Thus, if the source spectrum is,

$$H_\phi\left(\omega,a,\theta\right) = \sum_n \alpha_n(\omega)P_n^1(\cos(\theta)) \tag{5}$$

it follows that ,

$$D_n(\omega) = \frac{a\alpha_n(\omega)}{\hat{H}_n^{(2)}\left(\frac{\omega a}{c}\right)} \tag{6}$$

Now, assume that the far field spectrum constitutes a unipolar Gaussian pulse in the time domain, with some prescribed distribution on the sphere of radius r. Then, it too can be decomposed into spherical harmonics and we obtain (letting $\tau = t-r/c$):

$$H_\phi\left(r,\tau,\theta\right) = \exp\left(-p\tau^2\right)\sum_n \zeta_n P_n^1(\cos(\theta)) \tag{7}$$

It follows that equations (5) and (7) are connected through the inverse Fourier Transform, so that the coefficients α_n and ζ_n are related through the equation:

$$\alpha_n(\omega) = K\zeta_n \exp\left(-\frac{\omega^2}{4p}\right)\frac{\hat{H}_n^{(2)}\left(\frac{\omega a}{c}\right)}{\hat{H}_n^{(2)}\left(\frac{\omega r}{c}\right)}\exp\left(-j\frac{\omega r}{c}\right)\text{, with } K \text{ a constant.} \tag{8}$$

The term $\exp(-j\omega r/c)$ appears as a result of making the radius of observation the zero reference plane (by the definition of τ.)

Because in the far field r>>a, equation (8) has three distinct limiting behaviors in the spectrum. At high frequencies where $\omega a/c$ is greater than 1.0 and greater than n, then also $\omega r/c >> 1,n$ and both Hankel functions can be written as their large value approximations. At some intermediate frequencies $\omega r/c$ still remains $>>1,n$ but $\omega a/c < 1$. In that case the denominator of (8) still can be simplified using the large value approximation but the numerator tends to the small value approximation. Finally when ω is sufficiently small, both numerator and denominator can be approximated by the small value approximation. The result is,

$$\alpha_n(\omega) \approx K\zeta_n \exp\left(-\omega^2/4p\right)\exp\left(-j\omega a/c\right) \text{ for } \omega >> \frac{c}{a} \tag{9a}$$

$$\alpha_n(\omega) \approx K\zeta_n \exp\left(-\omega^2/4p\right)\left(\frac{c}{\omega a}\right)^n \text{ for } \frac{c}{r} << \omega << \frac{c}{a} \tag{9b}$$

$$\alpha_n(\omega) \approx K\zeta_n \exp\left(-\omega^2/4p\right)\left(\frac{r}{a}\right)^n \text{ for } \omega << \frac{c}{r} \tag{9c}$$

If the Gaussian pulse observed at r>>a is to be sharp, then p is to be very large, making the desired radiated pulse frequency domain spectrum very flat and broad. Equations (9) tell us that to obtain such a spectrum at the target, the source spectrum must be the opposite: very sharply peaked at low frequencies (as $(r/a)^n$.) This leads to the paradoxical result that if the pulse at r were to be shorter than the diameter of the source sphere (less than 2a/c in duration), the source pulse would have a duration greater than 2r/c. That is, the source would not even start "turning off" before the radiated pulse reached its target at r and the echo began its return. Such a situation violates the requirements for a range finding system to locate, much less identify, a target. The factor $(r/a)^n$ in equations (9) also tell us that the angular distribution of the energy in the pulse over the sphere of radius r is very different from the source distribution, and strongly dependent on the distance to the target. In other words, the pulse undergoes a severe metamorphosis on its trip to the target.

Neither of these results is unexpected. We saw precisely the same situation in two dimensions. In fact, the pulse of Figure 4 has all the same characteristics as the pulse we are trying to form in this three dimensional case. It may be unipolar but its source must be of extremely long duration, plus it changes continually with distance.

The final comment to be made is on the similarity of this problem to the problem of generating supergain beams from an electrically small radiator. Weeks' remark[4] on the practical limitations in that case apply here as well: The fact that the source spectrum for high orders n, goes as $(r/a)^n$, means that large amounts of reactive energy are being stored in the sphere of radius **a** in the form of large currents that oscillate strongly as a function of θ. Such currents imply large ohmic losses in the conductors of the radiator. For this and the previous reasons, it is not practical to expect sharp unipolar pulses from a finite radiator.

IV-CONCLUSION

It has been shown that the pulses generated by finite sources are in general bipolar pulses. The simplest way to come to this realization is to consider the phenomena from the standpoint of Heaviside. If there was no source before t=0 and there is no source after t=dt, then whatever fields were excited by the beginning of the pulse must be erased by the tail of the pulse. Otherwise at t=∞ there would be, left behind, fields in space without the support of sources. This erasing is done by the negative nature of the trailing pulse that is generated from the finite boundary of the source. The only conditions under which unipolar pulses can be generated involve either infinite sources or infinite source durations. Diffraction phenomena, field metamorphosis, Fresnel regions and Fraunhofer regions, well known trademarks of the behavior of fields from finite sources in the frequency domain, were shown to have direct analogies in the time domain.

The unipolar Gaussian pulse, so often used to represent ultra-wideband radiation, cannot be observed in the far field. It only exists, for a limited time, in the Fresnel zone of large radiators.

REFERENCES

1. O. Heaviside, "Electromagnetic Theory," Chelsea Publishing Co., New York, Vol.III, p. 6 (1971).
2. W. L. Weeks, "Electromagnetic Theory for Engineering Applications", Wiley, New York, ch. 7 (1964).
3. O. Heaviside, **op. cit.**, Vol.III, p. 95.
4. W. L. Weeks, **op. cit.**, p. 553-557.

SCATTERING UPON A MIXED DIELECTRIC-CONDUCTOR BODY: A TIME - DOMAIN APPROACH

Dan V. Gibson and Michel M. Ney

Department of Electrical Engineering
University of Ottawa
161 Louis Pasteur
Ottawa, Ontario
Canada K1N 6N5

ABSTRACT

A time-domain integral equation technique is presented for calculating the behaviour of a mixed dielectric - conductor body subjected to a transient plane wave. The incident pulse is a smooth Gaussian pulse. We develop an original formulation capable of calculating the electric and magnetic fields on or outside the surface of the scatterer from the moment the incident wave hits the target.

Scattering upon a body whose external surface consists of a dielectric part and a conductive one separated by a boundary line lying on that external surface, where the electrical and magnetic characteristics change abruptly, has not been dealt with before in a time-domain formulation. Numerical results are presented for (A): a single dielectric with a ground plane, covered by a conductive strip and (B): a combination of one dielectric and one solid conductor.

THEORY

Time-domain integral formulations for the scattering of dielectric and conductive bodies separately were proposed[1-6]. In the case of hybrid bodies the surface of the scatterer is not uniform in its electrical characteristics and a simple superposition of the treatments mentioned above is insufficient. The difficulty[7] is that the field vectors **E** and **H** are not continuous and do not have continuous first derivatives over the entire surface. There is a discontinuous change in their tangential components in passing from the conductive to the dielectric regions on the surface S. The surface is divided into two separate areas: A_1 (conductive) and A_2 (dielectric). An abrupt change in the surface current density can be accounted for by an accumulation of line charges on the

Ultra-Wideband, Short-Pulse Electromagnetics
Edited by H. Bertoni *et al.*, Plenum Press, 1993

contour. Let **ds** be an element of length along the contour in the positive direction as determined by the normal **n** to the surface. Also, $\mathbf{n}_1$ is the unit vector lying on the surface and normal to both **n** and **ds** and directed towards A_1 (Fig. 1). With the inclusion of terms taking into account the line densities, the frequency-domain equations are in the absence of sources:

$$\begin{aligned}\mathbf{E}(\mathbf{x},\omega) &= \mathbf{E}^{\mathbf{inc}}(\mathbf{x},\omega)+\\ &\frac{T}{4\pi j\omega\varepsilon}\oint_C \nabla'\varphi(\mathbf{x},\mathbf{x}',\omega)[\mathbf{H}_1(\mathbf{x}',\omega)-\mathbf{H}_2(\mathbf{x}',\omega)]\cdot\mathbf{ds}'-\\ &\frac{T}{4\pi}\int_{A_1\cup A_2}\Big[j\omega\mu[\mathbf{n}'\times\mathbf{H}_i(\mathbf{x}',\omega)]\varphi(\mathbf{x},\mathbf{x}',\omega)-\\ &[\mathbf{n}'\times\mathbf{E}_i(\mathbf{x}',\omega)]\times\nabla'\varphi(\mathbf{x},\mathbf{x}',\omega)-[\mathbf{n}'\cdot\mathbf{E}_i(\mathbf{x}',\omega)]\nabla'\varphi(\mathbf{x},\mathbf{x}',\omega)\Big]da'\end{aligned} \tag{1}$$

and

$$\begin{aligned}\mathbf{H}(\mathbf{x},\omega) &= \mathbf{H}^{\mathbf{inc}}(\mathbf{x},\omega)-\\ &\frac{T}{4\pi j\omega\mu}\oint_C \nabla'\varphi(\mathbf{x},\mathbf{x}',\omega)[\mathbf{E}_1(\mathbf{x}',\omega)-\mathbf{E}_2(\mathbf{x}',\omega)]\cdot\mathbf{ds}'+\\ &\frac{T}{4\pi}\int_{A_1\cup A_2}\Big[j\omega\varepsilon[\mathbf{n}'\times\mathbf{E}_i(\mathbf{x}',\omega)]\varphi(\mathbf{x},\mathbf{x}',\omega)-\\ &[\mathbf{n}'\times\mathbf{H}_i(\mathbf{x}',\omega)]\times\nabla'\varphi(\mathbf{x},\mathbf{x}',\omega)-[\mathbf{n}'\cdot\mathbf{H}_i(\mathbf{x}',\omega)]\nabla'\varphi(\mathbf{x},\mathbf{x}',\omega)\Big]da'\end{aligned} \tag{2}$$

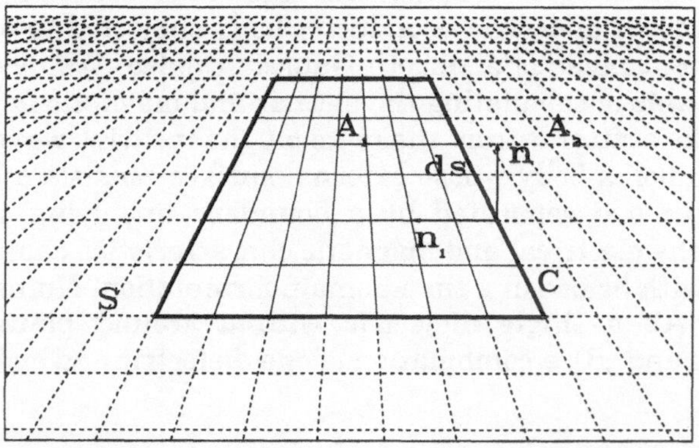

Figure 1. Dielectric A_2 and conductor A_1 separated by contour C

where subscripts 1 and 2 imply the values of the fields along the contour C, immediately inside areas A_1 and A_2. Subscript "i" in the surface integrals is assigned values 1 or 2, depending on whether we perform the integrals over A_1 or A_2. T is $(1-\Omega/4\pi)^{-1}$, where Ω is the solid angle at the observation point **x**. Primed quantities are in the source coordinates. On a smooth surface, $\Omega=2\pi$. $\mathbf{E}^{inc}$ and $\mathbf{H}^{inc}$ are the incident fields at the observation point. φ is:

$$\varphi=\frac{e^{-jk|x-x'|}}{|x-x'|} \tag{3}$$

Applying the inverse Fourier transform and the convolution theorem:

$$FT\left[\frac{1}{j\omega}\,\mathbf{H}(\omega)\,\nabla\varphi(\omega)\right]=\int_0^{t-\frac{R}{c}}\mathbf{R}\left[\mathbf{H}(t-\frac{R}{c}-t_1)\frac{1}{R^3}+\frac{1}{c}\frac{\partial}{\partial t}\mathbf{H}(t-\frac{R}{c}-t_1)\frac{1}{R^2}\right]dt_1 \tag{4}$$

where $\mathbf{R}$ = $\mathbf{x}$-$\mathbf{x'}$, we obtain after some manipulation including a technique proposed in [1], the time-domain equations. The fields are assumed causal time functions, $\mathbf{n'}$ is the normal to the surface at the source point $\mathbf{x'}$ and $\hat{\mathbf{R}}$ the unit vector from $\mathbf{x'}$ to $\mathbf{x}$. The fields and their derivatives on the surface and along the contour C are calculated at a time τ, where $\tau = t - R/c$. With the notations:

$$L_o=\frac{1}{R^2}+\frac{1}{Rc}\frac{\partial}{\partial t}\quad ;\quad L_i=\frac{1}{R^2}+\frac{1}{Rc_i}\frac{\partial}{\partial t} \tag{5}$$

we obtain the tangential component of the electric field on the dielectric:

$$\begin{aligned}\mathbf{n}\times\mathbf{E}(t) &= \mathbf{n}\times\mathbf{E}^{inc}(t)+\frac{1}{4\pi}\mathbf{n}\times\Big\langle\frac{1}{\varepsilon_o}\int_0^{\tau_o}dt_1\oint_C\hat{\mathbf{R}}L_o[\mathbf{H}_1(\tau_o-t_1)-\mathbf{H}_2(\tau_o-t_1)]\cdot\mathbf{ds}'-\\ &-\frac{1}{\varepsilon_i}\int_0^{\tau_i}dt_1\oint_C\hat{\mathbf{R}}L_i\mathbf{H}_2(\tau_i-t_1)\cdot\mathbf{ds}'-\int_{A_1}\left(\mu\frac{1}{R}\mathbf{n}'\times\frac{\partial\mathbf{H}(\tau_o)}{\partial t}-[\mathbf{n}'\cdot L_o\mathbf{E}(\tau_o)]\hat{\mathbf{R}}\right)da'+\\ &+\int_{A_2}\left(\mathbf{n}'\{L_o\mathbf{E}(\tau_o)-L_i\mathbf{E}(\tau_i)]\hat{\mathbf{R}}+[\mathbf{n}'\times[L_o\mathbf{E}(\tau_o)-L_i\mathbf{E}(\tau_i)]]\times\hat{\mathbf{R}}-\mu\frac{1}{R}\mathbf{n}'\times\frac{\partial[\mathbf{H}(\tau_o)-\mathbf{H}(\tau_i)]}{\partial t}\right)da'\Big\rangle\end{aligned} \tag{6}$$

the tangential component of the magnetic field on the dielectric:

$$\begin{aligned}\mathbf{n}\times\mathbf{H}(t) &= \mathbf{n}\times\mathbf{H}^{inc}(t)+\frac{1}{4\pi}\mathbf{n}\times\Big\langle\int_{A_1}[\mathbf{n}'\times L_o\mathbf{H}(\tau_o)]\times\hat{\mathbf{R}}\,da'+\\ &\int_{A_2}\left(\frac{1}{R}\mathbf{n}'\times\left[\varepsilon_o\frac{\partial\mathbf{E}(\tau_o)}{\partial t}-\varepsilon_i\frac{\partial\mathbf{E}(\tau_i)}{\partial t}\right]+\right.\\ &\left.\mathbf{n}'\cdot[L_o\mathbf{H}(\tau_o)-L_i\mathbf{H}(\tau_i)]\hat{\mathbf{R}}+[\mathbf{n}'\times[L_o\mathbf{H}(\tau_o)-L_i\mathbf{H}(\tau_i)]]\times\hat{\mathbf{R}}\right)da'\Big\rangle\end{aligned} \tag{7}$$

the normal component of the electrical field on the conductor:

$$\mathbf{n}\cdot\mathbf{E}(t) = 2\mathbf{n}\cdot\mathbf{E}^{inc}(t)+\frac{1}{2\pi}\mathbf{n}\cdot$$

$$\left\langle \frac{1}{\varepsilon_o}\int_0^{\tau_o} dt_1 \oint_C \hat{\mathbf{R}} L_o\left[\mathbf{H}_1(\tau_o - t_1) - \mathbf{H}_2(\tau_o - t_1)\right]\cdot \mathbf{ds}' - \int_{A_1}\left(\mu\frac{1}{R}\mathbf{n}'\times\frac{\partial \mathbf{H}(\tau_o)}{\partial t} - \left[\mathbf{n}'\cdot L_o\mathbf{E}(\tau_o)\right]\hat{\mathbf{R}}\right) da' - \right. \quad (8)$$

$$\left. - \int_{A_2}\left(\mu\frac{1}{R}\mathbf{n}'\times\frac{\partial \mathbf{H}(\tau_o)}{\partial t} - \left[\mathbf{n}'\cdot L_o\mathbf{E}(\tau_o)\right]\hat{\mathbf{R}} - \left[\mathbf{n}'\times L_o\mathbf{E}(\tau_o)\right]\times\hat{\mathbf{R}}\right) da' \right\rangle$$

and the tangential component of the magnetic field on the conductor:

$$\mathbf{n}\times\mathbf{H}(t) = 2\mathbf{n}\times\mathbf{H}^{inc}(t)+\frac{1}{2\pi}\mathbf{n}\times\left\langle \int_{A_1}\left[\mathbf{n}'\times L_o\mathbf{H}(\tau_o)\right]\times\hat{\mathbf{R}}\, da' + \right.$$

$$\left. + \int_{A_2}\left(\varepsilon_o\frac{1}{R}\mathbf{n}'\times\frac{\partial \mathbf{E}(\tau_o)}{\partial t} + \left[\mathbf{n}'\cdot L_o\mathbf{H}(\tau_o)\right]\hat{\mathbf{R}} + \left[\mathbf{n}'\times L_o\mathbf{H}(\tau_o)\right]\times\hat{\mathbf{R}}\right) da' \right\rangle \quad (9)$$

Two more equations similar to (6) and (7) express the normal components of the electric and magnetic fields on the dielectric side. They are not shown here for reasons of economy of space. Solving the above system of six equations does not require the inversion of a matrix. The reason is that the left-hand sides of the equations and their right-hand sides are not calculated at the same point in time. The unknown components on the left, at time t, are expressed in terms of the known quantities on the right, already calculated at earlier times.

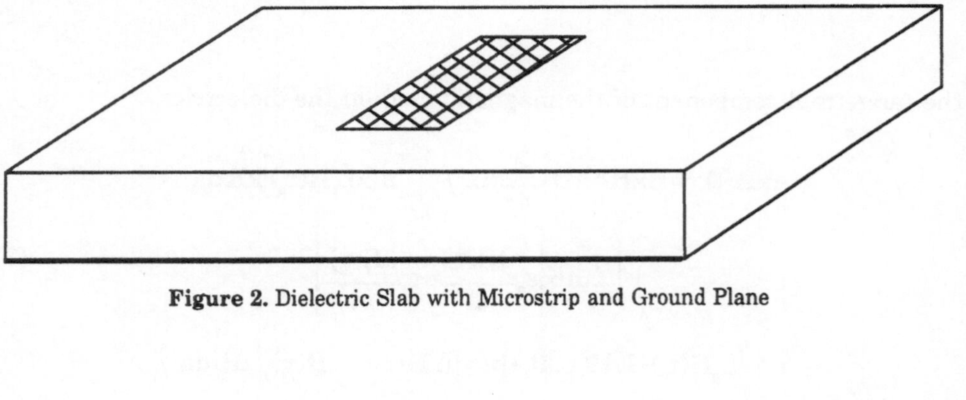

Figure 2. Dielectric Slab with Microstrip and Ground Plane

Bodies that have volume discontinuities, for example layered dielectrics, are not covered by the approach presented in this paper. There, the fields' discontinuities are accounted for by introducing a distribution of surface densities of charge and currents.

RESULTS

The surface of the scatterer was divided into square patches and the surface integrals were replaced by sums of integrals over patches. In turn, the integrals over patches were approximated by the product between the area of the patch and the average of the integrand. That average value was assigned to the center of the patch. For purposes of rough numerical implementation, the patch at which the observation point was located, was eliminated from calculation. A numerical accuracy check was based on the verification of Maxwell's second law.

The results for two configurations, i) dielectric slab with ground plane and a perfectly-conductive microstrip on the opposite face and ii) metallic box with thick walls, are discussed.

Dielectric Slab with Ground Plane and Microstrip

We calculated the current induced in the microstrip (Fig. 2) for a dielectric with $\varepsilon_r = 10.0$. The microstrip was 40 X 4 mm. The incident wave was normal to the microstrip and the electric field lay along it.

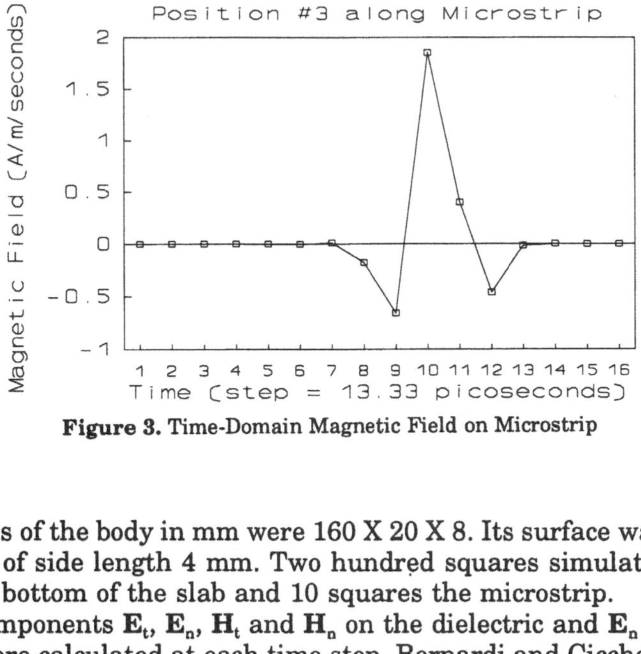

Figure 3. Time-Domain Magnetic Field on Microstrip

The sides of the body in mm were 160 X 20 X 8. Its surface was divided into 580 squares of side length 4 mm. Two hundred squares simulated the ground plate on the bottom of the slab and 10 squares the microstrip.

Field components $\mathbf{E}_t$, $\mathbf{E}_n$, $\mathbf{H}_t$ and $\mathbf{H}_n$ on the dielectric and $\mathbf{E}_n$ and $\mathbf{H}_t$ on the conductor were calculated at each time step. Bernardi and Cicchetti[8] studied a similar case. Their line was terminated into its characteristic impedance at both ends (ours was open-ended) and it had the same relative permittivity of the substrate.

For validation purposes we compared our results with theirs. The above authors calculated, using a transmission line model and assuming a quasi-TEM dominant mode, that a current in the range of a few microamperes was induced in the microstrip. We extrapolated their results, by taking the limit of those end impedances to infinity, and compared them to those obtained by us after the appropriate FFT.

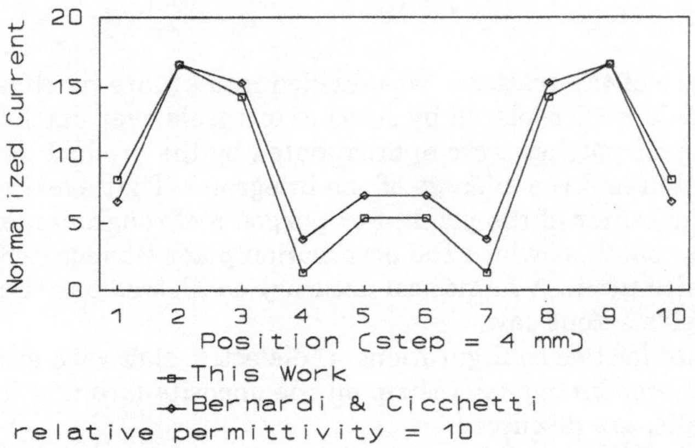

Figure 4. Normalized Current along Microstrip at 4.6875 GHz

Figure 3 displays an example of the induced magnetic field time-domain transform at all sixteen time steps. The graph in Figure 4 presents the relative behaviour of the current across the microstrip at the first frequency. FFT was performed on the magnetic field results along the microstrip at all 10 microstrip locations. Frequencies were multiples of the fundamental value of 4.6875 GHz. The values of the currents obtained in this work and in that of Bernardi and Cicchetti were normalized to a common maximum value.

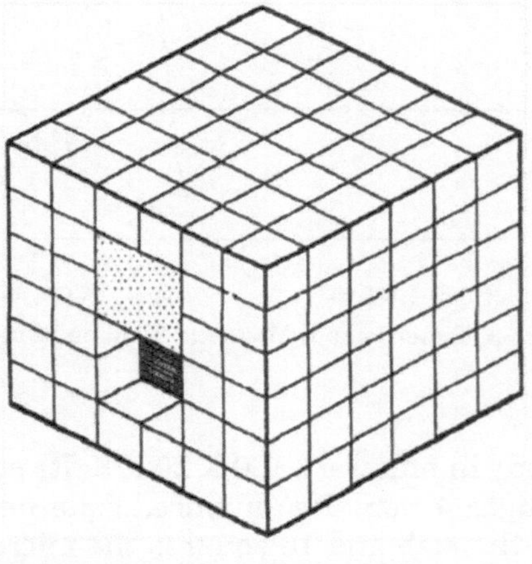

Figure 5. Conductive Box with Dielectric Plug

The integral present in Maxwell's second law was calculated over a surface comprising the body and a parallel surface just outside it. The closeness to a null value reflects the fact that the calculated fields respect that law. That integral, calculated at eight frequencies departed from zero by no more than 9 10^{-6} A·m.

Results of the Conductive Box with Dielectric Plug

The cubic box (Fig. 5) had a side of 30 cm and 5 cm thick walls. The opening in the front wall was 10 cm wide and 20 cm long and its length was varied between 5 and 15 cm by means of a dielectric plug ($\varepsilon_r = 10.0$).The graphs in Figure 6 show the behaviour of the electric field at the centre of the box in time-domain.

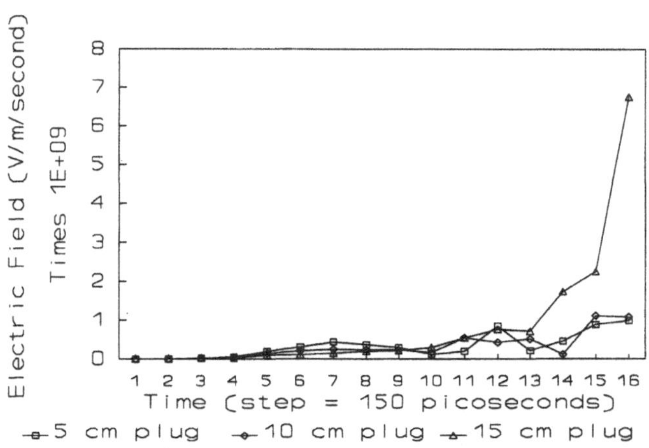

Figure 6. Time-Domain Electric Field at Center of Box

Figure 7 shows the magnetic field at multiples of f_0 = 417 Mhz. Like in the microstrip case, the integral present in Maxwell's Second Law continued to remain close to zero, with a maximum value of about 9 10^{-8}A·m.

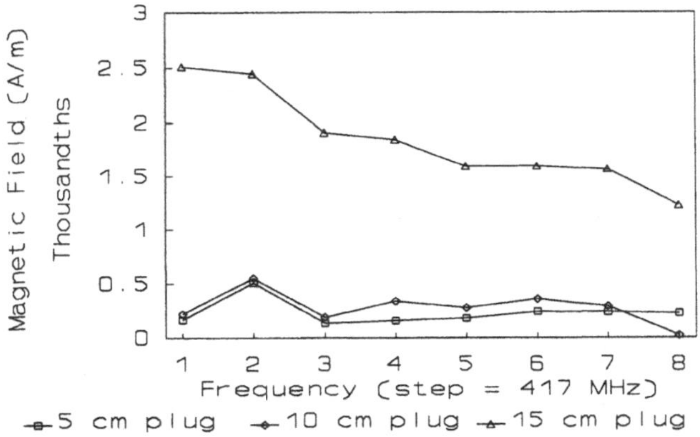

Figure 7. Frequency-Domain Magnetic Field at Center of Box

A study of the time behaviour of the electric and magnetic fields for several openings of the box, indicates that for the 5 or 10 centimetre plugs the results do not differ much. When the plug is 15 centimetres long however, the values of the fields increase significantly. A possible explanation would be that some type of constructive interference occurs in that situation.

CONCLUSIONS AND FUTURE WORK

The methods described herein and the results obtained are expected to provide a tangible means to model the behaviour of a mixed dielectric - conductor scatterer subjected to an incident plane wave. In the case of the microstrip the current induced can be calculated and the expected interference effects estimated. In the case of the metallic box with thick walls the fields at any location inside it could be estimated.

Such a box was originally conceived to serve as a testing environment for electronic equipment. The penetration of the outside field into the box is a measure of the expected interference upon the device or component undergoing testing.

The equations developed so far are new and are expected to provide a solid basis for performing numerical time-domain predictions on even more complicated hybrid structures. Future work may include the development of more general formulations that would describe the scattering of waves upon multi-dielectric bodies. Such integro-differential equations would take into account volume discontinuities, by introducing surface charges at the interface between the dielectrics. Such a development would greatly expand the number of structures which may be analyzed as possible scatterers.

REFERENCES

[1] Poggio, A.J. and Miller, E.K., "Solutions of three-dimensional scattering problems", in *Computer Techniques for Electromagnetics*, R. Mittra, Ed., 1973

[2] Bennett, C.L. and Weeks, W.L., "Electromagnetic pulse response of cylindrical scatterers", *G-AP Symposium*, Boston, Mass., 1968

[3] Bennett C.L. and Mieras H., "Time-domain scattering from open thin conducting surfaces", *Radio Science*, Vol. 16, 1231-1239, 1981

[4] Bennett C.L. and Ross, G.F., "Time-Domain Electromagnetics and its Applications", *Proc. IEEE*, Vol. 66, No. 3, 299-318, 1978

[5] Mieras H. and Bennett, C.L., "Space-Time Integral Equation Approach to Dielectric Targets", *IEEE Trans. Ant. Prop.*, Vol. AP-30, No. 1, 2-9, 1982

[6] Miller, E.K., "An Overview of Time-Domain Integral Equation Models in Electromagnetics", *J. Elec. Waves & Appl.*, Vol. 1, No. 3, 269-293, 1987

[7] Stratton, J.A., *Electromagnetic Theory*, McGraw-Hill, 1941

[8] Bernardi P. and Cicchetti R., "Response of a Planar Microstrip Line Excited by an External Electromagnetic Field", *IEEE Trans. Elec. Compat.*, Vol. 32, 98-105, 1990

SCATTERING OF ISO-DIFFRACTIVE PULSED BEAMS FROM RANDOMLY IRREGULAR SURFACES

Shimshon Frankenthal

Tel Aviv University
Ramat Aviv, Israel

Abstract

The paraxial wave equation is employed to track the bichromatic correlation function of signals scattered by a randomly-irregular surface modelled as a phase-screen, and hence to compute the intensity-pulse returns of elementary beamed pulses emitted by an iso-diffractive Gaussian aperture (across which spectral lines are Gauss-distributed, with spreads adjusted to achieve a common Rayleigh distance). In a suitable asymptotic regime, the returns consist of two pulses on radically different time scales, which reflect the surface roughness and its effective slope, respectively. They are compared here with returns from small deterministic targets.

1. Scope and Major Assumptions

Brief pulses are finding an ever-increasing domain of applicability in scenarios which require target-detection and identification and/or propagation-medium diagnostics. The present paper serves to illustrate the type of calculations that are required in such applications. It addresses the propagation of an elementary "iso-diffractive" pulse (a "complex-source" solution of the wave equation - see [1]), and particularly its scattering by a model randomly-irregular surface (e.g. the ocean) and by a model deterministic reflector. The results may be applied to the problem of characterizing the statistical properties of a randomly-irregular surface, or to the problem of identifying returns from deterministic reflectors against a background of radiation returned from such a surface.

In particular, we address the computation of the temporal correlation function, and the intensity, of brief pulses returned from the aforementioned model reflectors. The main assumptions are:

1. The pulse-beam is emitted from and returned to an aperture at a height z above the random surface, its axis is normal to that surface, and it is sufficiently narrow so that the paraxial approximation is adequate. Moreover, scalar modelling of the field is sufficient.

2. The scattering-reflection by the surface can be modelled by a random phase screen. The random phase process can be accounted for by a correlation function and/or a phase structure function which depends only on the separation-distance q between points on the surface.

3. Ths emitted pulse is Gaussian, in the sense that its spectral components are Gauss-distributed across the emitting aperture. In the iso-diffractive examples considered here, the spread of each spectral component varies inversely with the square-root of the frequency ω or the wavenumber k.

2. Formulation - The Bichromatic Coherence of the Screen-Return

The temporal correlation Γ and intensity I of a real stochastic time signal v(t)

$$\Gamma(t,\tau) = \langle v(t_1)\, v(t_2)\rangle \quad \text{with} \quad t_{1,2} = t \pm .5\tau; \quad I(t) = \Gamma(t,0) \tag{1}$$

may be represented in terms of the temporal Fourier-spectrum $\hat{v}$ of the signal thus

Ultra-Wideband, Short-Pulse Electromagnetics
Edited by H. Bertoni *et al.*, Plenum Press, 1993

$$\Gamma(t,\tau) = \frac{1}{\pi^2}\mathrm{Re}\left[\int_0^\infty d\omega \int_{-2\omega}^{2\omega} d\Omega + \int_0^\infty d\Omega \int_{-.5\Omega}^{.5\Omega} d\omega\right] e^{-j(\omega\tau+\Omega t)} \langle \hat{v}(\omega+.5\Omega)\, \hat{v}^*(\omega-.5\Omega)\rangle \tag{2}$$

$$I(t) = \mathrm{Re}\,\frac{1}{\pi}\int_0^\infty d\Omega\, e^{-j\Omega t}\,\frac{1}{\pi}\int_0^\infty d\omega\, \langle \hat{v}(\omega+.5\Omega)\, \hat{v}^*(\omega-.5\Omega)\rangle \tag{3}$$

where the $\langle\,\rangle$ denote ensemble-averaging. Since we shall use the wavenumber-spectrum

$$\hat{v}(k) = \int_{-\infty}^{\infty} dt\, e^{j\omega t}\, v(t) \qquad \text{with} \qquad \omega = kc \tag{4}$$

we next define the sum and difference variables corresponding to the wavenumbers k_1 and k_2

$$k_s = \tfrac{1}{2}(k_1 + k_2) \qquad k_d = \tfrac{1}{2}(k_1 - k_2) \qquad \text{so that} \quad \omega = ck_s \quad \text{and} \quad \Omega = 2ck_d \tag{5}$$

and therefore seek the bichromatic coherence for the signals of interest

$$\hat{\Gamma}(k_s,k_d) = \langle \hat{v}(k_1)\, \hat{v}^*(k_2)\rangle \qquad \text{where} \qquad k_{1,2} = k_s \pm k_d \tag{6}$$

To propagate field-spectra between range-planes,we shall employ the paraxial propagator

$$\hat{v}(k,z,\mathbf{u}) = -\frac{jk}{2\pi(z-z')}\, e^{jk(z-z')} \int_{-\infty}^{\infty} du' e^{j\frac{k}{2}\frac{(\mathbf{u}-\mathbf{u}')^2}{z-z'}}\, \hat{v}(k,z',\mathbf{u}') \tag{7}$$

where $\mathbf{u}$ denotes transverse location in a range-plane z.

Let the transmitting aperture and the phase-screen occupy the plane $z = -z_s$ and $z = 0$, respectively. We use Eq.(7) to propagate the spectrum $\hat{v}_a(k,\mathbf{u})$ of the aperture-signal from $z' = -z_s$ to $z = 0$, multiply the result by the exponential of the random phase $\phi(k,\mathbf{u})$ injected by the screen, and use Eq.(7) again to propagate the field to the generic field point $(z,\mathbf{u})$. The result is

$$\hat{v}(k,z,\mathbf{u}) = -\frac{jk}{2\pi z} e^{jkz} \int_{-\infty}^{\infty} du' e^{j\frac{k}{2}\frac{(\mathbf{u}-\mathbf{u}')^2}{z}}\, e^{j\phi(k,\mathbf{u}')} \left[-\frac{jk}{2\pi z_s}\right] e^{jkz_s} \int_{-\infty}^{\infty} du_a e^{j\frac{k}{2}\frac{(\mathbf{u}'-\mathbf{u}_a)^2}{z_s}}\, \hat{v}_a(k,\mathbf{u}_a) \tag{8}$$

The corresponding expression for the bichromatic coherence $\hat{\Gamma}(k_s,k_d)$ of Eq.(6) would thus require a four-fold, 2-dimensional vector integration over pairs of points $\mathbf{u}'_{1,2}$ and $\mathbf{u}_{a\,1,2}$ in the phase-screen and aperture planes, with the wavenumber k replaced by $k_{1,2}$ as required.

For fields whose spectra are Gauss-distributed across the aperture plane, the aperture-spectrum is

$$\hat{v}_a(k,\mathbf{u}) = V_a(k)\, \exp -\frac{u^2}{d^2(k)} \tag{9}$$

where the spread d of each spectral line may depend on the wavenumber. In this case, the inner integral of Eq.(8) yields an explicit expression for the spectrum of the field incident on the screen: it is the axial spectrum V_a, multiplied by the Gaussian aperture-to-screen transfer function

$$\hat{g}_s(k,\mathbf{u}') = -\frac{jb}{z_s - jb} e^{jkz_s} \exp\frac{jb}{z_s - jb}\frac{u'^2}{d^2} = -\frac{b}{\sqrt{z_s^2+b^2}} e^{j\theta_s} \exp jk\left[z_s + \frac{u'^2}{2(z_s^2+b^2)}\right] \exp -\frac{u'^2}{2R_s^2} \tag{10}$$

$$\text{where} \qquad \theta_s = \frac{\pi}{2} + \tan^{-1}\frac{b}{z_s} \qquad b = \frac{kd^2}{2} \qquad R_s^2 = \frac{z_s^2+b^2}{kb} \tag{11}$$

b is the Rayleigh distance about which a beam of wavenumber k transits from the near-field to the far-field behavior, and R_s is the e-folding distance of the intensity incident on the screen. The remaining integral extends over pairs of screen-points $\mathbf{u}'_{1,2}$, or over their sum-difference counterparts

$$\mathbf{u}' = \tfrac{1}{2}(\mathbf{u}'_1 + \mathbf{u}'_2) \qquad \mathbf{q}' = \mathbf{u}'_1 - \mathbf{u}'_2 \tag{12}$$

The integrand also includes the bichromatic ensemble-avereage of the random-phase exponentials. Our assumption regarding the screen now allows us to express this ensemble average in the form

$$\langle e^{j[\phi(k_1,u_1')-\phi(k_2,u_2')]}\rangle = e^{-S(k_1,k_2,q')} \tag{13}$$

where the scattering exponent S depends only on the separation $\mathbf{q}'$ between the correlated points $\mathbf{u}'_{1,2}$. The integration over $\mathbf{u}'$ then leaves just a single vector integration over the $\mathbf{q}'$ coordinate.

Further simplification occurs when we require that all the spectral components of the beam transit from the near-field to the far-field behavior at the same point b ("iso-diffractive" aperture). This happens - see Eq.(11) - when the beam-spread d(k) of each component of the aperture-spectrum varies as $1/\sqrt{k}$), and some care must be exercised in incorporating this relation in the present computation. In the domain of integration of Eqs.(2) and (3), the wavenumber $k_1 = k_s + k_d$ is always positive, but $k_2 = k_s - k_d$ may be positive or negative. In performing the computation, we therefore set $b_i = .5k_i d^2 = b\ \mathrm{sign}(k_i)$, where b is a positive number. As a result, the final expressions for the bichromatic coherence and the intensity vary according as k_s is larger or smaller than k_d.

Finally, when we specify that the receiver and the transmitter are co-located at a distance z from the phase screen (a randomly reflecting surface), the bichromatic coherence at the receiver is:

$$\hat{\Gamma}(k_s,k_d,z,\mathbf{u}) = G_s(k_s,k_d,z,\mathbf{u})\,\hat{\Gamma}_a(k_s,k_d) \tag{14}$$

where $\hat{\Gamma}_a$ is the Gaussian coherence of the transmitted signal

$$\hat{\Gamma}_a(k_s,k_d) = V_a(k_s+k_d)\,V_a^*(k_s-k_d) \tag{15}$$

and G_s is the transfer function

$$G_s(k_s,k_d,z,\mathbf{u}) = P\,e^{j4k_d z}\,e^{-\gamma^2 u^2}\,\frac{1}{\pi}\alpha^2\int_{-\infty}^{\infty} d\mathbf{q}' e^{-\alpha^2(\mathbf{q}'+j\beta\mathbf{u})^2}\,e^{-S(k_s,k_d,\mathbf{q}')} \tag{16}$$

$$P = \frac{b^2}{4z^2+b^2} * \begin{cases} 1 & \text{when } k_d < k_s \\ -e^{j\theta_p} & \text{when } k_s < k_d \end{cases} \quad (17);\qquad \alpha^2 = \frac{k_s^2-k_d^2}{4\eta b} * \begin{cases} (k_s\xi - jk_d)^{-1} & \text{when } k_d < k_s \\ -[k_d(\xi+j)]^{-1} & \text{when } k_s < k_d \end{cases} \tag{18}$$

$$\beta = 2\frac{bz}{4z^2+b^2} * \begin{cases} 1 & \text{when } k_d < k_s \\ 0 & \text{when } k_s < k_d \end{cases} \quad (19);\qquad \gamma^2 = \frac{1}{4z^2+b^2} * \begin{cases} bk_s - j2k_d z & \text{when } k_d < k_s \\ k_d(b-j2z) & \text{when } k_s < k_d \end{cases} \tag{20}$$

$$\theta_p = 2\tan^{-1}\frac{b}{2z} \qquad \xi = \frac{bz}{b^2+2z^2} \qquad \eta = \frac{z}{b}\frac{2z^2+b^2}{4z^2+b^2} \tag{21}$$

3. Screen Scattering Model

The simplest way to model scattering by a random surface is to assume that it injects a phase which is proportional to the wavenumber k and to the deviation of the surface-elevation from its mean value

$$\phi(k,\mathbf{u}) = k\langle\mu^2\rangle^{1/2}\bar{\mu}(\mathbf{u}) \tag{22}$$

where $\langle\mu^2\rangle$ is the mean-square deviation of the elevation and $\bar{\mu}(\mathbf{u})$ a normalized random function. With this assumption, the scattering exponent S defined in Eq.(13) becomes (see, e.g., [2])

$$S(k_1,k_2,\mathbf{q}) = \tfrac{1}{2}(k_1-k_2)^2 D_\infty + k_1k_2 D(q) \qquad D(q) = \langle\mu^2\rangle[1-B(q)] \tag{23}$$

where B(q) is the normalized correlation, and D(q) is the structure function, of the random process. Using Eq.(6) for $k_{1,2}$ in Eq.(23) - both apply throughout the range of integration of Eq.(3) - we find

$$S(k_s,k_d,\mathbf{q}) = 2D_\infty k_d^2 + (k_s^2-k_d^2)\,D(q) \tag{24}$$

for all positive k_s and k_d. As will be seen shortly, this causes some difficulty: since D is always positive, the exponential exp-S decreases with q when $k_s > k_d$, but increases with q otherwise.

In contradistinction to the monochromatic case, where the first term of S is zero, it is necessary to provide some information (D_∞) about the overall variance of the process (or about its outer scale) in order to fully specify the bichromatic coherence. Moreover, since $k_s^2 - k_d^2$ changes sign over the range of integration of Eqs.(2) and (3), we *cannot* employ the "quadratic-screen" approximation

$$D(q) \sim C_s^2 q^2 \quad \text{where} \quad C_s^2 = \frac{\langle\mu^2\rangle}{L^2} \tag{25}$$

to simplify the integral for the bichromatic transfer-function G_s in Eq.(16): Eq.(25), which retains only the leading term of the series-expansion of D, is justifiable only when $k_s - k_d >> 1$.

4. The Aperture Spectrum and The Emitted Intensity Pulse

The width of spectral lines, which are Gauss-distributed across the aperture, varies as $1/\sqrt{k}$. Thus

$$\hat{V}_a(\omega,\mathbf{u}) = V_a(\omega)\exp -\frac{u^2}{d^2} \quad \text{with} \quad d^2 = \frac{2bc}{\omega} \tag{26}$$

and it suffices to specify the axial (aperture-center) spectrum V_a. We shall consider two exponential spectra, which correspond to the analytic-delta beam treated in [1], and to its Hilbert-transform

$$V_a(k) = \exp-\frac{|k|}{k_a} * [1 \text{ or } j\,\text{sign}(k)] \tag{27}$$

where k_a (or the corresponding $\omega_a = ck_a$) is the aperture-bandwidth.

The bichromatic-coherence of these pulses is obtained using Eq.(15), with $k_{1,2} = k_s \pm k_d$. We find

$$\hat{\Gamma}_a(k_s,k_d) = \begin{cases} \exp -2k_s/k_a & \text{when} \quad k_d < k_s \\ \pm\exp -2k_d/k_a & \text{when} \quad k_s < k_d \end{cases} \tag{28}$$

where the + and - correspond to the two choices in [ ] in Eq.(27). Using dimensionless versions of the sum and difference wavenumbers k_s and k_d (and corresponding frequencies) and of the time t

$$\bar{\omega} = 2\frac{k_s}{k_a} = 2\frac{\omega}{\omega_a} \qquad \bar{\Omega} = 2\frac{k_d}{k_a} = \frac{\Omega}{\omega_a} \qquad t_a = ck_a t = \omega_a t \tag{29}$$

we integrate Eq.(3) to find the positive-frequency spectrum and the shape of the intensity-pulses

$$\hat{I}_a(\Omega) = \frac{c}{\pi}\left[\pm\int_0^{k_d} dk_s \exp-2\frac{k_d}{k_a} + \int_{k_d}^{\infty} dk_s \exp-2\frac{k_s}{k_a}\right] = \frac{\omega_a}{2\pi}[1 \pm \bar{\Omega}]\exp-\bar{\Omega} \tag{30}$$

$$I_a(t) = \text{Re}\frac{1}{\pi}\int_0^{\infty} d\Omega\, e^{-j\Omega t}\frac{\omega_a}{2\gamma}(1\pm\bar{\Omega})e^{-\bar{\Omega}} = \frac{\omega_a^2}{2\pi^2}\text{Re}\int_0^{\infty} d\bar{\Omega} e^{-\bar{\Omega}(1+jt_a)}(1\pm\bar{\Omega}) = \frac{\omega_a^2}{\pi^2(1+t_a^2)^2}*[1 \text{ or } t_a^2] \tag{31}$$

which evolve on the time scale $(ck_a)^{-1}$ established by the aperture bandwidth.

5. The Return from a Deterministic Reflector

This section computes the intensity returned from a deterministic reflector in the far-field of the aperture. The reflector is modelled by a circular perfectly-reflecting plate of radius R, located on the screen at a transverse distance $\mathbf{u}_r$ from the aperture nadir-point, and oriented normal to the line of sight from the aperture-center. The coordinates of a point on this plate will be denoted by (z-z′,$\mathbf{u}_r$+$\mathbf{u}'$), where z′ is the height of the point above the screen.

The field incident on the plate is obtained from Eq.(10) by setting

$$z_s = z - z' \qquad \mathbf{u} = \mathbf{u}_r + \mathbf{u}' \qquad \frac{1}{z - jb} \sim \frac{1}{z}\left(1 + j\frac{b}{z}\right) \qquad z_r = z + \frac{u_r^2}{2z} \qquad z' + \frac{1}{z}\mathbf{u}_r\cdot\mathbf{u}' = 0 \tag{32}$$

(the last relation means that the plate is normally-oriented). Using Eq.(7), the returned field is

$$g_r = -\frac{bk}{2\pi z_r(z-jb)} e^{j2kz_r} \exp - \frac{kb}{2z^2} u_r^2 \int_0^R du'u' \exp\left[j\frac{k}{2z} - \frac{kb}{z^2}\right]u'^2 \int_0^{2\pi} d\theta \exp\left[j\frac{kb}{z^2}u_r u'\cos\theta\right] \quad (33)$$

Assume that the aperture is located in the far-field of the plate, which is small and located within one e-folding distance R_s of the surface intensity - Eq.(11) - for all frequencies of interest:

$$\frac{kR^2}{2z} << 1 \qquad R << u_r \qquad u_f^2 < \frac{2z^2}{kb} \quad (34)$$

All exponents in the integrand of Eq.(33) are then negligible, and the integration produces the cross section A of the plate. The two-way transfer-functions for the field and the coherence are

$$g_r = -\frac{bk}{2\pi z_r(z-jb)} A\, e^{j2kz_r} \exp - \frac{kb}{2z^2} u_r^2 \quad (35)$$

$$G_r = \frac{b^2A^2}{4\pi^2 z^4} e^{j4k_d z_r} (k_s^2 - k_d^2) * \begin{cases} \exp -k_s b(u_r^2)/z^2 & \text{when } k_d < k_s \\ -e^{j\theta_r} \exp -k_d b(u_r^2)/z^2 & \text{when } k_s < k_d \end{cases} \qquad \text{with } \theta_r = 2\tan^{-1}\frac{b}{z} \quad (36)$$

The product $G_r\hat{\Gamma}_a$, which must be used in Eq.(2) to produce the spectrum and the shape of the returned intensity pulse, has the form of Eq.(36), except that the quantity bu_f^2/z^2 in the exponents is replaced by k_r^{-1}, defined (along with the corresponding frequency ω_r) by

$$\frac{2}{k_r} \equiv \frac{2}{k_a} + b\frac{u_r^2}{z^2} \qquad \omega_r \equiv ck_r \quad (37)$$

Dimensionless variables can now be defined as in Eq.(29), except with the reference frequency and wavenumber of Eq.(37). The spectrum and pulse-shape of the reflected intensity are

$$\hat{I}_r(\Omega) = \frac{c}{4\pi}\frac{b^2A^2}{4\pi^2z^4} e^{j2k_d z_r} \exp -\Omega_r \left[1 + \Omega_r \mp e^{j\theta_r}\frac{1}{3}\Omega_r^3\right] \quad \text{where} \quad \Omega_r = 2\frac{k_d}{k_r} \quad (38)$$

$$I_r(t) = \frac{c^2}{\pi^2}\frac{b^2A^2k_r^4}{4\pi^2z^4}\frac{1}{(1+t_r^2)^4}\left[(1-t_r^2)^2 \text{ or } 4t_r^2\right] \quad \text{where} \quad t_r = ck_r(t - 2z_r/c) \quad (39)$$

Beside the delay, the time-scale is stretched (by k_a/k_r) relative to the that of the aperture-pulse.

6. The Screen Return

We shall focus on the intensity-return at the center $u = 0$ of the receiving aperture, and in the limit where the screen lies in the farfield region of the aperture. We shall need the quantities

$$\delta \equiv 2C_s^2\eta bk_a \qquad \epsilon \equiv \xi\delta \quad (40)$$

and the far-field limiting forms of the various parameters that enter the discussion

$$\xi \sim \frac{b}{2z} << 1\ ; \quad \delta \sim zC_s^2k_a\ ; \quad e^{j\theta_p} \sim 1 + j2\xi\ ; \quad \epsilon = \xi\delta \sim \tfrac{1}{2}bC_s^2k_a\ ; \quad \nu^2 = \tfrac{1}{4}\langle\mu^2\rangle k_a^2\ ; \quad C_s^2 = \frac{\langle\mu^2\rangle}{L^2} \quad (41)$$

We next evaluate the coherence transfer-function G_s, Eq.(16), for the Gaussian screen whose structure-function is given in Eq.(24). We use polar coordinates for q', perform the angle-integration (which yields 2π), and employ the dimensionless parameters defined above, as well as the dimensionless variables defined in Eq.(29). We find that, except for multiplicative prefactors, G_s behaves as

$$\bar{G}_s = \tilde{\alpha}^2 \int_0^\infty dq_2 \exp -\tilde{\alpha}^2 q_2 \exp -\left[\bar{\alpha}(1 - \exp -q_2)\right] \quad \text{where} \quad \bar{\alpha} = \nu^2(\bar{\omega}^2 - \bar{\Omega}^2)\ , \quad q_2 = \frac{q'^2}{L^2} \quad (42)$$

and

$$\tilde{\alpha}^2 = \bar{\alpha} * \begin{cases} (\epsilon\bar{\omega} - j\delta\bar{\Omega})^{-1} & \text{when } k_s > k_d \text{ or } \bar{\omega} > \bar{\Omega} \\ -[\bar{\Omega}(\epsilon + j\delta)]^{-1} & \text{when } k_s < k_d \text{ or } \bar{\omega} < \bar{\Omega} \end{cases} \quad (43)$$

The expressions corresponding to $k_d < k_s$ and $k_s < k_d$ in Eqs.(28) and (43) must be used in the first and second integrals of Eq.(2), respectively, or in two separate ranges of the inner integral of Eq.(3). They produce two distinct contributions to the intensity return:

$$I_S(t) = \frac{\omega_a^2}{2\pi^2} \frac{b^2}{4z^2+b^2}(\mp J_1 + J_2) \tag{44}$$

$$J_1 = \mathrm{Re}\ e^{j\theta_p} \int_0^\infty d\bar{\Omega} e^{-\bar{\Omega}(1+jt_a)}\, e^{-2\nu^2\bar{\Omega}^2} \int_0^{\bar{\Omega}} d\bar{\omega}\, \bar{G}_S \quad (45);\quad J_2 = \mathrm{Re} \int_0^\infty d\bar{\Omega} e^{-j\bar{\Omega} t_a}\, e^{-2\nu^2\bar{\Omega}^2} \int_{\bar{\Omega}}^\infty d\bar{\omega}\, e^{-\bar{\omega}}\, \bar{G}_S \quad (46)$$

In general, these integrals must be evaluated numerically. However, in the far-field, high-frequency regime, the following approximate approach provides an idea of the pulse shapes and scales.

The $\bar{\Omega}$ dependence of the integrands of Eqs.(45) and (46) involves three frequency-scales: 1, δ^{-1} and ν^{-1}. The last two are related by the ratio

$$\frac{\nu^2}{\delta^2} = \frac{1}{4} \frac{L^4}{z^2\langle\mu^2\rangle} \tag{47}$$

which will be assumed very small (it is certainly so for a very wide range of altitudes z and ocean-surface conditions, for example). We shall also assume $\nu >> 1$ (typically, for an r.m.s. ocean wave-height of 1m, the transitional frequency is 70 Mhz - and for that example we shall consider much higher frequencies). Indeed, since the screen return will be found to decrease with increasing δ, we shall henceforth examine the regime $\nu >> \delta >> 1$, with $\xi \to 0$ (far field). In this regime, asymptotic expressions for J_1 and J_2 may be obtained.

For the region $\bar{\omega} < \bar{\Omega}$, the substitutions

$$\bar{\omega} = x\bar{\Omega} \qquad \Omega_\nu = \nu\bar{\Omega} \qquad \tilde{q} = \frac{\nu}{\delta} q_2 \tag{48}$$

convert the exponent in the integrand of Eq.(42) to the form

$$-\Omega_\nu(1-x^2)[\epsilon/\delta + j]^{-1}\tilde{q} + \Omega_\nu^2(1 - x^2)\left[1 - \exp-\frac{\delta}{\nu}\tilde{q}\right] \tag{49}$$

Since the expression for J_1, Eq.(45), contains a Gaussian term in Ω_ν^2, the dominant contribution to J_1 occurs when Ω_ν is of order 1. With $\nu << \delta$ - Eq.(47) - the exponential in [ ] in Eq.(49) vanishes very rapidly, implying that the structure-function is at saturation over most of the aforementioned range. Approximating this exponential by 0 over the entire range of integration permits the evaluation of J_1:

$$\bar{G}_S = \exp-\bar{\alpha} = \exp-\nu^2(\bar{\omega}^2-\bar{\Omega}^2) \tag{50}$$

$$J_1 = \frac{\sqrt{\pi}}{2}\frac{1}{\nu^2}\int_0^\infty d\Omega_\nu e^{-\Omega_\nu^2}\, e^{-\Omega_\nu/\nu}\, \mathrm{erf}(\Omega_\nu) \cos(\Omega_\nu t_\nu) \quad \text{where} \quad t_\nu = \frac{t_a}{\nu} \tag{51}$$

An equivalent simplification is not available in the range $\bar{\omega} > \bar{\Omega}$. In order to estimate J_2, we approximate the Gaussian structure function itself by the form

$$1 - e^{-q_2} \sim \begin{matrix} q_2 & \text{when} & q_2 < 1 \\ 1 & \text{when} & q_2 > 1 \end{matrix} \tag{52}$$

which contains both the quadratic approximation (the q_2-term - see Eq.(25)) and the "saturating" feature. The approximate form is everywhere larger than the actual function, and produces a lower bound on the contribution J_2 to the screen-return: its use in Eq.(42) produces

$$\bar{G}_S = \frac{\tilde{\alpha}^2}{\tilde{\alpha}^2 + \tilde{\alpha}}\left[1 - e^{-(\tilde{\alpha}^2+\tilde{\alpha})}\right] + e^{-(\tilde{\alpha}^2+\tilde{\alpha})} \tag{53}$$

and upon substituting the appropriate expressions for $\tilde{\alpha}^2$ and $\bar{\alpha}$, the leading terms read

$$\bar{G}_s = [1 + \epsilon\bar{\omega} - j\delta\bar{\Omega}]^{-1} + e^{-\nu^2(\bar{\omega}^2 - \bar{\Omega}^2)} \qquad \text{when} \qquad \bar{\omega} > \bar{\Omega} \tag{54}$$

When $\nu >> 1$, the contribution of the second term in Eq.(54) to J_2 is similar to Eq.(51), except that $\mathrm{erf}(\Omega_\nu)$ is replaced by $\mathrm{erfc}(\Omega_\nu)$. The contribution of the first term can be expressed as

$$J_\delta = \mathrm{Re}\int_0^\infty d\bar{\Omega} e^{-\bar{\Omega}(1+jt_a)}\ e^{-2\nu^2\bar{\Omega}^2} \int_0^\infty d\bar{\omega}'\ e^{-\bar{\omega}'}\ [1 + \epsilon\bar{\omega}' - \bar{\Omega}(j\delta-\epsilon)]^{-1} \tag{55}$$

This is integrated by ignoring ϵ relative to δ, and re-scaling the frequency and time variables thus

$$\Omega_\delta = \bar{\Omega}\delta \qquad\qquad t_\delta = \frac{t_a}{\delta} \tag{56}$$

With $\delta \rightarrow \infty$, the leading term of the result is

$$J_\delta = \frac{\pi}{\delta}\delta_{-1}(t_\delta)\ e^{-t_\delta}\ \frac{1}{1 + \epsilon t_\delta} \tag{57}$$

When these results are inserted in Eq.(44), the asymptotic expression for I(t) reads (see Figs.1 and 2)

$$I_s(t) = \frac{\omega_a^2}{2\pi^2}\frac{b^2}{4z^2}(J_\delta + J_\nu)$$

$$J_\delta = \pi\frac{1}{\delta}\delta_{-1}(t_\delta)\ e^{-t_\delta}\ \frac{1}{1 + \epsilon t_\delta}\ ;\ J_\nu = \frac{\pi}{4}\frac{1}{\nu^2}\mathrm{Re}\frac{2}{\sqrt{\pi}}\int_0^\infty d\Omega_\nu\ e^{-j\Omega_\nu t_\nu}\ e^{-\Omega_\nu^2}\ [\mathrm{erfc}(\Omega_\nu) \mp \mathrm{erf}(\Omega_\nu)] \tag{58}$$

$$1 << \nu << \delta \qquad \nu^2 = \tfrac{1}{4}\langle\mu^2\rangle k_a^2 \qquad \delta = zk_a\frac{\langle\mu^2\rangle}{L^2} \qquad \epsilon = \tfrac{1}{2}bk_a\frac{\langle\mu^2\rangle}{L^2}$$

We note that J_δ and J_ν(a lower bound!) are never negative: J_ν vanishes at t = 0 when the - sign is chosen ; it is a Gaussian pulse $(\pi/4\nu^2)\exp{-t_\nu^2/4}$ when the + sign is chosen. We also observe that the term J_δ still reflects the quadratic approximation. When the phase screen is employed to model the ocean-surface, J_δ is essentially sensitive to the slope, since it depends on $C_s^2 = \langle\mu^2\rangle/L^2$, while J_ν reflects the r.m.s. surface-height fluctuations.

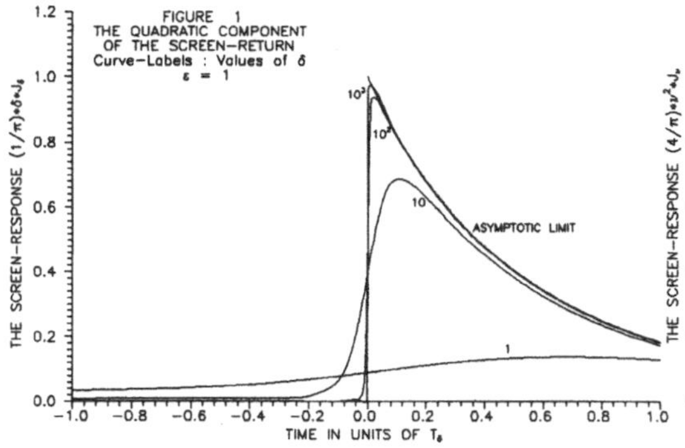

Figure 1

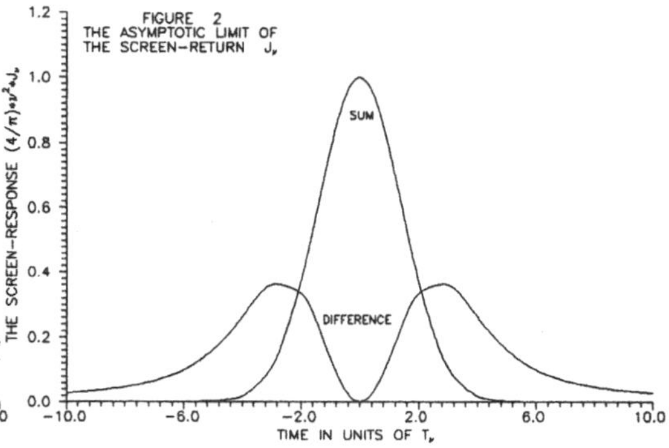

Figure 2

7. Discussion

We have illustrated the computation of the intensity-returns in response to a brief and narrow pulsed-beam (unit magnitude) emitted by an iso-diffractive aperture (Rayleigh-distance b), which varies on a time scale determined by the aperture-bandwidth k_a

$$T_a = \frac{1}{ck_a} \tag{59}$$

The computation entails the use of the paraxial wave equation to track the bi-chromatic coherence of the signal, and the use of a phase-screen to model interactions with a randomly-irregular surface. The results are effectively stated in a cylindrical coordinate system $(z,\mathbf{u})$ about the aperture, whose polar-axis z coincides with the beam axis.

A small, normally-oriented deterministic reflector of area A, centered on a point (z, u_r), will return a pulse of magnitude I_r, which evolves on a time scale T_r, such that

$$I_r = \frac{c^2}{\pi^2}\frac{b^2A^2k_r^4}{4\pi^2z^4} \qquad T_r = \frac{1}{ck_r} \qquad k_r = k_a\left[1 + \frac{u_r^2}{2R_S^2}\right] \qquad R_S^2 = \frac{z^2}{k_a b} \tag{60}$$

where R_S is the e-folding distance of the intensity incident on the plane z.

The return I_S from a randomly-irregular surface at z consists of two pulses, whose magnitudes are I_δ and I_ν, which evolve on the respective time-scales T_δ and T_ν, given by

$$I_\delta = \frac{c^2}{8\pi}\frac{b^2k_a}{z^3}\frac{L^2}{\langle\mu^2\rangle} \qquad I_\nu = \frac{c^2}{8\pi}\frac{b^2}{z^2}\frac{1}{\langle\mu^2\rangle} \qquad T_\delta = \delta T_a \qquad T_\nu = \nu T_a$$

$$\delta = zk_aC_S^2 \qquad \nu^2 = \frac{1}{4}\langle\mu^2\rangle k_a^2 \quad \text{with} \quad 1 << \nu << \delta \quad \text{and} \quad \epsilon = \tfrac{1}{2}ck_aC_S^2 \qquad C_S^2 = \frac{\langle\mu^2\rangle}{L^2} \tag{61}$$

where $\langle\mu^2\rangle$ is the roughness (mean-square wave-height of ocean, say) and L the correlation-length of the surface. The ν- and δ-pulses manifests the surface-roughness and slope, respectively.

To illustrate, consider the problem of detecting a deterministic "target" on the ocean against the background of the random ocean-return. A "design-goal" might be to reduce the ratios I_δ/I_r and I_ν/I_r over, say, a region of radius R_S (the incident intensity e-folding distance, or "footprint") to an acceptable level. The first of these two ratios is given by

$$\frac{I_\delta}{I_r} = \frac{\pi^3}{2}\frac{z}{A^2C_S^2k_a^3}\left[1 + \frac{u_r^2}{2R_S^2}\right]^4 = \frac{27}{16}\frac{z(\text{km})}{A^2C_S^2\, f_a^3(\text{Ghz})}\left[1 + \frac{u_r^2}{2R_S^2}\right]^4 \tag{62}$$

where the factor in [ ] represents the decrease in the target-reflected signal with increasing distance from "ground-zero", which amounts to a factor of 5 as the target moves from ground-zero to the edge of the footprint.

Numbers: With an altitude z = 300 km, an aperture-bandwidth f_a = 30 Ghz, and an ocean with an r.m.s. wave-height $\langle\mu^2\rangle$ = 1m and correlation-length L = 30m, a "ground-zero" signal-to-noise ratio of 5 can be attained. This drops to 1 at the "footprint-edge", which will be R_S = 10 km if b = 3 m, implying an aperture-diameter d = 5 cm at the aperture-bandwidth frequency. At this point, the I_ν-term contributes about as much as the I_δ-term.

References

1. E. Heyman and H. B. Felsen, "Complex Source Pulsed Beam Fields", J. Opt. Soc. Am. A, Vol. *6*, June 1989.

2. S. Frankenthal, "The Reflection of Radiation from Randomly Irregular Surfaces", J. Acous. Soc. Am. *85(1)*, pp. 110-121, 1989.

AN ANALYTIC CONTINUATION METHOD FOR THE ULTRA BROADBAND DETERMINATION OF THE ELECTROMAGNETIC PROPERTIES OF MATERIALS

Nicolaos G. Alexopoulos[1] and Rodolfo E. Díaz[2]

[1]Department of Electrical Engineering
University of California
Los Angeles, CA 90210

[2]Hexcel APD
2500 W. Frye Rd.
Chandler, AZ 85224

ABSTRACT

All dielectric materials, by nature, exhibit a frequency dependent permittivity. This property, termed dispersion, must be taken into account when materials are employed in applications that span very broad frequency ranges; such as ultra-wideband antenna applications and the design of Low Observable structures. Therefore, an accurate and compact representation of this frequency dependence over very broad bands of frequencies is needed. In addition, Time Domain Computational Electromagnetics computer codes can only realize their full potential when such a representation is used for modeling of material bodies, since only then will the pulse solution truly contain the proper solutions for the full spectrum of the pulse. Such a representation can be developed[1] by using the analytic function properties of the permittivity and a minimum set of physical assumptions. The result is a compact sum of special analytic basis functions. Because of their analyticity, dielectric data obtained over convenient portions of the Radio Frequency spectrum can be continued into other portions of the spectrum where measurements may be more difficult to perform. The usefulness of this technique is illustrated by generating the ultra broadband model of a carbon loaded absorbing foam of the type used in anechoic chambers. The model is then incorporated into a Dispersive Finite Difference Time Domain calculation of the absorption performance of a wall covered by wedges of that foam.

I - INTRODUCTION

The design of ultra-wideband systems introduces new challenges to the materials' engineer. Because of the broad spectrum of radio frequencies involved, the common narrow band information on the electromagnetic properties of materials is inadequate. This is particularly true of absorbing materials whose frequency dependent properties can vary widely from DC through the Gigahertz range. If real materials are to be used in the design and test of ultra-wideband radiating systems, their properties must be known over the entire spectrum. Furthermore, such knowledge should be cast in the form of a compact representation suitable for computational electromagnetics in order to obtain realistic performance expectations from the initial system design.

There have been two main obstacles to obtaining this representation: a lack of reliable ultra-wideband data on electromagnetic materials and a lack of a rigorous model valid for all physically realizable dielectrics. The first obstacle comes about because free space testing, the most reliable broadband test method for materials, is limited in practice by the requirement that the sample size be greater than or equal to one wavelength. Thus, testing materials at UHF and VHF frequencies becomes a challenge. The other test methods (waveguides, resonators) are by nature narrowband and have specific limitations on the kinds of materials they can accurately measure. (For instance, coaxial airline methods are not appropriate for anisotropic materials, while resonator perturbation methods are not appropriate for highly conducting materials.) The result is that our knowledge of the properties of engineering materials usually contains several holes in the spectrum. The second obstacle comes about because the traditional models of materials were derived on the basis of specific physical assumptions. (For instance, the Debye model of dielectric relaxation is derived by assuming the material is made up of permanent dipoles continually in collision with their neighbors.) Therefore these models have always been used with the caveat that they are just approximations and not truly rigorous[2].

It is shown in Reference 1 that this caveat is wrong. The classic physical models are terms of an expansion of the permittivity in special analytic basis functions. This expansion is founded on causality and a minimum of physical requirements and is rigorous within the given assumptions. Therefore it is not limited to specific physical models and can be used to represent all physically realizable dielectric materials. In thus removing the second obstacle this representation also removes the first, because its analytic function properties allow us to analytically continue band limited data into the regions of the spectrum where data is missing.

In Section II of this paper a summary of the derivation of this analytic representation is given. This is followed by a demonstration in Section III of the use of the analytic continuation method to obtain ultra broadband dielectric data for a typical absorbing foam material. This data is then used in a Dispersive FDTD computer program to calculate the broadband performance of this foam as an anechoic chamber wedge absorber, using a single pulse calculation.

II- THE ANALYTIC REPRESENTATION

If it is postulated that the experimentally observed phenomenon of dielectric Polarization is a macroscopic manifestation of the "polarizability" of the dielectric's microscopic structure by the applied field, then the following three physical requirements follow:

First, because the phenomenon involves the exchange and transport of energy between the field and the material, it must be causal. Second, because the phenomenon involves the reaction of microscopic "particles" to an applied force, it must be ultimately expressible in the form of an equation of motion. Third, because the dielectric is not a continuum but a granular ensemble of individual "particles", any model derived under the assumption of a continuum must break down at high enough frequencies. (Fortunately, for most materials, this latter limit of validity occurs beyond the ultraviolet.)

Given these requirements plus the assumption of invariance under translation in time and the assumption of linearity (to allow the use of superposition), the connection between the dielectric displacement vector (output) and the applied electric field (input) that caused it can be cast in the form[3]:

$$\mathbf{D}(t) = \mathbf{E}(t) + (2\pi)^{-1/2} \int_{-\infty}^{+\infty} T(t-t')\, \mathbf{E}(t')\, dt' \tag{1}$$

Where **T(t)** is the response of the system to a Dirac delta function input at t=0. And where we assume that both T(t) and **E**(t) are square integrable. (Square integrability for **E**(t) means that we are dealing with inputs that contain a finite amount of energy. This is reasonable for all practical cases of waves propagating through material media. For T(t) to be square integrable means that the response of the system after being "suddenly struck" eventually "dies out". In other words, there must always be a finite amount of loss in the system.) In equation 1, we have normalized to the permittivity of free space so that in the absence of any material, **D**=**E**. (Therefore, hereafter the term permittivity is used to denote relative permittivity.)

If the Fourier transform pair is defined by,

$$\mathcal{F}(\omega) = (2\pi)^{-1/2} \int_{-\infty}^{+\infty} F(t)\, e^{i\omega t} dt\ , \quad F(t) = (2\pi)^{-1/2} \int_{-\infty}^{+\infty} \mathcal{F}(\omega)\, e^{-i\omega t}\, d\omega\,. \tag{2}$$

Then, equation 1 transforms to:

$$\mathcal{D}(\omega) = (1+\mathcal{T}(\omega))\ \ \mathcal{E}(\omega) \tag{3}$$

From which we recognize the quantity $(1+\mathcal{T}(\omega))$ as the frequency domain permittivity, $\varepsilon(\omega)$. Therefore $\varepsilon(\omega)-1$ is the Fourier transform of $T(\tau)$, a causal, square integrable function (with $\tau = t-t'$).

$$\varepsilon(\omega) = 1 + \int_0^{\infty} T(\tau) e^{i\omega\tau} d\tau \tag{4}$$

The causality requirement means that $T(\tau)$ vanishes for $\tau<0$, making equation 4 an instance of Titchmarsh's Theorem 95[4]. From which it is concluded that the function $\varepsilon(\omega)-1$ is the boundary value (that is, as $\text{Im}(\omega)\rightarrow 0$) of a function which is analytic in the upper half of the complex frequency (ω) plane. This well known result leads to a thorough understanding of the analytic function properties of $\varepsilon(\omega)$ in the upper half-plane[5]: Namely, that the function $\varepsilon(\omega)$ must be a one-valued regular and zeroless function everywhere in the upper half-plane. Its imaginary part can only vanish on the imaginary ω axis and there can be no singularities on the real ω axis except possibly at the origin where the presence of DC conductivity allows a simple pole.

The problem is that despite all this information, none of this knowledge can be used to derive the desired compact representation of the permittivity. The closest we can get is the derivation of the Kramers-Krönig relations, which allow the completion of the function if we know either its real or its imaginary part; that is, if we already know half of the answer. So, where is the desired compact representation? In the lower half-plane, because there reside all the singularities of the function.

To see this, we must first consider what types of singularities are permissible in the lower half-plane. It has been shown in reference 1 that poles of first order are permissible but poles of second or higher order are disallowed by the requirement that $\mathrm{Im}(\varepsilon(\omega))$ only vanish at $\mathrm{Re}(\omega)=0$. Essential singularities are banned because at an essential singularity the function has no unique limiting value and it can come arbitrarily close to any assigned value an infinite number of times. If such a singularity (i.e. of the form $e^{1/\omega}$) were placed close to the real axis, the permittivity in its neighborhood could vary wildly for small variations of the frequency. Such behavior has never been reported in physical materials. Nonisolated natural boundaries as in the function $f(\omega) = \Sigma\, \omega^{n!}$ are also outlawed because such functions typically are not square integrable along the real axis. Logarithmic branch points can be similarly excluded. Finally, square (or higher order) root branch points turn out to be not permissible because the permittivity cannot be reduced to an equation of motion. The result is that the lower half-plane of the permittivity function may only have zeros or simple poles. And so, the complex permittivity, which was known to be holomorphic in the upper half-plane, is then meromorphic in the entire complex frequency plane. In particular, it is a rational function[6].

If $\varepsilon(\omega)-1$ is a rational function, we can express it as the product of all its zeros divided by the product of all its poles:

$$\varepsilon(\omega) - 1 = \frac{(\omega-\zeta_1)(\omega-\zeta_2)\ldots(\omega-\zeta_n)}{(\omega-\pi_1)(\omega-\pi_2)\ldots(\omega-\pi_n)} \tag{5}$$

This can be reduced further by invoking a physical fact: Since the microscopic structure of the dielectric is ultimately reduced to the response of its electrons to the incident wave, we know that its limiting behavior, as $\omega \to \infty$, must be that of the classical Solid State Physics model of the free electron gas. We then know that beyond a "plasma" frequency $\varepsilon(\omega)-1$ decays as $1/\omega^2$ and so there must be more terms in the denominator of equation 5 than in the numerator. Thus, we can do a partial fraction expansion. However, before proceeding with the expansion, we realize that equation 4 implies the fundamental causal symmetry $\varepsilon(-\omega^*) = \varepsilon^*(\omega)$. And so to every pole term $A/(\omega-\pi_1)$ there must be a term $A'/(\omega+\pi_1^*)$. Our partial fraction expansion takes the form of a sum of special functions:

$$\varepsilon(\omega) - 1 = \sum_n \frac{A_n}{\left\{(\omega - \pi_n)(\omega + \pi_n^*)\right\}} \tag{6}$$

The significance of these functions is grasped immediately upon further examination of one term. Let the pole be given by $\pi_1 = a + ib$. Then the first term,

$$C_1 = \frac{A_1}{(\omega\text{-}a\text{-}ib)(\omega + a\text{-}ib)} = \frac{A_1}{\left\{\omega^2 - (a^2 + b^2) - i2b\omega\right\}} \tag{7}$$

becomes, through the variable substitutions, $a = (1/LC - R^2/4L^2)^{1/2}$ and $b=-R/2L$,

$$C_1 = \frac{A_1}{\{1 - i\omega RC - \omega^2 LC\}} \tag{8}$$

This expression is special because, as the very suggestive variable substitutions denote, it is just the effective complex capacitance of a series LRC circuit. (With the subtle variation that to conform to the works in references 3, 4 and 5 we have used the

"Physicist's" convention $e^{-i\omega t}$ instead of the "Engineering" convention $e^{j\omega t}$ to denote a time harmonic field.) Therefore, the partial fraction expansion of the permittivity reduces to the effective complex capacitance of a parallel sum of series LRC circuit branches. (Where if the Admittance of the parallel sum is Y, the effective complex capacitance is defined as $C_{eff}=Y/j\omega$.) The circuit elements correspond to specific material properties: The capacitance C is the DC permittivity, $\varepsilon_0\varepsilon_r$, in farads/meter. The resistance R is the inverse of the bulk resistivity σ in mhos/meter. And the inductance L appears whenever the material has a natural resonance. In the absence of L, the circuit gives a pure Debye Relaxation. When L/R>>RC the circuit gives the classic Lorentz resonance.

This representation, just like the Kramers-Krönig relations, is a necessary consequence of the analytic function properties of the permittivity implied by causality. No "physical model" approximation has been made. We therefore expect equation 6 to be a complete representation of any physically realizable dielectric. In addition, the properties of completeness and analyticity mean that we can use analytic continuation to reconstruct the missing information from an incomplete set of permittivity data. As alluded to in the Introduction this is a very useful property for the determination of the electromagnetic properties of materials over ultra broadbands of frequency.

III- APPLICATION TO AN ABSORBER

The most convenient way to display ultra broadband data of materials is the Cole-Cole plot. In this plot, the abscissa is the real part of the permittivity, ε_r ,and the ordinate is the imaginary part, ε_i. For materials possessing Debye relaxation behavior, such a plot describes a semicircular arc. The rightmost end of the arc is the zero frequency value of the permittivity ($\varepsilon_r=\varepsilon_{DC}$, $\varepsilon_i=0$) and the leftmost end of the arc is the infinite frequency permittivity ($\varepsilon_r=\varepsilon_\infty$, $\varepsilon_i=0$.) In between, the permittivity may exhibit several peaks and valleys. For an absorber constructed by dispersing microscopic carbon particles into a dielectric foam matrix, it is expected that the behavior will be entirely of the Debye relaxation type. In other words, no resonances are expected because no conducting features of the order of the wavelength should exist inside the foam. Therefore the sum of equation 6 should reduce to a parallel sum of series RC (or $\sigma\varepsilon$) circuit branches. With this expectation, the measured data on a sample of absorber is plotted in Figure 1.

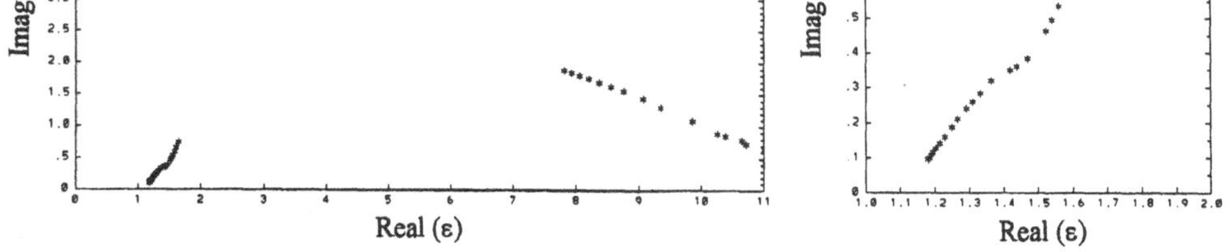

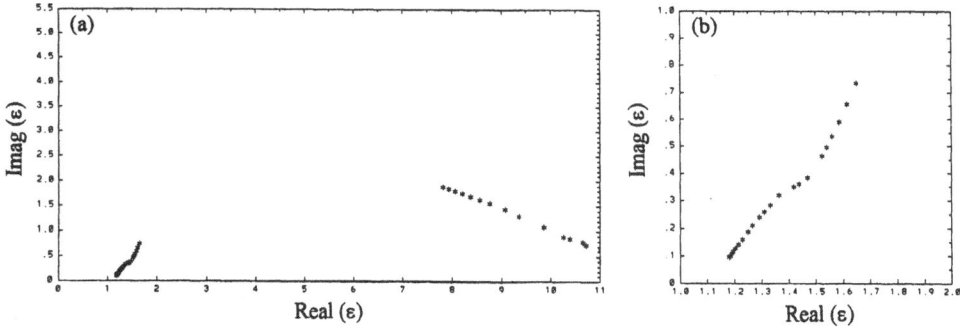

Figure 1. Cole-Cole plot of the permittivity data of a sample of carbon loaded foam absorber. (a) Full spectrum from 10 KHz to 20 GHz. (b) Detail of the high end of the spectrum from 500 MHz to 20 GHz.

The low frequency data in Figure 1 was obtained using a parallel plate capacitor test set-up and an impedance meter and it spans the frequency range from 10 KHz to 1 MHz. The high frequency data (highlighted by the dashed square in Figure 1a and detailed in Figure 1b) was obtained using two different free space test set-ups. The first set-up covered from 500 MHz to 2.5 GHz, the second, 1.5 GHz to 20 GHz. As is typical with this type of test set-up, the accuracy of the data degrades towards the band edges, so in the overlap region between 1.5GHz and 2.5 GHz the two sets of data were averaged together, resulting in the slight dip observed.

The question to be answered is, what is the value of the permittivity in the unknown frequency range from 1 MHz to 500 MHz? Can it take arbitrary values? The answer to the latter question is no, its values are uniquely determined as an analytic continuation of the values known in the rest of the spectrum. If the data had infinite precision (no error) then we could perform this analytic continuation by taking repeated derivatives (the Taylor expansion method.) However in the presence of noise, and a discrete number of data points, that approach is unwieldy. Instead, we just fit the sum of RC circuit branches to the data and express the presence of error as a bound on the possible variation of the unknown data.

The best way to perform this fit is a method derived from Grant[7,1] in which the high frequency data is used to determine the high frequency circuit branches plus the sum total of all the conductivities (σ_{tot}) and the low frequency data is used to determine the low frequency circuit branches plus the sum total of all the capacitances (ε_{tot}). The missing data then consists of some combination of capacitances and conductances whose combined values total: $\sigma_{missing} = \sigma_{tot} - \Sigma \sigma_n$, $\varepsilon_{missing} = \varepsilon_{tot} - \Sigma \varepsilon_n$.

Then the only span of variation allowable for the missing data is the range of values resulting from the way these missing quantities are distributed. The two limits of this distribution are: Either $\sigma_{missing}$ and $\varepsilon_{missing}$ are in series and constitute a single circuit branch or else they are themselves the sum of many other parallel $\sigma\varepsilon$ branches. The single term gives rise to the highest possible peak in ε_i, while the distributed terms tend to give a very flat ε_i profile. This range of variation is not arbitrarily large. If we know an estimate of the worst case error that can be expected from the edges of the known measured data, then the fit to the missing data must not to exceed this limiting error.

Applying the analytic continuation method to the absorber foam data and assuming that the data points near the edges of the missing frequency band are good to within 8%, gives the results of Figure 2.

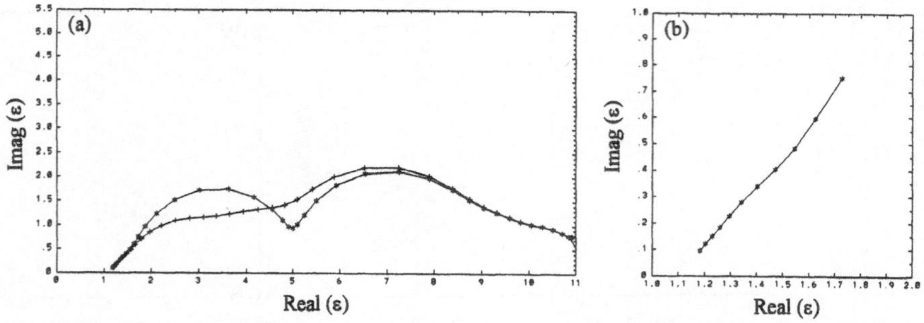

Figure 2. Result of the analytic continuation of the absorber foam data of Figure 1. (a) Full spectrum, (b) Detail of the high end of the spectrum.

The two curves in the full spectrum Cole-Cole plot of Figure 2 span the two extremes of the unknown data. When the missing circuit branches are collected into a single RC circuit, the curve with the sharp dip and large semicircle results. When the missing circuit branches are distributed into four separate terms with different time constants, the smoother curve results. Attempting to spread out the time constants of the missing circuit branches any farther exceeds the assigned 8% error of the measured data.

Another way to look at this exercise is to consider the ε_i versus logarithmic frequency plot of the measured data in Figure 3a. Given such a plot, we could not know what is the highest value of ε_i attained by this material in the VHF or UHF range. But by use of the analytic continuation method, we obtain Figure 3b which states that it is physically impossible, given the measured data, for the ε_i of this material to be much higher than 2.3. It also tells us that if we want a more accurate representation of the material properties in the missing range, and we wish to construct a special narrow band test set-up for this purpose, the test frequency should be chosen near either peak of the single RC branch model; since there the different circuit models will tend to differ the most. Figure 4 shows the circuit model of the analytically continued data; valid over a 2,000,000:1 bandwidth.

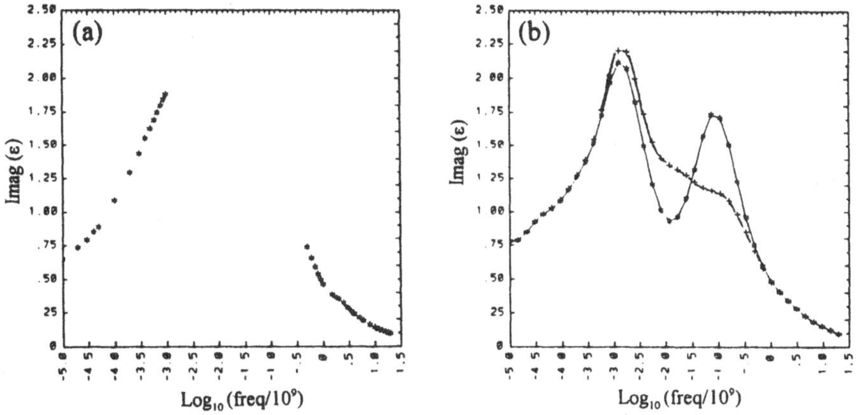

Figure 3. Imaginary part of the permittivity of the carbon loaded foam plotted against logarithmic frequency. (a) Measured data. (b) Analytic continuation to the full spectrum.

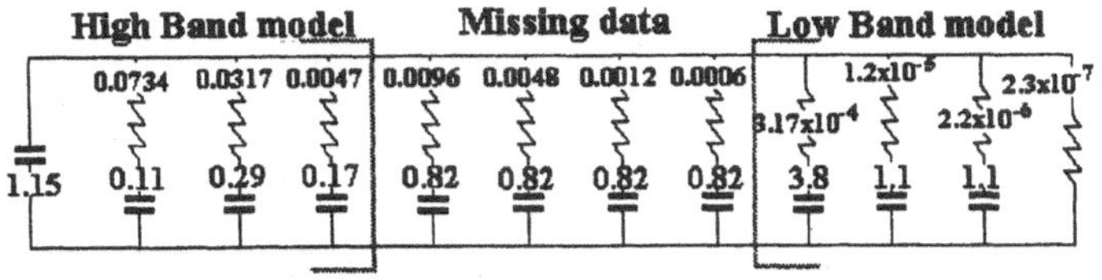

Figure 4. Circuit model for the permittivity of the carbon loaded foam, valid over a 2,000,000:1 bandwidth. Capacitors are given in units of ε_0, conductivities are in mhos/meter.

To illustrate the use of such an ultra broadband representation in computational electromagnetics, we consider the echo from an incident gaussian pulse onto an infinite ground plane covered by an infinite array of 24 inch deep absorbing foam wedges. The method of Joseph, et al[8] was used to introduce the material dispersion into FDTD. Their formulation requires the frequency domain permittivity to be represented as a sum in powers of ω. Such a sum is derived directly from the expression of equation 6 by putting all terms over the common denominator and performing the required algebra. Figure 5a shows the incident and reflected time domain pulses. The corresponding Reflection Coefficient of the wall of absorber wedges (Figure 5b) obtained by Fourier Transforming the time domain echo is valid over the entire bandwidth of the material model. Its only limitation is the frequency domain noise introduced by the finite time step size of the FDTD algorithm.

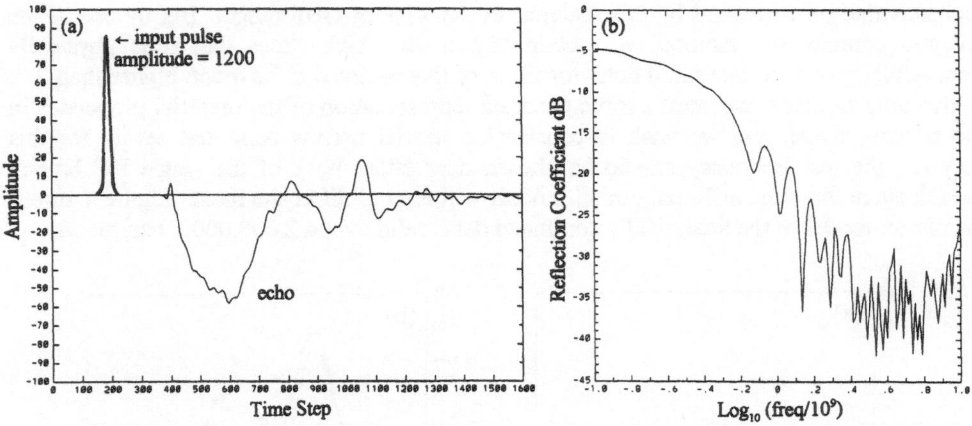

Figure 5. Dispersive FDTD computation of the performance of a wall of absorbing wedges. (a) Time domain pulse and echo. (b) Frequency domain reflection coefficient.

IV CONCLUSION

Because the permittivity is an expression of a causal phenomenon it has very special analytic properties in the complex frequency plane. One of the most useful consequences of this analyticity is that data measured over finite bands of the real frequency axis can be analytically continued into frequency bands where such data is not available. The value of this technique was demonstrated by obtaining a 2,000,000:1 bandwidth representation of a carbon loaded foam absorber and then using the representation to calculate the performance of wedges made of this absorber.

REFERENCES

1. R. E. Díaz, "The Analytic Continuation Method for the Analysis and Design of Dispersive materials," Ph.D. Dissertation, UCLA, Los Angeles (March 1992).
2. A. Von Hippel, "Dielectrics and Waves", Wiley, New York, p.178 (1954).
3. J. S. Toll, Causality and the dispersion relation: Logical foundations, *Phys. Rev.* 104:1760 (1956).
4. E. C. Titchmarsh, "Introduction to the Theory of Fourier Integrals," Clarendon Press, Oxford, p.128 (1967).
5. L. D. Landau and E. M. Lifshitz, "Electrodynamics of Continuous Media," Pergamon Press, New York, pp.249-268 (1960).
6. E. C. Titchmarsh, "The Theory of Functions," Oxford University Press, London, pp. 91-94 (1964).
7. F. A. Grant, Use of complex conductivity in the representation of dielectric phenomena, *J. Appl. Phys.* 29:76 (1958).
8. R. M. Joseph, S. C. Hagness, A. Taflove, Direct time integration of Maxwell's equations in linear dispersive media with absorption for scattering and propagation of femtosecond electromagnetic pulses, (Preprint provided to the author by Professor Taflove.)

DIFFRACTION AT A CONE WITH ARBITRARY CROSS-SECTION

V.A. Borovikov

Institute of Problems of Mechanics
pr. Vernadsky 101, Moscow, 117526 Russia

The non-stationary (time-dependent) problem of plane wave diffraction at a cone S is reduced to more simple problems which allow the numerical solution:

α.The solution of mixed (boundary and initial value) problem for hyperbolic equation with two space-like variables.

β.The solution of Dirichlet problem for Laplace equation in unite sphere with cut-out cone S.

The solution of the first problem has the explicit integral representation in the case of diffraction at polyhedral angle.

1.The posing of problem. Following [1,2], we consider the non-stationary diffraction at a cone S of the plane wave

$$v_t(t.r,\omega,\omega_0)= \frac{1}{\Gamma(\lambda+1)} (t+r\cdot\omega\cdot\omega_0)^{\lambda} \quad \text{if} \quad t+r\cdot\omega\cdot\omega_0>0$$
$$=0 \quad \text{if} \quad t+r\cdot\omega\cdot\omega_0<0 \qquad (1)$$

Here the parameter λ will be defined below; $\mathbf{r}=r\cdot\omega$; $r\geqslant 0$;

Ultra-Wideband, Short-Pulse Electromagnetics
Edited by H. Bertoni *et al.*, Plenum Press, 1993

$|\omega|=1$ are the spherical coordinates; $-\omega_0$ is the direction of incident wave propagation. We seek the solution $v(t.r,\omega,\omega_0)$ of wave equation

$$v_{tt}-\Delta v = v_{tt} - v_{rr} - \frac{2}{r}v_r - \frac{1}{r^2}\Delta_w v = 0 \qquad (2)$$

which satisfies on cone S the boundary condition

$$v\Big|_S = 0 \quad \text{or} \quad \frac{\partial v}{\partial n}\Big|_S = 0; \qquad (3)$$

the appropriate regularity conditions at the tip of S and coincides for sufficiently small t with the incident wave (1). Here Δ_w is the Beltramy-Laplace operator defined at unite sphere. In the spherical polar coordinates θ,φ

$$\Delta_w = \frac{\partial^2}{\partial\theta^2} + \cot\theta\,\frac{\partial}{\partial\theta} + \frac{1}{\sin^2\theta}\frac{\partial^2}{\partial\varphi^2}$$

We assume that the tip of cone S lies at the origin r=0.

At t=0 incident wave front hits the tip. At t>0 the diffracted spherical wave with front $t=r$ propagates. The main problem is the evaluation of this wave.

2.The reduction of problem. The solution $v(t,r,\omega,\omega_0)$ which we seek is a homogeneous function t,r of order λ:

$$v(t,r,\omega,\omega_0)= t^{\lambda}\, v(r/t,\omega,\omega_0)$$

Inserting it into the wave equation we receive ($\xi=r/t$):

$$\xi^2(1-\xi^2)v_{\xi\xi}+ 2\xi[1+(\lambda-1)\xi^2)v_\xi - \lambda(\lambda-1)\xi^2 v + \Delta_w v= 0 \qquad (4)$$

This equation is hyperbolic before the wave front $t=r$ (for $t<r$, i.e. for $\xi>1$) and became elliptic after front - for $t>r$ ($\xi<1$). For $\lambda=-1/2$ it reduced to the Laplace equation[1]: if $t^{-1/2}v(r/t,\omega)$ is the solution of wave equation (2), then

$$v(\xi,\omega)= [1+(1-\xi^2)^{1/2}]^{-1/2}w\left(\frac{\xi}{1+(1-\xi^2)^{-1/2}},\omega\right)$$

where $w(\rho,\omega)$ is the harmonic function:

$$w_{\rho\rho}+2\rho^{-1}w_\rho+\rho^{-2}\Delta_w w=0$$

We evaluate this function in domain Σ_S which is the part of sphere $\rho<1$ outside S. Let us define the boundary conditions on boundary of Σ_S. This boundary consists of two components - the part of S for which $\rho<1$ and the part N_S of unit sphere $\rho=1$ which lie outside S. On S the boundary condition (3) is posed. On N_S, i.e. for $r=t$ the values of w are defined by continuity as the limit of solution $v(t,r,\omega,\omega_0)$ taken at $t=1$ from the region before the front $t=r$.

3. The solution for $-r<t<r$. Let us define v for $t<r$. For $t<-r\cdot\omega\cdot\omega_0$, i.e. before the incident wave front this wave and the entire field $v(t,r,\omega,\omega_0)$ vanish. To find the solution for $-r<t<r$ it is convenient to let

$$v(t,r,\omega,\omega_0)=r^{-1/2}f(t/r,\omega,\omega_0) \tag{5}$$

The introducing of the independent variable s: $\cos s=-t/r$ carries the wave equation to the form[2]

$$f_{ss}+\frac{1}{4}f-\Delta_w f=0, \tag{6}$$

interval $-1<t/r<1$ into $0<s<\pi$ and the incident wave (1) with $\lambda=-1/2$ into the function

$$\begin{aligned} f_i &= \pi^{-1/2}(\cos\psi-\cos s)_+^{-1/2} \\ &= \pi^{-1/2}(\cos\psi-\cos s)^{-1/2} \qquad \text{if } \psi<s \\ &=0 \qquad\qquad\qquad\qquad\qquad\quad \text{if } \psi>s \end{aligned} \tag{7}$$

Here ψ is the angle between ω and ω_0: $\cos\psi=\omega\cdot\omega_0$.

For small values s function f_i identically vanishes outside a small neighborhood of w_0 and satisfies the boundary condition (2). Therefore f may be defined as a solution of eqn (6), satisfying (2) and coinciding with (7) for small s. It means that f satisfies the initial conditions:

$$f|_{s=0}=0;\quad \frac{\partial f}{\partial s}\Big|_{s=0}=2\sqrt{2\pi}\,\delta(\omega,\omega_0) \tag{8}$$

where δ is the Dirac δ-function defined on the unit sphere.

So, the problem is reduced to:

α. To find the solution $f(s,\omega,\omega_0)$ of eqn (6) in domain $0<s<\pi$; $\omega\in N_S$ with boundary and initial conditions (3),(7)

β. To find the harmonic function $w(\rho,\omega,\omega_0)$ in the domain Σ_S which equal to the $f(\pi,\omega,\omega_0)$ at $\rho=1$ and satisfies to the boundary condition (3) at S.

Let $f=f_i+f_r$, then f_r is defined as a solution of eqn (6) which vanishes for sufficiently small s and takes on S the inhomogeneous boundary conditions:

$$f_r=-f_i \quad \text{or} \quad \frac{\partial f_r}{\partial n} = -\frac{\partial f_i}{\partial n}$$

Because the incident wave (7) became infinite at the front $\psi=s$, the reflected wave has the same singularity on its wave front which hampers its numerical evaluation. The situation simplifies in two described below cases.

4. Diffraction at the smooth cone S which completely illuminated by the incident wave. Here we can extract the first term of ray expansion of function f_r:

$$f_r=\frac{[s-\sigma(\omega,\omega_0)]_+^{-1/2}}{\sqrt{\pi}}A_0(\omega)+f_r'$$

where $s=\sigma(\omega,\omega_0)$ is the equation of reflected wave front

$$\sigma(\omega,\omega_0)=\min_{\omega_S\in S}[\Delta(\omega_0,\omega_S)+\Delta(\omega_S,\omega)]$$

and $\Delta(\omega,\omega_0)=\arccos(\omega\cdot\omega_0)$ is the geodesic distance at N_S between ω and ω_0, i.e. angle between these vectors. The amplitude $A_0(\omega)$ is defined by appropriate transport equation (see [2], § 4.2, p.p. 3,4). The function f_r' is the continuous function and may be defined by the numerical methods (for example< by the finite differences method).

5. Diffraction by the polyhedral angle. We show that the function $f(s,\omega,\omega_0)$ is expressed here by the integrals.

The solution $v(t,r,\omega,\omega_0)$ in the domain $r>t$ is the sum of the following waves:

α. The incident plane wave and its reflections from the sides of S.

β.The primary conical edge waves excited by the incident wave at the edges OA_1, OA_2 etc of polyhedral angle S.

γ.The secondary conical edge waves excited by the primary waves at the edges of S.

δ.The edge waves of subsequent diffractions.

There is only finite amount of these waves.

The "cylindrical" edge wave which correspond in s,ω-space to the primary edge wave of some edge HA_n have the front $s=\Delta(\omega_0,\omega_n)+\Delta(\omega_n,\omega)$, where ω_n is the direction of edge OA_n. Accordingly, $s=\Delta(\omega_0,\omega_n)+\Delta(\omega_n,\omega_m)+\Delta(\omega_m,\omega)$ is the front equation of wave which correspond to the double diffraction at the edges OA_n, OA_m. There is the finite amount of such waves in the interval $0<s<\pi$.

6.The integral expressions for the edge waves. Let θ,φ be the spherical coordinates at N_S in which the direction ω_n of some edge OA_n has the equation $\theta=0$ and $\varphi=0$, $\varphi=\Phi$ are the equations of the sides of S incident with OA_n. Let θ,φ and θ_0,φ_0 define the directions ω and ω_0. Then $\Delta(\omega_n,\omega)=\theta$, $\Delta(\omega_0,\omega_n)=\theta_0$ and $s=\theta+\theta_0$ is the equation of wave front of primary edge wave excited at the edge OA_n. If $s<\theta+\theta_0$ then the function $f(s,\omega,\omega_0)$ is the sum of the incident wave (7) and its reflections from the sides of S. If $s>\theta+\theta_0$ then $f(s,\omega,\omega_0)$ expressed in terms of Sommerfeld integral:

$$f(s,\omega,\omega_0)=\int_{\gamma}\frac{H_{\pm}(\alpha+\varphi,\varphi_0,\Phi)d\alpha}{(\cos\theta\cos\theta_0+\sin\theta\sin\theta_0\cos\alpha-\cos s)^{-1/2}} \qquad (9)$$

Here denominator has cuts: $(\pi+i\sigma,\pi+i\infty)$, $(\pi-i\sigma,\pi-i\infty)$, $(-\pi+i\sigma,-\pi+i\infty)$ and $(-\pi-i\sigma,-\pi-i\infty)$ where $\sigma=\operatorname{arch}[(\cos\theta\cos\theta_0-\cos s)/\sin\theta\sin\theta_0]$. Contour γ consists of four loops bypassing clockwise these cuts. Function $H_{\pm}$ is the Sommerfeld kernel:

$$H_{\pm}(\alpha,\beta,\Phi)=\frac{\pi}{2\Phi}[\cot\tfrac{\pi}{2\Phi}(\alpha-\beta)\pm\cot\tfrac{\pi}{2\Phi}(\alpha+\beta)] \qquad (10)$$

sign + is taken for the boundary condition $\partial v/\partial n|_S=0$ and sign - for the condition $v|_S=0$.

The eqn (9) enables to find $f(s,\omega,\omega_0)$ for any ω,ω_0 but only for sufficiently small values s:

$$s \leqslant s_0 = \min_{n,m} [\Delta(\omega_0, \omega_n) + \Delta(\omega_n, \omega_m)]$$

where min is taken for all edges OA_n, OA_m of polyhedral angle S. Only the primary edge waves are excited for these s.

If $s_0 < s < 2s_0$ then the next expression is used

$$f(s_1+s_2, \omega, \omega_0) = \frac{1}{2\sqrt{2\pi}} \left(\frac{\partial}{\partial s_1} + \frac{\partial}{\partial s_2}\right) \cdot$$

$$\cdot \int_{N_S} f(s_1, \omega, \xi) f(s_2, \xi, \omega_0) d\xi \qquad (11)$$

This expression follows from (8). Only the finite amount of such steps needed for $0 < s < \pi$.

Note that the integral (9) is derived analytically and is equal to the half-sum of incident and reflected waves in the case of diffraction at plane screen with deleted sector[2].

7. Expansion by eigenfunctions of operator Δ_w at N_s. Let $\varphi_n(\omega)$ are these eigenfunctions:

$$\Delta_w \varphi_n + \lambda_n \varphi_n = 0; \ \varphi_n \Big|_S = 0 \ \text{ or } \ \frac{\partial \varphi_n}{\partial n}\Big|_S = 0; \quad \int_{N_S} \varphi_n^2 d\omega = 1$$

Expanding $\delta(\omega, \omega_0)$ by functions φ_n, we receive from (8):

$$f(s, \omega, \omega_0) = 2\sqrt{2\pi} \sum_n (\lambda_n + \tfrac{1}{4})^{-1/2} \varphi_n(\omega) \varphi_n(\omega_0) \sin s\sqrt{\lambda_n + 1/4} \qquad (12)$$

For $w(\rho, \omega, \omega_0)$ we receive:

$$w(\rho, \omega, \omega_0) = \qquad (13)$$

$$= 2\sqrt{2\pi} \sum_n (\lambda_n + \tfrac{1}{4})^{-1/2} \varphi_n(\omega) \varphi_n(\omega_0) \sin s\sqrt{\lambda_n + 1/4}\ \rho^{\sqrt{\lambda_n + 1/4}} - \frac{1}{2}$$

8. The solution of stationary problem. The integral

$$u(kr, \omega, \omega_0) = \exp\left(\frac{-\pi i}{4}\right)\sqrt{k} \int_{-\infty}^{\infty} \exp(ikt) v(t, r, \omega, \omega_0) dt \qquad (14)$$

is the solution of stationary problem diffraction at S of incident wave $\exp(-ikr\omega\cdot\omega_0)$. Letting $t=-r\cos s$ and using (12), (13) we receive Sommerfeld integral for u [3,4]:

$$u(kr,\omega,\omega_0) = \frac{\exp(\pi i/4)}{\sqrt{kr}} \int_C \exp(-ikr\cos s)\mathrm{Re}V ds$$

$$= \frac{\exp(\pi i/4)}{2\sqrt{kr}} \int_C \exp(-ikr\cos s) V ds \qquad (15)$$

where C consists of the branch $s=-i\xi$ $(0<\xi<\infty)$, interval $(0,\pi)$ and branch $s=\pi-i\xi$ $(0<\xi<\infty)$;

$$V= V(s,\omega,\omega_0)= \sum_n \varphi_n(\omega)\varphi_n(\omega_0)\exp(-is\sqrt{\lambda_n+1/4}) .$$

Short-wave asymptotic (at $kr \Rightarrow \infty$) of spherical diffracted wave excited at the tip of S is defined by stationary point $s=\pi$ in integral (15) and has the form [3,4]:

$$u_d = \sqrt{\pi/2}\ \frac{\exp(ikr)}{kr}\ V(\pi,\omega,\omega_0)$$

It is seen from (12),(13) that $V(s,\omega,\omega_0)$ and, therefore, the short-wave asymptotic u_d is expressed in terms of first derivatives of $f(s,\omega,\omega_0)$ at $s=\pi$ and $w(\rho,\omega,\omega_0)$ at $\rho=1$ [2].

The described approach enables to find the uniform asymptotic of solution in the vicinities of light-shadow boundaries. For diffraction at smooth completely illuminated cone the field in the vicinity of light-shadow boundary for reflected wave is expressed in terms of parabolic cylinder functions $D_{-1/2}, D_{-3/2}$ [2,5]. For diffraction at polyhedral angle the uniform asymptotic is expressed in terms of Fresnel integral in the vicinities of light-shadow boundaries for incident, reflected and edge waves and expressed in terms of generelized Fresnel integral (introduced in [6])in the vicinities of lines of merging of these boundaries, i.e. in the vicinities of rays of incident and reflected waves which pass through the tip of cone.

REFERENCES

1.Borovikov V.A. On the reduction of some three-dimensional problems of diffraction to the Dirichlet problem for Laplace equation. *Dokl. Akad. Nauk SSSR,* **144,** 527-530 (1961) (in Russian).

2.Borovikov V.A. "Diffraction by Polygons and Polyhedrons", Moscow, "Nauka" (1966) (In Russian).

3.Smyshljaev V.P. Diffraction of plane wave at conical bodies. *Matematich. Voprosy Teorii Rasprostranenija Voln,* **173** (18), 142-154 (1988) (in Russian)

4.Smyshljaev V.P. Diffraction by conical surfaces at high frequencies. *Wave Motion* **12** 329-339 (1990)

5.Brodskaja A.L., Popov A.V., Hosiossky S.A. Asymptotic of a wave reflected by cone in the penumbra zone. *VI All-Union Symposium on Diffraction and Propagation of Waves*. Moscow-Erevan, Vol 1,227-231 (1973) (in Russian)

6.Clemmow P.S., Senior T.B.A. A note on Generalized Fresnel integral. *Proc. Cambr. Phil. Soc.* 1953, v.9, N 3, 570-572.

SIGNAL PROCESSING TECHNIQUES

TIME-FREQUENCY-DISTRIBUTION ANALYSIS OF FREQUENCY DISPERSIVE TARGETS

E. K. Walton and A. Moghaddar

The Ohio State University ElectroScience Laboratory
1320 Kinnear Road
Columbus, Ohio 43212

INTRODUCTION

In the analysis of scattering targets, it is common to obtain time domain profiles by a Fourier transformation of the frequency response. In this process, wide-band frequency data are used to generate an approximation of the impulse response of the target via an inverse Fourier transform (IFT). The peaks in this impulse response then correspond to scattering centers of the target[1].

The process of associating the local maxima in the impulse response to the scattering center locations is based on the implicit assumption that the scattering mechanism is *not* frequency dispersive over the measurement bandwidth. For frequency dispersive targets, the propagation constant will be frequency dependent. The use of frequency dispersive media such as certain radar absorbing materials (RAM), or even a frequency dispersive structure such as an open waveguide duct or inlet can produce such behavior. For frequency dispersive situations, the impulse response alone cannot provide information about the dispersive characteristics of the individual scattering mechanisms of the radar target.

In this paper, we shall show that an analysis of scattering from frequency dispersive structures can be accomplished by time-frequency distribution (TFD) techniques. Three TFDs are investigated, and applied to the scattering analysis of a circular waveguide cavity. This paper comprises a brief description of the applied TFDs, and a detailed comparison of the TFD results for scattering analysis. We shall show that for some targets, such as the cavity, TFD is an effective method for description of the properties of the target. For the waveguide cavity, propagating modes and cutoff frequencies can readily be determined from the TFD, whereas neither time nor frequency representations will provide such information.

TIME-FREQUENCY DISTRIBUTIONS

Time-frequency representations (TFD) describe a signal in terms of its joint time and frequency content. Originally, the motivation for a time-frequency analysis was to devise a joint function of time and frequency, a distribution, which describes the energy density of a signal simultaneously in time and frequency[2]. Such a distribution can be

Ultra-Wideband, Short-Pulse Electromagnetics
Edited by H. Bertoni *et al.*, Plenum Press, 1993

used to give the energy of a signal contained in certain frequency and time intervals. If the TFD is represented by $P(t,\omega)$, the energy in any time-frequency interval (t_1,t_2) and (ω_1,ω_2) is given by the integral of the distribution over that time and frequency interval.

$$E = \int_{t_1}^{t_2} \int_{\omega_1}^{\omega_2} P(t,\omega) dt d\omega \quad (1)$$

At any time t_0, the summation of the $P(t,\omega)$ over the entire frequency support of the distribution should give the instantaneous energy of the signal at t_0, and summing up the distribution over all time at a single frequency ω_0 should give the energy of the signal at that particular frequency. These properties are called the marginals and are represented as

$$|s(t_0)|^2 = \int P(t_0,\omega) d\omega, \quad (2)$$

$$|S(\omega_0)|^2 = \int P(t,\omega_0) dt. \quad (3)$$

Time-frequency distributions were originally proposed for time-varying signals in the context of time-to-frequency transformations. For time signals, the Fourier transform will generate the frequency domain spectrum. For our present application, however, the original signal is in the frequency domain (the magnitude and phase of an electromagnetic scattered signal versus frequency), and its transform is the approximate impulse response. Therefore, in this paper, the TFDs (in contrast to the majority of previous TFD applications) are presented in the frequency-to-time transformation context.

The Wigner Distribution

The concept of a joint time-frequency distribution was originally developed in a quantum mechanics context. The original work of Wigner[3] was later advanced by Ville and Moyal[4]. Since then, several TFDs have been developed and a general framework for derivation and comparison of TFDs has been proposed. An excellent review of the TFDs is presented by Cohen[2].

Apart from quantum mechanics applications, TFDs did not attract much attention until the early 1980's. The surge in utilizing TFDs in other applications started with the work of Claasen and Mecklenbrauker[5,6,7]. In the last 12 years the Wigner distribution has attracted a wide range of applications from geophysical exploration, speech analysis, ECG and other time-varying signal analysis.

For a time signal $s(t)$, the Wigner distribution $WD_s(t,\omega)$ is defined by

$$WD_s(t,\omega) = \int_{-\infty}^{\infty} s(t+\frac{\tau}{2}) s^*(t-\frac{\tau}{2}) e^{-j\omega\tau} d\tau. \quad (4)$$

The Wigner distribution has a number of desirable properties, including satisfying the marginals, and high signal concentration in time-frequency[5]. It has been shown[8] that the Wigner Distribution has the highest signal concentration in the time-frequency plane. A more concentrated distribution would not satisfy the marginals, and would violate the time-frequency uncertainty principle[8].

Unfortunately, the Wigner distribution introduces cross-terms between multi-component signals which makes the interpretation of the Wigner distribution difficult in some applications. For example, for a two component signal $s(t) = x(t) + y(t)$, the

Wigner distribution can be written as:

$$WD_s(t,\omega) = WD_x(t,\omega) + WD_y(t,\omega) + 2Re[WD_{xy}(t,\omega)], \quad (5)$$

where the cross term $WD_{xy}(t,\omega)$ is defined as:

$$WD_{xy}(t,\omega) = \int_{-\infty}^{\infty} x(t+\frac{\tau}{2})y^*(t-\frac{\tau}{2})e^{-j\omega\tau}d\tau. \quad (6)$$

To circumvent the cross terms, several variations of the Wigner distribution have been proposed. A common point in all these variations is a filtering process in time or time-frequency domains. The pseudo-Wigner distribution (PWD) is defined[5] as

$$PWD_s(t,\omega) = \int_{-\infty}^{\infty} s(t+\frac{\tau}{2})s^*(t-\frac{\tau}{2})h(\tau)e^{-j\omega\tau}d\tau, \quad (7)$$

where $h(\tau)$ is a time-windowing function which has a frequency-convolution effect. In a more general case, the Wigner distribution can be convolved in both time and frequency with a two-dimensional filter $F(t,\omega)$. As an example, a Gaussian filtering function proposed by Choi and Williams has been shown[9] to effectively reduces the cross-terms. Reduction of the cross terms, however, is always at the cost of reducing the resolution in time-frequency plane. The smoothed Wigner distribution is always smeared in time and has less signal concentration compared to the WD.

Running-Window Fourier Transform

The most commonly used technique for analysis of time varying signals is the short-time Fourier transform (STFT). The STFT of a time signal $s(t)$ is defined as:

$$STFT(t,\omega) = \int_{-\infty}^{\infty} s(\tau)w(\tau-t)e^{-j\omega\tau}d\tau, \quad (8)$$

where $s(t)$ is the time signal and $w(t)$ is the window function. The function $w(t)$ can be considered as a window which selects a particular section of the signal centered around a given time location. The frequency content of the signal at any given time will then be determined by a Fourier transform of this windowed signal.

Analogous to the STFT, for frequency dispersive scenarios, we use a partitioning of the entire frequency band into smaller bands, and generate the impulse response for the sub-bands. In this method, we take a slice of the frequency data about a center frequency, and compute the inverse Fourier transform of the windowed data. By using a window of length $\Delta\omega$, centered about ω, and varying ω, after transformation of the windowed data to the time domain, we obtain a spectral density which is a function of both t and ω.

For the frequency domain data $S(\omega)$, we define the running-window Fourier transform (RWFT) as:

$$RWFT(t,\omega) = \int_{-\infty}^{\infty} S(\Omega)W(\Omega-\omega)e^{j\Omega t}d\Omega, \quad (9)$$

where $S(\omega)$ is the complex frequency data, and $W(\omega)$ is the windowing function. For a symmetric window function with non-zero weights for $|\omega| < \Delta\omega/2$, the integrand in Equation (9) will be a $\Delta\omega$ wide portion of frequency data centered about ω.

The term *running-window* in this paper is used to indicate that (1) only a subsection of the data is used to obtain the transform, (2) the center of this window is moved over the entire available data. The term *running-window* is preferred here, because for our frequency-to-time data analysis, the STFT will be a misnomer.

Running-Window Autoregressive TF Distribution

In the last decade a variety of high resolution spectral estimation techniques have been introduced. The major advantage of these techniques is their ability to resolve closely spaced spectral peaks for short data records. The high resolution method used in this study is an autoregressive (AR) spectral estimation of the sub-band frequency data.

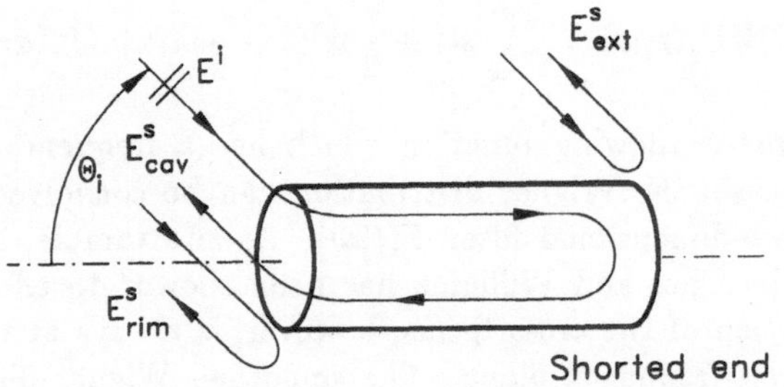

Figure 1. Geometry of the open-ended circular cavity and major scattering mechanisms.

The AR formulation of scattering data is based on the point-scatterer (non-dispersive) formulation of the radar target[10,11]. This point-scatterer assumption in the AR formulation is advantageous when the target can actually be modeled as a collection of scattering centers. Although point-scatter assumption is violated for dispersive features of a target, as we shall show, if a narrow window is used, reasonable distributions can be obtained.

TFD OF A CIRCULAR WAVEGUIDE CAVITY

In this section, the electromagnetic (EM) scattering from a waveguide cavity is analyzed using three time-frequency distribution techniques. A 2-ft (61 cm) long, 1.75-inch (4.44 cm) diameter circular cavity with an open end (Figure 1) is used as an example of a frequency dispersive scatterer.

Referring to Figure 1, one can write the total scattered field as

$$\vec{E}^s = \vec{E}^s_{rim} + \vec{E}^s_{cav} + \vec{E}^s_{ext}, \tag{10}$$

where $\vec{E}_{rim}$ is the field scattered by the edge of the aperture at the open end of the cavity, and $\vec{E}^s_{ext}$ is the scattering from the exterior of the cavity. The scattering from the interior of the cavity, $\vec{E}^s_{cav}$, is the frequency dispersive part of the total scattered signal. For frequencies below the cutoff frequency of the cavity, the $\vec{E}^s_{cav}$ does not contribute to the total scattering. For frequencies greater than cutoff, as the operating frequency increases, the group velocity of the propagating mode increases. The number of propagating modes also increases with frequency. For the circular cavity, the scattered signal versus frequency is measured for the case where the incident plane wave is normal to the open end of the cavity, and at 45° angle.

Normal Incidence

The EM scattered signal is measured from 2 to 18 GHz at normal incidence ($\theta = 0$). The band limited impulse response is computed using an inverse Fourier transform (IFT) of the windowed 2-18 GHz data. The magnitude of the measured frequency data, and the band limited impulse response are shown in Figure 2. From this figure, the scattering from the front and back ends of the cavity can be seen at -2.0 and 2.0 nano-seconds respectively. From the impulse response, we can also see a diffuse response after 2.0 nano-seconds. The diffuse feature in the impulse response can be a clue that scattering from the closed end has some variation with frequency, or that other dispersive features are present. However, neither the frequency nor the time-domain data of Figure 2 provides specific information about the frequency dispersive nature of the cavity.

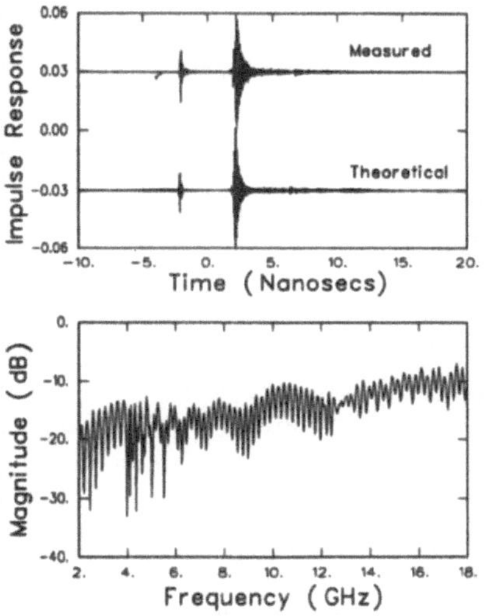

Figure 2. Magnitude of the measured frequency scan data, and the band limited impulse response for measured and theoretical data (0° incidence, horizontal polarization).

For comparison, theoretical frequency data are also computed using a high frequency ray approach[12] for an infinitely long, 1.75-inch diameter cylinder. In computing the theoretical data, it is assumed that a 1.75-inch disk is placed 2-ft deep inside the cylinder. The band limited impulse response for the theoretical data is shown in Figure 2.

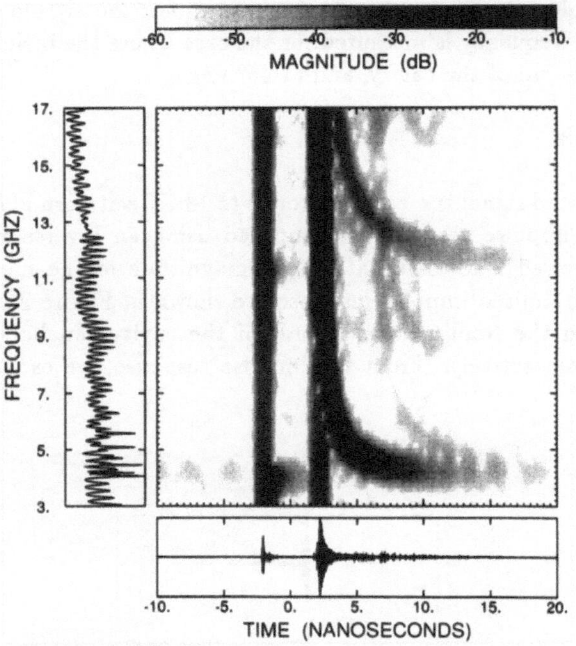

Figure 3. Running-window FT Time-frequency distribution using a 2-GHz wide Kaiser-Bessel window. Measured data for the 1.75-inch diameter circular cavity, horizontal polarization, 0° incidence.

To fully observe the frequency-dispersive behavior of the cavity, the time-frequency distribution is computed using a running-window-FT with a 2-GHz wide Kaiser-Bessel window. The magnitude of the distribution versus time and frequency is shown in gray-scale levels in Figure 3. In the same figure, along the time and frequency axes, the impulse response, and the magnitude of backscattered signal are also included. From the TFD, we can identify the leading and trailing ends of the cavity, and the two cutoff frequencies at 3.9 GHz and 11.5 GHz.

Since a horizontally polarized plane wave is incident along the axis of the cavity, only TE_{1x} modes have significant contributions to the propagating modes through the cavity. The cutoff frequencies for the TE modes[13] are obtained from

$$(f_c)_{np}^{TE} = \frac{x'_p}{2\pi a\sqrt{\epsilon\mu}}, \tag{11}$$

where a is the radius of the cavity, and x'_p is the pth zero of derivative of nth order Bessel function, $J'_n(x')$. Using the values for x'_p, we can compute the cutoff frequencies for TE modes. The computed cutoff frequencies for the 1.75-inch diameter circular cavity are:

$$(f_c)_{11}^{TE} = 3.90 \quad GHz \tag{12}$$

$$(f_c)_{12}^{TE} = 11.31 \quad GHz. \tag{13}$$

Referring to Figure 3, one can see that these values agree with the experimental results for the cutoff frequencies. The shape of the modes in the time-frequency plane also

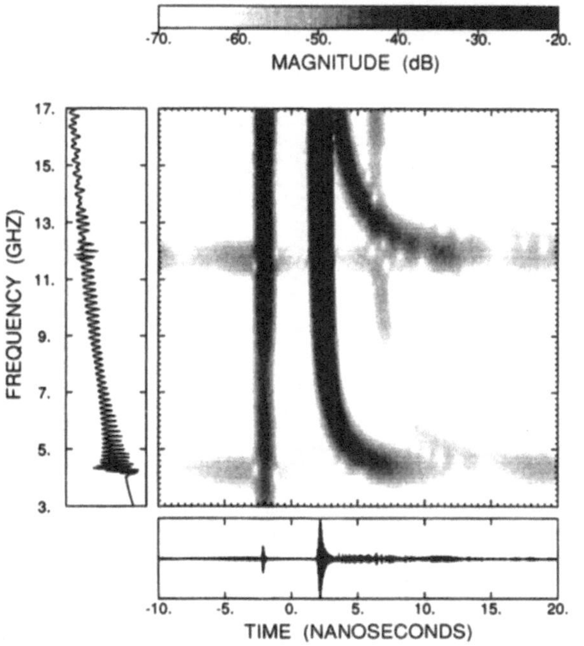

Figure 4. Running-window FT Time-frequency distribution using a 2-GHz wide Kaiser-Bessel window. Theoretical data for the 1.75-inch diameter circular cavity, horizontal polarization, 0° incidence.

indicate the rapid variation of the group velocity for frequencies greater than cutoff. As the operating frequency is increased, the group velocity of the propagating mode increases and the propagation time for the signal reflected from the closed end decreases. The "L" shaped curves in the T-F plane are a result of this variation in the propagation time.

The TFD for the theoretical data using the RWFT is shown in Figure 4. By comparing Figures 3 and 4, one can see that the cutoff frequencies, and the general shape of the distribution for the measured and theoretical data are very similar. The trailing end can be observed in the measured data for all frequencies. However, for the theoretical data, since an infinitely long cylinder is considered, the cutoff for the TE_{11} mode (3.9 GHz), is the starting frequency where the effect of the closed end can be observed. It is worth emphasizing that this information cannot be obtained by comparing the impulse responses in Figure 2.

Another difference between the TFD of the theoretical data and the TFD of measured data is the resonance behaviour at the cutoff frequencies for the theoretical data. At the two cutoff frequencies, an apparent non-causal response can be observed. This is actually due to the aliasing of the late time response which is folded into the early time region. This feature can be attributed to the lossless assumption in the theoretical data.

The second TFD technique applied to the cavity data is the Wigner distribution. The Wigner distribution for the measured data is shown in Figure 5. As expected, the Wigner distribution provides a higher resolution for the leading and trailing ends of the cavity as well as the propagating modes. The major disadvantage of the Wigner

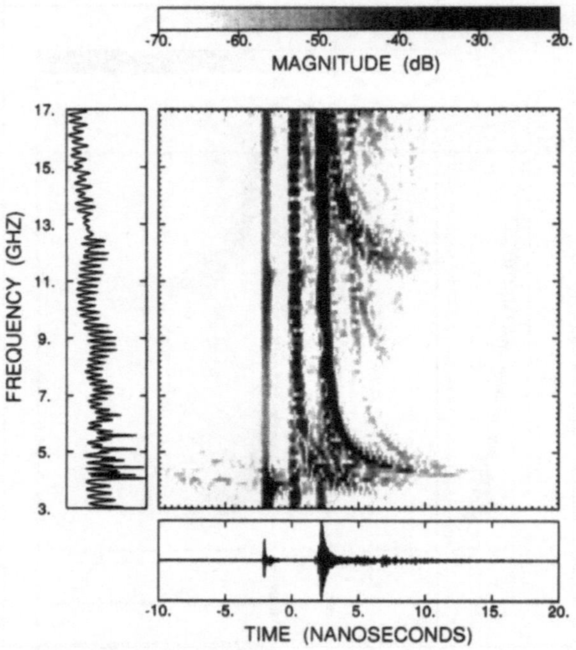

Figure 5. Wigner Time-frequency distribution for the measured data.

distribution, as seen in Figure 5 is the presence of cross-terms. For some sections of the distribution, the cross-terms obscure the distribution to a point where it is difficult to distinguish between the co- and cross-components.

The running-window autoregressive (RWAR) spectral estimation technique is the third TFD technique applied to the cavity data. The RWAR time-frequency distribution for the measured cavity data is shown in Figure 6. To obtain the TFD, a forward-backward linear prediction (FBLP) technique[14] is used to compute the AR parameters of the sub-band data. Then, the spectral density is computed from the AR parameters. A sequence of these densities for different center frequencies are used to generate the TFD.

The AR model is based on the point-scatterer (non-dispersive) assumption[15]. This assumption is violated for frequencies where transition from evanescent to propagating modes occur. As a result, the frequency resolution of the propagating modes is not as good as the RWFT distribution, whereas the resolution for the leading and trailing ends of the cavity is superior to that of RWFT or the Wigner-distribution.

45° Aligned Waveguide Cavity

In a second measurement, the EM scattering from the cavity was measured such that the angle between the axis of the cavity and the incident plane wave was 45°. The magnitude of the measured signal, and the band limited impulse response for horizontal polarization are shown in Figure 7. From the impulse response in this figure, we can identify the leading and trailing ends of the cavity. However, no specific informa-

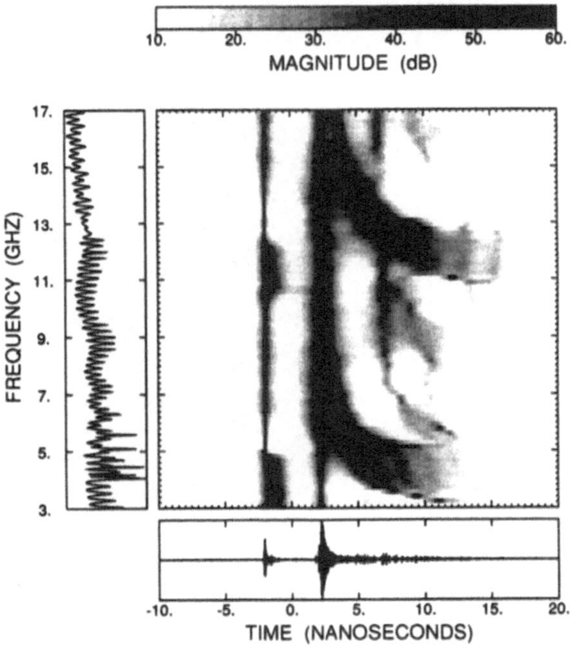

Figure 6. Time-frequency distribution for the measured data obtained from the running-window autoregressive spectral density.

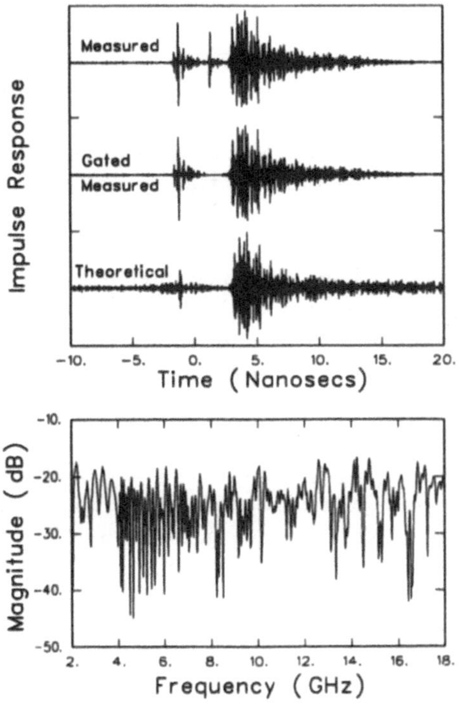

Figure 7. Magnitude of the measured frequency scan data, and the band limited impulse response for measured and theoretical data (45° incidence, horizontal polarization).

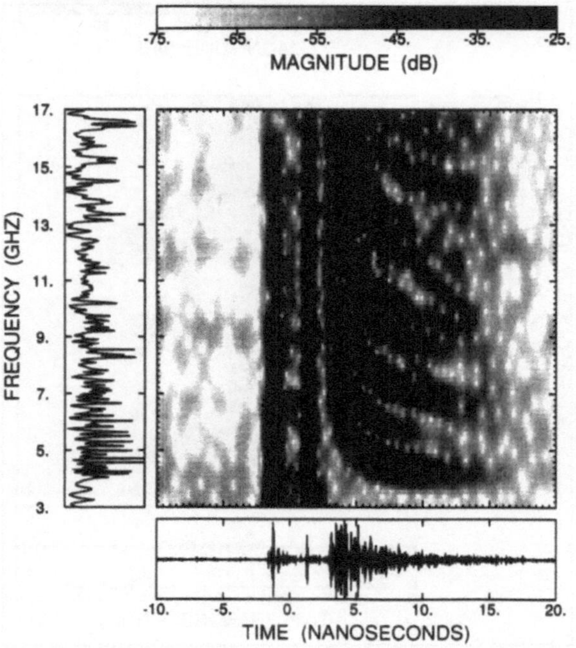

Figure 8. Running window FT Time-frequency distribution for the measured data. (1.75-inch diameter circular cavity, horizontal polarization, 45° incidence).

tion about the propagating modes or cutoff frequencies are available from the impulse response.

The theoretical frequency data are also computed for an infinitely long, 1.75-inch diameter cylinder at 45°. In computing these data, it is assumed that a 1.75-inch disk is placed 2-ft deep inside the cylinder. Since the theoretical data are computed for an infinitely long cylinder, scattering from the trailing end of the cylinder will be absent in these data (compare the impulse response at +1-nanosecond in for theoretical and measured data in Figure 7). To compare the measured data with the theoretical, the trailing end of the cavity is gated out of the measured data. In this process, the impulse response of the cavity is weighted by a window function which only selects the desired sections of the impulse response. The windowed impulse response is then transformed to the frequency domain by an FFT. The impulse response after this gating process is also shown in Figure 7.

The TFD obtained by the running-window FT technique for the measured data is shown in Figure 8. From this figure, the leading and trailing ends of the cavity as well as a number of propagating modes can be seen. The TFD for the measured data after gating the trailing end of the cavity is shown in Figure 9. For comparison, the TFD of the theoretical data is computed and the result is shown in Figure 10. The cutoff frequencies for the propagating modes of the cavity are also computed and the results are summarized in Table 1. By comparing the results form Figure 8 and Table 1, one can see that cutoff frequencies can be obtained from the TFD of the measured signal.

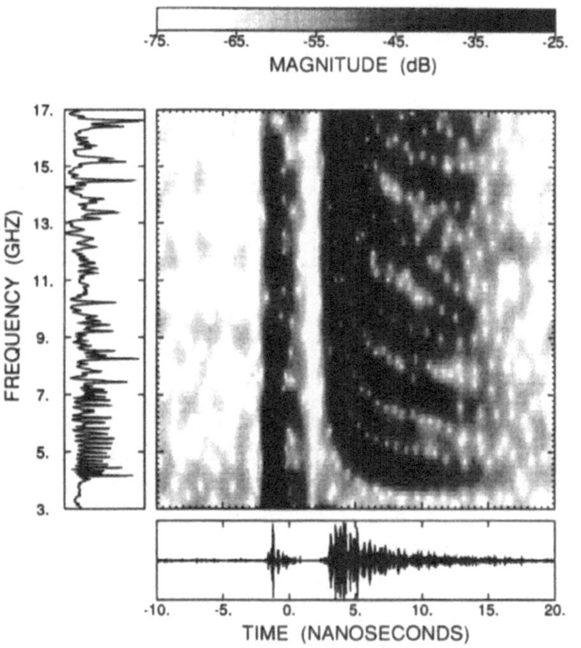

Figure 9. Running window FT Time-frequency distribution for the measured data after gating the trailing edge. (1.75-inch diameter circular cavity, horizontal polarization, 45° incidence).

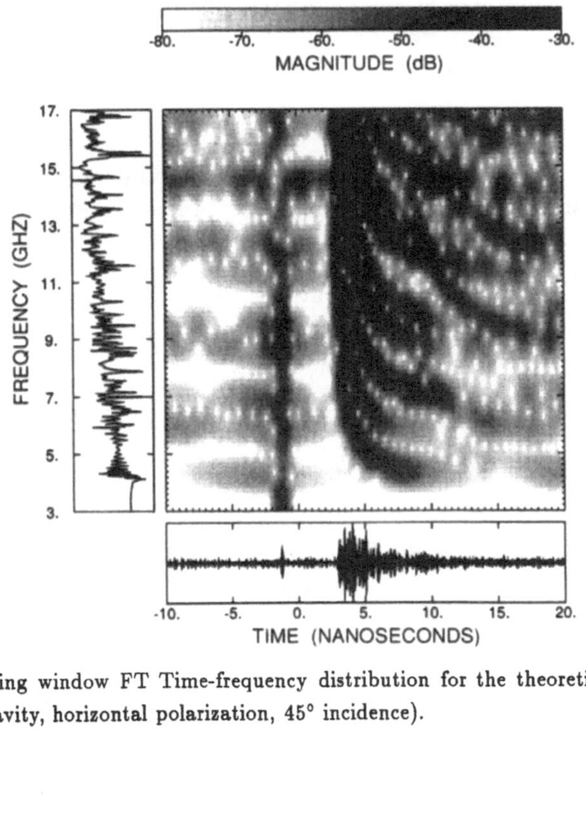

Figure 10. Running window FT Time-frequency distribution for the theoretical data. (1.75-inch diameter circular cavity, horizontal polarization, 45° incidence).

Table 1. Cutoff frequencies for a 1.75-inch diameter open-ended circular cavity.

Mode	TE_{11}	TM_{01}	TE_{21}	TM_{11} TE_{01}	TE_{31}	TM_{21}	TE_{41}
f_c (GHz)	3.95	5.17	6.56	8.23	9.02	11.03	11.42

Mode	TE_{12}	TM_{20}	TM_{13}	TE_{51}	TE_{22}	TE_{20} TM_{12}
f_c (GHz)	11.45	11.86	13.70	13.78	14.41	15.07

CONCLUSIONS

In this paper, an electromagnetic scattering analysis technique based on the computation of a time-frequency distribution (TFD) is presented. The TFD contains the signal density versus both time and frequency. As an example of a frequency dispersive target, measured and theoretical EM scattering data for a 1.75-inch (4.44 cm) diameter cylindrical waveguide cavity was used for TFD analysis. The TFD of the cavity were shown using three techniques: 1-the running-window Fourier transform (RWFT), 2-the Wigner distribution, and 3-the running-window autoregressive spectrum estimation. For the cavity, it was shown that the cutoff frequencies of the cavity modes can be determined from the TFD.

ACKNOWLEDGEMENT

This work was supported by the Ohio State University ElectroScience Laboratory Compact Range Consortium.

REFERENCES

1. E. K. Walton and J. D. Young, "The Ohio State University compact radar-cross section measurement range," *IEEE Trans. Antennas Propagat.*, 32:1218—1223 (1984).

2. L. Cohen, "Time-frequency distributions – a review," *Proc. IEEE*, 77: 941–981 (1989).

3. P. E. Wigner, "On the quantum correction for thermodynamic equilibrium," *Phys. Rev.*, 40:749–759 (1932).

4. J. E. Moyal, "Quantum mechanics as a statistical theory," *Proc. Cambridge Phil. Soc*, 45:99–124 (1949).

5. T. A. C. M. Claasen and W. F. G. Mecklenbrauker, "The Wigner distribution - a tool for time-frequency signal analysis; Part I: continuous-time signals," *Philips J. Res.*, 35: 217–250 (1980).

6. T. A. C. M. Claasen and W. F. G. Mecklenbrauker, "The Wigner distribution - a tool for time-frequency signal analysis; Part II: discrete-time signals," *Philips J. Res.*, 35:276–300 (1980).

7. T. A. C. M. Claasen and W. F. G. Mecklenbrauker, "The Wigner distribution - a tool for time-frequency signal analysis; Part III: relations with other time-frequency signal transformations," *Philips J. Res.*, 35:372–389 (1980).

8. D. Gabor, "Theory of communication," *J. Inst Elec. Eng.*, 93:429–441, (1946).

9. H.-I. Choi and W. J. Williams, "Improved time-frequency representation of multicomponent signals using exponential kernels," *IEEE Trans. Acoust., Speech, Signal Processing*, 37:832–871, (1989).

10. E. K. Walton, "Comparison of fourier and maximum entropy techniques for high-resolution scattering studies," *Radio Science*, 22:350–356, (1987).

11. E. K. Walton, "Far-field measurements and maximum entropy analysis of lossy material on a conducting plate," *IEEE Trans. Antennas Propagat.*, 37:1042–1047 (1989).

12. C. W. Chuang and P. H. Pathak, "Ray analysis of modal reflection for three-dimensional open-ended waveguides," *IEEE Trans. Antennas Propagat.*, 37:339–346 (1989).

13. R. F. Harrington, "Time-Harmonic Electromagnetic Fields," McGraw-Hill Inc., (1961).

14. D. W. Tufts and R. Kumaresan, "Estimation of frequencies of multiple sinusoids: making linear prediction perform like maximum likelihood," *Proc. IEEE*, 70:975–989 (1982).

15. E. K. Walton and A. Moghaddar, "High resolution imaging of radar targets using narrow band data," in *IEEE AP-S International Symposium*, (London, Ontario, Canada), (1991).

SCATTERING OF SHORT EM-PULSES BY SIMPLE AND COMPLEX TARGETS USING IMPULSE RADAR

G. C. Gaunaurd,[1] H. C. Strifors,[2] S. Abrahamsson,[3] and B. Brusmark[3]

[1]Naval Surface Warfare Center, Research Department (Code R-42)
White Oak, Silver Spring, MD 20903-5000, U.S.A.
[2]National Defense Research Establishment (FOA 2)
S-172 90 Sundbyberg, Sweden
[3]National Defense Research Establishment (FOA 3)
P.O. Box 1165, S-581 11 Linköping, Sweden

INTRODUCTION

The determination of radar cross-sections (RCSs) of scatterers of simple geometrical shapes is a well-studied problem area.[1-5] Traditionally, the analytical treatment has used continuous wave (CW) incidences on the targets, while in practice the interrogating waveforms are pulses, often of short duration, and the targets have complex shape. As we demonstrated earlier[4,5] a Fourier transform technique conveniently extends the steady-state cases to the transient situations when the incident waveform is a single pulse of arbitrary shape and duration.

We study the transient scattering of short ultra-wideband (UWB) electromagnetic pulses by targets of simple and complex geometrical shapes using an impulse radar. The complex-shaped targets are plastic scale-models of aircraft with metallized surface. The aircraft are selected to be a B-52 bomber (scale 1:72) and a C-130 transport (scale 1:48). For comparison we also use simple-shaped targets such as a metal sphere and a dielectric, air-filled, spherical shell. We display the various signature representations for these targets, both as theoretically predicted for the simple-shaped ones, and as observed from measured data. The actual shape of the interrogating pulses is theoretically modeled using a digital filter design technique together with pulse returns from a reference target, in this case the metal sphere. Using a second simple-shaped target, the dielectric spherical shell, we can then demonstrate that a good general agreement is obtained between theoretical predictions of pulse returns and measured data, both in the frequency domain and in the time domain.

The (monostatic) RCSs are measured when the aircraft models are illuminated from the front and the traditional frequency signatures are displayed. We then extend the signature representations of these targets to the combined time-frequency domain by computing and displaying pseudo-Wigner distributions (PWDs) of the transient responses obtained, when the targets are illuminated broadside as well as nose-on. The PWD is shown to be capable of extracting informative features about the targets within the frequency band of the incident pulses, in agreement with the general time-development of resonance features. It becomes clear that the analysis of backscattered pulses in the combined time-frequency domain by means of the PWD with an appropriate time-window will improve the target recognition capability furnished by the ordinary RCS of the considered scatterers.

Ultra-Wideband, Short-Pulse Electromagnetics
Edited by H. Bertoni *et al.*, Plenum Press, 1993

TRANSIENT SCATTERING

The backscattered electric field when a plane CW is incident on the North pole (defined by the spherical coordinate $\theta = 0$) of a spherical target of outer radius a at the distance r from the origin of the sphere is given by:[1,8]

$$\mathbf{E}_{sc}(0,t) = E_0 \frac{a}{2r} \mathbf{U}\, e^{i(\omega t - kr)} f_\infty(0,\omega), \tag{1}$$

where

$$\mathbf{U} \equiv \mathbf{e}_\theta \cos\phi - \mathbf{e}_\phi \sin\phi. \tag{2}$$

Here, E_0 is the amplitude of the incident field, ω the angular frequency, t the time, $k = \omega/c$ the wave number, c being the speed of light in free space. Moreover $\mathbf{e}_\theta$ and $\mathbf{e}_\phi$ are unit vectors in the colatitude direction θ and azimuth direction φ, respectively, and $f_\infty(\theta = 0, \omega)$ is the form-function in the backscattering direction ($\theta = 0$), here expressed as a function of the angular frequency ω. The normalized RCS in the backscattering direction then assumes the form

$$\frac{\sigma}{\pi a^2} \equiv \lim_{r \to \infty} \left(\frac{2r}{a} \frac{|\mathbf{E}_{sc}|}{E_0} \right)^2 = |f_\infty(0,\omega)|^2. \tag{3}$$

To generalize to pulsed incidences we follow our earlier procedure[4,5] by first introducing a Fourier transform pair:

$$G(\omega) = \int_{-\infty}^{+\infty} g(t) e^{-i\omega t} dt, \qquad g(t) = \frac{1}{2\pi} \int_{-\infty}^{+\infty} G(\omega) e^{i\omega t} d\omega, \tag{4}$$

where $g(t)$ here is the incident pulse and $G(\omega)$ is its spectrum. The backscattered electric far field then assumes the form

$$\mathbf{E}_{sc}(0,t) = E_0 \frac{a}{2r} \mathbf{U} \frac{1}{2\pi} \int_{-\infty}^{+\infty} G(\omega)\, f_\infty(0,\omega)\, e^{i(\omega t - kr)} d\omega. \tag{5}$$

The exponential $\exp(-ikr)$ in this equation is the travel time between the observation point and the back of the sphere (viz., the South pole), which implies that a value of $r/a > 2$ corresponds to a location of the observer outside the scatterer giving a positive value of the arrival time of the backscattered pulse.

When using the discrete Fourier transform (DFT) in numerical implementations where the incident pulse is given in the form of a discrete-time series being assumed periodic, the above formulation of the continuous-time Fourier transform pair is conveniently converted to:

$$X(k) = \sum_{n=0}^{N-1} x(n)\, e^{-i(2\pi/N)kn}, \qquad x(n) = \frac{1}{N} \sum_{k=0}^{N-1} X(k)\, e^{i(2\pi/N)kn}, \tag{6}$$

where the sequences $x(n)$ and $X(k)$ both contain N elements.[6]

COMBINED TIME-FREQUENCY DOMAIN SIGNATURE

While the standard Fourier analysis allows the decomposition of a signal into individual frequency components and the establishment of the relative intensity of each component, it does not describe *when* those frequencies occurred. However, representations of the signature of an object in the combined time-frequency domain could add information needed or desired for target identification purposes. To be useful such signature representations and the displays they produce should closely reflect their physical causes. From the vast repertoire of time-frequency distributions[7] we here choose the pseudo-Wigner distribution (PWD). The PWD of a function $f(t)$ is defined by[8]

$$\tilde{W}_f(\omega,t) = 2 \int_{-\infty}^{+\infty} f(t+t') f^*(t-t') w_f(t') w_f^*(-t') e^{-i2\omega t'} dt', \tag{7}$$

where an asterisk denotes complex conjugation. The corresponding discrete-time pseudo-Wigner distribution (DPWD) is given by

$$\tilde{W}_f(k,l) = 2\sum_{n=0}^{N-1} f(l+n)f^*(l-n)w_f(n)w_f^*(-n)\, e^{-i(4\pi/N)kn}, \tag{8}$$

where throughout this work a Gaussian window function is chosen:

$$w_f(t) = e^{-\alpha t^2}, \tag{9}$$

α being a positive real number that controls the width of the time-window. When applying the PWD, or DPWD, to pulses returned by a scatterer, a more or less detailed or smoothed-out signature can be obtained by controlling the width of the time-window through the choice of value of the parameter α. To avoid aliasing each PWD in this work is based on the analytic function[6,8] derived from the real-valued signal, and it is calculated using a fast Fourier transform (FFT) algorithm.

RESULTS FROM IMPULSE RADAR MEASUREMENTS

The transmitted pulses of the impulse radar system used for these measurements have length 1 ns with a pulse repetition frequency of 250 kHz and a peak power of 50 W. We use an antenna that consists of two parallel dipoles (transmitting and receiving) within a frequency band of 0.2-2 GHz. During the experiments the antenna and scattering object are located in a small anechoic chamber at the National Defense Research Establishment (FOA 3). Data are sampled and preprocessed by the aid of a (Tektronix) digitizing signal analyzer, which has a signal level resolution of 8 bits and a real-time sampling rate of 2 GS/s (gigasamples per second) while the equivalent-time sampling rate could be up to 1 TS/s. The analog bandwidth is 1 GHz, which imposes a restriction on the total bandwidth of the complete radar and signal analyzer system. To suppress noise, the time-series considered in this work are each composed of 256 (spherical targets) or 64 (aircraft models) ensemble averaged (time-overlayed) sampled waveforms using an equivalent-time sampling rate of 20 GS/s. Antenna coupling and clutter returns are eliminated by first measuring the background (without any target present) then subtracting it from each of the following time-series obtained with a target present.

Frequency Domain Signatures

We have demonstrated earlier[5] that we can conveniently model the very short and broad-band pulse of the employed impulse radar system using a digital filter design technique. We found that bandpass filtering an ideal (Dirac) impulse by using a sixth order Butterworth bandpass filter with cutoff frequencies 180 and 640 MHz and passband attenuation 3 dB resulted in a designed pulse that emulates the combined effect of the pulse actually transmitted by the impulse radar and the filtering properties of the other components of the system as the receiving antenna and digitizing signal analyzer. A metal sphere of outer radius of 254 mm is used as reference and the data determining the designed incident pulse are chosen to give a good agreement between the theoretically predicted spectrum of the backscattered pulse and the spectrum of the measured return pulse. Figure 1 (main plot) displays the comparison of these spectra, and Fig. 1 (insert plot) the comparison of the theoretically predicted waveform with the recorded experimental data. Solid lines are used for experimental data and dashed lines for theoretical predictions, and all data are normalized by setting the maximum absolute value equal to unity.

To examine the degree of agreement between theoretical prediction and experimental results obtained by using the designed incident pulse, Fig. 2 displays the corresponding comparison when the target is a dielectric (permittivity $\varepsilon = 2.6$) spherical shell of outer radius 320 mm and thickness 16 mm. Because of the good agreement between theoretical predictions and experimental results as shown in Figs. 1 and 2 we expect that the signal processing of measured return pulses also from complex-shaped targets gives accurate signature representations. Figure 3 displays the recorded pulse versus time in the interval $0 \leq t \leq 25$ ns (insert plot) and the spectrum of the exhibited waveform versus frequency in the interval $0 \leq f \leq 1.4$ GHz (main plot), when the target is the scale model of the B-52 bomber (scale 1:72, length 68 cm, wingspan 78 cm) illuminated from the front (antenna parallel to the wings). The corresponding results when the target is the scale model of a

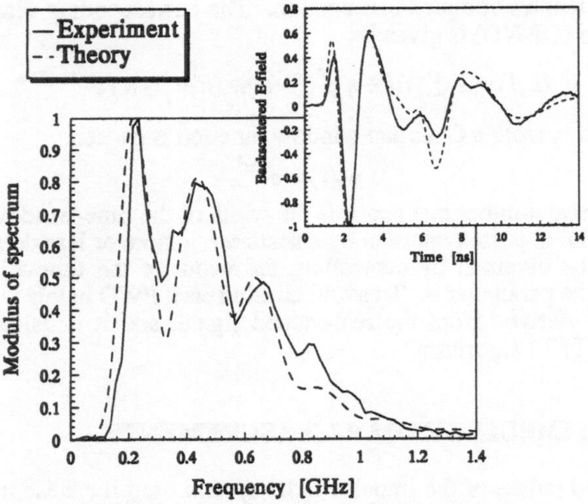

Figure 1. Comparison of the theoretically predicted spectrum (main plot) and waveform (insert plot) of the backscattered pulse (dashed lines) with the corresponding experimentally obtained data (solid lines). The target is a metal sphere with outer radius of 254 mm.

C-130 transport (scale 1:48, length 62 cm, wingspan 84 cm) are displayed in Fig. 4. Comparison of Figs. 3 and 4 discloses the degree of resemblance between the spectra of the two rather dissimilar targets.

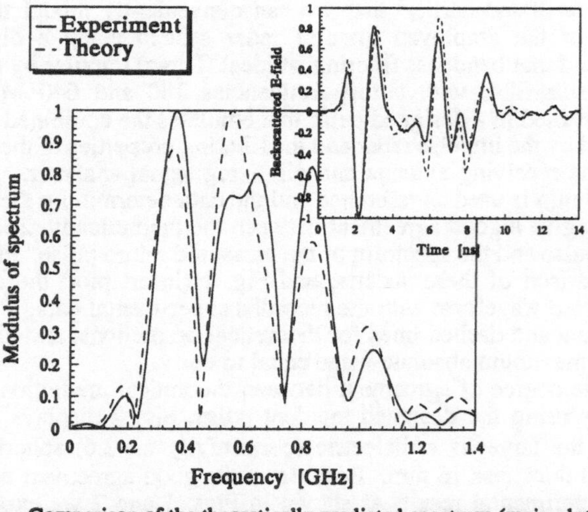

Figure 2. Comparison of the theoretically predicted spectrum (main plot) and waveform (insert plot) of the backscattered pulse (dashed lines) with the corresponding experimentally obtained data (solid lines). The target is a dielectric spherical shell with outer radius of 320 mm and thickness 16 mm.

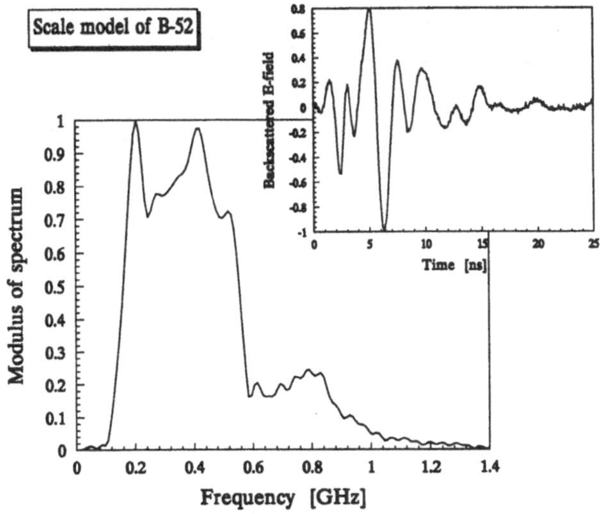

Figure 3. The spectrum (main plot) and waveform (insert plot) of the measured backscattered transient response when the target is a B-52 aircraft model being illuminated nose-on with the antenna parallel to the wings.

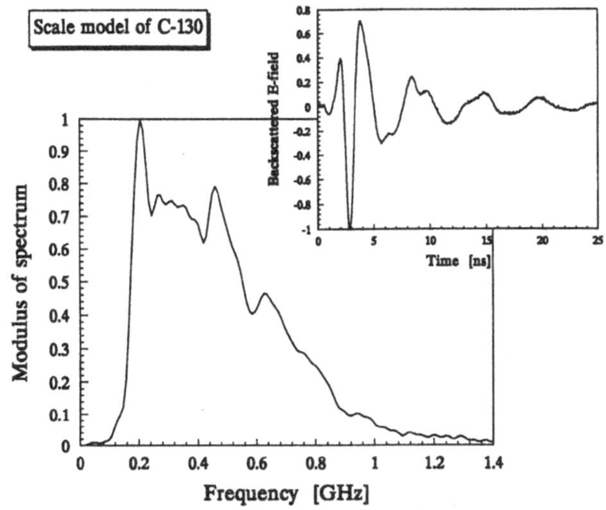

Figure 4. The spectrum (main plot) and waveform (insert plot) of the measured backscattered transient response when the target is a C-130 aircraft model being illuminated nose-on with the antenna parallel to the wings.

Time-Frequency Domain Signatures

We examine the signature representations in the combined time-frequency domain as furnished by the pseudo-Wigner distribution. The discussion will be confined to the complex-shaped targets for which only measured data are available, and we compare the signature features in the computed and displayed PWDs when the targets are illuminated at two different aspects. The parameter α in Eq. (9) controlling the width of the time-window in the PWD is chosen to be $\alpha = 0.10\,(\mathrm{ns})^{-2}$. Figure 5 displays a 3-D surface plot and its plane projection 2-D contour plot of the PWD of the backscattered pulse when the B-52 model is illuminated from the front (polarization parallel to the wings). The spectrum of the entire backscattered pulse as obtained by a DFT is shown in Fig. 3. The time-frequency domain signature as displayed in Fig. 5 exhibits features reflecting the extension of the target away from the antenna and, accordingly, gives more information of the shape of the target than does the ordinary RCS.

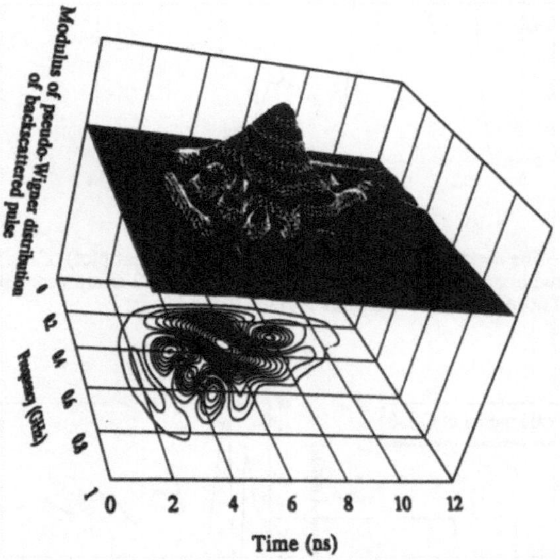

Figure 5. Surface plot and its plane projection contour plot displaying the PWD of the experimentally measured backscattered pulse when a B-52 aircraft model is illuminated by an impulse radar nose-on with the antenna parallel to the wings.

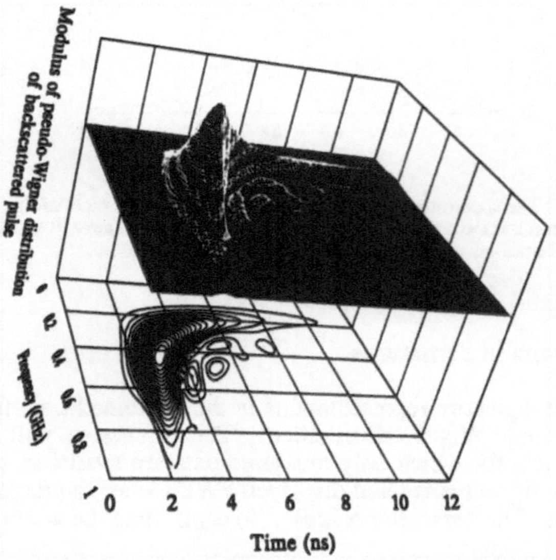

Figure 6. Surface plot and its plane projection contour plot displaying the PWD of the experimentally measured backscattered pulse when a C-130 aircraft model is illuminated by an impulse radar nose-on with the antenna parallel to the wings.

As a comparison we display in Fig. 6 the corresponding results when the target is the C-130 model illuminated as in the former case. The spectrum of the entire backscattered pulse is displayed in Fig. 4. We observe that while the waveform spectra of the pulse returns shown in Figs. 3 and 4 are rather similar to each other, the differences between the corresponding PWDs displayed in Figs. 5 and 6 are very noticeable.

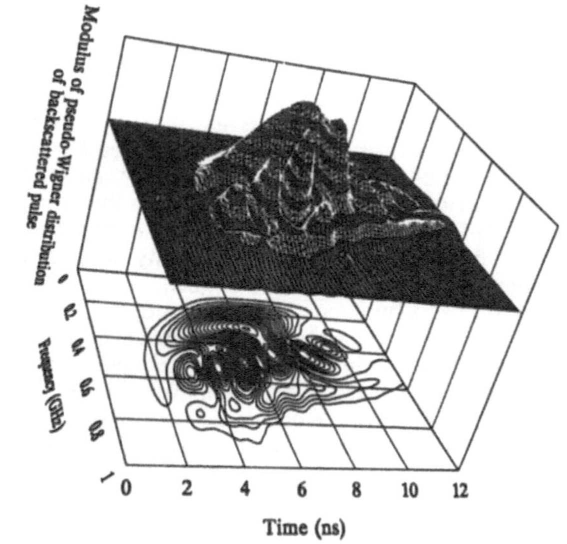

Figure 7. Surface plot and its plane projection contour plot displaying the PWD of the experimentally measured backscattered pulse when a B-52 aircraft model is illuminated broadside by an impulse radar (antenna parallel to the fuselage).

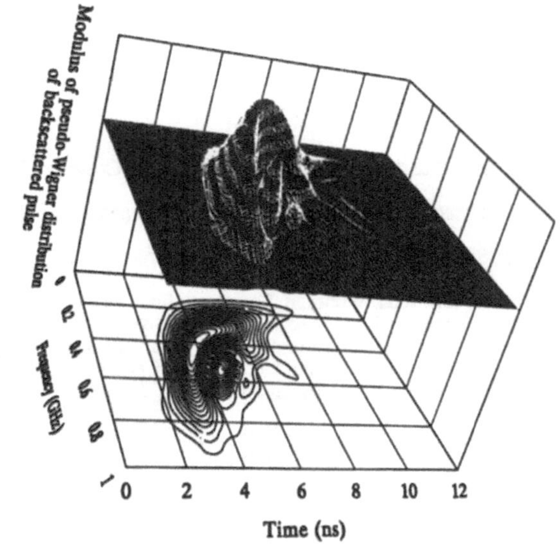

Figure 8. Surface plot and its plane projection contour plot displaying the PWD of the experimentally measured backscattered pulse when a C-130 aircraft model is illuminated broadside by an impulse radar (antenna parallel to the fuselage).

More evidence of the capability to distinguish one target from another that is inherent in the PWD is given by the comparison of the PWDs of the transient responses when the two aircraft models are illuminated from the side (antenna parallel to the fuselage). Figure 7 displays the PWD of the B-52 model and Fig. 8 the PWD of the C-130.

DISCUSSION

We have studied the transient interaction when a short, broad-band pulse is incident on a few targets, both theoretically and experimentally. These transient interactions are closely related to the performance of a specific impulse radar system to the extent that the broad-band pulse has been designed using a digital filter technique to model not only the pulse transmitted by the impulse radar but also to include the filtering effects of the complete radar and signal analyzer system. We have also computed and displayed PWDs using 3-D surface plots and 2-D contour plots to demonstrate the capability of the PWD to give informative signature representations in the combined time-frequency domain. The feasibility of the PWD becomes even more obvious when it is compared with the standard spectrum of the considered echo returns. We have shown that the PWD can extract and exhibit informative features, which improves the target-recognition capability ordinarily furnished by the standard RCS (or form-function) of the considered targets.

The Wigner distribution has a number of desirable properties, including high feature concentration in the time-frequency domain[8]. However, like any other bilinear distribution it also introduces cross-terms between multiple signal components in time-frequency, which in many applications makes the physical interpretation difficult. By introducing a suitable window function it is possible to control the occurrence of cross-terms, which in fact carry signature information with them when the Wigner distribution is applied to echo returns from targets being illuminated with appropriate waveforms. The Choi-Williams distribution[9] is but one of the many other approaches[7] to time-frequency signature analysis that has gained interest from many researchers in the field of signal processing. A desirable objective for future work would be to evaluate the relative merits of the many time-frequency distributions available for signature analysis of scattered pulses from appropriately illuminated targets.

ACKNOWLEDGEMENTS

The authors gratefully acknowledge the support of the Independent Research Boards of their respective Institutions. The work was supported in part by Defense Materiel Administration (FMV), Sweden.

REFERENCES

1. G.T. Ruck, D.E. Barrick, W.D. Stuart, and C.K. Krichbaum, "Radar Cross Section Handbook," Vol. 1, Plenum, New York (1970).
2. A.L. Aden, Electromagnetic scattering from spheres with sizes comparable to the wavelength, *J. Appl. Phys.* 22:601 (1951).
3. A.L. Aden and M. Kerker, Scattering of electromagnetic waves from two concentric spheres, *J. Appl. Phys.* 22:1242 (1951).
4. G.C. Gaunaurd, H.C. Strifors, and W.H. Wertman, Transient effects in the scattering of arbitrary EM pulses by dielectric spherical targets, *J.Electromagnetic Waves Appl.* 5:75 (1991).
5. S. Abrahamsson, B. Brusmark, G.C. Gaunaurd, and H.C. Strifors, Target identification by means of impulse radar, *in*: "Automatic Object Recognition," F.A. Sadjadi, ed., Proc. SPIE 1471, 130 (1991).
6. A.V. Oppenheim and R.W. Schafer, "Digital Signal Processing," Prentice-Hall, Englewood Cliffs, NJ (1975).
7. L. Cohen, Time-frequency distribution — A review, *Proc. IEEE* 77: 941 (1989).
8. T.A.C.M. Claasen and W.F.G. Mecklenbräuker, The Wigner distribution — A tool for time-frequency signal analysis, *Philips J. Res.* 35:217, Part I; 35:276, Part II; 35:372, Part III (1980).
9. H.-I. Choi and W.J. Williams, Improved time-frequency representation of multicomponent signals using exponential kernels, *IEEE Trans. Acoust., Speech, Signal Processing* 37:862 (1989).

ANALYSIS OF TIME DOMAIN ULTRAWIDEBAND RADAR SIGNALS[1]

S R Cloude[1], P D Smith[1], A Milne[1], D M Parkes[2] and K Trafford[2]

[1]Applied Electromagnetics Ltd, Scott Lang Building,
The Observatory, Buchanan Gardens, St Andrews,
KY16 9LU, Scotland, UK

[2]DRA Malvern, St Andrews Road,
Malvern, Worcs. WR14 3PS, UK

INTRODUCTION

Ultrawideband or Impulse Radar has for some time been postulated as having several advantages over conventional radar in that it has the potential for accurate target identification and discrimination. The fact that the transmitted pulse is of very short duration ensures that although the peak powers may be very high, the mean powers involved in the transmitters are low, even at relatively high repetition rates.

With the advent of very fast risetime, high voltage pulse generators it is now possible, with a suitable antenna, to radiate either a Gaussian like Electromagnetic field of very short duration, or a single cycle, very high frequency, sinusoid. For validation purposes these types of field are used to illuminate both generic and complicated structures. The scattered field is detected using very wideband sensors, and then analyzed in both the time and frequency domain.

The purpose of this paper is to describe the generation, calibration and analysis techniques for processing Ultrawideband (UWB) time domain radar signals and presents experimental measurements in support of these techniques[1].

UWB CALIBRATION FACILITY

The UWB range is located on an elevated earth plane such that the source of the radiated signal is sited at the centre of both planes of the test facility (Figure 1), to ensure that any reflected signals from the extremities of the enclosure are well outside the time window being used for the measurements.

The ground plane is 2m above floor level and is 10m x 20m in size.

As all the transmitted signals are above the plane, and the pulse generators are enclosed in screened boxes, sensitive detection equipment can be used under the ground without corruption. Figure 2 shows the experimental layout for the radiating and receive system.

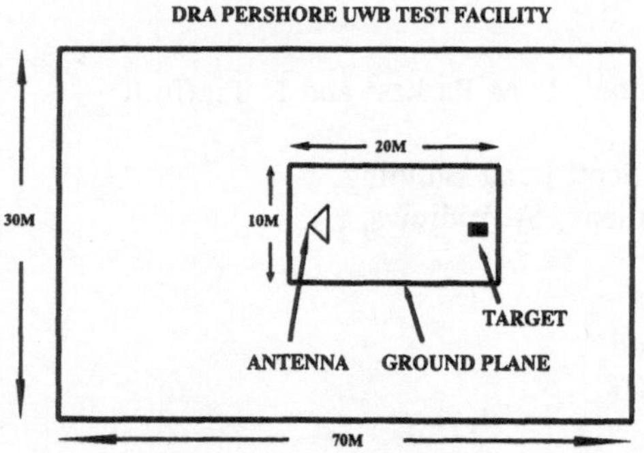

Figure 1 Configuration of the UWB Facility.

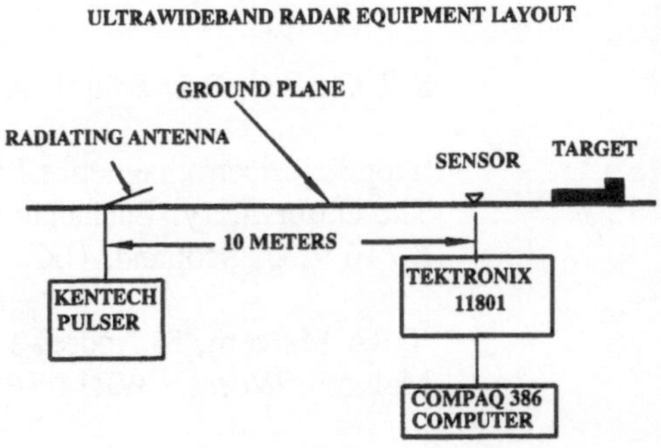

Figure 2 Configuration of UWB Radar system.

PULSE GENERATOR

The pulse generator used for the experiments was a Kentech HPM which has a pulse output voltage of 4kV and a 10 - 90% risetime of 90pS. The avalanche pulsers use a novel distributed trigger system to achieve fast risetimes coupled with high voltage. The pulsers use a series/parallel stack of devices each with a voltage switching capacity of a few hundred volts. These are triggered sequentially along a series string. Each device effectively drives an impedance of only a few ohms and the output is propagated through a stripline transformer to match into 50 ohms. The triggering scheme offers low jitter so several pulser boards can be combined to give higher peak power. The low jitter (typically <5pS) means that sampling technology can be adopted and used over a very long time window. The output of the pulse generator is shown in Figure 3.

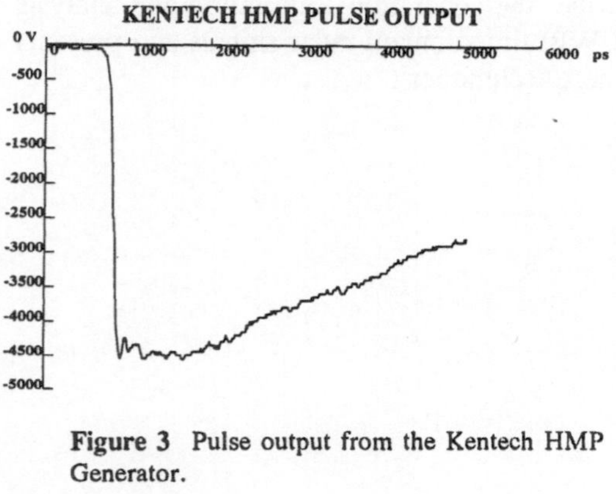

Figure 3 Pulse output from the Kentech HMP Generator.

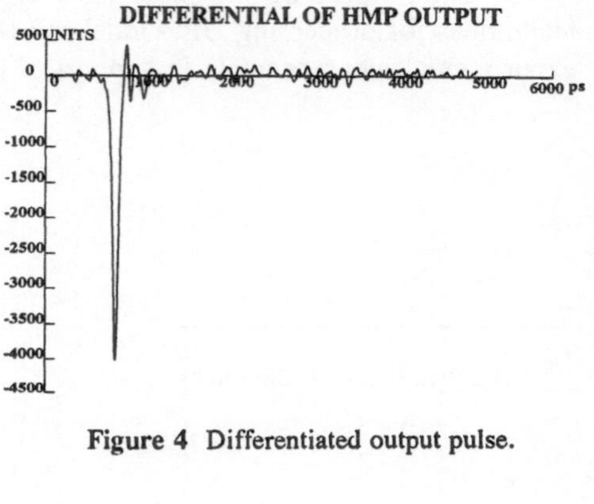

Figure 4 Differentiated output pulse.

This was obtained by feeding the pulser output to the Tektronix 11801 sampling system via suitable value Barth attenuators. The resulting waveform was then down loaded to a Compaq 386 computer using the DRA data acquisition programme (DATACQ)for storage and signal processing. The waveform that is expected to illuminate the target is the derivative of the pulser output and is shown in Figure 4.

RADIATED FIELD

It is necessary to have precise information about the field illuminating the target. A 9mm radius biconical sensor was used to detect the radiated E field at a range of 10m from the transmitting antenna, which in this instance was a "kipper" shaped horn[2]. The signal detected by the sensor was fed to the input of the Tektronix SD-26 sampling head and after averaging the signal was stored and then transferred to the computer for processing, as the detected signal was the derivative of the E field. Figure 5 shows the raw data and Figure 6 shows the time history and amplitude of the E field.

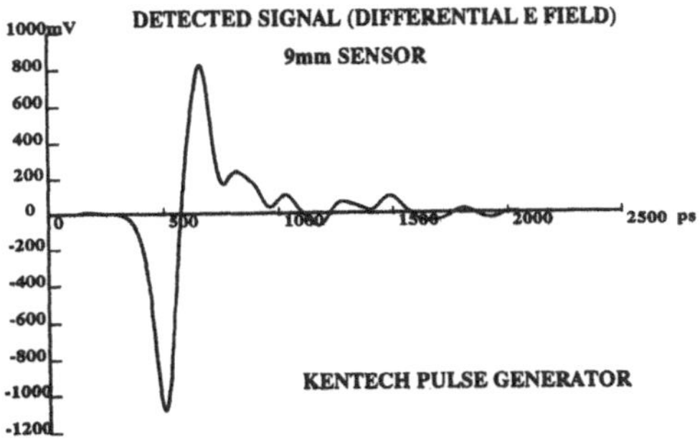

Figure 5 Detected signal from the 9mm sensor.

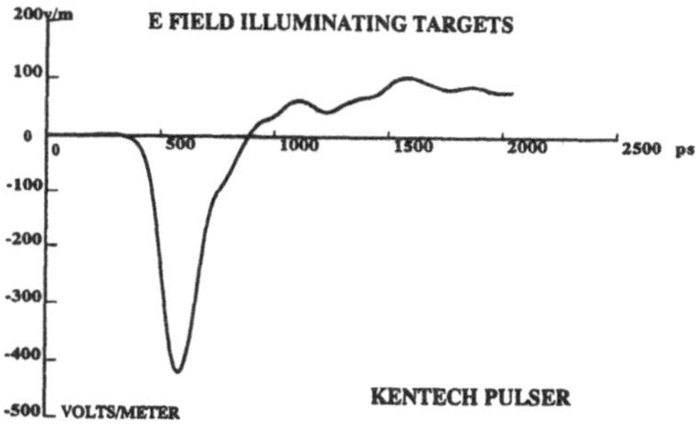

Figure 6 Amplitude and time history of the E field.

CANONICAL TARGETS

The targets used for the experiments were quickly manufactured on site using copper sheet, aluminium foil and available spun aluminium spheres. As the targets were going to be placed on the ground plane it is assumed that there will be a perfect image produced beneath the surface. It was therefore necessary to provide Half cylinders, half size boxes and hemispheres for validation of the theoretical predictions being undertaken.

HALF CYLINDER

The cylinder was constructed using thin aluminium sheet over a plastic former. The apertures were produced by cutting the desired shapes in the appropriate location. The fins were cut separately and then fixed in place at a later stage of the measurements. The dimensions of the cylinder with two sets of fins attached are shown in Figure 7.

HEMISPHERES

The hemispheres used in the experiment were the two halves of a standard Radar calibration sphere. The spheres are made of spun aluminium and the diameter in this instance was 30cm. The configuration when the two hemispheres were irradiated is shown in Figure 8.

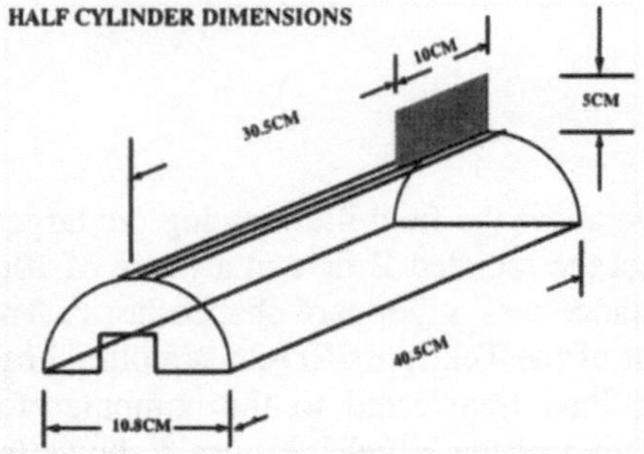

Figure 7 Dimensions of the "half" cylinder.

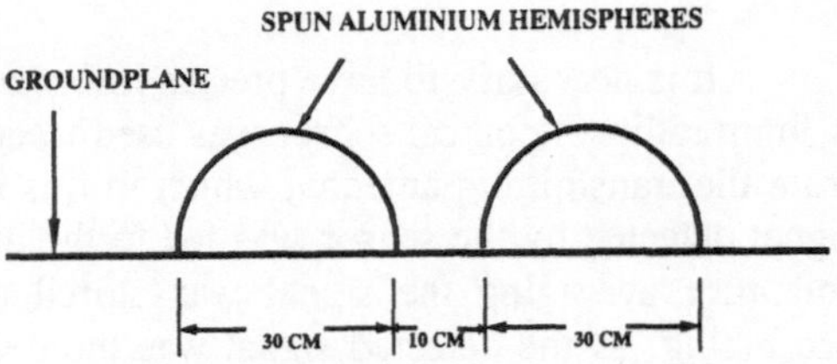

Figure 8 Configuration of the two hemispheres.

RECTANGULAR CAVITY

The cavity was constructed simply using aluminium foil to cover expanded polystyrene. The aperture size was in the "A" plane of the object and could be easily varied.

The dimensions of the cavity are 27cm x 18cm x 7.5cm and the position of the aperture is shown in Figure 9.

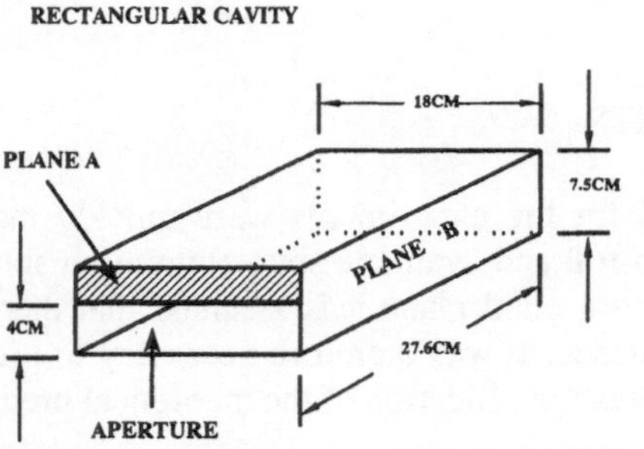

Figure 9 Dimensions of the rectangular cavity or box.

TIME DOMAIN ANALYSIS OF TARGET RETURNS

The time window spanning the proposed target location was identified and recorded ensuring that the data was not corrupted by any background signals. The targets were then placed at the predetermined location on the ground plane. The very low jitter associated with the pulse generator enables averaging to be accomplished over a long time period, hence improving the signal to noise ratio. The resulting signals were then subtracted from the background reference signal prior to transfer to the computer for storage and signal processing.

ULTRAWIDEBAND RETURN FROM HEMISPHERES

A sphere is a standard target for calibrating radars. The step function response of a sphere is known, and the "poles" can be calculated, therefore it is important to validate the theory on a target which can be used by both conventional and UWB radars.

The differential time domain response of a hemisphere on a ground plane is shown in Figure 10.

Figure 11 is the integrated signal and shows the classical response for a sphere to a step function (given experimental limitations) which are predicted by Quasi-Optical Analysis From this response the diameter of the sphere can be obtained. When a second hemisphere was added behind the initial target, the distance between the spheres (10cm) can be measured.

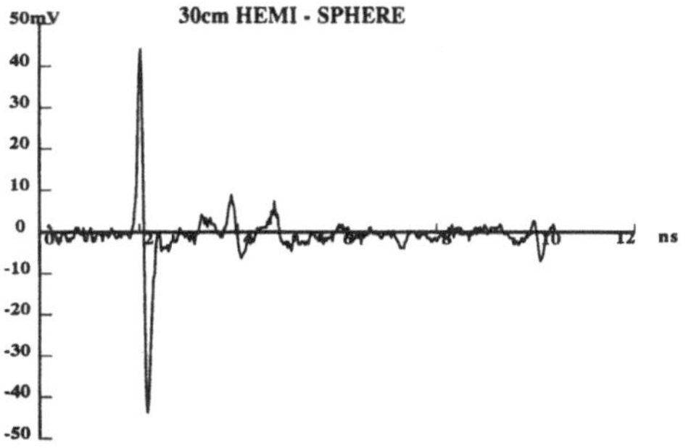

Figure 10 Return signal from a single hemisphere.

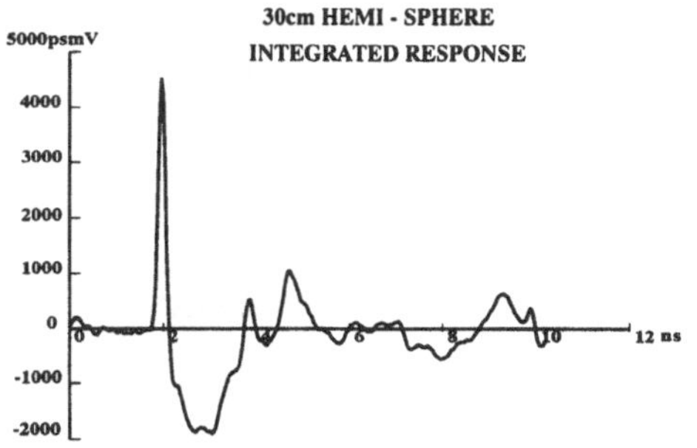

Figure 11 Integrated response from the hemisphere.

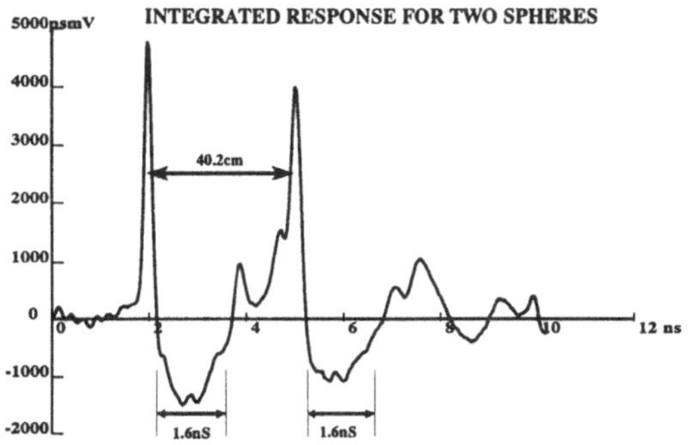

Figure 12 Integrated response for the return signal from two hemispheres.

Information is still available about both targets. Figure 12 shows the discrimination capability of the UWB system.

ULTRAWIDEBAND RETURNS FOR CYLINDER AND FINS

In the "real" world the majority of cylindrical targets of interest have complicated Radar Cross Sections (RCS) due to the complexities introduced by slots, fins and other surface features to the external surface. It is therefore essential that the UWB signatures of some basic structures are fully understood both in the time, as well as the frequency domain.

The response for the cylinder is given in Figure 13 and the fin at the rear is identified in Figure 14.

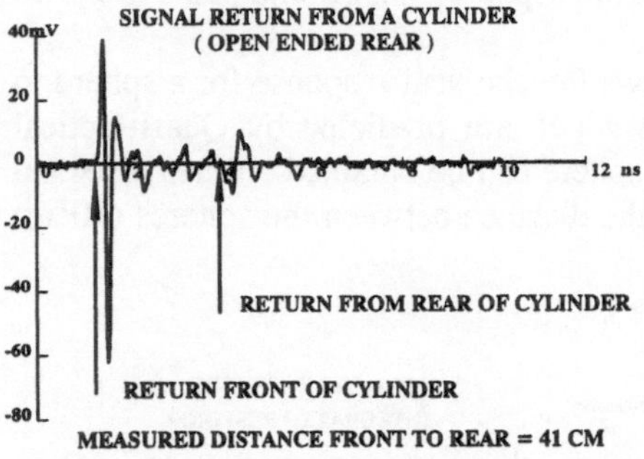

Figure 13 Response from the cylinder.

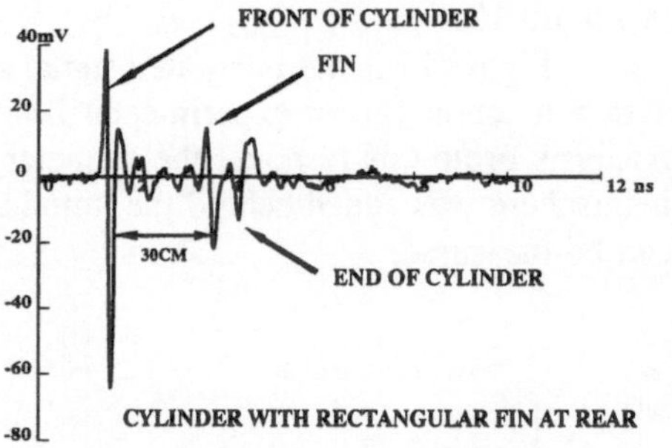

Figure 14 Response from the cylinder with the fin.

ULTRAWIDEBAND RESPONSE FOR A RECTANGULAR CAVITY

In many target systems there are rectangular apertures giving access to various cavities inside the main body. It is of interest to be able to accurately predict the internal resonances stimulated by the UWB field not only for the cavity in isolation, but also in the presence of a baffle which is large compared to the aperture. The results of this type of measurement are important when considering protection from UWB energy ingress. The differing time domain responses are shown in Figures 15 and 16.

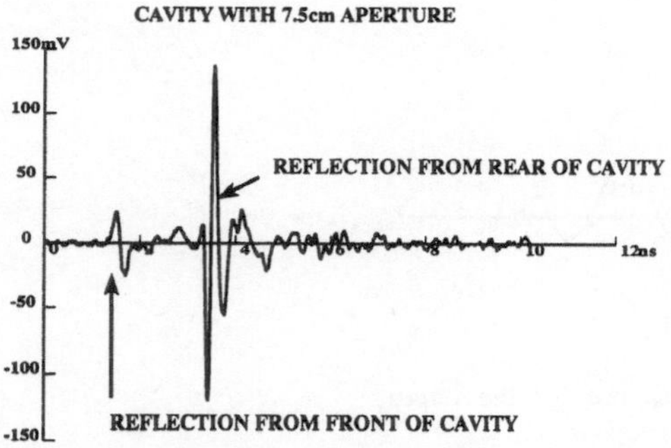

Figure 15 Response from the box with a large aperture.

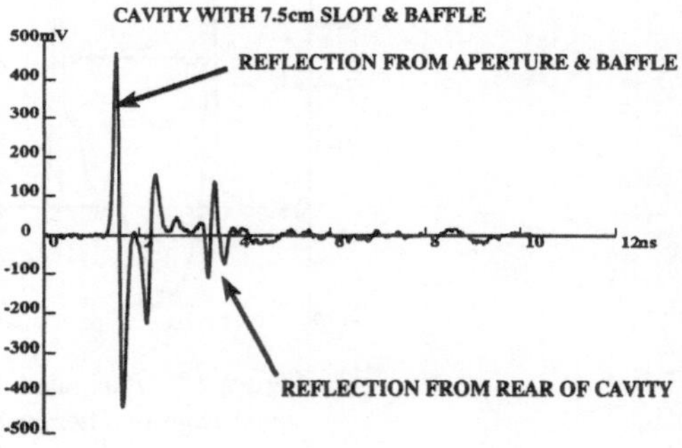

Figure 16 Effect of placing a baffle at the box aperture.

The reflection from the aperture and the rear of the cavity can be clearly identified, as can the late time resonances. The baffle exhibits a large "flash" response but the cavity parameters are still well pronounced.

A detailed examination of the results obtained experimentally, with time domain modelling and exploiting the benefits which incur from the transformation from the time domain to the frequency domain is described in the following sections.

FINITE DIFFERENCE TIME DOMAIN TECHNIQUES

We may use Finite Difference Time Domain (FDTD) techniques to investigate the effects of change of target geometry and incident pulse shape. This method utilises a FDTD discretisation of Maxwell's equations.

We discretise these equations in space and time with a rectangular (Cartesian) spatial grid with cell lengths Δx, Δy, Δz, and a time step Δt. Following Yee[3], central differencing in both space and time is used to replace derivatives. This is second order accurate. It requires a special ordering on the nodes of the components of the electric field **E** and magnetic field **H**. This ordering is the Yee cell, wherein the components are staggered, both in space and time, by half a step.

Arbitrary bodies are defined in this Cartesian space by their electrical properties and so curved bodies are represented by castellated surfaces. This representation of the bodies guarantees continuity of the tangential fields at the boundaries of different materials without the need for any special conditions there.

The resulting explicit difference is subject to a *Courant stability criterion*

$$c_{max}\Delta t < \left[(\Delta x)^{-2}+(\Delta y)^{-2}+(\Delta z)^{-2}\right]^{-1/2}$$

To avoid the introduction of spurious reflections at the edge of the computational mesh, a radiation boundary condition (Mur[4]) is imposed on the *scattered* wave. To model the incident field, a Huygens source representation[5] is introduced inside the mesh, splitting the problem into a total and a scattered field region.

The natural output of the FDTD method is the local field and current values. Thus near field values in the vicinity of scatterers are easily obtained. Surface currents are derived from the field quantities on the surfaces. A standard integral representation with suitable accounting for the retarded times enables the far field response of a scatterer to be obtained.

When this technique is applied to the cylinder[6], which is illuminated end on, we expect to see two main features (Figure 17). The first feature will be the front flash followed by a second flash corresponding to the pulse reaching the back end of the cylinder. The return from the end of the cylinder is at a later time with the delay being equal to twice the cylinder length. This return pulse should initially have the opposite polarity to the first flash, and as it has to transit over the front face again, the rear flash will be longer in duration than the front flash.

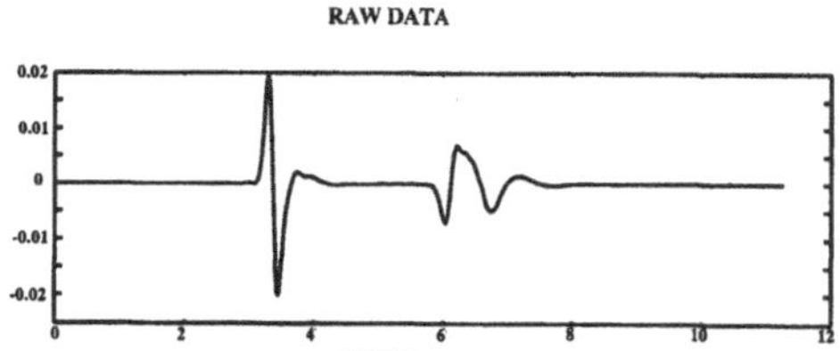

Figure 17 FDTD representation of the cylinder.

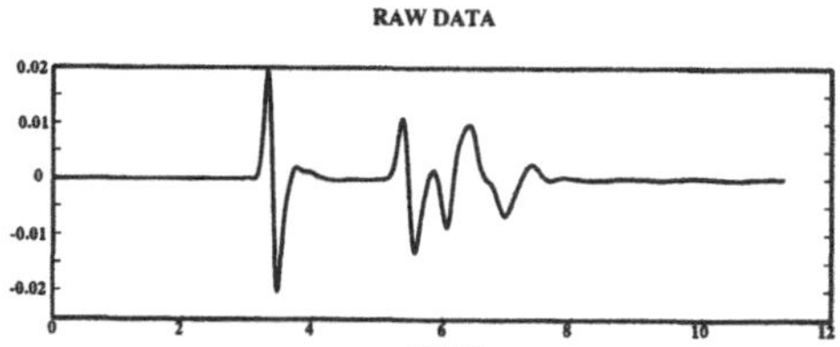

Figure 18 FDTD representation of the cylinder and fin.

With the rear fin in position (Figure 18) we can see when we compare Figure 17 with Figure 18 that both show the front flash unchanged but a new feature coming ahead of the rear flash. The location of this feature corresponds exactly with the front of the fin for both the experimental and calculated cases. The late time extension of the rear flash is also modified by the presence of the fin, but the fast transient, associated with the initial interaction of the pulse with the feature, is clearly evident. This is of great importance to target recognition. Examples of this technique are given in the following sections.

POLE EXTRACTION USING PRONY ANALYSIS

A classical approach to the description of impulsive radar signals models the late time backscattered response I(t) as a sum of decaying exponentials:

$$I(t) = \sum_{n=1}^{\infty} a_n e^{s_n t} \tag{1}$$

Here the "poles" s_n are independent of the direction of illumination, whereas the weighting coefficients an depend upon the measurement process. The target is uniquely identified by its (infinity of) poles. Many authors have devised techniques for locating these poles in the complex plane from experimental signals corrupted by noise. An excellent review of various approaches and problems is given elsewhere[7]. Most authors do not use a nonlinear least squares fit of the data to the model (1) because of the intensive computation involved, but rather prefer a modified form of Prony's algorithm.

If N poles are assumed to be present in the m data values $I_k=I(k\Delta t)$, sampled at intervals of Δt, then with $\lambda_n = e^{s_n \Delta_t}$, we have:

$$\sum_{n=1}^{N} a_n \lambda_n^k = I_k + \epsilon_k \quad (k = 0, 1, 2, ..., m-1)$$

where ϵ_k denotes the noise at time step k. The first step is to construct a polynomial

$$p(\lambda) = c_o \lambda^N + c_1 \lambda^{N-1} + ... + c_N \tag{2}$$

whose roots are $\lambda_1, ...\lambda_N$.

$$Let \quad A = \begin{bmatrix} I_0 & I_1 & \cdots & I_N \\ I_1 & I_2 & \cdots & I_{N+1} \\ . & . & \cdots & . \\ . & . & \cdots & . \\ I_{m-1+N} & I_{m-N} & \cdots & I_{m-1} \end{bmatrix}$$

In the absence of noise, the coefficient vector $c = [c_0, c_1, ...c_N]$ is the eigenvector corresponding to the zero eigenvalue of **A'A**; in the presence of noise, c is identified as the (normalized) eigenvector of the least eigenvalue of **A'A**. Having determined c, a polynomial rooting procedure yields the roots λ_n and the poles $s_n = \ln(\lambda_n/\Delta t)$. The full nonlinearity of the model fitting is concentrated in this step. That a sufficient number, N, of poles has been used to model the data is ascertained by examining the eigenvalues of A^tA. Ideally, there will be a sharp contrast between the magnitude of those eigenvalues corresponding to true poles and those which model the system noise, the latter being of magnitude proportional to the noise variance. (It is useful to over-estimate N somewhat although the use of too many spurious poles is undesirable). As the noise level increases, the fall-off is less pronounced and the choice of N correspondingly less certain.

The sphere and the box with a slot represent two extremes of success in this pole extraction technique. The late time response of the sphere is much smaller than that of early time, and so it becomes extremely difficult to reliably extract many poles. A detailed examination[8] concluded that "the identification of complex resonances ... with high radiation damping is a difficult proposition, even with relatively noise free data." Application of the technique described above to sphere scattering, measured at DRA Hanger II Facility, Pershore, confirms these results.

In contrast, a box with a 4 cm slot represents a class of object whose poles can be more reliably extracted. A box 7.5 x 18 x 27.6 cm was placed on the ground plane and illuminated broadside to the 7 .5 x 18 cm face, in which a 4 x 18 cm slot was opened. The algorithm was applied to the backscattered field data, both measured (see Figure 15) and calculated (by an FDTD code), as in an earlier section. Pole pairs so derived are listed in Table I and their corresponding (scaled) coefficients are shown in Figure 19.

Table I - Comparison of predicted and detected poles

Poles/2π - Ghz	
FDTD	10% Noise
	-0.0294 ± 0.5849i
-0.00358 ± 1.1701i	-0.0079 ± 1.1844i
-0.353 ± 1.4724i	-0.0583 ± 1.4843i
-0.0272 ± 4.0813i	-0.0171 ± 2.2092i
20% Noise	Experiment
-0.703 ± 0.4417i	-0.2007
-0.0730 ± 0.8383i	-0.1255 ± 0.6706i
-0.0113 ± 1.1825i	-0.0795 ± 1.1789i
-0.0979 ± 1.4855i	-0.2994 ± 1.5396i
	-0.0501 ± 2.3954i

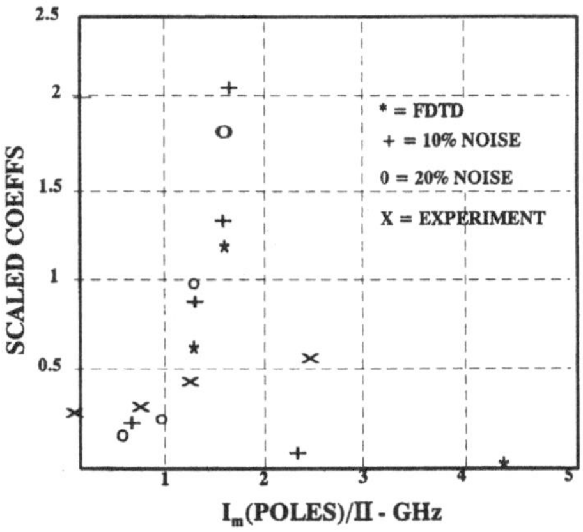

Figure 19 Comparison of pole coefficients.

The numerical results generate two dominant pairs corresponding to two internal resonances of the box (1.14 and 1.47 Ghz) excited by this particular field polarization when fully closed. Addition of Gaussian noise, (standard deviation = 10% of peak value), to the numerically computed results shows that the extracted poles are still readily identifiable. This required N=21, and only the lower frequency poles are shown. Notice that a low frequency fairly damped pole has been introduced. Increasing the noise to 20% makes recognition increasingly difficult, due to the larger number of poles needed to model the noise. In both the 10% and the 20% cases, the pole pair with smaller co-efficient is more susceptible to the added noise and beyond the 20% noise level, recognition is destroyed.

The algorithm was applied to the late time measured data, using every second data point, and N=41. Generally a choice of about one third of the number of data points gives reasonable results (even without filtering). The box resonances are recognizable, although the less dominant one stands out less well against the low frequency pole which is modelling a low frequency effect in the data measurement process.

MATCHED FILTER DETECTORS AND WAVELET PROCESSING

A range of techniques have been investigated for combined time/transform domain analysis of UWB radar signals. The main objective of such techniques is to enhance the detection and identification of features in the UWB response of complex composite targets.

The techniques investigated have included the early matched filter detection based on high frequency inverse scattering identities[9]. These have clearly demonstrated how asymptotic diffraction theory can be used to enhance detection of localised target features such as specular return from curved surfaces.

Gabor Transform analysis has been used to provide time/frequency localisation of UWB radar features for identification and detection[9,10]. These techniques are superior to straightforward matched filter implementation as they account for the relative timing of diffraction components in the UWB response.
An example of a Gabor plane transform for the cylinder with the fin is shown in Figure 20.

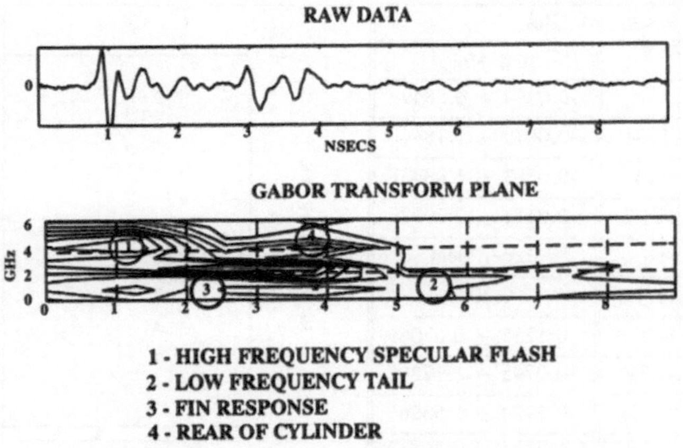

Figure 20 Gabor Transform analysis of the cylinder and fin.

Wavelet Transform techniques have recently found widespread application in many areas of signal processing[9]. They combine the time/frequency localisation properties of windowed Fourier Transform techniques with several additional features, including the concept of scale, which can be used to characterise composite UWB radar signals and as the basis for detection and identification in the presence of noise.

An example of the use of wavelet processing of FDTD predictions for the UWB radar return from the rectangular box is given in Figure 21, which shows a contour plot of the time/scale wavelet transform plane (using a Mallet wavelet) for the raw data (with decreasing scale down the page). Two important characteristics are identified in the data, the localisation in time (at fine scale) of the main specular radar returns from the front and rear of the box and the enhancement of the late time resonance behaviour at intermediate scales. This latter portion is of particular significance as it demonstrates the potential advantage of using wavelet analysis for combined early and late time processing. The scale parameter is also useful for the reduction of noise as illustrated in Figure 22. Here white noise has been added to the FDTD prediction for the box and then a wavelet analysis has been performed.

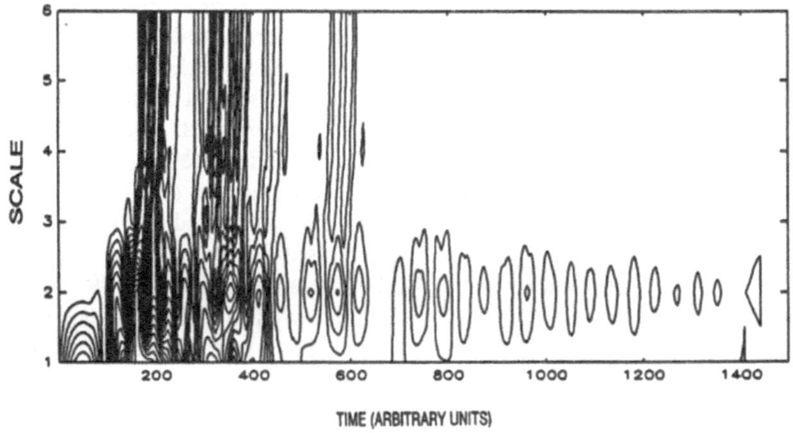

Figure 21 Wavelet Transform of FDTD Predictions for rectangular box.

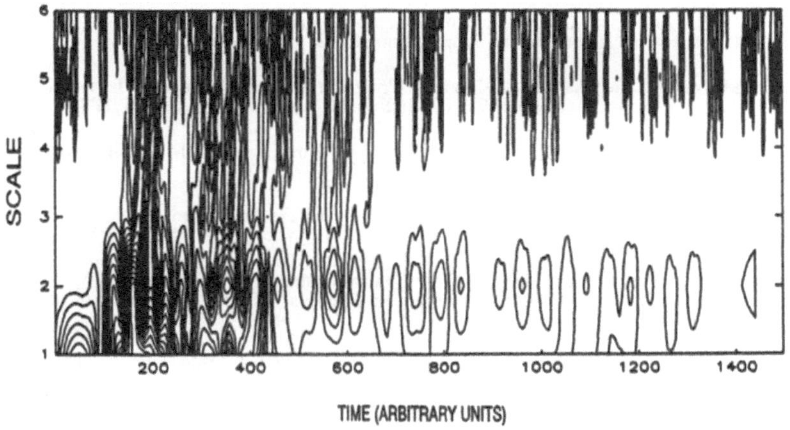

Figure 22 Wavelet Transform of box data with added gaussian noise.

We see that the resonance behaviour is still clearly apparent at intermediate scale values, while the noise is localised to fine scales. This observation may provide the basis for a noise filtering technique for target identification and detection. Future work will be aimed at quantifying this noise immunity as a function of signal to noise ratio.

CONCLUSIONS

It has been shown that with advanced Pulse Generator technology and modern detection equipment it is feasible to produce time domain signatures of simple and complex targets which can give precise details of the interrogated objects in the time domain, and produce results which approach the expected Quasi-Optic response for a step function illumination.

The FDTD modelling has been shown to be a useful aid to interpretation of the measured results. The computed results also provide a clean pulse for the signal processing studies where the effects of noise are studied.

A technique of target identification from later time backscattered responses was described and shown to be successful for a class of scatterers which posses several low radiation damped poles, at least in systems in which the noise level does not exceed 20%.

Combined time/transform domain processing has several attractive features for UWB radar: it permits localisation of important features of the target response as well as providing immunity to noise and interference. In particular, wavelet analysis seems to offer several distinct advantages in processing and interpretation of the radar returns.

The problem of detecting moving targets within a cluttered environment is being studied, and some of the techniques are proving to be effective.

Although there is still a long way to go before an operational UWB Radar can be completely specified, the results obtained so far are very promising. With the advances taking place in many technological areas, short range systems should be possible in the future.

REFERENCES

1. D M Parkes, P D Smith. "Transient Field Generation and Measurement" 1986 AP-S International Symposium Digest on Antennas and Propagation. Vol2, p 1015-18.

2. P D Smith, D M Parkes. "Compact and Directive Transient Antennas". Seventh International Conference on Antennas and Propagation ICAP 91. (Conf.Publ. No 333) Vol2 p.588-91,1991.

3. Yee, K.S. "Numerical solution of initial boundary value problems involving Maxwells equations in isotropic media" IEEE Trans. Antennas Propagat. AP-14,302-307, 1966.

4. Mur, G. "Absorbing boundary conditions for the Finite Difference Approximation of the time domain electromagnetic field equation" IEEE Trans . Electromagn. compat. EMC-23, 377-382, 1981.

5. Merewether, D. E., Fisher, R. and Smith, F. W., "On implementing a numeric Huygens source scheme in a finite difference program to illuminate scattering bodies" IEEE Trans. Nuclear Sci. NS-27, 1819-1833, 1980.

6. Milne, A. M., Interim Report on Impulse Radar Modelling (for DRA Malvern) CR6/92 Jan. 1992.

7. S.M.Kay, S.L.Marple, Spectrum Analysis - A modern Perspective, Proc.IEEE Vol.69, pp 1380-1419, 1981.

8. D.G.Dudley Progress in Identification of Electromagnetic Systems, IEEE Ant.Prop.Soc. Newsletter Vol.30, No.4, pp 5-11, 1988.

9. S R Cloude, P D Smith, A Milne, D M Parkes." Analysis of Time - Domain Ultrawideband Radar Signals". SPIE Technical Conference on Ultrawideband Radar Jan 1992 SPIE Proceedings Vol 1631 pp 111 - 122.

10. D.M.Parkes, S.R.Cloude, P.D.Smith,"Feasibility of Ultrawideband Radar", APMC Conference Proceedings. Vol.1. pp 357-360. August 1992.

ULTRA-WIDEBAND ELECTROMAGNETIC TARGET IDENTIFICATION

D.G. Dudley[1], P.A. Nielsen[2], and D.F. Marshall[2]

[1]Electromagnetics Laboratory
[2]Communications, Control, and Signal Processing Group
ECE, Bldg. 104, University of Arizona
Tucson, AZ 85721

BACKGROUND

Target identification is concerned with obtaining information about a target from input-output data. Specifically, in identification of electromagnetic targets, we are concerned with the interaction between transmitted electromagnetic waves and scattering objects. Examples can be found in such disciplines as electrical methods in geophysical exploration, monostatic and bistatic radar, medical hyperthermia, and microwave imaging.

Electromagnetic target identification is a synthesis procedure, rather than an analysis problem. In analysis, we are concerned with determining output(s), given input(s) and scatterer(s). This case is commonly called the *forward problem.* In synthesis, we are concerned with finding characteristics of scatterer(s), given input(s) and output(s). If the scattering system is completely determined, we call this case the *inverse problem.*[1] If we are able to determine parameters that give information, usually incomplete, about the scatterer(s), we call this case the *parametric inverse problem.* The inverse problem is one of formidable complexity; it has yielded usable solutions in only a very limited number of special cases. In the parametric inverse problem, we concede the overwhelming difficulty of the inverse problem and concentrate on adopting a parametric model and obtaining estimates of interesting parameters. Such parameters in the radar problem, for example, might be target range, nose-to-tail dimension and wingtip-to-wingtip dimension.

Parametric modeling applied to electromagnetic target identification had its beginnings in the early 1970's when several electromagnetic researchers began to consider the electromagnetic identification problem at a very basic level. These early researchers conceded the overwhelming difficulty in obtaining a quick solution to the practical problem of the user (target identification by a radar operator, for example) and returned to a fundamental examination of the basics. These researchers were able to build on mathematical scattering theory and its application to problems in modern physics. Lax and Phillips[2] had shown that for the scalar wave equation, the scattering from objects with Dirichlet boundary conditions is given by a *meromorphic function* of complex frequency. Essentially, what this result means is that the scattering can be expressed as the sum of the following two contributions: First, a temporal series of complex exponentials (poles in the s-plane) that describe the body resonances of the object; second, a temporal response, strictly limited to early time produced by an *entire function* of complex frequency (no s-plane poles). Following the mathematical theory, researchers quickly applied the results to electromagnetics. In 1972, Baum[3] postulated scattered wave expansions in a formulation he named the Singularity Expansion Method (SEM). In 1973, Marin[4], by a clever use of Fredholm determinants, showed that the meromorphic characteristic of scattering holds for the *vector* wave equation in electromagnetics, for scatterers with perfect conductivity ($\sigma \to \infty$). The stage was thus set for electromagnetic workers to examine how complex resonances could be extracted from scattered field data. Should it be possible to determine a set of body resonances, it was then hoped that the resonances would yield information about the scattering object. For example, if the object were an airplane, it was hoped that the resonances would determine such features as nose-to-tail length and wingtip-to-wingtip length.

The period from 1974 to 1976 was marked by efforts to adapt Prony's method to the determination of body resonances. Particularly strong programs appeared at Ohio State University,[5] University of Illinois,[6] Lawrence Livermore National Laboratory,[7] and the University of Arizona. In 1979, Dudley showed[8] that Prony's method is a rudimentary form of system identification, associated with single-input, single-output difference equation modeling. This result marks the beginning of studies aimed at understanding the basis of complex resonance modeling methods applied to electromagnetic scattering data.

Unfortunately, complex resonance modeling applied to target identification has met with only limited success. Based on our efforts at the University of Arizona over the past sixteen years, we have come to the following conclusions:

1. Successful identification of complex resonances from data requires that the resonances, considered in the frequency domain, be well above the noise and separately defined.

2. The identification of complex resonances associated with structures with high radiation damping is a difficult proposition, even with relatively noise-free data.

3. Complex resonance modeling ignores the information contained in early time in the scattered wave, an unfortunate fact, since early time usually contains the portion of the signal where the signal-to-noise ratio is high.

4. The complex resonances of major features of aircraft and missiles occur at frequencies so low that modern radar methods are difficult, if not impossible, to apply.

This view is supported by recent results in the Advanced Sensors Directorate, Research, Development, and Engineering Center, U.S. Army Missile Command. In discussing resonance modeling as applied to radar target identification (RTID), Smith and Goggans[9] conclude that although "many researchers have proclaimed that this approach would prove to be a panacea for RTID,...the small energy content and the low frequency of the resonance modes on aircraft-sized objects are seen as liabilities..."

The incompleteness of the complex resonance target description is now well known and well understood.[10] Indeed, in the early time, the complex resonance model is an improper model and must be supplanted, or supplemented, by techniques that consider local features of the scattering object. We have introduced[11] a method to extend single-input, single-output complex resonance models to include both multiple-look angles and early time information. We have also demonstrated a method for local feature identification in early time[12] based on asymptotic descriptions of electromagnetic scattering by concentrating on models involving the radius of curvature of an object.

Recent efforts in target identification[9] have recognized the necessity for combining more complete electromagnetic models with classical and modern methods of detection. The objective of a *detection system* is to determine accurately whether a particular signal (generated or emitted by a target) is present in an observation (the received electromagnetic wave). The simplest of detection problems can be formulated as a binary hypothesis test,[13,14] as follows:

> *There is no target versus there is a target.*

The goal of the detector is to determine which of these hypotheses is true by processing the observation. This simple test can be modified to account for multiple hypotheses,[13,15] as follows:

> *Target A is present, or target B is present, or target C is present, or, ..., or target Z is present.*

Thus the detector *classifies* the observation as being generated by one of the Z targets.

Precise specification of the decision rule requires that the observation be characterized under each hypothesis.[13,15] For the binary hypothesis problem, this requires that the noise and the noise-plus-signal be modeled. Due to the random nature of noise, the observation is described statistically. It is important to emphasize that characterization of the observation requires that the structure of the electromagnetic wave received by the antenna which is due to the target be defined.

In classification problems, the detector makes a choice between different known targets. Since the observation is known to be a function of the orientation of the target relative to the transmitting and receiving arrays, detailed information about the structure of the observation for each target at each orientation is necessary. When the orientation is included, the number of scattering models per target[9] is of the order of 10^5. Therefore, the direct method of calculating the scattering function of each aircraft for a range of different orientations and saving these results can quickly become computationally unmanageable.

An alternative to the classical detection described above is the identification of a target which is known to be present. This is similar to the classification problem, but the set of possible targets is not specified. Here, the objective is to determine meaningful or interesting parameters about the unknown target by processing the observation. We shall discuss this approach in greater detail in what follows.

There are several basic research issues in electromagnetic target detection and identification that we believe are important in ultra-wideband applications. A fundamental and difficult one concerns whether one can effectively trade off frequency for spatial position of sensors. We have recently being studying this tradeoff[16] in conjunction with geophysical targets. It is well known from applications such as medical imaging and geophysical tomography that efficient imaging can be obtained by moving a sensor completely around the target. In the usual radar scenario, however, one is usually limited to one, or at most a few, sensor positions. Indeed, by far the most usual situation is monostatic, where there is a single transmitter and receiver at one location. In such a situation, one gives up completely the three spatial "degrees of freedom." The question then becomes whether one can at least partially regain the loss in degrees of freedom with frequency bandwidth. Another issue involves selection of a model that in some way maximizes the obtaining of target information from the scattered electromagnetic return. It is still unclear what is the best electromagnetic model structure. A third issue involves bandwidth and center frequency (or frequencies) as a function of target size. There is a strong possibility that more than one segment of the frequency spectrum may be necessary to obtain enough information to identify, rather than simply detect, an object. A final issue is how one detects or estimates the information required by the model in the presence of noise.

MODEL-BASED DETECTION

Basic to the success of an electromagnetic target detection and identification procedure is the selection of a target model. Smith and Goggins[9] discuss the following scattering center model. The impulse response $h(t)$ of the target is modeled as a series of weighted delta functions, where each delta function contains a time delay, viz:

$$h(t) = \sum_{m=1}^{M} h_m \delta(t - \tau_m) \tag{1}$$

This model has the characteristic of dividing a target into *scattering centers* each of which make a distinct contribution to the scattered return. The model is easily extended to include multiple targets and/or multiple angles of observation. The model is limited, however, in that it neglects frequency dependence in the target echo. We presently have a model under study that incorporates frequency dependence. In addition, it is easy to show that (1) is a special case of the model we are considering.

Our model is due to Altes,[17] who introduced a transversal filter as a parametric model in underwater acoustics. What is novel is that the impulse response uses both integrals and differentials of delta functions, not simply delta functions as in (1) above. Our modeling studies have been motivated by results in electromagetic scattering by Kennaugh and Moffatt,[18] and in physical optics by Freedman.[19] Freedman's work shows that Rayleigh scattering is proportional to the square of the frequency, which is the second derivative of an impulse. Flandrin, Magan, and Zakharia[20] refer to the model of Altes[21] as a *Generalized Target Description.* They are able to relate it to physical situations using wavelets.

The Altes Model

The Altes Model is a channel identification model in which the following assumptions are made:

Additive noise is Gaussian and white.

The transmitted signal $x(t)$ is known completely. Its Fourier transform $X(\omega)$ has the property $X(\omega)_{\omega=0} = 0$.

The channel is linear with impulse response $h(t)$ and transfer function $H(\omega)$.

The measured echo or channel output is $y(t)$.

The modeled echo or channel output is $\hat{y}(t) = h(t) * x(t)$.

In the synthesis problem, we use the measured echo $y(t)$ and the transmitted pulse $x(t)$ to extract physical information about the scattering object. This information is contained in the impulse response $h(t)$. The difficulty with the synthesis problem is in modeling this impulse response so that the model parameters are physically meaningful. For example, having a parameter which corresponds to object size is desirable.

Altes has introduced the following transversal filter as a model of the scattering function (channel):

$$h(t) = \sum_{m=1}^{M} \sum_{n=-N}^{N} h_{nm}\delta^{(n)}(t - \tau_m) \tag{2}$$

where the $(2N + 1) \times M$ tap weights $\{h_{nm}\}$ and the M delays $\{\tau_m\}$ are estimated from the data. The superscript on the delta function determines whether the delta is integrated ($n < 0$), differentiated ($n > 0$), or is a standard delta function ($n = 0$). We call the model in (2) the *generalized target model.*

The Fourier transform of the modeled echo is

$$\begin{aligned}\hat{Y}(\omega) &= H(\omega)X(\omega) \\ &= \sum_{m=1}^{M} \sum_{n=-N}^{N} h_{nm}\left[(i\omega)^n X(\omega)\right] e^{-i\omega\tau_m} \\ &= \sum_{m=1}^{M} \sum_{n=-N}^{N} w_{nm} V_n(\omega) e^{-i\omega\tau_m}\end{aligned} \tag{3}$$

where

$$V_n(\omega) = (i\omega)^n X(\omega)/\sqrt{E_n} \tag{4}$$

$$w_{nm} = h_{nm}\sqrt{E_n} \tag{5}$$

and E_n is the energy of the nth component. Thus, the filters $\{V_n\}$, which represent the decomposition of the expected echo into to its energy components, are matched to the n-th derivative or integral (depending on the sign of n) of the transmitted signal $x(t)$.

If the delays $\{\tau_m\}$ are known, Altes suggests that the $(2N+1) \times M$ model parameters be chosen so that the mean square error (MSE) between the observation and the echo be minimized. These parameters can be calculated by solving a linear set of equations, $Aw = r$, where the kl^{th} entry of the correlation matrix A is given by

$$\mathrm{Re}\{v_k^*(-t) * v_l(t)\},$$

and the elements of the vector r are calculated as

$$\mathrm{Re}\{v_k^*(-t) * y_l(t)\},$$

where $v_k(t) \leftrightarrow V_k(\omega)$. For any fixed delay, τ_m the filter tap weights can be estimated by minimization of the MSE. (For the MSE criterion, for each fixed delay t_d, the estimation procedure requires that the received echo be correlated with a bank of matched filters, where each of the $2N \times 1$ filters is defined for one specific $\{n, t_d\}$ pair. It is well known that for white noise, matched filters maximize the output SNR at a fixed time of all linear filters. For Gaussian noise, the matched filter will minimize the error probability; this is not necessarily true for non-Gaussian noise.)

For the identification problem, the delays, which correspond to the time of arrival of the creeping waves, are useful for characterizing the size of the scatterer. However, in a realistic scenario, the delays are not known and must be inferred or estimated from the data. To compensate for lack of knowledge of the delays, Altes suggests using a structure that is a combination of the results of Middleton and Esposito[22] on Bayes optimal joint detection-estimation methods and generalized likelihood ratio detectors. Essentially, the parameters of the detector are estimated via maximum likelihood or some other technique.

The strengths of the model in (2) include its flexibility and its ability to model time delays. In addition, it models the complete meromorphic scattering function, as opposed to the simple pole locations. Finally, the state-space model used by Dudley[11] fits within this structure.

One of the weaknesses of the model is that a large number of filters are necessary to complete the estimation process. Additionally, the technique needs to be tested in the presence of noise. In cases of high SNR, we expect that the detection-estimation operation can be carried out with success.

Some Noise-Free Results

To determine the usefulness of the generalized scattering center model, we have performed the following study. To test the model, we use a hard acoustic sphere of radius a. The time-domain velocity potential for the sphere can be written as[23]

$$y(t) = \frac{1}{2\pi}\int_{-\infty}^{\infty} rV^s(\omega/c, r, 0)X(\omega)e^{i(r-2a)\omega/c}e^{i\omega t}d\omega \tag{6}$$

where

$$rV^s(\omega/c, r, \theta) \approx -\frac{i}{k}e^{ikr}\sum_{n=0}^{\infty}(-1)^n(2n+1)\frac{j_n'(ka)}{h_n'^{(2)}(ka)}P_n(\cos\theta) \tag{7}$$

and where r is the distance to the observer, θ is the angle between the observed and the z-direction, k is the wavenumber, P_n is an nth-order Legendre polynomial, j_n is a spherical Bessel function, $h_n^{(2)}$ is an n-th order Hankel function of the second kind, and $X(\omega)$ is the input.[23] By using a combination of Mathematica and Matlab tools, a bandlimited (5000 Hz) Fourier transform of the velocity potential has been evaluated for $x(t) = \delta(t)$. Figure 1 shows the magnitude of a bandlimited portion of $Y(\omega)$ assuming that the input is an impulse. Figure 2 is the inverse transform of the bandlimited $Y(\omega)$.

In matching the generalized scattering center model to the observation, an initial step is to select the model order N. To begin, we focus on the initial time portion of the backscatter (approximately the first 15 msec of $y(t)$), and set all $\tau_m = 0$. It is well-known that the asymptotic expansion of $Y(\omega)$ for the sphere contains only negative powers of n.[12] We therefore write the model for the transfer function as

$$w(t) = \sum_{n=-N}^{0} w_{n0}\delta^{(n)}(t) \tag{8}$$

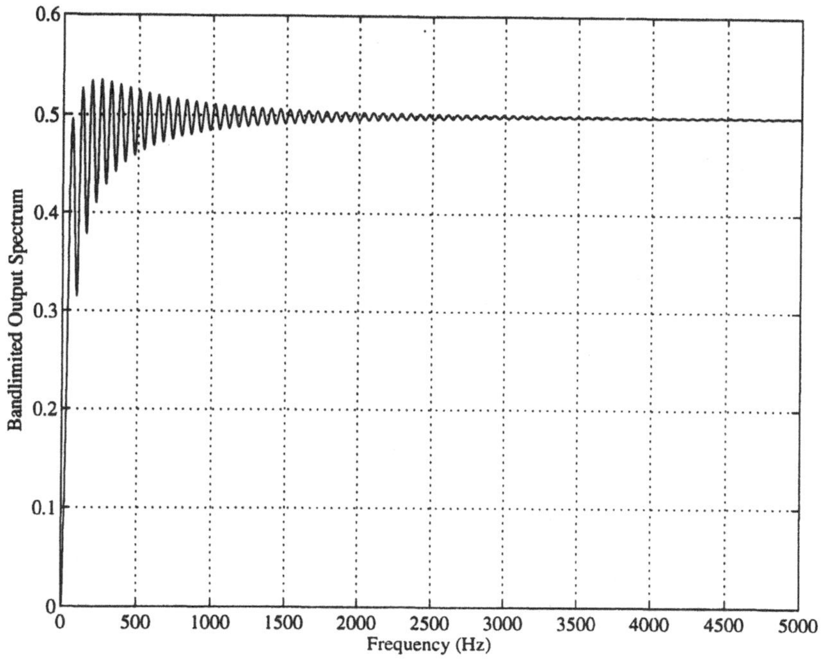

Figure 1. Bandlimited frequency response $Y_B(\omega)$ for a hard acoustic sphere.

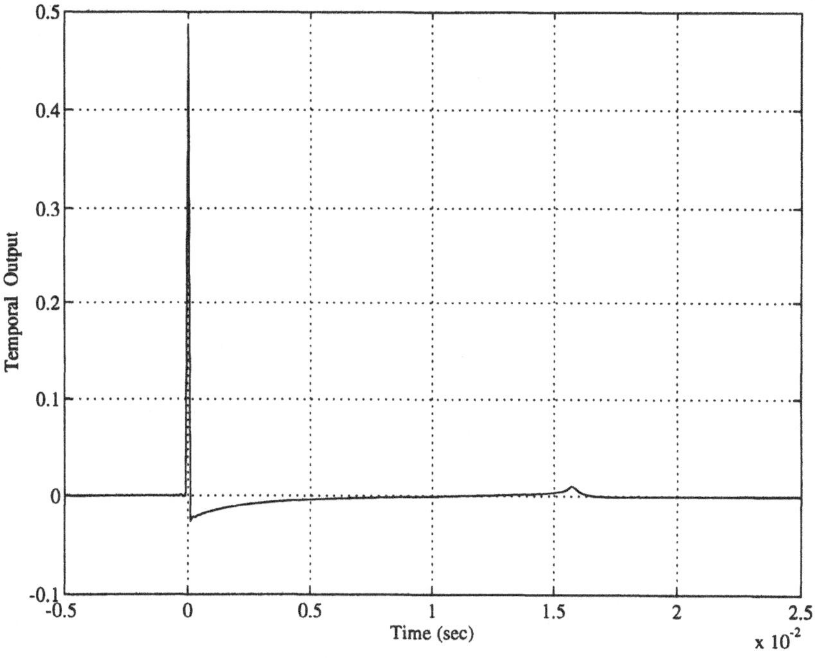

Figure 2. Inverse transform of $Y_B(\omega)$.

The model echo is then given by

$$\hat{y}(t) = \sum_{n=-N}^{0} w_{n0} v^{(n)}(t) \tag{9}$$

When $N = 0$, the generalized model simplifies to the scattering center model in (1), assuming only one time delay. As N increases, the echo is modeled as the transmitted pulse with integrated versions of the transmitted pulse. The model has been tested for $N = 0, 2$ and 4.

The weights w_{n0} are evaluated by minimizing the mean-square error (MSE) between the observation and the modeled response, viz:

$$\frac{1}{2\pi} \int_{-\infty}^{\infty} |Y(\omega) - \hat{Y}(\omega)|^2 d\omega \tag{10}$$

This operation results in a matrix calculation

$$Aw = r \tag{11}$$

where the elements of A are given by

$$a_{kl} = a_{kl}(t) = v_k^*(-t) * v_l(t)|_{t=0}$$

and the elements of the vector r are

$$r_k = r_k(t) = v_k^*(-t) * y(t)|_{t=0}$$

By inverting A (if A is square), or by using a least squares approach, we obtain estimates for the weights w_{n0}.

To excite the model, we use the transmitted pulse of Dudley and Weyker,[23] viz:

$$x(t) = \begin{cases} 0, & t < 0 \\ \sin 2\pi f_c t \sin^2 \alpha t, & 0 \leq t \leq \pi/\alpha \\ 0, & \pi/\alpha < t \end{cases} \tag{12}$$

The parameter f_c is the carrier frequency and α controls the duration of the pulse. The early time portion of the observation has a duration of approximately 15 msec for a sphere with a one meter radius. Hence, the pulse duration is constrained to be shorted than 15 msec.

Figure 3 illustrates the effect of increasing the model order on the MSE for fixed α. As the model order increases, the MSE decreases. Note the substantial difference between the MSE for the zeroth and second order models. The increase from a second to a fourth order model yields little further improvement. As the model order increases, the echo requires that progressively higher-frequency components be evaluated. For the implementation here, the maximum frequency which can be used is 5000 Hz. Eventually, increasing the model order will result in an increase in the MSE because

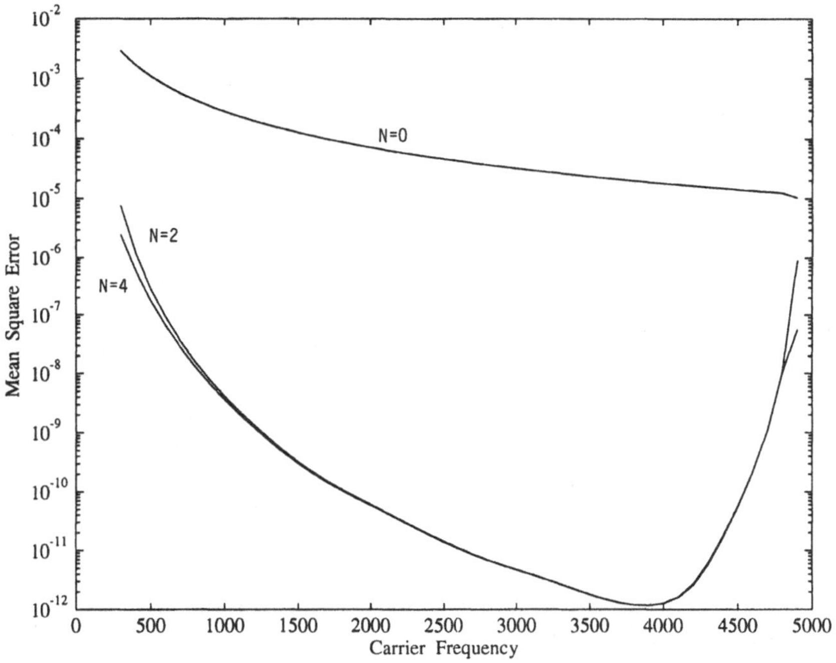

Figure 3. Mean-square error for models of order $N = 0, 2, 4$. Pulse duration $\alpha = 300$.

the necessary high frequencies are not available. We further note that the choice of α will affect the minimum MSE which can be achieved.

Another advantage of using the sphere as a test target is that the asymptotic expansion for the scattering function contains the sphere radius as an explicit parameter,[12] viz:

$$H(\omega) = \frac{a}{2}\left\{1 - \frac{3/2}{i\omega(a/c)} + \frac{5/2}{(i\omega)^2(a/c)^2} - \frac{25/4}{(i\omega)^3(a/c)^3} + \frac{22}{(i\omega)^4(a/c)^4} + O\left[\left(\frac{\omega a}{c}\right)^{-5}\right]\right\} \tag{13}$$

The response has been normalized for the distance to the observer and phase-adjusted so that the first observation occurs at time zero. The first item is the contribution due to geometrical optics and the higher order terms are corrections which improve with increasing frequency.[12] By comparing the asymptotic series expansion to the Altes model, we find that

$$a = 2h_{00} \tag{14}$$

That is, the radius is twice the value of the zeroth order coefficient in the asymptotic series. Similarly,

$$a = 5c^2/4h_{-20} \tag{15}$$

The radius of the sphere was estimated for models of order $N = 0, 2, 4$. Recall that when $N = 0$ the generalized scattering center model reduces to the conventional model. Figures 4 and 5 illustrate the results. Figure 4 shows the estimate for the radius using only (14) for the zeroth, second, and fourth-order models. Notice that the zeroth order model produces the worst estimate of the sphere radius for carrier frequencies below 4000 Hz. The second-order model produces the best estimate of the radius. The addition of two more terms in the model actually results in a degradation of the estimate of the spheres radius. The degradation is particularly noticeable when the carrier frequency is larger than 4000 Hz. These results suggest that the "best" model order to use is $N = 2$.

In Figure 5, the radius estimates computed using (15) for the second and fourth-order models are shown. The estimate from the second-order model is reasonably good for carrier frequencies in the range 1000 to 4000 Hz, although (14) produces better results. For the fourth-order model, (15) produces erroneous estimates. By increasing the model order, the filter coefficients h_{n0} for $n = 0, -1$, and -2 are perturbed from the values they had when the model order was $N = 2$. These perturbed coefficients are then used to estimate the radius. The reason the perturbation negatively affects the estimation procedure is that the determinant of the matrix A approaches zero as the model order increases. Hence, the relationship between the filter weights w_{n0} and the observation r is underdetermined.

Our conclusions from the noise-free analysis of the Altes model are as follows: The addition of terms in the parametric model will result in a decrease in the MSE. When implementing this on a computer, however, the MSE will eventually increase due to finite bandwidth. The accuracy of the radius estimate is degraded by the addition of higher-order terms. We conclude that the radius should be estimated from (14).

Some Preliminary Results with Noise

Using the results of the noise-free analysis, we have begun an investigation into the effects of adding noise to the echo from the target. The noisy observation z is modeled as:

$$z(t) = y(t) + N(t) \tag{16}$$

where $N(t)$ is a white, Gaussian noise process. The variance of the noise is defined as σ^2. White noise implies a constant spectral level, which is a reasonable assumption over a small range of frequencies.

We have preliminary results for two different cases: single measurement of the echo; multiple measurement of the echo. We first consider the single measurement of the echo, shown in Figures 6 to 13. In Figures 6 and 7, we plot the average MSE, denoted $\hat{mse}$, and the variance of the MSE, denoted $\hat{var}(mse)$, between the observation and the modeled response for a signal-to-noise (SNR) ratio of 20dB. In Figures 8 and 9, we show the average MSE and the variance for an SNR of 80dB. We refer to the zeroth-order model as the *conventional* model. We note that the addition of two-terms to the conventional model does result in a decrease in the $\hat{mse}$. Further, the $\hat{var}(mse)$ of the second-order model is smaller than that of the conventional model. Notice however, that the conventional model is more robust to the SNR than the second-order model.

In Figures 10 and 11, we plot the average radius estimate, $\hat{a}$, and the variance of the radius estimate, $\hat{var}(a)$, for the conventional and the second-order models with an

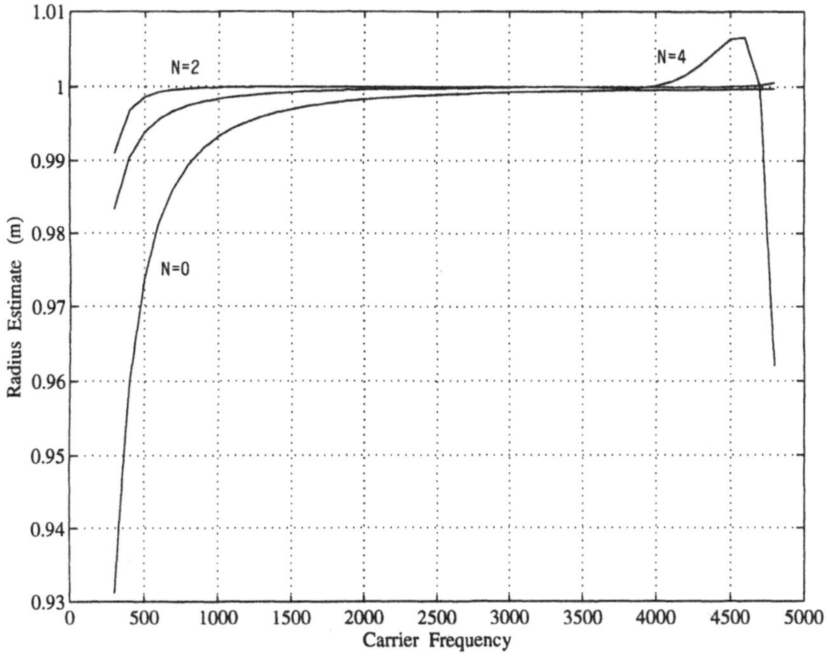

Figure 4. Estimate of the sphere radius using the estimator defined in (14) for model orders $N = 0, 2, 4$. Pulse duration $\alpha = 300$.

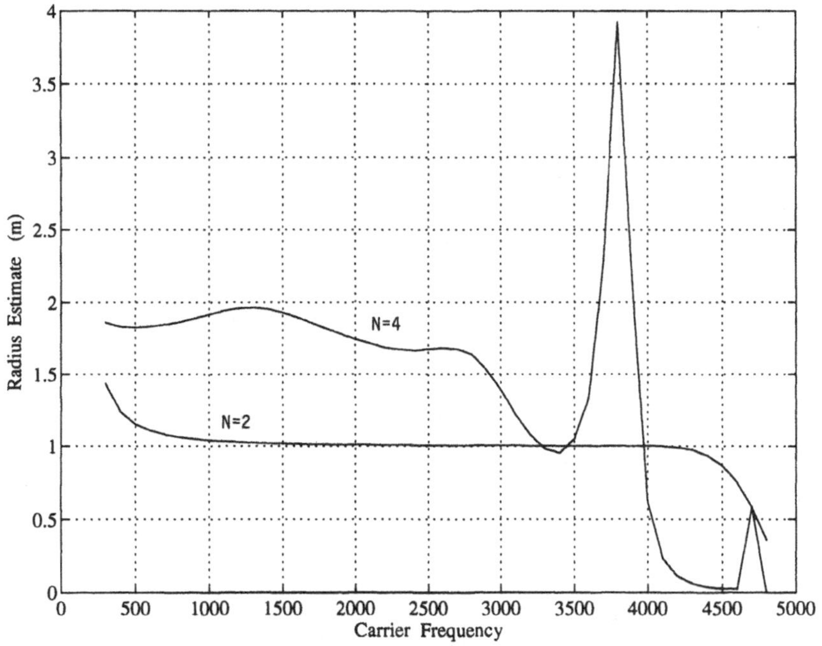

Figure 5. Estimate of the sphere radius using the estimator defined in (15) for model orders $N = 2, 4$. Pulse duration $\alpha = 300$.

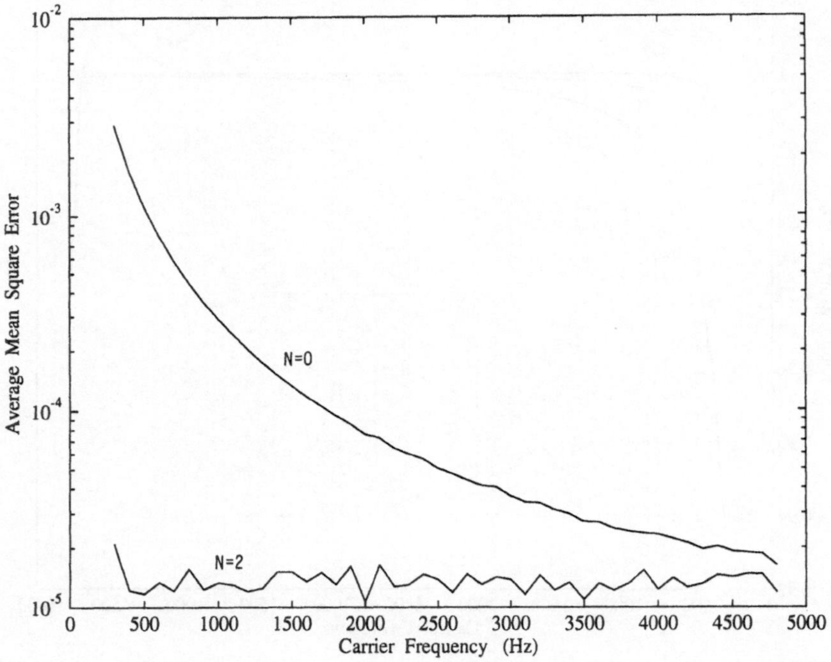

Figure 6. Average MSE between the observation and the model output for model orders $N = 0, 2$. $SNR = 20$.

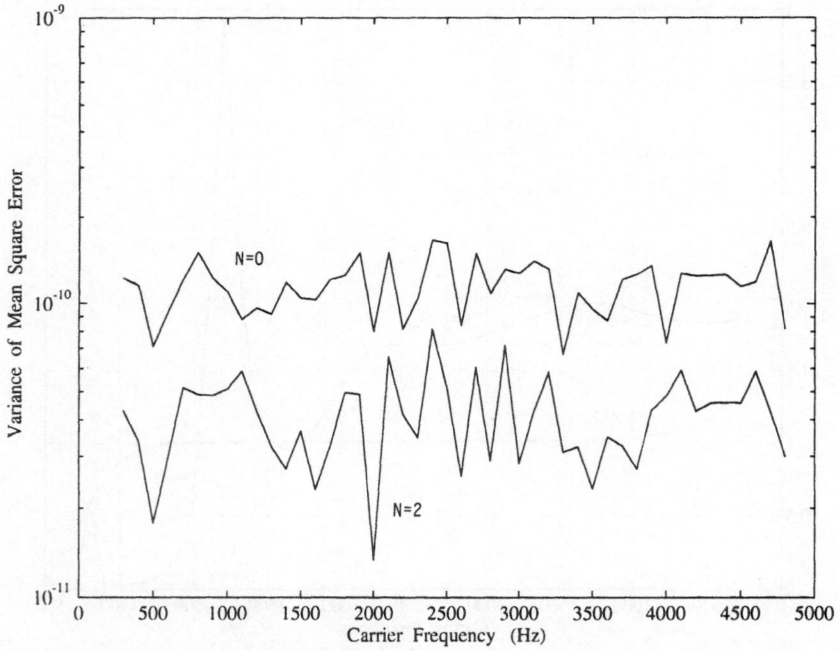

Figure 7. Variance of the MSE between the observation and the model output for model orders $N = 0, 2$. $SNR = 20$.

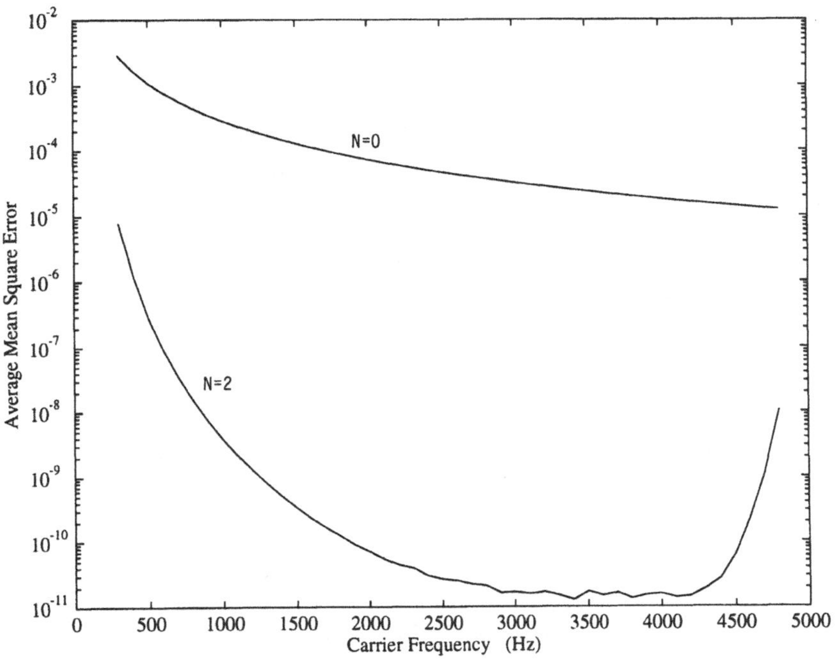

Figure 8. Average MSE between the observation and the model output for model orders $N = 0, 2$. $SNR = 80$.

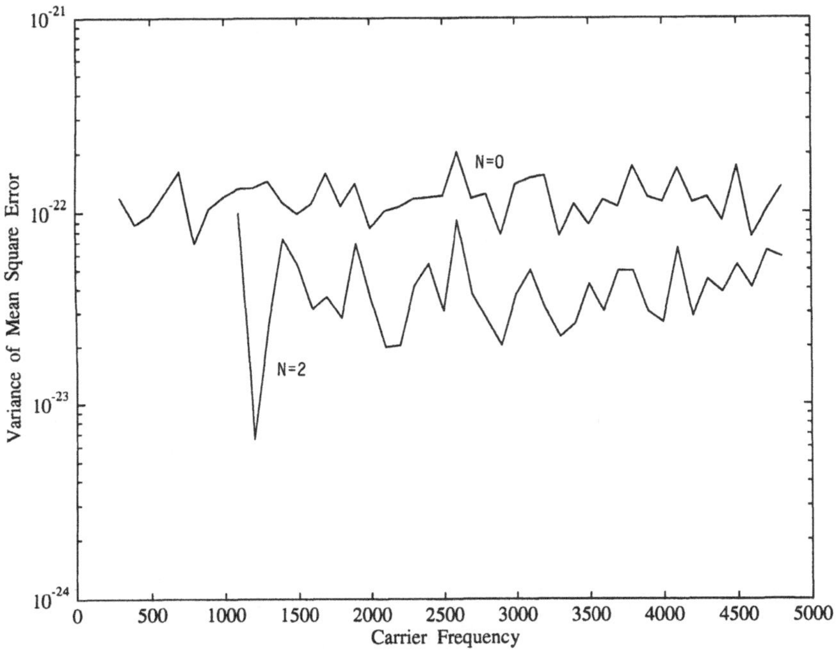

Figure 9. Variance of the MSE between the observation and the model output for model orders $N = 0, 2$. $SNR = 80$.

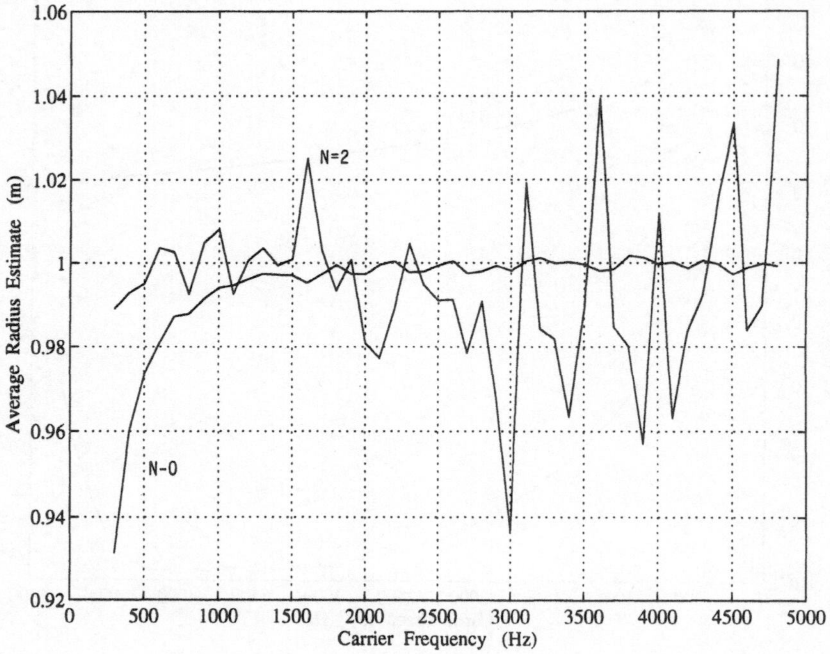

Figure 10. Estimate of the average radius for model orders $N = 0, 2$. $SNR = 20$.

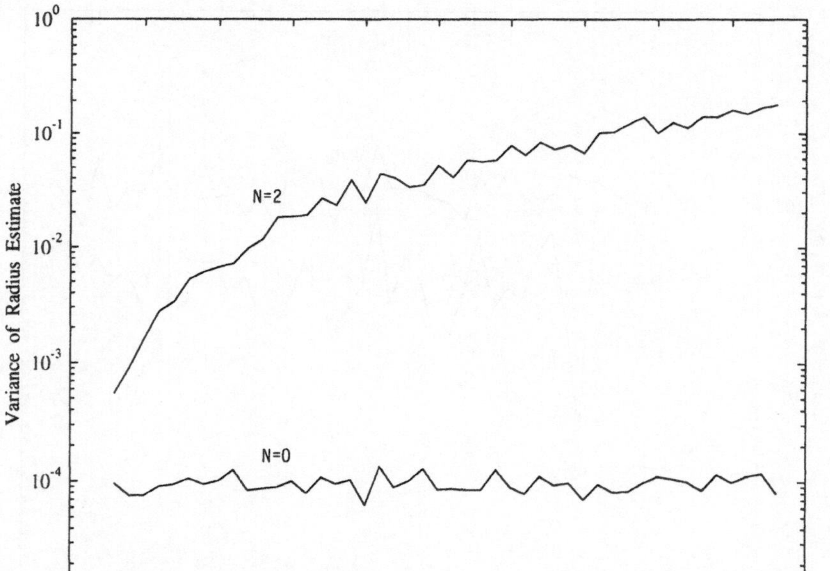

Figure 11. Variance of the radius for model orders $N = 0, 2$. $SNR = 20$.

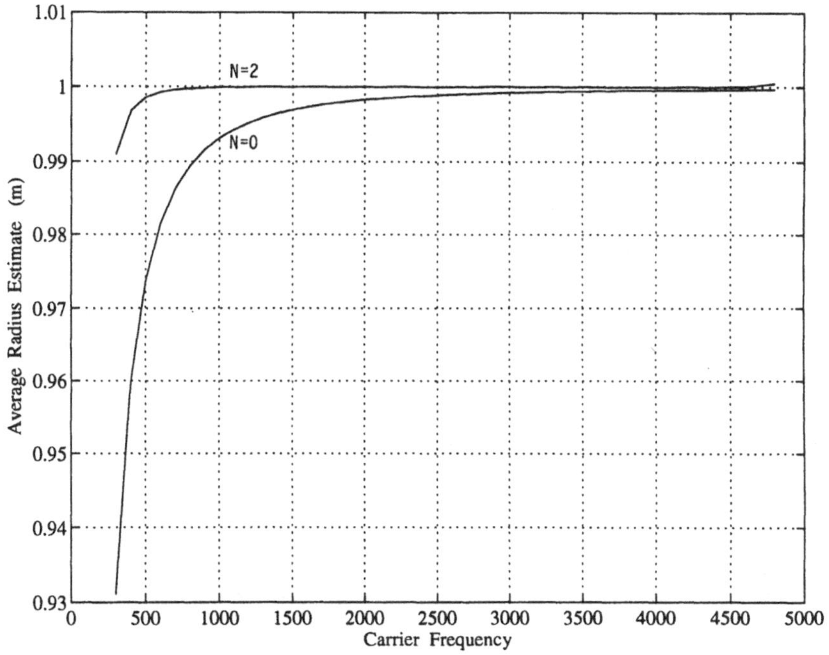

Figure 12. Estimate of the average radius for model orders $N = 0, 2$. $SNR = 80$.

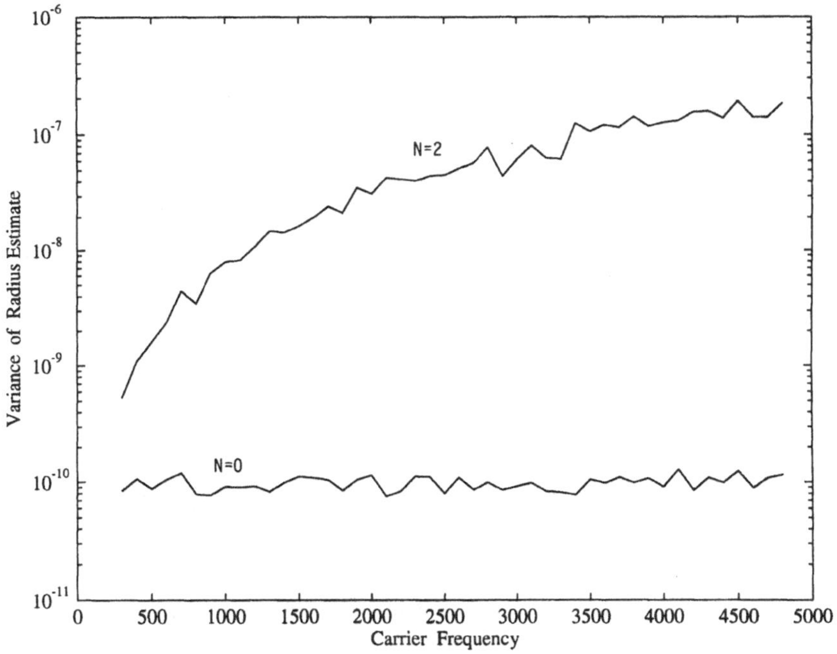

Figure 13. Variance of the radius for model orders $N = 0, 2$. $SNR = 80$.

SNR of 20 db. In Figures 12 and 13, we perform the same plots for an SNR of 80 db. The radius was estimated using (14). At low SNR's, the conventional estimate is very close to the true radius. For the second-order model, we find that the estimate is within 2% of the true value. At high SNR's, the radius estimate from the second-order model is consistently more accurate than that from the conventional model. One disadvantage of the estimate from the second-order model is that it has a high variance, relative to that of the conventional model.

For the conventional model, when multiple measurements of the echo are received and coherently added before estimation of the model parameters or estimating the sphere radius, essentially no improvement is observed. This is consistent with the robustness of the estimate to the SNR which was observed earlier. However, for the second-order model, even with only 10 measurements coherently added, the average MSE decreases substantially. Further, the variance of the radius estimate decreases by about an order of magnitude. Hence, if the receiving system is an array, these preliminary results suggest that the use of the second-order model allows the redundancy to be utilized.

Some Preliminary Results with Time-of-Arrival Estimation

For target identification, one of the most interesting parameters in the scattered signal is the time-of-arrival (TOA) of successive events. These times can often be related to physical parameters of the scattering object.[18] For the hard acoustic sphere, the TOA is determined by the event in the post-reflection part of scatter impulse response at approximately 16 msec (Fig. 1).

We have considered two methods to extract the TOA of the first creeping wave from the observation. The first method examines the magnitude of the envelope of $r_0(t)$ in the post-reflection portion of the observation. (Recall that r_n is a function of time and 0 indicates the correlation of the observation with the 0th-order component of $x(t)$). Preliminary noise free results indicate that the first local maximum occurs at about the same location as the "bump" in the scattering impulse response. The second method examines

$$\sum_{n=-N}^{0} r_n^2(t)$$

Again, the maximum of the envelope of this sum is examined in the post-reflection portion of the observation. Our results indicate that the maximum of the envelope of this sum occurs at about the same place as the "bump" in the scattering impulse response. Figure 14 illustrates the first creeping wave of the observation, the sum-or-squares of r, and the scattering impulse response.

OBJECTIVES IN FUTURE WORK

The results reported herein are the preliminary results from an on-going study of model-based detection methods for ultra-wideband applications. Our specific objectives are as follows:

1. To determine a parametric electromagnetic scattering model, or models, that will allow application of detection and identification techniques.

2. To determine a detection technique, or techniques, that will allow identification of parameters of interest associated with scattering objects of practical interest.

3. To determine robustness of the model and detection in the presence of noise.

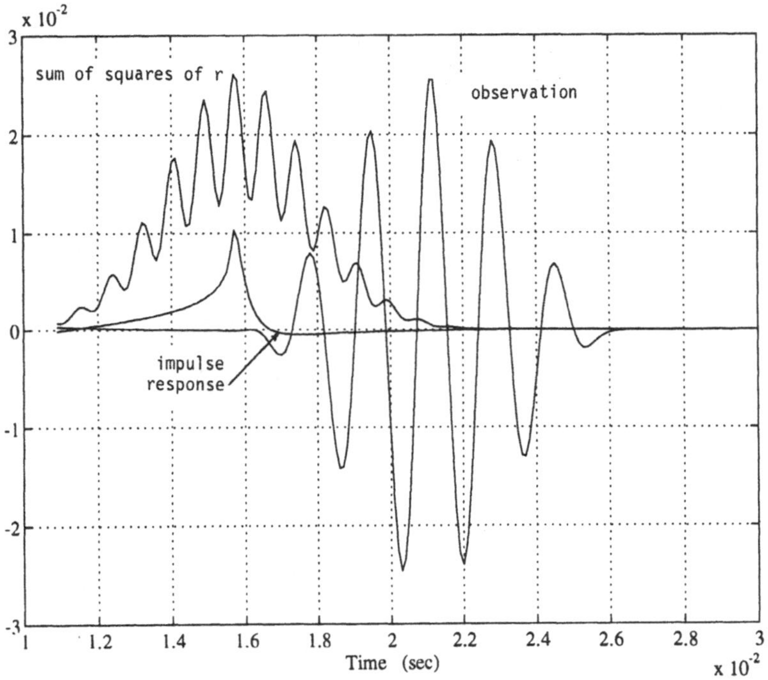

Figure 14. Localization of time-of-arrival (TOA) of the first creeping wave. Curves are of the observation, sums-of-squares of r, and impulse response of scatterer. Noise-free case. Pulse duration $\alpha = 300$.

4. To recommend a *Generalized Scattering Model and Detection Method* leading to partial or complete object reconstruction.

To accomplish our objectives, we are concentrating on several topics: First, we are studying the basic tradeoffs in target identification between spatial degrees of freedom and bandwidth. Included is a study of bandwidth extrapolators, such as the theory of alternating projections. Second, we are considering possible electromagnetic model structures with the goal of selecting a model or models containing parameters of interest associated with canonical scattering objects. Third, we are studying detection and processing methods to extract parameters (or information) of interest in broad-bandwidth signals scattered by canonical scattering objects. Fourth, we are interested in the robustness of model and detection methods in the presence of noise.

REFERENCES

1. Dudley, D.G. (1986), The parametric inverse problem in transient scattering, in *Review of Progress in Quantitative Nondestructive Evaluation*, D.O. Thompson and D.E. Chimenti, Ed., Plenum, New York, 317-326.
2. Lax, P.D. and R.S. Phillips (1967), *Scattering Theory*, Academic Press, New York.
3. Baum, C.E. (1971), On the singularity expansion method for the solution of electromagnetic interaction problems, *Air Force Weapons Laboratory Interaction Notes*, note 88.
4. Marin, L. (1973), Natural mode representation of transient scattered fields, *IEEE Trans. Antennas Propagat.*, vol. AP-21, 809-818.

5. Moffitt, D.L. and R.K. Mains (1974), Detection and discrimination of radar targets, *IEEE Trans. Antennas Propagat.*, vol. AP-23, 358-367.
6. Van Blaricum, M.L. and R. Mittra (1976), A technique for extracting the poles and residues of a system directly from its transient response, *IEEE Trans. Antennas Propagat.*, vol. AP-23, 777-781.
7. Lytle, R.J. and D.L. Lager (1976), Using the natural frequency concept in remote probing of the earth, *Radio Science*, vol. 11, 199-209.
8. Dudley, D.G. (1979), Parametric modeling of transient electromagnetic systems, *Radio Science*, vol. 14, 387-396.
9. Smith, C.R. and P.M. Goggins (1992), Radar target identification, Technical Report, Advanced Sensors Directorate, Research, Development and Engineering Center, U.S. Army Missile Command, Redstone Arsenal, AL 35898-5253, to appear (invited) in *IEEE Antennas and Propagation Society Magazine.*
10. Dudley, D.G. (1988), Progress in identification of electromagnetic systems, Feature Article, *IEEE Antennas and Propagat. Society Newsletter*, August issue, 3-11.
11. Dudley, D.G. (1985), A state-space formulation of transient electromagnetic scattering, *IEEE Trans. Antennas Propagat.*, vol. AP-33, 1127-1130.
12. Weyker, R.R. and D.G. Dudley (1987), Asymptotic model-based identification of a hard acoustic sphere, *Wave Motion*, vol. 9, 77-97.
13. Van Trees, H.L. (1968), *Detection, Estimation, and Modulation Theory, Part I*, John Wiley and Sons, New York.
14. Helstrom, C. W. (1968), *Statistical Theory of Signal Detection*, Pergamon Press, Oxford.
15. Thomas, J.B. (1969), *An Introduction to Statistical Communication Theory*, Wiley, New York.
16. Nabulsi, K.A. and D.G. Dudley (1992), A new approximation and a new measurable constraint for slab profile inversion, to appear, *IEEE Trans. Geosciences and Remote Sensing.*
17. Altes, R.A. (1976), Sonar for generalized target description and its similarity to animal echolocation systems, *J. Acoust. Soc. Am.*, vol. 59(1), 97-105.
18. Kennaugh, E.M. and D.L. Moffatt (1965), Transient and impulse response approximations, *IEEE Proceedings*, vol. 53, 893-901.
19. Freedman, A. (1962), A mechanism of acoustic echo formation, *Acoustica*, vol. 12, 10-21.
20. Flandrin, P., F. Magand, and M. Zakharia (1990) Generalized target description and wavelet decomposition, *IEEE Trans. Acoustics, Speech, and Signal Processing*, vol. 38(2), 350-352.
21. Altes, R.A. (1980), Detection, estimation, and classification with spectrograms, *J. Acoust. Soc. Am.*, vol. 67(4), 1232-1246.
22. Middleton, D.M. and R. Esposito (1968), Simultaneous optimum detection and estimation of signals in noise, *IEEE Trans. Infor. Theory*, vol. 14, 434-444.
23. Weyker, R.R. and D.G. Dudley (1987), Identification of resonances of a hard acoustic sphere, *IEEE Journal of Oceanic Engineering*, vol. OE-12, 317-326.

RADAR TARGET IDENTIFICATION AND DETECTION USING SHORT EM PULSES AND THE E-PULSE TECHNIQUE

E. Rothwell, K. M. Chen, D. P. Nyquist, P. Ilavarasan, J. Ross, R. Bebermeyer, and Q. Li

Department of Electrical Engineering
Michigan State University
East Lansing, MI 48824

INTRODUCTION

The E-pulse radar target discrimination scheme, a resonance cancellation technique based on the late-time behavior of the transient scattered field, has been successfully demonstrated[1,2,3,4] in the laboratory on a variety of occasions. The technique is based on the target natural frequencies and is inherently aspect-independent. Unfortunately, this approach ignores the early-time scattered field component, which is dominated by specular reflections from target scattering centers.

This paper presents an E-pulse technique which is usable with waveforms arising from specular scattering, such as the early-time response of a radar target. The technique uses resonance cancellation in the *frequency domain* to eliminate the sinusoidal functions arising from the aspect-dependent temporal positions of specular reflections. Target discrimination using early-time information is then possible using an algorithm identical to that used with late-time data, with the exception that discrimination is aspect dependent.

Cancellation of frequency-domain sinusoids is possible for any waveform which is specular in nature. This paper also shows how to enhance the detection of radar targets using transient pulses by eliminating clutter from specular structures such as the sea surface.

MODELLING OF ULTRA-WIDEBAND SCATTERING FROM RADAR TARGETS

Figure 1 shows the transient response of a 1:72 scale model of a B-52 aircraft to an excitation pulse with energy in the band 0.2-7.0 Ghz. It is typical of the return from an aircraft target, showing an early-time period (3.0-7.0 ns) dominated by localized specular reflections from scattering centers such as engine intakes and attachment points, followed by a late-time natural oscillation period consisting of global resonance information ($t > 7.0$

Ultra-Wideband, Short-Pulse Electromagnetics
Edited by H. Bertoni *et al.*, Plenum Press, 1993

ns). It is important to note that there can be no precise demarcation between the early-time and late-time portions of a scattered field response since substructure resonances are often established before the excitation signal clears the target, resulting in a resonance component to early time.

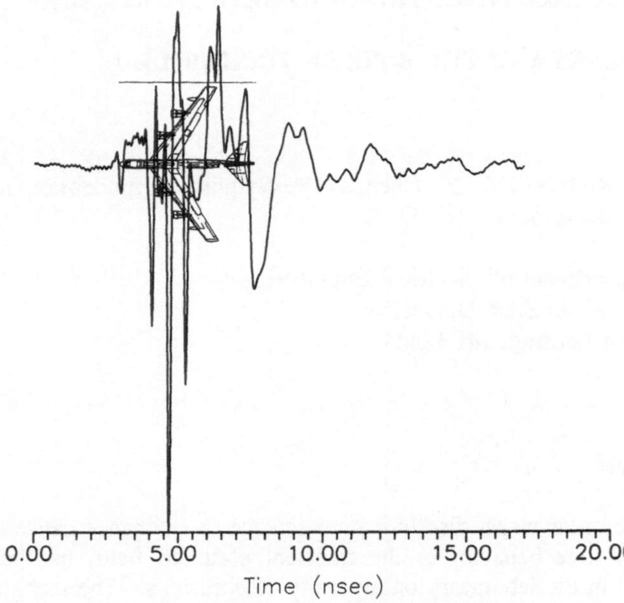

Figure 1. Response of 1:72 scale B-52 model measured at nose-on incidence in the frequency band 0.2-7.0 GHz. Polarization is in plane of wings.

Baum[5] has proposed a model of the late-time response using the aspect-independent natural resonance frequencies of the target $\{s_n=\sigma_n+j\omega_n\}$

$$r_L(t) = \sum_{n=-N}^{N} A_n e^{s_n t} \qquad t>T_L. \tag{1}$$

Here T_L designates the beginning of late time, N modes are assumed excited by the incident pulse, and the aspect-dependent complex amplitudes $\{A_n\}$, along with the natural frequencies, occur in complex conjugate pairs. This response has formed the basis for the aspect-independent late-time E-pulse technique.

Altes[6] has proposed a simple model for the early-time response

$$r_E(t) = \sum_{m=1}^{M} p(t)*h_m(t-T_m) \qquad t<T_L \tag{2}$$

where p(t) is the incident pulse and $h_m(t)$ is the localized impulse response originating at the mth scattering center at the time T_m. In the frequency domain this response becomes

$$R_E(\omega) = \mathscr{F}\{r_E(t)\} = \sum_{m=1}^{M} P(\omega) H_m(\omega) e^{-j\omega T_m} \tag{3}$$

where $H_m(\omega)$ is the transfer function of the mth scattering center and $P(\omega)$ is the spectrum of p(t). Hurst and Mittra[7] suggest that the transfer function can be approximated as an exponential function of frequency. Assuming that $P(\omega)$ is slowly varying then gives

$$R_E(\omega) = \sum_{m=1}^{M} B_m e^{\tau_m \omega} \tag{4}$$

where $\{\tau_m = \alpha_m - jT_m\}$ are complex times associated with the scattering center impulse responses. It is readily seen that there is a duality between the temporal late-time response (1) and the spectral early-time response (4). This duality allows the direct application of E-pulse cancellation to spectral early-time data.

THE E-PULSE TECHNIQUE FOR GENERAL EXPONENTIAL SIGNALS

The E-pulse technique has been formulated in detail for the case of late-time responses[1,2,3]. It is useful to review the basic premise of the technique in the context of a general complex exponential signal.

Consider a complex signal

$$f(x) = \sum_{k=1}^{K} C_k e^{Q_k x} \qquad X_L < x < X_F \tag{5}$$

where $\{C_k\}$ and $\{Q_k\}$ are complex numbers. An E-pulse e(x) is a real waveform of finite extent X_E which upon convolution with f(x) eliminates a preselected component of the exponential series. In the particular, the entire series can be eliminated resulting in

$$c(x) = e(x) * f(x) = \int_0^{X_E} f(x') e(x-x') dx' = 0 \qquad X_L + X_E < x < X_F . \tag{6}$$

The conditions for synthesizing such an E-pulse can be given in the context of resonance cancellation as[3]

$$E(s=Q_k) = E(s=Q_k^*) = 0 \qquad 1 \le k \le K \tag{7}$$

where E(s) is the Laplace spectrum of e(x).

Discrimination between waveforms having differing sets of complex frequencies $\{Q_n\}$ is accomplished by creating E-pulses for each frequency set. Upon convolution with an unknown waveform, the output with zero energy for $X_L + X_E < x < X_F$ identifies the frequency set. Application to the late-time response of a radar target results in an aspect-independent E-pulse waveform since the complex frequencies comprising the signal are the natural frequencies of the target. Application to the transformed early-time response of a radar target produces an aspect-dependent E-pulse since the frequencies are related to the temporal positions of the specular reflections.

To quantify discrimination an "E-pulse discrimination number" (EDN) is defined as

$$EDN = \frac{\int_{X_L+X_E}^{X_F} [c(x)]^2 dx}{\int_0^{X_E} [e(x)]^2 dx} . \tag{8}$$

In an ideal, noise-free situation the E-pulse convolution with zero EDN identifies the waveform. A more realistic scenario suggests that the convolution with the minimum value of EDN should identify the waveform. A measure of confidence is then given by the "E-pulse discrimination ratio" (EDR) defined as

$$EDR = 10\log_{10}\left\{\frac{EDN}{\min(EDN)}\right\} \quad dB \tag{9}$$

so that the identified waveform has an EDR of 0 dB, while the other values are greater than 0 dB.

DEMONSTRATION OF EARLY-TIME E-PULSE DISCRIMINATION

To demonstrate the feasibility of using E-pulses for early-time target discrimination, the scattered field response of 1:72 B-52 and 1:48 B-58 scale aircraft models have been measured using an HP 8720B network analyzer. (See Ross[8] for details on the MSU transient measurement system and calibration technique). Restrictions imposed by the anechoic chamber and the transmit/receive antenna system limit the lowest measurement frequency to 0.4 GHz. Unfortunately, this excludes the dominant resonances of larger targets. To overcome this problem, measurements were made of smaller 1:144 B-52 and 1:96 B-58 scale models over 0.4-4.4 GHz, and of larger 1:72 B-52 and 1:48 B-58 scale models over 1.0-7.0 GHz. The results were then scaled and combined to achieve an equivalent measurement of the larger targets with a bandwidth of 0.2-7.0 GHz. Figure 2 shows the time-domain pulse responses of the B-58 as a function of aspect angle, measured from nose-on (0°), as the model is rotated about its geometrical center in 5° increments, obtained by applying a cosine taper weighting function to the measured data and inverse-transforming. Note how the specular reflections rotate with the aircraft, and how the late-time period is aspect dependent, due to the changing coupling into the various natural modes. The rapid variation with aspect angle suggests that 5° increments are probably not fine enough to adequately describe the aspect dependence of early time.

E-pulse waveforms have been constructed for the early-time component of each measured waveform by truncating the time-domain waveform after the two-way transit time of the target along the aspect angle, transforming the result into the frequency domain, performing mode extraction[3] and applying the synthesis equations (7). A discrimination test is then performed by assuming one waveform arises from an unknown target of unknown aspect angle and convolving that waveform with each of the E-pulses. The smallest EDN value then identifies the target and its aspect angle with confidence given by the EDR.

Figure 3 shows the resulting values of EDR obtained from each of the E-pulse convolutions. Response and E-pulse numbers 1-7 represent the B-58 at aspects of 0° to 30° in 5° increments, and numbers 8-14 represent the B-52 at aspects 0° through 30°. Note that EDR values greater than 15 dB have been truncated for display purposes. It is easily seen that the correct waveform has been identified in all cases, with usually a high level of

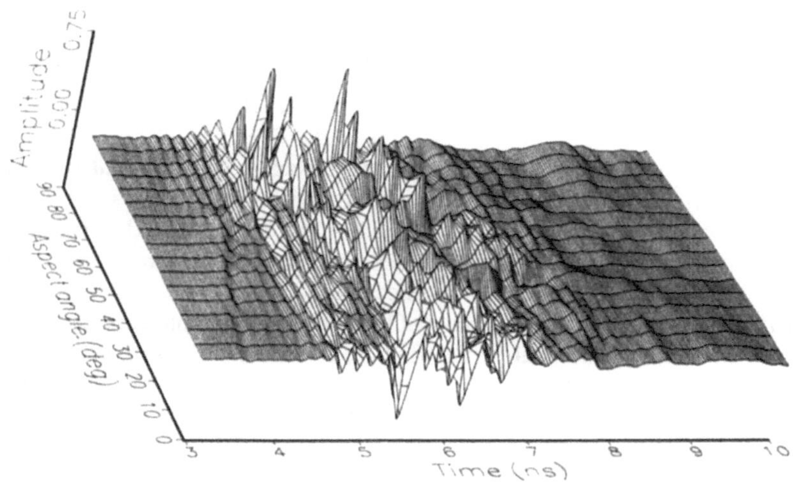

Figure 2. Response of 1:48 scale B-58 model measured at various incidence angles in the frequency band 0.2-7.0 GHz. Polarization is in plane of wings.

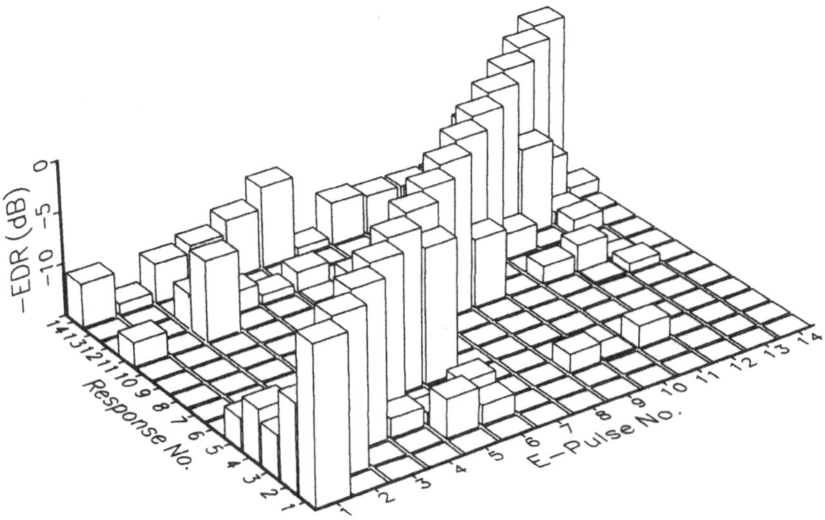

Figure 3. Early-time discrimination of B-52 and B-58 at aspects of 0° to 30° from nose-on.

confidence. Especially important is that any waveform from the B-52 is discriminated from any B-58 waveform with high confidence, and vice-versa. However, the fact that convolution of the 30° B-52 E-pulse with the 25° B-52 response is highly different than its convolution with the 30° B-52 response suggests that a much finer discretization on aspect angle is required. That is, a B-52 response at 28° might not be associated with the B-52 unless E-pulses are created for finer increments in aspect angle.

ENHANCEMENT OF TARGET DETECTION BY REDUCTION OF SEA CLUTTER

The detection of radar targets near the sea surface using transient signals is made difficult by the presence of a strong clutter return from a disturbed sea. However, if the scattering from water wave crests is primarily specular within the band of the interrogating signal, the E-pulse resonance cancellation technique can be used to eliminate the clutter return, thus increasing the probability of detection.

Assume that the sea surface consists of wave crests of nonuniform heights separated by the water wavelength λ_w. If the scattering from these wave crests is nearly specular, the transient back-scattered electric field response can be approximated precisely as in (2). At near-grazing incidence, the time between specular reflections is approximately

$$T_m = \frac{2m\lambda_w \cos(\theta_0)}{c} \tag{10}$$

where θ_0 is the incidence angle measured from grazing incidence, and M in (2) is the number of wave crest reflections within the time window of interest. Because of the form of the scattered field response, E-pulse cancellation can be used to eliminate the sea clutter response over part of its frequency band.

Figure 4 shows the time domain response of a simulated sea surface measured in the

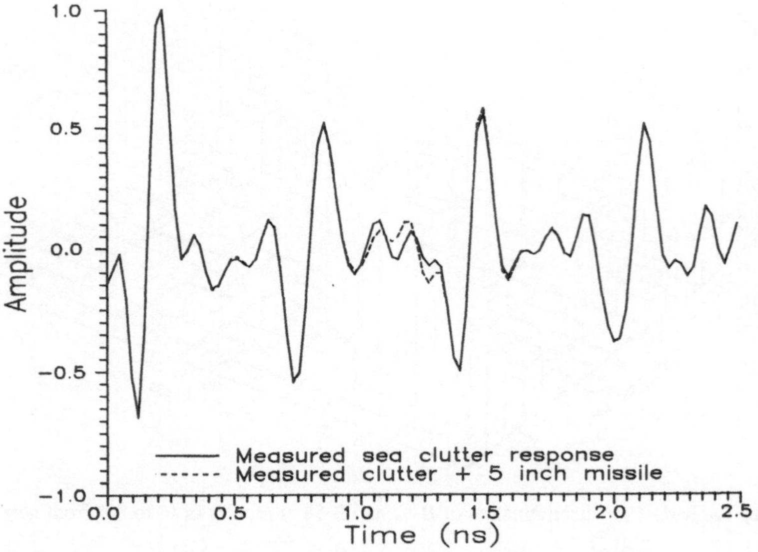

Figure 4. Measured response of simulated sea surface with and without addition of 5 inch missile response.

frequency band 1.0-7.0 GHz. The surface has been constructed from an aluminum sheet of width 12" adhered to styrofoam in a sinusoidal pattern, with six wave crests of height 1" and spatial wavelength 4". The measurement has been done with horizontal polarization at an incidence angle of 17° from grazing, with the reflections from the first four wave crests shown. Specular reflections from the wave crests are quite apparent, separated by about 0.65 ns, as predicted by (10).

To eliminate the sea clutter, an E-pulse of width 3.9 GHz has been constructed from the transform of the sea clutter response. Convolution of the E-pulse with the real part of the transform domain sea clutter, Figure 5, reveals that after a frequency of 3.9 GHz the convolved output is nearly zero, and the sea clutter above that frequency has been eliminated.

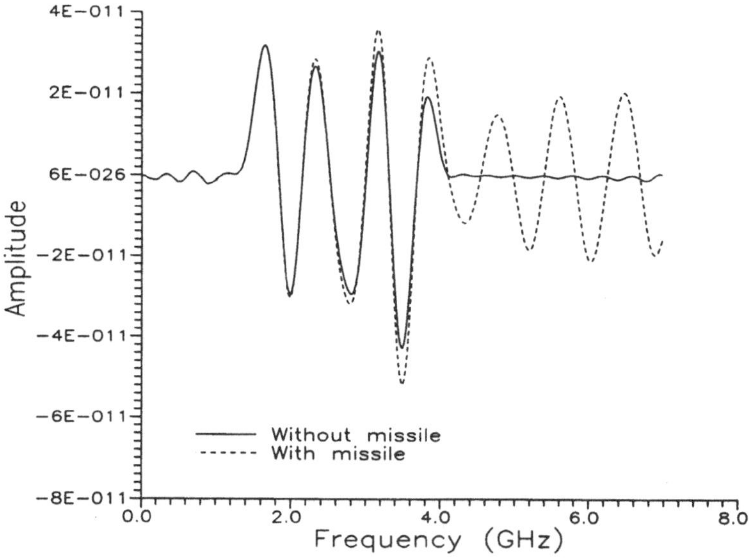

Figure 5. Convolution of frequency-domain E-pulse with transform of measured return from simulated sea surface, with and without addition of missile response.

As an example of target detection in clutter, the measured response of a 5" long aluminum missile model has been added to the sea clutter data, as shown in Figure 4, with a peak amplitude chosen to be 10% of the peak clutter amplitude. The convolution of the E-pulse with the target+clutter response, as shown in Figure 5, reveals that the missile response is still present. When the real and imaginary convolved outputs are windowed from 4.0-7.0 GHz and inverse transformed into the time domain a distorted, bandlimited version of the missile response is recovered, as shown in Figure 6. Thus, the missile can be easily detected even with the large sea clutter component present.

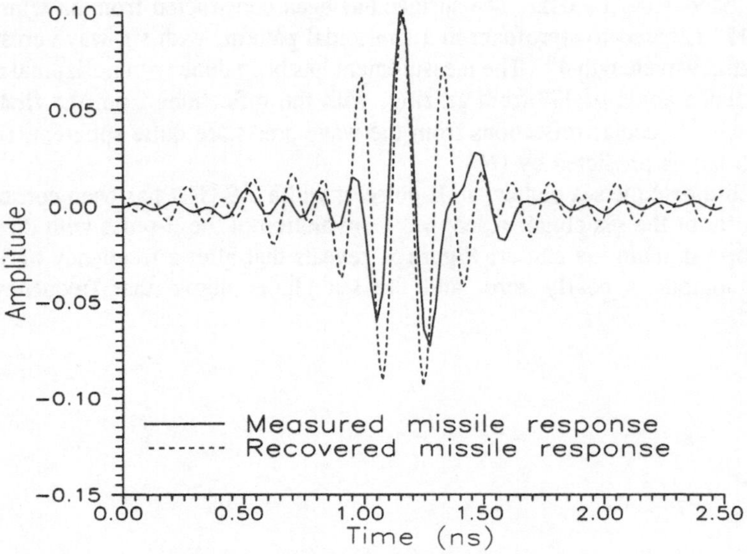

Figure 6. Response of 5 inch missile recovered from frequency-domain convolution.

REFERENCES

1. E.J. Rothwell, D.P. Nyquist, K.M. Chen and B. Drachman, Radar target discrimination using the extinction-pulse technique, *IEEE Trans. Antennas Propagat.* AP-33:929 (1985).
2. K.M. Chen, D.P. Nyquist, E.J. Rothwell, L. Webb and B. Drachman, Radar target discrimination by convolution of radar return with extinction-pulses and single-mode extraction signals, *IEEE Trans. Antennas Propagat.* AP-34:896 (1986).
3. C.E. Baum, E.J. Rothwell, K.M. Chen and D.P. Nyquist, The singularity expansion method and its application to target identification, *Proc. IEEE.* 79:1481 (1991).
4. K.M. Chen, D.P. Nyquist, E.J. Rothwell and W.M. Sun, New progress on E/S pulse techniques for noncooperative target recognition, *IEEE Trans. Antennas Popagat.* 40:829 (1992).
5. C.E. Baum, Emerging technology for transient and broad-band analysis and synthesis of antennas and scatterers, *Proc. IEEE.* 64:1598 (1976).
6. R.A. Altes, Sonar for generalized target description and its similarity to animal echolocation systems, *J. Acoust. Soc. Am.* 59:97 (1976).
7. M.P. Hurst and R. Mittra, Scattering center analysis via Prony's method, *IEEE Trans. Antennas Propagat.* AP-35:986 (1987).
8. J.E. Ross. "Application of Transient Electromagnetic Fields to Radar Target Discrimination," Ph.D. Dissertation, Michigan State University (1992).

STABLE POLE EXTRACTION FROM SCATTERING DATA*

S. Unnikrishna Pillai and Theodore I. Shim

Electrical Engineering Department
Polytechnic University
Five MetroTech Center
Brooklyn, New York 11201

INTRODUCTION

One way to express the input-output relations of a scattering system is in terms of its transfer function relating the incident and reflected waveforms. Due to the innumberable modes and complexities involved in such a phenomenon, the equivalent system can be at best represented by a nonrational model. The transfer function of such a nonrational system, unlike the rational systems, cannot be expressed as the ratio of two polynomials (of finite degree). Nonrational systems occur in practice in a variety of contexts such as in distributed structures, partial differential equations, systems regulated by differential equations with delays and scattering experiments. For example, e^{-z}, $(1-z)ln\,(1-z)$, $z/(e^z-1)$ all represent nonrational transfer functions, and in any rearrangement their numerators and/or denominators will contain an infinite number of terms. Thus, in general

$$H(z) = \frac{\sum_{k=0}^{\infty} \beta_k z^k}{\sum_{k=0}^{\infty} \alpha_k z^k} = \sum_{k=0}^{\infty} b_k z^k \tag{1}$$

where the power series is valid in some contour in the complex z-plane. If the system is known to be stable, for example, then $\sum_{k=0}^{\infty} |b_k| < \infty$, and the above power series is analytic in $|z| \leq 1$. The input-output relation of nonrational systems also can be expressed in terms of the infinite recursion

$$x(n) = -\sum_{k=1}^{\infty} \alpha_k x(n-k) + \sum_{k=0}^{\infty} \beta_k w(n-k) = \sum_{k=0}^{\infty} b_k w(n-k) \tag{2}$$

The problem now is how does one simulate such a system?

Obviously every term in (2) *cannot* be faithfully simulated in a practical setup and the straightforward choice of truncating the equivalent above infinite series in (1) is not

*This work was supported by the National Science Foundation under contract MIP–9020501 and the Office of Naval Research under contract N–00014–89–J–1512.

to be recommended for several reasons. First, truncation leads to severe round-off error. Even though approximating a nonrational function by truncating its power series (polynomial approximation) is not attractive, nevertheless, it may be possible to realize the same approximation by making use of a "simpler" rational function. This is, of course, the idea behind Padé approximations [1, 2], where the rational function

$$H_r(z) = \frac{\beta_0 + \beta_1 z + \cdots + \beta_q z^q}{\alpha_0 + \alpha_1 z + \cdots + \alpha_p z^p} = \frac{B(z)}{A(z)} \tag{3}$$

is said to be a Padé approximation to $H(z) = \sum_{k=0}^{\infty} b_k z^k$ in (1) if at least the first $p+q+1$ terms in the power series expansion of $H(z)$ and $B(z)/A(z)$ match exactly, *i.e.*,

$$H(z) - \frac{B(z)}{A(z)} = O(z^{p+q+1}). \tag{4}$$

Although Padé approximations are unique, unfortunately, since such approximations only involve a finite number of terms (b_0, b_1, ..., b_{p+q}) of the actual impulse response in determining the rational approximation, the crucial asymptotic characteristics of the original nonrational system can never be faithfully represented by this approach. Furthermore, such approximations need not be stable. For example, the AR(1) Padé approximation to the minimum phase function $ln\,(1+\epsilon+z)$ is not stable, if $0 < \epsilon < 0.75$ [4].

Following the approach developed in [3]–[5] for the rational case, our solution to this problem tries to impose Padé-like constraints on the magnitude square of the transfer function rather than on the transfer function itself by making use of a key result developed in [6].

PADÉ-LIKE APPROXIMATION

To be specific, let

$$S(\theta) = |H(e^{j\theta})|^2 = \sum_{k=-\infty}^{\infty} r_k e^{jk\theta} \geq 0 \tag{5}$$

represent the power spectral density $S(\theta)$ and the autocorrelations $\{r_k\}_{k=-\infty}^{+\infty}$ of the given nonrational function. Clearly,

$$r_k = 1/2\pi \int_{-\pi}^{\pi} S(\theta) e^{-jk\theta}\, d\theta. \tag{6}$$

Then a rational system as in (3) is said to be a Padé-like approximation to the above nonrational system if

$$|H_r(e^{j\theta})|^2 = \left| \frac{\beta_0 + \beta_1 e^{j\theta} + \cdots + \beta_q e^{jq\theta}}{1 + \alpha_1 e^{j\theta} + \cdots + \alpha_p e^{jp\theta}} \right|^2 = \sum_{k=-n}^{n} r_k e^{jk\theta} + O(e^{j(n+1)\theta}) \tag{7}$$

where $n \geq p+q$, *i.e.*, the Fourier coefficients of the ARMA(p,q) rational approximation must match with the Fourier coefficients of the given nonrational system *at least* up to the first $p+q+1$ terms.

More generally, any power spectral density $S(\theta) \geq 0$ that satisfies the integrability condition

$$\int_{-\pi}^{\pi} S(\theta) d\theta < \infty$$

and the Paley-Wiener Criterion

$$\int_{-\pi}^{\pi} ln\, S(\theta) d\theta > -\infty$$

can be factorized in terms of its minimum phase Wiener factor [7]

$$H(z) = \sum_{k=0}^{\infty} b_k z^k$$

that is analytic together with its inverse in $|z| < 1$, such that

$$S(\theta) = |H(e^{j\theta})|^2 = \sum_{k=-\infty}^{\infty} r_k e^{jk\theta} \,. \tag{8}$$

To make further progress in solving the problem posed in (7), it is best to make use of the class of all spectral extension formula developed in [6]. Given a set of autocorrelations $r_0, r_1, \cdots r_n$ the class of all power spectral density functions $K(\theta)$ that interpolate these autocorrelations can be parametrized by an arbitrary bounded function[1] $\rho(z)$, and, moreover, the associated Wiener factors are given by (up to multiplication by a constant of unit magnitude) [6]

$$K(\theta) = |H_\rho(e^{j\theta})|^2 \tag{9}$$

where

$$H_\rho(z) = \frac{\Gamma(z)}{A_n(z) - z\rho(z)\tilde{A}_n(z)} \,. \tag{10}$$

Here $\Gamma(z)$ represents the minimum phase factor associated with the spectral factorization

$$|\Gamma(e^{j\theta})|^2 = 1 - |\rho(e^{j\theta})|^2 \tag{11}$$

that is guaranteed to exist under the causality criterion [6]

$$\frac{1}{2\pi}\int_{-\pi}^{\pi} ln\,(1 - |\rho(e^{j\theta})|^2)d\theta > -\infty \,.$$

Further, $A_n(z)$ in (10) represents the Levinson polynomial at stage n generated from $r_0 \to r_n$ and $\tilde{A}_n(z) \triangleq z^n A_n^*(1/z^*)$. These Levinson polynomials can be recursively computed using

$$\sqrt{1 - |s_k|^2}\, A_k(z) = A_{k-1}(z) - z s_k \tilde{A}_{k-1}(z) \,, \tag{12}$$

that starts with $A_0(z) = 1/\sqrt{r_0}$, and the reflection coefficients s_k in (12) satisfy the relation

$$s_k = \left\{ A_{k-1}(z) \sum_{i=1}^{k} r_i z^i \right\}_k A_{k-1}(0) \,, \tag{13}$$

where $\{\ \}_k$ represents the coefficient of z^k in $\{\ \}$. Clearly, if $K(\theta)$ in (9) is nonrational, then it must follow from (11) for a specific choice of nonrational $\rho(z)$.

More interestingly, if an ARMA(p, q) rational function approximation as in (7) holds for some $H_r(z)$, then from the above argument, that too must follow from (9)–(10) for a specific *rational* $\rho(z)$ with $n \geq p + q$ [3]–[5]. In that case, since $\delta(\Gamma(z)) = q$, from (11), $\delta(\rho(z)) = q$. Let

$$\rho(z) = \frac{h(z)}{g(z)} \tag{14}$$

where $h(z)$ and $g(z)$ are polynomials of degree atmost equal to q. Then (10) simplifies into

$$H_r(z) = \frac{\psi(z)}{A_n(z)g(z) - zh(z)\tilde{A}_n(z)} = \frac{\psi(z)}{D(z)} \tag{15}$$

[1] A function $\rho(z)$ is said to be bounded if (i) it is analytic in $|z| < 1$ and (ii) $|\rho(z)| \leq 1$ in $|z| < 1$. Thus z, $(1 + 2z)/(2 + z)$, $e^{(z-1)}$ are all bounded functions.

where $\psi(z)$ represents the minimum phase polynomial resulting from the factorization

$$\psi(z)\psi_*(z) = g(z)g_*(z) - h(z)h_*(z) \tag{16}$$

and

$$D(z) = A_n(z)g(z) - zh(z)\tilde{A}_n(z)\,, \quad n \geq p+q\,. \tag{17}$$

However, for an ARMA(p,q) rational approximation to hold $\delta(D(z)) = p$ and since $\delta(\tilde{A}_n(z)) = n \geq p+q$, from (17), we must have $\delta(h(z)) \leq q-1$. As a result, $\delta(g(z)) = q$ and the bounded function $\rho(z)$ in (14) must have the representation

$$\rho(z) = \frac{h_0 + h_1 z + \cdots + h_{q-1}z^{q-1}}{1 + g_1 z + \cdots + g_{q-1}z^{q-1} + g_q z^q}\,. \tag{18}$$

The formal degree of $D(z)$ in (15) and (18) is still $n+q$, and to respect the ARMA(p,q) nature of $H_r(z)$, the coefficients of the higher order terms beyond z^p must be zeros there. This results in $n+q-p$ linear equations in $2q$ unknowns and since $n \geq p+q$, these equations are at least $2q$ in number. Clearly, the minimum number of these equations is obtained for $n = p+q$ and in that case the unkonwns, g_k, $k = 1 \to q$ and h_k, $k = 0 \to q-1$ can be obtained by solving the matrix equation [3]–[5]

$$\mathbf{Ax} = \mathbf{b} \tag{19}$$

where $\mathbf{A}$ is a $2q \times 2q$ matrix given by

$$\left[\begin{array}{cccc|cccc}
a_p & a_{p-1} & \cdots & a_{p-q+1} & -a_q^* & -a_{q+1}^* & \cdots & -a_{2q-1}^* \\
a_{p+1} & a_p & \cdots & a_{p-q+2} & -a_{q-1}^* & -a_q^* & \cdots & -a_{2q-2}^* \\
\vdots & \vdots & \cdots & \vdots & \vdots & \vdots & \cdots & \vdots \\
a_{p+q-1} & a_{p+q-2} & \cdots & a_p & -a_1^* & -a_2^* & \cdots & -a_q^* \\
\hline
a_{p+q} & a_{p+q-1} & \cdots & a_{p+1} & -a_0^* & -a_1^* & \cdots & -a_{q-1}^* \\
0 & a_{p+q} & \cdots & a_{p+2} & 0 & -a_0^* & \cdots & -a_{q-2}^* \\
0 & 0 & \cdots & a_{p+3} & 0 & 0 & \cdots & -a_{q-3}^* \\
\vdots & \vdots & \cdots & \vdots & \vdots & \vdots & \cdots & \vdots \\
0 & 0 & \cdots & a_{p+q-1} & 0 & 0 & \cdots & -a_1^* \\
0 & 0 & \cdots & a_{p+q} & 0 & 0 & \cdots & -a_0^*
\end{array}\right], \tag{20}$$

$$\mathbf{x} \triangleq [g_1\ g_2\ \cdots\ g_q\ h_0\ h_1\ \cdots\ h_{q-1}]^T\,,$$

and

$$\mathbf{b} \triangleq -[a_{p+1}\ a_{p+2}\ \cdots\ a_{p+q}\ 0\ 0\ \cdots\ 0]^T\,.$$

Here, a_k's denote the coefficients of the Levinson polynomial $A_{p+q}(z)$ in (12), i.e.,

$$A_{p+q}(z) = a_0 + a_1 z + a_2 z^2 + \cdots + a_{p+q}z^{p+q}\,.$$

However, (19) need not have a solution for every p, q. Even if there exists a solution in some cases, $g(z)$ so obtained need not be free of zeros in $|z| \leq 1$ and further $\rho(z)$ need not turn out to be a bounded function. If any of the above possibilities occur for some p and q, then there exists no such ARMA(p,q) rational approximation as in (7) to the given nonrational function. Interestingly, for some p and q, if $\rho(z)$ so obtained turns out to be a bounded function, then with the h_k's and g_k's so determined, the denominator polynomial $D(z) = g(z)A_{p+q}(z) - zh(z)\tilde{A}_{p+q}(z)$ in (15) takes the explicit form

$$D(z) = \tilde{a}_0 + \tilde{a}_1 z + \tilde{a}_2 z^2 + \cdots + \tilde{a}_p z^p \tag{21}$$

where

$$\tilde{a}_0 = a_0 \tag{22}$$

and (with $g_0 = 1$)

$$\tilde{a}_i = \begin{cases} \sum_{j=0}^{i} a_j g_{i-j} - \sum_{k=0}^{i-1} a^*_{p+q-i+k+1} h_k\,, & 1 \le i \le q \\ \sum_{j=0}^{q} a_{i+j-q} g_{q-j} - \sum_{k=0}^{q-1} a^*_{p+q-i+k+1} h_k\,, & q+1 \le i \le p\,. \end{cases} \tag{23}$$

The factorization in (16) can now be carried out in any number of ways and $H_r(z)$ given by (15) represents a minimum phase ARMA(p,q) system whose first $p+q+1$ autocorrelations are guaranteed to match with those of the given nonrational spectral density function $S(\theta)$. In this case, $H_r(z)$ is said to be a Padé-like approximation to the given transfer function $H(z)$. To sum up, if $\delta(D(z)) = p$, $\delta(g(z)) = q$ and $r_0 \to r_n$ represent the autocorrelations that match with those of the rational approximation $H_r(z)$, then the choice $\rho(z) = h(z)/g(z)$ in (18) is optimal if

1. $\rho(z)$ is bounded,
2. $p + q \le n$,
3. $p < n$ (If $p = n$, then $p+q \le n$ implies $q = 0$ and hence $g(z) =$ constant, $h(z) \equiv 0$. Thus $\rho(z) \equiv 0$ and it results in an AR(n) transfer function that satisfies the above conditions trivially.)

In that case the rational approximation $H_r(z)$ is automatically minimum phase. To summarize, we have the following theorem [3]–[5].

Theorem 1: Given the autocorrelations $r_0 \to r_{p+q}$ that form a positive definite sequence, the necessary and sufficient condition to fit an ARMA(p,q) model as in (7) is that the set of $2q$ linear equations in (19) obtained by equating the coefficients of the highest degree terms in (17), that involves the coefficients of the Levinson polynomial $A_{p+q}(z)$, yield a bounded solution for the function $\rho(z)$ defined in (18).

The next theorem suggests strongly that such optimal $\rho(z)$'s should always exist for every nonrational $S(\theta)$, at least for certain values of n. The numerical evidence presented in the next section in support of this conjecture appears to be overwhelming [3]–[5].

Theorem 2: Let $S(\theta)$ denote a nonnegative periodic function of θ of period 2π that satisfies the integrability condition as well as the Paley-Wiener criterion. Then, for all n, except a finite number at most, there exist polynomial solutions $g(z)$, $h(z)$ and $D(z)$ of (17) meeting the above last two requirements $q + p \le n$ and $p < n$ for optimality [3]–[5].

NUMERICAL RESULTS

If an ARMA(p,q) Padé-like in approximation $H_r(z)$ exists to a nonrational transfer function $H(z)$, the $2q$ linear equations (19) can be used to determine the desired bounded function $\rho(z)$ and the system parameters as in (19)–(23). The heavy dots on the curves in Figs. 1(b)–2(b) indicate the existence of such optimal rational approximations to $H(z)$, where $\rho(z)$ exists as a bounded function. Notice that the existence of $\rho(z)$ as a bounded function is necessary and sufficient to guarantee an ARMA(p,q) Padé-like approximation.

When $S(\theta) = |H(e^{j\theta})|^2$ is known in advance, the percent-error

$$\eta(n,m) \triangleq \sup_{\theta} \frac{|S(\theta) - K_r(\theta)|}{S(\theta)} \tag{24}$$

also functions as a trustworthy measure of fidelity. Here $r = n + m$, and

$$K_r(\theta) = |H_r(e^{j\theta})|^2 \tag{25}$$

represents the approximated ARMA(n, m) spectrum.

The transfer function $H(z)$ in Fig. 1 is clearly nonrational and defines a nonrational power spectral density $S(\theta) = |H(e^{j\theta})|^2$. By making use of r_k's obtained from (7), as shown in Fig. 1, the present approach produces an optimal ARMA(11,9) approximation $H_r(z)$ which is (necessarily) both stable and minimum-phase. Moreover, the first $11 + 9 + 1 = 21$ Fourier coefficients of $|H_r(e^{j\theta})|^2$ also coincide with the corresponding Fourier coefficients of $S(\theta)$. Hence, $K_r(\theta)$ is truly a Padé approximation to $S(\theta)$. Observe that $\eta(n, m)$ is displayed in the later half of each figure, while $K_r(\theta)$ and $S(\theta)$ appear superimposed in their former parts. Each heavy dot on Figs. 1(b) and 2(b) indicate the existence of a corresponding optimal Padé-like approximation to $H(z)$ such as ARMA(4,3), (6,2), (7,4), *etc.* This abundance of optimal approximations is also evident in Fig. 2.

Nonrational systems can involve complicated transcendental transfer functions $H(z)$ that satisfy the integrability condition and the Paley-Wiener criterion. However, the latter do not preclude either logarithmic or essential singularities on $|z| = 1$. Of these, Fig. 1 exhibits an example with a logarithmic singularity at $\theta = \pi$ and Fig. 2 shows an example with an essential singularity at $z = -1$. In Fig. 2

$$H(z) = 1 + (1 + z)e^{-(1-z)/(1+z)}\,. \tag{26}$$

Here, $H(z)$ is analytic in $|z| < 1$ and continuous everywhere on $|z| \leq 1$. Moreover, at $z = -1$, it also exhibits an essential singularity and naturally the behavior of the function in the neighborhood of $z = -1$ *outside* the unit circle will reflect this fact. Remarkably, as Fig. 2 shows, the ARMA(15,12) Padé-like approximation captures all essential features of this transcendental transfer function everywhere except possibly in the close neighborhood of $z = -1$.

An interesting application of this rational approximation procedure will be to explore ways in which this algorithm can be adapted to produce resonant frequencies present in scattering data. Fig. 3 shows the incident and reflected waveforms in a scattering experiment sampled at the rate of $T = 3.04\,psec$ [8], with Fig 3.a representing the incident waveform and Fig 3.b representing the back scatter waveform (low-Q data) from the center of a single PEC strip that is $30\,mm$ wide. Fig. 3.c represents the returns from a resonant structure formed by two such $30\,mm$ wide PEC strips that are arranged in a parallel configuration separated by $15\,mm$. The back scattered data (high-Q data) is observed from the center of the top strip and it shows the early retruns as well as the late returns.

Since the incident waveform is essentially an impulse, the outputs represent the impulse response $h(kT)$, or its noisy version and the problem here is to obtain a rational system that best approximates this data. For a causal system, since

$$H(z) \triangleq \sum_{k=0}^{\infty} h(kT)z^k\,,$$

represent the system transfer function and

$$S(\theta) = |H(e^{j\theta})|^2 = \sum_{k=-\infty}^{k=+\infty} r_k e^{jk\theta}$$

the spectrum, we have

$$r_k = \sum_{n=0}^{\infty} h(kT)h\left((n + k)T\right)$$

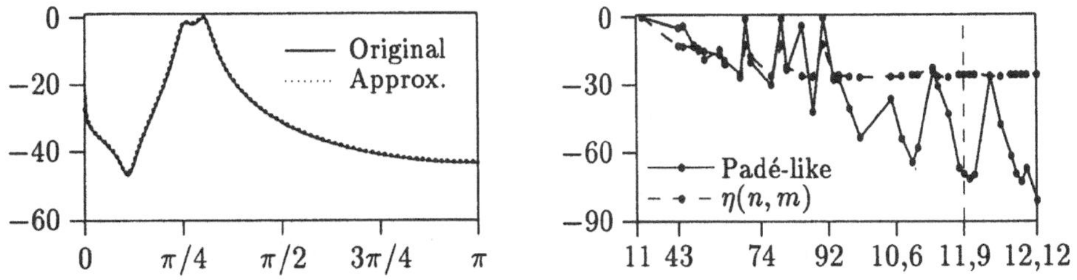

(a) Original and approximated spectra (b) Test criteria for model order selection

Fig. 1: Rational approximation of a nonrational system with a logarithmic singularity at $z = 1$. The original nonrational system transfer function in Fig. 1 is given by

$$H(z) = \frac{(1.1025 - 1.9734z + z^2)ln\,(1-z)}{1 - 2.4394z + 3.2856z^2 - 2.2126z^3 + 0.8227z^4}\,.$$

The approximated ARMA(11,9) model is given by $H_r(z) = B_m(z)/A_n(z)$ where

$$\begin{aligned} A_n(z) &= 1 - 4.56z + 9.07z^2 - 9.06z^3 + 2.07z^4 + 6.44z^5 - 9.43z^6 \\ &\quad +6.60z^7 - 2.65z^8 + 0.59z^9 + 0.062z^{10} - 0.0021z^{11}\,, \end{aligned}$$

and

$$\begin{aligned} B_m(z) &= 1.1 - 3.76z + 4.07z^2 + 0.209z^3 - 3.93z^4 + 3.35z^5 - 1.23z^6 \\ &\quad +0.203z^7 - 0.013z^8 + 0.0006z^9\,. \end{aligned}$$

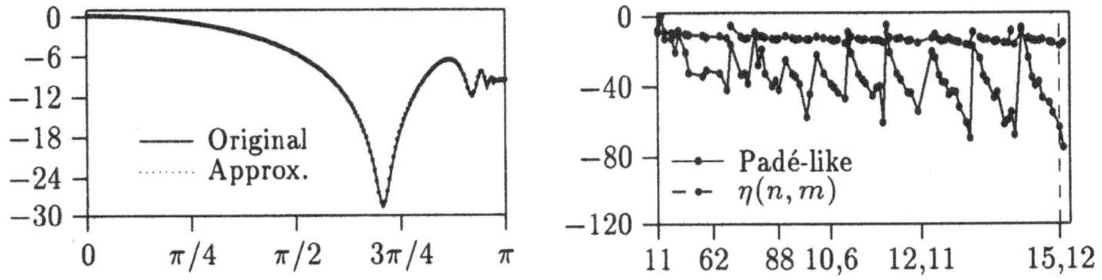

(a) Original and approximated spectra (b) Test criteria for model order selection

Fig. 2: Rational approximation of a nonrational system with an essential singularity at $z = -1$. The original nonrational system corresponds to $H(z) = 1 + (1+z)e^{-(1-z)/(1+z)}$ and the approximated ARMA(15,12) model is given by $H_r(z) = B_m(z)/A_n(z)$ where

$$\begin{aligned} A_n(z) &= 1 + 3.06z + 1.00z^2 - 5.70z^3 - 5.68z^4 + 2.57z^5 + 5.36z^6 \\ &\quad 0.65z^7 - 1.82z^8 - 0.61z^9 + 0.19z^{10} + 0.09z^{11} + 1.4 \times 10^{-7}z^{12} \\ &\quad -1.1 \times 10^{-6}z^{13} + 1.0 \times 10^{-6}z^{14} - 1.4 \times 10^{-7}z^{15}\,, \end{aligned}$$

and

$$\begin{aligned} B_m(z) &= 1.37 + 5.29z + 5.48z^2 - 4.69z^3 - 14.08z^4 - 7.10z^5 + 7.56z^6 \\ &\quad 9.92z^7 + 1.09z^8 - 3.64z^9 - 1.52z^{10} + 0.37z^{11} + 0.24z^{12}\,. \end{aligned}$$

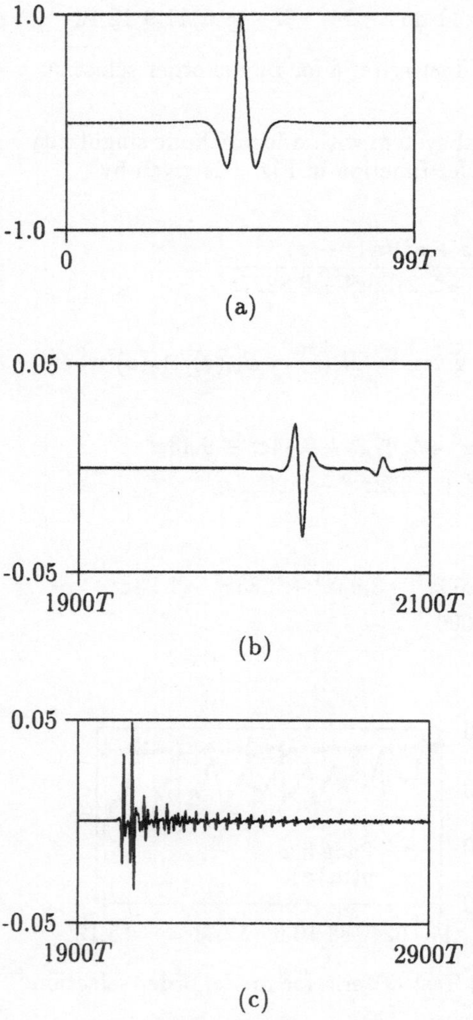

Fig. 3. Incident and reflected waveforms in a scattering experiment.
(a) Incident waveform (b) Return signal (Low-Q data) from a single strip (c) Return signal waveform (High-Q data) from a resonant structure. Sampling Period $T = 3.04$ *psec.*

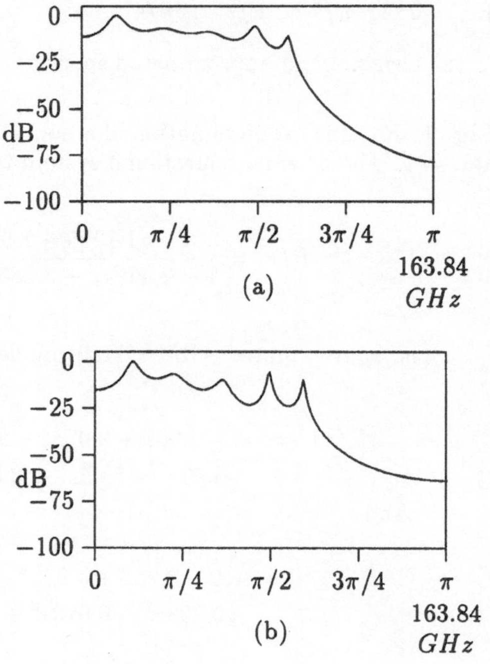

Fig. 4. Stable pole identification: Low- and High-Q spectra and their rational approximations. (a) ARMA(10, 1) approximation of the Low-Q data
(b) ARMA(10, 2) approximation of the High-Q data. (see text for details)
Actual resonant frequencies and the associated fractional power levels (in braces) in the low-Q case:
15.84 GHz (0.03), 38.46 GHz (0.013), 59.50 GHz (0.006), 80.63 GHz (0.002), 95.99 GHz (0.0002).
Actual resonant frequencies and the associated fractional power levels (in braces) in the high-Q case:
17.34 GHz (0.011), 36.88 GHz (0.006), 59.26 GHz (0.001), 80.84 GHz (0.0003), 96.74 GHz (0.0001).

and this formula can be used to obtain estimates of the autocorrelations $r_k,\ k = 0 \rightarrow n$ in both cases. The autocorrelations so obtained has been used to generate the stable rational approximations as described here, and the respective spectra are shown in Figs. 4.a and Figs. 4.b corresponding to the low- and high-Q cases respectively. The approximated stable transfer function $H_1(z) = B_1(z)/A_1(z)$ in the Low-Q case is a stable ARMA(10,1) model given by

$$\begin{aligned} A_1(z) &= 1.0 - 3.336z + 6.999z^2 - 11.034z^3 + 13.918z^4 - 14.376z^5 + 12.286z^6 \\ &\quad -8.553z^7 + 4.722z^8 - 1.871z^9 + 0.446z^{10} \end{aligned}$$

and

$$B_1(z) = 0.00093 + 0.00066z$$

and $H_2(z) = B_2(z)/A_2(z)$ in the high-Q case is a stable ARMA(10,2) model given by

$$\begin{aligned} A_2(z) &= 1.0 - 3.491z + 7.621z^2 - 12.364z^3 + 16.094z^4 - 17.136z^5 + 15.155z^6 \\ &\quad -10.944z^7 + 6.312z^8 - 2.644z^9 + 0.662z^{10} \end{aligned}$$

and

$$B_2(z) = 0.00154 - 0.00036z + 0.00084z^2 .$$

Since the poles of the rational system represent the normalized resonant frequencies with respect to $f_0 = 1/2T = 163.84\,GHz$, these poles together with their residues can be used to obtain an equivalent set of actual resonant frequencies and their associated power levels as shown in Fig. 4.

REFERENCES

[1] H. S. Wall, *Analytic Theory of Continued Fractions*, New York: Chelsea, 1973.

[2] G. A. Baker, Jr., *Essentials of Pade Approximants*, New York: Academic Press, 1975.

[3] S. U. Pillai, T. I. Shim, and D. C. Youla, "A new technique for ARMA-system identification and rational approximation," to appear in *IEEE Trans. Signal Processing*, vol. 42, no. 5, May 1993.

[4] T. I. Shim, *New Techniques for ARMA-System Identification and Rational Approximation*, Ph.D. Dissertation, Polytechnic University, Brooklyn, New York, June 1992.

[5] S. U. Pillai and T. I. Shim, *Spectrum Estimation and System Identification*, New York: Springer Verlag, Inc. 1993.

[6] D. C. Youla, "The FEE: A new tunable high-resolution spectral estimator," Part I, Technical note, no. 3, Department of Electrical Engineering, Polytechnic Institute of New York, Brooklyn, New York, 1980: also RADC Rep. RADC-TR-81-397, AD A114996, February 1982.

[7] N. Wiener and P. Masani, "The prediction theory of multivariate processes, part I: the regularity condition," *Acta Math.*, vol. 98, 1957.

[8] L. Carin and M. McClure," Single Pulse Scattering Measurements from a resonant structure," Technical note, Department of Electrical Engineering, Polytechnic University, Brooklyn, New York, 1992.

ON USING SCATTERING STATISTICS FOR ULTRA WIDEBAND ELECTROMAGNETIC TARGET CLASSIFICATION AND IDENTIFICATION

Edmund K. Miller

Group MEE-3
Los Alamos National Laboratory
Los Alamos, NM 87545

INTRODUCTION

An interrogating electromagnetic field has information encoded upon it by any target from which it scatters. This process presents the electomagneticest/signal processor with a problem that ranges across questions such as the following, in increasing order of specificity and information requirements:

1) Detection--Is there a target?
2) Classification--To which class does that target belong?
3) Identification--What specific member of that class is the target?
4) Reconstruction--What are the target's electrical and/or geometrical properties?

Radar systems have been analyzed primarily from the perspective of Question 1 above. Recently, increasing attention has been devoted to Questions 2-4, as more specific and complete information beyond target detection is desired. In the following discussion, we present conceptually a statistics, information-oriented approach to this general set of problems. Our goal is to assess the information requirements related to classification /identification (C/I). In particular, we consider the possibility of using statistical analysis as an intermediary between imaging targets, an approach that requires an excessive amount of data, and such feature-based techniques as using poles. We speculate that probability density functions (PDFs) of the aspect-angle-dependent radar cross section (RCS) provide an attractive opportunity for reducing the data requirements of imaging while avoiding some of the disadvantages of resonance techniques.

Our discussion on the use of scattering statistics for C/I is developed in three phases. The first phase, the least data-intensive approach, uses expected values of target RCS (<RCS>, averaged over aspect angle) as a function of frequency. In the second phase, knowledge of the aspect-angle-dependent PDFs is added to the process, with the goal of improving C/I performance at the expense of needing a more detailed data base and more computations. In the third phase, with a further increase in the amount of data and computation, a knowledge of the aspect-dependent scattering patterns is added in an attempt to reduce the ambiguity caused by aspect-angle variation. For brevity, we refer to the overall approach as statistical C/I or SCI.

Our discussion begins with some background, then continues with sections addressing each of these three phases. Two final sections consider the questions of observability and discriminability, followed by some concluding comments. Throughout this discussion, we assume the availability of the kinds of scattering data needed to "profile" a target by plotting scattering strength versus range, but we consider alternative uses for such data. Specifically, we assume data is available over an ultra-wide-bandwidth, though not necessarily obtained from impulse measurements but possibly instead being derived from discrete frequency

Ultra-Wideband, Short-Pulse Electromagnetics
Edited by H. Bertoni *et al.*, Plenum Press, 1993

sampling. Our approach is conceptual; no actual data are employed in this discussion except for "data images" as defined below. Our purpose is to suggest that SCI may be achievable by analyzing data and using stored-library information, a prerequisite of all C/I schemes, in a manner different from that of developing radar images using profiling.

BACKGROUND

The basic C/I problem considered here can be simply stated as follows:

PROBLEM--Find the target, T_i, from a set of T targets, that has produced the observed data.

STRATEGY--Minimize the data needed to accomplish C/I in terms of the required data base, the needed measurements, and the processing required of those measurements and data base.

ASSUMPTION--Discrete samples obtained across a wide enough bandwidth and range of viewing angles will permit C/I to be accomplished to some desired level of performance.

The overall problem involves data collection, information processing and decision making for which at least three distinct approaches can be identified. Perhaps most straightforward, but also most data intensive, is profiling and imaging, which require wide bandwidth (relative to wavelength change with respect to target size) and many aspect angles. This approach can yield target images of near photographic quality, but it is not considered further here. An apparently less data-demanding approach is one which is feature-based and derived from electromagnetic physics to reduce the number of frequencies and/or aspects that may be needed. Techniques that employ the idea of body resonances (SEM poles) would fall into this category. They offer the tantalizing possibility of realizing aspect-independent C/I, which is one reason why pole-based techniques have attracted much attention. The third approach, the one considered in most detail here, is also feature-based, but it employs more *ad hoc* features. Its data requirements might fall somewhere between the other two approaches, as will be illustrated conceptually in the discussion that follows.

Before presenting the details of our SCI approach, we first make some general observations concerning C/I or, as the problem is also known, automatic target recognition (ATR).

Experimental Design:

* As more confidence in C/I is desired, more information/data is required.
* As target numbers increase, more information is needed to maintain a specified confidence level.
* The amount of information available increases with bandwidth and the number of aspect angles available.
* If data are available across a wide-enough bandwidth, then fewer aspect angles are needed (i.e., bandwidth trades for aspect).
* The basic problem of ATR-C/I is one of developing more appropriate knowledge representation so that a better data-acquisition strategy can be defined and better use can be made of that data.

Target-Set Considerations:

* For a given set of targets, some set of observation variables will be optimum with respect to minimizing the measurement requirements for a specified confidence level to be achieved in C/I.
* The number of needed frequencies can be minimized if the frequencies sampled are optimized with respect to their information content.
* Higher spatial resolution requires wider bandwidth (i.e., $\Delta x \Delta f \sim$ constant).
* Imaging is overkill for most C/I applications.
* Just as images are not necessary for C/I, physical features are also optional.
* The more similar the targets, the more data that will be required to accomplish C/I to a desired level of confidence.

Data Representation and Processing:

* Target responses can be represented as feature vectors and intersecting regions in N-space.
* Target regions overlap to the extent that information is inadequate or unavailable.
* Feature-vector overlap can be reduced with more data and/or more data accuracy.

* Feature-vector overlap increases with increased noise and more targets.
* The C/I process begins with an imperfect data base and must be able to handle imperfect measurements.
* Redundant data can reduce noise effects and increase information, thereby increasing reliability and confidence.
* The quantity of information available increases with the data accuracy.
* Poles represent physics, so even if inaccessible they can provide insight concerning observables.
* Aside from differences in scattering physics, a 1GHz bandwidth centered at 1 GHz is equivalent in information concent to 1 GHz centered at 10 GHz.

We also note that all available *a priori* knowledge should be used to reduce data requirements. For example, knowing that the targets of interest are airborne, unlikely objects, such as tanks and ships, should not need consideration (unless they are being dropped by parachute!). Or, aspect angles can be limited if one knows a target's track and takes into account that airplanes don't fly sideways. Or, if only size is important for C/I, then only a few resonance peaks in the frequency-dependent RCS may be needed. Assuming that the same data are available whatever approach is to be used for C/I, the goal is to make optimal use of that data.

USING EXPECTED VALUES OF RCS FOR SCI

A commonly encountered question in inverse problems such as C/I is "At what frequencies should measurements be made, given a class of known targets, to maximize ATR reliability subject to certain constraints such as total bandwidth, number of sampling frequencies, number of aspect angles, signal-to-noise ratio, etc?" This question might be stated somewhat differently: "At what frequencies are the targets of interest most separable, or where are their frequency-dependent signatures most nearly orthogonal?"

A possible procedure for finding this set of frequencies can be outlined as follows:

1) For target i = 1, find $<R_i(\omega;\theta,\varphi)>$ over 4π steradian incident angles for a set of frequency samples ω_j, j = 1, . . . , F
2) Repeat for i = 2, . . . , T targets to obtain the matrix $M_{i,j}$.
3) Normalize the results for each target to get $\underline{M}_{i,j}$ (this may be unnecessary if the absolute value of the RCS can be measured).
4) Compute the distance (norm) measure d_i between $\underline{M}_{i,j}$ at each ω_j for each target pair for a total of $T^2/2$ - T values and sum to obtain D_j for j = 1, . . . , F.
5) Order the D_j from maximum to minimum values.
6) Select the set of frequencies F' for C/I from the largest of the D_j obtained.

Target C/I might then be attempted as follows:

7) Renormalize the $M_{i,j}$ for the new set of frequencies f'_j = 1, . . . , F'.
8) From the unknown target return T_x, compute $d'_{i,j}$, i = 1, . . . , T, j = 1, . . ., F', from the target library.
9) Sum the $d'_{i,j}$ over frequency to obtain D'_i, the target measures.
10) The library target with smallest value of target measure identifies unknown.

The goal is to identify the frequencies where the expected values of target RCSs are most different as illustrated graphically in Fig. 1. If the only information available were the aspect-angle-averaged RCS data, then the best measurement frequencies for C/I might be selected as indicated here. However, if the variances of the PDFs for the targets of interest are nonzero (assured for all but spherical targets) and their PDFs are also overlapping, the probability of a false identification can be higher than is acceptable. In that case, the target PDFs would also be needed, as is discussed in the next section.

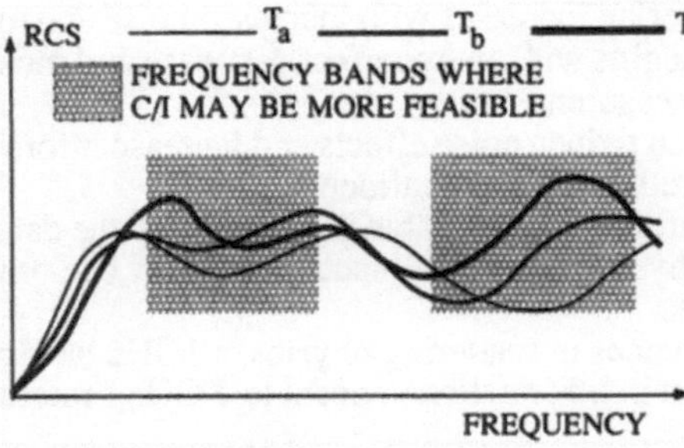

Figure 1. A conecptual frequency plot of the expected values for three radar targets illustrating the idea that some frequency bands might be more useful for target C/I.

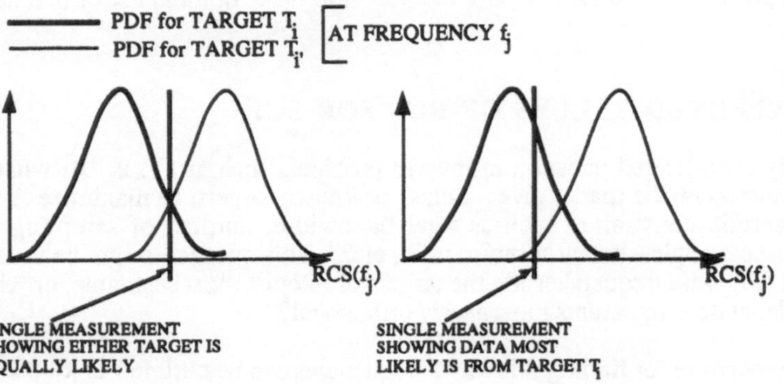

Figure 2. Plots of the PDFs for two targets to illustrate how a single RCS data value might be interpreted. If the measured value were to fall midway between the two PDFs at their crossover point, as in the left-hand plot, then neither target would be more likely to have produced that measurement. On the other hand, if the measured value were to be located as on the right-hand plot, then target T_i would be the more likely.

ADDING ASPECT-ANGLE STATISTICS FOR SCI

The RCS of large, complex targets is a rapidly varying function of aspect angle. Consequently, even though one measurement at a single frequency might enable discrimination among two, three, or even more targets with different <RCS> values, this possibility decreases with increasing target numbers because the RCS values will be distributed about the median and their PDFs are likely to overlap. This leads to the situation depicted in Fig. 2 in which the PDFs of two targets are seen to overlap significantly. Thus, even though their <RCS> values are distinctly different, they share a range of possible RCS values, making a decision about which target has produced the measured data less certain.

As a way to visualize the differences (or similarities) between different targets, including their aspect-angle dependence, we can develop a data "image" of the target's electromagnetic response (Fig. 3) for three simple wire models. These results, which were obtained from the computer model NEC [Breakall et al. (1985)], depict the aspect-dependent RCS as a function of frequency for a straight wire, a bent wire (having two right-angle bends one quarter of the distance from each end), and a crossed wire, all with the same overall length of 60 m.

Quantitative analysis of these kinds of data presentations for a variety of targets should help reveal the similarities (and differences) between the EM observables for such targets and should reveal the most suitable frequencies, across some band, for discriminating among

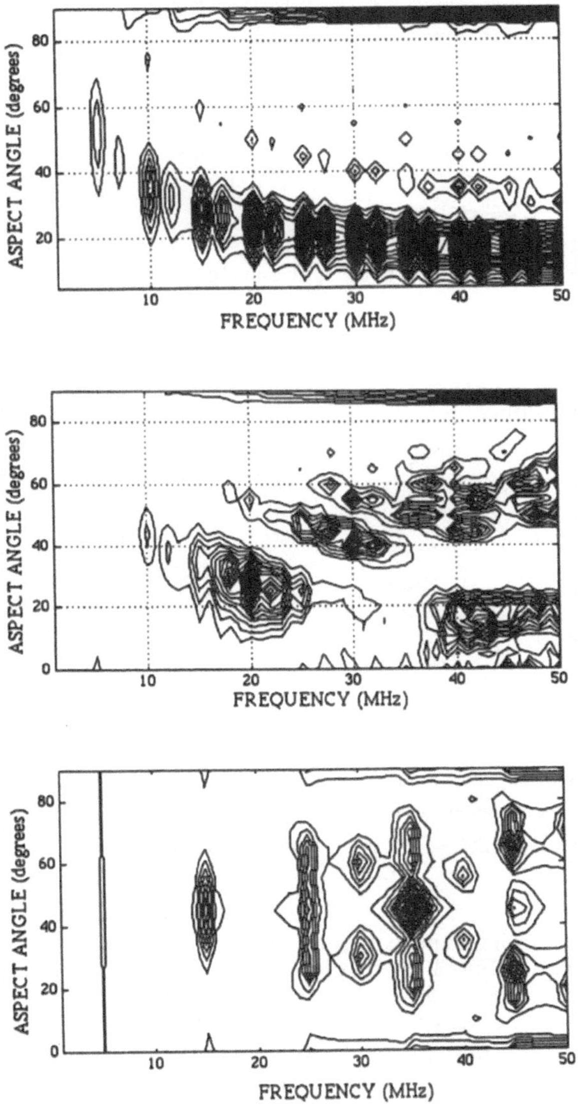

Figure 3. Data images for three 60-m wire targets, obtained from their backscatter RCS in the target plane as a function of aspect angle and frequency (top, straight wire; middle, bent wire; bottom, crossed wire). The "image," a contour plot in these two variables, graphically demonstrates the scattering characteristics of targets in a common format. A data image can provide much more information than the usual aspect angle-RCS plot. Although aspect angle is a function of two angles, elevation and azimuth, by transforming them to a single variable (for example, by measuring aspect angle on a downward spiral from one pole of a spherical coordinate system to the other), a single variable would result.

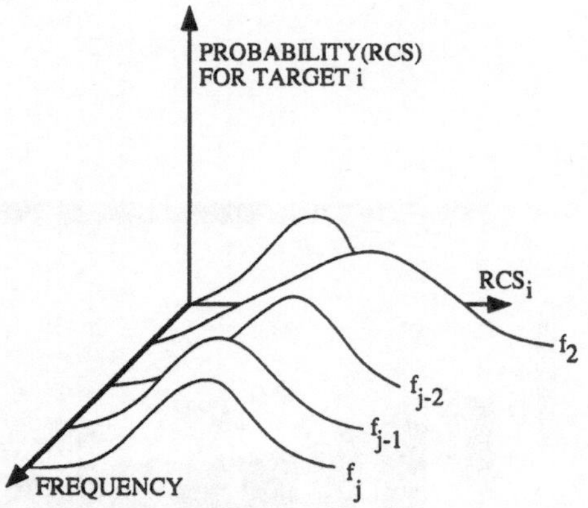

Figure 4. The PDFs of target T_i for a number of different frequencies.

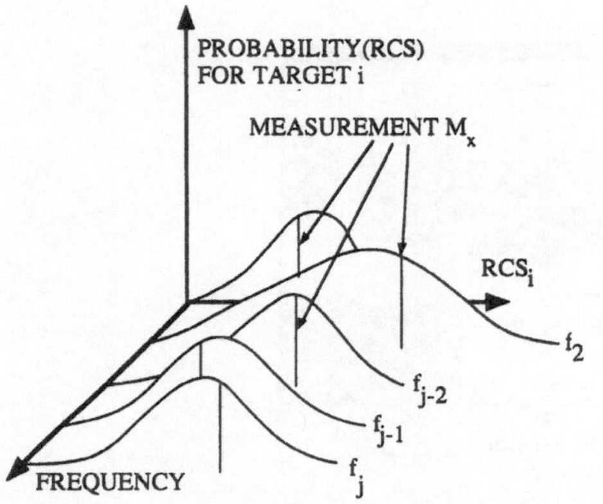

Figure 5a

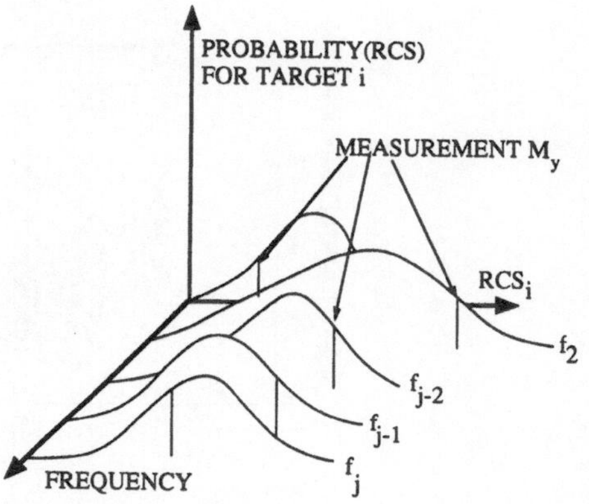

Figure 5b

Results from two different measurements plotted on the PDFs of target T_i. The data from measurement M_x, as depicted here, is more compatible with target T_i of the data base than is measurement M_y.

them. One way to do such an analysis, is to develop the PDFs for targets of interest from data such as that shown in Fig. 3. A set of PDFs like those shown in Fig. 4 might then be obtained (those shown here are for conceptual purposes only, and do not represent a real target). Given a set of measurements M_x at frequencies f_1, f_2, . . . , f_j, to be plotted on the PDF-frequency plot (Fig. 5a), we might conclude that the observed data have a high probability of coming from T_i. The basis for our conclusion is that the measured RCS values are generally found near the peaks of the PDFs for this target, or near the most probable RCS values. If the measurements produced the kind of plot seen in Fig. 5b, however, then it would be more likely that some other target generated the data because the measured RCS values in this case fall further into the tails of these PDFs. A quantitative measure of target-identification probability might simply be obtained by summing the PDFs of each target over the set of measured frequencies. Of course, a more rigorous measure would be preferred, using Bayesian or other statistical approaches to compute probabilities. Also note that specular-flash returns might significantly affect target PDFs, and thus require special consideration in the treatment outlined here, but that topic is deferred to a later discussion.

ESTIMATING ASPECT ANGLES FOR IMPROVED SCI

The PDFs alone cannot be expected to convey enough information to accomplish successful

SCI for target sets that are large-enough and/or similar-enough. As more confidence is required in target identification, more information will be necessary. As a logical next step beyond using only <RCS> or PDF information, we now consider the possibility that the angle-dependent scattering patterns are also available as part of the data base to which we have access. Assume that the RCS of an unknown target is available at F frequencies across some bandwidth for which an *a priori* data base has been established. The library data might be stored as x-y plots of RCS vs aspect angle, where the latter could be compressed into one dimension from the usual elevation-azimuth, two-dimensional coordinates by using an angle variable that spirals outward and downward from one pole of a spherical coordinate system towards the other. Alternatively, this information could be stored in the form of data images (such as those shown in Fig. 3) to capture the frequency-aspect information.

For any target other than a sphere and except at extrema, the backscatter pattern will contain two or more aspect angles at which the same RCS will be observed. The scattering pattern becomes more multiple-valued with increasing frequency, producing increasing ambiguity in aspect angle. However, the number of possible aspect angles at which a target might have been illuminated to produce a given RCS value at any frequency, although possibly large, will nonetheless be limited, assuming of course that the measurement falls within the range of values produced by that target. If RCS values are measured simultaneously (for a moving target), then they will all share a common aspect angle, providing an opportunity for reducing aspect-angle ambiguity or, possibly, of performing C/I itself.

The basic approach is to determine, for each target in the data base, that set of aspect angles consistent with the measured RCS at each of the available sampled frequencies as might be done using the data of Fig. 3. The result of this exercise could be visualized in a three-dimensional array, or template, of binary entries that contain x's where there is an angle match and no entry where there is none as in Fig. 6. Ideally, the correct target would be identified as the one whose backscatter patterns produce a single matching angle, to within some experimental error, at all observation frequencies.

ADDITIONAL CONSIDERATIONS

In just considering the possibility of using radar for C/I, the properties of observability and discriminability are crucial for success. In this discussion these terms are defined, for the target set of interest, as follows:

OBSERVABILITY--measures how accessible are the features that might differentiate the various targets from the observations available,

and

DISCRIMINABILITY--measures how dissimilar these features are.

An implicit assumption in even considering the possibility of successful C/I, is that both properties are present to the extent necessary, a somewhat imprecise, but quantifiable, measure. The more pronounced these properties are for the target set and measurements available, the amount of data and the data processing necessary for acceptable C/I performance, can be expected to be less. Conversely, when these properties are less pronounced, the amount of data, and processing, can be expected to increase.

As a specific example of how both observability and discriminability can depend on a specific choice of features, consider the use of resonances or poles for C/I. For transient data of given accuracy, the oscillatory components of the poles can be more accurately estimated than the damping components [Dudley and Goodman (1986)], i.e., damping rates are less observable than the oscillation rates in transient data. The implications for target discriminability directly follow when it is noted that the oscillation rate is more sensitive to target *size*, while the damping rate is more sensitive to target *shape* [Miller (1991)]. Therefore, if targets of similar size but different shape form the set of interest, that poles would probably not be a good choice for a feature set. Conversely, if target size is the discriminant, poles might be a more reasonable choice for discriminants or features.

The three phases of SCI considered here are intended to deal with increasing similarities of the target-set observables or with limitations in the available data. Expected values of the target RCSs might provide the simplest discriminant. If, however, that is not feasible, the

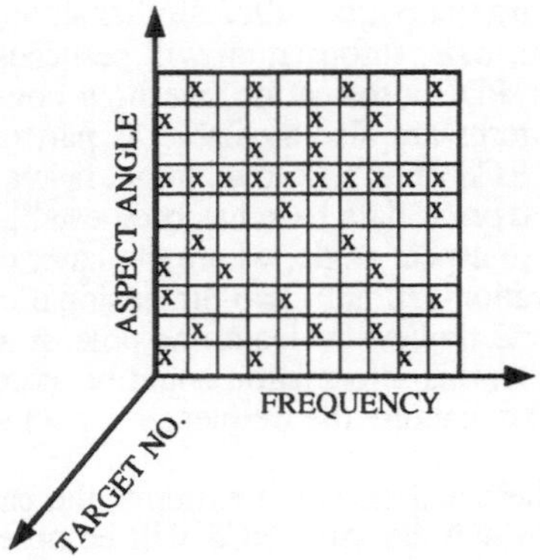

Figure 6. A matching angle-frequency template for one target from a set of T targts, showing how the aspect angle might be determined and target identificaton achieved, by finding that aspect angle consistent with a data base of scattering patterns for the target set of interest.

next level of discriminant would be to use the aspect-angle PDFs as a function of frequency. But if the PDFs and expected values of the RCSs are too similar, then the scattering patterns would need to be considered. If these also prove inadequate, the targets of interest could be too similar to be reliably identified using radar data and considering the constraints involved.

CONCLUDING REMARKS

We have discussed an approach to classification/identification (C/I) using statistical analysis of the aspect-angle-dependent RCS of a target over some frequency band. The basic idea is that, given the kind of data base needed for profiling and imaging prospective targets, where broadband data is needed over many aspect angles, alternative uses can be made of such data that might permit more parsimonious and efficient C/I to be realized. We suggest that, with such a data base, statistical analysis might permit us to attempt several stages of C/I, including the use of aspect-averaged RCS values; probability-density functions (PDFs) derived from using aspect angle as a random variable; and details of the angle scattering patterns themselves. This conceptual approach is speculative in that the ideas considered are not quantitatively tested here. The use of statistics as proposed for C/I, seems to be a logical extension of the basic detection problem, which, itself is a statistical problem. The approach outlined here might be profitably implemented in a pre-classification stage as a prelude to profiling and imaging, in those cases where the latter might be found to be necessary for more successful C/I. By following a "graded" approach, where the amount of data and processing progressively escalates only as C/I at a given stage is found inadequate, it should be possible to reduce the overall effort required to achieve a desired performance.

REFERENCES

Breakall, J. K., G. J. Burke, and E. K. Miller (1985), "The Numerical Electromagnetic Code (NEC)," in Proceedings of 6th Symposium and Technical Exhibition on Electromagnetic Compatibility, Zurich, March 5-7, pp. 301-308.

Dudley, D. G. and D. M. Goodman (1986), "Transient Identification and Object Classification," pp. 456-497 in *Time-Domain Measurements in Electromagnetics*, ed. E. K. Miller, Van Nostrand Reinhold, New York, NY.

Miller, E. K. (1991), "Model-Based Parameter-Estimation Applications in Electromagnetics," pp. 205-256 in *Electromagnetic Modelling and Measurements for Analysis and Synthesis Problems*, ed. B. de Newmann, Kluwer Academic Publishers, Dordrecht/Boston/London.

EVALUATION OF A CAUSAL TIME DOMAIN RESPONSE FROM BANDLIMITED FREQUENCY DOMAIN DATA

T. Sarkar, H. Wang, R. Adve, M. Moturi
Electrical and Computer Engineering Department
Syracuse University; Syracuse, NY 13203

M. Wicks
Rome Laboratory
Rome, NY 13441

INTRODUCTION

Broadband measurements may be performed in either the time domain or the frequency domain. Time domain measurements are easier to perform since the waveforms of interest are all real. However, one disadvantage is the limited dynamic range of available systems, whereas frequency domain measurement equipment benefits from large dynamic range. Also, it is difficult to recover the true time domain response from measurements made in a noisy environment. On the other hand, frequency domain measurements maybe carried out either over an entire range of frequencies or selectively over a band of frequencies, whichever is convenient and/or less susceptible to noise. From these measurements a time domain response can be extracted by an inverse Fourier transform. Since, bandlimited complex frequency domain data does not guarantee causality in the time domain, nor a real time domain response, measurements carried out in the frequency domain do not truly represent the transient response of the system.

The objective of this paper is to develop a technique for the extraction of a causal time domain response from bandlimited frequency domain data. Physical time domain signals are causal, by which we mean that the signal exists only for time greater than zero. We will show that it is possible to accurately extract a causal time response from bandlimited frequency domain data. However, bandlimited data does not necessarily produce a causal time domain signal when Fourier-transformed. Even so, we establish that it is possible to extract a bandlimited, causal response by extrapolating the complex frequency domain data under the premise that the time domain signal must be causal. In general, the real and imaginary parts of the complex frequency domain data are independent of each other. However, the causality of the time domain signal h(t), assures us that certain relationships exist between the real and imaginary components in the frequency domain. If we denote $H_{\Re}(j\omega)$ as the real part and $H_{\Im}(j\omega)$ as the imaginary part of the transfer function $H(j\omega)$ obtained from h(t), then, from causality, they have to be related by a special form of the Hilbert transform

Ultra-Wideband, Short-Pulse Electromagnetics
Edited by H. Bertoni *et al.*, Plenum Press, 1993

[1]. The property that the real and imaginary parts of the frequency domain data correspond to the even and odd parts of h(t) is exploited in extracting a causal response from complex bandlimited frequency domain data.

Since we process discrete frequency domain data, we handle frequency and time domain signals in the form of sequences. Simulation has been carried out for various waveforms. We obtained bandlimited frequency domain data by taking a portion of the data from the Fourier transform of these signals. Using this bandlimited frequency domain data we show that the causal time domain signal computed by our technique closely coincides with the true signal. Also, computer simulation has been included to show how noisy frequency data may be analyzed. We also obtained frequency domain data from a HP 8510 Network Analyzer using a Sliding load. The processing technique was applied to this data and the time domain response of the Sliding load was extracted. The performance of our technique is compared with the time domain response obtained by the inverse discrete Fourier transform of the measured data.

RELATIONSHIPS BETWEEN SEQUENCES AND THEIR FOURIER TRANSFORMS

This section briefly covers some of the properties of sequences and their Fourier transforms. Any complex sequence h[n] can be expressed as a sum of a symmetric sequence $h_e[n]$ and an anti-symmetric sequence $h_o[n]$. In the case of real sequences, these are called even and odd sequences [2].

Therefore,

$$h[n] = h_e[n] + h_o[n] \tag{1}$$

$$h_e[n] = h_e[-n] \tag{2}$$

$$h_o[n] = -h_o[-n] \tag{3}$$

The Fourier transform of any complex sequence h[n] is represented by H $(e^{j\omega})$, where

$$H(e^{j\omega}) = \sum_{n=-\infty}^{\infty} h[n]\, e^{-j\omega n} \tag{4}$$

Therefore,

$$H(e^{-j\omega}) = \sum_{n=-\infty}^{\infty} h[n]\, e^{j\omega n} \tag{5}$$

This implies,

$$H_R(e^{j\omega}) = \mathcal{F}\{h_e[n]\} \tag{6}$$

and

$$jH_{\Im}(e^{j\omega}) = \mathcal{F}\{h_o[n]\} \tag{7}$$

Also, $H(e^{j\omega}) = H^*(e^{-j\omega})$ for real h[n] and $H_{\Re}(e^{j\omega}) = H_{\Re}(e^{-j\omega})$ which is an even function, and $H_{\Im}(e^{j\omega}) = -H_{\Im}(e^{-j\omega})$ which is an odd function.

THE HILBERT TRANSFORM RELATIONSHIP

Consider a periodic, real, time domain sequence $h_p[n]$ with period N, that is related to a finite length sequence h[n] of length N by

$$h_p[n] = \sum_{i=-\infty}^{\infty} h[n+iN] \tag{8}$$

From Eq. (1) we have,

$$h_p[n] = h_{pe}[n] + h_{po}[n], \; n=0,1,2,\ldots,N-1 \tag{9}$$

and from Eq. (2) we have

$$h_{pe}[n] = \frac{1}{2}\{h_p[n] + h_p[-n]\} \tag{10}$$

and similarly from Eq. (3),

$$h_{po}[n] = \frac{1}{2}\{h_p[n] - h_p[-n]\} \tag{11}$$

If we have N=2r (where r is a positive integer)
then,

$$h_p[n] = \begin{cases} 2h_{pe}[n] & n = 1,2,3,\ldots,\frac{N}{2}-1 \\ h_{pe}[n] & n = 0,\frac{N}{2} \\ 0 & n = \frac{N}{2}+1,\ldots,N-1 \end{cases} \tag{12}$$

and also the odd part of the sequence can be expressed as,

$$h_{po}[n] = \begin{cases} h_{pe}[n] & n = 1,2,3,\ldots,\frac{N}{2}-1 \\ 0 & n = 0,\frac{N}{2} \\ -h_{pe}[n] & n = \frac{N}{2}+1,\ldots,N-1 \end{cases} \tag{13}$$

If we define $u_{pN}[n]$ as the periodic sequence

$$u_{pN}[n] = \begin{cases} 2 & n = 1,2,3,\ldots,\frac{N}{2}-1 \\ 1 & n = 0,\frac{N}{2} \\ 0 & n = \frac{N}{2}+1,\ldots,N-1 \end{cases} \tag{14}$$

then we can express $h_p[n]$ as

$$h_p[n] = h_{pe}[n]u_{pN}[n] \tag{15}$$

Equivalently the Fourier transform of $h_p[n]$ yields,

$$H_p[k] = \frac{1}{N}\sum_{m=0}^{N-1} H_{p\Re}[m]U_{pN}[k-m] = H_{p\Re}[k] + jH_{p\Im}[k] \tag{16}$$

Eq. (14) can be alternatively expressed as,

$$u_{pN}[n] = 2u[n] - 2u[n-\frac{N}{2}] - \delta[n] + \delta[n-\frac{N}{2}] \tag{17}$$

where $u[n]$ is the unit step sequence, and $\delta[n]$ is the unit sample sequence.
The Fourier transform of $u_{pN}[n]$ is,

$$U_{pN}[k] = \begin{cases} -2j\cot\left[\frac{\pi k}{N}\right] & k \text{ odd} \\ 0 & k \text{ even} \end{cases} \tag{18}$$

and $U_{pN}[0] = N$ as derived from the definition of $U_{pN}[k]$.
Defining

$$V_{pN}(k) = U_{pn}(k) - N\delta(k) \tag{19}$$

Therefore from Eq. (16),

$$H_p[k] = H_{p\Re}[k] + \frac{1}{N}\sum_{m=0,m\neq k}^{N-1} H_{p\Re}[m]V_{pN}[k-m] \tag{20}$$

Hence we get the imaginary part as,

$$jH_{p\Im}[k] = \frac{1}{N}\sum_{m=0,m\neq k}^{N-1} H_{p\Re}[m]V_{pN}[k-m] \tag{21}$$

This is the Hilbert transform relationship between the real and imaginary parts of the Fourier transform of a periodically causal sequence. If h[n] = 0 for n < 0 and for n > N/2 then the periodicity maybe removed and we have the same relationships between the real and imaginary parts of the Fourier transform of h[n].

$$jH_{\mathfrak{I}}[k] = \begin{cases} \frac{1}{N} \sum_{m=0,m\neq k}^{N-1} H_{\mathfrak{R}}[m] V_N[k-m] & 0 \le k \le N-1 \\ 0 & \text{otherwise} \end{cases} \tag{22}$$

In the previous expression the cotangent term is itself periodic with period N, so when we compute the N point discrete Fourier transform of the real causal time sequence, the relationship between the real and imaginary parts will be affected. Eq. (22) is the Hilbert transform relationship between $H_{\mathfrak{R}}[k]$ and $H_{\mathfrak{I}}[k]$. Thus with a knowledge of one the other can be evaluated. Alternatively, since $h_e[n]$ is the inverse Fourier transform of $H_{\mathfrak{R}}[k]$, it can be obtained by an inverse discrete Fourier transform (IDFT) of $H_{\mathfrak{R}}[k]$. Eq. (13) expresses a relationship between $h_e[n]$ and $h_o[n]$, thus $h_o[n]$ is known. $H_{\mathfrak{I}}[k]$ can be obtained as the Fourier transform $h_o[n]$, by utilizing a discrete Fourier transform (DFT) algorithm: This procedure forms the basis of our technique for the extraction of a real, causal time domain response from bandlimited complex frequency domain data. The theoretical development assures us that by computing the DFT's and IDFT's the original real time sequence will not lose its causal nature.

Extrapolation of frequency domain data for a causal system response

A method to generate a causal, finite duration time domain sequence h[i], (h[i] = 0 for i < 0 and i ≥ N) given a bandlimited data set of its Fourier transform is described. The iterative technique discussed in the subsequent part of the section describes a sequence of transformations between the time and frequency domains. To describe the algorithm utilized in our technique we shall assume that the bandlimited frequency domain data is available between frequencies f_1 and f_2 and sampled at n frequency points in between f_1 and f_2. This data can be expressed as a row vector of length n, represented as

$$H(1{:}n) = [H_1\ H_2\ H_3\ \ldots\ H_n] \tag{23}$$

where the notation H(1:n) indicates that samples 1 through n are elements of the row vector H. In general H(p:q) indicates samples H_p through H_q.

In order to extract the real and causal time domain signal for which we have the complex frequency domain data in the frequency band $[f_1,f_2]$, we first extrapolate the available frequency domain data between frequencies $[0,f_1]$ and from $[f_2,f_3]$, where f_3 is a frequency chosen until which extrapolation is considered necessary. Since $H(e^{j\omega}) = H^*(e^{-j\omega})$ the extrapolated data between frequencies $[-f_3,0]$ will be symmetrical to that between frequencies $[0,f_3]$. Therefore, extrapolation in the negative frequency domain will not require any additional calculation. The steps involved to extract the time domain sequence from a bandlimited frequency domain data are explained below.

1. The available bandlimited frequency domain data H(1:n) is padded with zeros to the extent necessary. To extract a casual, time domain sequence of finite length n samples the frequency data has to be padded with zeros to a minimum of 2N points, providing a sequence of even length. Therefore the available frequency domain data has to be padded

with 2N - n zeros. A zero padded vector HZ(1:2N) is formed from H(1:n) as

$$HZ(1{:}2N) = [H_1\ H_2\ \ldots\ H_n\ 0\ 0\ \ldots\ 0\ ]$$

2. The zero padded complex frequency domain data is now split into corresponding real and imaginary parts.

$$HZ_{\Re} = \mathrm{Real}(HZ)$$

$$HZ_{\Im} = \mathrm{Imag}(HZ)$$

3. An inverse discrete Fourier transform of the real part of the frequency data is performed which results in an even sequence Eq. (6). This is the even part of the actual time domain sequence.

$$h_e(1{:}2N) = \mathrm{idft}(HZ_{\Re})$$

This sequence is a row vector of dimension 2N.

4. The odd sequence is obtained from the even sequence by inverting the sign on the second half of the even sequence Eq. (13). In the odd sequence the first sample and the N+1th sample must be made zero. Hence,

$$h_o(1{:}2N) = [0\ \ h_e(2{:}N)\ \ 0\ -h_e(N+2{:}2N)]$$

5. The odd sequence is subjected to a discrete Fourier transform so as to create the imaginary part of the spectrum Eq. (11).

$$HZ_{\Im}new = \mathrm{Imag}[\mathrm{dft}(h_o)]$$

6. The imaginary part of the bandlimited frequency data extracted in step 2 is substituted into the data generated by the discrete Fourier transform of the odd sequence in the previous step.

$$HZ_{\Im}subs = [\ HZ_{\Im}(1{:}n)\ \ HZ_{\Im}new(n+1\ :\ 2N)\ ]$$

7. The new imaginary data $HZ_{\Im}subs$ thus formed is subjected to an inverse discrete Fourier transform, which will return an improved version of the odd sequence.

$$h_o new = \mathrm{idft}(jHZ_{\Im}subs)$$

8. The even sequence is obtained from the odd sequence by inverting the sign on the second half of the odd sequence. In the even sequence the first sample and the N+1th sample are substituted by the first and N+1th samples of the previous version of the even sequence. Thus,

$$h_e new = \left[\ h_e(1)\quad h_o new(2{:}N)\quad h_e(N+1)\ -\ h_o new(N+2\ :\ 2N)\ \right]$$

9. The even sequence of samples are subjected to a discrete Fourier transform so as to recreate the real part of the spectrum.

$$HZ_{\Re} new\ =\ Real[dft(h_e new)]$$

10. The real part of the bandlimited frequency data extracted in step 2 is substituted into the data generated by the discrete Fourier transform of the even sequence in the previous step.

$$HZ_{\Re} subs\ =\ \left[\ HZ_{\Re}(1{:}n)\quad HZ_{\Re} new(n+1\ :\ 2N)\right]$$

11. $HZ_{\Re}$subs thus created is subjected to an inverse discrete Fourier transform as in the third step above to result in an even sequence.

$$h_e(1{:}2N)\ =\ idft(HZ_{\Re} subs)$$

12. Subsequent processing is an iteration of the steps 2-11.
13. The time domain sequence is the sum of the even sequence and the odd sequence which have been created. The iteration process is terminated when a causal time domain sequence with sufficient resolution is generated.

$$h(1{:}2N)\ =\ h_e(1{:}2N)\ +\ h_o(1{:}2N)$$

14. If it is necessary to extrapolate the frequency data further, it is possible by zero padding the new frequency domain data, available between frequencies $[-f_3,f_3]$ until the frequency required say f_4 and the same procedure as above can be applied again.

Subsequent sections illustrate the efficacy of this technique. We have provided several simulation results. We have also implemented this technique on actual frequency domain measurements performed on a Hewlett Packard network analyzer HP 8510B.

SIMULATION RESULTS

Two Exponentially Decaying Sinusoidal Signals

In this example we consider two damped sinusoidal waves, giving a time signal of the form

$$f(t)\ =\ e^{-at}\sin(2\pi f_1 t)\ +\ e^{-bt}\cos(2\pi f_2 t) \tag{24}$$

where a = 0.6, b = 0.5, f_1 = 60.5Hz and f_2 = 32.5Hz. The signal is represented by a set of 32 samples. A 96 point DFT is performed on the signal to obtain the frequency domain data. In this example we assume knowledge of 16 samples out of the 96 samples lying in an intermediate band of frequencies. The frequency spacing between the samples is $2\pi/96$ rad/sec. The bandlimited data considered is in between $\omega = \pi/3$ rad/sec and $2\pi/3$ rad/sec. This bandlimited data was zero padded with 80 zeros for the spectral components between $2\pi/3$ and π and between $-\pi$ and $\pi/3$ rad/sec. The zero padded complex frequency data was processing using our technique. In Fig. 1(a) and Fig. 1(b) plots of the extrapolated real and imaginary parts of the frequency data are shown and they coincide closely with the actual real and imaginary parts of the DFT of the time domain response. A distinctive feature of this example is that the bandlimited data did not contain the spectral peak. However it becomes visible after processing. Fig. 2(a) contains the plot of the time domain response which is a result of the inverse discrete Fourier transform of the bandlimited data. It is non-causal and inaccurate. Compare it to the signal which results from the processing. The signal obtained after processing is found to coincide closely with the original signal. In Fig. 2(b) as in the previous example the imaginary part of time domain signals is found to be of considerable amplitude in the case of a direct transformation of the bandlimited data into the time domain. Whereas, after processing the frequency data, the time domain response has no imaginary component.

Rectangular Pulse

The signal considered in this example is a rectangular pulse of duration 2 ns. The amplitude of the pulse is 1 unit. To simulate the bandlimited frequency domain data for this pulse we perform a 64 point FFT on it. The real and imaginary parts of the 64-point FFT of the pulse are shown in Fig. 3(a) (and Fig. 3(b)) and corresponds to samples of $H_{\Re}(\omega)$ (and $H_{\Im}(\omega)$) at a frequency spacing of $2\pi/64$ rad/sec. The bandlimited data considered for this example is for $\omega = 0$ to $\pi/2$ rad/sec. The real and imaginary parts of the bandlimited data are also plotted in Fig.3(a) and Fig. 3(b). The bandlimited data was extrapolated using the iterative technique. The extrapolated data is observed to approximately coincide with the original data as in Fig. 3(a) and Fig. 3(b) and this does not improve with the number of iterations. The reason for this can be observed from Fig. 4(a). Though the time domain response after processing is more accurate than the one derived from the bandlimited data, the presence of Gibbs' effect limits this accuracy. However causality is maintained in the response derived after extrapolation. The imaginary part in the time domain data is completely attenuated by the processing, as seen in Fig. 4(b).

Experimental Data from an HP 8510B Network Analyzer

In this example we have used experimental data acquired from measurements performed on a Sliding load standard using HP 8510B network analyzer [3]. The load element of the Sliding load can be placed at five positions and after calibration at these positions the frequency response of the Sliding load is measured. The HP 8510B network analyzer was used to measure the S_{11} parameters from 45 MHz to 9.045 GHz using 401 samples. The frequency spacing between samples is .0225 GHz.

As a first step, the real and imaginary components of the frequency domain data, as plotted in Fig. 5 are padded with zeros to accommodate spectral components between frequencies -9.045 GHz and 45 MHz. Processing this zero padded complex frequency data using our technique, the data is extrapolated between frequencies -9.045 GHz and 45 MHz. This is plotted in Fig. 6(a) and Fig. 6(b). Fig. 7 shows the magnitude of the frequency domain data after extrapolation in the frequency domain. The time domain response extracted by directly performing an IDFT on the zero

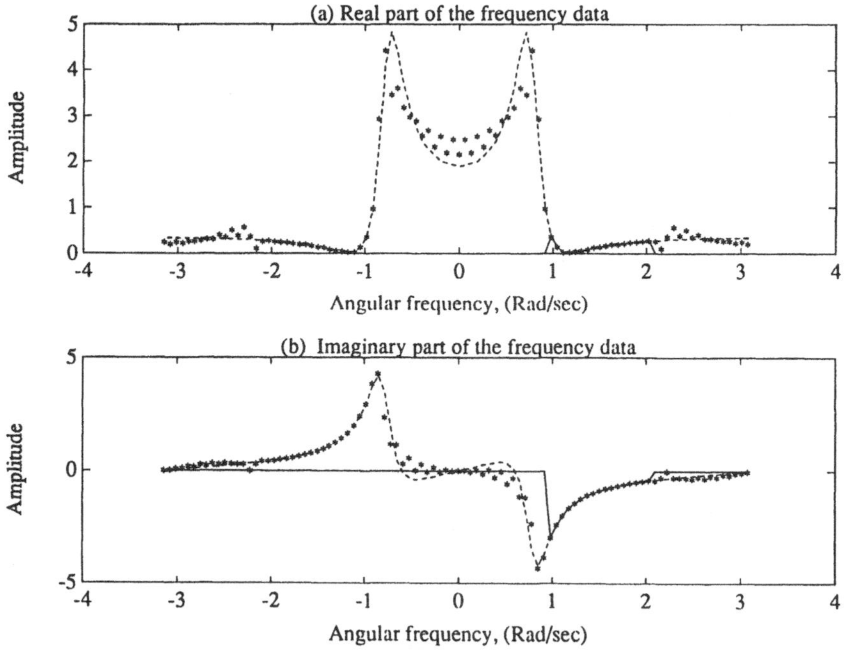

Figure 1. (a)Real part of the frequency response, the band limited data is between frequencies $\omega = \pi/3$ rad/sec and $2\pi/3$ rad/sec. (b)Imaginary part of the frequency response. legend: - Bandlimited frequency data, ★★★ Extrapolated frequency data, --- Simulated actual frequency response.

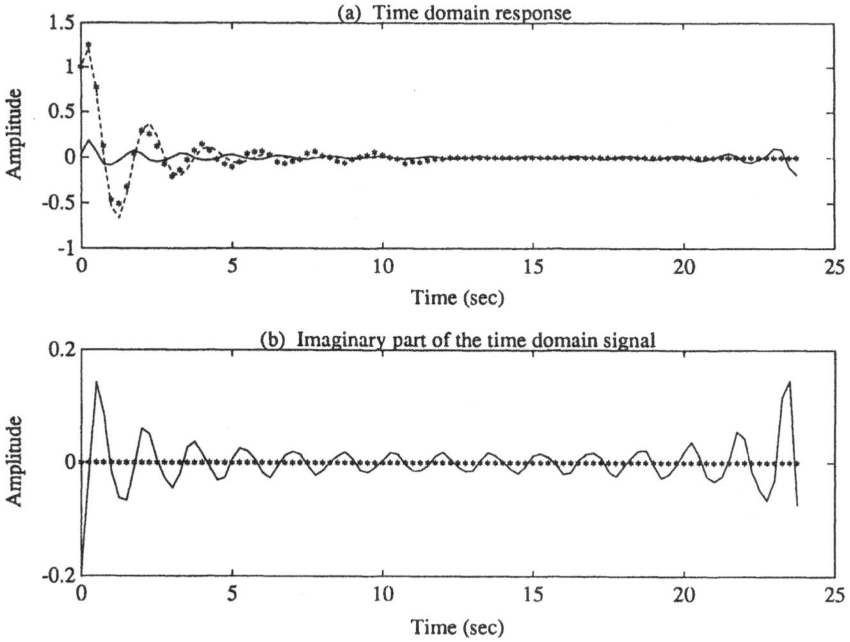

Figure 2. (a)The time domain responses extracted from the bandlimited frequency data and the actual simulated time domain response. (b)Imaginary part of the time domain response which appears as a result of the IFFT of the bandlimited complex frequency data. legend: - The time domain response extracted directly from the bandlimited data, ★★★ The time domain response extracted from the bandlimited data after processing, ---The simulated time domain signal.

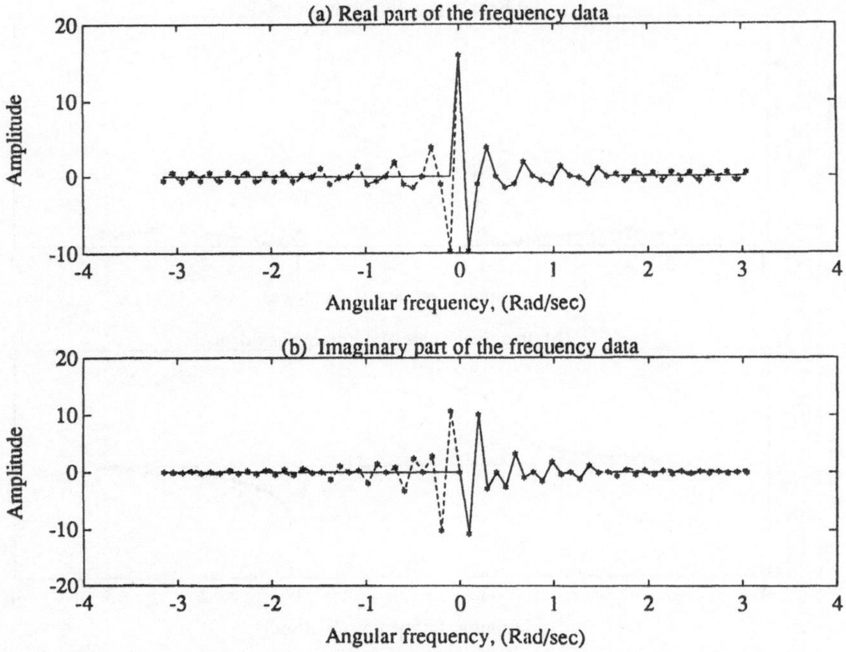

Figure 3. (a)Real part of the frequency response, the band limited data is between frequencies 0 and $\pi/2$ rad/sec. (b)Imaginary part of the frequency response. legends: - Bandlimited frequency data, ★★★ Extrapolated frequency data, --- Simulated actual frequency response.

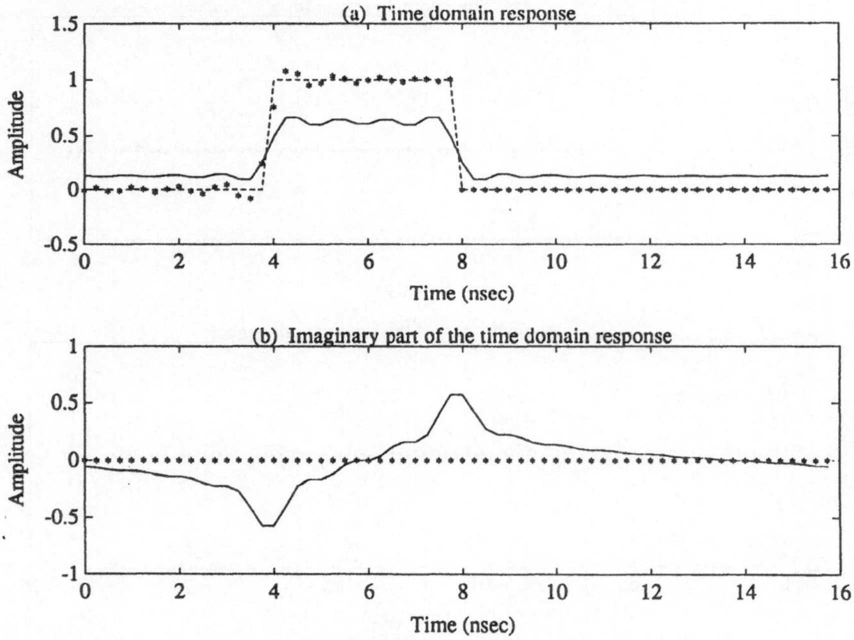

Figure 4. (a)The time domain responses extracted from the bandlimited frequency data and the actual simulated time domain response. (b)Imaginary part of the time domain response which appears as a result of the IFFT of the bandlimited complex frequency data. legends: - The time domain response extracted directly from the bandlimited data, ★★★ The time domain response extracted from the bandlimited data after processing, --- The simulated time domain signal.

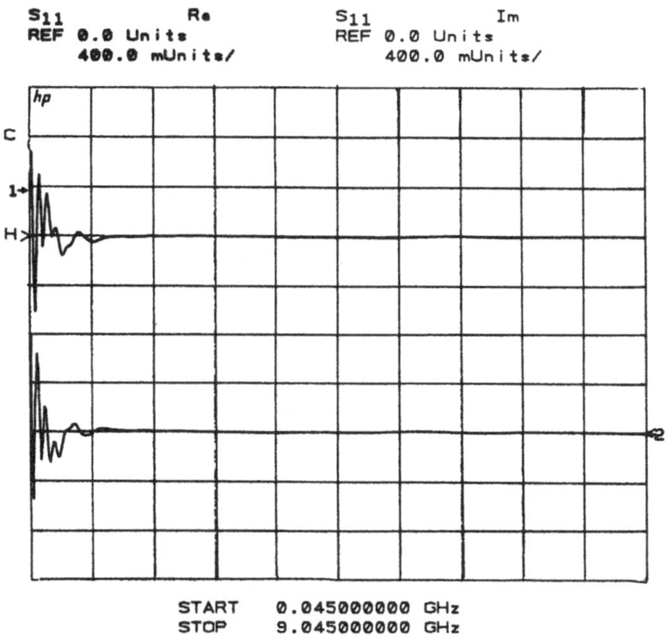

Figure 5. Real and imaginary parts of the frequency domain data measured on a HP 8510B Network Analyzer for a Sliding load.

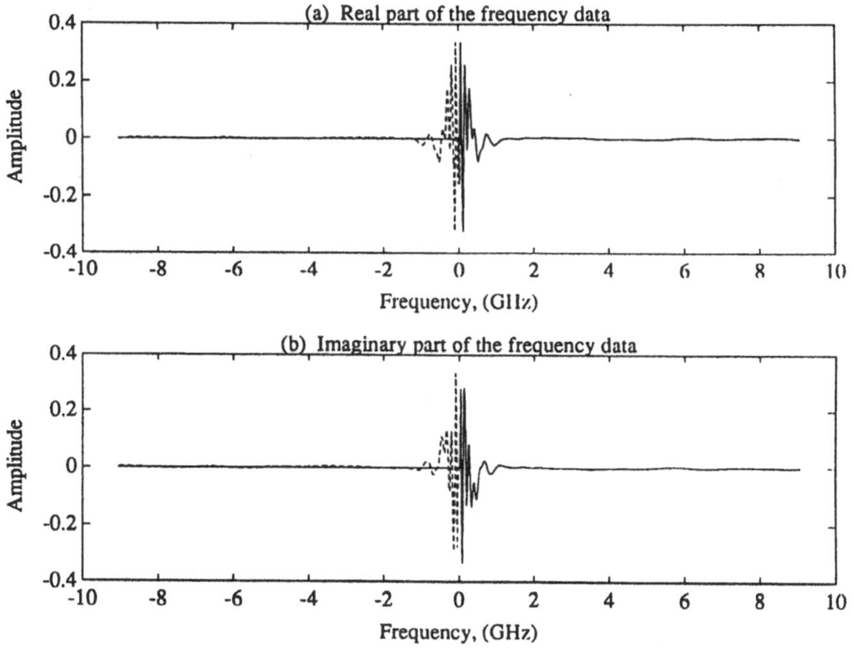

Figure 6. (a)Real part of the frequency response, the bandlimited data is between frequencies 0.045 GHz and 9.045 GHz. (b)Imaginary part of the frequency response. legends: - Bandlimited frequency data, --- Frequency data after extrapolation.

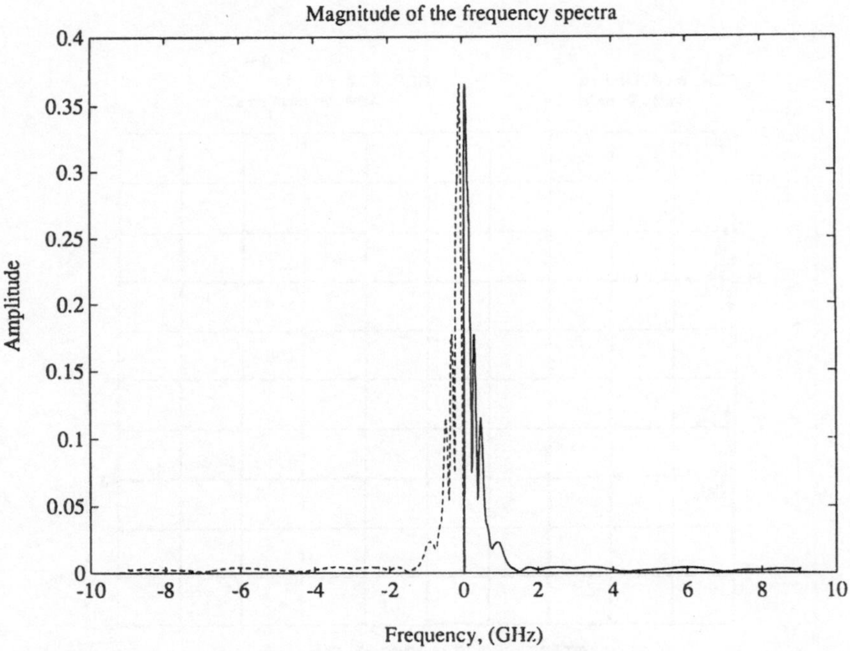

Figure 7. The magnitude of the frequency domain response prior to and after processing. legends: - Bandlimited frequency response, --- Frequency data after extrapolation.

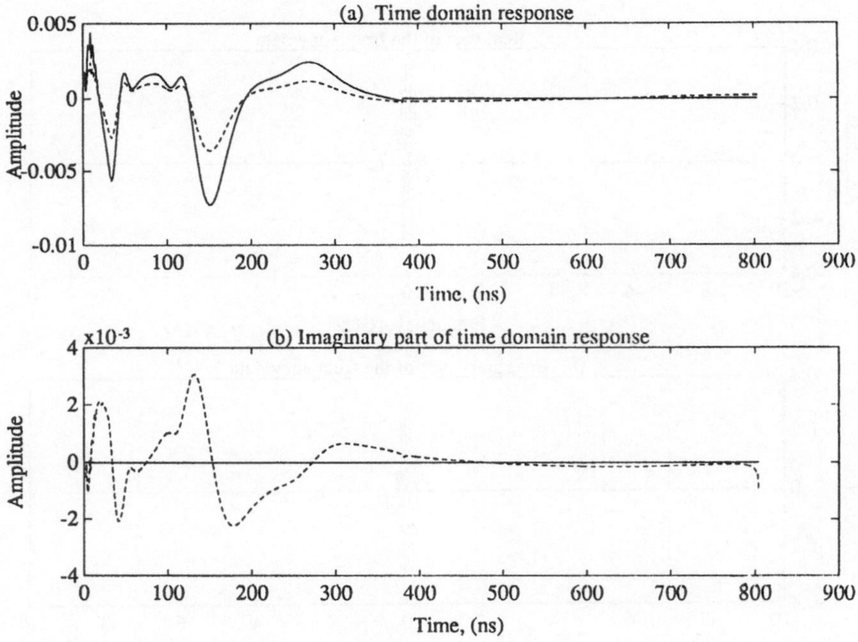

Figure 8. (a)The time domain responses extracted from the bandlimited frequency data. (b)Imaginary part of the time domain response which appears as a result of the IFFT of the bandlimited complex frequency data. legends: - The time domain response extracted from the bandlimited data after processing by the technique, --- The time domain signal extracted directly from the frequency domain data by performing an IDFT on the bandlimited data.

padded (bandlimited) frequency domain data and the time domain response extracted after processing the data are plotted in Fig. 8(a) and Fig. 8(b). We plotted the imaginary part of the time domain response in Fig. 8(b) to show that by processing the data the imaginary part of the time domain response becomes zero, as it should be. Whereas, by merely performing an IDFT of the bandlimited data creates a considerably large imaginary part of the time domain response. The time domain response can be seen in Fig. 8(a) to be casual.

Next, we have used data between .045 GHz and .3825 GHz, which is represented by 16 samples and applied our technique after padding with 789 zeros to accommodate spectral components between -9.045 GHz and .045 GHz, and spectral component between .3825 GHz and 9.045 GHz. The extrapolation of the real and imaginary parts of the frequency data are plotted in Fig. 9(a) and Fig. 9(b). The magnitude response of the Sliding load prior to and after processing are shown in Fig. 10. Observe the presence of a third peak in the magnitude of the frequency domain response created as a result of the extrapolation. The time domain response is extracted and the real and imaginary parts of the time domain response are plotted in Fig. 11(a) and Fig. 11(b). The time domain response extracted using the technique is a causal response and has no imaginary parts.

CONCLUDING REMARKS

The Hilbert transform relationship between the real and imaginary parts of the frequency domain response for causal time domain signals proved to be a sound basis for the extrapolation of the frequency domain data and extraction of a causal and real time domain response. The reconstruction of the time domain signal is excellent given a sufficient number of iterations. Specifically, the peak value of the impulse response and the shape of the time domain response are accurately represented. The presence of an imaginary component in the time domain response as a result of the inverse Fourier transform of bandlimited frequency domain data is nullified as a result of the processing. Rather than using the direct relationship between the real and imaginary parts of the frequency domain response, we used the fact that the real part of the frequency response is the Fourier transform of the even part of the time domain response, and the imaginary component of the frequency response is the Fourier transform of the odd part of the time domain response.

Application of this technique in the case of noisy frequency domain data has been discussed. With an approximate knowledge of the bands of frequency data which have been corrupted by noise it was shown possible to extract the true time domain response. The technique was applied to measurements performed on the HP 8510B vector network analyzer for a Sliding load standard and the results have been discussed.

The results may not be absolutely convergent if sufficient amount of bandlimited data is not available, however this also depends on the energy content of the spectral components in the bandlimited data. This is seen in the example of the noisy frequency data where four samples of the frequency data in the band $\omega = 0$ to $\pi/4$ rad/sec was sufficient to reconstruct the time domain response. Further investigation is necessary to find out what characterizes the quality of the extracted time domain response with regards to amount of bandlimited data available, and the energy of the spectral components present within the bandlimited data. The entire extrapolation procedure was performed by zero padding at the frequencies where the data is absent. Though this is ideal in dealing with any general frequency response,

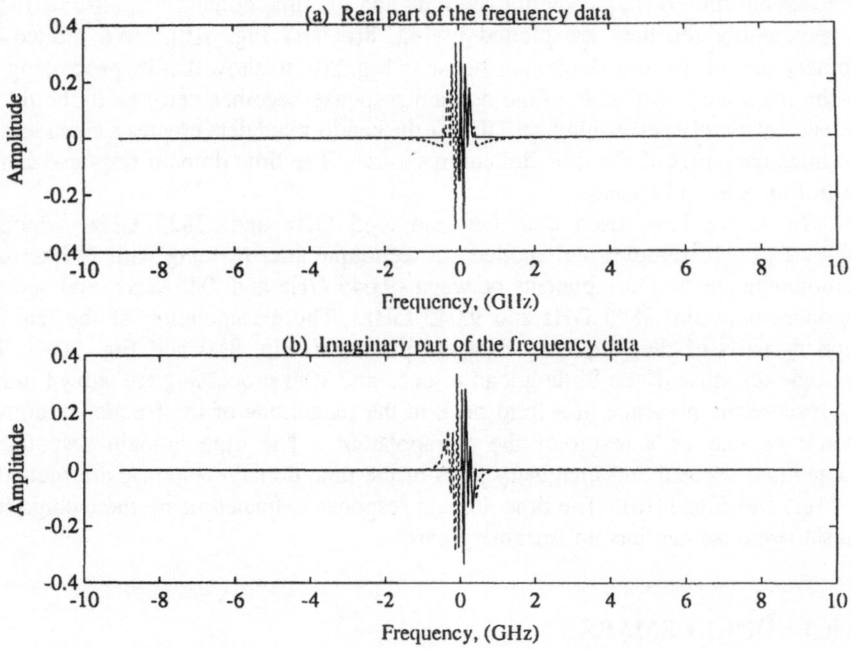

Figure 9. (a)Real part of the frequency response, the bandlimited data is between frequencies 0.045 GHz and 0.3825 GHz. (b)Imaginary part of the frequency response. legend: - Bandlimited frequency data, --- Frequency data after extrapolation.

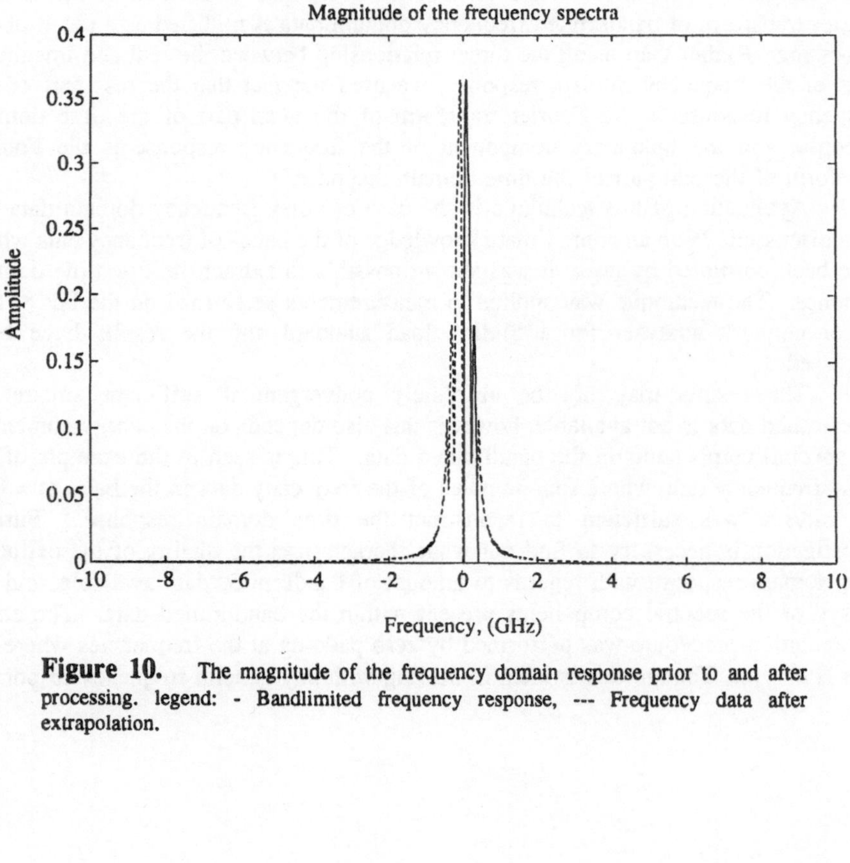

Figure 10. The magnitude of the frequency domain response prior to and after processing. legend: - Bandlimited frequency response, --- Frequency data after extrapolation.

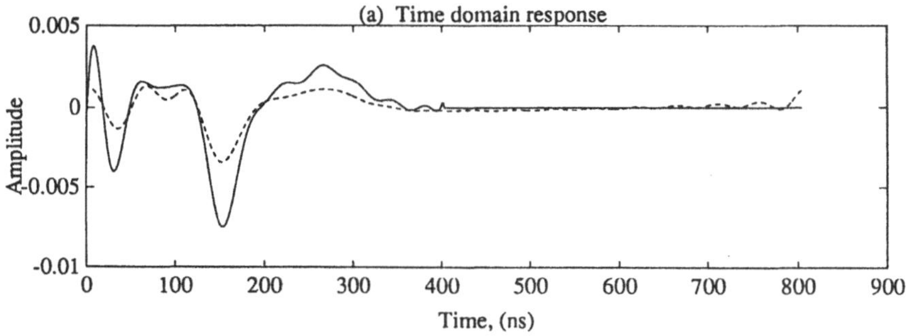

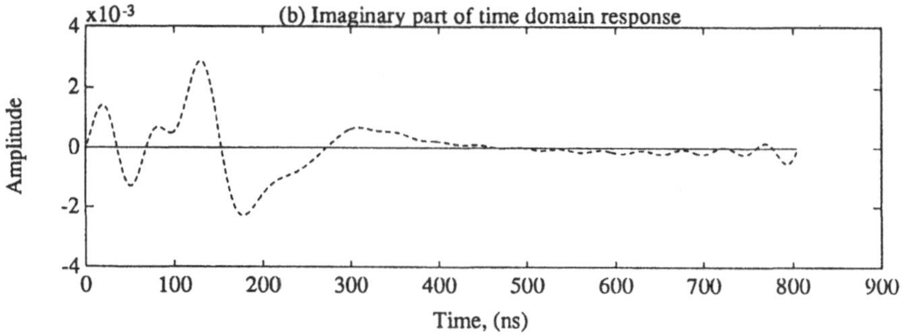

Figure 11. (a)The time domain responses extracted from the bandlimited frequency data. (b) Imaginary complex frequency data. legend: - The time domain response extracted from the bandlimited data after processing by the technique, --- The time domain signal extracted directly from the frequency domain data by performing an IDFT on the bandlimited data.

it will probably be advantageous in terms of reduced number of iterations and also a more accurate time domain response if it is possible to find a correlation between adjacent data on either side of the bandlimited complex frequency data and fit a curve to predict the frequency data. The entire extrapolation procedure was performed by zero padding at the frequencies where the data is absent. Though this is ideal in dealing with any general frequency response, it will probably be advantageous in terms of reduced number of iterations and also a more accurate time domain response if it is possible to find a correlation between adjacent data on either side of the bandlimited complex frequency data and fit a curve to predict the frequency data. The extrapolation procedure should remain the same as before.

Acknowledgments

The authors are grateful to Mr. Zoran Marićević for providing data measured on the HP 8510B network analyzer.

Bibliography

1. Discrete Hilbert transform, V. Čizěk, IEEE Trans., Audio and Electroacoustics, vol. AU-18, No. 4, pp. 340-343, Dec. '70.
2. Discrete-time signal processing, A. V. Oppenheim and R. W. Schafer, Prentice Hall, Englewood Cliffs, NJ, 1989.
3. Hewlett-Packard Company HP 8510B Operating and Programming manual, HP Part No. 08510-90070, Hewlett-Packard, Santa Rosa, CA, 1987.

POLARIZATION PROCESSING FOR ULTRA WIDEBAND TARGET DETECTION[1]

Chul Jai (CJ) Lee
Massachusetts Institute of Technology Lincoln Laboratory
244 Wood Street, Rm. KB-300
Lexington, MA 02173-9108

Hong Wang
Department of Electrical & Computer Engineering
Syracuse University
Syracuse, NY 13244-4100

Clifford Tsao
Directorate of Surveillance and Photonics
Rome Laboratory
Griffies AFB, NY 13441-4514

abstract - This paper reveals a detection performance advantage of polarimetric Ultra Wide Band (UWB) systems over polarimetric Narrow Band (NB) systems. With UWB to resolve the target of interest into its Multiple Dominant Scatterers (MDSs), the so-called signal cancellation problem associated with polarimetric NB systems is essentially eliminated via randomization of MDS polarization, resulting in a much more reliable detection performance improvement than what NB systems can offer with polarization processing.

I. Introduction

Polarization domain processing has been used for years in conventional Narrow-Band (NB) surveillance radar systems to improve detection of slowly and/or tangentially moving targets in clutter, and it has been an increasingly important research area whose potential is yet to be fully understood [1].

The Clutter Polarization Canceler (CPC) is among the simplest methods of polarization processing as it requires dual polarization channels on receive only and does not need to know the target polarization vector [2]. With NB systems, however, CPCs have a serious performance degradation problem of the so-called signal cancellation [3], which occurs when the polarization vector of the target to be detected is close to that of the surrounding clutter.

A Ultra Wide Band (UWB) system is characterized by its fine range resolution with the use of a large bandwidth, and it usually has a large relative bandwidth (even over 100%). Recently research and development of UWB radar systems has again received much attention, as their potentials as a tool for surveillance, target identi-

[1]This work was partially supported by the Defense Advanced Research Projects Agency under Air Force Contract F30602-91-C-0035 managed by the AFSC Rome Laboratory. The work of CJ Lee was also supported by the Graduate Assistant in Areas of National Need Fellowship from the US Dept. of Education, and it was conducted while he was with Syracuse University.

fication/ recognition, remote sensing, etc., are gradually recognized. A UWB based wide-area surveillance, for example, can offer such good features as foliage/ ground penetration, much better moving-target detection, low probability of interception, etc. Although most of UWB research and development has only involved a single "polarization" channel so far, the extension of the NB concept of polarization to UWB has been under investigation with an emphasis on the modeling of target-clutter polarization over a wide frequency band [4].

The purpose of this paper is to show and quantify an additional performance advantage of a UWB based surveillance system that employs a CPC. With UWB to resolve targets of interest into Multiple Dominant Scatterers (MDSs), we will show that the signal cancellation problem is essentially eliminated, resulting in much more reliable detection performance improvement than what NB systems can offer with polarization processing.

II . UWB POLARIMETRIC DATA MODEL

For the purpose of this paper stated in the previous section, we will be using the simplest UWB target and clutter models as outlined below. More sophisticated polarimetric UWB target and clutter modeling can be found in [4].

We consider a UWB system with a single polarized transmitter and dual orthogonal polarized receiver, which may have a large relative bandwidth (> 10%). The bandwidth of the system is large enough to break the target of interest into well separated Multiple Dominant Scatters (MDSs) each of which is modeled by a perfect point target with a flat frequency response.

Let $\mathbf{u}(t)$ be the waveform vector received by an UWB radar system at the front end of its dual-polarized receiver. It consists of the returns from the target, clutter as well as the internal receiver noise, the measurement from the MDS which can be expressed as

$$\begin{aligned}\mathbf{u}(t) &= \begin{bmatrix} \mathbf{u}_1(t) & \mathbf{u}_2(t) & \cdots & \mathbf{u}_m(t) & \cdots & \mathbf{u}_M(t) \end{bmatrix} \\ &= \begin{bmatrix} u_{h1}(t) & u_{h2}(t) & \cdots & u_{hm}(t) & \cdots & u_{hM}(t) \\ u_{v1}(t) & u_{v2}(t) & \cdots & u_{vm}(t) & \cdots & u_{vM}(t) \end{bmatrix}\end{aligned} \tag{1}$$

where the two subscripts in the last equation refer to the receiver polarization state and multiple dominant scatterer indices, $m = 1, 2, \cdots, M$, respectively. It indicates that the $2M$ waveform data are available to process.

The mth dominant scatterer waveform data $\mathbf{u}_m(t)$, 2×1, can be expressed in terms of received waveform $\mathbf{u}(t)$, i.e.,

$$\mathbf{u}_m(t) = \mathbf{u}(t + (m-1)T_M); \quad 0 < t < T_M \tag{2}$$

where T_M is the observation interval for a single dominant scatterer. Here it is assumed that the dominant scatterers are well separated in range, and the range resolution length must be large compared to its dominant scatterer range extent.

The attenuation factors for both h and v channels represent the target backscatter, progagation loses, antenna responses, and the radar cross section of the mth dominant scattering center. The attenuation vector may be expressed with an unit normalized polarization state vector $\mathbf{p}_{sm}$ which is deterministic

$$\begin{bmatrix} b_{shm} \\ b_{svm} \end{bmatrix} = \alpha_m \begin{bmatrix} p_{shm} \\ p_{svm} \end{bmatrix} = \alpha_m \mathbf{p}_{sm} \tag{3}$$

A probabilistic target model is employed such that the magnitude, $a = \alpha_m$ for all m, is assumed to be a random variable which is the resultant sum of all the scatterers within a range cell assumed to consist of the one-dominant scatterer plus Rayleigh fading, and its probability density function is given in [5].

The mth scattering target waveform can be written as

$$\mathbf{u}_{sm}(t) = \alpha_m f_m(t - \tau_m)\mathbf{p}_{sm} = \alpha_m \mathbf{p}_{sm}(t - \tau_m) \tag{4}$$

The target portion of received waveform, finally, given by

$$\mathbf{u}_s(t) = \mathbf{P}_s(t)\mathbf{a} \tag{5}$$

where

$$\mathbf{P}_s(t) = [\mathbf{p}_{s1}(t-\tau_1), \cdots, \mathbf{p}_{sm}(t-\tau_m), \cdots, \mathbf{p}_{sM}(t-\tau_M)]; \quad (2 \times M) \tag{6}$$

and

$$\mathbf{a} = \begin{bmatrix} \alpha_1 \\ \vdots \\ \alpha_m \\ \vdots \\ \alpha_M \end{bmatrix}; \quad (M \times 1) \tag{7}$$

The colored noise vector due to clutter is obtained by convolving (with respect to range) the transmit UWB signal $f(t)$ with the range-doppler variant scattering vector of the clutter process $\mathbf{b}_c(t-\lambda/2, \lambda)$, where

$$\mathbf{b}_c(t-\lambda/2,\lambda) = \begin{bmatrix} b_{ch}(t-\lambda/2,\lambda) \\ b_{cv}(t-\lambda/2,\lambda) \end{bmatrix} \tag{8}$$

The colored noise vector becomes

$$\mathbf{u}_c(t) = \sqrt{e_t}\int_{-\infty}^{\infty} \mathbf{b}_c(t-\lambda/2,\lambda) f(t-\lambda)\, d\lambda \tag{9}$$

where t is time, λ is radar range expressed in units of time, and $\sqrt{e_t}$ is the energy of the transmitted signal. This expression represents the colored noise vector due to the clutter.

The clutter covariance functional matrix is defined as a 2×2 real symmetric matrix,

$$\mathbf{K}_c(t,z) = E\{\mathbf{u}_c(t)\mathbf{u}_c^T(z)\} \tag{10}$$

Substituting $\mathbf{u}_c(t)$ into Eq.(10) and assuming the returns from different range intervals are statistically independent and that the return from each interval is a sample vector function of a stationary zero-mean Gaussian random process, the the waveform covariance matrix $\mathbf{K}_c(t,\tau)$ can be approximated by [5]

$$\mathbf{K}_c(t,\tau) = \frac{1}{2B}\mathbf{R}_c \cdot \delta(t-\tau) \tag{11}$$

where $\mathbf{R}_c$, size 2×2, is the clutter covariance matrix. Therefore, the total covariance matrix of the clutter and receiver noise is given by

$$\mathbf{K}_n(t,\tau) = (\frac{1}{2B}\mathbf{R}_c + \frac{N_0}{2}\mathbf{I}) \cdot \delta(t-\tau) = \mathbf{R}\cdot\delta(t-\tau) \tag{12}$$

where $\mathbf{R}$ is the total noise covariance matrix, and $\mathbf{I}$ is an 2×2 identity matrix. It is assumed that each element in $\mathbf{K}_n(t,z)$ is square-integrable and that the receiver noise and clutter components are independent. The *apriori* clutter plus receiver noise model is described by

$$\mathbf{R} = \sigma_c^2 \cdot \begin{bmatrix} \varepsilon & \varrho\sqrt{\varepsilon} \\ \varrho^*\sqrt{\varepsilon} & 1 \end{bmatrix} + \begin{bmatrix} \sigma_n^2 & 0 \\ 0 & \sigma_n^2 \end{bmatrix} \tag{13}$$

The power ratio between co-polarized and cross-polarized channels is specified by ε, with $0 \leq \varepsilon \leq 1$, and the complex correlation between the voltages x_h and x_v is specified by a coefficient ϱ with $0 \leq |\varrho| \leq 1$. Together with degree-of-polarization parameter, the complex correlation coefficient provides a useful description of the time-varying polarized waves.

III . UWB Target Detecion Based on Optimum Clutter Polarization Canceler

The probability density function of the received waveform is given as [5], under H_0,

$$f(\mathbf{u}(t) \mid H_0) = d_0 exp(-\frac{1}{2}\int \mathbf{u}^H(t)\mathbf{R}^{-1}\mathbf{u}(t)dt)\ , \tag{14}$$

and, under H_1, the likelihood function can be written similar way

$$f(\mathbf{u}(t) \mid H_1) = d_0 \cdot exp\{-\frac{1}{2}\int [\mathbf{u}(t) - \mathbf{u}_s(t)]^H \mathbf{R}^{-1}[\mathbf{u}(t) - \mathbf{u}_s(t)]dt\} . \quad (15)$$

with constant d_0.

The MDS target modeling which we developed in in waveform-based UWB modeling section presents the returns from all the scattering centers

$$\mathbf{u}_s(t) = \sum_{m=1}^{M} \alpha_m \mathbf{p}_{sm}(t - \tau_m) = \mathbf{P}_s(t) \cdot \mathbf{a} \quad (16)$$

where τ_m is arrival time of mth scatterer, and $\mathbf{P}_s(t)$ and $\mathbf{a}$ are specified by Eq.(6) and Eq.(7), respectively. The exponent part of Eq.(15) after substituting Eq.(16) becomes

$$-\frac{1}{2}\int [\mathbf{u}(t) - \mathbf{u}_s(t)]^H \mathbf{R}^{-1}[\mathbf{u}(t) - \mathbf{u}_s(t)]dt \quad (17)$$

$$= -\frac{1}{2}\{\int \mathbf{u}^H(t)\mathbf{R}^{-1}\mathbf{u}(t)dt - \int \mathbf{u}^H(t)\mathbf{R}^{-1}\mathbf{a}dt \quad (18)$$

$$-\int \mathbf{a}^H \mathbf{P}_s^H(t)\mathbf{R}^{-1}\mathbf{u}(t)dt + \int \mathbf{a}^H \mathbf{P}_s^H(t)\mathbf{R}^{-1}\mathbf{P}_s(t)\mathbf{a}dt\} \quad (19)$$

Let us define

$$\mathbf{G} = \int \mathbf{P}_s^H(t)\mathbf{R}^{-1}\mathbf{P}_s(t); \;\; (M \times M) \quad (20)$$

$$\mathbf{u} = \int \mathbf{P}_s^H(t)\mathbf{R}^{-1}\mathbf{u}(t)dt; \;\; (M \times 1) \quad (21)$$

Then Eq.(19) becomes

$$-\tfrac{1}{2}\{\textstyle\int \mathbf{u}^H(t)\mathbf{R}^{-1}\mathbf{u}(t)dt - \mathbf{u}^H\mathbf{a} - \mathbf{a}^H\mathbf{u} + \mathbf{a}^H\mathbf{G}\mathbf{a}\}$$

$$= -\tfrac{1}{2}\{\textstyle\int \mathbf{u}^H(t)\mathbf{R}^{-1}\mathbf{u}(t)dt + [\mathbf{a} - \mathbf{G}^{-1}\mathbf{u}]^H\mathbf{G}[\mathbf{a} - \mathbf{G}^{-1}\mathbf{u}] + \mathbf{u}^H\mathbf{G}^{-1}\mathbf{u}\} \quad (22)$$

As we are assuming the amplitude of the scatterers are unknown, the likelihood ratio test is formulated as following

$$\eta = \max_{\mathbf{a}}\{\frac{f(\mathbf{u}(t) \mid H_1)}{f(\mathbf{u}(t) \mid H_0)}\} \underset{H_0}{\overset{H_1}{\gtrless}} \eta_o \quad (23)$$

Note that when $\mathbf{a} = \mathbf{G}^{-1}\mathbf{u}$, the maximum to $f(\mathbf{u}(t) \mid H_1)$ can be achieved, then

$$\eta = \exp\{\mathbf{u}^H\mathbf{G}^{-1}\mathbf{u}\} \underset{H_0}{\overset{H_1}{\gtrless}} \eta_o \quad (24)$$

Since $\exp(\cdot)$ is a monitonically increasing function, the equivalent test becomes

$$\ell = \{\mathbf{u}^H\mathbf{G}^{-1}\mathbf{u}\} \underset{H_0}{\overset{H_1}{\gtrless}} \ell_o \quad (25)$$

The optimum UWB dual-polarized channel receiver is designed for detecting a target in clutter as well as white noise. The optimum receiver follows the threshold comparison test accordingly with Eq.(25) where $\mathbf{G}$ and $\mathbf{u}$ satisfy the Eq.(20) and Eq.(21),respectively. A block diagram of the optimum UWB polarization processor is presented in Fig.(1).

The detection and false alarm probabilities for the optimum UWB polarization processing, corresponding to the test statistic of Eq.(25), are [5]

$$P_{fa} = e^{-\ell_o} \cdot \sum_{m=1}^{M/2-1} \frac{\ell_o^{M/2-m}}{(M/2-m)!} \quad (26)$$

and

$$P_d \;=\; \int_{\ell_o}^{\infty} \frac{\ell^{(M/2)-1} \cdot e^{-\ell/(1+\Delta/4)} \cdot (M/2)!}{(1+\Delta/4)^M} \cdot$$

$$\sum_{m=1}^{(M/2)} (\frac{\Delta/4}{1+\Delta/4})^m \cdot \frac{\ell^m}{m!((M/2)-m-1)!((M/2)-m)!} dt \qquad (27)$$

The optimum UWB polarization processing developed so far is directly based on the received waveform without any preprocessing, and is assumed to have received M dual-polarized echoes scattered by a target illuminated by single polarized transmittion. The problem is to process in an optimum way the M dual-polarized echoes to detect a target against clutter background and receiver noise. This optimum processor requires a priori information about not only approximate locations of MDS centers but also the radar cross section distribution, which are not usually available to the processor

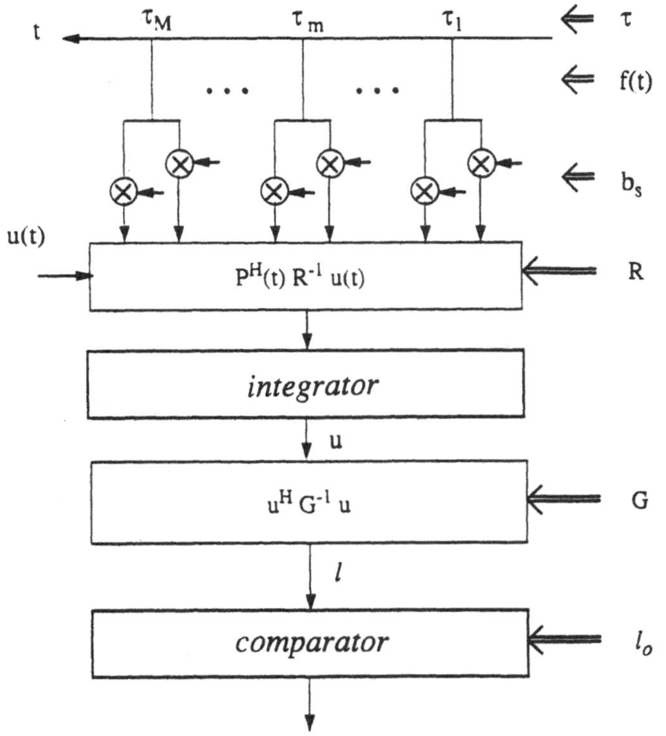

Figure 1: A Block Diagram of the Optimum UWB Polarization Processing.

in practice. The consequence of non-coherent integration approach without knowing these information will yield some performance degradation. Furthermore, the above waveform-based optimization procedurs involve integral and tensor equation which are very difficult to solve analytically. In general even the scalar case is not trivial. No further attempt was made to work out an specific example, since the main goal of this study is to compare the detection performance improvement of UWB system over the NB system in a polarization canceler based detector. However, the nature of these derived equations renders their solution a subject in itself.

IV. A Detecion Performance Comparison of Polarimetric NB and UWB Systems

With the optimum UWB CPC derived in the last section, we are to compare the probabilities of detection of polarimetric NB and UWB systems. We assume that the NB system also uses its optimum CPC. As any UWB system has the well-known clutter power reduction gain over on NB system due to its reduced range resolution cell, we will purposely specify the NB system in the following way which allows us to concentrate on the additional UWB gain over the NB with polarization domain processing.

Consider a NB system whose bandwidth is slightly smaller than the so-called critical bandwidth required to break a target into its MDSs. In other words, the size of the

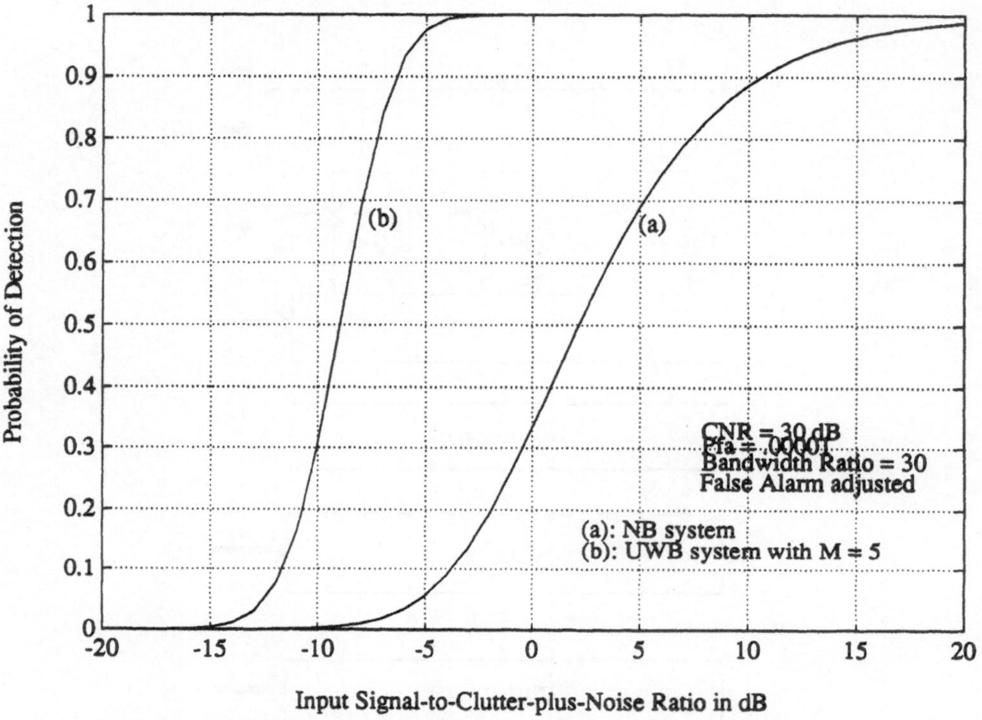

Figure 2: Optimum Polarimetric Processing UWB vs. NB : (a) NB polarimetric system and (b) UWB polarimetric system without the clutter power reduction gain.

range resolution cell of the NB system is nearly the same as the target size. Therefore any system with a larger bandwidth must process the data over multiple range cells to offer its best performance, i.e., the UWB range processing window must be as large as the range resolution cell size corresponding to the critical bandwidth. By setting the bandwidth of the NB system at the critical bandwidth and still approximating its target return by that of a perfect point target, we have equivalently deducted the well known UWB clutter power reduction gain in our performance comparison of NB and UWB systems. In addition, the detection thresholds of the NB and UWB systems are set to give the same probability of false alarm over the range processing window (equal to the NB system's range resolution cell size).

To make a comparison feasible while still keeping the basic conclusion valid, the following simplifications have been imposed on the target and clutter models.

(1) The MDSs of the UWB target have the same strength and in the comparison they are modeled by independent and identically distributed ($i.i.d$) complex Gaussian vectors.

(2) The point target model of the NB is also probabilistic and it is consistent with the above MDS model.

(3) The clutter return is frequency-independent.

Fig.(2) shows the detection probabilities of the above defined polarimetric NB and UWB systems, with the false alarm probabilities both set at $P_{fa} = 10^{-5}$ over the same range proessing window, and the number of the MDSs is 5. Clearly the polarimetric UWB system offers a significant detection performance gain, in addition to its clutter power reduction gain which has already been deducted from this comparison. Intuitively, this additional UWB gain comes mainly from the randomization of the target MDS polarization, as the MDS magnitude randomization has been shown in [6] to contribute a performance gain of only few dBs. One should note that, because of the MDS polariztion randomization, the polarimetric UWB performance gain is much more reliable than that of a polaimetric NB system which suffers from the signal cancellation effect.

V. CONCLUSION AND REMARK

The polarimetric UWB processor derived in this paper offers a much more reliable performance gain than the corresponding polarimetric NB processor which suffers from the signal cancellation effect. In [5,7] we showed that the signal cancellation problem can also be eliminated by using a multiple (narrow) band system. However multiband systems cannot offer many other features UWB systems have, such as a very fine range resolution.

REFERENCES

[1] *Proc. of the 2nd International Workshop on Radar Polarimetry*, Nantes, France, Sept. 8-10, 1992.

[2] D. Giuli, "Polarization Diversity in Radars," *Proc. IEEE,* vol. 74, No. 2, pp. 245-269, February 1986.

[3] M.C. Wicks, V.C Vannicola, K.C. Stiefvater, and R.D. Brown, "Polarization Radar Processing Technology," *Proc. IEEE International Radar Conference*, pp. 409-416, May 1990.

[4] W. M. Boerner, Chuan-Li Liu, Xin Zhang, and Vivek Naik, "Polarization Dependence in Ultra Wideband Impulsive Radar Target vs. Clutter Discrimination," *Proc. SPIE - The International Society of Optical Engineering*, pp. 87-110, Los Angeles, California, Jan. 22-23, 1992.

[5] H. Wang, and CJ Lee, "Adaptive Multiband and Optimum UWB Polarization Cancelers," *Pt. II of Phase I DARPA/RL Contract Report (F30602-91-C- 0035)*, June 1992.

[6] P. Swerling, "Detection analysis for range-extended targets," *TSC Course Notes*, April 1977.

[7] CJ Lee, and H. Wang, "An Adaptive Multiband Clutter Polarization Canceler," *Proc. of 1992 Joint Symposia, IEEE-APS/URSI/NEM*, Chicago, IL., July 18-25, 1992.

A TOMOGRAPHY-BASED APPROACH FOR IMAGING RADAR BACKSCATTER IN RANGE DELAY AND DOPPLER*

Marvin Bernfeld[1], F. Alberto Grünbaum[2]

[1]A.J. Devaney Associates, Inc.
355 Boylston Street
Boston, Massachusetts 02116
(508) 877-5995
[2]Mathematics Department
University of California
Berkeley, California 94720

ABSTRACT

High resolution imaging of radar backscatter distributed over range and Doppler can be achieved with a novel technique based on tomography. The technique employs tomographic imaging in conjunction with a chirp diversity radar that can view such distributions through a match filter system incorporating a suite of linear FM pulses which vary in their chirp rate. The motivation for this marriage conists of an analogy that exists with respect to post match filter signals and certain projections that are combined in tomography. An initial investigation has revealed that the point spread characteristic that can be achieved with this approach resembles the thumbtack-like ambiguity function long sought by radar scientists. The objective of this paper is to provide an explanation regarding the underlying analogy motivating this this new range-Doppler imaging technique.

INTRODUCTION

High resolution radar imaging can serve any combination of purposes in support of the fundamental radar modes, surveillance, tracking and classifying radar objects. One such purpose consists of mapping radar backscatter intensity that is jointly distributed over range delay and Doppler shift. Rotating objects such as the lunar surface are often candidates for this mapping which in this particular case is an intermediate step to obtaining high resolution physical images. However, potential applications occur in many other

*This research was supported by the Advanced Research Agency of the Department of Defense, and was monitored by the Air Force Office of Scientific Research under Contract No. F49620-89-C-0116.

fields such as reentry vehicle wake classification, weather forecasting, radar clutter mapping, microburst detection, traffic control, ionosphere sounding, medical imaging, astronomy, etc. In general, incorporating range-Doppler imaging in radar sytems can provide useful real-time knowledge concerning backscattering sources.

To extract these images via classical signal processing techniques, can be accomplished by actively correlating the backscatter with respect to range-Doppler displacements of the transmitted signal. Equivalent results can be obtained, however, from a match filter system with center frequency displacements covering the range of possible Doppler offsets. The images that are constructed in either case reflect a characteristic point spreading represented by the radar ambiguity function.

Practical implementations that have been developed are the pulse Doppler radars. These systems incorporate coherent signal processing in which the backscatter signals from a sequence of phase coherent pulse transmissions are first resolved in range, then transformed to the spectral distribution as a function of range. The coherent integration that is performed, during the latter processing, normally is preceded by linear FM pulse compression. This provides the resolution in range and, by employing linear FM for the pulse compression, the output amplitude is effectively normalized with respect Doppler shift.

Regarding pulse Doppler radars, the resolution in range is determined by the swept bandwidth contained in the linear FM pulses transmitted, while resolution in Doppler is determined by the total time duration that the backscatter signals are coherently integrated to derive their spectral representations. Stringent phase stability, rigid pulse schedules, range walk and possible deterioration of Doppler coherence during extended dwell times, as well as characteristic "bed-of-nails" ambiguities are several problems that can arise, ultimately limiting the applications for which the pulse Doppler radar can be an effective imaging approach.

An alternative technique to extract these images is currently being investigated[1]. The technique employs tomographic imaging in conjuction with a chirp diversity radar that can effectively view range-Doppler scattering distributions through a match filter system incorporating a suite of linear FM pulses which vary in their chirp rate. This marriage is motivated by an analogy that exists with respect to post match filter signals and certain projections that are combined in tomography. Since tomographic imaging does not depend on coherent signal processing, this novel imaging approach is fundamentally immune to the preceding problems confronting pulse Doppler radar developments. Perhaps, most significant among the benefits that the investigation has revealed consists of a point spread characteristic which resembles the classical thumbtack-like ambiguity function long sought by radar scientists. The objective of this paper is to provide an explanation regarding the underlying analogy motivating this new range-Doppler imaging technique.

TOMOGRAPHY

Tomography was initially applied in the field of medical imaging (c.1970). In this application, it is called CT -

an abbreviation for Computerized Tomography -, and it has revolutionized clinical radiology by supplying x-ray images which heretofore were not possible without invasive and often life threatening surgery.

Today, tomographic imaging methods have been developed for diverse fields such as geophysical probing, astronomy, and non destructive testing at industrial laboratories. Moreover the principles are not unfamiliar to radar engineers who have perceived that synthetic aperture radars ——spotlight SAR as well as ISAR——incorporate similar imaging methods.

Tomographic imaging methods are supported mathematically by the 2D Radon transform. Thus, tomography is based on the idea of reconstructing some 2D distribution from one dimensional projections of the distribution. In the classical implementation, the projections are individually obtained for uniformly sampled viewing angles covering 180 degrees, while individual projections are comprised of uniform samples consisting of parallel straight-line integrations across the distribution for one of these viewing angles.

Stating this idea in mathematical terms for a distribution $g(r,\psi)$, where r and ψ are polar coordinates that denote radial distance and angle, respectively, the 2D Radon transform can be described as follows

$$p(x',\theta) = \int_{-\infty}^{\infty}\int_{-\infty}^{\infty} f(x,y)\ \delta(x' - (x\ COS\ \theta - y\ SIN\ \theta))\,dxdy \qquad (1)$$

where $f(x,y)$ corresponds in cartesian coordinates to the distribution $g(r,\psi)$ and $\delta()$ denotes a line-type δ-function with linear projections on the x,y plane that are aligned as follows $x' = x \cos\theta - y \sin\theta$.

Equation (1) is a mathematical definition for the projections employed in tomography. To reconstruct $g(r,\psi)$, there are two possible approaches. One approach consists of the following inverse Radon transform.

$$g(r,\psi) = \frac{1}{4\pi^2}\int_0^{\infty} d\theta \int_{-\infty}^{\infty} dx'\ \frac{\partial p(x',\theta)/\partial x'}{r\cos(\psi - \theta) - x'} \qquad (2)$$

Alternatively, algebraic reconstruction techniques can be employed and enjoy equal popularity for imaging $g(r,\psi)$.

RADAR TOMOGRAPHIC PROJECTIONS

In x-ray tomography, the tomographic projections are acquired by physically realigning the direction that x-rays penetrate the distribution $g(r,\psi)$, where $g(r,\psi)$ represents a two-dimension distribution of x-ray attenuation. Another example of tomography in medicine consists of magnetic resonance imaging (MRI). In this case, the projections are acquired by electrically steering the straight-line integrations across $g(r,\psi)$ which now represents a distribution of nuclear magnetic resonance. Similarly, it is possible to acquire the tomographic projections of radar backscatter intensity jointly distributed over range and Doppler. This can be accomplished by viewing such distributions through a match filter system incorporating a suite of linear FM pulses that vary in their chirp rate. The technique is more reminiscent of MRI than x-ray tomography since the projections are acquired without physically rotating the direction of the parallel-line integrations. As explained below, this is determined by the chirp rate that is transmitted.

For transmitted radar pulses that contain linear frequency modulation, the match filter response coresponding to a point scatterer will appear at a time delay that depends not only on the range to the scatterer but also the scatterer's range-rate. The relationship can be expressed as follows

$$\left(\frac{C}{2}\right)\mathit{RANGE\ DELAY} = \mathit{RANGE} \pm \left(\frac{T}{\nabla f} f_0\right)\mathit{RANGE\text{-}RATE} \tag{3}$$

where T, ∇f and f_0, respectively, denote the time duration, the swept frequency, and the center frequency of the transmitted pulse. The addition or subtraction in eq.(3), respectively, holds for either up chirp or down chirp linear frequency modulation.

The graphic representation of the preceding equation is a straight line in the range and range-rate plane. The possible solutions for eq.(3) line up on a perpendicular to this line, intersecting it at a given range delay. Any scatterers distributed in range and range-rate so that they fall on a particular perpendicular line will contribute to the composite amplitude at the corresponding time delay measured at the output of the match filter. This sumation as a function of time delay is similar to the sequence of line integrations that are performed in tomography and represented by the 2D Radon transform expressed in eq.(1).

For a radar pulse with linear frequency modulation, therefore, the instantaneous match filter output can be described as a tomographic projection of the joint radar backscatter intensity in range delay and Doppler. Referring to (3), the line-of-view for which this projection is derived possesses a slope that is proportional $\nabla f/T$, the chirp rate. A suite of linear FM pulses that vary in their chirp rate, will yield a family of these projections. These can be combined, as in x-ray tomography or MRI, to produce an image of the joint radar backscatter intensity in range delay and Doppler.

Mathematically, the radar tomographic projections are defined by

$$\overline{\gamma(t)\gamma^*(t)} = \int_{-\infty}^{\infty}\int_{-\infty}^{\infty} \overline{|C(\tau,\phi)|^2}\ |\chi(\tau-t,\phi)|^2\, d\tau\, d\phi \tag{4}$$

where $\gamma(t)$, which is expressed below in complex notation, denotes the match filter output amplitude corresponding to the scattering amplitude, $C(\tau,\phi)$, distributed in range delay (τ) and Doppler (ϕ); the horizontal bar denotes ensemble averaging.

$$\gamma(t) = \int_{-\infty}^{\infty}\int_{-\infty}^{\infty} C(\tau,\phi)\ \chi(\tau-t,\phi)\ \exp\{-j2\pi f_0(\tau-t)\}\, d\tau\, d\phi \tag{5}$$

By utilizing the notation $|\chi(\tau,\phi)|^2$ to denote the radar ambiguity function, where $\chi(\tau,\phi)$ is the correlation function in complex notation expressed below, corresponding to a waveform $\mu(t)$, equations 4 and 5 are actually the general description of the match filter output for any waveform that might be transmitted.

$$\chi(\tau,\phi) = \int_{-\infty}^{\infty}\mu(t-\frac{\tau}{2})\mu(t+\frac{\tau}{2})\exp\{-j2\pi\phi t\}\,dt \tag{6}$$

The correlation function $\chi(\tau,\phi)$ corresponding to $\mu(t)$ possesses the important property that it becomes $\chi(\tau, \phi - a\tau/\pi)$ if $\mu(t)$ is multiplied by the quadratic phase function $\exp\langle jat^2\rangle$ or, equivalently, is modulated by linear FM with a chirp rate equal to $a/2\pi$. Thus, the correlation function is sheared in a perpendicular direction with respect to the τ axis as a result of this multiplication.

When the ambiguity function surface possesses a ridge-like appearance as is characteristic with the wideband linear FM pulses, the effect of this shearing is to rotate the ridge. For example, if the orignal waveform is an unmodulated pulse, the ridged surface is rotated to an alignment which coincides with the line $\phi-a\tau/2\pi=0$ where $a/2\pi$ is the slope of this line.

For pulses with sufficiently large bandwidth, the ridge will be sufficiently sharp that $|\chi(\tau,\phi)|^2$ can be approximated by a line-type δ-function. Equation 4 in that case describes a line integration reminiscent of the projections in tomography. With a suite of linear FM pulses that vary in their chirp rate providing different rotations of the linear FM ambiguity function, the corresponding match filter outputs can be combined as in tomography to derive high resolution images of backscatter intensity distributed over range delay and Doppler.

CONCLUSION

An analogy exists between the projections that are employed in tomography and the outputs that are obtained with a match filter radar incorporating a suite of linear FM pulses that vary in their chirp rate. Based on this, an alternative to coherent Doppler processing has been proposed for high resolution imaging in range delay and jointly in Doppler. The name that has been given to this idea is Chirp Diversity Radar.

The implementation of this idea has been approached so far by simulating the projections of some basic range-Doppler distributions. This has produced a point spread characteristic that resembles the the classical thumbtack-like ambiguity function long sought by radar scientists. As the investigation continues, with more realistic simulations introduced and with experiments involving real data, it is anticipated the results will support predictions that the preceding tomography-based imaging offers at least a viable alternative to pulse Doppler radar with respect to range-Doppler imaging.

REFERENCE

1. Bernfeld, M., Final Technical Report(1992): Tomographic Mathematical Ideas Applied to Radar Detection, DARPA Order No: 7090, AFOSR Contract F49620-89-C-0116.

A SIMPLE METHOD FOR OVER-RESOLVED MOVING TARGET DETECTION IN CLUTTER *

Lujing Cai[1], Hong Wang[1], Clifford Tsao[2], and Russell Brown[2]

[1]ECE Department, Syracuse University
Syracuse, NY 13244-1240

[2]Surveillance and Photonics Directorate, Rome Laboratory
Griffiss AFB, NY 13441-4514

ABSTRACT

This paper addresses the problem of the multipulse Moving Target Detection (MTD) with ultra wideband (UWB) radar systems. A simple UWB detection algorithm is developed to process the multipulse range walk data with large relative bandwidth. The new processing algorithm, which can "coherently" integrate the range-walk target components in multipulse returns without a predetection "line-up" operation, is shown by simulation to be able to deliver reliable detection performance close to the optimum MTD when the clutter spectrum spread is small.

I . INTRODUCTION

In [1], the performance potential of the moving target detection (MTD) with ultra-wideband/high-resolution radar systems is systematically evaluated and compared with that of conventional narrowband systems. The performance gains of the ultra wideband (UWB) systems are mainly identified as the clutter power reduction gain, the fluctuation reduction gain, and the extra discrimination gain, each of which shows up under different bandwidth ranges and target models. The overall performance improvement due to increasing system bandwidth suggests a potentially promising future of the UWB moving target detection in an increasing hostile surveillance environment.

To achieve the performance potential provided by the UWB waveforms, however, further research efforts are still necessary since the required implementation complexity becomes a major challenge to the application of the UWB technology to the moving target detection. With the use of the waveforms of very large relative

*The work of L. Cai and H. Wang was supported by the Defense Advanced Research Projects Agency under Air Force Rome Laboratory Contract F30602-91-C-0035.

bandwidth, the conventional processing methods, such as passband-to-baseband conversion and orthogonal receiver structure, are no longer suitable for the UWB processing. The optimum UWB processor derived in [1] performs its interference cancellation directly based on the received continuous waveform, which requires exact knowledge of the interference covariance. To maximally exploit the discriminations offered by the range walk and doppler shift, the target velocity is also required to be known to the multiple pulse integration part of the optimum processor. The optimum UWB processor is not practically realizable in the sense that it requires the *a priori* knowledge of the interference statistics and target characteristics, which are usually not available to the system. Therefore, how to perform the waveform-based interference suppression and moving target detection with the above practical limitations will be an interesting topic among the UWB applications.

As an extension of [1], this paper addresses the practical aspects of the multiple pulse moving target detection with UWB radar systems. Under the assumptions that the clutter is not moving and its doppler spectrum is relative simple, we propose a "realizable" UWB detection algorithm, which applies the simple pulse cancellation principle to the waveform-based interference suppression, and utilizes a "velocity filter bank" to cover the target velocity region of surveillance interest.

II . DATA MODELING AND OPTIMUM UWB PERFORMANCE

Consider an ultra wideband radar system which transmits a train of M pulses with constant pulse repetition interval T_p. Let $\mathbf{x}(t), M \times 1$ be a received multiple pulse vector waveform. It consists of

$$\mathbf{x}(t) = \mathbf{s}_r(t) + \mathbf{n}_c(t) + \mathbf{n}_n(t) \qquad 0 < t < T_p, \tag{1}$$

where $\mathbf{s}_r(t)$, $\mathbf{n}_c(t)$, and $\mathbf{n}_n(t)$ are the target signal return, interference/clutter return, and receiver noise, respectively. With a sufficient large signal bandwidth, the received target signal is composed of many multiple dominant scattering (MDS) centers, i.e., it can be expressed as [1]

$$\mathbf{s}_r(t) = \mathbf{G}(t)\mathbf{a}, \tag{2}$$

where the elements of the column vector $\mathbf{a}$ represent the amplitudes of the scatterers of the target, and $\mathbf{G}(t)$ is the MDS waveform matrix specified by

$$\mathbf{G}(t) = [\mathbf{g}(t-\tau_1) \quad \mathbf{g}(t-\tau_2) \quad \cdots \quad \mathbf{g}(t-\tau_J)], \tag{3}$$

with $\tau_j, j = 1, 2, ..., J$, being the arrival time of the scattering centers. Under the assumption that the propagation channel is non-dispersive, the single scatterer waveform vector $\mathbf{g}(t)$ received from a moving target can be written as

$$\mathbf{g}(t) = [g(t) \qquad g(t - \frac{2v_s}{c}T_p) \qquad \cdots \qquad g(t - \frac{2v_s}{c}(M-1)T_p)]^T, \tag{4}$$

where v_s and c are the target velocity and wave propagation velocity, respectively, and $g(t)$ is the transmitted single pulse waveform.

For the interference and noise components in Eq.(1), we assume that they are multivariate Gaussian random process with zero mean and waveform covariance function

$$\mathbf{K}(t,\tau) = \mathrm{E}\{(\mathbf{n}_c(t) + \mathbf{n}_n(t))(\mathbf{n}_c(\tau) + \mathbf{n}_n(\tau))^H\} = \mathbf{R}\delta(t-\tau). \tag{5}$$

where $\mathbf{R}$ is the total covariance matrix of interference and noise. The covariance function of this type presumes that the interference is not moving.

The data model described above is derived and reported in [1].

When the interference statistics and the target characteristics are known to the system, the optimum UWB processor, derived in [1], has the following decision rule

$$\eta = \mathbf{x}^H \mathbf{C}^{-1} \mathbf{x} \underset{H_0}{\overset{H_1}{\gtrless}} \eta_0, \tag{6}$$

where $\mathbf{C}$ is the inter-scatterer decoupling matrix defined by

$$\mathbf{C} = \int \mathbf{G}(t)^H \mathbf{R}^{-1} \mathbf{G}(t) dt \qquad J \times J, \tag{7}$$

and $\mathbf{x}$ is calculated from the test data

$$\mathbf{x} = \int \mathbf{G}(t)^H \mathbf{R}^{-1} \mathbf{x}(t) dt \qquad J \times 1. \tag{8}$$

Obviously, the *a priori* knowledge of the interference-plus-noise covariance $\mathbf{R}$, target velocity v_s, and scatterer arrival time $\tau_j, j = 1, 2, ..., J$, are required to form the test statistic. The optimum UWB processor possesses the maximum performance potential of the UWB moving target detection, which serves as the performance standard we attempt to approach when developing new realizable detection algorithms.

Under the assumption that the scatterer amplitudes $a_j, j = 1, 2, ..., J$ are non-fluctuating constants, the closed-form expressions of the probabilities of false alarm and detection are found as

$$P_f = \int_{\eta_0}^{\infty} \frac{\eta^{J/2-1}}{\Gamma(J/2)} e^{-\eta} dt \tag{9}$$

and

$$P_d = \int_{\eta_0}^{\infty} e^{-(\gamma+t)} \left[\frac{t}{\gamma}\right]^{\frac{J/2-1}{2}} \mathrm{I}_{J/2-1}(2\sqrt{\gamma t}) dt \tag{10}$$

where

$$\gamma = \mathbf{a}^H \mathbf{C} \mathbf{a}/2, \tag{11}$$

and $\mathrm{I}_{J/2-1}(\cdot)$ is the $(J/2-1)$th order modified Bessel function of the first kind.

III . THE SIMPLE UWB PROCESSOR

This section will present the simple UWB processor. It is considered as a realizable algorithm since it doesn't requir *a priori* knowledge about the interference covariance, scatterer separation and target velocity.

Since the clutter covariance is unknown to the processor, it is desired that the $\mathbf{R}$ in Eq.(7) and Eq.(8) can be replaced by a fixed matrix independent of $\mathbf{R}$. In a narrowband system, the conventional pulse canceler with binomial coefficients is usually successful in canceling the zero-velocity clutter with a relatively simple spectrum. For the UWB processing, we will use the similar pulse-to-pulse cancellation principle.

Let $\mathbf{T}$ be the transform matrix representing the operation of the pulse canceler, we operate it on the received data vector $\mathbf{x}(t)$

$$\mathbf{x}'(t) = \mathbf{T}\mathbf{x}(t). \tag{12}$$

This operation performs the pulse to pulse clutter cancellation directly based on the received continuous waveform, which can also be understood as a whitening procedure. It is easy to find that the signal component after the cancellation is given by

$$\mathrm{E}(\mathbf{x}'(t)) = \mathbf{T}\mathbf{G}(t)\mathbf{a} = \mathbf{G}'(t)\mathbf{a}, \tag{13}$$

where $\mathbf{G}'(t) = \mathbf{T}\mathbf{G}(t)$. Note that $\mathbf{x}'(t), M' \times 1$ is still a multiple pulse waveform and the useful information induced by target motion (i.e., range walk and doppler shift) is not employed yet. To maximally exploit the multiple pulse integration gain for target/clutter discrimination, the target velocity v_s should come into role in combining these multiple pulses. Thus the multiple pulse correlator, integrated with matched filtering, is formed in the following way

$$\mathbf{x} = \int \mathbf{G}'^H(t)\mathbf{x}'(t)dt, \qquad J \times 1, \tag{14}$$

where $\mathbf{G}'(t)$ is clearly a function of target velocity. As we will show in next section, the above multiple pulse correlator is useful when the clutter residue after the interference cancellation is still much larger than the target component.

The data vector $\mathbf{x}$ in Eq.(14) contains the signal components from each of the dominant scattering centers. If $\mathbf{G}(t)$ is completely known, i.e., if the target velocity and scatterer arrival time are known, the final test statistic can be formed by the inter-scatterer integration

$$\eta = \mathbf{x}^H \mathbf{C}^{-1} \mathbf{x} \underset{H_0}{\overset{H_1}{\gtrless}} \eta_0 \tag{15}$$

where $\mathbf{C}$ is the scatterer decoupling matrix

$$\mathbf{C} = \int \mathbf{G}'^H(t)\mathbf{G}'(t)dt, \qquad J \times 1. \tag{16}$$

The structure of the UWB processor is illustrated in Fig. 1. Similar to the optimum UWB processor, the processor consists of several major parts: the interference canceler, the multiple pulse correlator integrated with the matched filter, and the inter-scatterer integrator. It is not difficult to see that replacing $\mathbf{R}^{-1}$ by $\mathbf{T}^H\mathbf{T}$ in the optimum processor specified by Eq.(7) and Eq.(8) will lead to the above canceler-based UWB processor.

If scatterer separation is unknown, the range interval under test is evenly divided into N non-overlap range cells. Each of the range cells, no matter it contains a scatterer component or not, enters into the test. Thus the signal waveform matrix $\mathbf{G}(t), M \times N$, is assumed to include the signal vectors from all the range cells

$$\mathbf{G}(t) = [\mathbf{g}(t-\tau_1) \quad \mathbf{g}(t-\tau_2) \quad \cdots \quad \mathbf{g}(t-\tau_N)]. \tag{17}$$

where $\tau_l, l = 1, 2, ..., N$, are the centers of the range cells. As a result, the dimension of $\mathbf{x}$ becomes $N \times 1$. One can include all the data in the test by using Eq.(15) with $\mathbf{G}(t)$ being modified by Eq.(17). Since some of the range cells contain no target returns, the above "integration-of-all" scatterer integration scheme will introduce the so-called collapsing loss.

Other inter-scatterer integration schemes, such as the m-out-of-N and integration-of-m schemes, are studied and reported in our previous work [3]. The integration-of-all scheme is the special case of the integration-of-m if $m = N$. It is shown that the integration-of-m scheme outperforms the m-out-of-N scheme in general. Particularly, the integration-of-all is favorable as it shows a good performance over a wide range of conditions, and doesn't require the *a priori* knowledge about the number of scatterers. For this reason, we will use the integration-of-all scheme in our performance evaluation performed in next section.

Target velocity is another unknown factor we need to consider in practice. To achieve a reliable detection performance, the signal waveform vector $\mathbf{G}(t)$ implied in Eq.(14) is required to be identical to the actual one contained in the receive data. Thus the underlying range walk rate and doppler frequency shift in $\mathbf{G}(t)$ should be

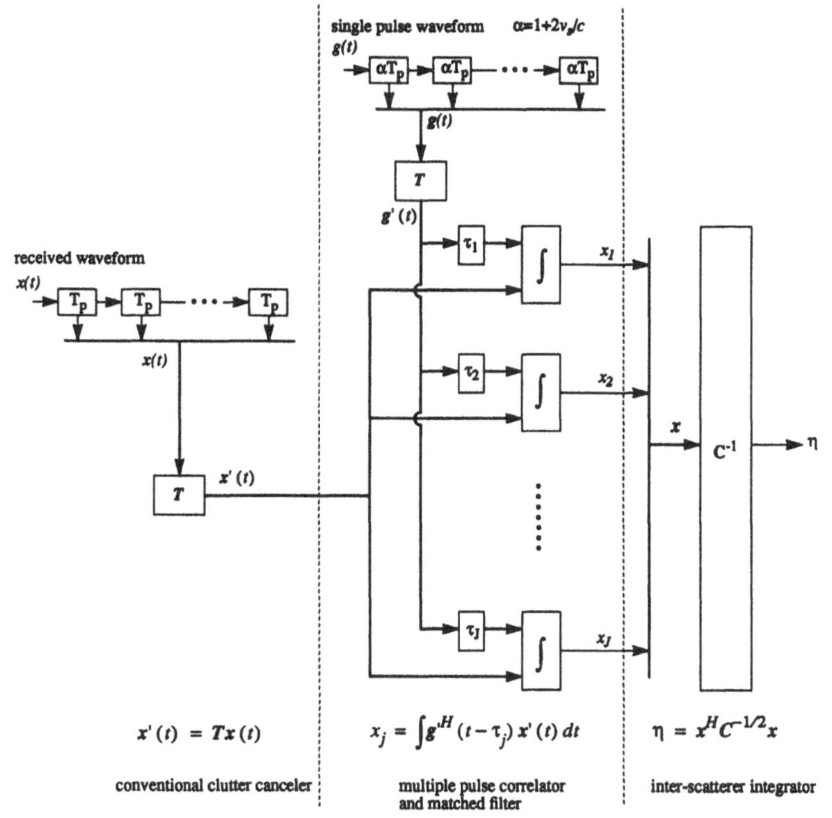

Figure 1. Structure of the simple UWB processor.

finely tuned at the corresponding target velocity. If the target velocity is unknown, one straightforward way is to place a set of the velocity-based correlators to cover the velocity range of interest, with each of the correlators being tuned at different velocity. The whole set of the velocity-based correlators can be considered as a "velocity filter bank" in the sense that they have the capability to reject the signal component with a velocity other than the tuned ones. One should note the such a strategy may largely increase the implementation complexity.

IV . PERFORMANCE EVALUATION

This section evaluates the detection performance of the canceler-based UWB processor. We will compare the processor with the UWB optimum to see how it performs with unknown interference statistics and target characteristics.

Even though it is possible to find a close-form expression for the detection performance of the canceler-based UWB processor, the result is tedious and inconvenient to perform numerical evaluation. Instead, we will conduct the performance evaluation based on computer simulation.

Under the assumption that the received waveform $\mathbf{x}(t)$ is a Gaussian process, it is not necessary to start the computer simulation from the uneasy job of generating the continuous waveform, since the data vector $\mathbf{x}$ is found to be a Gaussian distributed random vector with its covariance matrix and mean specified by

$$\mathrm{E}(\mathbf{x}\mathbf{x}^H|H_0) = \int \mathbf{G}^H(t)\mathbf{T}^H\mathbf{T}\mathbf{R}\mathbf{T}^H\mathbf{T}\mathbf{G}(t)dt = \mathbf{P}, \qquad N \times N \tag{18}$$

and

$$E(\mathbf{x}) = \begin{cases} \mathbf{0} & \text{under } H_0 \\ \mathbf{Qa} & \text{under } H_1 \end{cases}, \tag{19}$$

where

$$\mathbf{Q} = \int \mathbf{G}^H(t)\mathbf{T}^H\mathbf{T}\mathbf{G}(t)dt, \qquad N \times N. \tag{20}$$

Obviously, $\mathbf{Q}$ is a function of the single pulse waveform, the range walk rate $\rho = 2v_sT_pB/c$, and relative bandwidth RBW. In this paper, we choose the commonly used rectangular linear FM waveform [4].

For the inter-pulse covariance matrix $\mathbf{R}$, we assume

$$\mathbf{R} = \sigma_c^2\mathbf{R}_{c0} + \sigma_n^2\mathbf{I} \tag{21}$$

where σ_c^2 and σ_n^2 are the clutter and receiver noise power, respectively, and

$$\mathbf{R}_{c0} = \{\exp -2(\pi\sigma_f(m-n))^2\}, \tag{22}$$

with σ_f being the parameter controlling the interference spectrum spread. We note that the interference covariance setting given in Eq.(22) corresponds to a Gaussian shaped spectrum.

As we mentioned before, large portion of $\mathbf{a}$ will be zero as there are only a few scatterers distributed in the range extent of interest. We are interested in the general behaviors of the processors with respect to the scatterer range distribution. Thus, given the number of the scatterers, we assume each scatterer is randomly scattered in the N range cells with equal probabilities and energy. Since the waveform is real under the large relative bandwidth assumption, the phases of the scatterers is represented only by the sign of the scatterer amplitudes, which is assumed to change between plus and minus with equal probability.

Fig. 2 plots the detection performance curves as a function of signal-to-receiver-noise ratio (SNR=$\sum_{j=1}^{J}|a_j|^2/\sigma_n^2$) for the simple and optimum UWB processors. Due to computer resource limit, the false alarm rate is set to 10^{-3}. M, the total number of pulses in the processing, is chosen large (M=28) in order to acquire sufficient multiple pulse integration gain. A five pulse canceler with binomial coefficients is used for the interference canceler. The performance of the suboptimum processor with known interference covariance and integration-of-all scatterer integration scheme is also included for reference. The performance difference between the optimum and suboptimum processors is considered as the performance loss due to unknown number and separation of the scatterers, while that between the suboptimum and the simple ones indicates the loss caused by unknown interference covariance. It is seen from Fig. 2, with a interference-to-recever-noise ratio INR=σ_c^2/σ_n^2=50dB, the simple UWB processor delivers a reasonably good performance close to the optimum.

In order to show the importance of the velocity-based correlator, a UWB processor without the correlator, formed by setting $M = N_p$ and directly applying the scatterer integrator at the output of the pulse canceler, is compared with the previous one with the correlator. An equal total signal energy constraint is imposed over the received pulse train for fair comparison. Fig. 3 shows the comparison results, where the detection performance is depicted as a function of residual signal-to-interference-pluse-noise ratio at the output of the canceler. The performance difference between the two processor is clearly seen in Fig. 3, which reflects the multiple pulse integration gain obtained by the velocity-based multiple pulse correlator.

The interference-to-receiver-noise ratio is increased to INR=70dB in Fig. 4. It is seen that the canceler-based processor will experience more performance degradation under more severer interference conditions.

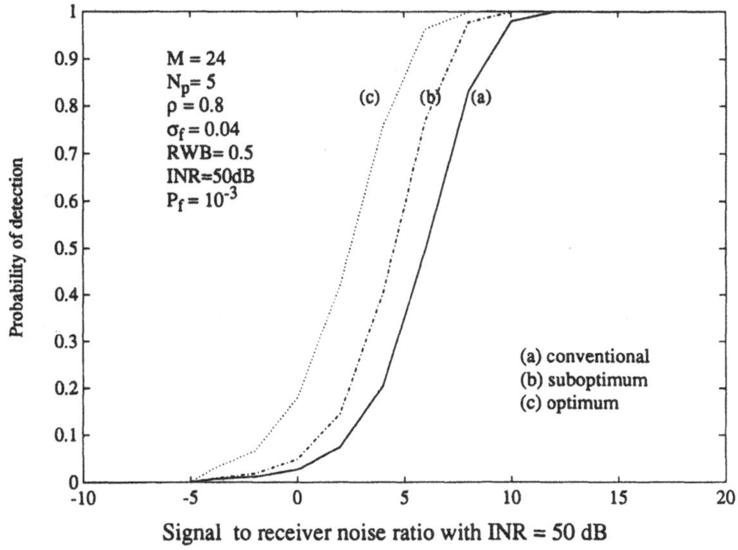

Figure 2. The detection performance of the simple UWB processor as compared to the optimum.

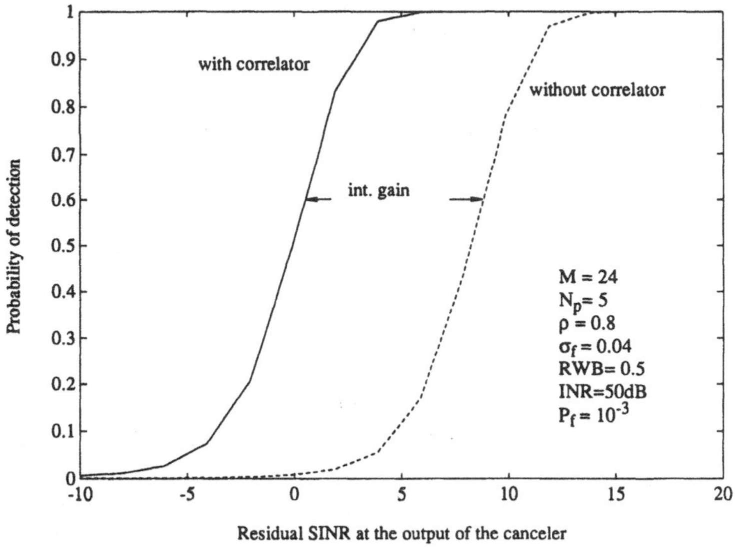

Figure 3. Performance gain provided by the velocity-based multiple pulse correlator.

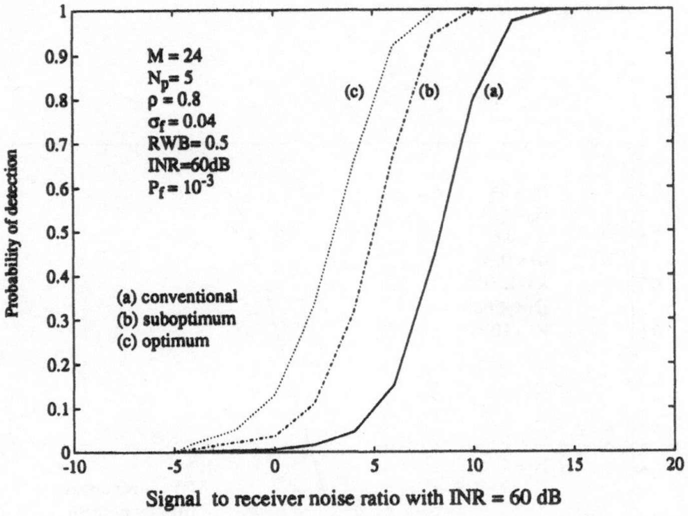

Figure 4. The detection performance of the simple UWB processor in a more severe interference enviroment.

V . CONCLUSION

To approach the performance potential provided by the UWB waveforms for weak moving target detection in strong clutter, a simple realizable UWB detection algorithm is developed to process the multipulse range walk data with large relative bandwidth. In the practical applications where both interference statistics and target characteristics are unknown to the system, the new processing algorithm proposes the use of the conventional nonadaptive cancellation method for clutter suppression, and the velocity filter bank for target discrimination against clutter. The computer simulation shows that the new MTD processor is capable to deliver a reliable detection performance close to the optimum in a non-moving clutter environment. It is also shown that when interference residue is still large due to insufficient cancellation, the velocity-based multipulse integration part of the processor, which jointly exploits range walk and doppler discriminations, becomes essentially necessary for further performance improvement.

This algorithm can be used for ground-based systems whose clutter spectrum is usually quite simple. Whether the UWB waveform is impulse or frequency/phase modulated, its implementation is well within the reach of currently available processor hardware.

REFERENCES

[1] L. Cai and H. Wang, "Moving target detection with UWB radar systems. Part I: overall performance assessment of performance potential," submitted to *IEEE Trans. Aerosp. Electron. Syst.*

[2] M. Skolnik, *Introduction to Radar Systems,* New York: McGraw-Hill, 1980.

[3] H. Wang and L. Cai, "Multiple pulse processing for over-resolved moving target detection," Part I of Phase I Contract Report to DARPA/RL, F30602-91-C-0035, June 1992.

[4] N. S. Tzannes, "Communication and Radar Systems", Englewood Cliffs, NJ: Prentice-Hall, p. 174, 1985.

INDEX